T0180520

Intelligent Decision Support Systems

Miquel Sànchez-Marrè

Intelligent Decision Support Systems

 Springer

Miquel Sànchez-Marrè
Dept. of Computer Science
Universitat Politècnica de Catalunya
Barcelona, Spain

ISBN 978-3-030-87792-7   ISBN 978-3-030-87790-3   (eBook)
https://doi.org/10.1007/978-3-030-87790-3

This Springer imprint is published by the registered company Springer Nature Switzerland AG
The registered company address is: Gewerbestrasse 11, 6330 Cham, Switzerland

To my beloved wife Anna for her continuous support and for the borrowed time from her that allowed me to devote time to write this book

Foreword

Each individual makes zillions of decisions every day, most of them unconsciously, without much reasoning. How hot to turn the shower, what to wear today, what to have for breakfast, which street to take on the way to the bus stop, and many more. Most of these decisions are embedded in some routine, reasonable and effective, and if not, there is usually no big problem in recovering from wrong or suboptimal choices. Decision making becomes more difficult and critical, for instance, when it is not clear, what the options are and how they are related to our objectives, when these objectives are different for the involved stakeholders, when actions decided upon cannot be undone, when they have a serious impact on entire companies, communities, states, and even at a global scope, and when bad or wrong decisions have a drastic impact on the living conditions and lives of people. The state of the world demands for many decisions of this kind, for instance, related to economy, international politics, military conflicts, ecology, and climate change.

In the last century, increasing efforts were undertaken to improve decision making through the development of more or less formal scientific theories, mathematical and algorithmic approaches, and the development of computer tools, initially mainly related to economy- and management-related decision processes, originally under the name of Information Management Systems. How and to what extent can these mathematical and computational means provide support or even automate the derivation of good decisions? The simplest scenario is being given a set of available options and a single criterion to evaluate and compare them and select the one that meets the interest of one stakeholder best. Unfortunately, such scenarios do not exist in the real world. There are always multiple criteria and multiple stakeholders, and one might hope for algorithms and formulas that assign some (quantitative or qualitative) value to different features and combine them using some weights to capture their importance to stakeholders in order to obtain the best result for all. Unfortunately, this does not work. A formula will make any of the options the winner when parameters are picked accordingly. And there is no best method in combining the measurements of criteria under reflection of values for the stakeholders; the "Paradox of multi-criteria decision analysis" reported some 40 years ago provides evidence for this.

The task of balancing the numerical or qualitative assessment and importance of different characteristics according to various objectives and interests cannot be delegated to mathematical expressions and algorithms that are bare of any meaning of the subject and impact of the decision. It will remain the duty of human decision makers. There is no universal and correct way of transforming the mesh of criteria and interests existing in the real world into a mathematical expression.

But then, what can be and should be the support in decision making that can be provided by computer systems? The answer is that we need to develop tools that capture and exploit knowledge about the domain of the decision—we need Intelligent Decision Support Systems. If we go back to the very core of decision making, we can identify which steps in the process can benefit from knowledge-based methods and what this support can look like. A decision means selecting one option of an action or plan out of a set that promises to satisfy particular goals if the respective action or plan is carried out. Apart from trivial cases, where an action is guaranteed to directly achieve a goal, such as turning a glass upside down yields an empty glass (if—solely—gravity forces apply), the goals we pursue are related to desirable states of some external system (like keeping global warming within certain limits), and suggested actions are meant to modify some part of this system to trigger an evolution of the system that, ultimately, approaches the goal, such as subsidizing energy production from renewable sources. This works only if we—or the decision support systems—dispose over knowledge about how the system functions and how it will respond to the actions taken.

This does not only mean that the system can predict this response with some certainty; even for deriving the set of promising options to choose from (which in many approaches is taken for granted), knowledge about the domain has to be present and applied. Furthermore, generating suitable options and assessing their potential to change the state of the system towards the intended goals requires to have an understanding of the state of the system one starts from (Which human activities, physical and chemical processes are present, and how do they influence global warming?). Even if lots of data (from earthbound measurements, satellites, lab experiments, . . .) are available, this alone does not help and may, on the contrary, pose a problem; interpreting them for assessing the current state and expected evolution of the system requires knowledge and is a challenge to human decision makers and to Intelligent Decision Support Systems.

Building such systems that, in one way or another, capture existing knowledge and experience related to a domain of decision making and can apply it to support different tasks in the process does not only face the fundamental challenge of formally representing knowledge and applying automated reasoning based on it. A number of secondary problems need to be addressed. For complex systems, knowledge from different sources needs to be integrated into a coherent representation, such as different research areas or the combination of scientific and experiential knowledge. How to avoid building IDSS for each individual region and problem instance of a class, and instead build solutions for a whole class that, for each instance, enable the transfer and exploitation of the relevant elements of knowledge? How to exploit uncertain and missing data, qualitative information or knowledge, and plausible assumptions, and, even more fundamentally, how to determine which information is essential or helpful? How to explain and justify results to the decision makers, which includes the problem of linking the model of the system domain, say, global warming, to the realm the decision makers can influence, e.g. the political, social, and economic means. How to cope with the uncertainty of predicting the system evolution triggered by actions, and how to support decisions that concern actions properly distributed over time? How to represent the spatial distribution of phenomena and perform automated reasoning about them?

Over the last decades, many contributions to the field of IDSS have been made, and very sophisticated and successful solutions have been produced, almost all of them facing real-world problems. Numerous formalisms and techniques from Computer Science and Artificial Intelligence have been used and further developed. When scanning the relevant literature, one gets the picture of a very vivid and creative field, eager to build useful tools—but also of a somewhat confusing landscape of quite diverse approaches and techniques, which reflects the variety of applications areas, problem characteristics, kinds of available knowledge, and, of course, favourite technologies. We are far from having a general theory of IDSS, standard architectures, or general guidelines of "How to build an IDSS". But in order to take steps in these directions, a comprehensive survey of the different concepts, formalizations, algorithms, and techniques present in the field is a prerequisite.

A comprehensive survey of this maze? I was sceptic when the author of this book disclosed his plan to me. On the other hand, I was certain that if anybody could achieve this, it would be him, being a researcher in different AI fields related to the topic, deeply involved in building IDSS, and teaching on this subject. And so, here is the book. It is more than a superficial list of various approaches, but thoroughly explains their theoretical and technical background. As such it will be both a very helpful assistant for anyone who intends to build an Intelligent Decision Support System and a valuable contribution to systematizing the field and bringing it closer to unifying perspectives, principles, and architectures.

Peter Struss
Munich, Germany
June 2021

Preface

There were three reasons that pushed me to write this book on Intelligent Decision Support Systems (IDSS). First, because I have been working within this area of research, probably sometimes even without knowing it or without knowing its current name, since my Graduate Thesis (*Tesi de Llicenciatura* or *Tesina*, in its original Catalan name) in 1991. That introductory research work led me into the complexity of real-world systems, and into the difficulty of applying the theoretical knowledge in Artificial Intelligence (AI) to solve a practical problem in real life. That problem was the diagnosis of a Wastewater Treatment Plant solved with the use of a knowledge-based system. Since then, I have been working in this field, making both research in the field and participating in the deployment of some IDSSs in several areas such as in Environmental Science, as in my Ph.D. thesis in 1996 about a distributed and integrated AI multi-level architecture for Wastewater Treatment Plants, or in Industrial Manufacturing field, such as an IDSS for optimizing the Textile Industry Manufacturing or a recommender system for improving Industrial Symbiosis (Circular Economy). More recently, in Medicine I have been participating in projects for deploying an IDSS for the prediction of Antimicrobial Resistance in Intensive Care Units in a hospital, or formerly in a project to develop an IDSS for the Analysis of the Evaluation of Adherence to Nutritional Intervention or in another one to develop an IDSS for Personalized Diet Recommendation.

Therefore, I felt that all my experience could be valuable and somehow interesting for other researchers in the field.

The second reason was because during these years I have noticed that there is not a general theory about IDSS, and a general framework for the development of IDSSs is missed. Usually, each IDSS is built up in an ad hoc manner, using diverse knowledge sources, using diverse approaches and techniques for different applications without any apparent standard protocol. All these facts outline there is some kind of disorientation among researchers in the field, and that there is neither a common agreed framework nor a general architecture and standard guidelines for deploying IDSSs, for selecting the most suitable techniques and models involved in an IDSS, for integrating different models in an IDSS, for evaluating IDSSs, etc.

Thus, a second issue of the book is to propose some general architecture and a cognitive interoperable framework for the development of IDSSs, providing guidelines on how to build an IDSS, trying to contribute to setting up Intelligent Decision Support Systems as an established and systematized field in Artificial Intelligence area.

Thirdly, because some years ago, in 2005 I started to teach, with my colleague Prof. Karina Gibert, a new Ph.D. Course on Intelligent Decision Support Systems in the Doctoral programme in Artificial Intelligence at Universitat Politècnica de Catalunya (UPC), transformed a few years later into a Master's course on Intelligent Decision Support Systems in an Inter-University (UPC-URV-UB) Artificial Intelligence Master. When I was browsing through the literature for related books on the field, I found out that there were no books covering the whole IDSSs topic, because most of available ones were focused on classical Decision Support Systems (DSSs) with a focus on Management Science and Business, most of them superficially treating the intelligent methods from Artificial Intelligence. Some others were collections of isolated chapters, commonly application-oriented articles, but none of them analysed how to build an IDSS, or proposed a general architecture for IDSSs, or explained and suggested evaluation techniques for IDSSs, or extensively described the integration of intelligent techniques in IDSSs, etc. This book aims at emphasizing the Computer Science and Artificial Intelligence point of view about IDSS, focusing more attention on the algorithmic aspects of the different machine learning and statistical techniques or model-driven methods than on the Management and Business aspects.

Therefore, I felt there was a lack of such a text both from the editorial and academic points of view.

To Whom

The book is oriented to a double spectrum of audience: both to professional practitioners and to academics (students, teachers, and researchers). On the other hand, the audience can belong to Business and Management area, and mainly to the field of, but not restricted to, Artificial Intelligence, Computer Science, Data Science and Engineering, and Information and Communication Technologies (ICT).

First, it can be useful to *Business Managers* to identify when a given business scenario is really a complex one and how many and of which nature are the decisions involved, how they can be analysed and what is the right or rational decision to be made in the management of the business scenarios, and how to solve that complex scenario with an IDSS. In addition, the basic principles of AI systems will be interesting for them. I suggest them to read Chaps. 1–4, 7, and 8 and optionally all the other ones.

In addition, it can be surely useful to *Computer Scientists* or/and *AI scientist* professional practitioners to learn how to analyse, to design, to implement, and to validate Intelligent Decision Support Systems (IDSSs) in complex real-world systems. The recommended roadmap for them is to read Chaps. 1–8 and 10, and optionally Chap. 9, with advanced topics.

Regarding the academic world, I have written the book assuming that the main audience for the book will be teachers/researchers and advanced undergraduate students or master's students in Computer Science, Artificial Intelligence, or Data Science and Data Engineering. For an undergraduate course on Intelligent Decision Support Systems, I suggest the reading of Chaps. 1–4, 6, 7, and 10. Complementary material would be for more advanced IDSS undergraduate courses or advanced students in Chaps. 5 and 8.

For a Master course on Intelligent Decision Support Systems, I recommend Chaps. 1–8 and 10. I would reserve the material of Chap. 9 for other advanced Master course/s or for eager students that want to know more on advanced topics in the field.

Furthermore, the whole Chap. 6 can be used as the basis for a specific course on Data Science, and the whole Chap. 5 may be used as the reference material for a course on Artificial Intelligence Reasoning Paradigms. The advanced topics material in Chap. 9 can be employed for an advanced course on Uncertainty Management in AI or for an introductory course on Spatial-Temporal Reasoning. The final section of Chap. 9 about Recommender Systems can be the basic material for an introductory course or seminar on Recommender Systems.

The book assumes a minimum knowledge of basic concepts in Computer Science, in programming or algorithmic, and basic concepts of logic, such as the notation of logical operators. Notwithstanding, when some not usual concept or definition appears, it is usually explained either in the text or in a footnote. Knowledge about Artificial Intelligence is not assumed, and thus, many parts of the book can be considered as an introductory text to AI.

Structure of the Book

The book is organized in three parts:

I—Fundamentals: it comprises an introduction to the motivations for the appearance of Intelligent Decision Support Systems. The complexity of real-world systems and its corresponding decisions are analysed (Chap. 1), and an introduction to the decision theory and to the decision process modelling is presented (Chap. 2). A historical evolution of classic Management systems, passing through Decision Support Systems until the emergence of Intelligent Decision Support Systems, is detailed (Chap. 3).

II—Intelligent Decision Support Systems: this part is the core of the book. The main characteristics of Intelligent Decision Support Systems (IDSS) are described. An introduction of Artificial Intelligence field and its main reasoning paradigms is explained. Intelligent Decision Support systems are defined, classified in different typologies and its conceptual components are analysed. Main requirements for an

IDSS are detailed and a proposal for a general IDSS architecture is presented, jointly with a cognitive-oriented approach for the IDSS development process (Chap. 4). Afterwards, a whole chapter is devoted to each major kind of IDSS according to the models used: *model-driven IDSSs* and *data-driven IDSSs*. Most common *model-driven methods* are explored in Chap. 5, and *data-driven models* within the context of Data Science are explored in Chap. 6. Finally, the use of Intelligent models in Decision Support tasks is illustrated with several case studies for both *model-driven* approaches and *data-driven* models (Chap. 7).

III—Development and Application of Intelligent Decision Support Systems: last part of the book is related to the practical development and application of IDSSs, and to analyse some advanced IDSS topics. Main useful software tools for the development of IDSS are presented, both including specific tools for given modelling techniques and general development tools (Chap. 8). The advanced topics analysed are the *Uncertainty Management* problem, the *Temporal Reasoning* aspects, and the *Spatial Reasoning* management. An advanced application of Intelligent Decision Support Systems is presented: *Recommender Systems* (Chap. 9). Finally, a summary of the book is synthetized; some open challenges in IDSS research such as the interoperability of models in IDSS or a general protocol for IDSS evaluation are presented. The book finishes with some concluding remarks (Chap. 10).

Contributor

Prof. Franz Wotawa, from Technische Universtät Graz, Institute for Software Technology (Graz, Austria), has contributed to Chap. 5 (Sect. 5.4) of the book, where he explained Model-based Reasoning methods, and to Chap. 7 (Sect. 7.1.3), where he described and analysed one case study for illustrating the use of Model-based Reasoning methods. He is a well-known and reputed researcher in Model-based Reasoning area, with a wide expertise in model-based diagnosis, model-based software debugging, model-based fault localization in hardware designs, and model-based diagnosis and reconfiguration of mobile robots.

Miquel Sànchez-Marrè

Barcelona

Acknowledgements

I am in debt to many people who helped me, in different ways, during the long process of writing this book. I want to acknowledge their contribution to finally let me reach the goal. It has not been an easy task due to many reasons. Since March 2016, during the last 5 years I have been writing the book. During this period, a lot of things have happened in my life that influenced on the writing process. Among all of them, I outline the appearance of the COVID-19 pandemic situation.

Along those years I have confirmed that it is almost impossible to write a book and at the same time fulfil my obligations as a university teacher, researcher, editorial board member of some journals, advisor of many students, etc. I have mostly advanced at the weekends, by nights, and at holiday periods. It has been a stressing period of my life, because I was constantly conscious of my delays with the editorial, but at the same time I have been doing the best I could to advance. Thus, I experimented a great frustration and stress: I wanted to progress, but at the end, the daily academic life, and its corresponding unexpected events, prevented me to make it. Finally, I am really exhausted but happy!

Being an Artificial Intelligence scientist, I will group my acknowledgements into two groups or clusters: the people from the academic world and the people from my personal environment,[1] who have helped me in this process. I am grateful to all of you for your help. In addition, I ask for your pardon if I forget to mention somebody. Sure in my mind, I am also grateful to you!

Regarding the scientific world, I want to sincerely thank a lot of people. First, to my friend and colleague Prof. Franz Wotawa for having accepted to contribute to this book with his well-known and reputed expertise in Model-based Reasoning field.

Thanks Franz for your great contribution, for your friendship, and for being my host in Graz in April 2016 when I explained and convinced you for helping me!

---

[1] Probably, the restless minds are already thinking that the two clusters can overlap. Yes, you are right, but the clusters can be fuzzy!

Next, I want to thank Prof. Peter Struss for his inspiring foreword. I am grateful for this nice text. I appreciate your text in a double way: as a friend and colleague, but even more as a reputed scientist in Artificial Intelligence and in Intelligent Decision Support Systems field, which makes me feel very happy with your kind words. If I contribute with just a small grain to the systematizing of IDSS area, I will be more than happy.

Thank you Peter for your nice words and for your friendship!

There is another group of people, which I will name as the "influencers"[2] of my research line to whom I am grateful for having met them in my academic and personal life. They led me to make research in Intelligent Decision Support Systems, even though none of us probably knew that we were working and building real applications in this new emergent field. Surely, without being conscious at all, you have influenced me and my research to be focused on Intelligent Decision Support Systems, with your collaboration along many years.

First, to my Ph.D. advisor, colleague, and friend Prof. Ulises Cortés, who led me to cope with real-world problems and intelligent systems to solve them. In early years with my Graduate Thesis, afterwards with my Ph.D. thesis, and in other more recent research projects.

Thank you Ulises for your friendship and invaluable and constant scientific pressure!

The other "influencers" are Chemical and Environmental Engineers, who introduced my Ph.D. supervisor and myself to actual environmental problems at the beginning of the 1990s, specially related to water treatment processes. They have fostered my interest to tackle complex real-world domains affording intelligent systems to support stakeholders to make decisions on those domains. They are Prof. Manel Poch, Prof. Javier Lafuente, Prof. Ignasi Rodríguez-Roda, and Dr. Joaquim Comas. During many years our scientific and also gastronomic meetings and research projects have motivated me and created a good atmosphere to make research in Intelligent Decision Support Systems.

Thank you Manel, Javier, Ignasi, and Quim for your friendship and your ever stimulating meetings!

In addition, I want to thank the narrow scientific collaboration of my colleague and friend Prof. Karina Gibert, who shares with me the teaching of three Master's courses on Intelligent Decision Support Systems and Data Science. Since many years ago, we had numerous vivid scientific discussions, specially around IDSS and Data Science, which surely have impacted my view on the field.

Thank you Karina for your friendship and for our interesting common research projects and numerous meetings which are always fruitful and profitable!

Furthermore, I strongly think that I have learned a lot from the scientific interaction with the students I have advised in their Ph.D. thesis and master's thesis. Certainly, the research works and our discussions and meetings have clearly

---

[2] Of course, influencers in the right sense of the word, and nothing to do with nowadays idolatrized characters in social networks!

provided me with new fresh ideas, points of view, problems, solutions, applications, etc., which have nurtured and influenced my perception and research on IDSSs field, which finally has ended in this book. Without your collaboration, all would have different.

Thank you for your friendship and your research collaboration Dr. Anna Gatzioura, Dr. Majid Latifi, Dr. Beatriz Sevilla Villanueva, Dr. Fernando Orduña Cabrera, Dr. Héctor Núñez Rocha, Dr. Luigi Cecaroni, Natalia Sarmanto, Àlvar Hernàndez Carnerero, Maritza Prieto Emhart, Gerard Reig Grau, Andrei Mihai, Nil Sanz Bertran, Maria Teresa Chietera, Jaume Martí i Ferriol, Thania Rendón Sallard, and Joel Segarra Rúbies!

Finally, in the academic environment I want to express my gratitude to all my colleagues in our Knowledge Engineering and Machine Learning Group (KEMLG), and to all our Intelligent Data Science and Artificial Intelligence Research Centre (IDEAI-UPC) colleagues for all the fruitful meetings and seminars which surely have influenced my research work with new ideas. Thank you all!

From a personal point of view, I also want to thank several people. First to the people from the Springer editorial for their continuous support and assistance in all the steps, and specially for their infinite patience with my delays in writing the book. As I mentioned above, I always have tried to advance faster, but actually I could not. I am sorry for that.

Thank you Mary James, Brian Halm, and Zoe Kennedy for believing in me and for your support!

Thank you Murugesan Tamilselvan and Sanjana Sundaram for your support and for the book coordination!

Last but not least, I want to thank all my direct family for having supported me along these years. I have felt that you were actually worried about the book, especially when you asked me with certain frequency: how is the book work progressing? You have pushed me not to relax on the task.

Thank you Anna, Paquita, Carme, Maria Reina, Àlex, Carme, Josep Maria, Laura, and Marta for being there!

Miquel Sànchez-Marrè
Barcelona, July 2021

Contents

Part II   Intelligent Decision Support Systems

Author and Contributor

About the Author

Miquel Sànchez-Marrè received a B.Sc. in Computer Science in 1988 and a M.Sc. in Computer Science in 1991 both from Barcelona School of Informatics (FIB), and a Ph.D. in Computer Science from Universitat Politècnica de Catalunya (UPC) in 1996. He is an Associate Professor in the Computer Science (CS) Dept. of UPC since 1997 (tenure). He is a member of the Intelligent Data Science and Artificial Intelligence Research Centre (IDEAI-UPC). He received an Accesit of the Oms i De Prat 1991 Prize, for his master's thesis. He co-founded the spin-off company Sanejament Intel·ligent S.L. (SISLtech, 2003–2017), which received the Energy Excellence Award 2013, and the Bioenergy Silver Award 2012. He is a pioneer member of Catalan Association of Artificial Intelligence (ACIA) and in 1994–1998 was a member of its board of directors. He is a Fellow of the International Environmental Modelling and Software Society (iEMSs) since 2005. He is mentioned as a contributor to the "Timeline of Artificial Intelligence" in the Wikipedia (https://en.wikipedia.org/wiki/Timeline_of_artificial_intelligence#1990s) for co-organizing the first Environment and AI Workshop in Europe: Binding Environmental Sciences and Artificial Intelligence (BESAI 98) at ECAI conference in 1998. He is a member of Spanish Association for Artificial Intelligence (AEPIA) and a pioneer member of iEMSs and a board member of iEMSs, too. He is also a pioneer member of the MIDAS (Spanish Network on Data Mining and Learning) and ÁTICA (Spanish network on Advance and Transference of Applied Computational Intelligence), and he was the main coordinator of the ÁTICA network. His main research topics are Intelligent Decision Support Systems, Case-Based Reasoning, Machine Learning, Data Science, Recommender Systems, Knowledge Engineering, Integrated AI architectures, and AI applied to Environmental, Industrial, and Health systems. He is the author of 220 peer-reviewed publications, including six other books.

Contributor

Franz Wotawa received a M.Sc. in Computer Science (1994) and a Ph.D. in 1996 both from the Vienna University of Technology. He is currently professor of software engineering at the Graz University of Technology (Graz, Austria) and the head of the Institute for Software Technology. His research interests include model-based and qualitative reasoning, theorem proving, mobile robots, verification and validation, and software testing and debugging. Besides theoretical foundations, he has always been interested in closing the gap between research and practice. For this purpose, he founded Softnet Austria in 2006, which is a non-profit organization carrying out applied research projects together with companies. Starting from October 2017, Franz Wotawa is the head of the Christian Doppler Laboratory for Quality Assurance Methodologies for Autonomous Cyber-Physical Systems closely working together with industrial partners. During his career, Franz Wotawa has written more than 400 peer-reviewed papers for journals, books, conferences, and workshops. He supervised 93 master's and 37 Ph.D. students. For his work on diagnosis, he received the Lifetime Achievement Award of the Intl. Diagnosis Community in 2016. Franz Wotawa has been a member of a number of programme committees and organized several workshops and special issues of journals. He is a member of the Academia Europaea, the IEEE Computer Society, ACM, the Austrian Computer Society (OCG), and the Austrian Society for Artificial Intelligence and a Senior Member of the AAAI.

Abbreviations

ACC	ACCuracy of a classifier/discriminant model
ACL	Agent Communication Languages
AdaBoost	Adapting Boosting
ADSS	Advanced Decision Support Systems
AHP	Analytic Hierarchy Process
AI	Artificial Intelligence
ANN	Artificial Neural Network
API	Application Program Interface
BTT	Backpropagation Through Time
Bagging	Bootstrap Aggregating
BDI	Belief-Desire-Intention
BER	Balanced Error Rate
BPMN	Business Process Model Notation
CB	Case Base
CBClas	Case-Based Classifier/Discriminant
CBPred	Case-Based Predictor
CBR	Case-Based Reasoning
CDPS	Cooperative Distributed Problem Solving
CF	Certainty factor
CH	Caliński-Harabasz index
CLD	Causal Loop Diagram
CLIPS	C Language Integrated Production System
CNN	Convolutional Deep artificial Neural Network
CPT	Conditional Probability Table
CRAN	Comprehensive R Archive Network
CRC	Constant Error Carrousel
CVD	Class Value Distribution
CVI	Cluster Validity Index
DA	Diet Algorithm
DAG	Directed Acyclic Graph

DBMS	Data Base Management System
DBN	Dynamic Bayesian Network
DIANA	Divisive ANAlysis Clustering
DLNN	Deep Learning artificial Neural Network
DSS	Decision Support Systems
EBL	Entropy-Based Local weighting method
EBR	Episode-Based Reasoning
EDSS	Environmental Decision Support System
EIS	Executive Information System
Ep	Hierarchical Episode Base
EPL	ECLIPSE Public License
FB	Fact Base
FFNN	Feed Forward artificial Neural Network
FIPA	Foundation for Intelligent Physical Agents
FIPA-ACL	Foundation for Intelligent Physical Agents ACL
FLAME	Flexible Large-scale Agent Modelling Framework
FN	False Negative instances
FP	False Positive instances
FRNN	Fully Recurrent artificial Neural Network
GA	Genetic Algorithm
GDPR	General Data Protection Regulation (EU 2016/679)
Gecode	Generic Constraint Development Environment
GESCONDA	Knowledge Management in Environmental Data
GIS	Geographic Information System
GPL	GNU General Public License
GQR	Generic Qualitative Reasoner
IBM	International Business Machines
IC	Integrated Circuit
IDE	Integrated Development Environment
IDLE	Python Integrated Development and Learning Environment
IDSS	Intelligent Decision Support Systems
IDSS-DD	IDSS Design Document
IDSS-IDD	IDSS Interface Design Document
IDSS-RSD	IDSS Requirements Specification Document
IDSS-IRSD	IDSS Interface Requirements Specification Document
IEDSS	Intelligent Environmental Decision Support Systems
ILD	Intra-List Diversity
IG	Information Gain
IPython	Interactive Python
IQR	Inter-Quartile Range
Jess®	Java Expert System Shell
JPD	Joint Probability Distribution
KB	Knowledge Base
KBR	Knowledge-Based Reasoning

KBS	Knowledge-Based Systems
KDD	Knowledge Discovery from Databases
KEEL	Knowledge Extraction based on Evolutionary Learning
KIF	Knowledge Interchange Format
KNIME	Konstanz Information Miner
KQML	Knowledge Query Manipulation Language
k-NN	k-Nearest Neighbour
LHS	Left-Hand Side
LR	(simple) Linear Regression
LSTM	Long Short-Term Memory network/model
MAE	Mean Absolute Error
MAPE	Mean Absolute Percentage Error
MAS	Multi-Agent Systems
MB	Measure of Belief
MBA	Model Builder Algorithm
MBR	Model-Based Reasoning
MCA	Multiple Correspondence Analysis
MCDA	Multi-Criteria Decision Analysis
MCMC	Markov Chain Monte Carlo method
MD	Measure of Disbelief
MEpB	Meta-Episode Base
MI	Mutual Information
ML	Machine Learning
MLP	Multiple-Layer Perceptron
MLR	Multiple Linear Regression
MSE	Mean Square Error
NLP	Natural Language Processing
NMSE	Normalized Mean Square Error
NoSQL	Non-SQL
NPV	Negative Predictive Value
ODE	Ordinary Differential Equation
OLAP	On-Line Analytical Processing
PCA	Principal Component Analysis
PDT	Prometheus Design Tool
PFA	Portable Format for Analytics
PHCT	Propositional Horn Clause Theory
PMML	Predictive Model Markup Language
PNN	Probabilistic artificial Neural Networks
PPV	Positive Predictive Value, Precision
PRO	Projection of attributes
PSF	Python Software Foundation
PyDev	Python IDE for ECLIPSE
PyPI	Python Package Index
QDE	Qualitative Differential Equation

QPE	Qualitative Process Engine
QPT	Qualitative Process Theory
QSIM	Qualitative SIMulation framework
Q_1	First Quartile
Q_3	Third Quartile
Rattle	R Analytic Tool to Learn Easily
RBFNN	Radial Basis Function artificial Neural Network
RBR	Rule-Based Reasoning
RDBMS	Relational Data Base Management System
RHS	Right-Hand Side
RMSE	Root Mean Square Error
RNN	Recurrent artificial Neural Network
ROC	Receiver/Relative Operating Characteristic
RS	Recommender System
RTRL	Real-Time Recurrent Learning
R2	Coefficient of determination
R^2	R-Squared, Corrected Coefficient of Determination
SA	Simulated Annealing
SCA	Simple Correspondence Analysis
SLP	Single-Layer Perceptron
Spyder	Scientific PYthon Development EnviRonment
SQL	Structured Query Language
SRNN	Simple Recurrent artificial Neural Network
SSE	Sum of Squares of Errors
SSR	Sum of Squares of the Regression model
SST	Total Sum of Squares
SVM	Support Vector Machine
SWF	Social Welfare Function
$TClR_k$	True Class Rate of class k
TP	True Positive instances
TPR	True Positive Rate, Recall, Sensitivity, Hit Rate
TN	True Negative instances
TNR	True Negative Rate, Specificity, Selectivity
TDNN	Time-Delay artificial Neural Network
UEB-1	Unsupervised Entropy-based weighting method 1
UEB-2	Unsupervised Entropy-based weighting method 2
UML	Unified Modelling Language
VDM	Value Difference Metric
VE	Variable Elimination algorithm
Weka	Waikato Environment for Knowledge Analysis
WWTP	Wastewater Treatment Plant
XML	eXtensible Markup Language
2TBN	Two-Time slice Bayesian Network

Part I
Fundamentals

Chapter 1
Introduction

Intelligent Decision Support Systems (*IDSS*) is an emergent field arising from the late 90s and beginning of the twenty-first century in Artificial Intelligence crossing with classical *Decision Support Systems* (*DSS*), which emerged in the Management Science and Business areas in the 60s and 70s.

Human beings are continuously facing many decisions every day in their life. Since they wake up in the morning, they make decisions about what they should wear, about what transport means they should select to go to work, about what to eat at lunch, and about many matters in their jobs within an organization.

The decision-making process depends on several *features*. For instance, in the former example above, usually, the clothes selected may depend on the foreseen activities for the day, on the weather conditions, on the available clean clothes, and on the personal state of mind. The number of possible *alternatives* for the combination of the clothes could be high. As the problem wrapping the decision is getting more difficult, the *number of features* that must be considered for making the decision and the *number of available alternatives* are higher. In addition, as the problem is more complex usually the human on charge of the decision-making, i.e. *the decisor*, is not able to make the right decision by himself or herself. Either the number of features and/or alternatives could be very large or their inherent complexity could be very high to be effectively manageable by a human. Here is where appears the idea of a system to help in the decision-making process of humans, i.e. *a Decision Support System* (*DSS*) (Marakas, 2003; Power, 2002).

Early Decision Support Systems, as it will be analyzed in Chap. 3, emerged in Management Science. One of the typical problems that a company must cope with is, for instance, the price of a new product. Of course, depending on the price, and many other related features like the economic status of the market, the existence of similar products in competitor companies, etc., the benefit for the company could range from very high to very low. It would be very interesting for the company managers they had some system or tool to help them in the assessment of the potential revenue depending on the price chosen for the new product. This *forecasting* ability of this kind of system will be one of the main characteristics of Decision Support Systems.

© Springer Nature Switzerland AG 2022
M. Sànchez-Marrè, *Intelligent Decision Support Systems*,
https://doi.org/10.1007/978-3-030-87790-3_1

As it will be shown in this book, this prognostic task as well as other tasks are carried out with some models which can be originated from experts' knowledge or first-principles theories (*model-driven DSS*) and/or some other models obtained from data analysis (*data-driven DSS*).

Intelligent Decision Support Systems (IDSS) (Dhar & Stein, 1997; Turban et al., 2005; Sauter, 2011) are the natural evolution of DSS when they incorporate some intelligent model or/and some intelligent data analysis method or/and some learning method. In other words, when some model or technique coming from the field of Artificial Intelligence[1] (Russell & Norvig, 2010; Schalkoff, 2011; Hopgood, 2012) is integrated in a DSS. Usually, these intelligent components make the IDSS more reliable to cope with a higher degree of complexity of the systems.

1.1 Complexity of Real-World System

In previous works, several researchers have tried to analyze and classify the degree of complexity and associated uncertainty of different kinds of systems and related problem-solving strategies. One of the most well-known is the classification proposed by Funtowicz and Ravetz (1993). They state that the two fundamental dimensions for classifying the strategies, and in fact, the systems are (see Fig. 1.1): the *system uncertainty* (*x*-axis) and the *decision stakes* (*y*-axis). They distinguish the uncertainty of systems among technical, methodological, and epistemological levels, ranging from low level to high level as you move along the axis. On the vertical axis, decision stakes can be divided into simple purposes, complex purposes, and conflicting purposes among stakeholders. It ranges from low to high complexity and risk. They propose to classify the problem-solving strategies of the real-world systems in three levels, each one including the precedent one, as shown in Fig. 1.1:

1. Applied Science Level: Simple systems, with low uncertainty, mainly technical uncertainty and with very limited risk associated with the decisions due to simple decision stakes purposes. The technical uncertainty will be managed by standard routines and procedures. The decision stakes are simple and small, probably because there is some external function for its results. This kind of systems can be managed with some scientists/engineers and some first-principles knowledge (equations, etc.).

   Environmental Example: The design of a wastewater treatment operation with a fixed inflow water, where an environmental or chemical engineer can solve the problem from the application of basic scientific principles.

---

[1] Artificial Intelligence is a subfield of Computer Science devoted to the study of the possible or existing mechanisms—in human or other beings—providing such behaviour in them that can be considered as *intelligence*, and the emulation of these mechanisms, named as *cognitive tasks*, in a computer through the computer's programming.

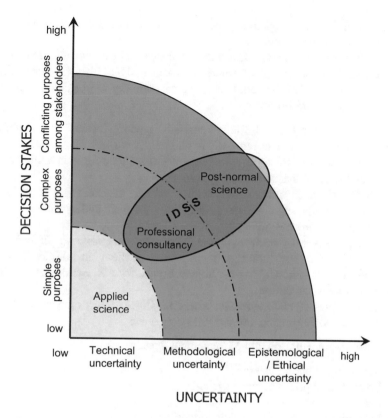

Fig. 1.1 Classification of problem-solving strategies for systems. (Adapted from Funtowicz and Ravetz 1993)

2. Professional Consultancy Level: Systems with a certain degree of uncertainty and the decision stakes are more complex. Usually, the professional work is performed for a client, whose requirements and purposes are to be satisfied. The reliability of theories and information makes the uncertainty be at the methodological level, which can be applied to several situations, where the experience and the human experts start to be needed. This kind of system can be managed with the knowledge and information provided by experts.

    Environmental Example: The supervision/management of a wastewater treatment plant, where the knowledge and the experience coming from experts are required to manage the uncertainty of the system.

3. Post-normal Science Level: Complex systems with very high level of uncertainty, and probably, with several contradictory goals among the stakeholders. Several kinds of expertise (knowledge, experience, historical data, etc.) are required to solve the conflicting purposes of stakeholders, and probably several groups of the society are needed (government, local authorities, citizens, etc.) to solve this kind

of problem. Epistemological uncertainty is involved because uncertainty is at the core of the problem itself.

Environmental Example: Environmental aspects of wastewater treatment at a global policy of one company, where environmental scientists, the plant manager, the executive chief of the company, the local authority responsible for the water, and other actors are involved.

As it is depicted in Fig. 1.1, the systems falling within levels 2 and 3, which are the more complex, are the systems where an *Intelligent Decision Support System* is envisioned to be able to cope with both the corresponding uncertainty and the complexity degree of these kinds of systems.

Complex real-world systems or domains exist in the daily real life of human beings, and normally show a *strong complexity* for their understanding, analysis, management, or problem-solving.

They imply several *decision-making tasks* very *complex* and *difficult*, which usually are faced up by human experts. Some of them, in addition, can have *catastrophic consequences* either for human beings, for the environment, or for the economy of one organization.

Some examples of such systems, where Intelligent Decision Support Systems seems to be a good approach, are the following ones:

- Business Administration and Management Systems:

  - Marketing support system
  - Products and prices support system
  - Human resources support system
  - Strategies and company policy support system

- Medical Systems

  - Medical system for heart disease diagnose support
  - Medical system for breast cancer diagnose support
  - Medical system for patient-assisted breathing control

- Industrial Process Systems

  - Industrial process management support system
  - Industrial design aided support system

- Environmental Systems

  - Wastewater treatment plants supervision support system
  - Air pollution control support system
  - Forest Management support system
  - Agricultural crop protection support system
  - River water quality supervision support system
  - Ecological population management support system

1.2 The Need of Decision Support Tools

Where comes the need for using Intelligent Decision Support Systems? When a user can start thinking of IDSS as the possible appropriate solution for a problem? As we have previously stated above, there are several reasons and features leading to its need and use, which can be summarized in the following ones:

- Cognitive limitations of the Decisor: Accurate evaluation of multiple alternatives in the decision process
- Problem structure at hand: Complexity of the decision-making process
- Need for forecasting capabilities
- Uncertainty management
- Data Analysis and exploitation (data-driven models)
- Need for including expertise (knowledge) (model-driven techniques)
- Need for using past experiences to improve the decision-making process

Typically, when several of these characteristics are present in a system, it is a good idea to design, implement, and use an Intelligent Decision Support System. Of course, by now, the reader has no idea how to design or implement an IDSS! However, in the next chapters of the book, we will analyze which are the different decisions that can appear in these kinds of systems, how to model these decisions, how to make easier the evaluation of the several alternatives for each decision involved, how to support users with forecasting abilities, etc. Summarizing, in the next three chapters the main elements for being able to understand how to analyze, design, develop, and validate an IDSS will be explained. In addition, we will outline the historical perspective of Decision Support Systems from its beginning with Management Information Systems, its evolution to Decision Support Systems, the integration of the advanced use of computers in DSS, until the final integration of intelligent methods, which made the DSS become actual Intelligent Decision Support Systems.

References

Dhar, V., & Stein, R. (1997). *Intelligent decision support methods: The science of knowledge work.* Prentice-Hall.

Funtowicz, S. O., & Ravetz, J. R. (1993). Science for the post-normal age. *Futures, 25*(7), 739–755.

Hopgood, A. A. (2012). *Intelligent systems for engineers and scientists* (3rd ed.). CRC Press.

Marakas, G. M. (2003). *Decision support systems in the twenty-first century* (2nd ed.). Prentice-Hall.

Power, D. J. (2002). *Decision support systems: Concepts and resources for managers*. Greenwood
    Publishing Group.
Russell, S. J., & Norvig, P. (2010). *Artificial intelligence: A modern approach* (3rd ed.). Pearson
    Education.
Schalkoff, R. J. (2011). *Intelligent systems: Principles, paradigms and pragmatics*. Jones and
    Bartlett Publishers.
Turban, E., Aronson, J. E., & Liang, T.-P. (2005). *Decision support systems and intelligent systems*
    (7th ed.). Pearson/Prentice Hall.

Further Reading

Gupta, J. N. D., Forgionne, G. A., & Mora, M. (Eds.). (2006). *Intelligent decision-making support
    systems: Foundations, applications and challenges (decision engineering)*. Springer.
Kuhn, T. S. (1962). *The structure of scientific revolutions*. University of Chicago Press.
Power, D. J. (2013). *Decision support, analytics, and business intelligence* (2nd ed.). Business
    Expert Press.
Sauter, V. L. (2011). *Decision support systems for business intelligence* (2nd ed.). Wiley.

Chapter 2
Decisions

2.1 Fundamentals About Decisions

A *decision* is the resolution or judgement or determination of an agent reached after consideration (a deliberative process) of a given *scenario* or *situation at hand*, which involves several *alternatives* or *options* or *courses of action*. Each alternative has several possible *consequences or outcomes*. A *consequence* or *outcome* for a given alternative is formed by the values of some *states of world/nature*, which are states out of the control of the decisor, and which are unknown at the time of making the decision. Thus, the states of the world provide the *uncertainty* in the decision process. Sets of related states of the world are called *events*.

In general, in decision theory, it is assumed that each alternative has one *consequence or outcome*. In the case that there is more than one *consequence* or *outcome*, usually from a different nature and compared through a different scale, it is commonly referred as *multi-attribute consequence/outcome* and *single-attribute consequence/outcome* when there is just one consequence or outcome.

This resolution of the decision-making implies the choice of one alternative, and the refusal of the other alternatives. In decision theory is commonly assumed that the set of alternatives is *closed* (i.e. the set of alternatives is fixed a priori) and that the alternatives usually are *mutually exclusive* (i.e. just one of them could be realized). The *decisor* can rank the alternatives using some *criterion* or *preferences* over the alternatives according to her/his *objectives*.

Decision-making is a process where a human agent (*decisor*) or a group of humans (*committee of decisors*) selects one possible alternative or course of action, among the available alternatives, being rational. *Rationality* means that the agent will use his/her beliefs to satisfy his/her desires or objectives. For instance, if you are thinking on how to spend the evening of a Saturday and your main objective is to be entertained and enjoying the free time, and you have two options: to stay at home cleaning and ordering your house or to go to the cinema, it seems more rational to *decide going to the cinema*. However, depending on the external factors, like

© Springer Nature Switzerland AG 2022
M. Sànchez-Marrè, *Intelligent Decision Support Systems*,
https://doi.org/10.1007/978-3-030-87790-3_2

whether you will have to make a queue for getting entrance tickets or not, or depending on how good will be the film, perhaps that is not the *right decision*. A *right decision* is a decision, which satisfies that its actual outcome is at least as good as that of every other possible outcome. The main drawback with decision theory is that you only know if the *rational decision* was the *right decision* after you have made your choice and you get the outcomes of your choice!

At the time of making the decision, we cannot know surely, which will be the right decision. Instead, we can make the rational choice, with the hope that it will be the right one.

2.2 Decision Typologies

As it has been outlined previously, there are many decisions in our everyday life, in all organizations, and companies all over the world. Of course, not all decisions are of the same type. Decisions can be grouped into different categories according to different dimensions or points of view. Here, we will describe the most relevant classifications proposed by several researchers in the area of Decision Support.

Herbert A. Simon proposed (Simon, 1960) the first classification of decisions' structure based on the programmability feature of the decisions. He distinguished decisions categorized according to the extent they are programmed:

- *Programmed Decisions*: This kind of decisions are repetitive decisions, routinely decisions, and commonplace decisions. Decisions commonly happening in the day-to-day of an organization, for which the organization and/or any decision support system, is prepared to make the final choices. They usually are simple decisions and correspond to well-*structured problems*.
- *Non-programmed Decisions*: This kind of decisions are novel, unique, and consequential. For these decisions, the systems have no specific procedure to deal with, as the decisions are very rare and only happens a few times. They usually are hard decisions, and correspond to *very poor structured problems* or *unstructured problems*.
- *Semi-programmed Decisions*: They are intermediate decisions in the continuum from programmed to non-programmed decisions, which probably happen some-times in the life of an organization. Usually, they are of a middle-level complex-ity. They correspond to *semi-structured problems*.

This typology of decisions by Simon is depicted in Fig. 2.1.

According to R. N. Anthony (1965), the different decision levels in an organiza-tion that usually cope with the managers can be classified in three groups:

- *Strategic Decisions*: These decisions are focussed on the strategical issues of the organization. These kinds of decisions are very rare and they happen just a few times in the life of an organization. These decisions are usually very complex and sometimes could involve several people in the decisions. For example, the decision whether a company must launch an international expansion to foreign markets.

Type of Decisions

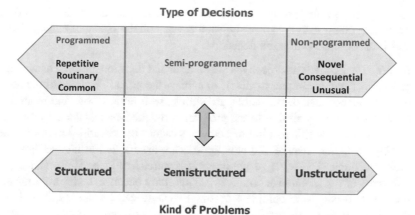

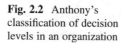

Kind of Problems

Fig. 2.1 H.A. Simon's classification of decision structure and associated problems

Fig. 2.2 Anthony's
classification of decision
levels in an organization

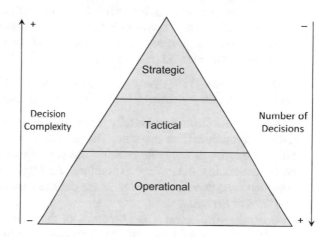

- *Tactical* or *Management Decisions*: These decisions are more related to the effective management and deployment of the assets of the organization. This kind of decision is not so frequent like the operational decisions but are more complex. For instance, some decisions about the production of a new product, and also the decision about what price should be the optimum to guarantee the maximum revenue for the organization.
- *Operational Decisions*: These are decisions appearing in the day-to-day operational activities of an organization. Usually are not very complex decisions. For instance, weekly decisions about the stock control, daily decisions about the production planning, etc.

This classification of this kind of decision can be depicted as in Fig. 2.2 with a triangle form. This means that as we move to the top of the triangle, the complexity of the decisions is higher but also the number of decisions of that type is decreasing.

A. L. Delbecq (1967) proposed another classification of decisions. He proposed the following classification in three types, focussing on the aspect of *negotiation*, and how is it undertaken within the decisions:

- *Routine Decisions*: The objectives are clear for the decision-makers, as well as the procedures and techniques available to achieve the goals. The level of negotiation is very low because the decisions are simple, well understood, and repetitive in time. They are very similar to the programmed decisions proposed by Simon.
- *Creative Decisions*: This kind of decisions cannot be solved by routinely methodologies. Other creative and new approaches are needed to handle these complex decisions. The level of negotiation required could be high. Usually, the outcomes of the possible alternatives are uncertain because the information about the scenario can be uncomplete or some strategy can be missed. This kind of decisions could be thought of as related to the non-programmed decisions or semi-structured decisions proposed by Simon.
- *Negotiated decisions*: Usually in this kind of decision, there are conflicts on the goals or in the strategies for the decision scenario. In addition, the manager is not the only decision-maker involved in the decision process. Thus, the final choice is obtained through a great negotiation effort made by a group of decisors to try to solve main differences in their criteria.

J. D. Thompson (1967) proposed a classification of four types of decisions according to the primary strategy used in making the final choice:

- *Computational Strategies*: Usually, the outcomes of the states of the world are rather deterministic than uncertain. The problem is well structured and the cause/effect relationships are quite certain. The preferences for some outcomes over others are clearly determined. Therefore, the final choice can be made using *computer programmed strategies*. These decisions are equivalent to programmed decisions of Simon.
- *Compromise Strategies*: The outcomes of the states of the world are rather deterministic and the cause/effect relationships are highly certain. The preferences for some possible alternatives are weak or unclear. The final choice must be agreed through some *compromise strategies* among the different alternatives, which probably are a little bit overlapping and contradictory. These decisions correspond to the semi-programmed decisions of Simon.
- *Judgemental Strategies*: The outcomes of the alternatives are highly uncertain. The problems are not well structured and the cause/effect relationships are also highly uncertain. Notwithstanding, the preferences for some possible alternatives are quite strong. Thus, the final choice must be obtained after some *judgemental strategies* have been done. These decisions are similar but more complex than the compromise strategy decisions and are within the semi-programmed decisions of Simon too, but more sliding to the right end, near to non-programmed decisions.
- *Inspirational Strategies*: these decisions are the hardest ones. The outcomes of the alternatives are highly uncertain, the problems are very bad structured, and the cause/effect relationships are highly uncertain. The preferences for some possible alternatives are weak or unclear. These decisions are very complex due to the high

level of uncertainty on all the elements of the decision process. The final choice can be obtained through some *inspirational* or *experiential-based strategy* from the decision-maker. These decisions correspond to the non-programmed decisions of Simon.

Henry Mintzberg (1973) proposed a new typology of decisions, paying attention to the *main activities* related to the decision:

- *Adaptive Activities*: These kinds of decisions usually have high levels of uncertainty. They are produced many times as the result *of reactive* activities to problems that happen regularly in the organization *and commonly are centred on the short-term* issues. They can be assimilated to operational decisions.
- *Entrepreneurial Activities*: These types of decisions are generated by several *entrepreneurial* activities in the organization, which are characterized also by high levels of uncertainty. The selection of the alternatives and the final choice is the result of *proactive considerations*, and focussed more on *near-term growth* over *long-term issues*. They can be considered as similar to management or tactical decisions.
- *Planning Activities*: In this kind of decisions, usually there are high-risk levels, and the decisions usually are generated by planning activities using both proactive and reactive thinking. The focus is on the *long-term growth* and efficiency. They can be thought as similar to strategic decisions.

All the different typologies described above match with a three-level classification of decisions and the corresponding complexity of the problems. In Fig. 2.3, there is an integration of the several types of decisions, including the main characteristics of the decisions. As soon as we move from left to right, the complexity of the decisions is increasing, in a continuum space of complexity. The amount of support needed for each kind of decision clearly depends on how the decision is positioned in this chart. As we move to the right, each time more support is needed. Therefore, the realm of *Intelligent Decision Support Systems* is located between the middle and rightmost decision group, which are the most complex and difficult decisions that will need a lot of models, expertise, knowledge, information, experience to be able to efficiently support the users to make the best choice.

2.3   Decision Theory

2.3.1   Origins of Decision Theory

Decision theory (Steele and Stefánsson, 2016) is not a unified field, and several approaches and theories have nurtured the field. The general term *Decision Theory* includes the standard decision-making analysis of a single person (*Individual decision-making Theory*), the decision-making analysis of a group of people (*Collective decision-making Theory* or *Social Choice Theory*) and the decision-making

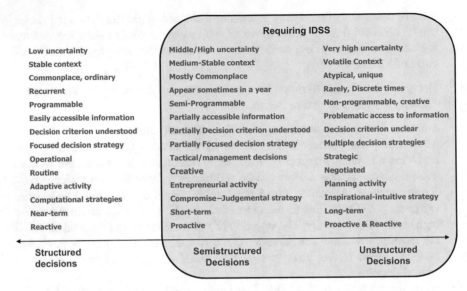

Requiring IDSS		
Low uncertainty	Middle/High uncertainty	Very high uncertainty
Stable context	Medium-Stable context	Volatile Context
Commonplace, ordinary	Mostly Commonplace	Atypical, unique
Recurrent	Appear sometimes in a year	Rarely, Discrete times
Programmable	Semi-Programmable	Non-programmable, creative
Easily accessible information	Partially accessible information	Problematic access to information
Decision criterion understood	Partially Decision criterion understood	Decision criterion unclear
Focused decision strategy	Partially Focused decision strategy	Multiple decision strategies
Operational	Tactical/management decisions	Strategic
Routine	Creative	Negotiated
Adaptive activity	Entrepreneurial activity	Planning activity
Computational strategies	Compromise–Judgemental strategy	Inspirational-intuitive strategy
Near-term	Short-term	Long-term
Reactive	Proactive	Proactive & Reactive
Structured decisions	Semistructured Decisions	Unstructured Decisions

Fig. 2.3 Continuum of the different types of decisions and localization of IDSS application area

process where the decisor must take into account the decisions of other people, which are influencing his/her own decisions (*Game Theory*). Martin Peterson, in (Peterson, 2009) distinguish three major periods in the history of decision theory:

- *The Old Period*: The foundations of decision theory have arisen since ancient Greek civilization with Greeks philosophers, which proposed some definitions regarding *right* or *rational acts* (Herodotus) or seeding the origins of the *logic of rational preferences* (Aristotle).
- *The Pioneering Period*: It started in 1654 with the origin and foundations of *probability theory*, which can be dated by the exchange of letters between Blaise Pascal and Pierre de Fermat over *games of chance*. Other outstanding events in the period were:

  - The publication of pioneering books in *expectation concept and expectation values* by Christian Huygens, the famous astronomer and physicist, in 1657: *De ratiociniis in ludo aleae* (Huygens, 1657).
  - The publication of another pioneering book in *principles of decisions* (*maximising the expected value*) by Antoine Arnauld and Pierre Nicole in 1662, in anonymous way, because they were Jansenist members of the Catholic Church: *La Logique, ou l'art de penser*, book known as *Port-Royal Logic* (Arnauld and Nicole, 1662).
  - Finally, Daniel Bernoulli, in 1738, introduced the notion of *moral value*, corresponding to the modern term of *utility*, which measures how good or bad is an outcome from the point of view of the decision-maker.

- *The Axiomatic Period*: It started at 1926, where *modern decision theory* emerged and was targeted to *axiomatize the principles of rational decision-making*. The outstanding events of the period were:

  - The publication of the article *Truth and Probability* by philosopher Frank Ramsey, written in 1926, but published in 1931 (Ramsey, 1926), where he proposed a set of eight axioms for *how rational decision-makers should choose among uncertain actions*.
  - Moreover, the publication of the book *Theory of Games and Economic Behavior* by von Neumann and Morgenstern, in 1944, and especially the second edition in 1947 (von Neumann and Morgenstern, 1944), where they wrote about *individual decision-making under risk*, coined the term *utility* for the same concept of *moral value* from Bernoulli, and presented their work on *game theory*. In the second edition, they proposed a set of axioms (*maximizing expected utility*) for how rational decision-makers should choose among lotteries.[1]
  - Finally, in 1954, Leonard Savage published the book *The Foundations of Statistics* (Savage, 1954), where another axiomatic analysis of the principle of maximising expected utility appeared.

In addition to these three periods, we want to outline an important period for the collective decision-making process, usually named as *Social Choice Theory* or *Social Decision Theory*. This period could be named as *The origins of Social Choice*.

- *The Origins of Social Choice Period*: It would start after *the old period*, in the Middle Ages until the French revolution period in 1789. It starts with the works of the lay philosopher and theologian Ramon Llull, born in Mallorca (1232–1316). Llull's contributions have been completely ignored until the last decade of the twentieth century by *Social Decision Theory* literature, probably due to the fact of his great number of written works in diverse languages: His native Catalan language, Latin, and Arabic, and some of them where rediscovered in 1937 (Honecker, 1937) and disseminated since 2001 (Hägele and Pukelsheim, 2001). He anticipated important work on *election theory* and he can be considered the pioneer of *voting theory* and *social choice theory* (Colomer, 2011), and he wrote, at least, two seminal works on voting and elections: *Artificium electionis personarum*, i.e. the method for the elections of persons (Llull, 1274), and *De Arte electionis*, i.e. on the method of elections (Llull, 1299). Both texts translated in English are available at (Hägele and Pukelsheim, 2001) and (Drton et al., 2004). Major outstanding contributions in this period were:

  - The appearance of *Artificium electionis personarum*, (Llull, 1274) which was Llull's basic proposal. It was a system based on the *binary comparisons of candidates*, where the winner is the candidate winning by the majority in the greatest number of comparisons. This system turns out to be more effective in

---

[1] A lottery is a probabilistic mixture of outcomes, according to von Neumann and Morgenstern.

preventing cycles and producing a winner than the well-known *Condorcet system,* which requires the winner to win all binary comparisons, for elections with five or more candidates.

The North American mathematician Arthur H. Copeland rediscovered this system in the mid-twentieth century (Copeland, 1951), with the very minor difference that Llull proposed to sum 1 point for each candidate in the voting when there is a tie, and Copeland proposed to sum half a point for each one.

- The writing of *De Arte electionis* (Llull, 1299), where he presented a variant of the voting system able to avoid ties. This system was a *non-exhaustive binary comparison system* by which after every round of voting the loser candidate is eliminated, the winner by majority is then compared with another candidate, and so on, and the elected is the candidate winning the last comparison. This system for four o less candidates can prevent ties and be more efficient than the initial Llull's system, and find a winner when the *Condorcet system* cannot. However, for a higher number of candidates, the system can produce different winners, depending on the order than the candidates are compared.
- Nicholas of Cusa (1401–1464), proposed the *rank-order count system* which requires the voters to rank all the candidates and give them ordinal points (0, 1, 2, etc.) from the least to the most preferred, being the winner, the candidate obtaining more points. Jean Charles de Borda (1733–1799) rediscovered this system during the French revolution (1789).
- Finally, this period ends with the proposal of Jean Antoine Nicolas de Caritat, marquis of Condorcet (1743–1794), who proposed a very similar system (*Condorcet system*) to Llull's first proposal of selecting the candidate winning all binary pairwise comparisons. His proposal was done during the French revolution (1789).

2.3.2 Modern Decision Theory

Modern Decision theory is an interdisciplinary subject influenced by several disciplines as Economy, Philosophy, Statistics, Psychology, Social Sciences, Political Sciences, and Computer Science. Researchers coming from these different disciplines are more interested in slightly different aspects of the decision process but the addition of all researchers' efforts have developed Decision Theory as it is.

As said above, Decision Theory is grouping several subfields in the area. The major subfield is *Individual Decision-making Theory*, which is the main field of study. Here, and in the rest of the book, we will focus on this subfield and by default when we will talk about Decision Theory, we will mean Individual Decision-Making Theory, if it is not differently specified.

Usually in Decision Theory, there is the distinction between *Normative Decision Theories* and *Descriptive Decision Theories*. Normative Decision Theories aim at providing norms about how the decisor *should rationally make* the decisions (Hansson, 1994). Descriptive Decision Theories aim at analyzing and explaining how people *actually make* the decisions. In this chapter, we will focus on Normative Decision Theories.

Probably, the first general theory of how to model the decision process, within the context of social choice theory, was proposed by Condorcet, because of his participation in the elaboration of the French Constitution of 1793. He proposed (Condorcet, 1847) that the decision process should be divided in three steps:

1. In the first step, each person should individually analyze and discuss the general principles, which will constitute the basis for the decision. Each one should analyze the different aspects and consequences of the different alternatives to make the decision.
2. In the second step, there should be a discussion among all decisors to identify the different opinions and try to get a consensus on them to get a small number of general opinions. This way, the set of alternatives is reduced to a manageable number.
3. In the final step, the actual choice among the alternatives must be done.

In the next subsection, several models proposed in the literature will be analyzed.

2.3.2.1 Analyzing the Decision Process

In the twentieth century, several researchers in the field have proposed different modern theories. As the decision process flows along time, it seems quite natural that the process should be modelled as a sequential process of different phases.

Sequential Models

The most influential models proposed in the literature are the models proposed by Dewey, Simon, and Brim. Dewey (1910) proposed in 1910 that five consecutive steps should compose the decision process, as follows:

1. The detection problem or difficulty
2. The definition of the character of that problem or difficulty
3. The suggestion of possible solutions
4. The evaluation of the suggestions
5. Further observation and experimentation leading to the acceptance or rejection of the suggestions.

Simon (1960) proposal was simpler to make it more adequate for the context of decisions in Management Science. His pioneering work research in the decision-making process within economic organizations, for which he received the Nobel Prize in 1978, contributed to establish both the field of Decision Support Systems on the one hand, and Artificial Intelligence field, in collaboration with Allan Newell, on the other, as in their seminal work on a *General Problem Solver* (Newell and Simon, 1961). According to Simon, decision-making consists of three principal phases:

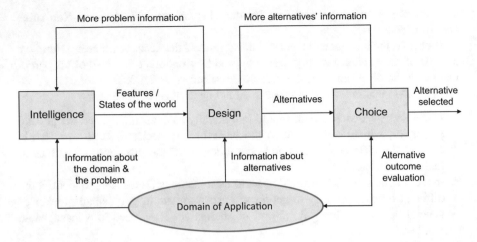

Fig. 2.4 Decision making model by H. Simon

1. *Intelligence*[2] Phase: At this level, there is the search for information and knowledge, in order to *identify a problem* requiring a decision-making task.
2. *Design* Phase: At this stage, the elicitation and building of *possible alternatives* or strategies to solve the detected problem is done.
3. *Choice* Phase: The decisor evaluates the possible alternatives and *chooses one* of them analyzed and generated in the previous step. The selected alternative satisfies the rational behaviour criterion. This means the selected alternative is the best one having in mind the fact that it provides the decisor with the maximum utility or benefit.

A representation of the model is depicted in Fig. 2.4.

Brim Jr. (1962) proposed another outstanding model of the decision process. He proposed to split the decision process into the following five steps:

1. Identification of the problem
2. Obtaining the necessary information
3. Production of possible solutions
4. Evaluation of such solutions
5. Selection of a strategy for performance

It could be easily checked that the steps 1 and 2 of Dewey's and Brim's proposal match the intelligence phase of Simon's proposal. In addition, both third step is the same than the design phase of Simon, and finally, the steps 4 and 5 of Dewey's and Brim's proposal match the choice phase of Simon's proposal. Thus, clearly, this

---

[2]Herbert Simon borrowed the military meaning of *intelligence* for naming the first phase, as he wrote in his book (Simon, 1960).

alignment of stages in main sequential models indicates that all of them are quite reasonable.

Non-sequential Model

Although the previous sequential models of the decision process are reasonable, one could think that there could be some kind of interaction among the different parts, and that possibly the different stages could happen in a different order. The most well-known model was proposed by Mintzberg, Riasinghani, and Théorêt in (Mintzberg et al., 1976). They state that the decision process is composed also by the same major phases than Simon's proposal, but they do not have a sequential relationship. The three phases according to them are:

1. *Identification* phase, which groups two tasks or routines:

    (a) *Decision recognition*, where the problem and the opportunities are identified from the large verbal data that decision-makers receive.
    (b) *Diagnosis*, which consists in using the available information and knowledge to clarify and define the issues.

2. *Development* phase, which comprises the definition and clarification of the options. Also, it includes two routines:

    (a) *Search* routine aiming at finding ready-made solutions
    (b) *Design* routine which aims at developing new solutions or modifying the ones generated by the previous task

3. *Selection* phase, which wrap three routines:

    (a) *Screen* routine is only activated when search is expected to generate more ready-made alternatives than can be intensively evaluated. In this routine, the suboptimal alternatives are eliminated.
    (b) *Evaluation-choice* routine implements the actual choice among the alternatives, using different evaluation schemes.
    (c) *Authorization* routine aims to obtain the approval for the selected alternative higher in the hierarchy.

The relationship among these routines is rather circular than sequential. The decision-maker could cycle and go back to other previous routines. In fact, this property of going back is also outlined in the Simon's proposal, as depicted in Fig. 2.1. Anyway, this model can also be clearly aligned with the Simon's model.

2.3.2.2   Managing the Decision Process

We will adopt the Simon's model, as the generic model for the decision process, from now on, due to the observation that main other models can be satisfactorily aligned with the three phases of Simon's model, and hence, are rather equivalent. Simon explained in (Simon, 1960) that executives spend most of the time in the intelligence and design phases rather than in the choice phase. This fact is a little bit surprising because the decision theory literature has been greatly focussed on the *choice* phase. Probably, it is because the choice phase is the one that can be mathematically modelled, and on the other hand, its result, the final choice, is the important output of the decision process. In the next subsections, the three phases will be analyzed. Of course, all three stages are very important to get a rational, and if possible, right final decision, but indeed, from a complexity point of view, the choice phase is really the more complex.

Facing the Intelligence Phase

At this first stage, the decision-maker should carefully analyze the problem or domain at hand. He/she must try to identify the decision or decisions involved in the problem at hand. For each decision, the *objectives* or issues must be clearly stated. In addition, the different relevant features, which can influence the decision process, should be searched and elicited. These features could be of different nature: *deterministic* or *stochastic*. Usually these features, especially in the case of stochastic nature are known as the *states of the world*, which reflects the effects of what happens outside the control of the decision-maker or what other persons decide.

For instance, let us suppose that one person wants to buy a new car. The problem is that in the market, there are many cars available, and then a decision should be made regarding *which car to buy*. Let us suppose that *main objective* is that the new car should be *a "good" c*ar, from the subjective point of the decisor. From a further analysis, let us suppose that a "good" car property can be decomposed on several dimensions or features: the elected car should have a price as low as possible, a high performance (velocity, acceleration, etc.), having no mechanic problems and having no electronic/electrical problems after buying it. At this point, the decision-maker can split the features into two types:

- Deterministic Features: good design, low price, high performance.
- Stochastic Features: no mechanic problems and no electrical/electronic problems.

Facing the Design Phase

In the second stage of the decision-making process, the set of possible *alternative options* for the actual decision must be enumerated. This step is a synthetic task, where all possible options must be considered. This step is a formalization step where all elements in the scenario must be outlined:

- The set of possible alternatives $A = \{a_1, a_2, \ldots, a_n\}$
- The set of *deterministic features* (p features) and/or *stochastic features* (m-p features), usually named as the states of the world $S = \{s_1, \ldots, s_p, s_{p+1}, \ldots, s_m\}$.

Table 2.1 General scheme of a decision matrix

	s_1	...	s_p	s_{p+1}	...	s_m
a_1	o_{11}	...	o_{1p}	o_{1p+1}	...	o_{1m}
a_2	o_{21}	...	o_{2p}	o_{2p+1}	...	o_{2m}
...	...	...	...	...	...	...
a_n	o_{n1}	...	o_{np}	o_{np+1}	...	o_{nm}

- The *outcomes* of each state of the world for each alternative (a_i), $O_i = \{o_{i1}, \ldots, o_{ip}, \ldots, o_{im}\}$.

Outcomes can be expressed in different ways, and this matter is an important topic in decision theory. They can be described in a binary way {high, not-high}, in a set of ordered qualitative values like {low, medium, high}, in a textual representation or in a numerical representation.

This formalization could be done in different ways like a decision matrix, a collection of sets or vectors, a decision tree, etc. Usually, in the decision theory literature, the formalization has been done using a *decision matrix*. Decision trees, as it will be shown in Sect. 2.4, have been commonly used for modelling a complex scenario with several decisions. Therefore, using a *decision matrix*, the result of this step is a matrix like the one depicted in Table 2.1:

In this matrix, each row (a_i) corresponds to each alternative or action or act, and each column (s_j) to a corresponding state of the world or combination of states. For each alternative and a given state, the outcome is specified. In our previous example, the obtained decision matrix for the problem could be such as the one described in Table 2.2.

Let us suppose that five possible alternatives have been formulated: *to buy car A, to buy car B, to buy car C, to buy car D and to buy car E.* Some features or states are deterministic like *price* and *performance.* The other ones are stochastic, like *mechanical problems* and *electrical problems,* which can happen or not when using the car. At each cell of the matrix, there is the corresponding outcome generated by the state corresponding to that column, for the alternative of that row. For instance, the alternative of *buying a car A* would get a not-high outcome for the price state, a low outcome for the performance state, a completely satisfied outcome for the no mech. Probs. and no elect. Probs. state, etc.

Facing the Choice Phase

The last step of the decision process aims at evaluating the different alternatives available to the decision-maker, and selecting the alternative producing the best outcome. The question is how to evaluate the best outcome for the decision-maker? This best outcome should be evaluated according to the objectives of the decision-maker.

In decision theory, there are some approaches to measure the degree of desirability of one alternative. One of them is based on the use of *preferences.* A preference of

Table 2.2 Decision matrix for the which car to buy decision problem

	Price	Performance	No mech. probs & no electr. probs	No mech. probs & electr. probs	Mech. probs & no electr. probs	Mech. probs & electr. probs
Buy car A	Not-high	Low	Completely satisfied	Partially satisfied	Partially satisfied	Partially unsatisfied
Buy car B	High	High	Completely satisfied	Partially unsatisfied	Partially unsatisfied	Completely unsatisfied
Buy car C	High	Medium	Completely satisfied	Partially unsatisfied	Partially unsatisfied	Completely unsatisfied
Buy car D	Not-high	Low	Completely satisfied	Partially satisfied	Partially satisfied	Partially unsatisfied
Buy car E	Not-high	Medium	Completely satisfied	Partially satisfied	Partially satisfied	Partially unsatisfied

one alternative over another one means the decision-maker perceives the former as producing a more beneficious or desirable outcome than the later, according to his/her objectives. Thus, it is a comparative method, which express a relation between two alternatives. The idea is that the decision-maker makes comparisons between any two alternatives, with the hope that a *preference ordering* could be established among all alternatives. Let us make some definitions:

Weak Preference Definition $A \succcurlyeq B$ defines a *weak preference* relation, which means that A is "at least as good as" B.

Therefore, $A \succcurlyeq B$ represents that the decision-maker considers option B is not preferred to A.

Strict/Strong Preference Definition From the weak preference relation, it can be defined the *strict preference* relation, $A \succ B$, which means that A "is better than" B, as follows:

$$A \succ B \Leftrightarrow A \succcurlyeq B \text{ and } \neg(B \succcurlyeq A)$$

Indifference Definition The *indifference* relation, $\sim$, which means that A and B are equally preferable, is defined as:

$$A \sim B \Leftrightarrow A \succcurlyeq B \text{ and } B \succcurlyeq A$$

It can be stated that the *weakly preference relation* $\succcurlyeq$, *weakly orders* a set S of options, *without any cycle of preferences*, whenever it satisfies the following two conditions:

Completeness Axiom For any $A, B \in S \Rightarrow A \succcurlyeq B$ or $B \succcurlyeq A$ or $A \sim B$
It means that all options are comparable.

Transitivity Axiom For any $A, B, C \in S \Rightarrow$ if $A \succcurlyeq B$ and $B \succcurlyeq C$ then $A \succcurlyeq C$

Therefore, if we have a weakly ordered set of alternatives, then a *preference ordering* could be made explicit, and then, the *most preferred alternative,* which is as good as all the others should be the *rational choice*.

Let us suppose that in our previous example about which car to buy, the decision-maker states these set of preferences:

to buy car E $\succcurlyeq$ to buy car D
to buy car E $\succcurlyeq$ to buy car C
to buy car E $\succcurlyeq$ to buy car B
to buy car E $\succcurlyeq$ to buy car A
to buy car C $\succcurlyeq$ to buy car D
to buy car B $\succcurlyeq$ to buy car D
to buy car A $\sim$ to buy car D
to buy car B $\succcurlyeq$ to buy car C
to buy car C $\succcurlyeq$ to buy car A
to buy car B $\succcurlyeq$ to buy car A

After a detailed analysis, it can be outlined that all pair of preferences can be compared, and that the preferences satisfy the transitivity property. Hence, a *preference ordering* can be drawn:

to buy car E $\succcurlyeq$ to buy car B $\succcurlyeq$ to buy car C $\succcurlyeq$ to buy car A $\sim$ to buy car D

Therefore, the selected choice, which is the alternative preferred to all the other ones, is the alternative "*to buy car E*".

Unfortunately, these two assumptions of *completeness* and *transitivity*, not always are satisfied, and then, does not exist one alternative being preferred to all the other ones. This situation can happen when some *preference cycle* appears, and no alternative is preferred to all the other ones. Sometimes this can happen due to the complexity of the alternatives, which are composed of many states or features, and it is difficult for the decision-maker to evaluate the alternatives and to state the preferences.

To make easier the comparison among alternatives, an approach that appeared in decision theory is to use a *numerical representation* of the values of the alternatives, i.e. of the values of the outcomes. These numbers are commonly named as *utilities*. In fact, a utility value is the transformation of an outcome to a numerical value. This assignation is the result of the application of a *utility function* to an outcome.

A *utility function* (u) can be defined as the following mathematical function:

$$u : O \rightarrow \mathbb{R}$$

$$o_i \mapsto u(o_i)$$

where O is the set of all outcomes,

o_i is a particular outcome, and

$u(o_i)$ is the utility value of outcome o_i.

The numerical values can be of two main types: *ordinal utilities* and *cardinal utilities*. The numerical representation of outcomes (utilities) is especially suitable in Management and Business Sciences, where the goodness or badness of some alternatives is usually expressed in monetary terms, indicating the *cost* or *benefit* of the alternatives. In addition, the use of numbers, allows using mathematical computations associated to probability theory and to additive or multiplicative aggregation functions of utilities, as it will be explained in Sect. 2.4.

What ensure that *ordinal utility functions* can be used for selecting the best alternative are the following two conditions:

Ordinal Representation Theorem Let S be a finite set of alternatives, and $\succcurlyeq$ a weak preference relation on S. Then, if the *weak preference relation $\succcurlyeq$ is complete and transitive*, then there is an *ordinal utility function* representing $\succcurlyeq$.

Representability of a Weak Preference Order Condition Any *weak preference order* among the options of a set of alternatives S can be represented by an *ordinal utility function (u)*, if and only if:

$$\text{For any } A, \ B \in S \Rightarrow A \succcurlyeq B \Leftrightarrow u(A) \geq u(B)$$

Thus, it can be concluded that if we have *a weak preference relation $\succcurlyeq$* which is *complete* and *transitive*, there is an *ordinal utility function representing the relation $\succcurlyeq$*. Moreover, as a weak preference order can be obtained, the utility function searched should verify the representability condition described above. Therefore, the alternative with the *highest value of the utility function* should be the rational choice. Thus, when working with *ordinal utility functions*, the choice *maximizing the utility function* is selected.

For instance, the same example regarding the different alternatives for *buying of a car* could be formalized again *using numerical measures* as the utility values for each outcome. See the Table 2.3.

The outcomes of *price state* are numerical values expressed in an *ordinal scale*: $2 > 1$. Thus, according to this state or feature, the alternatives "*buy a car A*", "*buy a car D*", and "*buy a car E*" would be the best and alternatives "*buy car B*" and "*buy car C*" would be the worst. The *ordinal utility* values of *price state* indicate than the value 2 is higher than value 1, but it is just an ordinal measure. We do not know if the worse alternatives are very near to the best or not, we just know the ordering. The other features are expressed with *cardinal utility* values, which will be explained below.

Ordinal scales have the inconvenient that it is not known how distant one utility value of one alternative is from the other. The number just express the positioning of the alternative according to the utility values. Think in some contest in a university for covering a position of full professor where an Evaluation Committee, after evaluating the curriculum of all the candidates in the contest, finally outputs the

Table 2.3 Decision matrix for the which car to buy decision problem using numerical values

	Price	Performance	no mech. probs & no electr. probs	no mech. probs & electr. probs	mech. probs & no electr. probs	mech. probs & electr. probs
Buy car A	2	3	+50	+25	+25	−25
Buy car B	1	10	+50	−25	−25	−50
Buy car C	1	7	+50	−25	−25	−50
Buy car D	2	3	+50	+25	+25	−25
Buy car E	2	7	+50	+25	+25	−25

ordered names of three people. The first one in the scale is the proposed candidate; the second one and the third one are the other best candidates, following the winner. However, you do not know if really the second one is almost as good as the proposed candidate or not, or if really the third one was much worse than the second one or not. In addition, the requirement that the weak preference relation should be complete and transitive is difficult to be satisfied in most of decision scenarios.

To solve this problem, *cardinal utility measures* can be used. Using a cardinal scale, the utilities express a numerical magnitude value, which is comparable among the different utilities of the different alternatives. Cardinal scales can be *interval scales* or *ratio scales*. Utilities in an *interval scale* express values which can be compared through a *difference operation*. For example, a grading system with a numerical scale between 0 and 10, being a 0 the worst grade and 10 the best grade. With this scale, if we have the scenario as depicted in Table 2.4.

Comparing Student 3 with a grade of 4 and Student 2 with a grade of 9, we can compute the difference between them, 9–4 = 5, and this difference is meaningful because the grade values 4 and 9 are really quite different. The 4 would be a grade corresponding to failing grade, and a 9 would be an excellent grade.

Utilities in a *ratio scale* express value, which can be compared with a *division operator*. For instance, in a weight ratio scale of some packages, where the reference value is 2 kg, the ratio scale can include several values like ¼, ½, 1, 1.5, 2, 5, 10, etc. as shown in the Table 2.5 (the last column with the absolute weight is added for better understanding the ratio scale utility).

These values mean that if we have the value 0.5, which means that this value is ½ times the reference value of 2 kg, i.e. 1 kg, the value 1.5 means that this value is 1.5 times the reference value, i.e. 3 kg, and so on. Although the utilities are not expressed in absolute values, the values can be compared through a *division (ratio) operation*.

Table 2.4 Using a cardinal
interval-value scale for a
grading decision scenario

	Grade
Student 1	7
Student 2	9
Student 3	4
Student 4	3
Student 5	8

Table 2.5 Using a cardinal
ratio scale for a package
weighting decision scenario

	Weight (ratio)	Absolute Weight (kg)
Package 1	5	10
Package 2	1.5	3
Package 3	½	1
Package 4	1	2
Package 5	2	4
Package 6	¼	0.5
Package 7	10	20

Thus, the package 7, which has a value of 10 if compared with the package 5, which
has a value of 2, gets the value of 10 / 2 = 5, meaning that the package 7 is 5 times
heavier than the package 5. Also, the package 7, if compared with the package
3, which has a value of ½, gets the value of $10/_{1/2} = 20$, meaning that the package 7 is
20 times heavier than the package 3.

Until now, we have not provided a general procedure for selecting the rational
choice, when several states or features should be compared, as in the case of the
"which car to buy" scenario. We have supposed until now that only one state or
feature is describing the decision scenario. Of course, in the practice, this simplification is not possible and most decision scenarios must take into account several
states of the world, ones being deterministic and the other ones being stochastic. This
general situation, which depends on the degree of stochasticity of the states, on the
existence of only deterministic features or not, etc., will be discussed in the next Sect.
2.4.

2.4   Decision Process Modelling

In this section, the problem of how to model and effectively make the final choice in
the decision process will be afforded from a practical point of view. We will
distinguish between a *single decision scenario* and a *multiple decisions' scenario*.

2.4.1 Single Decision Scenario

As seen in previous Sect. 2.3, we know how to formalize a decision scenario, and make the choice when we have a simple situation with just one state or feature: using *preferences, ordinal preferences,* or *cardinal preferences.* The *rational choice* is *the most preferred alternative* or *the alternative with a higher utility value.* As we have explained, in previous section, the use of numerical representation for the outcomes, i.e. *utilities,* and especially *cardinal utilities* seems to be the best option for solving the decision problems. Therefore, we will assume than all the outcomes of the alternatives, for each state in the different problems are expressed with cardinal utilities, preferably in an *interval-value scale.* In this section, we will analyze how to proceed to select the best alternative when in general, there are several states of the world influencing the decision process. Depending on the different nature of the states of the world, several scenarios can be distinguished. Mainly, one can differentiate between *decision-making under certainty* and *decision-making under no-certainty.* See Knight (1921) and Luce and Raiffa (1957).

1. *Decision-making under certainty* means that the *states of the world* are *deterministic,* and thus, the outcomes of a given alternative are invariably known.
2. *Decision-making under no-certainty* is the opposite case when *the states of the world are stochastic.* In this situation, a further classification can be made in the no-certainty situation: *decision-making under risk,* and *decision-making under uncertainty.*

   (a) *Decision-making under risk* happens when each alternative leads to the outcomes, and *each outcome is occurring with a known probability value.* The *states of the world are stochastic,* and they have an *associated probability distribution, which is known* by the decision-maker.
   (b) *Decision-making under uncertainty* or *ignorance* includes two situations:

   • *Decision-making under classical ignorance:* The situation when the alternatives and outcomes are known, the states are stochastic, and have an *associated probabilistic distribution, which is unknown* by the decision-maker.
   • *Decision-making under unknown consequences:* The situation when *there are unknown states and/or outcomes* by the decision-maker. Some consequences could be catastrophic.

2.4.1.1 Decision-Making Under Certainty

In this scenario, the outcomes of each state for a given alternative are deterministic. The outcomes are expressed in an interval-value cardinal utility scale. There are two possibilities: only one state or attribute describes the decision process (*single-attribute approach*) or several states or attributes are describing the decision process (*multi-attribute approach*).

Single-Attribute Approach

The rational choice is to choose *the alternative with the highest utility value for the attribute or state*. Hence,

$$a_i \text{ is the best alternative} \Leftrightarrow \forall a_j \in A \ u(a_i) \geq u(a_j)$$

Main problem is that usually, there are more than one state or attribute. Then, the unique way to apply this procedure is that the different states or attributes must be combined in a *unique state* and in a *common scale*. That means that the utility function should evaluate the desirability of the different states/attributes in a unique value, and possibly combining different cardinal scales, which could be difficult or even having not much sense.

For instance, considering again the "what car to buy" decision problem from Sect. 2.3, and taking just the two deterministic attributes (*price, performance*), and let us suppose that the outcomes are measured in an *interval-value scale* [1,3], for the price, and [0,10] for the *performance*, like depicted in Table 2.6(a). In addition, we consider the same problem but just with one attribute/state which has to be the *combination/summarization of the utility values of the two attributes*. Supposing that both attributes have the same importance, we could express the utility of the new unique attribute (*Price* and *Performance*) as the mean of the two values, like depicted in the first column of the Table 2.6(b). Of course, as the two attributes have different scales; to get a combined attribute could be done in a *rather intuitive* and difficult way integrating two values in different scales in a final value in one scale, or more elegantly, transforming the values of one scale into values on the other preserving the magnitudes (*scale transformation*). Here, we made a *scale transformation* to get the values of *price*, which are in the scale [1,3] into the same scale than *performance* ([0,10]). Thus, the *price* values [2,1,1,2,2] are transformed in

Table 2.6 Decision matrix for the which car to buy decision problem with two deterministic attributes (a) and with just one combined deterministic attribute (b)

(a)		
	Price	*Performance*
Buy car A	2	3
Buy car B	1	10
Buy car C	1	7
Buy car D	2	3
Buy car E	2	7

(b)		
	Price & Performance	*Price & Performance₂*
Buy car A	4	11/3
Buy car B	5	**20/3**
Buy car C	3.5	14/3
Buy car D	4	11/3
Buy car E	**6**	19/3

[5,0,0,5,5] in the same scale than *performance*. Afterwards, the average utility value is computed for the new unique attribute *Price* and*Performance*.

This way, considering the first column of the Table 2.6(b), and applying the criteria of selecting the alternative with the highest utility, we will select the alternative "*buy car E*" which have a utility of 6. In the case that, it was supposed that the importance of *price* is about 1/3 and the importance of *performance* is about 2/3, then the corresponding combined utilities would be the ones depicted in the second column of Table 2.6(b). Then, applying the same criteria, the alternative "*buy car B*" would be the candidate, because it has the highest utility value of 20/3. The highest utilities are marked in boldface in the Table 2.6(b).

From this simple example, can be inferred that if the decision problem at hand has several states, the strategy to combine them into just one attribute and one single scale, could be difficult, especially if you have many attributes to be combined. For those cases where there are several attributes, some researchers in the literature suggest the multi-attribute approach.

Multiple-Attribute Approach

The multiple-attribute approach has the advantage that the *different attributes/states can be compared in their own different scale*, which could make more sense, when the attributes and their scales are measuring very different things, like monetary units, subjective personal characteristics of people, etc. In this approach, each attribute can be measured in the most suitable numerical unit.

The *utility value of each alternative* is *the aggregation of the utilities of all the attributes*. Usually, this aggregation criterion can be classified in *additive criteria* and *non-additive criteria*.

The *additive approach* assigns a relevance factor (weight) to each attribute, and then compute the *aggregated utility value* of each alternative as the weighted sum of the utilities of each outcome of the attribute for the alternative. The weights are usually real values normalized in the interval [0,1], and their sum is up to one. More formally:

$$aggu(a_i) = \sum_{j=1}^{m} w_j * u(o_{ij})$$ (2.1)

$$\text{where } 0 \leq w_j \leq 1 \text{ and } \sum_{j=1}^{m} w_j = 1$$

Given that,
o_{ij} is the outcome for the attribute j in the alternative a_i
w_j is the relevance/weight of the attribute j
m is the number of attributes
$aggu(a_i)$ is the aggregated utility of the alternative a_i

Of course, a difficulty in this additive aggregation is the estimation of the weights, which should be made by the decision-maker.

Non-additive approaches do not compute the aggregated value of an alternative as a weighted sum of utility values, but instead they propose to use other mathematical functions. For instance, one that has been proposed in the literature of decision theory is to use the *multiplication of the utility values of each attribute*. Thus, it can be computed as follows:

$$aggu(a_i) = \prod_{j=1}^{m} u(o_{ij}) \tag{2.2}$$

This multiplicative aggregation penalizes the low values of the utilities of the attributes.

Coming back to the example of the decision problem of *"which car to buy"* described in Table 2.6(a), if we compute the aggregated utility values using an *additive approach* for all the alternatives, assuming an equal relevance situation (weights equal to ½) and a situation where the importance of *performance* is higher (2/3), we have:

$$aggu(a_1) = ½ * 2 + ½ * 3 = 5/2 \qquad aggu(a_1) = 1/3 * 2 + 2/3 * 3 = 8/3$$
$$aggu(a_2) = ½ * 1 + ½ * 10 = \mathbf{11/2} \quad aggu(a_2) = 1/3 * 1 + 2/3 * 10 = \mathbf{21/3}$$
$$aggu(a_3) = ½ * 1 + ½ * 7 = 8/2 \qquad aggu(a_3) = 1/3 * 1 + 2/3 * 7 = 15/3$$
$$aggu(a_4) = ½ * 2 + ½ * 3 = 5/2 \qquad aggu(a_4) = 1/3 * 2 + 2/3 * 3 = 8/3$$
$$aggu(a_5) = ½ * 2 + ½ * 7 = 9/2 \qquad aggu(a_5) = 1/3 * 2 + 2/3 * 7 = 16/3$$

Thus, the selected alternative would be the second one, *"buy car B"* in both scenarios. The best options are marked in boldface.

Otherwise, if we use a *multiplicative* approach to compute the aggregated utility values, we have:

$$aggu(a_1) = 2 * 3 = 6$$
$$aggu(a_2) = 1 * 10 = 10$$
$$aggu(a_3) = 1 * 7 = 7$$
$$aggu(a_4) = 2 * 3 = 6$$
$$aggu(a_5) = 2 * 7 = \mathbf{14}$$

According to this multiplicative approach, the best alternative now would be the fifth alternative *"buy car E"*, which was the second best one when using the additive approach. It can be seen that the low value of alternative a_2 in the price attribute has penalized it in front of the alternative a_5. The best option is marked in boldface.

2.4.1.2   Decision-Making Under Risk

Decision-making under risk is the scenario when the states of the world are stochastic, and the decision-maker knows the probabilities of the outcomes for the different alternatives. The outstanding method in decision theory to cope with this probabilistic situation is the *Expected Utility Theory*, formulated by von Neumann and Morgensten (1944, 1947). The concept of utility, and especially expected utility was an evolution of older concepts developed in mathematics in seventeenth century (Arnauld and Nicole 1662), but the term *expected utility*, was coined by von Neumann and Morgensten with this precise meaning.

The *Expected Utility*, as thought by von Neumann and Morgensten, was defined for a lottery, which is a probabilistic mixture of outcomes. Using lotteries, an interval-value utility measure over the alternatives can be constructed introducing lottery options. The basic idea is that the judgment about one alternative can be measured by the riskiness of a lottery, which have as prizes the other alternatives. When there is an indifference between an alternative and a lottery with other two alternatives given certain probabilities, the utility of the alternative can be computed.

Expected Utility Definition   Given, L_i that is a lottery from a set L of lotteries, O_{ik} is the outcome or prize of lottery L_i, p_{ik} is the probability of occurrence of outcome O_{ik}.

The expected utility of L_i is defined as follows:

$$EU(L_i) = \sum_{k=1}^{m} p_{ik} * u(O_{ik}) \tag{2.3}$$

von Neumann and Morgenstern Representation Theorem   If the preferences of an agent satisfy these principles: Transitivity, Completeness, Continuity, and Independence, then there is *an expected utility function* that represents the agent's preferences, and thus, the agent's preferences can be represented as *maximising expected utility*. Formally,

$$\text{For any } L_i, L_j \in L, \ L_i \succcurlyeq L_j \Leftrightarrow EU(L_i) \succcurlyeq EU(L_j)$$

The continuity axiom and independence axiom are described below, and the former two were explained in Sect, 2.3.2.2.

Continuity Axiom   Suppose that $A \succcurlyeq B \succcurlyeq C$, then
there is a $p \in [0, 1]$ such that $\{pA, (1 - p)C\} \sim B$
where $\{pA, (1 - p)C\}$ represents a lottery that results either in A, with probability p, or C, with probability $1 - p$.

Independence Axiom   Suppose $A \succcurlyeq B$, then
for any C, and any $p \in [0, 1] \Rightarrow \{pA, (1 - p)C\} \succcurlyeq \{pB, (1 - p)C\}$.

Table 2.7 Decision matrix for the *which car to buy* decision problem under risk

	no mech. probs & no electr. probs	no mech. probs & electr. probs	mech. probs & no electr. probs	mech. probs & electr. probs
Buy car A	+50	+25	+25	−25
Buy car B	+50	−25	−25	−50
Buy car C	+50	−25	−25	−50
Buy car D	+50	+25	+25	−25
Buy car E	+50	+25	+25	−25

Let us see an example of the application of the *principle of maximizing expected utility*. Let us consider the decision problem of "*what car to buy*" from Table 2.3, and let us just consider the stochastic states. Thus, our decision problem is defined in Table 2.7.

Let us assume that the probability of having mechanical problems is 0.5, the probability of no having mechanical problems is 0.5, the probability of having electrical problems is 0.5, and the probability of no having electrical problems is 0.5, and that they are independent. Thus, the probability of the intersection is the product of probabilities (0.25).[3] Let us compute de expected utility of all alternatives:

$$EU(a_1) = 0.25 * (+50) + 0.25 * (+25) + 0.25 * (+25) + 0.25 * (−25) = \mathbf{18.75}$$
$$EU(a_2) = 0.25 * (+50) + 0.25 * (−25) + 0.25 * (−25) + 0.25 * (−50) = −12.5$$
$$EU(a_3) = 0.25 * (+50) + 0.25 * (−25) + 0.25 * (−25) + 0.25 * (−50) = −12.5$$
$$EU(a_4) = 0.25 * (+50) + 0.25 * (+25) + 0.25 * (+25) + 0.25 * (−25) = \mathbf{18.75}$$
$$EU(a_5) = 0.25 * (+50) + 0.25 * (+25) + 0.25 * (+25) + 0.25 * (−25) = \mathbf{18.75}$$

Therefore, the alternatives maximizing the expected utility are a_1, a_4, and a_5. Thus, "*buy car A*", "*buy car D*", and "*buy car E*" are the alternatives maximizing the expected utility. The best options are marked in boldface.

Although the principle of maximize the expected utility has been deeply used in the practical application of decision theory in risk situations, several *controversial literature* and *paradoxes* (San Petersburg, Allais, Ellsberg, etc.), have appeared in the decision theory literature. In addition, some other concepts appeared in the decision theory literature like *subjective utility* and *subjective probability* concepts, which originated a subfield of decision theory called *Bayesian Decision Theory*. See further information in (Peterson, 2009).

---

[3] $p(A \cap B) = p(A) * p(B|A)$, but if A and B are independent, then $p(A \cap B) = p(A) * p(B)$, because $p(B|A) = p(B)$.

2.4.1.3   Decision-Making Under Uncertainty

As described above there are two different scenarios according to the uncertainty or ignorance situation. Both scenarios will be analyzed.

Decision-Making Under Classical Ignorance

In this situation, the decision-maker does not know the probability distribution of the states of the world considered in the decision problem. Therefore, the expected utility cannot be used for selecting the best choice. In the literature of decision theory, there are several decision criteria for making the choice of the best alternative. The most commonly accepted are the following:

Maximin Decision Rule   This decision rule is based on the idea of maximizing the minimal utility of each alternative. Thus, the minimal utility of each alternative should be computed, and from those ones, the maximum value is the one which identifies the best alternative. It is a *pessimistic criterion*, because it examines the worst cases of each alternative, and among them, it selects the best one. It considers how bad the course of action could be, and selects the best one among the worst expectations. Von Neumann proposed this criterion in adversarial game theory, but was popularized by Wald (Wald, 1950). It is commonly named as *Wald's criterion*.

Let us formalize the rule:

$$a_i \text{ is the best alternative} \Leftrightarrow \forall a_j \in A, \quad min\,(a_i) \geq min\,(a_j)$$

Given that,

$A = \{a_1, a_2, \ldots, a_n\}$ is the set of possible alternatives
$min(a_j)$ is the minimal utility/value of the alternative a_j

Let us suppose that we have the decision problem expressed in the following decision matrix of Table 2.8, with five alternatives and four states.

The best alternative according to the *maximin rule* is a_3, because the value 4 is the maximum of the minimum values of all alternatives $\{3, 2, 4, 3, 2\}$. The minimum values are in cursive font and the maximin value is in cursive and bold.

Leximin (Lexicographic Maximin) Decision Rule   This decision rule was originated to solve some not very rational behaviour of the *maximin rule* (Sen, 1970). When applying the *maximin rule* there is a tie between two or more alternatives, which have the same worst value, whatever of these alternatives can be selected as the best one, but usually it is not true that all are similarly preferable. According to *leximin rule*, when there is a tie among the worst value in some alternatives, the second worst value of those alternatives must be compared, and the maximum value should be selected. In the case that a new tie is produced, then the third-worst values should be compared and so on. Finally, the maximum value is the one, which

	s_1	s_2	s_3	s_4
a_1	6	4	*3*	9
a_2	10	2	5	6
a_3	*4*	8	7	12
a_4	5	4	9	*3*
a_5	8	11	5	*2*

Table 2.8 Applying the *maximin decision rule* to a 5-alternative and 4-state problem

identifies the best alternative. This criterion is also a pessimistic decision rule like the maximin criterion, but a little bit more reasonable.

Let us formalize the rule:

$$a_i \text{ is the best alternative} \Leftrightarrow \forall a_j \in A \Rightarrow \exists n \in \mathbb{Z}, n > 0, \ min^n(a_i) \geq min^n(a_j)$$
$$\text{and } \forall m \in \mathbb{Z}, 0 < m < n, \ min^m(a_i) = min^m(a_j)$$

Given that,

$A = \{a_1, a_2, \ldots, a_n\}$ is the set of possible alternatives
$min^1(a_j)$ is the worst utility/value of the alternative a_j
$min^2(a_j)$ is the second worst utility/value of the alternative a_j
$min^k(a_j)$ is the kth worst utility/value of the alternative a_j

Let us suppose that we have the decision problem expressed in the following decision matrix of Table 2.9, with 5 alternatives and 4 states.

The best alternative according to the leximin *rule* is a_4, because when comparing the worst utilities of all alternatives $\{3, 2, 3, 3, 2\}$ there is a tie among a_1, a_3, a_4. After comparing the second worst utilities of these three alternatives $\{4, 4, 4\}$ there is another tie among them. Finally, when comparing the third-worst utility/value $\{6, 8, 9\}$, the maximum value is 9, which corresponds to alternative a_4. The worst values are in cursive font, the second-worst values are in cursive an underlined and the leximin value is in cursive and bold.

Maximax Decision Rule This decision rule proposes to choose the best option as the alternative that maximizes the maximum utility values of all alternatives. Thus, this criterion is radically different from *maximin* and *leximin* criteria. It is completely optimistic, as it is focussing on the best possible outcomes of the alternatives, and selects the maximum one among the bests.

Let us formalize the *maximax rule*:

$$a_i \text{ is the best alternative} \Leftrightarrow \forall a_j \in A, \ max(a_i) \geq max(a_j)$$

Given that,

$A = \{a_1, a_2, \ldots, a_n\}$ is the set of possible alternatives
max(a_j) is the maximal utility/value of the alternative a_j.

Table 2.9 Applying the *leximin decision rule* to 5-alternative and 4-state problem

	s_1	s_2	s_3	s_4
a_1	6	4	3	10
a_2	10	2	5	6
a_3	3	8	4	12
a_4	11	4	9	3
a_5	8	11	5	2

Table 2.10 Applying the *maximax decision rule* with 5 alternatives and 4 states

	s_1	s_2	s_3	s_4
a_1	6	4	3	*9*
a_2	*10*	2	5	6
a_3	4	8	7	***12***
a_4	5	4	*9*	3
a_5	8	*11*	5	2

Let us suppose that we have the same decision problem than in Table 2.8 expressed in the decision matrix of Table 2.10, with 5 alternatives and 4 states.

The best alternative according to the *maximax rule* is a_3, because the value 12 is the maximum of the maximum values of all alternatives {9, 10, 12, 9, 11}. The maximum values are in cursive font and the maximax value is in cursive and bold.

Optimism-Pessimism Decision Rule or Alpha-Index Rule This decision was proposed by Hurwicz (Hurwicz, 1951) and it is commonly known also as *Hurwicz's criterion*. This proposal considers both the worst outcome and the best outcome of each alternative, and according to the degree of optimism and pessimism of the decision-maker, the best alternative is selected. The evaluation of each alternative a_i is done through a weighted formula of the best and worst values:

$$\alpha * max\,(a_i) + (1 - \alpha) * min\,(a_i) \qquad (2.4)$$

where $\alpha \in \mathbb{R}$, $0 \leq \alpha \leq 1$, represents the degree of optimism of the decision-maker, which should be determined for a concrete decision problem. As high optimism feels the decision-maker a higher value of α is set, and as pessimistic feels, a lower value of α is set. Of course, it is an intermediate criterion between the pessimistic maximin criterion and the optimistic maximax criterion:

if $\alpha = 1$, then what is maximized is $max(a_i)$ [*maximax criterion*]
if $\alpha = 0$, then what is maximized is $min(a_i)$ [*maximin criterion*]

Let us formalize the *optimism-pessimism rule*:

$$a_i \text{ is the best alternative} \Leftrightarrow \forall a_j \in A,$$

$$\alpha * max\,(a_i) + (1 - \alpha) * min\,(a_i) \geq \alpha * max\,(a_j) + (1 - \alpha) * min\,(a_j)$$

Given that,

Table 2.11 Applying the
*optimism-pessimism decision
rule* to 5-alternative and
4-state problem and the results
with two values of α *(0.8 and
0.2)*

	s_1	s_2	s_3	s_4	$\alpha = 0.8$	$\alpha = 0.2$
a_1	6	4	*3*	*10*	8.6	4.4
a_2	*10*	2	5	6	8.4	3.6
a_3	2	8	4	*12*	**10**	4
a_4	*11*	4	9	*3*	9.4	**4.6**
a_5	8	*11*	5	*2*	9.2	3.8

$A = \{a_1, a_2, \ldots, a_n\}$ is the set of possible alternatives
$\alpha \in \mathbb{R}, 0 \leq \alpha \leq 1$, expresses the degree of optimism/pessimism
max(a_j) is the maximal utility/value of the alternative a_j
min(a_j) is the minimal utility/value of the alternative a_j

Let us suppose that we have the decision problem expressed in the following
decision matrix of Table 2.11 with 5 alternatives and 4 states.

Let us to find out what happens with different settings of α parameter. If we
suppose that the decision-maker is a rather optimistic person, and sets the parameter
to 0.8 ($\alpha = 0.8$) and if the decision-maker is a rather pessimistic person and the
parameter is set to 0.2 ($\alpha = 0.2$):

The evaluation is the following:

$$\text{Evaluation of } a_1 : \quad (\alpha = 0.8) \quad 0.8 * 10 + 0.2 * 3 = 8.6$$
$$(\alpha = 0.2) \quad 0.2 * 10 + 0.8 * 3 = 4.4$$
$$\text{Evaluation of } a_2 : \quad (\alpha = 0.8) \quad 0.8 * 10 + 0.2 * 2 = 8.4$$
$$(\alpha = 0.2) \quad 0.2 * 10 + 0.8 * 2 = 3.6$$
$$\text{Evaluation of } \mathbf{a_3} : \quad (\boldsymbol{\alpha = 0.8}) \quad 0.8 * 12 + 0.2 * 2 = \mathbf{10}$$
$$(\alpha = 0.2) \quad 0.2 * 12 + 0.8 * 2 = 4$$
$$\text{Evaluation of } \mathbf{a_4} : \quad (\alpha = 0.8) \quad 0.8 * 11 + 0.2 * 3 = 9.4$$
$$(\boldsymbol{\alpha = 0.2}) \quad 0.2 * 11 + 0.8 * 3 = \mathbf{4.6}$$
$$\text{Evaluation of } a_5 : \quad (\alpha = 0.8) \quad 0.8 * 11 + 0.2 * 2 = 9.2$$
$$(\alpha = 0.2) \quad 0.2 * 11 + 0.8 * 2 = 3.8$$

The best alternative being optimistic ($\alpha = 0.8$) according to the *optimism-
pessimism rule* is a_3, because the value 10 is the maximum of the weighted values
of all alternatives {8.6, 8.4, **10**, 9.4, 9.2}. The best alternative being pessimistic
($\alpha = 0.2$) according to the *optimism-pessimism rule* is a_4, because the value 4.6 is the
maximum of the weighted values of all alternatives {4.4, 3.6, 4, **4.6**, 3.8}. The
maximum and minimum values are in cursive font and the *optimism-pessimism*
value is in bold in the Table 2.11.

Minimax Regret Rule Savage (Savage, 1951) proposed this decision rule. The
rationality of this criteria is related with the human feeling of *regret* that some people
experiments when after having made an action, for instance a choice, and given new

information available, starts to regret the action (choice) done. For instance, after deciding that you will not take your swimsuit for the next weekend, and while you are in your car going to the destination of the weekend, you get new information about the hotel you are staying: it has a beautiful swimming pool. That moment you start to regret not having taken your swimsuit with you! Moreover, you start to think that possibly your choice was not the best one. Savage and other researchers found that this concept could be relevant for making a rational decision. If your choice does not generate any regret or a minimal one, that is the best choice. Other alternatives generating much regret are not good alternatives. The idea of the *minimax regret criterion* is that the best alternative is the one minimizing the maximum amount of regret of the alternatives.

The usual procedure for computing which is the best alternative is generating a *regret matrix*. The values of the *regret matrix* are the result of subtracting the value of each outcome from the value of the best outcome of each state. This way the *regret values* (distance of the outcome to the best outcome of each state) are computed. The regret values are always negative or zero when are equal to the best outcome. Then, from the *regret matrix*, the maximum regret values for each alternative can be computed. These maximum regret values (in absolute value) are the lower values, because all are negative values. Finally, all these maximum regret values for each alternative are compared, and the minimum regret value (in absolute value), which is the high value, among them identifies the best alternative. Therefore, the outcomes must be expressed in a cardinal interval scale.

Let us formalize the rule:

$$a_i \text{ is the best alternative} \Leftrightarrow \forall a_j \in A,$$
$$min\{o_{i1} - max(s_1), \ldots, o_{im} - max(s_m)\} \geq$$
$$min\{o_{j1} - max(s_1), \ldots, o_{jm} - max(s_m)\}$$

Given that,

$A = \{a_1, a_2, \ldots, a_n\}$ is the set of possible alternatives
o_{ip} is the outcome of alternative a_i for the state of the world s_p
$max(s_p)$ is the maximal value of the state of the world s_p across all alternatives

Let us suppose that we have the decision problem expressed in the following decision matrix of Table 2.12 with 5 alternatives and 4 states.

We will compute the corresponding *regret matrix*. To that end, first the regret values must be computed. They are the difference between each outcome and the best outcome of each state. The best outcome of each state is marked in boldface {11, 11, 9, 12}. The resulting *regret matrix* is depicted in Table 2.13.

Then, in the *regret matrix*, we must compute for each alternative the maximum regret value (minimum value), and after all are computed for all alternatives, then the alternative with the minimum regret (maximum value) is the selected. The maximum regret values of each alternative are marked in bold {−7, −9, −9, −9, −10}. Then,

Table 2.12 Decision matrix
for applying the *minimax
regret rule* to a 5-alternative
and 4-state problem

	s_1	s_2	s_3	s_4
a_1	6	4	3	10
a_2	10	2	5	6
a_3	2	8	4	12
a_4	11	4	9	3
a_5	8	11	5	2

Table 2.13 *Regret matrix*
corresponding to the decision
matrix of Table 2.11

	s_1	s_2	s_3	s_4
a_1	−5	−7	−6	−2
a_2	−1	−9	−4	−6
a_3	−9	−3	−5	0
a_4	0	−7	0	−9
a_5	−3	0	−4	−10

the minimum regret is −7, which corresponds to alternative a_1, which is the
selected one.

The Principle of Insufficient Reason Decision Rule Jacques Bernoulli
(1654–1705) formulated this criterion. This principle states that if the decision-
maker has *no reason* to believe that one state of the world is more probable to
occur than the other states, then equal probabilities should be assigned to all the
states. In general, if there are m states, the probability assigned to each state will be
$1/m$.

In fact, what we are doing is transform an initial problem of decision under
ignorance to a problem of decision under risk. That means the decision under risk
problem could be solved with any technique for solving decisions under risk. As the
most common decision strategy in decision under risk is the use of the Expected
Utility Theory, we can formalize the rule as follows:

$$a_i \text{ is the best alternative} \Leftrightarrow \forall a_j \in A,$$

$$\sum_{k=1}^{m} \frac{1}{m} * u(o_{ik}) \geq \sum_{k=1}^{m} \frac{1}{m} * u(o_{ik})$$

Given that,

$A = \{a_1, a_2, \ldots, a_n\}$ is the set of possible alternatives
$u(o_{ip})$ is the utility of the outcome of alternative a_i for the state of the world s_p

For example, coming back to the example of Table 2.12, to select the best
alternative, we must compute the Expected Utility of each alternative, assuming
that the probability of each state is ¼. If we compute them, we get:

$$EU(a_1) = 0.25 * 6 + 0.25 * 4 + 0.25 * 3 + 0.25 * 10 = 23/4$$
$$EU(a_2) = 0.25 * 10 + 0.25 * 2 + 0.25 * 5 + 0.25 * 6 = 23/4$$
$$EU(a_3) = 0.25 * 2 + 0.25 * 8 + 0.25 * 4 + 0.25 * 12 = 26/4$$
$$EU(\mathbf{a_4}) = 0.25 * 11 + 0.25 * 4 + 0.25 * 9 + 0.25 * 3 = \mathbf{27/4}$$
$$EU(a_5) = 0.25 * 8 + 0.25 * 11 + 0.25 * 5 + 0.25 * 2 = 26/4$$

Therefore, the best alternative according to the *principle of insufficient reason decision rule* is alternative a_4, which is the one getting the highest expected utility value.

Decision-Making Under Unknown Consequences

This scenario implies a higher degree of uncertainty or ignorance. The decision-maker ignore what the possible consequences (and the corresponding outcomes) are. That means that for some consequence of a certain alternative the decision-maker does not know if the probability is zero or not. In other words, some consequences of one alternative could not be known when making the choice.

This is really a complex and extremely difficult scenario because some consequences are not known and could not be taken into account when evaluating the alternatives. The scenario can be even worse when some unknown consequences could lead to catastrophic outcomes. These catastrophic consequences can be of different nature. Most common can be listed here:

- *Economic Scope*: Unforeseen catastrophic negative monetary outcomes, which can lead a company or organization to bankruptcy. For instance, when estimating the costs of a company expansion in other foreign markets, etc.
- *Environmental/Chemical/Ecological Scope*: Unforeseen catastrophic outcomes for the nature and human beings. For instance, when deciding too much high levels of allowed contaminants in a city or a region, when supervising a chemical plant in an inaccurate way, use of dangerous pesticides for the crops, etc.
- *Pharmaceutical Scope*: Unforeseen catastrophic outcomes of launching a new experimental drug in the market, if the perceptive clinical studies on the effect of the drug on patients, has not been carried on in a fair and accurate way, and some negative effects have not been detected.
- *Biological Scope*: Unforeseen catastrophic outcomes of new mortal viruses emerging when doing some genetic engineering activity, transgenic vegetables manipulation, etc. are causing negative effects on human beings.
- *Medical Scope*: Unforeseen catastrophic outcomes caused by a wrong medical decision (diagnosis, treatment, etc.) on a patient due to unknown characteristics or symptoms of the patient (allergies, etc.).

With the higher degree of uncertainty and ignorance, we are moving here, there are not decision criteria available to cope with this scenario. Of course, what is suggested from a rational point of view is to make a rational analysis of these possible uncertain consequences, and decide whether they could be unconsidered or not.

Sometimes, even having unknown consequences, some risky decisions should be made, like when sending the first manned mission to explore the space, or when testing a new drug on humans without being completely sure about possible negative effects: Science has advanced sometimes selecting options with high uncertain consequences!

When they should be considered, a rational rule is to avoid the alternatives most related with higher degrees of ignorance or uncertainty.

On the other hand, these very complex decision scenarios are the focus of Intelligent Decision Support Systems (IDSS), where most of the decision criteria reviewed in this chapter cannot be uniquely applied. Then, a lot of expert knowledge, experiential data models, and reasoning techniques coming from Artificial Intelligence field must be integrated to cope with complex real-world scenarios.

2.4.1.4   Decision-Making Under Hybrid Scenarios

Major scenarios analyzed in the literature are decision-making under certainty conditions, decision-making under risk conditions, and decision-making under uncertainty/ignorance conditions. However, a natural question emerges here. What to do when facing a hybrid scenario where there are *some deterministic/certain states or features*, but other ones are *stochastic* either with known probabilities or with unknown ones?

These scenarios are complex situations where uncertainty and deterministic knowledge must be used to solve the involved decision problems. Again, Intelligent Decision Support Systems appear as the potential solution for these complex scenarios.

Notwithstanding, we can draw some procedures to manage this situation:

An *integration of the deterministic states, the stochastic ones with unknown probability distributions with the stochastic states with a known probability distribution*. Main idea here is that deterministic states can be easily transformed into stochastic states with the probability of one, as they are completely certain states, and the stochastic states with unknown probability distribution, can be converted to known probabilistic states applying the *principle of insufficient reason decision rule* (i.e. each state can be considered as equally probable). The main problem is that each group of states forms a different *event*,[4] and all the *events* must be combined in a new one event, where the new states are all the possible combinations through intersection operations of the old states of the old events. Another minor problem is that all the utilities should be transformed to a unique scale to allow the

---

[4] An *event* is a random variable, which have as possible values the different states. It is formed by the union of all its constituent states. The sum of all probabilities of the states of an event sum one.

aggregation of the utilities of the old states to compute the new utilities of the new states of the new event.

Given:

$S_d = \{s_1, \ldots, s_d\}$ is the set of the *deterministic states*, and each deterministic state is an event,

$S_k = \{s'_1, \ldots, s'_k\}$ is the set of the *known probabilistic states*, and the union of all these states forms an *event*,

$S_u = \{s''_1, \ldots, s''_u\}$ is the set of the *unknown probabilistic states*, and the union of all these states forms an event

Therefore, the *old events* that should be combined are:

$$\text{Old-Events} = \{\{s_1\}, \ldots, \{s_d\}, \{s'_1 \cup \ldots \cup s'_k\}, \{s''_1 \cup \ldots \cup s''_u\}\}$$

In addition, the new states of the *new event* will be the intersection of the Cartesian product of the old states:

$$\text{New-Event-states} = \{\{s_1 \cap \ldots \cap s_d \cap s'_1 \cap s''_1\}, \ldots, \{s_1 \cap \ldots \cap s_d \cap s'_1 \cap s''_u\},$$
$$\ldots, \{s_1 \cap \ldots \cap s_d \cap s'_k \cap s''_1\}, \ldots, \{s_1 \cap \ldots \cap s_d \cap s'_k \cap s''_u\}\},$$

where,

$$\#\text{New-Event-States} = k * u$$

The new utilities of the outcomes of the new states of the new event will be computed as follows:

$$u(o_{ij}) = \sum_{l=1}^{d} u(s_{il}) + u\left(s'_{i,ik(j,k)}\right) + u\left(s''_{i,iu(j,u)}\right) \qquad (2.5)$$

where,

o_{ij} is the outcome of the alternative a_i for the new state s_j

$u\left(s'_{i,ik(j,k)}\right)$ is the utility of the *old known probabilistic state* $s'_{i,ik(j,k)}$ of the alternative a_i

$ik(j,k)$ is a function that computes the index of the *known probabilistic states* (k), depending on the values of j and k, and is defined as follows[5]:

---

[5]The operator *div* provides the quotient of the integer division and the operator *mod* provides the remainder of the integer division.

$$ik(j,k) = \begin{cases} j \ div \ k & if \ j \ mod \ k = 0 \\ \\ j \ div \ k + 1 & if \ j \ mod \ k \neq 0 \end{cases}$$

$u\left(s''_{i,iu(j,u)}\right)$ is a function that computes the index of the *unknown probabilistic states* (*u*), depending on the values of *j* and *u*, and is defined as follows:

$$iu(j,u) = \begin{cases} u & if \ j \ mod \ u = 0 \\ \\ j \ mod \ u & if \ j \ mod \ u \neq 0 \end{cases}$$

Finally, the Expected utility of all the alternatives for the new event and associated new states will be computed as in the case of decision-making under risk (2.3), as all the involved probabilities are known.

$$EU(a_i) = \sum_{j=1}^{k*u} p_{ij} * u(o_{ij}) \tag{2.6}$$

where,

p_{ij} is the probability of outcome o_{ij} for alternative a_i and for each new state *j*, which can be computed as the product of the probabilities of each old states as follows:

$$p_{ij} = \prod_{l=1}^{d} p(s_{il}) * p\left(s'_{i,ik(j,k)}\right) * p\left(s''_{i,iu(j,u)}\right) \tag{2.7}$$

Taking into account in (2.7) that the probabilities of the old deterministic states are equal to one, $p(s_{il}) = 1$, then (2.7) can be simplified as:

$$p_{ij} = p\left(s'_{i,ik(j,k)}\right) * p\left(s''_{i,iu(j,u)}\right) \tag{2.8}$$

Note that we are assuming that the probabilities of the old known and unknown probabilistic states are independent, and thus, p_{ij} can be computed as the product of their probabilities. The unknown probabilities, as explained above, can be converted to known probabilistic states with equal probabilities applying the *principle of insufficient reason decision rule*.

Another possible approach can be to use an *additive aggregation of all three (or two) groups of states*. Each resulting utility of each group will be weighted according to the degree of relevance or reliability that the decision-maker considers for each group.

First, a *scale transformation* will be needed for having all the utilities in all the states for all alternatives on the same scale, and this way, making possible the additive aggregation of utilities.

For the group of *deterministic states* or *features*, the aggregated utility can be computed as specified in decision-making under the certainty section, in the general case of the multi-attribute situation as stated in (2.1).

The utility for the *group of stochastic states with known probabilities* can be computed through the Expected utility theory as specified in the decision-making under risk section in Eq. (2.3).

The utility for the *group of stochastic states with unknown probabilities* can be computed, using some of the decision rule criteria available specified in the decision-making under uncertainty section. Specifically, *maximin, maximax,* and *optimism–pessimism* decision rules seem reasonable. As explained before the optimism–pessimism decision rule wraps both *maximin* and *maximax.* Therefore, the idea is that the decision-maker will decide their degree of optimism or pessimism related to the unknown probabilities of those states. The best outcome according to the *alpha-rule decision* can be computed as detailed in (2.4).

Finally, *the additive aggregation of the different utilities and outcomes* will be weighted by three parameters β, γ, and $\beta + \gamma - 1$, which denote the degree of reliability of each group of states according to the judgment of the decision-maker. The *mixed aggregated utility for the hybrid situation* is defined as follows:

$$
Mixaggu(a_i) = \beta * \sum_{j=1}^{d} w_j * u(o_{ij}) + \gamma * \sum_{j=d+1}^{k+d} p_j * u(o_{ij}) + (\beta + \gamma - 1)*
$$
$$
(\alpha * \max{}_{j>k+d}(u(o_{ij})) + (1 - \alpha) * \min{}_{j>k+d}(u(o_{ij})))
\tag{2.9}
$$

where,

$\alpha, \beta, \gamma \in \mathbb{R}$
$0 < \alpha, \beta, \gamma < 1$
$m = d + k + u$ *is the number of total states/attributes*

2.4.2 Multiple Sequential Decisions

Previously, in this chapter, we have described the process of how to formalize and model a decision scenario, when we have just one single decision. As seen, it is not an easy task to understand and characterize the decision problem, to generate all the alternatives, and finally to make a rational choice.

Real-world problems are even more complex because usually there a more than one decision that must be coped with. These decisions are processed sequentially, one after another, because some decisions can depend on the previous decisions made as well as other random or stochastic events that can happen. Therefore, if one

Table 2.14 Decision matrix for the "*selecting to connect the alarm of your home or not before going out for the weekend*" decision problem

	attempted robbery ($p = 0.3$)	No attempted robbery ($p = 0.7$)
Connecting the alarm	No robbery and $- 5$ €	No robbery and $- 5$ €
Not connecting the alarm	Robbery and $- 50{,}000$ €	No robbery and 0 €

needs to model all decisions in a concrete problem, some formalism is needed to model and visualize sequentially all the decision processes. Most commonly used tools are *decision trees* and *influence diagrams*. In the next subsections, these formalisms will be analyzed.

2.4.2.1   Decision Trees

A *decision tree* is a formalism that can represent and visualize one or more sequential decisions in a concrete scenario. They were popularized in Decision Theory and Game theory for modelling a visualization of the decision scenario. In fact, the most common formalism we have been using in previous sections, a *decision matrix*, can be represented and visualized in an equivalent form, with a decision tree.

For instance, the single decision problem of "*selecting to connect the alarm of your home or not before going out for the weekend*" is described in the following decision matrix depicted in Table 2.14.

If you connect the alarm, you have an estimated electric cost of 5 €. If thieves come to your home and attempt to make a robbery, and you do not have connected the alarm, they get in and pick up 50,000 € in money and other valuable assets. If thieves do not come to your home, you do not have any monetary loss. Let us suppose that the estimated probability that thieves attempt to make a robbery at your home is 0.3. Thus, this decision is formalized as *decision-making with risk* scenario. This decision scenario represented in the decision matrix can be represented in the following decision tree, as shown in Fig. 2.5.

As it can be seen in the example of Fig. 2.5, a decision tree is a graphical model of a decision process. The first node (squared node) represents the decision ("*connect the alarm at home?*"), and the two available alternatives of the decision are the branches outgoing the node (*Yes, No*). The two states of the decision matrix form an event ("*Attempted robbery?*"), which is a random success. In decision trees, the events are represented with nodes (circled nodes), and its branches are the set of its corresponding states (*Yes, No*). The end nodes (round-shaped rectangle nodes or sometimes no specific nodes) are the outcomes for each specific state of the event.

Decision trees can be used for the formalization of a decision problem under certainty conditions, under risk conditions with known probabilities of the states of each event, or under uncertainty or ignorance conditions. Decision trees have been extensively used in Economics and Management science, because in addition to the meaningful graphical representation of the problem at hand, it can be very useful in scenarios under risk conditions. In risk conditions scenario, for each state of an

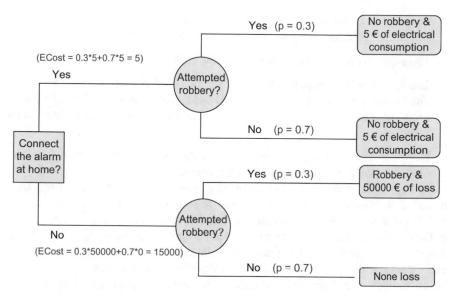

Fig. 2.5 Decision tree for the *"selecting to connect the alarm of your home or not before going out for the weekend"* decision problem

event, a probability value can be depicted in the corresponding branch of the tree. If the outcomes at the leaves of the tree are expressed in a numerical form, this way the Expected Utility for each one of the alternatives can be easily computed and *propagated backwards* in the tree, from the leaves (outcomes) to the branches of the decision node (alternatives). In this example, using the formula of the Expected utility (2.3), we have:

$$EU\left(\text{``Connecting the alarm} \equiv \text{Yes''}\right) = 0.3 * (-5 \text{ €}) + 0.7 * (-5€) = -5 \text{ €}$$

$$EU\left(\text{``No Connecting the alarm} \equiv \text{No''}\right) = 0.3 * (-50,000) + 0.7 * 0$$
$$= -15,000 \text{ €}$$

Therefore, the alternative maximizing the Expected Utility is the alternative of "Connecting the alarm". These expected utility values could also be depicted in the tree associated to each alternative, as shown in Fig. 2.5. This is extremely useful for the managers or economists to make a decision.

Decision trees are trees that graphically model a decision-making process, which can be formed of several decisions and random events with their corresponding states.

The *nodes* in the decision tree represent *decisions* or *uncertain/random/stochastic events:*

- *Decision nodes* are represented in the tree chart by *rectangles*
- *Uncertain event⁶nodes* are represented in the tree chart by *circles*

The *edges* connecting the nodes are named as *branches*:

- The *branches leaving a decision node* are the *set of available alternatives*
- The *branches leaving an uncertain event node* are *the possible results of the event or states of the world*. Probability values, when known, can be associated to the branches.

The *final leaves* of the tree are the *outcomes* of the decision path. They are represented usually within *round-shaped rectangles*.

In Fig. 2.6 a new example of a decision tree with more than one decision is depicted, to illustrate that this formalism is adequate for a sequential decision-making scenario. In the example, the decision-maker is wondering "where to go in holidays" and afterwards has to face another important decision, "what to do once there". In this scenario, we have two sequential decisions, the nodes "Zone to visit?" and "Touristic activity?" Thus, after deciding what tourist zone to visit (candidate alternatives are: Aquitaine in France, the Swiss Alps, or Lapland), and depending on the weather that could be where it has been decided, the decisor must select the touristic activity to be done, among "Beach and pool", "Monument visit", "Countryside excursion", and "Staying at the hotel". Finally, depending on the degree of satisfaction, the outcomes are reached ("very happy", "partially satisfied", and "completely unhappy"). In the decision tree, there are the estimations of the probabilities of the random events, and hence, we are facing decision-making under risk scenario. However, the outcomes are not represented numerically. Notwithstanding, transforming the qualitative outcomes to numerical ones could make that the Expected Utilities could be computed for all the decisions involved in the decision tree following a backward procedure. In the chart, not all possible combinations are shown because the complete chart was very huge.

In fact, this is one of the problems of decision trees. When adding events, the size of the tree and specifically the number of branches grows exponentially.

Decision trees are a very interesting modelling technique, especially for Intelligent Decisions Support Systems, because from a decision tree, a set of inference rules⁷ can be derived automatically, from the root of the tree to each one of the leaves. For example, in a diagnosis step in an IDSS. In addition, the sequential appearance of the decisions in the decision process in the tree can guide the development of an IDSS to give support to the decision scenario described in the tree.

---

⁶An *event* can be assimilated to the statistical concept of random variable, and its states or possible results are the possible values of the random variable.

⁷*an inference rule* is a rule with this format: *if <conditions> then <conclusions>*

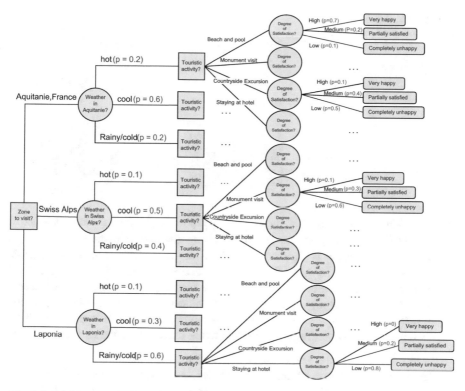

Fig. 2.6 Decision tree for the *"preparing a leisure trip"* decision problem

2.4.2.2   Influence Diagrams

The other formalisms to model the decision process making when there are sequential decisions involved are the *influence diagrams* or *relevance diagrams*. They can be considered as an extension of *path diagrams* (Wright, 1921). On the other hand, as we will explain at the end of this subsection, they are somehow a generalization of *Bayesian Networks* or *Belief Networks*. They were proposed by a Stanford research group working in Decision Analysis (Howard et al., 1976; Howard and Matheson 1984).

Influence Diagrams are Directed Acyclic Graphs (DAGs) which represents graphically a decision-making process. A graph is a structure, which is composed of nodes and edges or arcs. Directed means that the edges or arrows have a determined direction associated with them. Acyclic means that no cycles are allowed in the graph, i.e. there is no way to get back to the same node after outgoing it and following the directed edges.

The nodes represent *decisions, uncertain events or deterministic events,* and *objectives/values.* Let us see the usual conventions for describing these elements in an influence diagram chart:

Fig. 2.7 Graphical
representation of the nodes

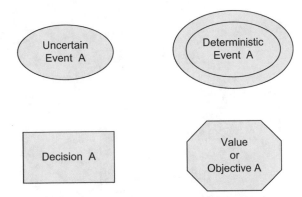

- A *decision node* is drawn as a rectangle. It represents variables under the control of the decision-maker and models the decision alternatives available.
- An *uncertain node* is drawn as a circle or oval. It is a random variable representing uncertain quantities that are relevant to the decision problem.
- A *deterministic node* is drawn as a double circle or double oval. It represents constant values or algebraically determined values from the states of their parents.
- A *value/objective node* is drawn as an octagon or diamond. It represents utility, i.e. a measure of desirability/satisfaction of the outcomes of the decision process.

In Fig. 2.7 there is the graphical representation of the different kinds of nodes.

There are three kinds of relations between the nodes: *relevance/dependence relation* or *conditional dependence*, *precedence relation* or *informational influence*, and *functional dependence*. Thus, the edges connecting the nodes could be of three different types:

- *Conditional Arcs* (solid edges): They indicate that the preceding node is *relevant* for the assessment of the value of the following component. Always are directed to *events*. They represent a *conditional dependence*.
- *Informational Arcs* (dashed edges): They indicate that a decision has been made *knowing the result of the preceding node*. Always are directed to *decision nodes*. They represent a *precedence relation* or *informational influence*.
- *Functional Arcs* (solid edges): They indicate that one of the components of the additively separable utility function *is a function of all nodes at their tails*. Always end in a *value/objective node*. They represent a *functional dependence*.

In Fig. 2.8, there is the graphical representation of the different kinds of relations and its representation through the different kinds of arcs.

As an example, let us model with an influence diagram the scenario when a citizen of a big city, each morning must decide which transportation mean will select: by bus, by metro, by tram, with a bicycle, with a car, with a motorbike, walking, a combined transport, etc. After the transportation mean is selected, the

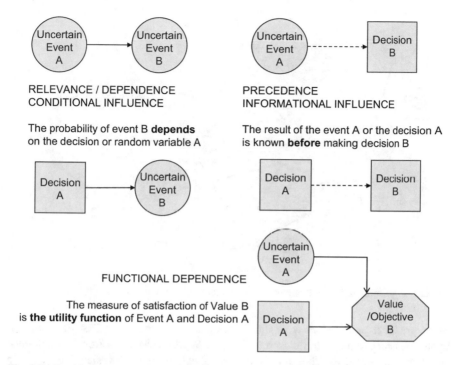

RELEVANCE / DEPENDENCE
CONDITIONAL INFLUENCE

The probability of event B **depends**
on the decision or random variable A

PRECEDENCE
INFORMATIONAL INFLUENCE

The result of the event A or the decision A
is known **before** making decision B

FUNCTIONAL DEPENDENCE

The measure of satisfaction of Value B
is **the utility function** of Event A and Decision A

Fig. 2.8 Graphical representation of the relations and the kind of arcs used

route should also be decided. Our citizen is a very conscious citizen, and she/he desires to arrive at the destination point maximizing the degree of satisfaction with the transportation selected, but at the same time wants that the transportation is as environmental friendly as possible, as healthiest as possible, and as efficient as possible. These are her/his objectives. To model this scenario, we should take into account the decisions involved, the objectives, and the different events, which can influence both the decisions and other events. For instance, it can be thought that the traffic status, the distance to the goal, the health status of the citizen, the air pollution level in the city, the actual weather, the private traffic, the public traffic, and the weather forecast can have some incidence on the decision problem. In Fig. 2.9, there is a picture of a possible influence diagram for this scenario.

In the influence diagram, there are three deterministic nodes (health status, distance to goal, and actual weather); five uncertain events (private transport status, public transport status, traffic status, weather forecast, and air pollution level in the city); two decision nodes (transportation system? and which route?); and four objective nodes (the global transport satisfaction, the transport efficiency, the healthiest transport quality, and the environmental-friendly degree of the transport).

Influence diagrams can be very useful for the analysis of the decision scenario, for making explicit several conditioning events in the decision process, but also from a

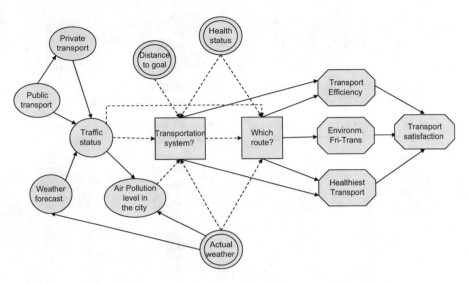

Fig. 2.9 A possible influence diagram for the *transportation mean* problem

more practical point of view in the development of Intelligent Decision Support
Systems, because they can be used to derive a *Bayesian Network* from the influence
diagram. An influence diagram is a generalization of a Bayesian Network. In
Bayesian Networks, all nodes are random variables (events), and the relations in
the graph are always conditional dependence relations. Therefore, the decision nodes
should be avoided. The decision nodes, in fact, can be considered as random variables, with the possible values, the different alternatives that are available. In
addition, the precedence relations can be transformed into conditional dependence
relations. In the same way, the objective nodes can be transformed into random
variables, and the possible values could be obtained from a discretization process of
the numerical cardinal utilities used as outcome or used directly if they form a
qualitative set of values.

2.5   Group Decision-Making

Until now, we have been focussing the decision problem under the assumption that
there is just one person involved in the decision-making process, i.e. *individual
decision-making*. From the analysis made, it can be concluded that individual
decision-making is a complex process. Then, you can imagine that when the same
process is under the responsibility of several people, then the complexity of the
process can be extreme. This situation, where more than one person has to make a

decision is commonly known as *collective decision-making* or *group decision-making* or *social decision-mak*ing. The subfield of Decision Theory studying this situation is named as *Social Decision Theory* or *Social Choice Theory*.

Of course, there are multiple situations in collective decision-making. The most complex scenario in collective decision-making is when the people in charge of the decision could have conflicting objectives. In addition, the decisors could have different relevant focus on different features or states of the world or different points of view on how the objectives could be achieved. Different preferences for the set of possible alternatives or different evaluations of the probability values of the stochastic states of the world could happen in this scenario. A typical example is *social election processes* in a democratic institution.

In other scenarios, the group of decision-making people could share the main objectives, as it is usual in *teamwork decision-making* in an organization, but each of them could have a particular opinion about which should be the best choice for the organization. Particularly important in organizations is the situation where sometimes the group could have a *leader* or *manager* of a supporting group, who take into account the opinion of the support group, but finally, the manager is responsible for making the final choice. This scenario can be named as *leaded group decision-making*, and indeed, could be considered as individual decision-making.

The concern of collective decision-making is to find a rational procedure to combine the individual preferences or choices into a collective social preference or choice. Common methodologies proposed in social decision theory for this combination are:

- *Consensus Decision-Making*: Strategy based on the negotiation among the collection of decision-makers in order to be able to arrive at a consensual preference among the alternatives, and produce a consensual social choice. This strategy probably could be feasible when all the decision-makers share the main objectives and it could be easier to reach a consensus on the best choice (*teamwork decision-making scenario*).
- *Aggregation-Based Decision-Making:* Strategy based on the finding *a Social Welfare Function (SWF)*[8], i.e. using any decision rule to aggregate a set of *individual preference orderings* over social states into a *social preference ordering* over those states. This general strategy is the most suitable for combining conflicting objectives and different preferences over the alternatives (like in a *social election process scenario*). A *social state* is a state of the world, which includes a complete description of alternatives for society.

In social decision theory, there have been two major findings about the *cycling of preferences* and about the *impossibility to convert the ranked preferences of*

---

[8]According to Arrow (1951), who was awarded the Nobel prize in economics in 1972, a social welfare function maps a set of individual orderings (ordinal utility functions) for everyone in the society to a social ordering, i.e. a rule for ranking alternative social states.

individuals into a social ranking, confirming the complexity of the *collective decision-making*: the *Voting paradox* and the *Arrow's impossibility theorem*:

The Voting Paradox (also known as the Condorcet paradox) is a situation noted by the Marquis de Condorcet in the late-eighteenth century, in which collective preferences can be cyclic, even if the preferences of individual voters are not cyclic.

The Arrow's Impossibility Theorem (Arrow, 1951) Arrow's impossibility theorem, demonstrated by Kennet J. Arrow in his doctoral thesis, is an impossibility theorem stating that when voters have three or more distinct alternatives (options), there is no social welfare function (SWF) while also meeting a specified set of reasonable criteria: ordering, non-dictatorship, Pareto efficiency, and independence of irrelevant alternatives. See Martin Peterson's book (Peterson, 2009) for further details.

As described in Sect. 2.3.1, the origins of social decision theory and especially of the voting theory were due to Ramon Llull. He proposed at least two voting schemes: *majority of binary comparisons voting* and *successive pairwise alternatives voting*. Next, we will review these voting schemes, as well as other ones, commonly used for *aggregation-based decision-making*, and especially in elections voting theory:

- *Majority of binary comparisons voting* (Llull, 1274) is a system based on the *binary comparisons of candidates* where the winner is the candidate winning by the majority in the greatest number of comparisons. This system turns out to be more effective in preventing cycles and producing a winner than the well-known *Condorcet system*, for elections with five or more candidates.
  Arthur H. Copeland rediscovered this system in mid-twentieth century (Copeland, 1951), with the very minor difference that Llull proposed to sum 1 point for each candidate in the voting when there is a tie, and Copeland proposed to sum half a point for each one, as described in (Colomer, 2001).
- *Successive pairwise alternatives voting* (Llull, 1299) presented a variant of the previous voting system able to avoid ties. This system is a *non-exhaustive binary comparison system* by which after every round of voting the loser candidate is eliminated, and the winner by majority is then compared with another candidate, until the last comparison which provides the winner. This system for four o less candidates can prevent ties and can be more efficient than the initial Llull's system, and find a winner when the *Condorcet system* cannot. However, for a higher number of candidates the system can produce different winners, depending on the order than the candidates are compared, as detailed in (Colomer, 2001).
- *Range voting* (Nicholas of Cusa, 1401–1464), proposed the *rank-order count* system which requires the voters to rank all the candidates and give them ordinal points (0, 1, 2, etc.) from the least to the most preferred, being the winner, the candidate obtaining more points. Jean Charles de Borda (1733–1799) rediscovered this system during the French revolution (1789) as explained in (Colomer, 2001).

- *Binary comparisons voting* (*Condorcet system*) was proposed by Jean Antoine Nicolas de Caritat, marquis of Condorcet (1743–1794). He proposed a very similar system to Llull's proposal of selecting the candidate winning *all binary pairwise comparisons* among all the candidates (de Caritat, 1785). His proposal was done during the French revolution (1789) as described in (Colomer, 2001).
- *Majority rule* or *majority voting*: is a voting system in which the candidate obtaining the highest number of votes, from all the electors, is the winner if the number of votes is higher than half of the votes cast. Thus, the winner should obtain two votes more than the sum of all the votes of the other candidates.
- *Plurality voting*: is a voting system in which the candidate obtaining the highest number of votes from the electors wins, with no requirement to get a majority of votes. Also is known as *simple majority voting*.
- *Weighting voting*: is a voting system where some individuals can cast several votes. Thus, the system is giving a higher weight to some specific electors in comparison to a normal voting scheme, where each elector has the same weight, i.e. all electors just cast one vote.

References

Anthony, R. N. (1965). *Planning and control systems: A framework for analysis.* Harvard University Graduate School of Business Administration.

Arnauld, A. and Nicole, P. (1662). La logique ou l'art de penser. Logic or the art of thinking. 5th ed. (J. V. Buroker, Trans. and ed.), Cambridge University Press. (1996).

Arrow, K. J. (1951). *Social choice and individual values* (2nd ed.). Wiley. Yale University Press, 1963.

Brim, O. G., Jr. (1962). *Personality and decision processes: Studies in the social psychology of thinking.* Stanford University Press.

Colomer, J. M. (2011). From *De arte electionis* to Social Choice Theory. Chapter in: *Ramon Llull: From the Ars Magna to Artificial Intelligence* (Eds. Alexander Fidora and Carles Sierra). Institut d'Investigació en Intel·ligència Artificial (IIIA-CSIC). Available at http://www.iiia.csic.es/library

Copeland, A. H. (1951). *A 'Reasonable' Social Welfare Function.* University of Michigan, Seminar on Applications of Mathematics to the Social Sciences.

de Caritat, J.-A.-N. [marquis de Condorcet] (1785). *Esai sur l'application de l'analyse â la probabilité des decisions rendues à la pluralité des vois.* Paris: Imprimerie royale. (English translation in McLean and Urken in 1995).

Delbecq, A. L. (1967). The management of decision-making within the firm: Three strategies for three types of decision-making. *Academy of Management Journal, 10*(4), 329–339.

Dewey, J. (1910, 1933, 1978). *How We Think*, 1910. Substantially revised edition in 1933. Also in *The Middle Works of John Dewey, Volume 6: Journal Articles, Book Reviews, Miscellany in the 1910–1911 Period, and How We Think*, vol. 6, pp. 177–356, 1978.

Drton, M., Hägele, G., Hanenberg, D., Pukelsheim, F., & Reif, W. (2004). A rediscovered Llull tract and the Augsburg web edition of Llull's electoral writings. *Le Médiéviste et l'ordinateur, 43.* Available at http://lemo.irt.cnrs.fr/43/43-06.htm

Hägele, G., & Pukelsheim, F. (2001). Llull's writings on electoral systems. *Studia Lulliana, 41,* 3–38. Available at https://www.math.uni-augsburg.de/emeriti/pukelsheim/2001a.html

Hansson, S. O. (1994). *Decision theory: A brief introduction*. Dept. of Philosophy and the History of Technology. Royal Institute of Technology (KTH), Stockholm, Sweden.

Honecker, M. (1937). Lullus-Handschriften aus dem Besitz des Kardinals Nikolaus von Cues - Nebst einer Beschreibung der Lullus-Texte in Trier und einem Anhang über den wiederaufgefundenen Traktat De arte electionis. Spanische Forschungen der Görresgesellschaft. *Erste Reihe, 6*, 252–309.

Howard, R.A. and Matheson, J.E. (1984). Influence diagrams. In R. A. Howard & J. E. Matheson (Eds.), *Readings on the principles and applications of decision analysis* (Vol. *II*). Strategic Decisions Group.

Howard, R. A., Matheson, J. E., Merkhofer, M. W., Miller III, A. C., Rice, T. R. (1976). *Development of automated aids for decision analysis*. DARPA contract MDA 903-74-C-0240, SRI International.

Hurwicz, L. (1951). Some specification problems and application to econometric models. *Econometrica, 19*, 343–344.

Huygens, C. (1657). De ratiociniis in ludo aleae. Appendix to Frans van Schooten's *Exercitationum Mathematicarum Libri Quinque.*

Knight, F. H. (1921). *Risk, uncertainty and profit*. Hart, Schaffner & Marx/Houghton Mifflin Co.

Llull, R. (1274). *Artificium electionis personarum* (English title: The method for the elections of persons).

Llull, R. (1299). *De Arte electionis* (English title: On the method of elections).

Mintzberg, H. (1973). *The nature of managerial work* (p. 1973). Prentice Hall.

Mintzberg, H., Raisinghani, D., & Théorêt, A. (1976). The structure of 'Unstructured' decision processes. *Administrative Sciences Quarterly, 21*, 246–275.

Newell, A., & Simon, H. A. (1961). In L. Automaten & H. Billing (Eds.), *GPS, a program that simulates human thought* (pp. 109–124). R. Oldenbourg KG.

Peterson, M. (2009). *An introduction to decision theory*. Cambridge University Press.

Ramsey, F. P. (1926). Truth and probability. In *Ramsey, 1931, The Foundations of Mathematics and other Logical Essays, Ch. VII, p.156–198*, edited by R.B. Braithwaite, London: Kegan, Paul, Trench, Trubner & Co. Harcourt, Brace and Company.

Savage, L. J. (1951). The theory of statistical decision. *Journal of the American Statistical Association, 46*, 55–67.

Savage, L. J. (1954). *The foundations of statistics*. John Wiley. Second revised edition, 1972.

Sen, A. K. (1970). *Collective choice and social welfare*. Holden-Day Inc..

Simon, H. A. (1960). *The new science of management decision*. Harper & Row.

Steele, K. & Stefánsson, H. O. (2016). Decision Theory. In The Stanford Encyclopedia of Philosophy (Winter 2016 Edition), Edward N. Zalta (ed.), https://plato.stanford.edu/archives/win2016/entries/decision-theory/

Thompson, J. D. (1967). *Organizations in action*. McGraw-Hill.

von Neumann, J., & Morgenstern, O. (1944, 1947). *Theory of games and economic behaviour* (2nd ed.). Princeton University Press.

Wald, A. (1950). *Statistical decision functions*. Wiley.

Wright, S. (1921). Correlation and causation. *Journal of Agricultural Research, 20*, 557–585.

Further Reading

Fidora, A. and Sierra, C. (Eds.) (2011). *Ramon Llull: From the Ars Magna to Artificial Intelligence*. Institut d'Investigació en Intel·ligència Artificial (IIIA-CSIC). Available at http://www.iiia.csic.es/library

Gilboa, I. (2009). *Theory of decision under uncertainty*. Cambridge University Press.

Studia Lulliana journal. Collection of articles since 1957. Càtedra Ramon Llull. Universitat de les Illes Balears. Available at http://ibdigital.uib.es/greenstone/collect/studiaLulliana/

Llull, R. (1283). *En qual manera Natana fo eleta a abadessa* (English title: In which way Nathana was elected abbess), 1283. This was a divulgation text about the election of the abbess of a convent, included in chapter 24 of one of the first novels written in Romance language: *Blaquerna*.

Luce, R. D., & Raiffa, H. (1957). *Games and decisions: Introduction and critical survey*. Wiley.

The Augsburg Web Edition of Llull's Electoral Writings. (2016). Available at www.uni-augsburg. de/llull/

Chapter 3
Evolution of Decision Support Systems

3.1 Historical Perspective of Management Information Systems

Management Information Systems (MIS) is a general term including different kinds of computer systems. All of them are centred on the management of data and information by means of electronic computer processing. Therefore, both older Transactions Processing Systems, as well as Decision Support Systems and Intelligent Decision Support Systems can be considered as MISs.

Decision Support Systems (DSS) arc the natural evolution of Transactions Processing Systems. In this section, we will analyze the main characteristics of Transactions Processing Systems, to outline the main differences with the Decision Support Systems.

Transaction Processing Systems were originated in the late 60s and early 70s, when the first high-level programming languages (FORTRAN, COBOL, ALGOL, etc.) were standardized and available in the computers in that era, usually named as *mainframes*. Mainframes were high powerful computers like the classical IBM 360 series. New high-level programming languages appeared as LISP[1] in 1962, PL/I, and BASIC in 1964. This age is commonly named as the *third computer generation*. In Transaction Processing Systems, the main tasks were electronic data processing. Usually, this processing involved large volumes of business transactions, which were processed in *batch mode*. The batch mode is a mode when all the processing is done in batches of similar works that the users of the computer system have been submitting to a queue and the processing is done later, sometimes by

---

[1] John McCarthy (1927–2011) designed LISP (LISt Processing) at MIT in 1959–1960. In 1962, the first stable version, Lisp 1.5, was delivered. It was a declarative functional programming language, based on the *lambda calculus*, oriented to the computation of symbolic expressions, and both data and programs (functions) were coded as lists. It has been the classical programming Language for Artificial Intelligence.

© Springer Nature Switzerland AG 2022
M. Sànchez-Marrè, *Intelligent Decision Support Systems*,
https://doi.org/10.1007/978-3-030-87790-3_3

night. That means this processing is *not interactive* but deferred. For instance, think of a system to manage an inventory with all the orders modifying the stocks, or a billing system to manage the offered utility services. In that era, the data was provided to the computer through punched cards and magnetic tapes.

From the early 70s, systems started to be distributed. The computation was distributed among several medium–small computers, instead of having a unique mainframe. At the same time, computer systems were moving to be *interactive*. In this period appeared the *microprocessor*, which integrated in just one circuit the unit control and the data path of the processor. The computers in this era were named as *microcomputers* or *workstations*. These workstations were fully interactive and the user could execute the commands online in an *interactive mode*. Usually, they were connected on a local area network and to the internet. New high-level programming languages appeared like Pascal in 1971, Prolog[2] in 1973, Smalltalk in 1976, C in 1978. Data was stored in direct-access devices like magnetic disks. This era is usually named as the *fourth computer generation*. In the middle 70s started to appear a new kind of processing systems which were called *Decision Support Systems* (DSS). They were *interactive systems* especially devoted to *support business decision-making* for the managers and executives of different firms. With the increasing power and connectivity of the microcomputers, it was started to be clear the power of the data and information recorded and available from a workstation computer.

Here, we can distinguish between the early Decision Support Systems (DSS) that were developed in the late 70s from others more Advanced Decision Support Systems (ADSS) developed in the 80s.

Early Decision Support systems were very simple and elementary. They were just basic systems developed using *spreadsheet software*. Spreadsheet software is software that emulates the classical "spread" of usually two "sheets" of paper, like an open newspaper, that accountants of firms have been using for hundreds of years (Power, 2004). The spreadsheet is a matrix of rows and columns. Usually, the columns correspond to different kinds of expenditures and the rows to different invoices. Each cell has the amount of a concrete invoice for a concrete expenditure. The great advantage of the spreadsheet software is that it can make basic arithmetic operations (sum, add, multiply, divide, etc.), and other more complex functions (average, min, max, etc.) and formulas provided by the user. This way, the data can be analyzed in multiple ways, and the software can outline valuable knowledge. This way, simple decisions, like which categories of expenditures should the firm try to decrease the costs to improve the benefit for the company can be supported.

Advanced Decision Support Systems (ADSS) were the DSS developed in the 80s, which incorporated an extensive use of the capabilities of the computer. Some authors named them as Computer-Assisted Decision Support Systems. They

---

[2] Alain Colmerauer (1941–2017) designed Prolog (Programmation Logique) in 1972. Prolog is a logic programming language based on the automation of the resolution principle of Robinson (1965), where facts and rules (relations) are represented by Horn clauses. It is a declarative logic programming language, which has been extensively used in Artificial Intelligence, especially in logic programming, and natural language processing.

integrated some *models* coming from *operations research* field and *management science* area. For instance, they integrated some *linear programming model* to optimize an objective function, usually the costs or benefit for an organization, subject to some constraints. In addition, some *business model* could be integrated to obtain financial information relevant for the decision-making process.

Another important feature was the ability to manage the so-called *"what if" analysis*. This kind of "what if" analysis means that the interface of the system allowed the user to run several models or the same models with different parameterization, and evaluate what would happen if the user selected some possible alternative. This was a very powerful mechanism, because, these simulation models enable to highly support the user before making the final choice, evaluating the possible consequences of each alternative.

The final step in the evolution of Advanced Decision Support Systems are the Intelligent Decision Support Systems (IDSS). They can be situated in the late 80s, and the 90s. The main novelty is the use of *models* coming from the *Artificial Intelligence* field. These models, as we will see in detail in the next sections and chapters of the book, are of different nature. Most of them are *machine learning methods*, which are induced from the data (data mining models), where others are biologically inspired methods, etc. Thus, many of them are data-driven models, but also others are model-driven methods. Another important feature of IDSSs is the integration of artificial intelligence models with other kind of models, like statistical models, etc. Therefore, even though the main characteristic of IDSSs is the inclusion of some artificial intelligence model, the integration of several kind of models is a frequent characteristic. In addition, the development of IDSSs has outlined the *importance* and *power of the data*, and the knowledge behind the data.

Some authors name the DSSs using artificial intelligence methods as *Knowledge-Based Decision Support Systems* as Klein and Methlie (1995) or *Knowledge-driven Decision Support Systems* as Power (2002). We guess that they use the knowledge-based term as a way to indicate that the models used in these DSSs are focussed on the exploitation of the knowledge obtained from the data or from the experts. We share partially that view, but the problem is that for artificial intelligence scientists, a knowledge-based system has a clearly and well-defined meaning, as a paradigm for problem-solving relying in a knowledge base, usually expressed by inference rules, which represents the "expert knowledge" in a concrete domain. Therefore, is not including other artificial intelligence techniques, which can be used in DSSs. Hence, in this book, we always will talk about *Intelligent Decision Support Systems*.

In Dhar and Stein (1997), a taxonomy for Management Information Systems was proposed. We have modified the classification to situate the IDSSs in that taxonomy. Therefore, in Fig. 3.1, there is our proposed classification of MIS. We have described the characteristics of Transaction Processing Systems and Decisions Support Systems. Just as a summary, the main characteristics are:

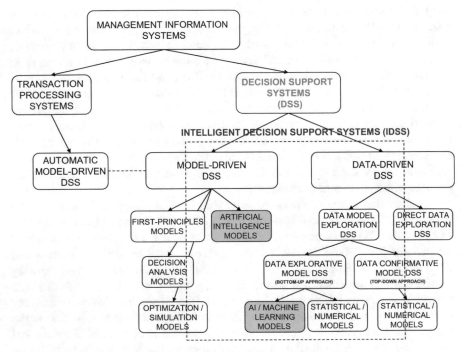

Fig. 3.1 Management Information Systems classification. (Adapted from Dhar and Stein 1997)

- Transaction Processing Systems:

  - Large volume of data processing
  - Simple decisions
  - Automated decisions

- Decision Support Systems

  - Supporting business managers
  - More complex decisions
  - Decisions made by humans
  - "what if analysis" ability

We share the distinction between Model-driven DSS and Data-driven DSS as proposed by Dhar and Stein (1997), but explicitly adding the Automatic Model-driven DSS, and some further classification of both Model-driven DSS and Data-driven DSS, with the following features:

- Model-driven DSS

  - Using one or more strong models to support decisions. There are different kinds of models. The classic models of DSS, such as First-Principles models,

Decision Analysis Models, Optimization models, or Simulation models; and the Artificial Intelligence models.
- The quality of the DSS relies on the quality of the model/s
- Friendly interface to use the model/s

- Data-driven DSS

  - Using data to extract relevant information for the decision-maker. There are different levels of data extraction, as we will analyze later: Direct data exploration DSS and Data model exploration DSS. Furthermore, Data model exploration DSS can be divided into Explorative Model DSS and Confirmative Model DSS.
  - The quality of the DSS relies directly in the data or in the models induced from data.
  - Model/s, when available is/are simpler than in model-driven DSS

- Automatic Model-driven DSS

  - A mixture kind of DSS between Transaction Processing Systems and Model-driven DSS
  - Automated decision without a human
  - Relying in an efficient model/s

As mentioned above, we propose an additional subdivision of Data-driven DSS into Direct data exploration DSS and Data model exploration DSS as depicted in Fig. 3.1.

Direct data exploration DSSs will be analyzed in Sect. 3.3.2.1, but can be classified into Data-reporting DSS, Data-analytic or Data-intensive DSS, Database Querying Model DSS, Executive Information System DSS, and Data Warehouse and OLAP System DSS. All Direct data exploration DSS are characterized by a direct use of data through several techniques for analyzing and querying data.

On the other hand, Data model exploration DSSs are characterized to exploit the information hidden in data, but not in a direct way, but indirectly by means of models induced from data. They can be further split into *Data confirmative model DSS* and *Data explorative model DSS*:

- *Data Confirmative Model DSS*: These DSSs use *confirmative models*. These models use a top–down approach. Some kind of model is hypothesized, and it should be confirmed with the available data. For instance, a *linear regression model*. The hypothesis is that the response variable is linear regarding the other explicative variables, that the noise in the model follows a normal distribution, and so on. This hypothesis should be confirmed once the model is obtained (the coefficients of the linear regression are obtained), and the approximation errors (residuals) should satisfy several conditions, etc. The models are obtained from the data, but in a confirmative way. Several statistical and numerical methods belong to this group.
- *Data Explorative Model DSS*: These DSSs use *explorative models*. These models use a bottom–up approach. In this situation, the models are obtained from the

exploration and direct mining of the data. For instance, a *classification rule model*. From the available data, an inductive algorithm explores and mine that data to obtain a set of classification rules, which form a discriminant model able to predict the class label of new data examples or instances. The models are obtained from the data, but in a bottom–up way. Several machine learning methods and artificial intelligence methods belong to this group, as well as some explorative statistical methods.

Finally, in Fig. 3.1 it is marked with a rectangle with dashed lines the zone where we situate the Intelligent Decision Support Systems. As it is shown in Fig. 3.1, the required kind of models (*artificial intelligence models* or *AI/machine learning models*) needed in order that a DSS can be considered as an IDSS are marked with a light-red/pink colour.

3.2   Decision Support Systems

As described in the previous section, the concept of Decision Support Systems was originated in the early 70s, and it is commonly agreed among the management science literature that two articles written by then were the pioneering starting points. The two articles are:

- "Models and Managers: The Concept of Decision Calculus" written by J.D. Little (1970). He described the concept of decision calculus as a "model-based set of procedures for processing data and judgements to assist a manager in his decision making" (p. B470). He proposed that this kind of system should be simple, user-friendly, easy to control, robust, and adaptive to the needs of the user.
- "A Framework for Management Information Systems" written by Gorry and Scott Morton (1989). In the article, they coined the term *Decision Support System* and developed a two-dimensional framework for computer support of managerial activities. One dimension was similar to Simon's classification (1960) structured vs. unstructured decisions, and the other one following Anthony's classification (1965) of managerial activity in operational, tactical, strategic.

Let us review and analyze different definitions of the concept under the broad term of Decision Support System:

- J. D. Little defined: "DSS are a *model-based* set of procedures for *processing data and judgments* to *assist a manager in his/her decision*" (Little, 1970).
- Keen and Scott Morton wrote that "DSS couple the intellectual resources of individuals with the capabilities of the computer to *improve the quality of decisions*. It's a *computer-based support* for *management decision* makers who deal with *semi-structured problems*" (Keen & Scott Morton, 1978).
- Moore and Chang said that "DSS is a system that is *extendable*, capable of *supporting ad hoc analysis and decision modelling*, oriented towards *future*

planning, and of being used at irregular, unplanned intervals" (Moore & Chang, 1982).

- Beulens and Van Nunen stated that "DSS enable managers to *use data and models* related to an entity (object) of interest to *solve semi-structured and unstructured problems* with which they are faced" (Beulens & Van Nunen, 1988).
- Emery and Bell wrote that "Main feature of DSS rely in the *model component. Formal quantitative models* such as statistical, simulation, logic and optimisation models are used to represent the decision model, and their solutions are alternative solutions" (Emery, 1987; Bell, 1992).
- McNurlin and Sprague stated that "DSS are systems for *extracting, summarising* and *displaying data*" (McNurlin & Sprague, 1993).
- George M. Marakas wrote that "a Decision Support System (DSS) is a system under the control of *one or more decision makers* that *assists in the activity of decision making* by providing an organised *set of tools* intended to impart structure to portions of the decision-making situation and to *improve the ultimate effectiveness of the decision outcome*" (Marakas, 2003).

We can make the following definition: A Decision Support System (DSS) is a system, which lets *one or more people* to *make decision/s* in a concrete domain, in order to manage it the best way, selecting at each time the best alternative among a set of alternatives, which generally are contradictory.

From these definitions, and the historical analysis from Sect. 3.1, it can be summarised the main characteristics of Decision Support Systems:

- Provide the best support to assist the decision-makers
- Use of computer-based tools and models to support making the best choice
- Improve the outcome of the final decision
- Useful for not simple decisions
- Coping with semi-structured and unstructured problems
- Used at irregular and unplanned intervals
- Not routinely decisions
- Decisions made by humans
- "What if analysis" ability by means of some tools, models, etc. to assess the available alternatives

This scenario is matching clearly with complex problems and decisions falling between the second and third kind of problems both in the classification from Simon (see Chap. 2, Sect. 2.2) and in the classification from Funtowicz and Ravetz (1993) (see Chap. 1, Sect. 1.1). Decision Support Systems have been used in several domains of application. As they emerged in management science and business, they have been widely applied in this area. Let us review the main areas where they have been applied:

- DSS in Business and Management

  - Decisions on prices
  - Decisions on products
  - Decisions on marketing
  - Decisions on employees and resources
  - Decisions on strategic planning

- DSS in Engineering

  - Decisions on product design
  - Decisions on process supervision/control

- DSS in Financial Entities

  - Decisions on loans
  - Decisions on investments
  - Decisions on portfolios

- DSS in Medicine

  - Decisions about presence/absence of illness
  - Decisions on medical treatments
  - Decisions on specialized medical machine control

- DSS in Environmental Systems (EDSS)

  - Decisions about reliable and safe environmental system management
  - Decisions on control/supervision of chemical plants
  - Decisions on control/supervision of wastewater treatment plants
  - Decisions on control/supervision of waste plants

Common tasks that DSS should perform are the following ones:

- Communication with the user
- Data Analysis
- Monitoring/Control
- Prediction
- Planning
- Management
- Data Recording/Retrieval

As we have described previously, DSS integrate some mathematical, statistical, and operations research models. Most usual models integrated are:

- Statistical/Numerical Models

  - Linear Regression Models
  - Logistic Regression model
  - Markovian Models
  - Multivariate Statistical Data Analysis

Principal Component Analysis (PCA)
Simple Correspondence Analysis (SCA)
Multiple Correspondence Analysis (MCA)

- Simulation Models
- Control Algorithms
- Optimization Techniques

  – Linear Programming
  – Queue Models
  – Inventory Models
  – Transport Models
  – Multiple Criteria Decision Analysis/Making (MCDA/MCDM)

3.3   Classification of DSS

In the literature, there are several classifications of Decision Support Systems, according to different criteria. As all of DSSs share several common components, we think that the *type of knowledge* they are focussing and using can be a good criterion to classify them. Attending to this criterion, DSSs can be classified into two major types: *model-driven decision support systems* and *data-driven decision support systems*. Notwithstanding this classification of DSSs, in practice, any DSS could be a mixture of different kinds of DSS types merging some model-driven DSS method and some data-driven DSS method. Next, these two general types and the models they are using will be described.

3.3.1   Model-Driven DSSs

As described in Sect. 3.1, model-driven DSSs, or model-centric DSSs as some researcher names them, use one or more strong models to support decisions. The quality of the DSS relies in the quality of the model/s. They usually have a friendly interface to use the model/s. Most common kind of models used in model-driven DSSs are:

- First-Principles models or Mechanistic models
- Decision Analysis Tools and models

  – Analytical Hierarchy Process model
  – Decision Matrix and Decision Table models
  – Decision Tree model

- Optimization models

  – Multi-Criteria Decision Analysis (MCDA) models

- Simulation models
  - Monte Carlo simulation models
  - Discrete-event simulation models

In the following subsections, these different kinds of models are described.

3.3.1.1   First-Principles Models or Mechanistic Models

First-Principles models or mechanistic models are based on an understanding of the behaviour of a system's components, analyzing the system from its first-principles in a given concrete domain. They are composed of basic and self-evident propositions, which are translated in some formal representation. Usually, these mechanistic models are algebraic models (a set of mathematical formulas) and/or equations' models (a set of differential equations, etc.). This kind of models exist in many fields like Economic models, Econometric models, Climate models, Ecological Models, Hydrological models, Geological models, Chemical models, Physical models, Biological models, Molecular models, Statistical models, Social Science models, Political Science models, etc.

3.3.1.2   Decision Analysis Tools and Models

In classical Decision Theory, several tools and models can be used to analyze, formalize, and visualize a decision-making scenario. Most commonly used are the following ones:

Analytic Hierarchy Process Model

The Analytic Hierarchy Process (AHP) is a technique for organizing and analyzing complex decisions in a hierarchical way, based on the mathematics and psychology. Thomas L. Saaty proposed AHP in the 1970s (Saaty, 1980). It is especially suitable for group decision-making. The method consists in first decomposing the decision problem (the goal node) into a hierarchy of easier sub-problems (sub-goals or criteria), each of which can be analyzed independently. The process goes on (criteria can be subdivided into new sub-criteria) until arriving at the leave nodes which are the alternatives or options of the decision problem. Once the hierarchy is built, the decision makers systematically evaluate the various elements on a tree level (siblings' nodes) by pairwise comparisons with respect to their impact on an element above them in the hierarchy, to derive numerical priority values (between 0 and 1). Priorities are organized in a priority matrix, and finally after some mathematical computations (adding and multiplying) the final priority value is computed for each alternative. The alternative with the highest priority is the suggested alternative.

Decision Matrix and Decision Table models

In Chap. 2, Sect. 2.3.2.2 *Decision matrices* have been detailed. They are a formalizing and visualization tool for a decision problem, which has been commonly used in Decision Theory.

Decision tables are a compact way to model complex decision rule sets and their corresponding actions. Decision tables, like if-then-else rules, associate conditions with actions to perform. In a decision table, the first rows are the conditions of the rules, and the next rows are the actions of the rules. The columns are the different possible decision rules, and each cell of the table are the values of the conditions or the actions corresponding to a concrete rule.

In the 60s and 70s some "decision table based" languages were designed and get some popularity for business programming.

Decision Tree Model

Decision trees were explained in detail in Chap. 2, Sect. 2.4.2.1. They are formalisms especially suitable for modelling a sequential decision problem.

3.3.1.3   Optimization Models

Optimization models (Luenberger & Ye, 2008) are a set of different models to optimize (maximize or minimise) a defined function (objective function) under some restrictions (usually expressed as inequalities). It is a general problem in operations research, which can be solved through several methods like linear programming, non-linear optimization, inventory models, transport models, queue theory models, etc. These models have been used to maximize the benefit or minimize the costs of an organization. A specific kind of models is the so-called multi-attribute and multi-criteria models.

Multi-Criteria Decision Analysis (MCDA) Models

Multiple-criteria decision-making (MCDM) or multiple-criteria decision analysis (MCDA) is a sub-discipline of operations research that explicitly evaluates multiple conflicting criteria in decision-making. Usually, in such problems, there is no a unique optimal solution, as the different criteria are usually contradictory. For each alternative some criteria are favourable and other ones are unfavourable. To solve these problems, it is necessary to use preferences of the decision-makers to set an order among the alternatives. The preferences are modelled through constraints on the objective function, which usually describes a cost/benefit function. Commonly, it can be solved through mathematical programming.

3.3.1.4   Simulation Models

Simulation models (Law & Kelton, 2000) are the methods used to provide the answer to the "what-if" scenario analysis. Most classical simulation models are numerical methods. A simulation model is a system, which emulates the behaviour of a real system based on different methodologies: discrete-event simulation models, activity-based simulation models, interaction-process simulation models, Monte Carlo simulation methods, etc. Most common simulation models used in decision Support Systems are Monte Carlo methods and discrete-event models.

Monte Carlo Simulation Models

Monte Carlo simulation models provide the decision-maker with a range of possible outcomes and the probabilities they will occur for any possible alternative. It computes results repeatedly, each time using a different set of random values from the probability functions assumed for each input random variable of the system. They are used in modelling scenarios with significant uncertainty in inputs such as the calculation of risk in business, in complex optimization, and numeric problems.

Discrete-Event Simulation Models

A *discrete-event simulation model* is a time-dependent simulation model, which is governed by the occurrence of events. The logical sequence of events is modelled, like for instance, in a queue system, which models the arrival time, the waiting time for a service, the time of receiving a service, etc. Different simulation runs let know the user valuable information about different possible alternatives in a decision-making scenario.

3.3.2   Data-Driven DSSs

Data-driven DSSs or data-centric DSSs as named by some authors in the literature use data to extract relevant information for the decision-maker. The quality of the DSS relies in the data. Models, when available are usually not so complex than models in model-driven DSS. There are different levels of information extraction from data. Others DSSs makes an indirect exploration of data through a model obtained from the data (data model). This main classification leads to the next classification level:

- Direct Data Exploration DSSs

  - Data-reporting DSSs
  - Data-analytic DSSs or Data-intensive DSSs

    Database Querying model DSSs
    Executive Information System model DSSs
    Data Warehouse and OLAP system model DSSs

- Data Model Exploration DSSs

  - Data Confirmative model DSSs
  - Data Explorative model DSSs

3.3.2.1   Direct Data Exploration DSSs

There are some DSSs, which make a direct data exploration, through different methods for analyzing and querying data. Some of them are based on simple summarizing and reporting tasks and other use more data-intensive information extraction.

Data-Reporting DSSs

The first data-driven DSSs computed information from the data available. Main goal is to condense and summarize large amounts of data, through the computation of averages, totals, data distributions, and the corresponding visualization of the information, to be helpful to the business managers of an organization. Usually, some *spreadsheet software* does this kind of reporting about the data. The extraction level of the data information is rather low.

Data-Analytic or Data-Intensive DSSs

The Data-analytic DSSs or Data-intensive DSSs show a high extraction level of data information. Data is analyzed in a more intensive way than just reporting information from the data.

Database Querying Model DSSs

In the evolution of Management Information Systems, and DSS in the 70s, appeared the database management systems (DBMS), which were designed to organize, store, and retrieve data as fast as possible. Moreover, the Relational DBMS dominated the market very fast. They provided a separated structure of the data in tables, avoiding data redundancy. This kind of model let the users to write database queries for their specific needs. The outstanding feature of these systems is that they provided a flexible query language called SQL (*structured query language*). This way, the user could consult complex information from the database of the organization, but the drawback is that the user had to know, both the structure of the database and the syntax and semantic of SQL. For instance, the manager of a firm could consult from the database, which are the international loyal customers with a high expenditure, to focus a new marketing campaign on a new product being commercialized only for international customers:

```
SELECT name, country FROM customer c, order o
WHERE  c.id = o.id
  AND    SUM(o.id, o.amount) > 4000
  AND    c.type = "international"
  AND    c.loyalty = "yes"
```

Executive Information System (EIS) Model DSSs

These kind of models appeared in DSSs as a solution to the business managers who did not know very well how to make the queries to a database. Executive Information Systems were a front-end for making fixed predefined database queries and generating usual reports for the executives of a firm. The front-end interface translated the input information data from a usual format to the necessary SQL code to make the appropriate query or queries to the RDBMS, and provide the useful information to support decisions. In EIS, all the database queries and reports were predefined. The most common queries were programmed, but the problem was that if the manager of a company wanted to query something different from the predefined queries he/she could not get the needed information. Furthermore, another problem was that it was really difficult or impossible to obtain and combine data from different databases of an organization.

Data Warehouse and OLAP System Model DSSs

These kinds of data-driven models in DSSs appeared to solve the problems EIS models showed, based on the idea to allow making open database queries and reports to the whole databases of an organization. This kind of models is based on the use of data warehouses and On-Line Analytical Processing (OLAP) systems. These two components provided DSS with high flexibility and easiness to use for the business managers. A *data warehouse* is a specific database designed to answer business queries. It is a repository of data coming from different data sources in a company. The different data sources are moved to the warehouse, where they could be accessed in different forms, because they are indexed and combined for having a fast access. *OLAP systems* are powerful front-ends able to process interactively data in a very fast and flexible ways. As it is commonly said, an OLAP system can "slice and dice" data in almost any way you can imagine. In fact, a useful way to think about OLAP systems is imagine that the data warehouse is like an n-dimensional hypercube of determined database features that you can index, cut, and obtain information. For instance, one dimension could be the different products of one firm, the other the customers, a third one could be the orders from the different customers over the different products of the company. With this structure, the user can query the data warehouse taking into account any dimension depending on his/her interest.

3.3.2.2   Data Model Exploration DSSs

Data Confirmative Model DSSs

Data Confirmative model DSSs are data model exploration decision support systems which capture the information from data through a data confirmative model. These kind of DSS are data-driven, but indeed also use a model obtained from data. Data confirmative models, as described in Sect. 3.1 use a top–down approach. Some kind of model is hypothesized, and it should be confirmed with the available data. Several statistical and numerical methods belong to this group:

- Linear Regression Models
- Logistic Regression Model
- Markovian Models

Data Explorative Model DSSs

Data Explorative model DSSs are model exploration decision support systems, which capture the information from data through a data explorative model. These kind of DSS are data-driven, but indeed also use a model obtained from data. Data explorative models, as described in Sect. 3.1 use a bottom–up approach. The models are obtained from the exploration of the data. Some explorative statistical method like multivariate statistical data analysis models belong to this category:

- Multivariate Statistical Data Analysis

  - Principal Component Analysis (PCA)
  - Simple Correspondence Analysis (SCA)
  - Multiple Correspondence Analysis (MCA)

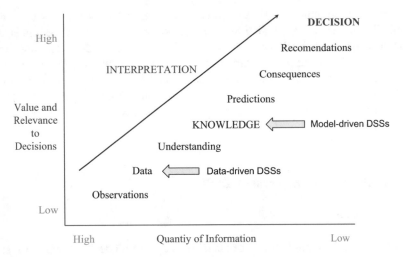

Fig. 3.2 Interpretation process in a DSS: from observations to decisions. (Adapted from A. D. Wittaker 1993)

3.4   The Interpretation Process in Decision Support Systems

As the complexity of Decision Support Systems is increasing, the complexity of the models increases too. Both model-driven DSSs and data-driven DSSs requires an *interpretation process*, which can start at the *data* level for data-driven DSSs or at the level of *knowledge* for model-driven DSSs. This interpretation process (see Fig. 3.2) must transform *observations* into data, when just only the necessary data and its corresponding meta-data are considered. Data should be extracted *and* understood, as some data-driven models do. The result is some important knowledge for the decision process. This knowledge coming from the data (data-driven DSS) or directly coming from a model-driven DSS is the basis for the next steps in the interpretation process. The next step is usually the *prediction* step, which can be implemented through a simulation model to let the user know the possible consequences of each one of the alternatives. This way the DSSs can make the appropriate recommendations to the decision-maker for making the final decision.

As far as we move in the interpretation process in any DSS, the quantity of information is reduced, as indicates the *x*-axis in Fig. 3.2, and the value and relevance for the final decision increases, as indicates the *y*-axis in Fig. 3.2.

References

Anthony, R. N. (1965). *Planning and control systems: A framework for analysis*. Harvard University Graduate School of Business Administration.

Bell, P. C. (1992). Decision support systems: Past, present and prospects. *Revue des systèmes de décision, 1*(2–3), 126–137.

Beulens, J., & Van Nunen, J. A. (1988). The use of expert system technology in DSS. *Decision Support Systems, 4*(4), 421–431.

Dhar, V., & Stein, R. (1997). *Intelligent decision support methods. The science of knowledge work.* Prentice-Hall.

Emery, J. C. (1987). *Management information systems. The critical strategic resource.* Oxford University Press.

Funtowicz, S. O., & Ravetz, J. R. (1993). Science for the post-normal age. *Futures, 25*(7), 739–755.

Gorry, G. A., & Scott Morton, M. S. (1989). A framework for management information systems. *Sloan Management Review, 13*(1), 49–62.

Keen, P. G. W., & Scott Morton, M. S. (1978). *Decision support systems: An organizational perspective.* Addison-Wesley.

Klein, M. R., & Methlie, L. B. (1995). *Knowledge-based decision support systems* (2nd ed.). Wiley.

Law, A. M., & Kelton, W. D. (2000). *Simulation modelling and analysis* (3rd ed.). McGraw-Hill.

Little, J. D. (1970). Models and managers: The concept of a decision calculus. *Management Science, 16*(8), B466–B485.

Luenberger, D. G., & Ye, Y. (2008). *Linear and nonlinear programming* (3rd ed.). Springer Science & Business.

Marakas, G. M. (2003). *Decision support systems in the twenty-first century* (2nd ed.). Prentice-Hall.

McNurlin, C., & Sprague, R., Jr. (1993). *Information systems management in practice* (4th ed.). Prentice-Hall.

Moore, J. H., & Chang, M. G. (1982). Design of decision support systems. *Database, 12*(1–2), 8.

Power, D. J. (2002). *Decision support systems: Concepts and resources for managers.* Greenwood Publishing Group.

Power, D. J. (2004). *A brief history of spreadsheets.* DSSResources.COM. version 3.6. http://dssresources.com/history/sshistory.html, Photo added September 24, 2002

Saaty, T. L. (1980). *The analytic hierarchy process: Planning, priority setting, resource allocation (decision making series).* McGraw-Hill International Book Company.

Simon, H. A. (1960). *The new science of management decision.* Harper & Row.

Wittaker, A. D. (1993). Decision support systems and expert systems for range science. In J. W. Stuth & B. G. Lyons (Eds.), *Decision support systems for the management of grazing lands: Emerging lands* (pp. 69–81).

Further Reading

Power, D. J. (2013). *Decision support, analytics, and business intelligence (Information systems)* (2nd ed.). Business Expert Press.

Turban, E., Aronson, J. E., & Liang, T.-P. (2005). *Decision support systems and intelligent systems* (7th ed.). Pearson/Prentice Hall.

Part II
Intelligent Decision Support Systems

Chapter 4
Intelligent Decision Support Systems

Intelligent Decision Support Systems are an evolution of the Advanced Decision Support Systems. IDSS are decision support systems, which integrate some artificial intelligence techniques. In addition, commonly there is the integration with other kinds of models (numerical, statistical, mechanistic, etc.), which confers them higher reliability to support the decision-maker.

Very probably the term *Intelligent Decision Support System* was used for the first time by Clyde W. Holsapple in 1977, in his Ph.D. Thesis entitled "Framework for a generalized intelligent decision support system" (Holsapple, 1977). The framework developed was a generalized and intelligent query processor for decision support searching in an information base, based on the concept maps. The concept of IDSS had been evolving since then until nowadays. Let us try to define what an Intelligent Decision Support System is. To do that, we analyze a few definitions made in the literature. Even though the name is not always IDSS and they are named as DSS or EDSS for specifically environmental domains, they refer to the same concept:

- Gottinger and Weimann (1992) said that "An Intelligent Decision Support System (IDSS) is an *interactive tool* for decision making for well-structured (or well-structurable) decision and planning situations that uses *expert system techniques* as well as *specific decision models* to make it a *model-based expert system* (integration of information systems and decision models for decision support)".
- Haagsma and Johanns (1994) stated, "An EDSS is an intelligent information system that *reduces the time in which decisions are made* in an environmental domain, and *improves the consistency and quality of those decisions*".
- Fox and Das (2000) wrote that "A DSS is a computer system that *assists decision-makers in choosing between alternative beliefs or actions* by *applying knowledge* about the decision domain to arrive at recommendations for the various options. It incorporates an explicit decision procedure based on a set of theoretical principles that justify the rationality of this procedure".
- Cortés et al. (2000) stated that "One of the essential points of the application of AI techniques to this area relies on the *knowledge-based facilities that they provide*

© Springer Nature Switzerland AG 2022
M. Sànchez-Marrè, *Intelligent Decision Support Systems*,
https://doi.org/10.1007/978-3-030-87790-3_4

to accelerate the problem identification. Another point is the *integration of several AI techniques with numerical and/or statistical models* in a single system providing higher accuracy, reliability and usefulness".

- Poch et al. (2004) said that "The *use of AI tools* and *models* provides direct access to *expertise*, and their flexibility makes them *capable of supporting learning* and *decision making* processes".
- Phillips-Wren et al. (2009) wrote that "IDSS add *artificial intelligence* (AI) *functions* to traditional DSS with the aim of *guiding users* through some of the *decision making phases and tasks* or *supplying new capabilities*".

We propose the following definition:

- An IDSS is a *highly reliable* and *accurate computer-based system* that commonly uses *several multidisciplinary methods*, either *data-driven* or *model-driven*, being at least one of them from Artificial Intelligence field, to *support* and to *improve* the *decision-making process* of a user or users through *analytical*, *synthetic*, and *prognosis tasks* in an *unstructured complex domain*, and being able to *learn from past decisions*.

Before analyzing in detail Intelligent Decisions Support Systems, let us review what the aims of Artificial Intelligence field are, and what are the main approaches or paradigms of Artificial Intelligence for a better understanding of the AI models used in Intelligent Decision Support Systems.

4.1 Artificial Intelligence

Artificial Intelligence is considered to be born as a field of research in Computer Science in the meeting at Darmouth College, a private university in New Hampshire in the USA, in 1956. John McCarthy (from Massachusetts Institute of Technology, MIT) convinced Marvin Minsky and Claude Shannon (from MIT), and Nathaniel Rochester (from International Business Machines, IBM) to help him bringing other USA researchers interested in the study of intelligence.

Finally, most regularly attended Allen Newell and Herbert A. Simon (from Carnegie Mellon University, CMU, by then Carnegie Tech), Trenchard More (from Princeton), Ray Solomonoff and Oliver Selfridge (from MIT) and Arthur Samuel (from IBM) to that 2-month workshop in the summer of 1956. In addition to these ten researchers, other ten ones attended at some time, which according to Ray Solomonoff notes were: Julian Bigelow, W. Ross Ashby, W.S. McCulloch, Abraham Robinson, Tom Etter, John Nash, David Sayre, Kenneth R. Shoulders, a Shoulders' colleague, and Alex Bernstein. Especially, the first ten including the four organizers, who attended the meeting most of the weeks, became the founders and pioneers of AI research. In the meeting, they started to think about what could be considered as "intelligent" with the aim to reproduce such an "intelligent" behaviour, and they thought that some activities, like for example, playing a game, solving word problems in algebra, proving logical theorems and speaking English, could be considered as highly intelligent. They set it as a great challenge to produce programs emulating the human behaviour of playing a game like chess, etc. They and their students produced such programs.

The name "*Artificial Intelligence*" is agreed that was coined by John McCarthy in that meeting. AI is an interdisciplinary science nurtured by Computer Science, Mathematics, Cognitive Psychology, Linguistics, Philosophy, etc. Notwithstanding, Ramon Llull in the middle ages (1305) was the first to propose and conceive a mechanism to automate the reasoning processes (the method of *Ars Magna*, which was named as *ars combinatoria* by Leibniz), which influenced Leibniz to construct the first mechanical calculator with the four basic arithmetic operations (Universal calculator) in 1694. Therefore, he can be considered as an earlier pioneer of the field. Many researchers in the literature have defined Artificial Intelligence (AI). Let us review some of them:

- E. Charniak and D. McDermott (1985) said "*Artificial Intelligence is the study of mental faculties through the use of computational models*".
- E. Rich and K. Knight (1991) defined "*Artificial Intelligence (AI) is the study of how to make computers do things, which, at the moment, people do better*".
- L. Steels (1993) stated "*Artificial Intelligence is a scientific research field concerned with intelligent behaviour*".
- H. A. Simon in June 1994 was interviewed by Doug Stewart from OMNI magazine (OMNI, 1994) and to the specific question "What is this the main goal of AI?", he answered: "AI can have two purposes. One is *to use the power of computers to augment human thinking*, just as we use motors to augment human or horsepower. Robotics and expert systems are major branches of that. The other is *to use a computer's artificial intelligence to understand how humans think*. In a humanoid way. If you test your programs not merely by what they can accomplish, but how they accomplish it, they you're really doing cognitive science; you're using AI to understand the human mind".

We propose the following definition of AI:

Artificial Intelligence is the study of the possible or existing mechanisms—in humans or other beings—providing such behaviour in them that can be considered as *intelligence*, and the simulation of these mechanisms, named as *cognitive tasks*, in a computer through the computer programming.

Another important element of AI is to know what the main AI areas are. Depending on the point of view, different areas can be distinguished. Taking into account the *application field*, there are many areas, but the following ones are chronologically the most important in AI:

- Puzzle Resolution
- Automatic Theorem Proving/Logic Programming
- Game Theory
- Medical Diagnosis
- Machine Translation

- Symbolic Mathematics and Algebra
- Robotics
- Fault Diagnosis
- Text Understanding and Generation
- Monitoring and Control Systems
- e-Commerce
- Business Intelligence
- Intelligent Decision Support Systems, Recommender Systems
- Intelligent Web Services
- Social Networks Analysis

According to the *cognitive tasks*, the following AI areas can be enumerated:

- Vision and Perception
- Natural language understanding
- Knowledge acquisition
- Knowledge representation
- Reasoning
- Search
- Planning
- Explanation
- Learning
- Motion
- Speech and Natural language generation

To analyze the different methodologies and paradigms of AI, which have been proposed in the literature, the second classification is better because focus on the main *cognitive tasks* of humans. Therefore, the ideal AI system would be what is named as an *intelligent agent* (i.e. an intelligent robot) showing *all the cognitive tasks* above enumerated. Thus, the goal of AI is to construct such *intelligent systems* or *intelligent agents*.

That means that the *agents* must show an *intelligent behaviour* like:

- Autonomy
- Learning skills
- Communication abilities
- Coordination abilities
- Collaboration abilities with other agents, either humans or artificial ones.

The general scheme of humans or intelligent agents' behaviour is summarized in Fig. 4.1.

According to M. Wooldrige and P. Jennings, "an *intelligent agent* is a computer system that is capable of *flexible* autonomous action in order to meet its design objectives" (Wooldrige & Jennings, 1995).

Agents must be:

- *Responsive*: Agents should perceive their environment and respond in a timely fashion to changes that occur in it.

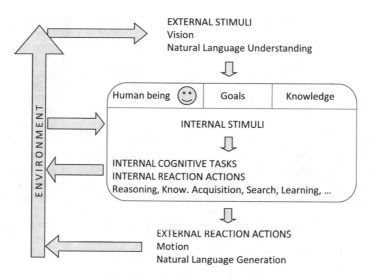

Fig. 4.1 Human beings or intelligent agents' behaviour chart

- *Proactive*: Agents should be able to exhibit opportunistic, goal-directed behaviour, and take the initiative where appropriate.
- *Social*: Agents should be able to interact, when they deem appropriate, with other artificial agents and humans in order to complete their own problem-solving capabilities and to help others with their activities.

In Table 4.1, there are several examples of different kinds of intelligent systems or intelligent agents.

Of course, to exhibit all cognitive tasks in a computer program or a robot is extremely difficult. What usually happens is that the intelligent systems that have been historically built, emulate some of them. For example, the first *game playing intelligent systems* for chess (see the Chess Playing System in Table 4.1) in the 60s and 70s were emulating *human search processes* and some kind of *human reasoning*, to explore the possible moves of the adversary in the game to select the best move for the player. One of the most successful intelligent systems were the *expert systems* in the 80s, which solved complex problems using specific *expert knowledge* in a concrete domain, commonly coded as inference rules (see the Medical Diagnosis System in Table 4.1). Expert systems emulated several *human reasoning* mechanisms, as well as *knowledge representation* abilities, and showed *explanation* skills too. Each one of the intelligent systems and the AI paradigms tries to emulate as much of these cognitive tasks as possible to solve complex problems of the real world. Specifically, reasoning, search, planning and learning cognitive tasks are the main issues of the different AI paradigms. However, what are these AI paradigms?

Table 4.1 Examples of intelligent systems or agents. Adapted from Russell and Norvig (2010)

Agent type	Goals/ performance measure	Environment	Sensors	Actuators
Chess playing system	To win, number of won games	Chess board, chess pieces	Input of opponent movements	Depict the movement of chess pieces, Display of the chess board
Medical diagnosis system	Healthy patient, Reduced costs	Patient, hospital, staff	Keyboard entry of symptoms, findings, patient's answer	Display of questions, tests, diagnoses, treatments, referrals
Satellite image analysis system	Correct image categorization	Images from orbiting satellite	Colour pixel arrays	Display of scene categorization
Internet Softbot	Summarize relevant information	Users, the web	Web pages, text, links	Interesting topic detection in web pages, going to other web pages
Machine Translation System	Text translation from one language to another	Text, Linguistic knowledge	Text	Sentence translation, Semantic interpretation of paragraphs
MARS Pathfinder	Maximum information gathering from Mars	Mars surface, Earth signals	Telemetry, Mars environment images, Velocity, and balance information	Forward, backward, accelerate, braking, turning, take one photo
Refinery Controller System	Purity, yield, safety	Refinery, operators	Temperature, pressure, chemical sensors	Valves, pumps, heaters, displays
Interactive English tutor	Student's score on test satisfactory	Set of students, testing agency	Keyboard entry	Display of exercises, suggestions, corrections

4.1.1   AI Paradigms

The different AI paradigms solve the same complex problems, but using a different approach, a different way of representing the *knowledge* required, a different *reasoning mechanism*, etc. They are usually classified in the AI literature as:

- The *cognitive* or *deliberative* paradigms

  - They are concerned on the *processing of symbols* rather than numerical values.
  - They use a *latent reasoning mechanism*.
  - Most of them are *cognitive-inspired* approaches.

- The *reactive* or *behavioural* paradigms

  - They are concerned about more *numerical computations* and providing nice and intelligent optimizations schemes or function approximation schemes.

- *No evident reasoning mechanisms* are used.
- Most of them are *bio-inspired* approaches.

In addition, there are some techniques oriented to manage the uncertainty originated in the data or in the knowledge or the reasoning processes involved in the real-world problems, which will be reviewed below. These approaches are usually named as *uncertainty reasoning models*.

4.1.1.1 Deliberative Approaches

Logic Paradigm

It is based on representing the knowledge about the problem and the domain theory through *logical formulas*.

The main reasoning mechanism is the *automatic theorem proving* using the automated resolution process set by Robinson (1965). It is a very general mechanism, which can be applied to any domain. Major techniques are based on Logic Programming.

For instance, in the domain of the "family theory", logic programming using Prolog, can be used to formalise the problem. The knowledge is expressed with facts (logical propositions) and with inference rules (relations among first-order logic predicates and/or facts). In the example, we have some facts like "John is a man", "Ann is a woman", "Michael is the father of Peter":

```
man(john).
man(peter).
man(michael).
woman(ann).
woman(eliza).
woman(ada).
father(michael, peter).
father(john, michael).
mother(ann, peter).
```

Moreover, the inference rules, which express the definition of a first-order logic predicate, in terms of other facts and predicates like "X is the grandfather of Y if X is a man, X is the father of Z, and Z is the father of Y":

```
grandfather(X,Y) :- man(X), father(X,Z), father(Z,Y).
grandfather(X,Y) :- man(X), father(X,Z), mother(Z,Y).
grandmother(X,Y) :- woman(X), mother(X,Z), father(Z,Y).
grandmother(X,Y) :- woman(X), mother(X,Z), mother(Z,Y).
son(X,Y) :- man(X), (father(Y,X); mother(Y,X)).
daughter(X,Y) :- woman(X), (father(Y,X); mother(Y,X)).
brother(X,Y) :- man(X), ((mother(Z,X), mother(Z,Y);
        (father(Z,X), father(Z,Y))).
sister(X,Y) :- woman(X), ((mother(Z,X), mother(Z,Y);
        (father(Z,X), father(Z,Y))).
```

Then, the system can answer questions like:

```
? - grandfather(john, X).
 Yes
 X = peter
```

The operator "*A, B*" is a conjunctive logical operator (*A* and *B*), and the operator "*A; B*" is the logical disjunctive operator (*A* or *B*). The operator "*A :- B*" is equivalent to the inference rule (logical implication operator, $B \Rightarrow A$) "If *B* then *A*".

Heuristic Search Paradigm

It is based on *searching* within a space of possible states, starting from the initial state to a final state, where the problem has been solved. Each *state* is the representation of the current state of the problem-solving process. The *state space* is a graph structure to be intelligently explored. The ordered list of the different operators that must be applied, from the initial state to the final state is the *plan* that must be followed to solve the problem.

Most commonly used techniques are *the beam search algorithm, the A* algorithm* and other heuristic search approaches. See Rich and Knight (1991) or Russell and Norvig (2010) for further information. The initial node represents the initial situation of the problem. For instance, a *numerical puzzle* is to be solved with the initial position of all the puzzle pieces. See the Fig. 4.2 for the representation of the initial state in the numerical puzzle example.

Starting from the initial node (s_0), the different methods to find a solution (the solved puzzle with all the pieces in its right position) generate all the possible successor nodes from the current node, applying all the possible *operators*. In the case of the puzzle problem, the *operators* are the possible movements of the pieces, and the nodes are the different puzzle board situations. After the possible successor nodes are generated, they are evaluated according to some cost function ($f = g + h'$). The cost function f computes the cost of arriving from the initial state s_0 to a final state (s_f) passing through the current node s_i. The function g is the real cost from the initial state s_0 to the current node s_i, and the heuristic function h' is an estimation of the cost to arrive from the current node s_i to a final state (s_f) following that path. In the example of the puzzle, h' is an estimation of the number of movements required

Fig. 4.2 An initial state of a possible numerical eight-puzzle problem

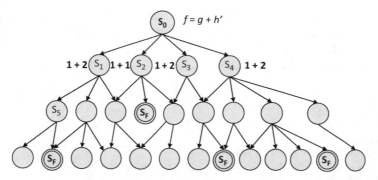

Fig. 4.3 The state space graph of a possible eight-puzzle problem

to arrive at the final position with all the pieces in the right position. The path indicating the minor cost is selected to try to reach a solution as fast as possible among the entire node graph. In the example of a puzzle node in Fig. 4.3, at the initial node, four successor nodes are generated (the four possible movements). Then applying the *A* algorithm* it evaluates all the possible nodes computing the h' heuristic, and selects the node with a lower estimated cost (node s_2 with a h' (s_2) = 2), which effectively with just a new more movement can reach a final state (s_f).

Knowledge-Based Paradigm

This kind of approach tries to get benefit from the particular *knowledge* of a concrete domain, which is normally used by experts when facing the problems to be solved.

This knowledge is encoded in what has been named as a *Knowledge Base* or *long-term memory*. Most common knowledge bases are implemented as *inference rules* (IF <conditions> THEN <actions>). The knowledge is explicitly represented by the inference rules. In addition, the *short-term memory* or *Fact Base* represents the knowledge about the current problem being solved. Sometimes, some researchers name the *Knowledge Base* as the *Rule Base* and use the term *Knowledge Base* to refer to both the *Fact Base* and the *Rule Base*.

This paradigm is based on the cognitive theory that states experts have their knowledge in the brain coded as inference rules. When solving a new problem, the rules are fired according to the current information (facts) of the problem at hand. Main examples of this paradigm are the *Expert Systems* and the *Intelligent Tutoring Systems*.

The reasoning mechanisms are the *forward reasoning* and *backward reasoning* engines. The technique used here is commonly known as *Rule-Based Reasoning* (*RBR*), or sometimes also known as *Knowledge-Based Reasoning* (*KBR*), and the system is usually called as a *Knowledge-Based System* (*KBS*). This paradigm will be deeply explained in Chap. 5, Sect. 5.3.

For instance,[1] suppose a Rule-Based Reasoning system (expert system) to decide whether a loan can be given to a customer or not for launching a new software company for developing IDSSs!

The Fact Base can be formed by:

```
Assets-value (AV)
Amount-required (N)
Financial-support (FS)
Financial-history (FH)
Amount-already-pending (AAP)
Reliability-of-devolution (RD)
Company-viability (CV)
...
Loan-given (I)
Loan-not-given ()
Loan-given-with-preferential-interest (PI)
```

Moreover, the Rule Base can be formed by several inference rules, coming from the expertise of the bank staff related to loan assessment, like:

```
Assets-value < 5*10⁵ → Insufficient-assets-value
Assets-value ≥ 5*10⁵ ∧ Assets-value < 3*10⁶ → Sufficient-assets-value
Assets-value ≥ 3*10⁶ → Excellent-assets-value
...
Financial-support=low ∧ Insufficient-assets-value → Loan-not-given
Reliability-of-devolution=low → Loan-not-given
. . .
Financial-support=good ∧ Sufficient-assets-value ∧
Financial-history=good ∧ Company-Viability=good → Loan-given (I=3)
. . .
Financial-support=high ∧ Excellent-assets-value ∧ Financial-
history=good ∧ Company-Viability=very-good → Loan-given-with-PI
(I=2.25)
. . .
```

Using a forward reasoning mechanism, the intelligent system can arrive to the adequate loan concession or not conclusion, applying the suitable inference rules.

Model-Based Paradigm

This approach is very similar to the Knowledge-based because the knowledge of a particular domain is used. The difference relies on the fact that the *knowledge* is *implicitly encoded* in some kind of model.

---

[1]This example is borrowed from my colleague Dr. Javier Béjar, Dept. of Computer Science, Universitat Politècnica de Catalunya.

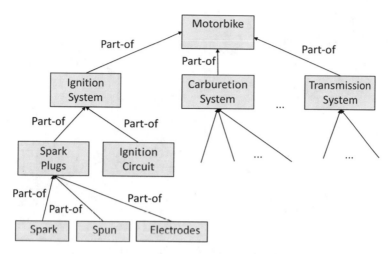

Fig. 4.4 A component-oriented model for the motorbike diagnosis problem

Most common approaches are *causal models* or *component-models* reflecting the causal or composition relationships among several components of a system, or *qualitative models* reflecting the qualitative relationships among several attributes, which are characterizing the domain.

The reasoning process is done through some kind of interpreter of the model and its component relationships.

Model-Based Reasoning (MBR) (Reiter, 1987) and *Qualitative Reasoning* (de Kleer & Brown, 1984; Forbus, 1984; Kuipers, 1994) are major techniques using this approach. Both approaches will be explained in Chap. 5, Sects. 5.4 and 5.5, respectively.

For instance, a *diagnose system for identifying problems in a motorbike*, could be based in the model about the different components of the motorbike, and making the corresponding inferences implicitly signalled by the model component chart. If there is a problem in the motorbike, this problem should be in one or more of its sub-components, like the ignition system, or the carburetion system or in the transmission system, etc. Figure 4.4 depicts a component-oriented model for the motorbike diagnose problem.

Experience-Based Paradigm

This approach tries to solve new problems in a domain by reusing the previous solution given in the past to a similar problem in the same domain (*analogical reasoning*). Thus, the solved problems constitute the "knowledge" about the domain.

As more experienced is the system better performance achieves, because the experiences (cases or solved problems) are stored in the *Case Base*. This way the system is *continuously learning* to solve new problems. The technique used in this

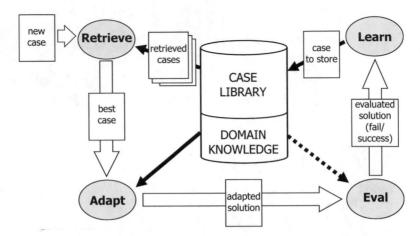

Fig. 4.5 Basic reasoning cycle of Case-Based Reasoning (CBR)

approach is known as *Case-Based Reasoning (CBR)* (Riesbeck & Schank, 1989; Kolodner, 1993; Richter & Weber, 2013). It is based on the *theory of dynamic memory* of Roger Schank (1982), which states that human memory is dynamic and change with its experiences along with their lives. Humans learn new things and forget others. The acts of humans are recorded as scripts in their memory (Schank & Abelson, 1977). In addition, the process of learning, understanding, reasoning, and explaining are intrinsically bind together in human memory.

The main reasoning cycle in Case-Based Reasoning is depicted in Fig. 4.5. This paradigm will be described in Chap. 6, Sect. 6.4.2.

4.1.1.2   Reactive Approaches

Connectionism Paradigm

This approach is inspired by the biological neural networks, which are in the brain of many living beings. The model of an *Artificial Neural Network (ANN)* mimics the biological neural networks with the interconnection of artificial neurons.

The ANN will produce an output result, as an answer to several input information from the input layer neurons, emulating the neural networks of the brain, which propagate signals among all the neurons interconnected.

The ANNs can have several intermediate layers (hidden layers) between the input layer and the output layer.

ANNs are *general approximation functions* very useful in non-linear conditions. Therefore, they can be used as a prediction system for a variable of interest.

In Fig. 4.6, there is a chart of an ANN with one hidden layer of five neurons, six neurons in the input layer and two neurons in the output layer. This paradigm will be explained in Chap. 6, Sect. 6.4.2.2.

Fig. 4.6 A chart of an ANN
with one hidden layer

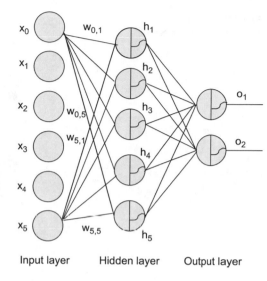

Input layer   Hidden layer   Output layer

Evolutionary Computation Paradigm

Evolutionary computation, and especially the most commonly used approach, *the genetic algorithm* paradigm, is a bio-inspired approach mimicking *selection natural process* in biological populations (Holland, 1975; Goldberg, 1989). It uses iterative progress, such as growth or development in a population. This *population* is composed of a set of *individuals*. Individuals represent, usually with binary code, entities, which are the objects to be explored in order to optimize a *fitness function*. The *fitness function* encodes the target to be optimized. This target could be a set of rules, a classification process, a variable prediction, etc.

The population is being evolved by means of several *genetic operators*. Most used operators are the *selection*, *mutation*, and *crossover* operators. All of them are inspired by the equivalent genetic processes in live beings. Starting from an initial population, new generations are built after applying the genetic operators until a satisfactory solution is obtained or the number of generations exceeds a maximum threshold.

It is possible to use parallel processing to fasten the optimization process. Such processes are inspired by biological mechanisms of evolution. Evolutionary computation provides a *biological combinatorial optimization* approach. This paradigm will be described in Chap. 6, Sect. 6.4.3.

A general scheme of a genetic algorithm process is depicted in Fig. 4.7.

Fig. 4.7 A genetic
algorithm general scheme

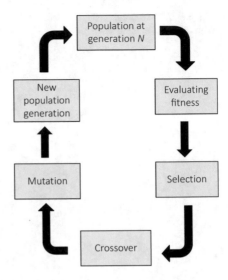

Other Optimization Paradigms

There are other several optimization techniques, named also as *metaheuristics*,[2]
which mostly emerged by the observation of nature or human activity. They can
be grouped in two major types:

- *Collective problem-solving* techniques named as *Swarm Intelligence optimization*, which are inspired by the biological groups of insects (ant colony, bee
  colony), birds (flock birds), etc. Most well-known techniques are:
  - *Ant Colony Optimization* (Dorigo, 1992): It is a technique, which mimics the
    method used by an ant colony to find the shortest path to interesting food for
    the colony. They use some pheromone quantities to signal the length of the
    path explored. After a short time, all the ants go on the shortest path to the
    food. This kind of optimization technique is useful to find the *shortest paths* in
    a graph.
  - *Swarm Particle Optimization* (Kennedy & Eberhart, 1995): It is a method,
    which optimizes a defined function or measure through the iterative movement
    of particles. Each particle is a candidate solution to the problem. Each particle
    changes its position and velocity, according to some mathematical formulas
    combining a random component with the information about the local optimum
    of the particle and the global optimum found by all the particles. The idea is
    that the swarm moves towards the optimum solutions.

---

[2] A metaheuristic is a high-level heuristic designed to find, generate, or select a heuristic (a criterion
or function to decide the best alternative to search in an optimization problem), to provide the best
possible solution to an optimization problem.

- Other techniques are specific *local search optimization techniques using several meta-heuristics*. Most commonly used are:

  - *Tabu Search* (Glover, 1986, 1989, 1990) is a local search technique to optimize a certain function by trying to improve a candidate solution through moving to some neighbour solutions. It enhances the performance of local search by allowing some relaxation on it. First, at each step worsening moves can be accepted if no improving move is available. Moreover, prohibitions, hence the term *tabu*, are introduced to discourage the search from coming back to previously visited solutions in a short-time period or solutions having violated some rule. The method uses a *tabu* list of prohibited candidate solutions to be visited.
  - *Simulated Annealing* (*SA*) Method (Khachaturyan et al., 1979; Kirkpatrick et al., 1983): is a probabilistic technique to approximate the global optimum of a given function. The name of the technique comes from *annealing process in metallurgy*, which is a technique involving heating and controlled cooling of a material. Heating and cooling the material affects both the temperature and the thermodynamic free energy. The function $E(s)$ to be minimized, is analogous to the internal energy of the system in that state. The goal is to bring the system, from a randomly initial state, to a state with the minimum possible energy, and temperature. At each step, the SA method considers some neighbour state, s', of the current state, s, and probabilistically decides between moving the system to state s' or staying in state s. These probabilities ultimately lead the system to move to states of lower energy. This iterative step is repeated until the system reaches a good state for the application, or until an allowed time has been spent.

4.1.1.3 Uncertainty Reasoning Models

Bayesian Networks

A *Bayesian network* or *belief network* (Pearl, 1988) or directed acyclic graphical model is a *probabilistic graphical model* that represents a set of random variables and their *conditional dependencies* via a Directed Acyclic Graph (DAG).

A Bayesian network is composed of a set of nodes (the random variables) and the associated conditional probabilities to each node for each possible value of the random variable.

They provide an *inference reasoning mechanism* to obtain the new probability values of any variable within the network after some new evidences are known. When new evidence are available, the conditional probabilities must be updated according to the Bayes theorem and conditional probability definition. This uncertainty reasoning model will be described in Chap. 9, Sect. 9.2.4.

A classic example (Pearl & Russell, 2000) based on a former example by K. Murphy (1998) is the scenario where one wants to model the conditional dependences between the event of slippery when going out of your home, where

Fig. 4.8 A classic example
described in Pearl and
Russell (2000) of a Bayesian
network for the slippery
causality problem

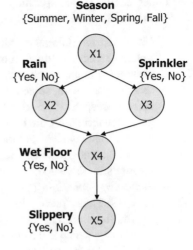

Fig. 4.9 Possibilistic
functions for the sets "Cool
Temperature" and "Young
Age"

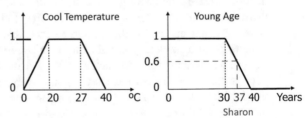

you have a garden, which can be wet, and other related events. Moreover, you can
take into account that it could have been raining during the night or perhaps you had
your sprinkler open during the night, or the current season of the year. This simple
Bayesian network is depicted in Fig. 4.8.

Fuzzy Logic Systems

Systems based on *fuzzy logic* and *possibilistic theory* (Zadeh, 1965) model the
vagueness and imprecision concepts. They were risen to model the inherent vagueness of natural language. For instance, think in the statement "Sharon is young",
which is completely vague, and the age of Sharon is not clearly determined.

The mathematical possibilistic model assigns a possibility value for each element
to belong to a fuzzy set. Values are evaluated in terms of logical variables that take
the value on a continuous range between 0 and 1. If the age of Sharon is 37, then her
possibility value of being young, according to the fuzzy set "Young Age" is 0.6, as
depicted in Fig. 4.9. Moreover, the possibilistic function to evaluate the degree of
possibility of any temperature to be a "Cool Temperature" is also represented in
Fig. 4.9 with a trapezoidal function. Fuzzy logic has been successfully applied in

rule-based reasoning systems to model the uncertainty of the real world. This uncertainty reasoning model will be described in Chap. 9, Sect. 9.2.5.

4.2   IDSS Typology

Intelligent Decision Support Systems can be divided, according to the *nature of the problem or process* they are giving support, into:

- Static IDSS
    This kind of IDSS give punctual support to decision-making, usually offline, and are mainly used to support multi-criteria decisions of policy-makers more than to make real decisions on a day-to-day basis. They are commonly used to support the selection of the *best design* or *configuration of a system*. The user of the IDSS can analyze the different "What-if" scenarios, to explore the response surface and the stability of the solution.
    Usually, the decision-maker needs to know how sensitive are the possible alternatives to small variations in the weight or/and value of the relevant variables in the decision scenario.
    These IDSSs are very specific and sometimes are only built to make or justify one decision. However, the complexity of these decisions could be very high. Some decisions could be tactical or strategic for an organization. Therefore, the *experiential knowledge* in these systems is not available, and many times what is most important is the *expert knowledge* to know how to cope with a complex problem, which has been not faced in the past.
    Commonly, some *mechanistic* or *first-principles models* are used in this kind of systems.
    Examples: Industrial process design IDSS systems, Training systems for emergency management IDSS systems, Launching a new product to the market IDSS systems.
- Dynamic IDSS
    This kind of IDSSs aim at controlling, managing, or supervising a process or system in real-time (online), facing similar situations on a regular basis (Sànchez et al., 1996). As they are managing usually an online process, they must be able to guarantee robustness against noise, to manage missing data, and to work with any combination of input data.
    These kinds of systems are very complex, as they have to manage and support the decision-makers to several kinds of decisions. Some decisions are routine decisions, which appear commonly in the daily operation of the systems managed. Other decisions could be more strategical and complex. For the routine decisions, the past experience can be used to help in the decision-making problem, as the decision problems could be repetitive in time. Thus, the *experience-based paradigm* here appears to be a good solution for this kind of system.

In addition, as many data can be recorded in this kind of system, data-driven, and specifically, data model exploration methods can be used.

Usually, the end user is responsible for accepting, refining, or rejecting proposed system solutions, but as this kind of system evolves in time, and there is an increase in the IDSS confidence because the IDSS is solving similar situations to those successfully solved in the past.

Examples: Wastewater Treatment Plant Supervisory IDSS System, Patient Disease Management IDSS System

- Hard-Constrained Dynamic IDSS

   These kinds of systems are Dynamic IDSS facing *hard real-time constraints* or *possible hard catastrophic consequences*. Therefore, they share the same features described above for the Dynamic IDSS. Moreover, some of them have the added difficulty of the hard real-time constraints. This means that all the support provided to users must be computed in a fast and accurate way to satisfy the constraint of a very fast response of the system. Therefore, the different models integrated into these systems should be computationally fast.

   Examples: Automatic Supervisory IDSS in a Medical Intensive Care Unit, Environmental Emergency Management IDSS

4.3   Classification of IDSS

As done with DSS, we propose to group the different kinds of Intelligent Decision Support Systems, attending to the *type of knowledge* used. Attending to this criterion, DSSs can be classified in the two major types: *model-driven intelligent decision support systems* and *data-driven intelligent decision support systems*. Of course, as happened with DSS, in practice any IDSS could be a mixture of different kind of IDSS types merging some model-driven IDSS method and some data-driven IDSS method. Moreover, some kinds of methods are usually data-driven methods, like the induction of a decision tree model (see Chap. 6), but also a decision tree could be constructed just as a model coming from an expert, thus being a model-driven method. On the contrary, some methods which usually are model-driven, like a qualitative reasoning model, obtained from the experts and the first-principles of a theory or domain (see Chap. 5), could be also obtained directly from data (a data model exploration model), thus being data-driven. Next, these two general types and the models they are using will be described, grouped in the most common way of use.

4.3.1   Model-Driven IDSSs

Model-driven IDSSs use one or more models coming from the artificial intelligence field to support decisions. The quality of the IDSS mostly relies on the quality of the model/s. They usually have a friendly interface to use the model/s. All these methods will be described in detail in Chap. 5. Most common kinds of models used in model-driven IDSSs are:

- *Agent-Based Simulation Models*: Agent-based simulation models are simulation models using an emerging paradigm of artificial intelligence, the multi-agent systems (Wooldrige & Jennings, 1995). Agent-based models consist of dynamically interacting rule-based agents. Each one of the elements of the system to be simulated is associated with an agent: the user, the different components of the system, etc. These agent-based simulation models can simulate different scenarios for evaluating the consequences of critical processes in decision-making. These models have been developed especially in natural systems (biology, environment), in medicine area, in management, and social systems (game theory, social networks, etc.). These models will be described in Chap. 5, Sect. 5.2.
- *Expert-based Models*: These kinds of models are models derived from the *expert knowledge* about a concrete task. Experts usually have a precise knowledge they use to solve successfully complex tasks. These models code this knowledge patterns in some kind of representation formalism. The most common way of representing this knowledge is the use of *inference rules*. These *inference rules* form the *Knowledge Base* of this kind of systems. Thus, usually, the reasoning mechanism used is rule-based reasoning. This kind of models have been used both for diagnosis problems and for design or configuration problems. These models will be explained in Chap. 5, Sect. 5.3.
- *Model-based Reasoning Methods*: These kinds of models are models derived from the *observation of a compositional model* of the system object to study. The *knowledge is implicitly coded in the model of the system.* They allow finding explanations for certain observed situations of a system. Thus, they have been successfully applied for *diagnosis tasks*. Most common approaches are *causal models* or *component models* reflecting the causal or composition relationships among several components of a system. These kinds of models are also named as *reasoning from first-principles* models (Reiter, 1987), as they only rely on the models for the problem-solving process. They are much related to the first-principles or mechanistic models explained in Chap. 3, Sect. 3.3.1.1. Main difference is that here the models are not expressed in algebraic or differential equation models but usually in logical rule models, and they use some formal reasoning mechanisms. These models will be described in Chap. 5, Sect. 5.4.
- *Qualitative Reasoning Models*: In addition, there are *qualitative models* reflecting the qualitative relationships among several attributes, which are characterizing the domain. They are also related to the first-principles or mechanistic models explained in Chap. 3, Sect. 3.3.1.1. These models were emerged from the *qualitative physics* domain (de Kleer & Brown, 1984) and *qualitative process theory* (Forbus, 1984). *Qualitative reasoning* (Kuipers, 1994) can be used as a tool both for modelling and for the simulation of a system or process. Given some observation about variables' values, it can be predicted how will evolve the values of other related variables (*prognosis task*), or how these values could have been produced (*diagnostic task*). These models will be explained in Chap. 5, Sect. 5.5.

4.3.2 Data-Driven IDSSs

Data-driven IDSSs use one or more data-driven models from artificial intelligence, and specifically from the inductive machine learning field. Data is used to extract relevant information, i.e. models, for the decision-maker. The quality of the IDSS relies on the *data*, which these IDSSs *indirectly explores using a model obtained from the data* (data model). Therefore, all are Data Model Exploration IDSSs. Furthermore, all are Data Explorative model IDSSs. This main classification leads to the next classification level:

- Data Model Exploration IDSSs

  - Data Explorative model IDSSs: *Data Explorative model IDSSs* are data model exploration intelligent decision support systems, which capture the information from data through a data explorative model. These kinds of IDSS are data-driven, but indeed, they use a model obtained (induced) from data. Data explorative models, as described in Chap. 3, Sect. 3.1 use a bottom–up approach. The models are obtained from the exploration of the data. Most intelligent models induce the model through some *machine learning* techniques. All these methods will be deeply described in Chap. 6. These methods can be divided in:

    Unsupervised Models: These models aims to get some knowledge from the data (*cognitive process of discovering new knowledge patterns*), searching for relationships among elements. There is no target or response variable being the object of interest:

    - Descriptive Models: These models search relationships among data instances or examples. Some authors in the literature name them as profiling models.
    - Associative Models: These models search relationships among random variables.

    Supervised Models: These models aim to get some model able to produce or predict the values of a target or response variable of interest (*recognition process of previous knowledge patterns*).

    - Discriminant Models: The variable of interest is a *qualitative variable*, which can take one of a finite set of different qualitative values (class labels). The models will be able to *discriminate* to which qualitative/class label belongs a new unseen instance, for this qualitative variable. In the literature, these models are also referred to as *classification models*.
    - Predictive Models: The variable of interest is a *quantitative variable*, which can take a numerical value, integer or real, among an infinite set of values. The models will be able to *predict* the quantitative value of a new unseen instance for this quantitative variable. In the literature, these models are also referred to as *regression models*.

4.4   Conceptual Components of an IDSS

All IDSSs usually share the same conceptual components (as depicted in Fig. 4.10): The artificial intelligence techniques component, some statistical/numerical methods component, an ontological component, an economical cost component, and some other complementary components. Let us describe each one of these components:

- The *artificial intelligence techniques component*: main feature of an IDSS is the use of, at least, one method coming from artificial intelligence field.

  As it has been explained in previous Sect. 4.3, these methods could be *data-driven*, *model-driven*, or including both kinds of methods. This component is essential in an IDSS, and confer it the label of "intelligent".

  These models in the IDSS can give support to different steps of the problem at hand.
- The *Statistical/Numerical Methods Component*: as it has been previously explained usually IDSSs integrates some statistical or numerical methods. These methods could be *data confirmative models* like a multiple linear regression model or *data explorative models* like a principal component analysis method. In addition, they can be *model-driven methods* as some optimization or

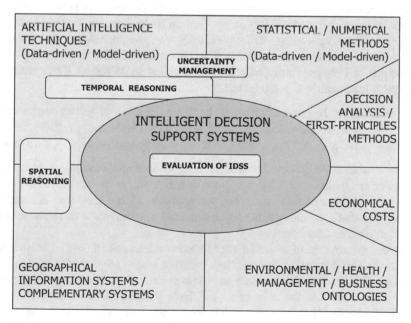

Fig. 4.10 Conceptual components of an IDSS

simulation model. These methods could complement the methods from artificial intelligence in different ways, providing IDSSs that are more reliable.

- The *Decision Analysis or First-Principles Methods Component*: An IDSS can integrate some model-driven methods regularly used in classic DSS such as some *decision analysis* methods like a decision matrix, a decision table, a decision tree model, or an analytic hierarchy process model. Furthermore, some *first-principles models* from de application domain can be incorporated into the IDSS.
- The *Ontological Component*: From a conceptual point of view, each IDSS must have an ontological component, which takes care of the necessary and important *knowledge* about a concrete domain (business, management, health, environmental, etc.). An *ontology* is a description of the important characteristics and relations of a domain. This knowledge can be represented with different kinds of formalisms, and not just an ontology, but conceptually it can be referred as the ontological component.
- The *Economical Cost Component*: In most IDSSs, there are some *economic constraints*, which must be taken into account and must be satisfied as much as possible. Usually, most organizations want to make the best possible decision but are constrained to the fact that some benefit function must be maximized or some cost function must be minimized.
- *Other Complementary Systems* component: Depending on the concrete domain to which the IDSS is giving support, some other components can be needed. For instance, a common component in environmental domains could be a Geographical Information System (GIS) component. In those domains, it is very useful to have a GIS component providing the IDSS with geographic-related information in several layers of a located zone, which is usually divided in a grid of cells. In addition, a GIS provides powerful visualization tools for an IDSS with geographic localization requirements.

In addition, in the chart of Fig. 4.10 there are depicted, in yellow round-shaped rectangles, four of the main *advanced* and *open challenges* in IDSS research (Sànchez-Marrè et al., 2006), which will be thoroughly analyzed in Chaps. 9 and 10:

- *Temporal Reasoning*: Usually in most IDSS, the temporal component of available data or available knowledge is not considered to simplify the complexity of the problem at hand. Most reasoning mechanisms do not take into account the temporal relations among some important entities, which make them less accurate and reliable as they could be.
- *Spatial Reasoning*: In a similar way to temporal reasoning, spatial reasoning is not considered in most of IDSS when there are spatial relationships among some spatial entities, like for instance in environmental domains where the different spatial regions, i.e. the cells of a grid, have an intrinsic relation of closeness. Taking into account these spatial relations could improve the reasoning mechanisms of these spatial IDSSs.

- *Uncertainty Management*: In real-world problems, both the data available and the knowledge can be uncertain. Sometimes, the online sensors are not working properly and the values generated are not completely true, other times the values are subjective values observed by humans, and finally, the different models encoding the knowledge should take into account that the knowledge patterns could not be completely certain. Thus, some uncertainty models to manage the inexactness in IDSSs should be used.
- *Evaluation of an IDSS*: The evaluation of a whole IDSS is a complex task because the whole IDSS can be composed of several subsystems (different kinds of models) and components. We will analyze in this chapter the main features to be considered in the evaluation of an IDSS, but there is not a unified protocol to validate IDSSs. This issue is an open challenge not yet solved. In final Chap. 10, a general protocol for IDSS evaluation will be provided.

The location of the yellow round-shaped rectangles in the picture is significant. Thus, they are located within the conceptual components which should cope with these challenges. For instance, the *temporal reasoning challenge* is fully related to the *artificial intelligence techniques component*, and a little bit related to the *statistical/numerical technique component*. Both components should be analyzed and modified to be able to cope with this temporal reasoning component.

4.5   Considerations and Requirements of an IDSS

Now, we have analyzed the different typologies of IDSS, a possible classification of IDSS according to the type and origin of the knowledge used (data-driven or model-driven), and the conceptual components of an IDSS. It is time to think about which should be the general requirements of an Intelligent Decision Support System. What features must be considered to build an IDSS? The following list enumerates the main requirements and features to be taken into account for the construction of an IDSS:

- *Analysis of the Domain and Scope of the Decision Problem*: The domain and the scope of the decision problem must be known, understood and analyzed. Depending on the domain at hand, some first-principles or mechanistic models could be used. It is very important to determine the clear scope of the problem. The decision problem must be analyzed, and the important features characterizing the decision problem must be discovered.
- *Defining the Objectives of the IDSS*: Many software systems, including the IDSS fail because the aims of the systems were not clearly defined. The objectives of the IDSS must be clearly stated. It is not the same to give support for a static IDSS than a dynamic IDSS. It is not the same to give support for a diagnostic task than

for a design task, etc. Each one of the tasks requiring support must be identified and the degree of support needed must be assessed.

- *Identifying the Type/s of Decision/s Problem/s*: First, the different decision problems involved should be identified before constructing the IDSS. Then, for each decision problem, its type, and characteristics should be determined. From the point of view of one company, the decision could be an operational decision, a tactical decision or a strategic decision. Depending on the type, the degree of complexity is different, and the frequency of the decision-making process is different.

- *Analysis of Decision-Making Process/es*: The IDSS should provide the user with the analysis of each one of the decision-making processes, following the basic three steps proposed by Simon (1960): *intelligence*, *design*, and *choice*. At intelligence step, the search for information about the decision process is done. At the design step, the set of all possible alternatives is built. Finally, at choice step, all alternatives are evaluated and the best one is selected.

- *Evaluation of the Expected Consequences of Decision Execution*: One important feature, which is fundamental for most IDSS is the ability to forecast the possible consequences of each one of the alternatives to give full support to the decision-maker. These prognostics skills are an essential requirement for most IDSSs. Usually, the prognosis is obtained by some simulation models.

- *Exploration of Data and Knowledge Availability*: it is very important to explore how the *knowledge* about a decision problem can be obtained. If there are data available, then it makes sense to set up *data-driven models*. If data are not available, but there are knowledge coming either from experts of the decision problem or from first-principles theories, then it make sense to set up *model-driven methods*.

- *Inspection of the Organizational and Structural boundaries*: It is necessary to inspect which are the boundaries of the organization. For instance, it is needed to know whether some data merging from different databases in the organization will be possible or not. Alternatively, whether some knowledge from the company can be used in the construction of the IDSS or not, due to confidentiality issues of the company regarding its customers.

- *Impact on and Synergy with the Existing Systems*: It is very important to assess the impact on the existing systems, especially software systems of the organization. If the IDSS can generate new data from solved problems, it is worth to search how these data will affect in the other systems, like the databases of the organization, etc. In addition, it is worth to explore the synergy with other systems in the organization, like how to integrate the IDSS to be constructed with an existing econometric simulation model in the organization.

- *Defining the Profiles of the Users of the System*: It is important to define the different profiles, if needed, of the users of the system. Of course, the final user, which is the decision-maker, is a fundamental exploitation profile, but other profiles could be needed. For example, an administrator user role with different privileges and functionalities available, or an AI scientist user profile with other different authorized privileges and functionalities.

- *Exploring the External Constraints and Contexts*: Usually, some external constraints like environmental laws or regulations in an environmental domain problem, some medical rules in a medical/health system, or some economical costs budget or boundaries need to be taken into account. In addition, some other conditions or facts could be imposed by the specific context.
- *Learning from Past Solved Problems*: This can be a very interesting requirement for an IDSS, which really makes it an intelligent system. If the IDSS can receive feedback about the final decision made by the user and the final outcome obtained, all this information can be used in the future to improve and make more efficient and reliable the IDSS. Notwithstanding, not always this requirement is met by all IDSSs.

Dhar and Stein (1997) propose to group the requirements in four categories depending on what they are related to: *quality of the system, quality of available resources, engineering of the system,* and *logistic constraints.* These requirements can be further classified as belonging to the *system requirements* or to the *organization requirements* related to the environ dimension of the IDSS, and belonging to *execution requirements* or to the *development* requirements related to the deployment stage dimension of the IDSS, as depicted in Fig. 4.11.

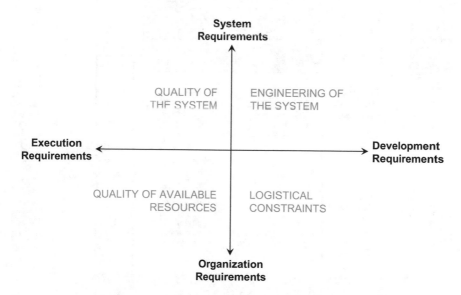

Fig. 4.11 Two-edge dimensions for IDSS requirements. (Adapted from Dhar and Stein 1997)

4.6 IDSS Architecture

In the literature, to our best knowledge, there is no a general architecture proposal to build Intelligent Decision Support Systems. Usually, each particular IDSS is constructed depending on the type of problem, the static or dynamic nature of the IDSS, the type of information, and knowledge available. We proposed in Cortés et al. (2000), an architecture for dynamic IDSS based in fifth levels or steps. Here, we propose a modification to get a six-step levels (see Fig. 4.12):

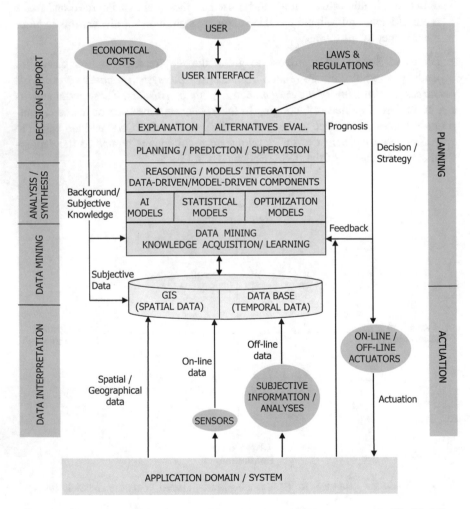

Fig. 4.12 Dynamic Intelligent Decision Support System architecture proposed. (Modified from Sànchez-Marrè's proposal in Cortés et al. 2000)

- The first step of the IDSS (*data interpretation*) encompasses the tasks involved in data gathering and registration into databases. Original raw data are often defective, requiring a number of pre-processing procedures before they can be registered in an understandable, and interpretable way. Missing data and uncertainty must be also considered at this level.
- In the second step, the *data mining level*, which usually is an offline step, where the *data-model data exploration models* are induced from the available data. This level is the knowledge discovery step providing the IDSS with the data-driven knowledge about the domain or process.
- The third step, *the analysis/synthesis level*, includes the reasoning models, data-driven or model-driven components that are used to either infer the state of the system or process (analytical problem solving) or construct the possible solutions (synthetically problem solving) so that a reasonable proposal of actuation can be reached. This is accomplished with the help of statistical, numerical, and artificial intelligence models, which will use the knowledge previously acquired from data, from experts or from first-principles theories.
- The fourth step, *decision support level*, establishes a supervisory task that entails gathering and merging the conclusions derived from AI knowledge models and other models. This level also raises the interaction of the users with the computer system through an interactive graphical user interface. When a clear and single alternative cannot be reached, a set of alternatives ordered by their probability or certainty degree should be presented to the user. Usually, the user takes profit from the prediction abilities of the IDSS using some kind of simulation method to assess the possible consequences of each alternative, prior to selecting the best alternative.
- In the fifth level, *planning level*, plans are formulated and presented to users or managers, as a list of general actions or strategies suggested to solve a specific problem. Plans are generated for the alternative or alternatives selected by the user.
- The set of actions to be performed to solve problems in the domain considered are the sixth and last step, the *actuation level*. The system recommends not only the action, or a sequence of ordered actions (i.e. a plan), but values or parameters that have to be accepted by the decision-maker. This is the final step in the architecture closing the loop. After the actuation, there is the possibility to give feedback to the IDSS on how was the actuation and its evaluation. This *learning ability*, is a very nice opportunity for experience-based reasoning paradigms and can increase the accuracy and reliability of the IDSS along time.

For static IDSS, the architecture could also be applied, but just using the necessary steps. In static IDSS, usually, there is no need for an actuation step.

4.7 IDSS Analysis, Design, and Development

There is a need for a common framework to develop an IDSS even it is either a static IDSS or a dynamic IDSS. Analyzing the main cognitive tasks of human beings within the decision-making environment, the common underlying tasks in most of IDSSs can be extracted.

A *possible cognitive framework for the development of IDSSs* was proposed by Sànchez-Marrè et al. in (2008), which is based on a three-layer architecture, as depicted in Fig. 4.13:

- *Analysis Layer*: In this layer is where most of analytical and interpretative processes are run. At this stage, the data gathering processes, as well as the knowledge discovery process by means of some data mining techniques, are undertaken to get *diagnostic models*. These models will provide the IDSS with hard analytical power to get an insight into the system/process being supervised in real-time or managed in an offline basis. Most of these techniques are *data-driven*. However, other *model-driven methods* can be employed using expert or first-principles knowledge models.
- *Synthesis Layer*: This layer wraps all the work necessary to synthesize possible alternative solutions for the different diagnostics found in the previous step. This synthetization can be done through several *solution generation methods* based on statistical techniques, artificial techniques, or numerical techniques. Of course,

Fig. 4.13 Cognitive-oriented framework for IDSS development

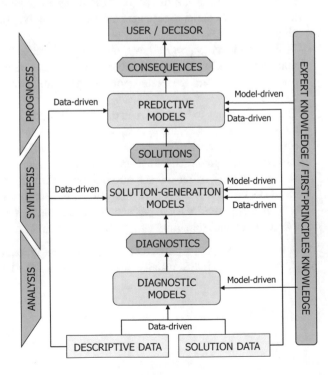

the integration of different nature methods could enhance the problem-solving ability of the IDSS.

- *Prognosis Layer*: At this upper layer, rely on the inherent ability of IDSS to decision support tasks. At this level, the several *predictive models*, which can be numerical (mostly simulations) or rather qualitative (qualitative reasoning or qualitative simulations), are used to estimate the consequences of several actions proposed in the previous step by the solution generation methods. These "what if" models let the final user make a decision based on the evaluation of several possible alternatives. At this stage, the temporal and spatial features could be very important for a good modelling practise.

Finally, for the design and the development of an IDSS, we propose the flow chart depicted in Fig. 4.14. This process is a particularization of the classic software engineering development cycle. Main phases in the development of an IDSS are:

- *Domain Analysis*: The first step in building an IDSS is the analysis and study of the domain problem we are coping with. Main issues of the IDSS should be outlined. The characterization of the most important features to be taken into account must be done. The different decisions and decision problems must be identified. Thus, the different subcomponents of the system should be identified. In order to explore whether some *data-driven models* can be constructed, the availability of historical data about the domain must be done. To assess the possibility of building *model-driven components*, the availability of expert knowledge or first-principles knowledge should be explored.
- *Data Collection* and *Knowledge Acquisition*: If some data-driven explorative model must be built, the corresponding data information, usually represented in data matrices, must be set-up. From the data, some knowledge acquisition methods should be applied (data mining techniques) to get the data models.
- *Model Selection*: Depending on the different issues needed in the domain, on the data available, on the experts' knowledge, or first-principles knowledge available, the most suitable models should be selected. The models could be models from artificial intelligence field, from statistics field, etc.
- *Model Implementation*: According to the models selected, each one of them must be implemented in the IDSS system. Depending on the nature of each model, different kinds of reasoning mechanisms must be used or implemented.
- *Model Integration*: In the usual case that several models are used in the construction of the IDSS, they must be integrated to form the whole IDSS. There are several kinds of integration schemes.
- *Evaluation of the IDSS*: The evaluation of an IDSS is a complex open challenge in IDSS as explained in Sect. 4.3. Anyway, a general practical procedure to evaluate the IDSS is to make an experimental evaluation of the system against some testing scenarios. If the tests give good results, the IDSS can be used. If not, there is a wrong IDSS development. The errors must be located in some of the preceding steps: model integration, model implementation, model selection, data collection, and knowledge acquisition or in the domain analysis step as shown in Fig. 4.14.

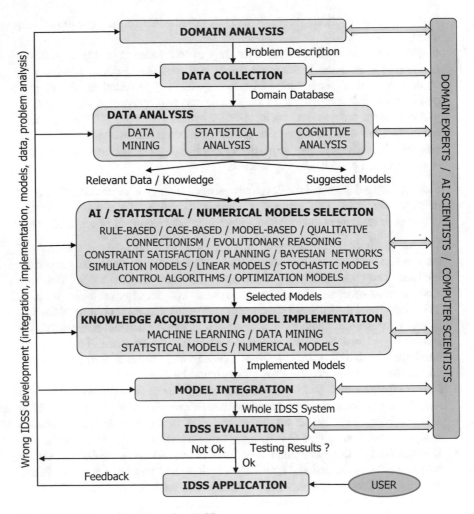

Fig. 4.14 Design and Building of an IDSS

4.8   IDSS Evaluation

As explained in previous Sects. 4.4 and 4.8, the evaluation of an IDSS is a complex task and an open research challenge in IDSS field. We will address this challenge in Chap. 10. Nevertheless, some criteria can be taken into consideration for the assessment of the IDSS. This checklist was proposed by Klein and Methlie (1995) for evaluating a DSS. Main criteria can be grouped in four categories: *system performance criteria*, *task performance criteria*, *business opportunities criteria*, and *evolutionary aspects criteria*.

- System Performance Criteria:

  - Efficiency and Response Time: This criterion is a very important one to ensure the efficiency and accuracy of the IDSS. The response time will be important in dynamic IDSS and hard-constrained dynamic IDSS, but not in static IDSS.
  - Data Entry: In general, the communication of data (online, offline, etc.) must be checked to assess the data transferring functionality within the IDSS.
  - Output Format: The output of the IDSS, which usually is the set of possible alternatives for the decision ordered according to some degree of trust or certainty, must be evaluated to check whether the output is the right one for each possible input decision problem or not.
  - Hardware: All the hardware subsystems and components must be assessed to ensure full connectivity and operation, like database systems, local area networks, operating systems, backup systems, etc.
  - Usage: The IDSS should be tested from the point of view of the end user to check for the right output information, right explanations when required, right user functionalities implemented, and other ergonomic criteria.
  - Human–Machine Interface: The interface should be as friendly as possible, easy to use, with a certain degree of customization for the user, supporting different kinds of users if needed, etc.

- Task Performance Criteria:

  - Decision-making time, alternatives, analysis, quality and participants: the main task of an IDSS is to give the best support to the decision-maker. To evaluate that, the formation of alternatives, the analysis done by the system, the quality of the support should be assessed to evaluate the degree of reliability and accuracy of the support provided by the IDSS.
  - User perceptions of trust, satisfaction and understanding: this criterion is a common and important one for any software system. The end user must understand and trust on the results provided by the system. If the end user does not trust on the system, the system will not be used. Thus, it is very important to evaluate the degree of trust and satisfaction of the user in an IDSS.

- Business Opportunities Criteria:

  - Costs of development, operation, and maintenance: to evaluate the business opportunity of a deployed IDSS, the costs of development, operation and maintenance must be computed.
  - Benefits associated with increased income and reduced costs: another interesting item to assess the business opportunity of an IDSS is the calculation of the benefit produced by the use of the system. The increased income and the reduced costs should be computed and evaluated.
  - Value to the organization of better service, competitive advantage, and training: another important aspect is the business opportunity derived from the competitive advantage of one organization due to having a powerful IDSS,

which other competitors could not have. Moreover, the better service to the organization, and the training value of the deployed tool, must be evaluated.

- Evolutionary Aspects Criteria:
  - Degree of flexibility, ability to change: Any software system should be as flexible as possible to changes and maintenance operations. IDSSs are not an exception, and they would better if they were flexible enough to be adapted to changes in several of their components (user interface, data, models, integration of components, etc.).
  - Overall functionality of the development tool: The development tool should provide the adaptability and flexibility issues. As much flexible be the development tool, much better will be the IDSS built.

4.9   Development of an IDSS: A First Example

In this section, a first example of how to develop an IDSS will be presented. Along the book, and especially in Chap. 7, several examples or case studies will be analyzed to illustrate the application of intelligent methods in the deployment of IDSSs. We will follow the methodologies for deployment of IDSSs proposed in previous sections of this chapter. Most of the case studies are based on the true stories about some companies and organizations, and others are specifically designed for illustrating the concrete use of some models. The names of the companies and other details have been changed to preserve confidentiality issues, and sometimes adapted to better illustrate fundamental aspects of the deployment of IDSSs.

Case Study: Customer Loyalty Analysis

The Scenario

All companies in the world are interested to manage in a proper way their relationships with their customers. One interesting point in these relationships is the loyalty analysis of their customers. The companies are very interested in know which customers are the loyal ones, to set-up special relations with them. They want to focus their efforts on analyzing the purchase behaviour of their loyal customers, on increasing the purchase level of their loyal customers, on sending their marketing campaigns addressed to their loyal customers. Our company of study is a European company selling books by internet: INTER BOOK. This company sells books to all European countries by internet. They have grouped all the countries in several regions with different managers: Western-Europe, Northern-Europe, Central-Europe and Eastern-Europe.

INTERBOOK is interested to know who their loyal customers are, which the main characteristics of their loyal customers are and which their purchasing behaviour are. Finally, they are interested in being able to know whether a new customer will be a loyal one or not, after her/his first purchase.

(continued)

Can an Intelligent Decision Support Systems help them providing adequate support and information to them for making the appropriate decisions?

Problem Analysis
Of course, an IDSS can help them. Let us analyze the main issues of the company:

- INTERBOOK wants to *study the buying behaviour* of their customers, and *to discover factors* which make a customer to be *loyal* or to be an *occasional* buyer
- INTERBOOK priority issues:

  - Need to characterize and distinguish the *loyal customers* from *occasional customers* to focus their marketing efforts on the right audience (*loyal customers*).
  - Analyze the purchase behaviour of their loyal customers
  - Get the most of their *loyal customers*
  - Being able to distinguish loyal customers from occasional ones *as soon as possible for new customers*

- INTERBOOK general issue: *minimize* the costs (marketing, etc.) to get a *maximum* benefit

Main problem features are the different behaviour of customers. One customer can order a large number of books or not in an order. A customer can pay an order with instalments or not. The amount of an order can be high or not, and so on.

Decisions Involved
It seems clear that the company has to face at least two main decisions:

- The first one is to decide which features *characterize and identify a loyal customer* and which ones not. With that decision solved, the actual customers can be classified as loyal or occasional depending on the degree they fulfil the necessary constraints to be loyal customers.
- Secondly, they need to have *the ability to decide given a new customer whether she/he will be or not a loyal customer* as soon as possible, and with the minimum information.

Requirements
The IDSS should have the following functional requirements:

- The IDSS must be able to give accurate support to analyze all the actual customers of INTERBOOK, in order that the general manager can define the characteristic features of a loyal customer.

(continued)

- Another functionality of the IDSS must allow the manager to easily classify each customer as loyal or occasional
- The last functionality of the IDSS must enable INTERBOOK to have some advanced support to decide whether a new costumer will be loyal or not with the minimum possible information and as soon as possible.

Data Availability

In all the scenarios, prior to the building of the IDSS, it must be carefully analyzed whether there are data available or not to get some *data-driven model*. In our case study, INTERBOOK had some historical data about its customers and about their orders. The company was rather new. Therefore, it did not have many years of historical data. They had just 5 years of history. They had *a database with the demographic and basic information about the customers*: "name", "address", "region", "country", "sex", "age", etc. They had about 1820 customers.

In another database, they had the orders of all their customers. For each order, they had "the customer identifier", "the date of the order", "the number of books ordered", "the amount of the order", "the payment method (credit card, bank transfer, payment with instalments)", etc. The database had 3570 orders.

In order to apply data explorative models, a *unique database* with the appropriate record type should be formed. Thus, both available databases should be merged to have just one record for each customer. That implies "the number of total orders" of one customer must be computed from one database to be integrated as a new feature in the final database. Other attributes like "date of first order", "date of last order", "total amount paid", "total amount ordered", and so on, must be equally processed. In addition, as it will be detailed in Chap. 6, the database should be previously pre-processed including data errors management, data missing management, data outliers' detection, data exploration analysis, bivariate plots, etc., to apply a machine learning method to get a data explorative model.

Expert or First-Principles Knowledge Availability

Another usual analysis in every scenario is to check whether there are experts available that can provide the system with expert knowledge. More-over, the possibility of having some first-principles knowledge coming from some theoretical principles should be also searched.

In our case study, there was not first-principles knowledge available regarding the loyal customers. However, the company has several people, including the general manager, the regional managers, the sales manager, etc. who had some expert knowledge coming from some years of experience in that business. This lead to consider the possibility of constructing some *model-driven component* for the IDSS. They had some knowledge about how to

(continued)

identify the loyal customers, which could be used in the building of the IDSS. They know some relationship between the loyalty condition and some other variable available in the database.

A Possible Solution

After the scenario and concrete problem analysis, a prototype IDSS started to be envisioned. Clearly, the IDSS should give support to both decisions involved: identification and classification of each customer as loyal or not, and to discriminate new costumers into loyal or occasional buyers. This fact suggests that probably the IDSS should have two subsystems or components.

Type of IDSS

Probably each component would belong to a different type of IDSS. The identifying component can be considered as a static IDSS component because it would be executed just one time to make the grouping of customers, and to classify them in the two groups. On the contrary, the discriminant IDSS component, should be used repeatedly each time a new costumer comes in business relation with INTERBOOK to classify the new buyer, as soon as possible, as a loyal or occasional costumer. Hence, this component should be a dynamic IDSS subsystem.

Kind of Tasks Involved (Analysis, Synthesis, Prognosis)

The main tasks are related to the identification of loyal customers and to the prediction of the customer type for new customers. The *identification task* is an *interpretative/analytical process*, which from a set of customer features should analyze which are the common features identifying loyal customers. The *prediction task* to decide whether a new customer is a loyal one or not is a *prognostic process*, which given the data describing a new customer and her/his behaviour should forecast the kind of customer he/she would be.

Model Selection

The *analytical task* could be supported in two ways: One using the *expert knowledge*, which leads to suggest that if "the number of orders" or "the number of books ordered at the first order" or "the total amount paid" are higher than some boundary, then the customer is a loyal one according to the knowledge of the experts. This way, an *expert-based model* can be used (*model-driven intelligent component*), implemented through a rule-based reasoning, as it will be explained in Chap. 5. Alternatively, given that there are *data available* about the costumers and their purchasing behaviour, a data explorative model can be used (*data-driven intelligent component*). Concretely, an *unsupervised descriptive model* (a *clustering method*) can be used

(continued)

to obtain two clusters of different data grouping themselves similar customers. From the two clusters obtained, through an interpretation process the common characteristics of its elements (customers) can be identified. The hope is that the two groups should express the difference regarding the buying behaviour of the customers. These methods will be detailed in Chap. 6.

After the *identification task* is done, each one of the customers can be easily classified (by expert rules or by its belonging to one cluster or the other) as loyal or occasional customers.

The *prognosis task* could be undertaken getting profit from the data available. Thus, again a data explorative model (*data-driven intelligent component*) can be used. However, now the methods need to be *supervised discriminant models*. Supervised, because we need a qualitative variable of response "kind-of-customer" (with possible labels being loyal-customer or occasional-customer), which will be the qualitative variable to be predicted by the model. Several machine learning techniques can be used to induce a discriminant data mining model. In our case study, *a decision tree approach* and a *k-nearest neighbour* approach (case-based reasoning paradigm) were proposed. These methods will be described in Chap. 6.

Model Integration
In this case study, the different models used should be integrated in a sequential way, because the two decisions involved are sequential. However, for each decision support, more than one model could be used as explained above. Thus, some of them could be executed in parallel, if needed, because they are alternative models. This integration scheme is depicted in the functional architecture shown below.

IDSS Functional Architecture
The functional architecture of the IDSS is composed of several subsystems, which are described in Fig. 4.15.

Implementation and Software Tools
Unfortunately, as it will be discussed in Chap. 8, there is no a general IDSS tool for the general development of IDSSs. However, some tools can help in the different kinds of models. Especially in data-driven models (data explorative models), there are some general tools which can be used to obtain the data models from the data. Moreover, some reasoning engine shells[3] can be used for rule-based reasoning (Drools, CLIPS, etc.).

(continued)

---

[3] Drools and CLIPS are open rule-based reasoning shells, which support forward and backward reasoning with inference rules.

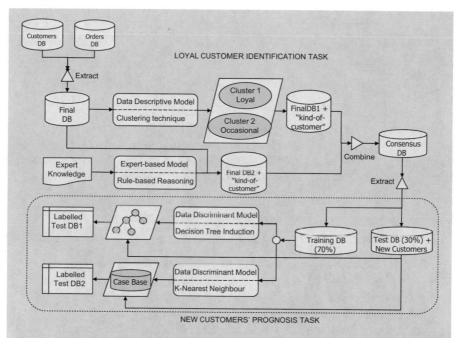

Fig. 4.15 Functional architecture of the loyalty analysis IDSS

For the deployment of the prototype in this case study, some R[4] packages (stats, cluster, etc.) or Python[5] libraries (scikit-learn, etc.), as well as RapidMiner[6] operators (kMeans, Decision Tree, NearestNeighbors, etc.), can be used.

Results and Evaluation

After the deployment of the different subsystems, a prototype of the IDSS for INTERBOOK company was built. Regarding the identification task, after applying the clustering technique, two clusters or groups were obtained. After an analysis, the experts of the company realized that mostly the elements in one cluster (721 customers) corresponded to what they considered as "loyal customers", and the other group (1099 customers) corresponded to "occasional customers". The group of the loyal customers could be characterized by

(continued)

---

[4]R is a language and environment for statistical computing and graphics, which can be enlarged with other packages for many data mining tasks.

[5]Python is a high-level general-purpose programming language. It is an interpreted, interactive, object-oriented programming language. It can be extended with packages and libraries for machine learning and data mining tasks.

[6]RapidMiner is a platform for developing data science projects. It supports data preparation, machine learning, and model deployment.

Table 4.2 Accuracy results of both discriminant methods

Discriminant model	Accuracy on training set	Accuracy on test set
Decision tree (C4.5 method)	0.775	0.746
K-NearestNeighbour ($k = 5$)	0.749	0.728

several variables, which have frequently similar values. Variables like "the number of orders" >2, "the total amount paid" >700 €, or "the number of products at first order" >5 were characterising the loyal customers, and the experts agreed on this identification. In addition, this characterization matches the expert-based model using rules, which inferred the same profile for the loyal customers.

Accuracy and Efficiency

Regarding the discriminant models used for predicting the "type-of-customer" variable, decision trees and k-nearest-neighbour ($k = 5$ gave the best results), they were trained using the 70% of the consensus database, and the remaining 30%, plus some new costumer data were used to test the accuracy of the predictive system. In the Table 4.2, there is the accuracy results of both approaches in the training and the validation test. From the results, it seemed that both methods were rather good, but the decision tree approach was slightly better with the training tests, but with the validation test, the k-nearest neighbour seemed to generalize a little bit better than the decision tree method. Of course, other methods could have been tested.

Quality of the Output Generated (Alternatives, Support Given)

The experts evaluated the general behaviour of the IDSS prototype as very good. They felt that this prototype could be evolved into a very useful support tool for the company. The prognosis part was considered as very useful, both quantitatively regarding the accuracy obtained, and qualitatively, as they mostly agree with the classification of costumers proposed by the IDSS. Moreover, the identification task supported by the clustering algorithm was very similar to the expert-based model obtained with the expert knowledge. Summarizing, the expert evaluation of the IDSS prototype was very good.

Business Opportunities (Benefit/Cost)

The managers of INTERBOOK thought that the IDSS could be very fruitful for the company, and provide them with great benefits. The IDSS can help them to focus the marketing campaigns on the loyal customers. This way, the company can increase its benefits, getting the most of their customers. In addition, they thought that the IDSS could be used as a training tool for new people in the company.

References

Charniak, E., & McDermott, D. (1985). *Introduction to artificial intelligence*. Addison-Wesley.

Cortés, U., Sànchez-Marrè, M., Ceccaroni, L., R-Roda, I., & Poch, M. (2000). Artificial intelligence and environmental decision support systems. *Applied Intelligence, 13*(1), 77–91.

Dhar, V., & Stein, R. (1997). *Intelligent decision support methods*. Prentice-Hall.

Dorigo, M. (1992). *Optimization, learning and natural algorithms*. Ph.D. thesis, Politecnico di Milano, Italy.

Forbus, K. (1984). Qualitative process theory. *Artificial Intelligence, 24*, 85–168.

Fox, J., & Das, J. (2000). *Safe and sound. Artificial intelligence in hazardous applications*. AAAI Press/The MIT Press.

Glover, F. (1986). Future paths for integer programming and links to artificial intelligence. *Computers and Operations Research, 13*(5), 533–549.

Glover, F. (1989). Tabu search - Part 1. *ORSA Journal on Computing, 1*(2), 190–206.

Glover, F. (1990). Tabu search - Part 2. *ORSA Journal on Computing, 2*(1), 4–32.

Goldberg, D. E. (1989). *Genetic algorithms in search, optimization and machine learning*. Addison-Wesley Publishing Company.

Gottinger, H. W., & Weimann, P. (1992). Intelligent decision support systems. *Decision Support Systems, 8*, 317–332.

Haagsma, I. G., & Johanns, R. D. (1994). Decision support systems: An integrated approach. In P. Zannetti (Ed.), *Environmental systems* (Vol. II, pp. 205–212).

Holland, J. H. (1975). *Adaptation in natural and artificial systems*. University of Michigan Press. New edition, Cambridge, MA, USA: MIT Press.

Holsapple, C. W. (1977). *Framework for a generalized intelligent decision support system*. Ph.D. Thesis, Purdue University, USA.

Kennedy, J., & Eberhart, R. (1995). Particle swarm optimization. In *Proceedings of IEEE International Conference on Neural Networks* (Vol. IV, pp. 1942–1948).

Khachaturyan, A., Semenovskaya, S., & Vainshtein, B. (1979). Statistical-thermodynamic approach to determination of structure amplitude phase. *Soviet Physics. Crystallography, 24* (5), 519–524.

Kirkpatrick, S., Gelatt, C. D., Jr., & Vecchi, M. P. (1983). Optimization by simulated annealing. *Science, 220*(4598), 671–680.

de Kleer, J., & Brown, J. (1984). A qualitative physics based on confluences. *Artificial Intelligence, 24*, 1–83.

Klein, M. R., & Methlie, L. B. (1995). *Knowledge-based decision support systems* (2nd ed.). Wiley.

Kolodner, J. (1993). *Case-based reasoning*. Morgan Kaufmann.

Kuipers, B. (1994). *Qualitative reasoning: Modeling and simulation with incomplete knowledge*. MIT Press.

Murphy, K. (1998). *A brief introduction to graphical models and Bayesian networks*. http://www. cs.ubc.ca/murphyk/Bayes/bintro.html

OMNI. (1994, June). Herbert Simon on the mind in the machine. Herbert A. Simon's interview by Douglas Stewart. *OMNI Magazine*. https://web.archive.org/web/20161019182013/, http:// www.omnimagazine.com/archives/interviews/simon/index.html

Pearl, J. (1988). *Probabilistic reasoning in intelligent systems*. Morgan Kaufmann.

Pearl, J., & Russell, S. (2000). *Bayesian networks*. Technical Report R-277. UCLA Cognitive Systems Laboratory.

Phillips-Wren, G., Mora, M., Forgionne, G. A., & Gupta, J. N. D. (2009). An integrative evaluation framework for intelligent decision support systems. *European Journal of Operational Research, 195*, 642–652.

Poch, M., Comas, J., Rodríguez-Roda, I., Sànchez-Marrè, M., & Cortés, U. (2004). Designing and building real environmental decision support systems. *Environmental Modelling & Software, 19* (9), 857–873.

Reiter, R. (1987). A theory of diagnosis from first principles. *Artificial Intelligence, 32*(1), 57–95.

Rich, E., & Knight, K. (1991). *Artificial intelligence* (2nd ed.). McGraw-Hill.

Richter, M. M., & Weber, R. O. (2013). *Case-based reasoning: A textbook.* Springer.

Riesbeck, C. K., & Schank, R. S. (1989). *Inside case-based reasoning.* Lawrence Erlbaum Associates Publishers.

Robinson, J. A. (1965). A machine-oriented logic based on the resolution principle. *Journal of the ACM, 12*(1), 23–41.

Russell, S. J., & Norvig, P. (2010). *Artificial intelligence. A modern approach* (3rd ed.). Prentice-Hall, Pearson Education.

Sànchez, M., Cortés, U., Lafuente, J., R-Roda, I., & Poch, M. (1996). DAI-DEPUR: An integrated and distributed architecture for wastewater treatment plants supervision. *Artificial Intelligence in Engineering, 10*(3), 275–285.

Sànchez-Marrè, M., Gibert, K., Sojda, R., Steyer, J. P., Struss, P., & Rodríguez-Roda, I. (2006). Uncertainty management, spatial and temporal reasoning and validation of intelligent environmental decision support systems (AITENV). In *Proc. of iEMSs 3rd Biennial Meeting on Environmental Modelling and Software, II IEDSS Workshop, CD-Rom, Burlington, Vermont, USA.*

Sànchez-Marrè, M., Comas, J., Rodríguez-Roda, I., Poch, M., & Cortés, U. (2008). Towards a framework for the development of intelligent environmental decision support systems. In *Proc. of 4th International Congress on Environmental Modelling and Software (iEMSs '2008). iEMSs 2008 Proceedings* (pp. 398–406).

Schank, R. (1982). *Dynamic memory: A theory of learning in computers and people.* Cambridge University Press.

Schank, R., & Abelson, R. (1977). *Scripts, plans, goals and understanding.* Lawrence Erlbaum.

Simon, H. A. (1960). *The new science of management decision.* Harper & Row.

Steels, L. (1993). *Ten years.* VUB Artificial Intelligence Laboratory. Vrije Universiteit Brussel.

Wooldrige, M., & Jennings, P. (1995). Intelligent agents: Theory and practice. *The Knowledge Engineering Review, 10*(2), 115–152.

Zadeh, L. A. (1965). Fuzzy sets. *Information and Control, 8*(3), 338–353.

Further Reading

Curwin, J., & Slater, R. (2001). *Quantitative methods for business decisions.* Thomson Business Press.

Dewhurst, F. (2002). *Quantitative methods for business and management.* McGraw-Hill.

Fayyad, U. M., Piatetsky-Shapiro, G., Smyth, P., & Uthurusamy, R. (1996). *Advances in knowledge discovery and data mining.* AAAI Press/MIT Press.

Huhns, M. N., & Stephens, L. M. (1999). Multi-agent systems and societies of agents. In G. Weiss (Ed.), *Multi-agent systems* (pp. 79–120). MIT Press.

McNurlin, C., & Sprague, R. H., Jr. (1993). *Information systems management in practice* (4th ed.). Prentice-Hall.

Mitchell, T. (1997). *Machine learning.* McGraw-Hill.

Moore, J. H., & Chang, M. G. (1982). Design of decision support systems. *Database, 12*(1–2), 8.

Morris, C. (2002). *Quantitative approaches in business studies.* Financial Times/Prentice Hall.

Power, D. J. (2013). *Decision support, analytics, and business intelligence (Information systems)* (2nd ed.). Business Expert Press.

Sprague, R. H., Jr., & Carlson, E. D. (1982). *Building effective decision support systems.* Prentice Hall.

Turban, E., Aronson, J. E., & Liang, T.-P. (2005). *Decision support systems and intelligent systems* (7th ed.). Pearson/Prentice Hall.

Witten, I. H., Frank, E., Hall, M. H., & Pal, C. J. (2016). *Data mining. Practical machine learning tools and techniques* (4th ed.). Morgan Kaufman.

Chapter 5
Model-Driven Intelligent Decision Support Systems

Miquel Sànchez-Marrè
Contributor: Franz Wotawa

5.1 Introduction

In the next sections the main kinds of Model-driven methods that can be used in Intelligent Decision Support Systems are explained. First, *Agent-based Simulation models* are detailed. They are based on the multi-agent system approach to model complex scenarios. There are several different methodologies for multi-agent system simulation modelling. Here, a general methodology for agent-based simulation is abstracted from particular methodologies. Next, *Expert-based models* are described. This kind of expert models are usually expressed as a set of inference rules, and therefore, rule-based reasoning is the major technique used. The general architecture of a *Rule-Based Reasoning system* is outlined and its major components are analyzed. The knowledge base, the reasoning, and the meta-reasoning components are especially analyzed, and algorithms for forward reasoning and backward reasoning are explicated. Afterwards, *Model-Based Reasoning methods* are explained. Its foundations and motivations to be used in IDSS are detailed. One of the applications field where model-based reasoning methods have been successfully applied is in diagnosis. Two model-based reasoning methods have been applied in this area: consistency-based diagnosis and abductive diagnosis. The first one makes use of models describing the correct behaviour of systems comprising components and connections, the latter one relies on logical models comprising knowledge of faults and their consequences. Algorithms for both kinds of methods are detailed. Finally,

The original version of this chapter was revised. The correction to this chapter is available at https://doi.org/10.1007/978-3-030-87790-3_11

Prof. Franz Wotawa has written the Sect. 5.4 about Model-Based Reasoning Methods.

F. Wotawa
Institute for Software Technology, Technische Universität Graz, Graz, Austria
e-mail: wotawa@ist.tugraz.at

Qualitative Reasoning models are introduced. The basic principles of qualitative reasoning techniques are explained: qualitative modelling, including model formulation and model representation, and qualitative simulation. The basic flowchart for deploying a qualitative reasoning model is presented. Most common qualitative reasoning approaches in the literature are reviewed.

All these models are illustrated with some examples to provide a better understanding of them.

5.2 Agent-Based Simulation Models

Agent-based simulation models are simulation models using an emerging paradigm of artificial intelligence: the *multi-agent systems* (Weiss, 1999; Wooldridge & Jennings, 1995). A multi-agent system implies more than one *agent* interacting with each other within an underlying communication infrastructure and without a procedural control mechanism; and, the individual agents often are distributed and autonomous (Huhns & Stephens, 1999). An *agent* is a computer system that is capable of independent action on behalf of its user or owner. It can decide what are the needs which satisfy its design objectives, rather than being explicitly told what to do at any moment (Wooldridge & Jennings, 1995; Wooldridge, 2002).

A multi-agent system consists of a number of agents, which *interact* with one another, typically by exchanging messages through some computer network infrastructure. In order to successfully interact, these agents will require the ability to *cooperate*, *coordinate*, and *negotiate* with each other, in the same way that humans cooperate, coordinate, and negotiate with other people in our everyday lives. (Wooldridge, 2009).

5.2.1 Multi-Agent Systems

Multi-Agent Systems (MAS) are based on the idea that a cooperative working environment can cope with problems that are hard to solve using the traditional centralized approach to computation. Intelligent agents are used to interact in a flexible and dynamic way to solve problems more efficiently (Mangina, 2002).

Wooldridge's work distinguishes between an agent and an intelligent agent, which is further required to be autonomous, reactive, proactive, and social (Wooldridge, 2002):

- *Autonomous*: Agents are independent and make their own decisions without the direct intervention of other agents or humans and agents have control over their actions and their internal state.
- *Reactive*: Agents need to be reactive, responding in a timely manner to changes in their environment.
- *Proactive*: An agent pursues goals over time and takes the initiative when it considers it appropriate.
- *Social*: Agents very often need to interact with other agents to complete their tasks and help others to achieve their goals.

A key issue in agent architecture is balancing reactiveness and proactiveness (Padgham & Winikoff, 2004). On the one hand, an agent should be reactive so its plans and actions should be influenced by environmental changes. On the other hand, an agent's plans and actions should be influenced by its goals. The challenge is to balance the two, often conflicting, influences: if the agent is too reactive, then it will be constantly adjusting its plans and not achieve its goals. However, if the agent is not sufficiently reactive, then it will waste time trying to follow plans that are no longer relevant or applicable.

Agents tend to be used where the domain environment is challenging; more specifically, typical agent environments are dynamic, unpredictable, and unreliable (Padgham & Winikoff, 2004):

- *Dynamic*: These environments are dynamic in that they change rapidly. Thus, the agent cannot assume that the environment will remain static while it is trying to achieve a goal.
- *Unpredictable*: It is not possible to predict the future states of the environment; often this is because it is not possible for an agent to have perfect and complete information about their environment.
- *Unreliable*: The actions that an agent can perform may fail for reasons that are beyond an agent's control.

Multi-agent systems are based on the idea that a cooperative working environment can cope with problems that are hard to solve using the traditional centralized approach to computation. Intelligent agents are used to interact in a flexible and dynamic way to solve problems more efficiently.

5.2.1.1 The Belief-Desire-Intention Model

Rational agents have an explicit representation of their environment and of the objectives they are trying to achieve. *Rationality* means that the agent will always perform the most promising actions (based on the knowledge about itself and the world) to achieve its objectives. As it usually does not know all of the effects of an action in advance, it has to deliberate about the available options.

Among the numerous deliberative agent architectures found nowadays the most widespread one is the *Belief-Desire-Intention* (*BDI*) proposed by Bratman as a model for describing rational agents (Bratman, 1987). The concepts of the BDI-model were later adapted by Rao and Georgeff to a more formal model that is better suitable MAS in the software architectural sense.

The BDI architecture uses the concepts of *belief, desire,* and *intention* as mental attitudes. *Beliefs* capture informational attitudes, *desires* motivational attitudes, and *intentions* deliberative attitudes of agents (Rao & Georgeff, 1995). Its central concepts are:

- *Beliefs*: Information about the environment.
- *Desires/Goals*: Objectives to be accomplished.

- *Intentions*: The currently chosen course of action.
- *Plans*: Means of achieving certain future world states.
- *Actions*: Ways the agent can operate on the environment.

The advantage of using mental attitudes in the design and realization of agents and multi-agent systems is the natural (human-like) modelling and the high abstraction level, which simplifies the understanding of systems (McCarthy 1979).

5.2.1.2 Agent Architectures

Agent architectures represent the move from theoretical specification to software implementation (Mangina, 2002):

- *Deliberative Architectures*: Agents should maintain an explicit representation of their world, which can be modified by some form of symbolic reasoning. The Belief-Desire-Intention (BDI) model (Rao & Georgeff, 1995) is the most widespread model used on this kind of architecture. The most typical architecture is the *Procedural Reasoning System* (Georgeff & Lansky, 1987).
- *Reactive Architectures*: They aim to build autonomous mobile robots, which can adapt to changes in their environment and move in it, without any internal representation. The agents make their decisions at run time, usually based on a very limited amount of information and simple situation-action rules:

$$\text{if} < \text{situation} > \text{then} < \text{action} >$$

  Decisions are based directly on sensory input. The most well-known reactive example is the *subsumption architecture* (Brooks, 1991).
- *Hybrid or Layered Architectures*: Many researchers suggested that a combination of the deliberative and reactive approaches would be more appropriate, as it would combine the advantages of both kinds and avoid the disadvantages. Layered architectures represent the way different subsystems are arranged into a hierarchy of interacting layers, which involve the different types of behaviours. *TouringMachines* (Ferguson, 1992) and the *Stanley architecture* (Thrun et al., 2007) are two examples of these architectures.

Agent-based models consist of dynamically interacting agents. The agents could be of any kind: deliberative, reactive, or hybrid. The power of agent-based simulation relies on modelling a complex system with different components, with probably different goals. These different components of a complex system, can be modelled with an agent. The agents *cooperate* to solve the problem by communicating among them. The *communication* is made upon sending and receiving *messages*.

5.2.1.3 Communication

In agent-based systems, all agents cooperate to implement a simulation of the complex system in an easier way. The different agents must communicate with each other, which means that some standard communication languages must be used. Agents have to communicate their knowledge for achieving their goals. Sometimes they just want to *inform* other agents about some of their beliefs, and other times they *request* for information or *request* another agent to perform an action. Communication can be accomplished through the *Agent Communication Language (ACL)*, by using communication protocols including TCP/IP, SMTP, and HTTP to exchange knowledge (Mangina, 2002).

Agents need a standard language for parsing and understanding the content of the messages. To that end, the *Knowledge Interchange Format (KIF)* (Genesereth & Fikes, 1992) was designed. KIF provides a syntax for message content, which is essentially first-order predicate calculus with declarative semantics. It is a language aiming to allow the representation of knowledge about some domain. It was intended to form the content parts of KQML messages.

The best-known ACLs are the *Knowledge Query Manipulation Language (KQML)* (Finin et al., 1993), and the *Foundation for Intelligent Physical Agents ACL (FIPA-ACL)* (FIPA, 1999):

- KQML: It is both a message format and a message-handling protocol to support runtime knowledge sharing among agents for cooperative problem-solving. KQML defines a common format for messages. It employs a layered architecture of communication, where at the bottom the functionality for message transport or communication occurs and at the top, the contents are specified by the application. A KQML message is like an object in object-oriented programming. Each message has:

  - A *performative*, which identifies the class of the message (achieve, ask-if, ask-all, evaluate, forward, tell, reply, etc.).
  - Some *parameters*, which are pairs of attribute values (:content, :force, :reply-with, :in-reply-to, :sender, :receiver, :language, :ontology, etc.) defining the message.

An example of a KQML message is described in Fig. 5.1.

Fig. 5.1 An example of a KQML message

```
(tell
    :sender    Settler-Agent
    :receiver  BioReactor-Agent
    :in-reply-to q1
    :content   (= (status settler1) normal)
)
```

Fig. 5.2 An example of a
FIPA-ACL message

```
(inform
   :sender  agent1
   :receiver  agent2
   :content (price good2 150)
   :language sl
   :ontology  hpl-auction
)
```

- FIPA-ACL aims to set general standards for agent interoperability. FIPA has defined an ACL (FIPA-ACL), which includes basic communicative actions (inform, request, propose, and accept) together with interaction protocols.

    Figure 5.2 depicts an example of a FIPA-ACL message, extracted from (FIPA, 1999).

The *communications among agents* let them to exchange knowledge and request other agents to perform some actions.

5.2.1.4   Cooperation

Agents must act autonomously, and hence, they should make decisions at runtime, rather than having all its behaviour hard-coded at design time. They should be able to dynamically coordinate their activities and cooperate with the other agents to solve the problem at hand (Jennings et al., 2001).

    The degree of *cooperation* can depend on whether the agents *share a common goal*, and there are no potential conflicts among them, or on the contrary, this assumption cannot be ensured because the agents are *self-interested agents*. The former situation is usually given when all agents are designed or owned by the same organization or individual, and the latter situation is given on the opposite case.

    Cooperative Distributed Problem Solving (*CDPS*) can be analyzed as a three-stage activity, according to Smith and Davis (1980):

- *Problem Decomposition*: The problem to be solved is decomposed in smaller subproblems. The decomposition process is usually a hierarchical process, where subproblems are further decomposed into smaller subproblems, and so on, until the final subproblems are of an appropriate granularity to be solved by individual agents. This decomposition task could be performed just by one agent, but this way there is the assumption that the agent has the necessary knowledge and expertise about the *task structure*. Usually, other agents have valuable knowledge about the task structure to improve the decomposition process.
- *Subproblem Solution*: The subproblems identified in the previous phase are individually solved. Typically, this step involves sharing of information between agents. They can help other agents providing information which can be useful to them, for solving a subproblem.

- *Solution Synthesis*: At this final stage, the different solutions of the different subproblems must be integrated into the final global solution. This synthetic process can be hierarchical, merging the different partial solutions at the different levels of abstraction of the hierarchy of tasks.

Two specific cooperating activities are usually present in a CDPS:

- *Task Sharing*: This task could be easier if all agents have the *same capabilities*, and then, any task can be assigned to any agent. However, in the general case when agents may have *different capabilities*, the task assignment process can be more difficult involving agents reaching agreements with others, using negotiation or auction techniques.
- *Result Sharing*: Agents usually must share relevant information about its solved subproblems. This sharing could be done *proactively* when an agent believes that its result could be interesting for other agents, or *reactively* when it sends the information to some other agent which requested that information.

5.2.1.5   Coordination

The *coordination* problem relies on managing the interdependencies between the activities of the agents. The agents develop the activities signalled by some plans. According to von Martial (1990), there can be a distinguished typology of coordination relationships:

- *Negative Relationships*: Relationships between two agents' plans, when accomplishing one plan implies a negative effect on the other agent's plan satisfaction.

  - *Resources Relationships*: Relationships in which the achievement of one agent's plan implies some resource allocation that prevents from achieving the other agent's plan goal.
  - *Incompatibility Relationships*: Relationships that following one agent's plan to reach a goal it is incompatible with achieving the goal of the other agent following its plan.

- *Positive Relationships*: They are relationships between two agents' plans from which some benefit can be obtained, for one or both of the agents' plans, by combining them.

  - *Requested Relationships*: Relationships when one agent explicitly requests help from another agent with its activities.
  - *Non-requested Relationships*: Relationships when by working together, the agents can achieve a better solution, at least for one agent, without worsening the other agent.

Action Equality Relationships: Both agents plan to perform an identical action, and detecting this situation, one of them perform the action alone, saving the other agent to make the same action.

Consequence Relationships: The actions of one agent's plan have the consequence (side effect) of achieving one of another agent's goal. Thus, the other agent is not needed to reach this satisfied goal.

Favour Relationships: Some action/s in the plan of one agent favourably contribute to the achievement of one of the other agent's goals. For instance, satisfying the precondition of one action of the plan of the other agent.

5.2.1.6 Design and Development

Software-engineering methodologies assume the existence of a set of concepts that it builds upon. For example, object-oriented notations such as UML (Unified Modelling Language) assume certain concepts, such as object, class, inheritance, and so on. With agent-oriented methodologies, we also need an appropriate set of concepts, and it turns out that the set of concepts is different from the object-oriented set.

Since the selection of the right methodology is crucial for any software project (Sudeikat et al., 2004), a special emphasis is put on the choice of a suitable methodology for the agent platform selected. Most common agent-oriented methodologies in the literature are:

- MAS- COMMONKADS (Iglesias et al., 1998) due to its origins in CommonKADS (Tansley and Hayball, 1993), it is a methodology oriented to the development using practical experience in expert systems.
- Gaia (Wooldridge et al., 2000) studies the views definitions in a methodology and tries to integrate a software life cycle. It is intended to allow a designer from a statement of requirements to an enough detailed design that can be implemented in a direct way. Gaia provides an agent-specific set of concepts through which a software engineer can understand and model a complex system. It encourages a developer to analyze the process as an organizational design.
- MaSE (DeLoach, 2001) has its own tool that supports the methodology. The primary focus of MaSE is to help a designer take an initial set of requirements and analyze, design, and implement a working multi-agent system.
- MESSAGE (Caire et al., 2002) defines the elements to take into account in the development of a MAS by means of the specification of meta-models. It proposes the adoption of a standard process, the RUP.
- The Tropos Methodology (Bresciani et al., 2004) aims to provide an agent-oriented view of software throughout the software development life cycle. It offers a conceptual framework for modelling systems based on the concepts of actors, goals, plans, resources, dependencies between actors, capabilities of the actors, and believes of the actors.

- Prometheus (Padgham & Winikoff, 2004) is a mature and well-documented methodology; it supports BDI-concepts and additionally provides a CASE-Tool, the Prometheus Design Tool (PDT), for drawing and using the notation. The Prometheus methodology consists of three phases that can be done simultaneously: system specification, architectural design, and detailed design.

5.2.1.7 Multi-agent Applications

Multi-agent systems have been applied in many domains like in business process management, in distributed sensing, in information retrieval, in electronic commerce, in human–computer interfaces, in industrial system management, in air-traffic, and in system simulation.

One of the most useful fields where multi-agent systems have been applied is in the *simulation of natural or artificial societies*. Usually, these applications include education systems, training systems, scenario exploration and analysis, policy systems, and entertainment systems.

The simulation of scenario analysis has been entitled as *social simulation* field. Thus, multi-agent systems have arisen as excellent experimental tools in the social sciences. This field of agent-based social simulation when facing complex systems, with the aim to see how are the joint behaviour of several individual agents interacting in society, has been named as *agent-based modelling*.

5.2.2 Agent-Based Simulation

Multi-agent systems provide a model for representing real-world environments with an appropriate degree of complexity and dynamism. Agent-based simulation is at the intersection of three scientific fields: agent-based computing, social sciences, and computer simulation.

Social sciences study the interaction among social entities and include social psychology, management, policy, and certain areas of biology. Computer simulation comprises techniques for simulating different scenarios of a phenomenon on a computer, such as discrete-event simulation, object-oriented simulation, or equation-based simulation. The joint area of computer simulation and multi-agent systems is named *multi-agent-based simulation* or simply, *agent-based simulation*.

In *agent-based simulation*, agents can be used to simulate the behaviour of human societies. Individual agents can be used to represent not only individual people but also organizations and similar entities. Agent-based simulation of social processes has some benefits, as described Conte and Gilbert (1995):

- Computer simulation allows the observation of properties of a model that may be analytically derivable but could have not yet been established.
- Several alternatives to a phenomenon can be discovered.

- Difficult properties to observe in sensitive complex domains like nature may be studied in a safety mode exploring the possible dangerous consequences of several alternatives.
- Scenarios can be replayed in the simulated model.
- Social relationships among agents can be analyzed.

The *scenario analysis* by the users of the agent-based simulation model (business managers, social policy analysts, air-traffic controllers, industrial system managers, environmental policy managers, etc.) is the main benefit of these kinds of models. Each different scenario is characterized by a different configuration of several parameters. After some running time of the simulation, the results of that scenario configuration can be reported and analyzed. This information is very valuable to help in the decision-making process of the user of an IDSS.

Each one of the elements or entities of the system to be simulated is associated to an agent: the user, the different components, etc.

5.2.2.1 Deployment and Use

For the *deployment of an agent-based simulation model*, one must follow some concrete methodology to design the multi-agent system. Notwithstanding the different methodologies, there are some major abstraction steps in all of them that can guide to deploy an agent-based simulation model:

1. Analyze the complex system to be modelled
2. Identify the different entities composing the system
3. Determine the different types of agents needed
4. Associate the entities with some kind of agent
5. For each type of agent determine:

    (a) The agent's goals: The main issue of the agent
    (b) The agent's perceptions

    - Online sensors
    - Other data

    (c) The agent's Actions: possible actions that an agent can perform
    (d) The agent's Roles: the different roles that an agent could have

6. Identify the interesting different scenarios and/or strategies
7. Characterize the different scenarios and/or strategies

    (a) Name them with a meaningful identifier
    (b) Select the different global parameters of the system characterizing the scenario and/or strategy
    (c) Select the different parameters for the different agents of the model characterizing the scenario and/or strategy

8. Run the different simulation scenarios and/or strategies

   (a) Try different simulation runtime lengths

9. Evaluate the simulation scenario and/or strategy results

   (a) Analyze the different simulation runs from the different simulation runtime lengths

10. Extract the valuable knowledge from the assessment of the consequences of each scenario and/or strategy

11. Make the rational decision (most profitable or useful decision from the decisor's point of view)

These *agent-based simulation models* can simulate different scenarios for evaluating the consequences of critical processes in decision-making. Agent-based simulation lets the user of an IDSS to perform dynamic simulations of the system to evaluate the operation of the whole system. Furthermore, by experimenting with possible "what if" scenarios, the end user can benefit from exploring the response surface and the stability of the solution from the safety of a simulated environment where no actual damage can be caused to people's property or the environment because of incorrect decisions. These models have been developed especially in natural systems (biology, environment), medicine, management, and social systems (game theory, social networks, etc.).

5.2.2.2   An Example of an Agent-Based Simulation Model

In the literature, there are many examples of application of agent-based simulation models in different areas. Here, we will review an application of agent-based simulation for cargo routing, implemented for Southwest Airlines, and described in (Thomas & Seibel, 1999). The application will be analyzed using our proposal methodology described in the previous section. This application revealed several missing opportunities to load cargo in the company. Final improvements enabled an average 75% reduction in the handling of freight, and an average increase in revenue of US\$ ten million (Luck et al., 2003).

Analysis of the System

Southwest Airlines both moved passengers and cargo from one city to another in the USA. The company analyzed its business cargo for some years, and they felt that this business could be expanded. Especially, the cargo capacity was thought that could be enlarged (by then just a, 7% of bin space was used). An inspection revealed that there were some bottlenecks in their system. Some aircrafts were scheduled to carry a large load of cargo but lacked the bin space to accommodate this load, and sometimes, some departure delays were originated. In addition, their *ramp agents* (people moving the cargo), were experimenting some frustration, because they were taking the cargo from one aircraft, and throwing it to the next plane, without a clear objective issue of optimizing the loading/unloading processes.

Therefore, their main issues were: improve the lives of their ramp agents, try to improve the unused bin space capacity, and improve the business cargo to increase its profit.

Identification of the Entities

To reproduce the current operations of the company they created an agent-based simulation model. After an extensive historic data analysis, main components or entities of the models were identified: the *shipment descriptions*, the *flight schedules,* and the *freight logs*. Another important feature detected was the *amount of cargo that had to be stored overnight* at each of the stations/airports, which had a considerable cost.

Types of Agents

In the simulation model, they identified several kinds of particular agents:

- *Freight forwarders*, whose main goal was to decide how to route a package to its destination, i.e. assignment of a shipment to a flight.
- *Ramp agents*, which were responsible to load/unload the cargo from planes.
- *Shipments*, which had the information about the origin and destination, the number of pieces, the weight, etc. to be managed.
- *Flights*, which were scheduled for the company to the different destinations.

  Other important entities considered were:

- Ramp operations
- Stations/airports
- Routes
- Overnight stores
- ....

Identification of Scenarios/Strategies

They identified several scenarios: One was how the company thought it was running its operations, the other was how the company was actually working, and the other was how the company should run its cargo-loading operations.

They envisioned two basic strategies for the ramp agents:

- The *"hot potato" strategy* of loading the cargo onto the next plane, which was the common strategy used by the ramp agents.

- The *same-plane strategy*, which was proposed after an accurate analysis of data, which basically consisted in leaving the cargo in the plane, even the plane had to fly to other cities, because it was known that the same plane would come back the cargo to the corresponding station/airport.

Evaluation of the Simulation Results

They ran the agent-based model for a one-week duration simulating the network behaviour. The evaluation of the possible different strategies was very clear and useful. The proposed *same-plane strategy* dramatically reduced the amount of cargo being transferred (loaded/unloaded), as well as the amount of cargo that should be stored overnight. And as a consequence, the work of the ramp agents was reduced, because the amount of cargo transfer was diminished.

The company, in its sixth-busiest cargo stations, experimented with a reduction of 50–85% in cargo transfer. Regarding the overnight transfer rate, it was reduced from approximately 240,000 pounds per week to 50,000 pounds per week.

Therefore, the IDSS they constructed, which was using the agent-based simulation model let the company to increase their benefits, and improve its cargo management strategies. See Thomas and Seibel (1999) for further details.

5.3   Expert-Based Models

These models are originated by the knowledge that experts have about some concrete tasks. Experts have acquired a specific knowledge about how to successfully solve *complex tasks*. This *expert knowledge* must be coded in some kind of knowledge representation formalism. The most used formalism to represent those knowledge patterns is the use of *inference rules*. These rules constitute the *Knowledge Base* of these systems, which represent the knowledge provided by the experts. As the knowledge is expressed with rules, the reasoning mechanism is named as *rule-based reasoning* or *knowledge-based reasoning* (Jackson, 1999; Gonzalez & Dankel, 1993; Buchanan & Duda, 1983). In early rule-based reasoning systems, also named as *Expert Systems*, these kinds of systems were also named as *Production Systems*, because the inference rules produce new knowledge.

The architecture of a rule-based reasoning system is depicted in Fig. 5.3.

The main components of a rule-based reasoning system are: the fact base, the knowledge base, the inference engines (reasoning component), the meta-reasoning component, the user interface, the knowledge acquisition module and the explanation module, and the knowledge engineer interface.

Now, these components will be described and analyzed.

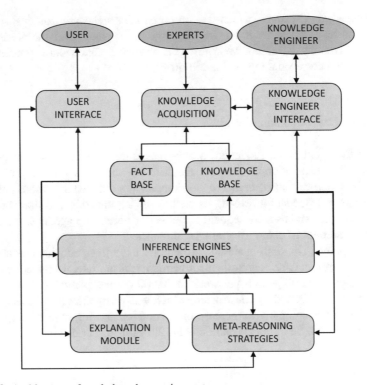

Fig. 5.3 Architecture of a rule-based reasoning system

5.3.1 Fact Base

The fact base describes the different relevant features (facts) characterizing the domain. Facts have some *static properties* and some *dynamic properties*. The static properties, sometimes are named as the *dictionary* of the rule-based reasoning system. The dynamics properties, which represent the state of the current problem being solved, are sometimes named as *short-term memory* or *working memory*.

Among the usual *static properties*, there are:

- *<Identifier of the fact>*: it is the name given to the fact.
- *<Type of the fact>*: describes the type of the values that the fact can handle. The type can be numerical, Boolean, string, enumerated ordered, enumerated non-ordered, etc.
- *<Questionable?>*: indicates whether the value of the fact can be obtained by interactively asking the user or not. The possible values are yes or not.
- *<Question>*: is a natural language string containing the appropriate question to be asked to the user to get the value of the fact.

Among the common *dynamic properties*, there are:

- *<Value>*: the current value of the fact, at a given moment of the reasoning process. The value can be *unknown,* when the value is not known, or must be a compatible value with the type of the fact.
- *<Certainty-of-the-value>*: it is the believed degree of certainty of the value of the fact, at a given moment of the reasoning process. The certainty values can be expressed with different uncertainty management models.

  - Without modelling uncertainty at all. Thus, all values are sure.
  - Using the *possibilistic theory* (values belonging to [0,1] real interval),
  - Using the *certainty factor model* of MYCIN (Shortliffe & Buchanan, 1975) (values belonging to $[-1, +1]$ real interval),
  - Using *qualitative probabilistic labels* (values belonging to an ordered enumerated label set like {completely-improbable, very-improbable, improbable, unknown, probable, very-probable, completely-sure}.

Some examples of facts are depicted in Fig. 5.4.

In the rule-based reasoning inference process, the dynamic properties of value and certainty need to be consulted often. Therefore, the facts are usually implemented in fast-accessing data structures like hash tables or similar ones.

5.3.2   Knowledge Base

The Knowledge Base (KB) is the core component where the knowledge coming from the experts is stored. As previously mentioned, the most common way of representing the knowledge in the Knowledge Base is using *inference rules* or *production rules*. This is motivated for the computational efficiency in the inference reasoning process. However, other formalisms to represent the Knowledge Base could be more structured. Common structured formalisms used are frames, semantic networks, (Jackson, 1999), and more recently ontologies.[1]

The KB organizes the knowledge available of the domain and the knowledge of the expert problem-solving process.

The inference rules have the following form:

$$\textbf{IF} < conditions > \textbf{THEN} < actions >$$

Each rule is normally composed of several elements:

- *<Rule-Identifier>*: It is an identifier assigned to each rule.
- *<Conditions or Premises>*: They are the conditions that must be satisfied in order that the rule can be fired, and the actions be executed. The set of conditions of a

---

[1] An ontology is the formal representation of a determined domain of discourse, by means of the specification of the relevant concepts, properties and their interrelationships. Usually can be a hierarchy or network of related concepts.

(Outside-Temperature
  :type Numeric
  :questionable? yes
  :question "What is the outside temperature (in ° C)?"
  :value 18.5
  :certainty   probable
)

(Weather-Forecast
  :type Enumerated-non-ordered {sunny, cloudy, windy, rainy, stormy}
  :questionable? no
  :question ""
  :value unknown
  :certainty
)

(Sludge-Waste-Primary-Settler
  :type Boolean
  :questionable? yes
  :question "Is the primary settler wasting sludge?"
  :value no
  :certainty 0.8
)

(Sludge-Concentration-Primary-Settler
  :type Enumerated-ordered {very low, low, normal, high, very-high}
  :questionable? yes
  :question "What is the concentration level of sludge in the primary settler?"
  :value high
  :certainty 1
)

Fig. 5.4 Examples of facts

rule are named in general as the *antecedent* of a rule, or the *left-hand side* (LHS) of a rule. The conditions can be combined with several *logical connectives* or *operators*:

– *Negation operator*: $\neg <$ cond$>$, with the following semantics:

<cond>	$\neg$ <cond>
True	False
False	True

– *Conjunction operator*: <cond1> $\wedge$ < cond2>, and the semantics of the operator is:

<cond1>	<cond2>	<cond1> $\wedge$ <cond2>
False	False	False

(continued)

\<cond1\>	\<cond2\>	\<cond1\> ∧ \<cond2\>
False	True	False
True	False	False
True	True	True

– Disjunction operator: \<cond1\> ∨ \< cond2\>, and the semantics of the operator is:

\<cond1\>	\<cond2\>	\<cond1\> ∨ \<cond2\>
False	False	False
False	True	True
True	False	True
True	True	True

The conditions or premises can be expressed as *propositions* (zero-order predicates) or *first-order predicates*:

– *Propositions*: It is a simple statement which can be true or false. They are simple predicates without any variable.

For instance: it-rains ≡ R, it-is-cloudy ≡ C

– *First-order Predicates*: They are predicates which can have variables for expressing the same predicate with other instantiations, and the formulas contain variables that can be quantified, and functions can be used.

For instance: man(x), father(y, x), it-rains(today), etc.

Note, that the predicates man(x) and father(x,y) are not yet instantiated, but the predicate it-rains(today) is already instantiated.

- *\<Rule-Certainty\>*: It is a measure of the certainty of the co-occurrence of the conditions and the conclusions of a rule. The certainty values can be expressed with the same uncertainty management models detailed above for the facts' values.
- *\<Actions or Conclusions\>*: The actions or conclusions of a rule are named in general as the *consequent* of a rule, or the *right-hand side* (RHS) of a rule. They can be actions or conclusions.

  – Actions: Which can be just sending messages to the user, making some computations, etc.
  – Conclusions: Which are new deductions (propositions or predicates) over the truth or not of some fact of the Fact Base.

Some examples of rules are depicted in Fig. 5.5.

The usual format of the rules is that in the antecedent only can appear the negative and the conjunction connectives, and in the consequent of the rules, commonly just one action or conclusion appears. The first rule in the examples of Fig. 5.5 describes that if the outside temperature is higher than 18 °C but lower than 25 °C, and the weather forecast for today is sunny, then it is suggested to play golf. The second rule states that if the sludge concentration of a primary settler in a wastewater treatment

Fig. 5.5 Examples of some
rules

(RPLAY-002
   IF
     Outside-Temperature > 18
     Outside-Temperature < 25
     Weather-Forecast = sunny
   THEN
     Play = golf
   ...)

(RPRIM-SET03
   IF

     Sludge-Concentration-Primary-Settler = high
     No Sludge-Waste-Primary-Settler
   THEN
     <u>0.8</u>
     Clean-sewage-pipe-Prim-Settler
   ...)

(RMOD8-007
   IF
     No patient-has (neutropenia)
     Associated-dermatology (ectima-gangrenosum)
   THEN
     <u>very-possibly</u>
     Origin-patient-disease (pseudomonas)
   ...)

plant is high, and the primary settler is not wasting sludge, then with a certainty of 0.8, you must clean the sewage pipe of the primary settler, because very probably it is plugged.

5.3.2.1   Modularization of the KB

In the early rule-based reasoning systems, named as *first-generation expert systems*, all the rules of one particular domain were organized together in the same KB, being able to reach more than one million rules in a KB. This approach had some drawbacks regarding the depuration of errors, the maintenance of this KB, and the meaningful understanding of the knowledge coded in the rules. In addition, this unique large KB can cause a high computational cost, when applying the inference engines.

Aiming to solve that, in latter rule-based reasoning systems, the so-called *second-generation expert systems*, the rules were grouped into some components named as *modules*. Each *module* grouped the rules according to some criteria. Usually, the rules guiding to deduce similar facts were put together in the same module. This way let us just consider one module at each reasoning step of the inference engine. This modularization of the KB, had several benefits. It reduced drastically the

Fig. 5.6 Hierarchical
modularization of the KB
for the diagnosis task in the
WWTP example

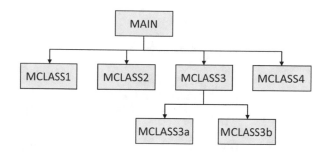

computational time for executing the reasoning step because each module had a lot
fewer number of rules to be considered. On the other hand, the different modules
gave a clear structuration of all the knowledge embodied in the rules, packaging the
"similar" rules into the same modules.

Modules usually are formed by:

- *<Identifier-of-the-module>*: An identifier of the module. The *main module* is the
  module where the execution is started at the beginning of the execution of the
  system.
- *<Conclusions/diagnostics>*: The list of possible conclusions or diagnostics that
  can be reached from the application of the knowledge contained in the module.
- *<Rules-of-the-module>*: The set of rules grouped together which are expressing
  the body of knowledge coming from the experts or the literature. All these rules
  are usually concluding similar conclusions or related conclusions about the same
  topic.
- *<Meta-rules-of-the-module>*: The set of meta-rules is representing the *meta-
  knowledge* on how to lead the reasoning process following some kind of strategy.
  There are different kinds of meta-rules, as it will be explained later.

Usually, this structuration of the KB has a hierarchical form, where the root
module is commonly the main module. In Fig. 5.6, there is an example of the
modularization of the KB using a hierarchical structure. This example corresponds
to a classical diagnostic process in a system. For instance, the diagnosis task in the
operation of a Wastewater Treatment Plant (WWTP) using the activated sludge as
the main treatment technology. The operation problems can be due to the presence of
heavy metals like zinc (*class1* problems), or to a high value of dissolved oxygen in
the *biological reactor* (*class2* problems), or to problems of bulking (*class3* problems), which can be due to two major causes (*class3a* or *class3b* problems) or to
overloading problems (*class4* problems). Thus, each module corresponds to the
possible diagnostic problems of that concrete class that can happen in the WWTP.

Thus, in the *MAIN* module, there are some general rules trying to identify which
is the *type* of problems occurring in the WWTP (see Fig. 5.7). For each possible type
of problem, there is, at least one module (*MCLASS1*, *MCLASS2*, *MCLASS3*, etc.),
which contains rules to continue discriminating which are the actual causes of the
operation problems, but just constrained to sub-problems and/or causes of that class
or type. All this process of exploration of the different modules is led by the
corresponding meta-rules.

```
(module MAIN                        ... )

    //Rules of the module
    ((RPR01
        IF                              //Meta-Rules of the module
            (high-COD-exit)             ((MRCL1
            (high-conc-Zn-inflow)           IF
            (high-conc-Zn-bioReactor)           (class1)
        THEN                            THEN
            (class1))                       (goto-module MCLASS1))
    (RPR02                          (MRCL2
        IF                              IF
            (normal-COD-exit)               (class2)
            (high-DO)                   THEN
        THEN                            (goto-module MCLASS2))
            (class2))                   ... ))
```

Fig. 5.7 Content of the *main* module for the diagnosis task in the WWTP example

5.3.3   Reasoning Component: The Inference Engine

The *reasoning module* is composed by the *inference engine*. Its main goal is to deduce/demonstrate new facts or/and to execute actions to solve the current problem, from a set of initial facts' values, and a given knowledge base, with the user interaction.

The inference engine is formed off a *rule interpreter system* and a *control strategy system*. The rule interpreter subsystem is able to interpret and understand the inference rules, and depending on the used control strategy to apply the rules.

5.3.3.1   Reasoning Cycle

The general cycle of the inference engine is the following:

1. *Detection*: This step aims to obtain the set of applicable rules. Usually, this step is named as the *Conflict Set Formation* step, because as commonly more than one rule could be applicable, there is a conflict regarding which rule has to be used.
2. *Selection*: This step consists of the selection of the rule to be applied or fired. Usually, it is named as the *Conflict Set Resolution* step, because at this point one of the candidate rules of the conflict set is selected, and thus, the conflict is solved.
3. *Application*: The last step is the application or firing of the selected rule. Commonly, it is referred to as the *Inference* step, because this step is where the new inferred facts are produced or new facts or sub-hypotheses must be validated. The new inferred facts' values are stored in the Fact Base or the new sub-hypotheses are added to the current stack of non-yet validated hypothesis or goals.

Detection of Candidate Rules

In this step of the reasoning cycle, the detection of the candidate rules to be applied is done. Rules will be the candidate or not depending on the control strategy used (i.e. *forward reasoning* or *backward reasoning*). Thus, the antecedent or the consequent of the rule must be checked for being satisfied.

The *rule interpreter* makes the computations and needed instantiations of the variables within the facts composing the rules, which are possible at each problem-solving state. In *forward reasoning*, if facts are expressed as *zero-order predicates* or *propositions* the matching process is easy because one rule can be only satisfied in a unique way. However, if the facts are represented by *first-order predicates*, then the pattern matching process between the available rules and the available objects is quite complex because one rule can be satisfied with many different instantiations of the variables of the facts.

One of the best known and efficient methods for the *rule pattern matching problem* is the *Rete*[2] *algorithm*, which was proposed by Charles L. Forgy (Forgy, 1982). The basic idea of the Rete algorithm is to save computation time in some naïve approaches to the pattern matching problem, which repeatedly evaluate the left-hand side of the rules against the working memory, in forward chaining. In addition, he observed that the working memory is only modified a little bit each time and that it is not needed to match all the patterns against all the working memory facts at each reasoning cycle. The Rete approach builds a network of nodes, where each node, except the root, corresponds to a pattern occurring in the left-hand side of a rule. Thus, Rete exploits the fact that usually the left-hand side of the rules in the working memory share conditions (like common prefixes in strings). The path from the root node to a leaf node defines a complete rule left-hand side (conjunction of conditions). Each node has a set of facts satisfying that pattern. This structure is a generalized *trie structure*.[3] As new fact values are known or modified, they are propagated along the network. This propagation causes nodes to be annotated when that fact matches that pattern. When a fact or combination of facts causes all of the patterns of a given rule to be satisfied, a leaf node is reached, and the corresponding rule is detected as a candidate to be triggered.

---

[2] Charles L. Forgy stated that he adopted the term "Rete" because of its use in anatomy to describe a network of blood vessels and nerve fibres.

[3] A *trie structure* is a data structure used in computer science. A trie is a set of nodes structured in a hierarchical way (a tree) used to store a hash table (a mapping function between a key and a value) where the keys are usually strings.

Selection of the rule to be applied

Once the *conflict set* has been formed with all candidate rules to be triggered, the cycle must continue with the selection of the best rule among the obtained ones in the previous step.

The selection depends on the *conflict resolution strategy* used by the inference engine. In the literature, there are many strategies, but the most commonly used criteria are:

- *Specificity*: The most specific rule (or the opposite, the most general rule) is selected. One rule is more specific than another one when it haves a greater number of conditions than the other. Therefore, it is considered a more difficult rule to be satisfied. These rules are supposed to be better because more data is taken into account.
- *Utility*: The more or less used rule is the one selected. This criterion can be used when working with rules using first-order predicates in the facts, and therefore, the rules can be reused with several instantiations. The rules should have a utility counter to measure the *utility* of the rule (number of times that the rule has been used).
- *Recency*: The rule matched by more recent facts is selected before other matched with older facts. To implement this criterion a time stamp system must be recorded for each fact of the Fact Base.
- *Certainty*: The rule with a higher certainty degree associated is selected before others with less certainty degree. Of course, this criterion only makes sense when rules have associated a certainty degree.
- *Ordering*: The first rule in order in the conflict set is the one selected. It is a very simple but efficient way of selecting the rule to be applied.
- *Randomness*: The rule selected is a random rule taken from the conflict set.
- *Information*: The most informative rule, which is the one giving the highest number of conclusions, should be selected before others. The rational is that for the same "price" you obtain more information with the same effort (a reasoning cycle). Of course, it only makes sense when rules can have more than one conclusion on the right-hand side of the rule.

Usually, more than one criteria are used, in an ordered way to decide which rule to select, when there are ties between rules just considering the previous conflict resolution strategies.

In addition, a common general strategy, with facts expressed as first-order predicates, is that a rule should not be allowed to fire more than once with the same instantiation of its facts. The reason is that if the rule were fired, it would produce the same conclusions than before, and thus, the problem-solving resolution would not advance.

Application of the Selected Rule

The *rule interpreter* applies or executes the selected rule in a different manner depending on the inference engine used. In *forward reasoning*, it updates the state of the Fact Base with new deductions, computations, or actions. The deductions are the consequent part of the rule. In *backward reasoning*, it adds new sub-goals in the stack of current sub-goals to be validated. The new sub-goals are the premises of the antecedent of the rule.

If the rules and facts are expressed using first-order predicates, then *instantiation propagation* of the same variables in the rule must take place.

If the rule-based system is managing the certainty degrees of facts and rules, then, *certainty propagation* from premises to conclusions, by means of different logic connectives must be done.

For instance, if we have this rule in the Knowledge Base:

$$R1 : IF \ human(x) \ THEN \ mortal(x)$$

and in the Fact Base we have:

$$human(Socrates)$$

$$human(John)$$

$$\ldots$$

using a forward reasoning engine, there would be two possible instantiations of R1:

$$R1 \ (x/Socrates) : IF \ human(Socrates) \ THEN \ mortal(Socrates)$$

$$R1(x/John) : IF \ human(John) \ THEN \ mortal(John)$$

where x/Socrates means the instantiation of the variable x by the atom Socrates, and x/John means the instantiation of the variable x by the atom John.

Depending on which rule instantiation was selected, a different conclusion could be reached (i.e. mortal(Socrates) or mortal(John)).

Otherwise, if backward reasoning were used, and we had the following state of the stack of hypotheses or conclusions to be validated:

$$Stack \ of \ hypotheses : mortal(Socrates)?$$

R1 would be a candidate rule because it is consequent (mortal(x)) can be matched with the hypothesis which needs to be validated (mortal(Socrates)), just making the instantiation of x/Socrates. Then, the application of R1:

$$mortal(Socrates) \ can \ be \ validated \ if \ we \ can \ validate \ human(Socrates)$$

Would produce a modification of the stack of hypotheses:

$$Stack \ of \ hypotheses : human(Socrates)?$$

Which could be validated in the next reasoning cycle, because human(Socrates) is a fact which is in the Fact Base (i.e. it is known to be true).

End of the Cycle

The cycle ends when no more applicable rules, within the current context (the whole KB or the current KB module), are found, or when the desired conclusion/s are reached, or when some unexpected exception happens.

Depending on the problem and on the control strategy, the reasoning chain could be cut (*backward reasoning*). Some back steps must be reconsidered. Therefore, a backtracking technique must be used.

5.3.3.2   Inference Engines

There are two main inference engines or control strategies in rule-based reasoning systems:

- *Forward reasoning* or deductive engine: This inference engine is a *data-driven* strategy. It starts with the evidence, or symptoms, or data, which are available in the Fact Base, and keep trying to apply rules to produce some new conclusions or hypotheses, with the hope that at some point in the reasoning steps, some of the expected conclusions or hypotheses will be deduced. Therefore, data are the elements guiding the reasoning process. It is also known as a *progressive chaining* or *forward chaining* of the rules because the rules are applied in a "forward" way (from left to right):

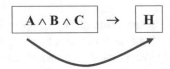

  If the antecedent of the rule is true ($A \wedge B \wedge C$, which depends on the truth of A, B, and C), then the consequent of the rule is also true (H).

- *Backward reasoning* or inductive engine: This inference engine is a *goal-driven* strategy. It starts picking up the goal or hypothesis on the top of the stack of hypotheses to be validated, and try to validate such hypothesis using the rules in the Fact Base that could conclude that hypothesis. This produces that the facts appearing in the antecedent of the rules should be also validated, and they are pushed into the stack of hypotheses to be validated, and so on, until all this backward chain can be validated with the available evidence, symptoms, or data, i.e. the facts of the Fact Base. Therefore, the goals, conclusions, or hypotheses are the elements guiding the reasoning process. It is also known as a *regressive chaining* or *backward chaining* of the rules because the rules are applied in a "backward" way (from right to left):

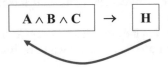

To validate the consequence of the rule (H), it must be validated also (A) and (B) and (C), which will be the new sub-goals or sub-hypotheses to be validated in the next reasoning steps.

Forward Reasoning

The *forward reasoning* mechanism is based on the application of *modus ponendo ponens* or *modus ponens*[4] (in Latin) inference rule in propositional logic[5]:

$$A \rightarrow B, A \vdash B$$

The inference rule states that if *A* implies *B* is true, and *A* is true, then we can conclude that *B* is also true. Therefore, it is a deductive method according to logic. See (Kleene, 1967) for further details on Logic.

The problem resolution runs searching from an initial state (data or symptoms or evidences) to the final state (goal), through the intermediate states generated by the chain of inferences derived from rule application. The idea is starting from evidences/symptoms/data try to deduce all possible conclusions by repeated application of *modus ponens* to all applicable rules.

The basic operation of a forward reasoning engine can be algorithmically described as follows:

```
Algorithm Forward Reasoning Engine (FRE)

Input: FB: The Fact Base; KB: The Knowledge Base,
       Criteria: Ordered list of conflict resolution criteria
Output: an updated FB

begin
  initialize_the_FB_with_known_facts'_values
  ConfSet ← compute_rule_conflict_set(FB, KB)
  while ConfSet ≠ ∅ & not exception_is_produced & problem_is_not_solved do
    Rule ← select_one_applicable_rule(ConfSet, Criteria)
        // Applicable rules will be those with antecedents being true
    Conclusion ← apply_rule(Rule)
    add_or_update_FB(FB, Conclusion)
    ConfSet ← compute_rule_conflict_set(FB, KB)
  endwhile
end
```

---

[4] *Modus Ponendo Ponens* or abbreviated *Modus Ponens* is the name of the Latin expression for a deductive argument form and inferential reasoning mechanism in propositional logic meaning: "mode that by affirming, affirms".

[5] The symbol ⊢ in logic means "then it is deduced or implied".

Main drawbacks of this inference mechanism are:

- It does *not focus on a concrete goal*. Thus, using this inference engine, the conflict resolution strategy can be very critical.
- Due to different possible instantiations of the premise predicates, in first-order predicates, a *combinatorial explosion* of possible instantiations of the same rule can appear. This could lead to multiple applications of the same rule with different instantiations that do not advance in the problem resolution, making the reasoning process very inefficient.

Main advantages of this inference engine are:

- It makes easier the knowledge formalization in form of inference rules because the application forward of a rule is more easily understandable.
- In addition, the modus ponens inference rule is very intuitive.

Let us describe a pair of examples to illustrate the use of forward reasoning engine.

Example 1

Let us suppose that we have the following Knowledge Base, Fact Base, and Goals to be deduced, with the facts described as propositions (zero-order predicates), as indicated in Fig. 5.8.

If forward reasoning engine is used, it will produce several different alternatives, depicted in Fig. 5.9, depending on the *conflict set resolution* criterion employed. Thus, if R4, R6, R5, and R2 are used, the goal will be verified in just four steps. However, if other rule orders (other paths) are used, it can be verified up to six steps.

Example 2

The Example 2, described in Figs. 5.10 and 5.12, shows that the selected *conflict resolution strategy* is very critical in a forward reasoning mechanism. As shown in Fig. 5.11, when it uses the *first rule in order* as the conflict resolution strategy (Fig. 5.10) it needs 50 steps to reach the goal A(f,b). However, when using the *most specific rule* as the conflict resolution strategy (Fig. 5.12) it only needs three steps, as shown in Fig. 5.13.

Fig. 5.8 Description of the Knowledge Base, initial Fact Base, and the goals in the Example 1 of the use of forward reasoning in propositional logic

Knowledge Base	Fact Base	Goal/s
R1: $A \wedge B \wedge C \rightarrow D$	A	G?
R2: $A \wedge E \wedge F \rightarrow G$	E	
R3: $B \wedge C \wedge D \rightarrow H$	B	
R4: $E \rightarrow C$		
R5: $A \wedge H \rightarrow F$	None fact is questionable	
R6: $A \wedge C \rightarrow H$		

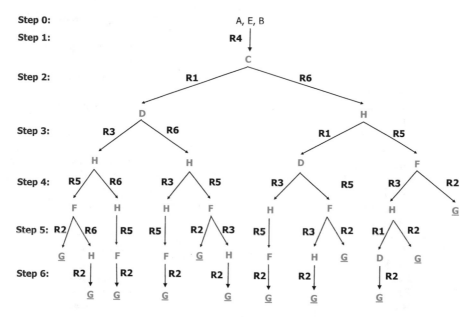

Fig. 5.9 All the different alternatives of rule applications depending on the *conflict set resolution* strategy in the Example 1 of the use of forward reasoning in propositional logic

Fig. 5.10 Description of the Knowledge Base, initial Fact Base, and the goals in the Example 2 of the use of forward reasoning in first-order logic, when using the first rule in order as the conflict resolution strategy

Knowledge Base	Goal
R1: B(X,Y) → A(X,Y)	A(f,b)?
R2: A(X,Y) → A(Y,X)	
R3: A(X,Y) ∧ A(Y,Z) → A(X,Z)	

Fact Base

Conflict resolution strategy
First rule In order

A(a,b)	B(a,c)	
A(c,b)	B(a,b)	
A(d,a)	B(b,c)	⟿ None fact is questionable
A(e,a)	B(d,a)	
A(f,a)	B(f,a)	

Backward Reasoning

The *backward reasoning* mechanism is not directly a deductive method. There is no direct inference rule in logic supporting it:

$$A \rightarrow B \nvdash B$$

However, taking into account that from *modus ponens* we know:

$$A \rightarrow B, A \vdash B$$

Comparing both logical expressions, we know that if the truth of the antecedent of the rule, *A* can be verified, then the *modus ponens* can be applied to that rule, i.e. a

R1 Application
A(a,c)
A(a,b) → ✓
A(b,c)
A(d,a) → ✓
A(f,a) → ✓
R2 Application
A(b,a)
A(b,c) → ✓
A(a,d)
A(a,e)
A(a,f)
A(c,a)
A(c,b) → ✓
A(a,b) → ✓
A(d,a) → ✓
A(e,a) → ✓
A(f,a) → ✓
A(a,c) → ✓
R3 Application
A(a,c) → ✓
A(a,a)
R2 Application
A(a,a) → ✓
R3 Application
A(c,c)
R2 Application
A(c,c) → ✓
R3 Application
A(c,a) → ✓
A(d,b)
R2 Application
A(b,d)
R3 Application
A(a,d) → ✓
A(c,d)
R2 Application
A(d,c)
R3 Application
A(d,c) → ✓
A(d,d)

R2 Application
A(d,d) → ✓
R3 Application
A(d,e)
R2 Application
A(e,d)
R3 Application
A(d,f)
R2 Application
A(f,d)
R3 Application
A(d,a) → ✓
A(e,b)
R2 Application
A(b,e)
R3 Application
A(a,e) → ✓
A(c,e)
R2 Application
A(e,c)
R3 Application
A(e,c) → ✓
A(e,d)
R2 Application
A(d,e)
R3 Application
A(e,e)
R2 Application
A(e,e) → ✓
R3 Application
A(e,f)
R2 Application
A(f,e)
R3 Application
A(e,a) → ✓
A(f,b) !!

In 50 steps !!

Fig. 5.11 Rule application steps using the *first rule in order* as the *conflict set resolution strategy* in Example 2 of the use of forward reasoning in first-order logic

forward interpretation of the rule must be done, and then it can be stated that B is true.

The *backward engine* states that to verify that B is true, given that A implies B is true, A should be validated to be true. The problem is that to validate A, if it is not already known as a true fact in the Fact Base, the same procedure must be done recursively with another rule to verify that A is really true. Thus, several sub-goals appear that must be verified. Even though this procedure does not match any direct

Fig. 5.12 Description of the Knowledge Base, initial Fact Base, and the goals in Example 2 of the use of forward reasoning in first-order logic, when using the most specific rule as the conflict resolution strategy

Knowledge Base

R1: B(X,Y) → A(X,Y)
R2: A(X,Y) → A(Y,X)
R3: A(X,Y) ∧ A(Y,Z) → A(X,Z)

Fact Base

A(a,b)	B(a,c)
A(c,b)	B(a,b)
A(d,a)	B(b,c)
A(e,a)	B(d,a)
A(f,a)	B(f,a)

Goal

A(f,b)?

Conflict resolution strategy
Most specific rule

⤳   **None fact is questionable**

R3 Application
A(d,b)
A(e,b)
A(f,b) !!

In 3 steps !!

Fig. 5.13 Rule application steps using the *most specific rule* as the *conflict set resolution strategy* in Example 2 of the use of forward reasoning in first-order logic

deductive logical inference, can be solved through the repeated application of the *modus ponens*, after a reverse chain of implications, involving several premises of the corresponding rules to verify the goal, has been constructed.

However, in practical implementations of a *backward chaining* inference, such as in logic programming, it is implemented using a *resolution* process to *refute* the negation of the goal to be validated. This *refutation process* uses iteratively the application of *modus tollendo tollens* or simply *modus tollens*,[6] to verify the refutations of all sub-goals. At the end, if all the refutations are validated, this means that the goal cannot be validated, but if the refutation process fails, then, it means that the goal is validated.

All in all, the repeated application of *modus tollens*, is at the end, the application of *modus ponens*, because *modus tollens* is defined as:

$$A \rightarrow B, \neg B \vdash \neg A$$

which is actually a specific application of *modus ponens*, because:

$$A \rightarrow B \equiv \neg B \rightarrow \neg A$$

---

[6] *Modus Tollendo Tollens* or abbreviated *Modus Tollens* is the name of the Latin expression for a deductive argument form and inferential reasoning mechanism in propositional logic meaning: "mode that by denying, denies".

In this inference mechanism, the problem resolution is clearly led by a goal. To validate a goal or hypothesis, the engine must validate a reasoning chain from the data or evidence or symptoms to the goal or hypothesis, but it must be constructed in reverse order. Thus, starting from the hypothesis, it must validate all the premises (facts of the antecedent) of the rule that conclude the hypothesis. Then, all these premises are new hypotheses that are required to be validated (*and-node* of the exploration graph). Therefore, each step implies new sub-goals or sub-hypothesis to be validated. This procedure is continued until some premises of a rule can be directly validated in the Fact Base or by the user. Sometimes, the reasoning chain cannot be validated and some backtracking steps must be done to consider other possible rules (reasoning chains) to validate the current hypotheses to be validated (*or-node* of the exploration graph).

The basic operation of a backward reasoning engine can be described as follows:

Algorithm Backward Reasoning Engine (BRE)

Input: *PendHyp*: a stack of hypotheses pending to be validated;
Output: *ValHyp:* a list of validated hypotheses

begin backward (**in** *PendHyp*: Stack(Hypotheses), **in out** *ValHyp*: List(Hypotheses))
    // Let be FB: the Fact Base; *KB*: the Knowledge Base;
     Criteria: Ordered list of conflict resolution criteria;
     CurrHyp: is the current hypothesis being validated

  if PendHyp $\neq \varnothing$ **then**
    CurrHyp $\leftarrow$ top(*PendHyp*)
    PendHyp $\leftarrow$ pop(*PendHyp*)
    if validated(*CurrHyp, FB*) **or** (CurrHyp **in** ValHyp) **then**
      backward(*PendHyp, ValHyp*)
      ValHyp $\leftarrow$ add(*ValHyp*, CurrHyp)
    else
      ConfSet $\leftarrow$ compute_ConfSet_rule_backward(*FB, KB, CurrHyp*)
        // Applicable rules will be those with the consequent being the CurrHyp
      while *ConfSet* $\neq \varnothing$ **and not** (*CurrHyp* **in** ValHyp) **do**
        Rule $\leftarrow$ select_one_applicable_rule_backward(*ConfSet, Criteria*)
        AntFacts $\leftarrow$ antecedent_facts(*Rule*)
        Ok $\leftarrow$ true
        while *AntFacts* $\neq \varnothing$ **and** Ok **do**
          Fact $\leftarrow$ get_Fact(AntFacts)
          push(*Fact, PendHyp*)   *// each fact is a new sub-goal to be validated*
          backward(*PendHyp, ValHyp*)
          pop(*PendHyp*)
          Ok $\leftarrow$ *Fact* **in** ValHyp
        endwhile
        if Ok **then** *ValHyp* $\leftarrow$ add(*ValHyp*, CurrHyp) **endif**
      endwhile
    endif
  endif
end

The problem-solving strategy is actually the exploration of an *and/or graph*. This means a graph where the nodes are alternatively of type *and* or type *or*. An *and-node* means that to validate the hypothesis of the node, *all* (the first and the second and . . . and the last) the hypotheses of the connecting nodes at the end of the output edges of the node must be validated. An *or-node* means that to validate the hypothesis of the node, *just one* (the first or the second or . . . or the last) of the hypotheses of the connecting nodes at the end of the output edges of the node is needed to be validated.

In the algorithm, before calling the function backward, the Fact Base must be initialized with the set of known facts' values. The initial call to the function will have as a parameter the hypothesis/goal to be verified, and an empty list of validated hypotheses: backward({H}, ∅). If there are more hypothesis to be validated, the function *backward* must be called for each one of the hypotheses to be validated.

Main advantages of *backward reasoning* engine are:

- The problem resolution is better directed than in forward reasoning because there is a clear goal to be concluded/validated. Only the necessary knowledge (i.e. inference rules) and facts are considered to solve the problem.
- Therefore, always that it is possible, it is better to use a backward reasoning engine instead of a forward reasoning.

Main drawback of this inference engine is that to employ backward reasoning, we must have a clear goal to be validated. Otherwise, it cannot be used, and a forward reasoning engine must be used.

Example 1

In Example 1 of the use of the backward reasoning, the conflict strategy resolution is the first rule in order (see Fig. 5.14). Thus, to verify whether hypothesis H is true or not, it first checks that H is not a currently known fact in the FB. As it is not there, then it sets H as the current hypothesis to be validated. Hence, it searches for rules where H is the consequent in KB, and it finds just one rule, R6. Thus, the premises of the antecedent of the rule R6 (A and G) are the new hypothesis to be validated, and they are added to the pending hypothesis stack (see Fig. 5.15a).

The next step is to take the first new hypothesis of the pending hypothesis stack (A), and apply the same procedure to validate A. The process continues going on trying to validate all the hypotheses appearing in the process.

Fig. 5.14 Description of the Knowledge Base, initial Fact Base, and the goals in Example 1 of the use of *backward reasoning in propositional logic*

Knowledge Base	Fact Base	Goal/s
R1: $A \wedge B \rightarrow C$	A	H??
R2: $C \rightarrow D$	B	
R3: $E \wedge F \rightarrow G$		
R4: $A \rightarrow E$ All facts are questionable		
R5: $D \rightarrow G$		
R6: $A \wedge G \rightarrow H$		

However, when validating G through rule R3, it has to validate F. This fact is not in the FB, and there is no rule that can conclude it. In this situation, the unique possibility to be validated is that this fact be askable to the user, and that the user answers that it is true. In this case, let us suppose that the user says no. Then G cannot be validated through R3 (see Fig. 5.15b).

Thus, it has to go back and reconsider other options, i.e. other possible rules to validate G. In this example, R5 can be used (see Fig. 5.15c). Finally, following the path through R5 and R2 all the necessary hypotheses can be validated (D, C, A, B), and thus, G is validated. As A is also validated, H is validated through R6 (see Fig. 5.15d). All these steps are depicted in Fig. 5.15.

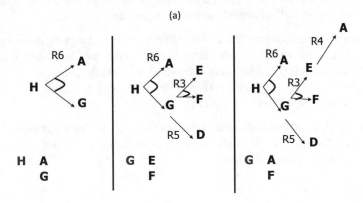

Fig. 5.15 Steps of the exploration of the and/or graph for Example 1 of the use of *backward reasoning in propositional logic*

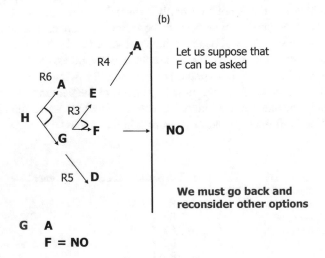

Fig. 5.15 (continued) Fourth step (**b**) where the fact F cannot be validated

(c)

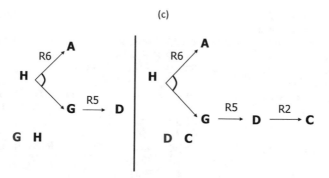

Fig. 5.15 (continued) fifth and sixth steps (**c**) where the fact G will be validated through rule R5, and fact D through rule R2

(d)

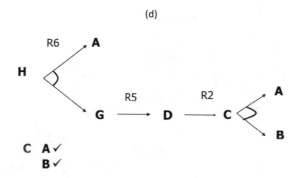

Fig. 5.15 (continued) Last step (**d**) where the fact C is validated through rule R2

Example 2a and 2b

The Examples 2a (Fig. 5.16) and 2b (Fig. 5.18) show that depending on the conflict resolution strategy, the exploration of the graph can be shorter or larger. In the Example 2a, the goal A(f,b) is trying to be validated through the use of R1 and R2 and R3, in this order because the conflict resolution strategy selects the first rule in order. In some steps there is a new hypothesis to be validated that is the same that is trying to validate, i.e. there is a cycle in the validation process. When this happens, this path must be cut, and reconsider other options (*backtracking process*). It needs 13 steps to finally validate the goal (see Fig. 5.17). However, in the Example 2b (Fig. 5.18), using the most specific rule as a conflict resolution strategy, it selects R3 as the first rule to try to validate the goal hypothesis, and just in 1 step can be validated (see Fig. 5.19). Thus, again, it can be stated that the conflict resolution strategy can be very critical in the resolution process. In addition, in the backward reasoning, the exploration of some paths in the and/or graph is cut, and some previous decisions must be reconsidered, i.e. a backtracking process must be applied.

Fig. 5.16 Description of
the Knowledge Base, initial
Fact Base, and the goals in
Example 2a of the use of
*backward reasoning in first-
order logic*, using the *first
rule in order* as the conflict
resolution strategy

Knowledge Base

R1: B(X,Y) → A(X,Y)
R2: A(X,Y) → A(Y,X)
R3: A(X,Y) ∧ A(Y,Z) → A(X,Z)

Goal

A(f,b)?

Fact Base

A(a,b)   B(a,c)
A(c,b)   B(a,b)
A(d,a)   B(b,c)
A(e,a)   B(d,a)
A(f,a)   B(f,a)

Conflict resolution strategy
First rule in order

⟿   **None fact is questionable**

(a)

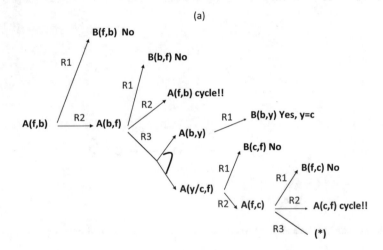

Fig. 5.17 Exploration of the first ten steps (**a**) of the and/or graph for Example 2a of the use of
backward reasoning in first-order logic, when using the *first rule in order* as the conflict resiolution
startegy

(b)

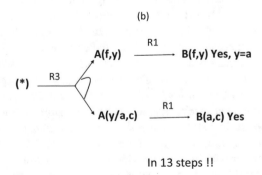

In 13 steps !!

Fig. 5.17 (continued) Exploration of the last three steps (**b**) of the and/or graph for Example 2a of
the use of *backward reasoning in first-order logic*, when using the *first rule in order* as the conflict
resiolution startegy

Fig. 5.18 Description of the Knowledge Base, initial Fact Base, and the goals in Example 2b of the use of *backward reasoning in first-order logic*, using the *most specific rule* as the conflict resolution strategy

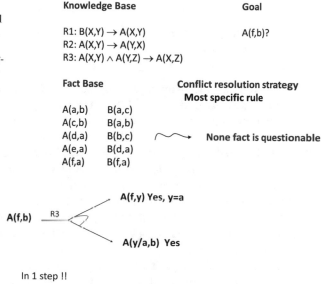

Knowledge Base

R1: B(X,Y) → A(X,Y)
R2: A(X,Y) → A(Y,X)
R3: A(X,Y) ∧ A(Y,Z) → A(X,Z)

Goal

A(f,b)?

Fact Base

A(a,b)	B(a,c)
A(c,b)	B(a,b)
A(d,a)	B(b,c)
A(e,a)	B(d,a)
A(f,a)	B(f,a)

Conflict resolution strategy
Most specific rule

None fact is questionable

A(f,b) —R3— A(f,y) Yes, y=a
                      A(y/a,b) Yes

In 1 step !!

Fig. 5.19 Exploration of the and/or graph for the Example 2b of the use of *backward reasoning in first-order logic*, when using the *most specific rule* in order as the conflict resolution strategy

5.3.4   Meta-Reasoning Component

The Meta-reasoning component is in charge to control how the reasoning process is done. Meta-reasoning means reasoning over the reasoning. Therefore, this component must control *how* and *when* to apply the reasoning processes. This control, in fact, is some knowledge on *how* and *when* we should use the knowledge on the concrete domain. Thus, we must be able to express meta-knowledge, i.e. knowledge over the knowledge.

The meta-knowledge can be implicit or explicit.

- *Implicit Meta-knowledge*: One way of controlling when to use a particular pattern of knowledge (i.e. a rule) is by means of the *conflict resolution strategy*. The set of ordered criteria that can be used, impose a particular ordering of the rules' execution.

  In addition, in the first-generation of rule-based reasoning systems or Expert Systems, an implicit mechanism appeared to control the rule applicability. They used some added *artificial premises* to the rules to control their applicability, like in the below example:

  Original KB:

  R1: A ∧ B → C
  R2: D → E
  R3: F ∧ G → H

Transformed KB to impose this order of application: R2, R3, and R1:

R1: $AP_1 \wedge AP_2 \wedge A \wedge B \rightarrow C$
R2: $D \rightarrow E \wedge AP_1$
R3: $AP_1 \wedge F \wedge G \rightarrow H \wedge AP_2$

Although this mechanism perfectly guarantees that R1 will be applied always after the application of R3 and R2, and that always R3 will be applied after the application of R2, it has a great drawback: the *complexity of the rules* is increasing as well as the *loss of meaning of the knowledge* coded in the rules. The use of this mechanism leads to merge in the rules *control knowledge* with *actual knowledge* over the domain. This mixture of both control and actual knowledge is not a good option. You can imagine what happens if you have just ten rules instead of three, and you want that one rule should be executed after the remaining ones! This rule will have nine *artificial premises* (AP_i) added in its precondition! This way the meaning of the rule is completely lost, and the complexity of the KB has increased very much.

- *Explicit Meta-knowledge*: To solve the problem of the artificial premises in the rules, the introduction of the concept of *meta-rules* by Randall Davis in (Davis, 1980) supposed a great advance in the use of explicit meta-knowledge mechanisms. These systems using explicit meta-knowledge mechanisms were named as the second-generation of rule-based reasoning systems.

  A *meta-rule* is a rule acting over rule/s. This mechanism provided a clear separation between *control* and *knowledge*. Rules express the knowledge about some domain, and the meta-rules express the control strategy over the knowledge. The use of meta-rules is fully linked to the use of a modular structure for the KB. In addition, two important features appeared with the use of meta-rules:

  - A *Unified Reasoning Mechanism*: The inference engine can be used both by rules and by meta-rules because they have the same syntactical structure.
  - *Strategy* Concept: the use of meta-rules in addition to the modular structure of the KB allows that the necessary modules can be ordered for the problem-solving process, building up the concept of strategy. A *strategy* can be thought as an ordered set of modules to be explored.

Thus, a *meta-rule* can be defined as an explicit control unit over the knowledge. There can be different *types of meta-rules*, from a semantic point of view:

- *Meta-rules over rules*: these meta-rules control the activation or not of some rules. They are useful when a set of rules only should be considered in the problem-solving process under some conditions. There are two subtypes of meta-rules:

  - Meta-rules activating rules: When applied, these meta-rules activate (i.e. make visible for the detection step) the indicated rules.
  - Meta-rules deactivating rules: When applied, these meta-rules deactivate (i.e. make invisible for the detection step) the indicated rules.

- *Meta-rules over modules*: thcsc mcta-rules have several subtypes, addressing different knowledge control purposes within a module:

  - Meta-rules for selecting the kind of reasoning mechanism to be used in the modules (forward, backward).
  - Meta-rules for setting the minimum certainty level of the rules in the module to be applied.
  - Meta-rules for rule subsumption.

- *Meta-rules over Strategies and Actuation Plans*: These meta-rules allow to work with the concept of *strategy*, i.e. an ordered set of modules to be explored, and an *actuation plan*, that could be a combination of several strategies. Most common types are:

  - Meta-rules for setting a strategy to be followed: These meta-rules let decide what modules and in what order should they be explored.
  - Meta-rules for determining an actuation plan: These meta-rules allow to select which strategy should be applied first when more than one is available.
  - Meta-rules for detecting exceptions in the problem-solving process.

From a syntactical point of view meta-rules are very similar to rules, but with the following characteristics:

- The *antecedent* will have the same form of a rule: A combination of premises with logical operators.
- The *consequent* has special propositions or predicates: The most used consequents are the following:

  - ACTIVATE-RULES <list of rule identifiers>: meta-rule that activate the list of rules indicated by the specified identifiers.
  - DEACTIVATE-RULES <list of rule identifiers>: meta-rule that deactivate the list of rules indicated by the specified identifiers.
  - FORWARD: meta-rule that specifies that the reasoning mechanism from now on, in the corresponding module, should be a forward chaining.
  - BACKWARD <list of facts to be validated>: Meta-rule that specifies that the reasoning mechanism from now on, in the corresponding module, should be a backward chaining with the given facts as the set of hypotheses to be validated.
  - GOTO-MODULE <list of module identifiers>: This meta-rule indicates that the new *strategy*, from now on is formed by the exploration, in order, of the given modules.
  - STOP-MODULE: This meta-rule indicates that the exploration of the current module must end, probably because, some important fact or facts have been already deduced within the module, and the current strategy must be continued as soon as possible.
  - HALT: This meta-rule stops the reasoning process of the system because some kind of exception or unexpected situation is detected in the problem-solving process.

Fig. 5.20 Examples of
different kinds of meta-rules

(MR-PRIMSET01
  IF
    OK-PUMP
  1.0
  THEN
    (DEACTIVATE-RULES RPRSET005
    RPRSET006 RPRSET007
    RPRSET008 RPRSET009
    RPRSET019 RPRSET020))

(MR-MAIN
  IF
    FEVER
  1.0
  THEN
    (BACKWARD-ENGINE FLUE))

(MR-STRATEGY01
  IF
    CLASS1
  POSSIBLE
  THEN
    (GOTO-MODULE MCLASS1))

(MR-03024
  IF
    AIDS
  POSSIBLE
  THEN
    (GOTO-MODULE ATYPICAL-
BACTERIAL
      PNEUMOCISTIS-CARINI TBC
      CITOMEGALOVIRUS CRIPTOCOCCUS
      NOCARDIA ASPERGILLUS
      PNEUMOCOCCUS
ENTEROBACTERIA))

(MR-02012
  IF
    AGE < 14
  SURE
  THEN
    (HALT))

Some examples of meta-rules are shown in Fig. 5.20. The first meta-rule "MR-PRIMSET01" belongs to the module of the Primary Settler of a WWTP that will be fired to deactivate some rules that do not apply when there is a pump in the Primary Settler. The second one "MR-MAIN" belongs to the main module that will be fired when it is true that a patient has fever. Then, the meta-rule changes the reasoning engine to a backward engine to validate a flue diagnostic. The third one "MR-

STRATEGY01" of a main module will be fired if it is true that the possible classification of an entity was of CLASS1, and then the strategy is directed to explore de module MCLASS1, which will be in charge of the accurate discrimination of the entity as belonging to the CLASS1. The next one "MR-03024" is a similar meta-rule. Finally, the fifth meta-rule "MR-02012" aims to detect an exceptional situation like the patient being younger than 14-years-old, and then, the medical diagnostic system is not able to make any diagnosis, and hence, it stops the reasoning process.

5.3.4.1 Reasoning Cycle with Meta-Rules

The general cycle of the inference engine needs to be modified when working with meta-rules. The new cycle is composed of the same three steps for detecting, selecting and applying a rule, followed with three new similar steps for the possible meta-rules:

1. *Detection of the candidate rules.*
2. *Selection of the rule to be applied.*
3. *Application of the selected rule.*
4. *Detection of the candidate meta-rules.*
5. *Selection of the meta-rule to be applied.*
6. *Application of the selected meta-rule.*

Now, the cycle ends in a concrete context (the whole KB or the current KB module) when neither anymore rule is applicable nor anymore meta-rule is applicable, within the current context, or when the problem has been solved, or when some exception happens.

5.3.4.2 Hybrid Reasoning

The appearance of meta-knowledge that can change the reasoning mechanism (forward, backward) provides the possibility of problem-solving in a *hybrid reasoning* way: i.e. a bidirectional reasoning to connect both the data and the conclusions space, in a more optimal way.

The general problem-solving process in rule-based reasoning is a process that tries to connect the data and the conclusion space in a concrete way:

- *Forward reasoning* tries to connect both spaces starting from the data available (data-driven) and arriving to some conclusions.
- *Backward reasoning* tries to connect both spaces starting from a concrete conclusion (i.e. a hypothesis) (goal-driven) and arriving to data.

In a hybrid reasoning way, both spaces can be connected traversing the path in the two directions (bidirectional search), as illustrated in Fig. 5.21, where the first piece

Fig. 5.21 Hybrid reasoning
paradigm

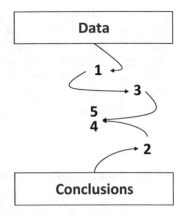

of the path is connected in a forward reasoning way, the second piece of path is connected in a backward reasoning mode, and so on, until both paths are finally connected. The change of the strategy is done through the corresponding meta-rules.

Main advantages of hybrid reasoning are that it solves the *impasses* that could appear in backward reasoning, when the reasoning chain cannot be continued, and there are no other options to explore. Then, a forward reasoning mechanism can be performed. In addition, hybrid reasoning can reduce the combinatorial explosion of forward reasoning, and focus on a more concrete fact to be validated with backward reasoning.

Example

Let us suppose that we have the following description of a KB, initial FB, and the corresponding goals, the specified meta-knowledge, and as a conflict resolution is defined the most specific rule (Fig. 5.22).

If we do not use the meta-knowledge, the forward reasoning engine using the most specific rule as conflict resolution strategy, and as a second criterion, the first rule in order, will reach the goals K and L, after 10 and 11 steps correspondingly, as described in Fig. 5.23.

On the other hand, if we use the meta-knowledge provided, which are the three meta-rules aiming to change the reasoning mechanism from forward to backward as soon as one of the facts, A or B or F were validated, the goal K will be reached in just seven steps. This happens because after the fact A is validated after the application of rule R5 in step one, the meta-rule MR1 is activated, and the reasoning engine is changed to backward with K as a hypothesis, just shortening the steps to the necessary ones. After that, the reasoning engine will be restored to forward reasoning, and it will need two steps more (R10 and R11) to reach the goal L (Fig. 5.24).

Knowledge Base
R1: $F \wedge G \rightarrow A$
R2: $A \wedge B \wedge F \rightarrow K$
R3: $D \wedge E \rightarrow B$
R4: $I \rightarrow A$
R5: $C \rightarrow A$
R6: $J \rightarrow D$
R7: $I \wedge D \wedge J \rightarrow F$
R8: $H \rightarrow I$
R9: $H \rightarrow E$
R10: $F \wedge G \wedge E \rightarrow M$
R11: $M \wedge J \wedge B \rightarrow L$

Facts Base
C, J, H, G

Goals
K??
L??
Conflict resolution strategy
Most specific rule

Meta-knowledge
MR1: $A \wedge$ Forward? $\rightarrow$ Backward (K)
MR2: $B \wedge$ Forward? $\rightarrow$ Backward (K)
MR3: $F \wedge$ Forward? $\rightarrow$ Backward (K)

⌇⟶   **None fact is questionable**

Fig. 5.22 Description of the Knowledge Base, the initial Fact Base, the goals, conflict resolution strategy, and the meta-knowledge for the example of using a hybrid reasoning

Fig. 5.23 Results of the example without using the meta-knowledge

There are no rules with 3 or 2 premises to be applied

(1) A (R5)
(2) D (R6)
(3) I (R8)
(4) F (R7)
(5) A (R1)✓

(6) A (R4)✓
(7) E (R9)
(8) M (R10)
(9) B (R3)
(10) K̲ (R2) !!
(11) L̲ (R11) !!

(1) A R5 $\rightarrow$ metarule $\rightarrow$ backward (K)

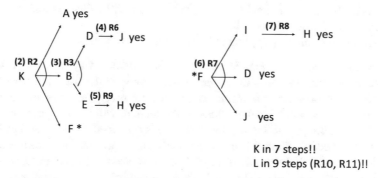

K in 7 steps!!
L in 9 steps (R10, R11)!!

Fig. 5.24 Results of the example when using the meta-knowledge

5.3.5   User Interface

The User Interface component is in charge to let the user interact with the system. Usually, this interaction is done through Natural Language (NL). Although this sounds very interesting, many times it is a "pre-fabricated" NL. This means that the questions the system make to the user are pre-stored in the definition of the Fact Base as explained in Sect. 5.3.1. And the answers of the user usually are an acceptable value belonging to a set of predefined values ({Yes, No}, {Low, Normal, High}, etc.) or a numeric answer depending on the nature of the fact.

For instance, the system can ask to the user: What is the outside temperature? to get the value associated to the fact Outside-Temp. And the user can answer 28.

Of course, it would be very nice that an actual Natural Language Processing (NLP) (Allen, 1987) step would happen, and then, the user could answer using whatever NL expression he/she wanted. For instance, in the above example, the user could answer using whatever NL expression like the following ones, and the system must understand and extract the valuable information, i.e. the value, for the fact:

- 28
- Twenty-eight
- 28 degrees
- 28°
- Twenty-eight degrees
- 28 ° C
- Twenty-eight Celsius degrees
- The temperature is 28
- The outside temperature is twenty-eight Celsius degrees

Furthermore, the system could generate the questions in NL, at each moment, instead of using the pre-stored LN questions. Some rule-based reasoning systems has used a truly NL approach, analyzing the text introduced by the user and generating the needed text for questioning the user.

The main functionalities provided by the user interface are:

1. *Introduce the problem data in advance*: The common way of running a rule-based reasoning system is in an *interactive way* with the user, but there is the possibility of no interaction. This means that the user gives in advance, all the information known about the values of the facts describing the problem. Commonly, this kind of *non-interactive execution* is produced when several problems must be solved, and thus, this mode is usually known as *batch execution mode*. In the batch mode, all the problems described are solved one after the other, without the interaction of the user. The results of the reasoning process are communicated to the user, at the end of the process, usually through a console and also in a text file.

2. *The system asks interactively questions to the user*: In the common interactive execution mode, the system can ask the user some information when it is required for the reasoning mechanism, either forward or backward. Main types of questions are:

   (a) About facts: When the reasoning mechanism requires to know some value of a fact, for determining whether a rule is a candidate rule in forward reasoning, and for the application of a rule in backward reasoning, when the fact is marked as a questionable in the Fact Base.
   (b) Requiring confirmations: Sometimes the system may require some kind of confirmation by the user, which is usually a binary one ({Yes, No}).

3. *The user asks questions to the system*: both in the middle of the interactive execution mode, when it is done in a step-by-step basis, or after the whole problem-solving reasoning process has ended, usually the user can ask several kinds of questions:

   (a) About the problem-solving process (Why?): these questions are related to the explanation component of the system. The user asks, for instance, why a determined conclusion have been made.
   (b) About the state of the Fact Base: These kinds of questions are useful when the user wants to know a concrete value of a fact, in the middle of a problem-solving process, or even though at the end of the reasoning process to check the values of several facts to fully understand why the conclusions reached by the system are those ones.

5.3.6 Explanation Module

The explanation module is the component of the system responsible for the credibility of the system. In any software engineering project, the user must trust on the resulting software product. If users do not trust on the software, they will not use them. This is especially true in the outputs of a reasoning process solving a complex problem. The user must be very convinced of the rationality of the conclusions of the system and the way the system has reached the conclusions. Thus, a rule-based reasoning system must provide explanations, or at least visualizations, in the reasoning process. This way, the system should explain or justify the *strategy* followed, or why some *module* has been explored, why some *rules* have been applied, and why some *meta-rules* have been also applied.

The typical functionalities offered by the explanation module in most rule-based reasoning systems are:

1. Why? explanations: These explanations aim to answer to the user why the system wanted to solve the intermediate goals it has solved to reach the final problem-solving state, i.e. the final goals have been reached. For instance, the system can justify that following a *strategy*, concrete *modules* have been explored and some

facts have been deduced. The chosen *strategy* has been chosen according to some *meta-rules*, which could be applied because some *facts* were verified, and so on.

2. How? explanations: Chain of reasoning until the current state. At any moment, if the reasoning process is going on step by step, or at the end of the process, the use can require explanations about the chain of reasoning that leads to satisfy a specified fact. For instance, given that the fact A is true, and as the rule R3 is selected. Its application provides the truth of the fact D, and then the rule R6 can be applied to finally validate the fact H.

Commonly, rule-based reasoning systems can offer two different layers of explanation, according to the degree of detail required by the user:

1. Visualization: At the visualization layer, it is enough to outline the trace of the reasoning. The trace of the reasoning means to show which is the chain of modules explored, and for each module which rules have been applied (R3, R6, R4, etc.) and the trace of all the facts that have been validated (A, D, H, etc.).
2. Justification: At this layer, the user requires more than just outlining the traces of the rules applied and facts deduced. It requires reasons for the reasoning strategy followed, justifications for the goals validated, why certain questions are asked, etc. For instance, the strategy must be justified because some meta-rule has suggested that line of reasoning, exploring some concrete modules, and the selected rules of one module has validated some facts.

The justifications and visualization can be communicated to the user in two ways:

1. Using prefixed or precomputed NL patterns of text. The different patterns of the text statements are precomputed to generate NL sentences, like in "Exploration of the module" + NameModule +"\nl" + "The rule", Rule, "has been applied, then The fact" + Fact + "was deduced. Next ...".
2. Generating NL depending on the context. That means that the syntactical and semantic structure of the statement must be generated, and then, depending on the context, the required semantic information must be added.

5.3.7 Knowledge Acquisition Module

The knowledge acquisition module is the experts' interface component. The experts can provide their knowledge to the system through the knowledge acquisition module. This component is not always present in all rule-based reasoning systems. The experts can provide the system with both factual knowledge and relational knowledge. Therefore, usually this component provides the experts with two different kinds of functionalities regarding the Fact Base or the Knowledge Base:

- Fact Base functionalities:

  - Definition of a new fact.
  - Edition of a fact of the Fact Base.

- Removing a fact from the Fact Base.
- Importation of the Fact Base.
- Exportation of the Fact Base.

- Knowledge Base functionalities:

  - Management of Modules.

      Definition of a new module of the KB.
      Edition of a module of the KB.
      Removing a module from the KB.

  - Management of Rules.

      Definition of a new rule of the KB.
      Edition of a rule of the KB.
      Removing a rule from the KB.

  - Management of Meta-rules.

      Definition of a new meta-rule of the KB.
      Edition of a meta-rule of the KB.
      Removing a meta-rule from the KB.

5.3.8 Knowledge Engineer Interface

The Knowledge Engineer/Engineering Interface module is the component which provides an interface to the knowledge engineer/AI scientist to access to the core technical functionalities of a rule-based reasoning system.

The knowledge engineer interface allows the knowledge engineer to access the knowledge acquisition module, the Reasoning component and the Meta-reasoning component. Thus, the knowledge engineer interface offers the following functionalities:

- Management of the Fact Base and the Knowledge Base. The knowledge engineer can access to the Fact Base and the Knowledge Base, through the knowledge acquisition module.
- Management of the Inference Engines module. The knowledge engineer can access the reasoning module to make adjustments and some tuning of the reasoning engines.
- Management of the Meta-reasoning module: The knowledge engineer can access the meta-reasoning module to adjust, modify and tune the meta-reasoning module. For instance, the order of the conflict resolution strategies can be defined or modified, and the meta-knowledge can be tuned for testing or maintenance purposes.

5.3.9 The Knowledge Engineering Process

The process of designing and building a rule-based reasoning is commonly named as the knowledge engineering process. This process has several steps to build a rule-based reasoning system as described in (Hayes-Roth et al., 1983), and depicted in Fig. 5.25.

In fact, these steps are a particularization of the general classical software engineering cycle. The general steps in the knowledge engineering process are the following:

1. *Identification*: in the identification step, the concrete problem to be solved must be clearly identified. Thus, task comprises the following items:

    (a) Analyze the *viability* of the rule-based reasoning system building. It is worth to build it or a more algorithmic solution could solve the problem? Availability of knowledge sources?
    (b) Search the available *knowledge sources* (experts, books, etc.)
    (c) Determine the *necessary data* to solve the problem.
    (d) Determine the *issues* (solutions) and the criteria which determine the solution.

2. *Conceptualization*: In this step, all the relevant elements must be characterized. This task involves the following sub-steps:

    (a) Find and enumerate the basic elements to characterize the domain (*relevant facts*) and their own *relations*.
    (b) Distinguish the *evidences*, the *hypotheses* and the *actions* to be done.
    (c) Enumerate the different *hypotheses/objectives*.
    (d) Decompose the problem into *sub-problems*, if it is possible.
    (e) Characterize the reasoning blocks and the reasoning flow.

Fig. 5.25 The Knowledge Engineering process according to Hayes-Roth et al. (1983)

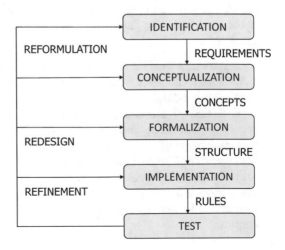

3. *Formalization*: in this step, the reasoning strategies must be identified for each one of the sub problems defined:

   (a) Determine the *reasoning strategies* required:

   - Classification, diagnosis, temporal planning, spatial design, configuration.

   (b) Identify *search space* and *search type*.
   (c) Identify the *problem-solving methodology*:

   - Heuristic classification, constructive resolution or a hybrid methodology. These practical methodologies will be explained in Chap. 7.

   (d) Analyze the *inexactness* (uncertainty, imprecision, or incompleteness) and the *completeness* of the information and knowledge available.

4. *Implementation*: In this step, the available knowledge about the domain and the problem to be solved must be implemented in the Fact Base and the Knowledge Base. Furthermore, the meta-knowledge must be implemented by means of meta-rules associated to the modules of the KB. Therefore, main sub-steps are the following ones:

   (a) *Knowledge* representation and implementation.

   - Fact Base definition.
   - Definition of the modular structure of the Knowledge Base.
   - Definition of inference rules for each module.

   (b) Decisions over the problem-solving control (*meta-knowledge*).

   - Definition of meta-rules associated to each module.

5. *Testing*: The last step in the building process of a rule-based reasoning system is its validation. To validate it, the usual procedure is the definition of representative cases with the experts, in order to be used as a test case for the system. The validation means to assess the system regarding several properties. Main steps are:

   (a) Determine with expert/s a set of representative test cases.
   (b) Assess the system operation regarding:

   - Correctness? Is the system providing correct solutions to the input test cases or not?
   - Completeness? Is the system able to solve a wide range of problems?
   - Inexactness? Is the computed inexactness estimation of the solutions reliable?
   - Credibility and explanations? Are the explanations, justifications provided by the system correct and credible?

If there are errors found in the testing, the process come back to previous steps, to refine the system, to redesign the system or to reformulate the characterization of the system. Afterwards, the updated prototype of the system is tested again, and the loop continues as far as no more errors are found.

5.3.10   An Example of an Expert-Based Model

To illustrate the application of an expert-based model, we will analyze a model for the identification of the most representative species of a Mediterranean tree, from those ones existing in Catalonia.[7]

5.3.10.1   Identification

One of the main tasks at the identification step is the clear definition of the problem we are addressing. In this case, the problem is that having a leave of a Mediterranean tree in Catalonia, and being able to see the tree, especially the top of the tree, the issue is to identify the species to which the tree belongs.

The problem seems enough complex to not simply apply classical algorithmic solutions. This kind of problems can be solved by an expert botanist and thus, an expert-based model seems a good solution. In addition, to ensure the viability of the system, the availability of the knowledge sources must be confirmed. In this case, there were some *available books on botanical taxonomy* of Mediterranean trees to be consulted.

The data needed for solving this problem are clearly the relevant features that characterise a tree. They can be obtained in the books about the taxonomies of the trees. The issues of the problem are to *identify* or *classify* a given tree, which we can see its top and one leave of it is available. The solution will be when we identify both the *genre* and the *species* of the tree.

5.3.10.2   Conceptualization

This step involves the characterization of the relevant facts to be taken into account for the identification process. From the books, we gathered the most *relevant facts* for the discrimination purposes:

- Regarding the tree

    - Durability-of-leaves (perennial, deciduous, unknown)
    - Top-of-the-tree (umbrella, conic-shape, round-shape, fusiform, unknown)
    - Tree-height (more-than-5 m, less-than-5 m)

---

[7] This example is expanded and adapted from a former example created by my colleague Dr. Ramon Sangüesa, Department of Computer Science, Universitat Politècnica de Catalunya.

- Tree-bark-colour (red, grey-black, shallow-grey)
- Tree-bark-espinosity (yes, no)
- Tree-bark-roughness (yes, no)
- Tree-bark-cork (yes, no)
- Tree-branch-disposition (opposed, sparse)
- Tree-branch-espinosity (yes, no)
- Tree-branch-rigidity (straight, arched)

• Regarding the leaves

- Leave-length (a positive number in cm or on average value)
- Leave-width (a positive number in cm)
- Leave-simplicity (a positive number). If it is one, it means that it is a simple leaf, and if it is greater than one, it is a compound leaf.
- Leave-colour (blue–green, black–green, grey–green, intense-green)
- Leave-morphology (needle, round-linear, pin-linear, elliptical, oval)
- Leave-lining (scaly, hairy)
- Leave-midrib-number (one, three, more-than-three)
- Leave-odour (yes, no)
- Leave-margin (entire, tooth-shaped, lobed)
- Number-sheathed-leaves (a positive number)

• Regarding the flower

- Flower-colour (white, pinked-white)
- Flower-clusters (grouped, isolated)
- Acute-angle-number-of-flower-disc (5, 10)
- Flower-petals-number (4, 5)

• Regarding the fruit

- Type of fruit (orange, woody cones, . . ., unknown)

• Regarding the bud

- Bud-colour (green, light-green, purple, unknown)

• Regarding the petiole

- Petiole-over-the-stem (yes, no)
- Petiole-length-related-to-leave (greater-than, smaller-than)
- Folioles-existence (yes, no)

• Regarding the folioles

- Foliole-number-per-leave (from-5-to-9, from-9-to-13) or (less-than-10, more-than-10)
- Foliole-teeth-disposition (small-and-regular, large-and-irregular)
- Foliole-parity-number (even, odd)

In this problem, the hypotheses are the possible genres and species of a Mediterranean tree, which are:

- Genre: There are 31 different genres taken into account, which are the most representative:

  - Pinus
  - Abies
  - Cedrus
  - Juniperus
  - Faxos
  - Punica
  - Eucaliptus
  - Citrus
  - Buxus
  - Olea
  - Laurus
  - Fagus
  - Salix
  - Diospyros
  - Quercus
  - Pyros
  - Cydonia
  - Mesplius
  - Cupressus
  - Tamarix
  - Phoenix
  - Acer
  - Fraxinus
  - Aesculus
  - Robinia
  - Sorbus
  - Pistacia
  - Juglans
  - Ceratonia
  - Ailanthus

- Species: There were 50 most representative different species taken into account to develop the prototype of the rule-based reasoning system:

  - Pinus pinea
  - Pinus sylvestris
  - Pinus mugo uncinata
  - Pinus halepensis
  - Pinus migra salzamannii (Scotch pine)
  - Pinus Pinaster
  - Abies picea

- Abies alba (Fir tree)
- Cedrus brevifolia
- Juniperus communis communis (Juniper)
- Juniperus oxycedrus
- Juniperus phoenicea (Savin)
- Taxus baccata (Yew)
- Punica Granatum (Pomegranate tree)
- Eucalyptus globulus (eucalyptus)
- Citrus aurantium sinensis (orange tree)
- Citrus limon (lemon tree)
- Citrus medica (citron tree)
- Buxus sempervirens (boxwood)
- Buxus balearica
- Olea europaea sylvestris (wild olive tree)
- Olea europaea euiropaea (olive tree)
- Laurus nobilis (laurel)
- Fagus sylvatica (beech)
- Salix caprea
- Salix atrocinerea catalaunica
- Salix alba (willow)
- Diospyros kaki
- Quercus ilex ilex
- Quercus suber (cork tree)
- Pyros comunis (pear tree)
- Cydonia oblonga
- Mesplius germanica
- Cupressus sempervirens (cypress)
- Tamarix Africana (tamisk)
- Tamaris canariensis
- Phoenix dactylifera
- Phoenix canariensis
- Acer negundo
- Fraxinus ornus (ash)
- Fraxinus angustifolia
- Fraxinus excelsior
- Aesculus hippocastanum
- Pistacia lentiscus
- Ceratonia silique (carob tree)
- Robinia pseudo-acacia
- Juglans regia (walnut tree)
- Ailanthus altissima
- Sorbus domestica (service tree)
- Sorbus Aucuparia

The problem of identifying the concrete species and the genre of a tree, can be divide, at least, in two sub-problems:

- Identification of the genre, or group of similar genres
- Identification of the concrete species

In principle, to make this identification, we do not know possible goals (species) to be identified. Thus, what makes sense is to use a *forward reasoning mechanism* to deduce, from the data available of a new tree to be identified, the genre and the species of the tree.

Notwithstanding, in the case that some facts can be identified as key fact to suggest a possible genre and/or species, some meta-rules can be associated in the corresponding modules to change the forward reasoning to a backward reasoning to validate the corresponding genre and/or species.

5.3.10.3   Formalization

The reasoning strategy to be used in this problem is clearly a *classification* problem. The system must classify a new tree into one of the available species. Thus, from the data obtained from the observation of the tree, and from the observation of one leaf, the system should assign the most adequate species that matches the characteristics.

To organize the knowledge available in the KB, some structuration of this knowledge in *modules* seems a good idea. To define the different modules, it seems very reasonable to use the *biological taxonomy* of the trees in genres, species, subspecies, etc.

Furthermore, in order not to get too much modules, it was thought that some similar genres could be grouped in the same module. Thus, some genres like Pinus, Abies, Juniperus, Taxus, and Cedrus, which share some characteristics like the length width of a leaf is lower than 5 cm, and the length of a leaf has an average value, can be joined in the same module. Thus, we obtain a module named PAJTC (the first letter of each genre). Afterwards, other modules will be defined to continue the classification/discrimination process to identify the corresponding genre and species. This way the hierarchical structure of the different modules proposed is described in Fig. 5.26.

The problem-solving methodology which is suitable for this problem is a *heuristic classification*, which comprises an abstraction process, a heuristic association and a refinement process, to get a classification label, from concrete values of the case.

In this system, the *inexactness* problem of the information and knowledge was not modelled, because, from the expert knowledge obtained, it seemed that the classification processes were very reliable.

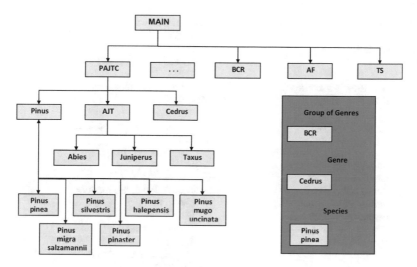

Fig. 5.26 Hierarchical modular structure of the KB for the tree identification problem

5.3.10.4   Implementation

The next step is the implementation of the *knowledge* available. These means to define the Fact Base, the Knowledge Base, and the meta-knowledge. From the data analyzed in the conceptualization step, the Fact Base can be easily implemented. The definition of the KB means to define the rules corresponding to each module of the KB, from the modular structure obtained in the previous step. In addition, the *meta-knowledge* is defined through the meta-rules associated to each module. In the next Fig. 5.27, there is a summary of the implementation of the different modules with the corresponding rules and meta-rules.

5.3.10.5   Testing

First, several test cases were defined according to some cases of species described in the literature. From the 50 species near 40 cases were selected to be assessed with the rule-bases reasoning system:

Four examples of genre Pinus
An example of Abies genre
An example of Juniperus genre
An example of Taxus genre
An example of Cedrus genre
Some example of Tamarix genre
Some example of Fraxinus genre
Some example of Sabus genre
Some example of Salix genre

(**module** Main

  // Rules of the module
  ((**R1**
    IF
      (leave-length average)
      (leave-width < 5)
    THEN
      (group PAJTC))
  (**R2**
    IF
      (leave-length average)
      (leave-width > 5)
      (leave-simplicity 1)
      (leave-midrib-number 1)
      (leave-margin entire)
    THEN
      (group VCR))
  ...)

  // Meta-Rules of the module
  ((**MR1**
    IF
      (group PAJTC)
    THEN
      (goto-module PAJTC))
  (**MR2**
    IF
      (group VCR)
    THEN
      (goto-module VCR))
  ...))

(**module** PAJTC

// Rules of the module
((**RPAJCT1**
  IF
    (number-sheathed-leaves 2)
  THEN
    (genre pinus))
(**RPAJCT2**
  IF
    (number-sheathed-leaves 1)
  THEN
    (group AJT)))
...)

// Meta-Rules of the module
((**MRPAJCT1**
  IF
    (genre pinus)
  THEN
    (goto-module pinus)
(**MRPAJCT2**
  IF
    (group AJT)
  THEN
    (goto module AJT))))

(**module** Pinus

  // Rules of the module
  ((**Rpinus1**
    IF
      (top-of-the-tree umbrella)
    THEN
      (species pinea))
  (**Rpinus2**
    IF
      (top-of-the-tree other)
      (length-leave > 3) & (length-leave < 6)
    THEN
      (species sylvestris mugo-uncinata))
    ...))

Fig. 5.27 Implementation of the modular Knowledge Base for the tree identification problem

Some example of Olea genre
Some example of Buxus genre
Some example of Citrus genre
Some example of Quercus genre
One example for each one of the following species: Cedrus, Picea abies, Punica granatum, Eucaliptus globulus, Laurus nobilis, Fagus sylvatica, Diospyros kaki, Pyros comunis, Cydonia oblonga
All the trees were correctly identified with the exception of the two Sabus genre species, which were confused. Thus, giving a 95% of accuracy in the identification.

The problems detected with the Sabus genre, were easily corrected given the modular structure of the KB. The problem was that the discriminating rules of the two Sabus species were swapped.

5.4   Model-Based Reasoning Methods

Model-based reasoning is a general methodology originating from research in artificial intelligence that allows to find explanations for certain situations only using models of systems. One application area where model-based reasoning has been successfully applied is diagnosis where the focus is only explaining a detected misbehaviour.

5.4.1   Introduction

The purpose of Intelligent Decision Support Systems (IDSS) is to support humans carrying out certain tasks like diagnosis. The object of diagnosis is to explain a certainly observed behaviour in terms of causes. In the context of IDSS someone, for example, might be interested in explaining why the population of a particular species is decreasing in a given environment, why a wastewater treatment plant is no longer working as expected, or why a computer program is not returning an expected value. In all these examples, we ask ourselves about the reasons for the behaviour aiming at identifying root causes and afterwards providing means for bringing the system back in the predefined wanted state. In contrast to other approaches of IDSS, we focus on methods utilizing system models directly for explaining an observed behaviour. One advantage of this method is that besides providing a system model and observations no other information is needed and no further additional tasks are required. Because of the property of the proposed model-based reasoning method only to rely on models directly for explanation purposes, it is also referred to as reasoning from first principles (Reiter, 1987).

In this section, we discuss two variants of model-based reasoning. The first one, which is also called *consistency-based diagnosis*, relies solely on models of the correct behaviour of a system for computing diagnoses, i.e. explanations for an observed behaviour. The second one uses models capturing known faults for explaining an unexpected behaviour. The second model-based reasoning technique is also known as *abductive diagnosis*. Besides discussing the underlying foundations of consistency-based diagnosis and abductive diagnosis, we also have a closer look at modelling standards someone should follow when applying model-based reasoning for IDSS. But before let us have a closer look at the basic concepts of model-based reasoning using a small example from the waste water treatment domain.

In Fig. 5.28 we outline a simplified schematics of a constructed wetland used for cleansing grey or black water in practice. The advantage of constructed wetlands is

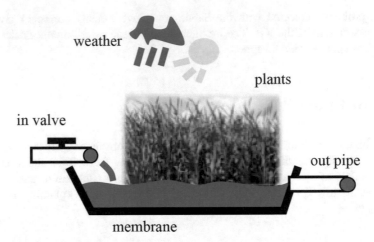

weather

plants

in valve

out pipe

membrane

Fig. 5.28 Simplified schematics of a constructed wetland

that due to its design the whole system is almost maintenance free. The idea behind constructed wetlands is to use plants, soil, filtering material, or organisms for treading water in order to remove unwanted substances. Depending on the required treatment effect the composition of the used parts and the overall treatment time is important. Like every engineered system constructed wetlands have to be designed to fit their purpose under given environmental conditions and the wanted effects.

For illustration purposes, we use a simplified model of a constructed wetland only. In our case, we have an input valve letting water to be treated into the wetland.

Providing that the water has enough phosphates and nitrates, and that there is always enough water in the basin, the plants should healthily grow. For simplicity we assume that the constructed wetland is designed such that the water remains in the basin as long as required for filtering unwanted ingredients and that the output valve has no influence on this duration. However, the weather conditions do influence the behaviour. If it is too sunny for a longer period of time, the water level might drop below a limit harming the planted vegetation. If it is raining too much, there might be a degradation of filtering, which is usually taken into account when designing such a system.

Let us consider now the case, where the vegetation of a constructed wetland is in bad shape and someone is interested in finding out the reasons behind. From the textual description of the behaviour of a constructed wetland we can conclude that only the water level and the phosphates and nitrate concentration of the incoming water influences the plants. The water level itself is influenced by the input valve and the weather. Hence, we obtain several causes explaining the current situation. For example, because of an unusual period of very hot and dry weather the water level decreased too much, finally causing the plants to die. In order to distinguish, the competing explanations someone has to further obtain knowledge, like the position of the input valve, the weather of the past month, or the water quality in terms of its phosphate and nitrate concentration.

What we can learn from this example is that a model of the behaviour of a system relying on its parts and potential interactions is sufficient to come up with explanations for an observed behaviour. In addition, we see that there might be many competing explanations requiring further information to allow distinguishing them. In the following, we make use of this running example for explaining the different methods behind model-based reasoning aiming at providing information about causes of certain observations.

We organize this section as follows. First, we discuss preliminary definitions where we focus on propositional horn-clause logic. Reader not familiar with logic for knowledge representation should consider reading Sect. 5.4.2. Afterwards, we discuss consistency-based diagnosis in Sect. 5.4.3 where we introduce the basic definitions and also the required modelling. In Sect. 5.4.4, we outline the foundations behind abductive diagnosis and also focus on the modelling aspects using our waste water treatment plant example. Finally, we conclude the section and compare the different diagnosis approaches based on their foundations and modelling requirements.

5.4.2 Preliminaries

In order to be self-contained, we briefly discuss some basic definitions and preliminaries in this section. Because both diagnosis methods of model-based reasoning are based on some sort of formal reasoning mechanism, we purely focus on models represented in form of logical rules. For simplicity reasons, we consider propositional models comprising only simple logical implications and facts. This type of models (or logical theories) is also referred to as propositional horn clause theories. For the interested reader, we refer to introduction literature to logic, e.g. (Smullyan, 1968), covering more details and foundations behind logical-based reasoning.

The basic concept behind propositional horn clause theories is a proposition variable, which represents a certain conceptual object of the real world. For example, in the domain of constructed wetlands in the introduction, we consider the level of the water in the basin. This level can be within the expected range, which can be expressed with propositional variable (or proposition for short) *water_in_range*. In case the water level fulfils this state in reality, we say that the proposition is true, and otherwise false. Note that, we are able to represent the contrary fact as proposition, e.g. *water_not_in_range*. Although, these two propositions have a meaning when considering the semantics of a natural language like English used for in this example, in classical logic the semantics is only given when assigning truth values to propositions. Hence, within propositional logical all propositions $p_1, p_2, \ldots$ do not have a meaning that corresponds to a certain representation as a string. Instead of writing *water_in_range* as a proposition representing this part of a model, we would also be able to write *wr*. However, in order to make a logical model readable, the recommendation is to use meaningful textual names for propositions.[8] For simplicity reasons, we further assume two specific propositional variables $\top$ and $\bot$ representing a proposition that is always true and always false, respectively. Hence, $\top$ represents a true proposition and $\bot$ a false proposition.

In propositional logic the propositional variables can be seen as atoms, i.e. things that cannot be divided into pieces. In order to allow combining such propositions we have to introduce operators. We start with the conjunction operator $\wedge$ representing the logical-and. For example, if we want to express that two propositions have to be true at the same time, we connect them using $\wedge$. In our running example, we might want to state that both the water level and the nitrate and phosphate concentration should be within a predefined range. This can be expressed as *water_in_range* $\wedge$ *nitrate_phosphate_in_range* where the proposition *nitrate_phosphate_in_range* represents the case where the nitrate and phosphate concentration of the water is within an expected level. In this example, we might also want to express that *water_in_range* $\wedge$ *nitrate_phosphate_in_range* are necessary for having a healthy vegetation. For this purpose, we need the second operator, i.e. the logical implication $\rightarrow$. For the implication, we allow a single proposition or a conjunction of proposition on the left side, and a single proposition only on the right side. Using the propositional variable *vegetation_healthy* for representing that the plants are in good shape, we can write the following logical rule:

$$water\_in\_range \wedge nitrate\_phosphate\_in\_range \rightarrow vegetation\_healthy$$

The semantics of the implication is rather straightforward. The proposition on the right side of the $\rightarrow$ has to be true if the formula of the left side, which is in this case a conjunction of proposition, is true. If the conjunction is not true, the proposition of the right side may be true or may be false. Using the operators, we are now able to formulate that a certain proposition has to be true, i.e. stating that this proposition is a fact. For example, we might have a look at the water level at our constructed wetland and classify this a being as expected. In this case, *water_in_range* is a fact and we can express this as rule:

$$\top \rightarrow water\_in\_range$$

For simplicity, we may also write $\rightarrow$ *water_in_range* and drop the $\top$ from the rule. We may also express such a fact as *water_in_range* ignoring even the $\rightarrow$ operator. From here on, we consider all these representations of facts someone finds in other texts on logic as equivalent. However, in order to avoid confusing the reader, we use $\top \rightarrow p$ for any proposition p in this section as facts only.

We also might want to express that a certain proposition is in contradiction with another proposition, i.e. stating that they cannot be true at the same time. For

---

[8]Logic has been originally developed to represent natural language in a formal way that allows to derive conclusions, which has to be generally accepted.

example, we only allow either *water_in_range* or *water_not_in_range* being true. In order to do that we can state this explicitly using only the ingredients we just introduced:

$$water\_in\_range \land water\_not\_in\_range \to \bot$$

In this rule $\bot$ would be required to be true if both propositions are true, which contradicts our initial definition of $\bot$.

After discussing the basic concepts of propositional horn clause models, we define propositional horn clause theories as follows:

Definition 1 (Propositional Horn Clause Theory (PHCT)) A tuple (P, Th) is a propositional horn clause theory (PHCT) if P is a set of propositions (comprising at least $\top$ and $\bot$), and Th is a set of rules of the form $p_1 \land \ldots \land p_n \to p_{n+1}$ with $p_i \in P$ for all $i = 1, \ldots, n + 1$.

We further define the semantics of PHCT using resolution, i.e. a rule that allows to infer facts from rules and other facts. For example, if a theory comprises the rules *water_in_range* $\land$ *nitrate_phosphate_in_range* $\to$ *vegetation_healthy*, $\top \to$ *water_in_range* and $\top \to$ *nitrate_phosphate_in_range*, then we want to derive the fact $\top \to$ *vegetation_healthy*. In terms of natural language, we want to state that if there is enough water in the basin of a constructed wetland, and the nitrate and phosphate concentration is fine, then we conclude the plants to be in good shape. This resolution principle behind is general and also reasonable. Formally, it can be stated as follows:

Definition 2 (Modus Ponens) Given a rule $p_1 \land \ldots \land p_n \to p_{n+1}$ and rules $\top \to p_1$, $\ldots$, $\top \to p_n$ where all the p_i's are propositions, then we can derive the fact $\top \to p_{n+1}$. Formally, we write:

$$\frac{p_1 \land \ldots \land p_n \to p_{n+1}, \top \to p_1, \ldots, \top \to p_n}{\top \to p_{n+1}}$$

or

$$\left(p_1 \land \ldots \land p_n \to p_{n+1}, \top \to p_1, \ldots, \top \to p_n\right) \vdash \left(\top \to p_{n+1}\right)$$

As already said, we might drop the $\top \to$ part from the rules and only state that p_{n+1} can be inferred from the available information. Note that the Modus Ponens (MP) rule is only one form of resolution but there are more. However, for our purpose considering PHCT and the MP inference rule are sufficient to explain model-based reasoning. Using MP, we are able to define:

Definition 3 (Closure) Given a PHCT (P, Th). The closure of resolution $Cl(Th)$ is the set of all facts that can be derived from Th when applying the Modus Ponens rule as often as possible. Formally, we define closure as fix point of the following recursive equation:

1. $Cl^0(Th) = Th$,
2. For $k > 0$:

$$Cl^k(Th) = \{\top \rightarrow p \mid \exists r_1, \ldots, r_k \in Cl^{k-1}(Th) \; s.t. (r_1, \ldots, r_k) \vdash (\top \rightarrow p)\}$$

The closure now is a fix point defined as:

$$Cl(Th) = Cl^k(Th) \text{ iff } \exists k \geq 0 \text{ with } Cl^k(Th) = Cl^{k-1}(Th) \qquad (5.1)$$

A fix point in the closure definition is the first set obtained when applying Modus Ponens where no further facts can be added. Note that because of the fact that every application of the MP rule adds one additional fact, and the restricted numbers of rules to be applicable, there must be a k where no further facts can be obtained, and thus a fix point exists. Let us illustrate, the definition of closure using a more elaborated example extending the model for our constructed wetland. We add a rule stating that healthy vegetation led to treated water at the out pipe and use the previously used rule. In addition, we assume that the nitrate and the phosphate concentration is within its intended range, and that the same holds for the level of the water in the basin. Hence, we obtain the following PHCT (P, Th):

$$P = \left\{ \begin{array}{c} water\_in\_range, nitrate\_phospate\_in\_range \\ vegetation\_healthy, out\_water\_treated \end{array} \right\}$$

$$Th = \left\{ \begin{array}{c} water\_in\_range \wedge nitrate\_phosphate\_in\_range \rightarrow vegetation\_healthy \\ vegetation\_healthy \rightarrow out\_water\_treated \\ \top \rightarrow water\_in\_range \\ \top \rightarrow nitrate\_phosphate\_in\_range \end{array} \right\}$$

In the first iteration, $Cl^1(Th)$ is set to Th itself. In the second iteration, we use the rules $water\_in\_range \wedge nitrate\_phosphate\_in\_range \rightarrow vegetation\_healthy$, $\top \rightarrow water\_in\_range$, and $\top \rightarrow nitrate\_phosphate\_in\_range$ to derive a new fact $vegetation\_healthy$, i.e. $Cl^2(Th) = Th \cup \{\top \rightarrow vegetation\_healthy\}$. In the third iteration, we use $vegetation\_healthy \rightarrow out\_water\_treated$ and $\top \rightarrow vegetation\_healthy$ to derive $\top \rightarrow out\_water\_treated$, i.e. $Cl^3(Th) = Th \cup \{\top \rightarrow vegetation\_healthy, \top \rightarrow out\_water\_treated\}$. In the fourth iteration, there are no further general rules to apply. Hence, $Cl^4(Th) = Cl^3(Th)$ leading to $Cl(Th) = Cl^3(Th)$.

This process of computing the closure can be applied to any arbitrary PHCT. The outcome is a set of rules and facts that can be inferred from a given PHCT using $\vdash$. There is one interesting case to consider. What should we do when we are able to infer $\top \rightarrow \bot$? In this case, the theory allows to derive a proposition, which has to be false. In this case, the proposition $\bot$ has to be true, which contradicts the definition of $\bot$. Hence, such a case should never happen in any PHCT.

Definition 4 (Satisfiable, Contradiction) Given a PHCT (P, Th). (P, Th) is said to be satisfiable if and only if $\top \rightarrow \perp \notin Cl(Th)$. Otherwise, (P, Th) is said to be a contradiction (or unsatisfiable).

Note that a PHCT might be a contradiction but when removing facts, it might become satisfiable again. It is also worth mentioning that a PHCT can never be a contradiction if there is no rule of the form $p_1 \wedge \ldots \wedge p_n \rightarrow \perp$. Checking for satisfiability of PHCT can be done very fast. There are algorithms available that deliver an answer in linear time depending on the size of the PHCT, where size is defined as the number of propositions used in all the rules (and not the number of propositions in P). See (Minoux, 1988) for such an algorithm. For our purpose, it is sufficient to assume that we have an algorithm TP (for theorem prover), which takes a PHCT as input, and returns $\sqrt{}$ in case of satisfiability and $\times$, otherwise. We use TP when describing the implementation basics behind the different model-based reasoning variants.

Before discussing the two diagnosis variants in detail, we briefly discuss their common foundations. In order to allow reasoning from models of systems, we require a model. In our case, we assume models to be formulated using PHCTs. Although, the models used for consistency-based diagnosis and abductive diagnosis are different, there are some aspects that are shared. For both variants, we do not only require logical rules for representing relevant diagnostic information about systems, we also need to know those parts of the model that correspond with root causes. For example, if the position of the input valve influences the water level in our constructed wetland example, we have to state this information explicitly in order to allow an algorithm to select this as a potential root cause if necessary. Therefore, we do not only capture the propositions and theories like in the definition of PHCT but also special propositions that serve as a connection to root causes. In the case of diagnosis, these propositions are called hypotheses and we define diagnosis models as follows:

Definition 5 (Diagnosis Model) A tuple (P, Th, Hyp) is a diagnosis model, where (P, Th) is a PHCT and $Hyp \subseteq P$.

In addition to models, we also require information about the current state of a system for diagnosis. For example, we need to know whether the vegetation looks healthy or whether the water level is within its expected range. This information is called observations and finally allows us to formulate the diagnosis problem.

Definition 6 (Diagnosis Problem) Given a diagnosis model (P, Th, Hyp) and a set of observations $Obs \subseteq P \setminus Hyp$. The tuple (P, Th, Hyp, Obs) is a diagnosis problem.

In the definition of diagnosis problems, the observations are restricted to propositions that are no hypotheses. This is reasonable because a hypothesis is a representation of a cause root that cannot be observed directly. A solution to a diagnosis problem is an explanation for given observations. This explanation is a subset of the set of hypotheses, but the formal definition of diagnosis is different for consistency-based diagnosis and abductive diagnosis. The reason is that both methods require a different kind of representation of systems. Whereas consistency-based diagnosis relies on models of the correct behaviour of a system, abductive diagnosis is based on knowledge about faults and their corresponding behaviour. We will discuss these differences in detail in the next two sections.

5.4.3   Consistency-Based Diagnosis

Consistency-based diagnosis, which has also been called model-based diagnosis, relies on a model of the system to derive explanations for unexpected observations. According to Reiter (1987), a model in this setting has to capture the correct behaviour of a system only providing that we have special propositions, i.e. the hypotheses that allow to reason about the correctness of certain system components when searching for a diagnosis. In Fig. 5.29, we depict the general principle behind consistency-based diagnosis. On the left, we have a system from which we obtain observations. On the right, we have a system model from which we are able to derive the expected behaviour in a certain situation. When comparing the expected behaviour with the observed behaviour, we might find a discrepancy, which is used to derive diagnoses using the model and the observations directly. These diagnoses can be mapped to the components of the real system thus providing root causes explaining the observed behaviour.

Before defining diagnosis formally, we first have to discuss the properties of models to be used in consistency-based diagnosis. From Definition 5, we only see that we need a set of propositions P, a theory Th, and a set of hypotheses Hyp to obtain a diagnosis model. This definition only considers the formal requirements regarding the involved parts of a model but does not consider certain requirements. The main model requirement is that the model should comprise interconnected components where for each component we define its correct behaviour. Components

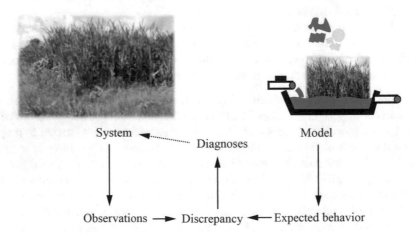

System ◄┄┄┄┄┄┄ Diagnoses   Model

Observations ➝ Discrepancy ◄━ Expected behavior

Fig. 5.29 The principle behind consistency-based diagnosis

in this setup change values of connections based on the values of other connections. Hence, some connections can be considered as outputs and others as inputs. However, there might be rules also stating that a particular input has a certain value based on other input and output values. Hence, the whole behaviour of a component is defined using rules mapping input and output values. In addition, we use a hypothesis in each rule on the left side that represents the fact that the component is working as expected. Hence, the set of hypotheses is for representing the–in this case–health status of components. In addition, to the component models, we require formal rules representing the interconnections and also inconsistencies of propositions like discussed in the previous section.

Let us illustrate, the modelling process for consistency-based diagnosis using the constructed wetland example from Fig. 5.28. The first step is to identify components. Components need not to be physical components. They can also represent certain quantities or other aspects of interest for diagnosis. For example, in a constructed wetland the nitrate and phosphate concentration of the water to be treated is important for the health status of the vegetation. Hence, this concentration needs to be with a certain range. Therefore, we have to introduce a component representing the state of the nitrate and phosphate concentration. In addition, to this component there are components necessary for diagnosis that have a physical counterpart like the input valve or the basin. The latter comprises of a membrane that holds the water preventing it from entering the surrounding soil. A leakage of the membrane might be a cause for a decreasing water level. Another component might be the weather condition of the past weeks. If the weather during summer is too hot without substantial rainfall, the evaporation may have a negative effect to the water level as well. Note, that too heavy rain has no real negative effect to the constructed wetland because the output pipe, which we do not consider in this model, would always assure that the water cannot go higher than a certain specified value. A last component may be the vegetation itself. In summary, we would consider the components as considered in Fig. 5.30.

In a second step, we have to identify the connections between components. Components communicate with another component using connections. For example, in case the nitrate and phosphate concentration are in the expected range the corresponding diagnosis component is providing this value, and the vegetation uses this value for growing in healthy way and thus treating the water. Because of the fact that in propositional logic, we cannot really state quantitative values like 0.1 we only represent qualitative values relevant for diagnosis, e.g. the nitrate and phosphate concentration is within its limits. For this purpose, we consider the correct behaviour for all components and make the health status of a component explicit, introducing a hypothesis Ok_C for all components C representing a root cause. This hypothesis is used at the left side of all rules describing the behaviour of a component. In the third step, we have to formalize the model. As an example, we formalize the model for the constructed wetland example. Note, that we write all hypotheses with a capitalized starting character to distinguish them from ordinary propositions.

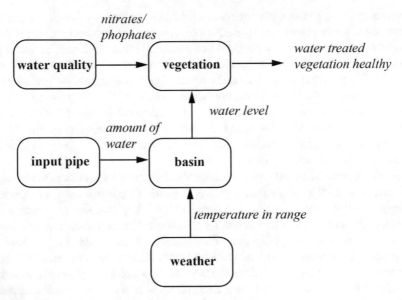

Fig. 5.30 A component-oriented model for constructed wetlands

Water Quality If the water quality is working as expected, then it is in its expected range.

$$Ok\_water\_quality \rightarrow good\_nitrate\_phosphate\_conc$$

Input Valve If the input valve is adjusted well (and assuming that there is always enough water flowing in), then enough water is passing the valve.

$$Ok\_input\_valve \rightarrow enough\_water\_passing$$

Weather If the weather within the last weeks has had not too much sunshine and high temperature, then there is no evaporation.

$$Ok\_weather \rightarrow no\_evaporation$$

Basin If the membrane of the basin is not leaking, and there is no evaporation and there is enough water passing from the input, then the water level has to be in its expected range. Note, that this model also makes use of the information provided by other components. Hence, there is a connection between these components. However, we do not model this explicitly as a separate connection because of simplicity.

$$Ok\_basin \wedge no\_evaporation \wedge enough\_water\_passing \rightarrow expected\_water\_level$$

Vegetation If the vegetation receives enough water as well as nitrates and phosphates, they are healthy and treating the water as expected. Note, that in this case, we do not assume the vegetation to be a potential root cause. Hence, we do not introduce a hypothesis in this case.

$$good\_nitrate\_phosphate\_conc \wedge expected\_water\_level \rightarrow vegetation\_healthy$$

$$good\_nitrate\_phosphate\_conc \wedge expected\_water\_level \rightarrow water\_treated$$

What is missing in this model are rules stating inconsistencies. For example, someone sees that the vegetation is not healthy. For this negated information, we have to introduce a proposition and add a rule stating that this proposition together with the one representing the opposite cannot occur at the same time. We do this for all qualitative values in the model used as connections between the components.

$$good\_nitrate\_phosphate\_conc \wedge bad\_nitrate\_phosphate\_conc \rightarrow \bot$$

$$expected\_water\_level\_conc \wedge low\_water\_level \rightarrow \bot$$

$$vegetation\_healthy \wedge vegetation\_not\_healthy \rightarrow \bot$$

$$water\_treated \wedge water\_not\_treated \rightarrow \bot$$

$$evaporation \wedge no\_evaporation \rightarrow \bot$$

$$enough\_water\_passing \wedge not\_enough\_water\_passing \rightarrow \bot$$

With these rules added to the theory *Th*, we conclude the model for our constructed wetland example. The described model only considers the correct case. There is no information formalized stating how components behave if they are faulty. Thus, consistency-based diagnosis is the best diagnosis method if the faulty behaviour of components is unknown. Note, that all of the propositions where we added a rule for stating inconsistency can be used as an observation for diagnosis. Observations are the starting point of diagnosis, because they have to be explained using the model. Before giving an example of diagnosis, we have to formally define what diagnosis means in the context of consistency-based diagnosis. As already discussed, a diagnosis has to explain the given observations. Explanation in this specific context is to identify a root cause (or root causes) for the observations. In the formal model, we use the hypotheses to represent root causes. Hence, we have to identify all hypotheses that when not be true lead to a satisfiable PHCT. Why satisfiable? Because, in this case, these hypotheses do not contradict the observations, which is exactly what we want. Formally, we define consistency-based diagnosis as follows:

Definition 7 (Consistency-Based Diagnosis) Given a diagnosis problem (P, Th, Hyp, Obs). A set $\Delta \subseteq Hyp$ is a consistency-based diagnosis if and only if the PHCT $(P, Th \cup \{\top \to o \mid o \in Obs\} \cup \{\top \to h \mid h \in Hyp \setminus \Delta\})$ is satisfiable. A diagnosis Δ is minimal if there is none of its subsets $\Delta' \subset \Delta$ being itself a diagnosis accordingly to this definition.

Having a closer look at Definition 7, we see that a diagnosis must be a subset of *Hyp* with the following additional constraint. Assuming all observations being valid and all hypothesis without the ones in *Hyp* being facts as well together with the theory *Th* is satisfiable and therefore, does not lead to a conflict. In terms of logic, we cannot derive $\bot$ from the provided logical rules.

Let us illustrate, the consistency-based diagnosis definition using our running example. We might visit a constructed wetland immediately seeing that the vegetation is not in a healthy state thus $Obs = \{vegetation\_not\_healthy\}$. Let us assume, an extremely hot summer and thus having $Ok\_weather$ as a diagnosis. We can check this diagnosis via taking *Th* for the constructed wetland and adding the facts $\top \to vegetation\_not\_healthy$, $\top \to Ok\_basin$, $\top \to Ok\_water\_quality$, and $\top \to Ok\_input\_valve$ to form a new PHCT accordingly to Definition 7. From $\top \to Ok\_input\_valve$ and rule $Ok\_input\_valve \to enough\_water\_passing$, we infer *enough_water_passing* accordingly to Definition 2. Using $\top \to Ok\_water\_quality$ and the rule $Ok\_water\_quality \to good\_nitrate\_phosphate\_conc$ the Modus Ponens allows us to obtain *good_nitrate_phosphate_conc*. Because we do not have the fact $Ok\_weather$, we cannot conclude *no_evaporation* and thus also not the fact *expected_water_level* preventing us from deriving finally *vegetation_healthy*. Therefore, we cannot infer a contradiction $\bot$ and $\{Ok\_weather\}$ is indeed a diagnosis accordingly to Definition 7. This diagnosis is alsominimalbecause the empty set is not a diagnosis, which can be proven easily assuming all hypotheses to be facts, which allows to infer $\bot$.

In practice, we are interested in minimal diagnoses accordingly to Definition 7. In addition, we are–in most cases–interested in computing diagnoses of the smallest cardinality. Hence, we prefer single fault diagnoses, i.e. diagnoses comprising only one hypothesis, over multiple fault diagnoses. So how can we utilize these preferences for computing diagnoses automatically? One simple – but maybe not the most efficient – algorithm is searching for diagnoses starting with diagnoses of the smallest size and increasing diagnosis size until the first set of diagnoses can be derived. The following algorithm **CBDiagnose** makes use of these assumptions and computes all minimal diagnosis of the smallest size starting with the empty diagnosis. The algorithm takes a diagnosis candidate and checks the condition given in Definition 7. If it is fulfilled, the candidate is added to the set of diagnoses. In the case at least one diagnosis of a certain size is found, the algorithm terminates giving back all found diagnoses.

Algorithm CBDiagnose

Input: Diagnosis model (*P, Th, Hyp*) and a set of observations *Obs.*
Output: All minimal diagnoses of smallest cardinality.

begin
1. Let *C* be the set comprising the empty set only, i.e., *C* = {∅}, and let *DS* be an empty set.
2. Let *C'* be an empty set.
3. **For all** elements Δ in *C* **do**:
   a. Construct a PHCT for the diagnosis problem (*P, Th, Hyp, Obs*) and the diagnosis candidate Δ accordingly to Definition 7.
   b. Call the theorem prover using the constructed PHCT. **If** the theorem prover returns √, **then** Δ is a diagnosis and added to the set *DS*. **Otherwise**, add the supersets Δ ∪ *h* for all hypothesis *h* ∈ *Hyp* \ (*Hyp* ∩ Δ) to *C'*.
4. **If** *DS* is empty **and** *C* is not equivalent to *C'*, **then** let *C* be *C'* and **go to 2**
5. **return** *DS*
end

Algorithm **CBDiagnose** obviously terminates, because in the worst case *C* becomes the set comprising the set of hypotheses only. In this case, no further hypothesis can be added and there is no further iteration. In this case, the algorithm returns the empty set stating that there are no diagnoses. The algorithm computes only minimal diagnoses. This is due to the fact that diagnoses are computed of increasing cardinality and computation is stopped whenever there is a diagnosis of a certain size. Hence, it is impossible to compute a diagnosis of larger cardinality and hence, all diagnoses must be minimal. The time complexity of **CBDiagnose** is obviously exponential because the number of diagnosis candidates is in the worst case $2^{|Hyp|}$. This is due to the fact that we have to check all subsets of *Hyp* in the worst case. However, in practice, there are usually single fault or double fault diagnoses, which makes the computation tractable. Moreover, someone can also introduce a limit for the diagnosis cardinality in order to stop computing diagnoses larger than a pre-specified cardinality.

There are other algorithms available for diagnosis. One of the diagnosis algorithms is based on the dual concept of diagnosis, which are conflicts. Conflicts are–similar to diagnoses–assumptions about truth hypotheses. But in case of conflicts, we are interested in hypotheses that when assumed to be true lead to an inconsistency. For more details, including the algorithm, we refer the interested reader to Reiter's seminal paper on diagnosis (1987) and Greiner et al. (1989), where the latter paper presented a corrected version of Reiter's original diagnosis algorithm. In addition, de Kleer and Williams (1987) presented an algorithm based on the de Kleer's ATMS (de Kleer, 1986) that allows to deal with fault probabilities and allows for computing the next best observation to be obtained for reducing the number of diagnoses. The underlying idea there is to use entropies for finding a particular point in the system where we want to apply a probe for gaining a new measurement that allows to distinguish diagnoses requiring a minimum number of additional observations. All the details can be obtained from the mentioned literature. In the following, we only briefly discuss the use of probabilities for allowing to distinguish diagnoses accordingly to their likelihood of being true.

Probabilities for distinguishing diagnosis can be introduced quite simple. We only need the fault probabilities for the hypotheses. A fault probability in this context is the probability for having the truth value of a particular hypothesis to be false. We now assume that we have the fault probability for every hypothesis $h \in Hyp : p_F(h) \in [0, 1]$. Such probabilities can be obtained from past experiences. For example, when we know that having a membrane leaking occurs only in 1% of the cases, we can set the fault probability of *Ok_basin* to 0.01. From the fault probabilities, we can easily obtain the probability of a certain diagnosis Δ when assuming that faults are stochastically independent, i.e., no fault triggers another fault. In this case, the probability of a diagnosis is the probability of the event that all hypotheses in Δ are faulty and all other hypotheses are true, i.e.:

$$p(\Delta) = \prod_{h \in \Delta} p_F(h) \cdot \prod_{h \in Hyp \setminus \Delta} (1 - p_F(h)) \tag{5.2}$$

For example, when using the following fault probabilities:

$$p_F(Ok\_weather) = 0.1 \quad p_F(Ok\_basin) = 0.01$$

$$p_F(Ok\_water\_quality) = 0.1 \quad p_F(Ok\_input\_valve) = 0.02$$

we obtain, the following probability of diagnosis {*Ok_weather*}:

$$p(\{Ok\_weather\}) = (Ok\_weather) \cdot (1 - (Ok\_basin)) \cdot$$
$$(1 - p_F(Ok\_water\_quality)) \cdot (1 - p_F(Ok\_input\_valve))$$
$$= 0.1 \cdot (1 - 0.01) \cdot (1 - 0.1) \cdot (1 - 0.02)$$
$$= 0.087318$$

Note, that in practice the real probabilities are not that important. Instead, the used probabilities should reflect the likelihood of some hypothesis to be true or false. Hence, sorting the hypotheses regarding their likelihood and setting the fault probability accordingly to reflect the obtained order, is usually good enough for practical application. In our example, we stated that the weather and the water quality are both similar likely to cause a fault but the input valve and the basin are less likely, stating that the membrane is the least likely reason for misbehaviour.

In the literature, the use of consistency-based diagnosis for gaining root cause information for environmental decision support has gained a lot of attention. Most recently, Struss et al. (2016) discussed a model for supporting wastewater treatment in India. In this paper, the focus is on the modelling part that allows for using consistency-based diagnosis in the particular application domain. The application of consistency-based diagnosis in the environmental decision support domain has been very well described in (Struss, 2012) and (Struss, 2011), where the author state preliminaries, challenges, advantages, and also discusses the architecture of such systems. Older work from the same author includes (Struss, 2009) and (Struss, 2008). Sànchez-Marrè et al. (2008) presented general account on the environmental decision support systems including model-based concepts.

Preliminary work on the application of consistency-based diagnosis to the environmental domain include (Struss et al., 2003; Heller & Struss, 2002, 2001; Struss & Heller, 1999). Heller and Struss (2001) are a paper of special interest because it shows how qualitative models can be utilized for consistency-based diagnosis. Qualitative models are of particular interest because they allow to generalize concepts without taking care of concrete numbers. The use of qualitative models in the environmental domains has been well described in (Bredeweg & Salles, 2009; Cioaca et al., 2009; Kansou et al., 2013). We refer the interested reader in this field to those papers for further studying.

Apart from the environmental domain, model-based reasoning has been used for diagnosing cars, space probes and even programs. Malik et al. (1996) provides several case studies where model-based diagnosis was used to analyze the root cause of failures detected in cars. Picardi et al. (2002) summarized their findings obtained for diagnosis designs of automotive systems. In many of these examples, abstract models, like qualitative models, e.g., see (Cascio et al., 1999), have been used for supporting the fault localization task. Pell et al. (1996) discussed the use of model-based reasoning to increase the autonomy of spacecrafts where the decision process itself is fully automated. For a general overview of the use of model-based diagnosis in the mobile robotics domain, refer to (Steinbauer & Wotawa, 2013).

In the context of fault localization in programs Console et al. (1993) introduced the basic foundations of debugging of logic programs relying on model-based diagnosis. Since then many others have provided models for supporting fault localization in programs written in different programming languages varying from Java (Mateis et al., 2000; Mayer et al., 2002; Mayer & Stumptner, 2004; Wotawa et al., 2012), functional languages (Stumptner & Wotawa, 1999), VHDL (Friedrich et al., 1999; Peischl & Wotawa, 2006), Verilog (Peischl et al., 2012), to even spreadsheets (Hofer et al., 2017a, b; Abreu et al., 2015). In the case of fault localization in programs model-based diagnosis provides potential fault candidates, e.g. the line of code that explains the fault behaviour, to the user and therefore, supports the user in finding and fixing bugs.

5.4.4 Abductive Diagnosis

Abductive diagnosis can be best explained when considering a patient coming to a medical doctor. The patient describes symptoms of his or her disease and the doctor's task is to identify the root causes of these symptoms using knowledge of the form: "causes imply symptoms". If there are more potential causes the doctor either relies on probabilities of diseases, asks questions about further symptoms, or alternatively obtains other information using physical examinations. Friedrich et al. (1990a) provided a formalization of this process relying on the abductive diagnosis.

In the following, we make use of Friedrich et al.'s outlined foundations and adapt them to fit our formal setting.

We start by explaining how models for abductive diagnosis can be obtained using the constructed wetland example depicted in Fig. 5.28. What we have to provide is a model of the constructed wetland providing rules leading from a cause to a set of symptoms. Hence, in the first step, we have to identify potential root causes. This can be done based on the components of a system and via assigning potential faults to these components. Let us start with the components given in Fig. 5.30. For water quality, we know that there is a concentration of nitrates and phosphates attached. In case of a too low concentration, there will be a negative effect on the vegetation. The position of the input pipe influences the water flow. Therefore, a wrong position might lead to a water flow that is not enough for sustaining the water level. The membrane of the basin might be leaking and thus also have a negative impact on the water level. A weather situation of a lot of sunshine and high temperatures lead to evaporation and thus again impacting the water level. A too low water level negatively impacts vegetation. The vegetation component itself might not has a corresponding fault mode and thus can be safely ignored. Using this brief discussion, we are able to come up with the following table of the components, their fault modes, and impacts (see Table 5.1).

Using the information provided in this table, we are able to come up with a set of rules. We form hypotheses for each pair of components and their fault modes and let them imply the symptoms.

$$Low\_np\_concentration \rightarrow low\_np\_concentration$$
$$Low\_np\_concentration \rightarrow vegetation\_not\_healthy$$
$$Wrong\_valve\_adjustment \rightarrow not\_enough\_water\_passing$$
$$Wrong\_valve\_adjustment \rightarrow low\_water\_level$$
$$Wrong\_valve\_adjustment \rightarrow vegetation\_not\_healthy$$
$$Too\_hot\_weather \rightarrow evaporation$$
$$Too\_hot\_weather \rightarrow low\_water\_level$$
$$Too\_hot\_weather \rightarrow vegetation\_not\_healthy$$
$$Membrane\_leaking \rightarrow water\_leakage$$
$$Membrane\_leaking \rightarrow low\_water\_level$$
$$Membrane\_leaking \rightarrow vegetation\_not\_healthy$$

Table 5.1 Components, fault modes and impacts

Component	Fault mode	Symptoms
Water quality	Low np concentration	Low np concentration, vegetation not healthy
Input valve	Wrong adjustment	Not enough water passing, low water level, vegetation not healthy
Basin weather	Membrane leaking too hot	Water leakage, low water level, vegetation not healthy Evaporation, low water level, vegetation not healthy

These rules can be reformulated using the information that either not enough water passing, evaporation, or water leakage leads to a low water level, and either a low nitrate and phosphate concentration or a low water level would harm vegetation. In this case, we would obtain a reduced set of rules that allows to draw the same conclusions:

$$Low\_np\_concentration \rightarrow low\_np\_concentration$$
$$Wrong\_valve\_adjustment \rightarrow not\_enough\_water\_passing$$
$$Too\_hot\_weather \rightarrow evaporation$$
$$Membrane\_leaking \rightarrow water\_leakage$$
$$water\_leakage \rightarrow low\_water\_level$$
$$not\_enough\_water\_passing \rightarrow low\_water\_level$$
$$evaporation \rightarrow low\_water\_level$$
$$evaporation \rightarrow low\_water\_level$$
$$low\_water\_level \rightarrow vegetation\_not\_healthy$$

Because of the fact that, we also want to incorporate information that a symptom is not visible, we have to add propositions stating the negated symptom in a second step. In addition, we add rules formalizing that the negated symptom together with the symptom cannot occur at the same time.

$$good\_np\_concentration \wedge low\_np\_concentration \rightarrow \perp$$
$$enough\_water\_passing \wedge not\_enough\_water\_passing \rightarrow \perp$$
$$evaporation \wedge no\_evaporation \rightarrow \perp$$
$$expected\_water\_level \wedge low\_water\_level \rightarrow \perp$$
$$vegetation\_healthy \wedge vegetation\_not\_healthy \rightarrow \perp$$

All propositions occurring in the rules and not being hypotheses can be used as observations. Some observations, i.e. the symptoms, are used in the rules on the right side. These observations need to be explained for diagnosis, i.e. for these observations we want to come up with their underlying causes, i.e. hypotheses supporting these observations. The other observations that occur only in rules implying $\perp$, like *good_np_concentration* bring in some sort of "positive" information. These observations need not to be explained using abductive diagnosis. Even stronger, these observations cannot be explained because there are no rules of the form "causes to symptoms" where such observations do not occur as symptoms. But how to integrate positive information? This can be easily done via adding these observations as facts to the PHCT when computing abductive explanations. Before illustrating the different handling of observations using our running example, we introduce formally the definition of abductive diagnosis.

Definition 8 (Abductive Diagnosis) Given a diagnosis problem (P, Th, Hyp, Obs), where $Obs = Obs^- \cup Obs^+$ with $Obs^- \cap Obs^+ = \emptyset$, Obs^+ are positive observations and Obs^- are the symptoms to be explained. A set $\Delta \subseteq Hyp$ is an abductive diagnosis if and only if the following two properties hold for the PHCT $(P, Th \cup \{\top \rightarrow o \mid o \in Obs^+\} \cup \{\top \rightarrow h \mid h \in \Delta\})$:

1. $Cl(Th \cup \{\top \rightarrow o \mid o \in Obs^+\} \cup \{\top \rightarrow h \mid h \in \Delta\}) \supseteq \{\top \rightarrow s \mid s \in Obs^-\}$
2. The PHCT is satisfiable, i.e. $(\top \rightarrow \bot) \notin Cl(Th \cup \{\top \rightarrow o \mid o \in Obs^+\} \cup \{\top \rightarrow h \mid h \in \Delta\})$.

A diagnosis Δ is minimal if no subset $\Delta' \subset \Delta$ is itself a diagnosis accordingly to this definition. Minimal diagnoses are also called parsimonious diagnoses.

In this definition, abductive diagnosis is defined as a subset of the set of hypotheses necessary to derive the symptoms to be explained. In addition, we require the corresponding PHCT to be consistent. This is a more technical detail because from inconsistent theories every proposition can be obtained from any arbitrary set of hypotheses. Such explanations would not be wanted.

It is worth noting that abductive diagnosis in comparison with consistency-based diagnosis relies on models capturing the faulty behaviour of components. The positive observations Obs^+ have therefore be separated from the negative ones Obs^- for which we want to find an explanation. Note also that there are relationships between consistency-based diagnosis and abductive diagnosis. Basically, when introducing fault models into consistency-based diagnosis we can also obtain diagnoses that would be derived using Definition 8. For more details, we refer the interested reader to (Console et al., 1991; Console & Torasso, 1990).

Let us illustrate Definition 8 using our constructed wetland example and the abductive diagnosis model developed before in this section. When using this model together with the observations $Obs = Obs^+ \cup Obs^-$ where $Obs^+ = \emptyset$ and $Obs^- = \{vegetation\_not\_healthy\}$, we obtain 4 parsimonious explanations: $\{Too\_hot\_weather\}$, $\{Wrong\_valve\_adjustment\}$, $\{Membrane\_leaking\}$, and $\{Low\_np\_concentration\}$. This number of explanations can be improved when adding additional information. Let us assume, that we are in front of the constructed wetland and see that the water level is fine. In this case, we change the positive observations $Obs^+ = \{expected\_water\_level\}$ and leave the negative observations, i.e. the symptoms as they are. For these observations, we finally would receive only one explanation, i.e. $\{Low\_np\_concentration\}$. This is obvious because the other explanations influence also the water level but the nitrate and phosphate concentration does not.

In order to show the case where we have multiple hypotheses explanations. For this purpose, assume $Obs^+ = \emptyset$ and $Obs^- = \{vegetation\_not\_healthy, low\_water\_level, low\_np\_concentration\}$. For this case we obtain again more than one diagnosis:

- $\{Too\_hot\_weather, Low\_np\_concentration\}$,
- $\{Wrong\_valve\_adjustment, Low\_np\_concentration\}$, and
- $\{Membrane\_leaking, Low\_np\_concentration\}$.

All of these explanations comprise the hypothesis *Low_np_concentration*. Further information about the setting of the valve, etc., may help to reduce the number of diagnoses.

The following **AbDiagnose** algorithm uses abductive diagnosis models together with observations to compute explanations for symptoms. The algorithm works very much similar to the **CBDiagnose** algorithm introduced for consistency-based diagnosis. Starting from the single fault hypotheses the algorithm searches for explanations of increasing size.

Algorithm AbDiagnose

Input: Diagnosis model (P, Th, Hyp) describing the faulty behaviour of a system, and
   a set of observations $Obs = Obs^+ \cup Obs^-$ where $Obs^+ \cap Obs^- = \emptyset$.
Output: All minimal abductive explanations of smallest cardinality.

begin
1. Let C be the set of sets each of them comprising exactly one hypothesis, i.e.,
   $C = \{\{h\} \mid h \in Hyp\}$, and let E be an empty set.
2. Let C' be an empty set.
3. **For all** elements Δ in C **do**:
   a. Construct a PHCT for the diagnosis problem (P, Th, Hyp, Obs) and the
      diagnosis candidate Δ accordingly to Definition 8, i.e.,
      $PHCT = (P, Th \cup \{\top \to o \mid o \in Obs^+\} \cup \{\top \to h \mid h \in \Delta\})$
   b. Check $Cl(Th \cup \{\top \to o \mid o \in Obs^+\} \cup \{\top \to h \mid h \in \Delta\}) \supseteq \{\top \to s \mid s \in Obs^-\}$.
   c. Check the satisfiability of the PHCT.
   d. **If** both checks are successful, **then** add Δ to E. **Otherwise**, add the supersets
      $\Delta \cup h$ for all hypothesis $h \in Hyp \setminus (Hyp \cap \Delta)$ to C'.
4. **If** E is empty **and** C is not equivalent to C', **then** let C be C' and **go to** 2.
5. **return** E
end

The **AbDiagnose** algorithm searches for diagnoses starting with a single hypothesis diagnosis first. At the end, all hypotheses are set to true for checking diagnosis. The algorithm returns a set of parsimonious explanations explaining all negative observations Obs^- returned, there are no explanations for the given observations. **AbDiagnose** obviously terminates. The runtime complexity is exponential on the size of hypotheses in the worst case. This algorithm is very much similar to **CBDiagnose** with differences coming from the different definitions of diagnosis. **AbDiagnose** also does not start with the empty set, because we need at least one hypothesis to explain observations. Hence, an empty diagnosis cannot be the outcome in the case of an abductive diagnosis.

It is worth noting that we are also able to add probabilities to abductive diagnosis. We simple can apply the same equation for computing probabilities for consistency-based diagnosis described in the previous section. This of course requires knowledge about the probability p_F for each hypothesis in the model.

The use of abductive diagnosis in the environmental domain is more recent (Wotawa et al., 2010, 2009) and has been mainly motivated when focussing on alternatives for other means for reasoning and in particular decision trees. Wotawa (2011) discussed the use of abductive diagnosis and showed improvements, e.g. regarding the ease of adaptivity in case of system changes, when compared to decision trees. Compared with consistency-based diagnosis, abductive diagnosis can be easily applied in all cases where fault modes for components and their consequences are available. In the case of the availability of component models describing the correct behaviour only, the recommendation would be to use consistency-based diagnosis instead.

The use of abductive diagnosis for other disciplines like medical diagnosis (Friedrich et al., 1990b) or obtaining explanations for technical systems like wind turbines have been reported. Wotawa (2014) showed that failure mode and effect analysis can be used for obtaining models for abductive diagnosis. In case of wind turbine diagnosis Koitz and Wotawa (2015) showed the feasibility of abductive diagnosis for providing root causes. Gray et al. (2015) discussed the use of abductive diagnosis for wind turbines in a more general setting.

5.4.5   Conclusions

In this section, we discussed two approaches for model-based reasoning to be used by an intelligent decision support system to advice user whenever explanations or diagnoses are needed. Both presented approaches are based on models of the system. In the case of consistency-based diagnosis, we need a component-connection model of the system where we have to formalize the correct behaviour of each component. A certain proposition indicating the correctness of the behaviour is used to allow an algorithm searching for those propositions to be set to false in order to get rid of any inconsistencies with given observations of the system. In the case of abductive diagnosis, we require a model capturing the faulty behaviour of components. There the idea is to have rules specifying the hypothesis to symptoms relation. The task of abductive diagnosis is to find those hypotheses that explain given symptoms.

Both presented diagnosis approaches can be fully automated. We discussed also two algorithms that compute minimal diagnoses of the smallest size, starting with diagnoses of size 1. Note, that in case of consistency-based diagnosis we start with the empty diagnosis in order to see whether there is a need for computing diagnoses. In case there is no contradiction with the observations, there is no need to find any single fault or multiple fault diagnosis. The algorithms for computing diagnoses are very much similar despite using different kinds of models. In addition, the time complexity of both algorithms is also the same. In the worst case, we have a time complexity of order exponential in the number of used hypotheses. However, in practice, it is very much likely to have single or double fault diagnoses, where all minimal diagnoses can be computed within polynomial time.

Table 5.2 Comparison of consistency-based and abductive diagnosis based on observations, models, complexity, and the modelling process

	Consistency-based	Abductive
Observations	All information regarding measured values but no hypotheses.	The set of observations distinguishes positive observations, i.e. proposition not used on the right side of a rule, and negative observations, i.e. observations that should be explained
Model	A component-connection oriented model formalizing the correct behaviour of the system	A model comprising rules of the form "hypothesis implies symptoms", where symptoms are the negative observations, and rules stating that symptoms and their negated propositions cannot occur at the same time
Time complexity	$O(2^{\|Hyp\|})$ in the worst case, and $O(\|Hyp\|)$ for computing single fault diagnoses	$O(2^{\|Hyp\|})$ in the worst case, and $O(\|Hyp\|)$ for computing single fault explanations
Modelling process	(1) Start with deciding on components and their connections. (2) For each component specify its behaviour considering inputs and outputs. (3) Formalize the behaviour of the components and the connections.	(1) Identify the components and their fault modes. (2) For each fault mode introduce a hypothesis and their effects. (3) For each hypothesis and the effects introduce a rule mapping the hypothesis to the effect. (4) Add a rule stating that the effects and their negating proposition cannot be true at the same time.
Best used	Whenever only the correct system's behaviour is known.	Whenever we have hypotheses and the caused symptoms.

In addition to the foundations, we explained all approaches using a small example coming from the environmental domain. In particular, we showed how models of constructed wetlands can be obtained for consistency-based diagnosis and abductive diagnosis. In both cases, the models are rather small and more or less easy to develop. Modelling of course is not always that easy and requires the right steps to be carried out. For this purpose, we also discussed modelling considering the different steps and tasks to be carried out in a sequential fashion.

In Table 5.2, we summarize the properties of both approaches to model-based reasoning. The table also includes a row discussing the different steps for constructing models. It is worth noting that both approaches have advantages over other methods used for diagnosis. They are very much flexible, and modelling used is close to models already under consideration when designing systems. In addition, in most cases models can be easily reused and adapted. This is hardly the case for other alternative models. However, modelling of course comes not for free and requires further effort to be spent when coming up with a diagnostic solution.

5.5   Qualitative Reasoning Models

In addition to previous model-driven methods, there are *qualitative models* reflecting the qualitative relationships among several attributes or variables, which are characterizing a domain. They are also related to the first-principles or mechanistic models explained in Chap 3, Sect. 3.3.1.1. These models were emerged from *qualitative physics* domain (de Kleer, 1977; de Kleer & Brown, 1984) and *qualitative process theory* (Forbus, 1984, 1990). *Qualitative reasoning* (Kuipers, 1986, 1994) can be used as a tool both for modelling and for the simulation of a system or process. Given some observation about variables' values, it can be predicted how will evolve the values of other related variables (prognosis task), or how these values could have been produced (diagnostic task).

5.5.1   Basic Principles of Qualitative Reasoning

The world is plenty of very complex information. Humans knowledge about the world is limited, and hence, incomplete. Notwithstanding, humans are able to manage that complexity and reason with incomplete information, and make deductions about it. One could argue that humans know how to reason because they own some *common sense knowledge* ability, or even some *expert knowledge*, gained with experience by solving the same domain problems. However, both kinds of knowledge are usually incomplete. We do not know all the details of complex mechanisms of the real world, but we can survive in it.

Qualitative reasoning basic approach is to provide some techniques being able to represent and reason with incomplete knowledge about physical devices or mechanisms. This field was originated from the qualitative physics subfield, which aid to reason about physical domains and the temporal evolution of these systems, in a different way than the classic numerical and quantitative one, mostly based in Ordinary Differential Equations (ODEs). According to Kenneth D. Forbus (Forbus, 1996), "*qualitative reasoning* is the area of AI which creates representations for continuous aspects of the world, such as space, time, and quantity, which support reasoning with very little information". The hypothesis of qualitative reasoning is that people can build and use *qualitative descriptions* of the mechanisms in the physical world.

A *qualitative description* is a description which captures the "qualitative" important parts of the physical world, avoiding the irrelevant ones. Each one of the important aspects becomes an important qualitative difference. For instance, when we are preparing an infusion and put our kettle with water on the stove to be heated, we do not care about the infinite temperatures the water will have, from the initial one to the boiling point temperature. Thus, the important temperature points are the *initial temperature*, and the *boiling-point temperature*. These are the qualitative important points.

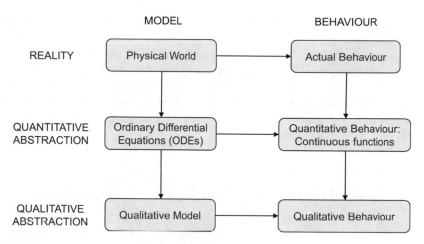

Fig. 5.31 The qualitative approach to modelling the real world. Adapted from Kuipers (1994)

On the other hand, the traditional quantitative approach in physics and engineering is to use Ordinary Differential Equations (ODEs) to describe the systems and their temporal behaviour, and then drawing inferences about it. Solving an ODE means to use some complex analytical mathematical methods or to use some numerical approximation methods for obtaining a solution. In this quantitative approach, much information is required. In the example of the kettle on the stove, the following information would be required: the temperature of the water, the temperature of the stove, the water volume, the boiling point, the rate of change water temperature, the heating power of the stove, all initial conditions, and so on.

Thus, making an abstraction of many values, and keeping just two qualitative values, people can reason and manage how to get an infusion with their kettles.

In Fig. 5.31, the qualitative modelling approach is shown. In the picture, the first row represents the reality: the *actual physical world*, and the observed *actual behaviour*. Moving to the second row, we need an abstraction process to simplify the complexity of the real world. This second row shows the classic *quantitative modelling* of physical systems: Ordinary Differential Equations (ODEs), which represents the dynamics of the world, and the *quantitative behaviour* observed, through some continuous functions, which are the solutions of the ODEs. Finally, the third row depicts the qualitative abstraction step to get a *qualitative model* of the physical world and the *qualitative behaviour* obtained from this modelling approach.

As it will be described below, what constitutes a *qualitative model* is slightly different within each one of the approaches or frameworks developed in the literature. Notwithstanding, we could name the qualitative model, in general, as a *Qualitative Differential Equation (QDE)*. The rational is that a QDE is an abstraction of an ODE, where numerical constants have been substituted by symbolic names, and a wide range of different initial conditions have been abstracted. The different approaches in the literature represent this QDE in different formalisms, but all have the same goal: describe in a qualitative way the behaviour of the physical scenario at hand, using some sets of entities (qualitative variables, concepts of ontology, etc.), and representing the relationship among the entities in some kind of formalism (confluences, influences, inequalities, mathematical functions, qualitative constraints, equations, etc.). In addition, the execution of the qualitative model, i.e. the *qualitative simulation*, to obtain the corresponding observed behaviour, is performed in slightly different ways, too. Some obtain a tree of behaviours, while others obtain a state-transition graph, where the behaviours are hidden in the different possible paths.

5.5.2   General Flowchart of a Qualitative Reasoning Model

Furthermore, although each approach developed in the literature to implement a qualitative reasoning system, uses its own terminology, formalisms, simulation algorithms, etc., a general flowchart for the deployment of a qualitative reasoning model can be depicted. Figure 5.32 shows this general flowchart.

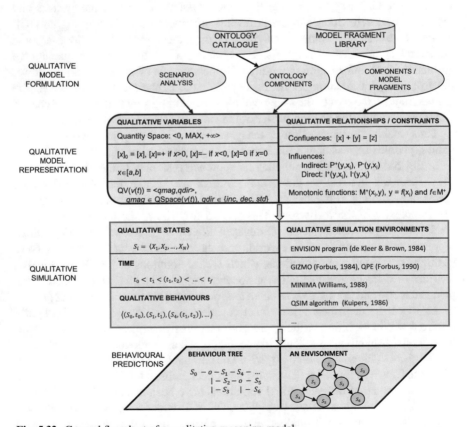

Fig. 5.32 General flowchart of a qualitative reasoning model

The flowchart is composed of three steps. The first two steps are related to the *qualitative model building*, and the later one to the *qualitative model simulation*.

The first one is the *qualitative model formulation*. At this step, a scenario analysis is undertaken to analyze the physical mechanism or device and its particular topology in order to get the important knowledge to be represented in the model. In addition, some parts of the new qualitative model can be reused from some existing ontology, within an existent catalogue of ontologies (liquids, solids, physical devices, electrical circuits, chemical compounds, etc.). Some model fragments or components can be included from an existing library of models or model fragments, to formulate the model.

The next step is the *representation of the qualitative model* previously formulated from the components and model fragments integrated, selecting the necessary qualitative variables, represented in a concrete formalism (quantity spaces, sign algebra, etc.), to express the important qualitative aspects of the given scenario. The qualitative relationships or constraints, expressed in the corresponding format (confluences, influences, etc.) are also added to the model.

The third step is the execution of the model, i.e. the *qualitative simulation*. The simulation will compute the temporal evolution of the different *qualitative states* of the model, from some initial state/s. The output of this step is the *behavioural predictions*. A qualitative behaviour is the sequence of consecutive qualitative states that can be reached from an initial state. Depending on the simulation environment, two kinds of output are generated: a behaviour tree or an envisionment. The qualitative behaviours are analyzed to get useful predictions of the temporal evolution of the system being modelled, and therefore, giving a very relevant knowledge for a decision-making process. This simulation can be done using some of the different existing environments in the literature like the ENVISION program (de Kleer & Brown, 1984), the GIZMO (Forbus, 1984) or QPE (Forbus, 1990) implementation, the MINIMA approach (Williams, 1988) or the QSIM algorithm (Kuipers, 1986, 1994).

5.5.3 Qualitative Model Building

As in any model-driven reasoning technique, there are two general steps. The first is the *building of the model*, which consists in selecting the appropriate information and knowledge to describe the real world, and in representing those information and knowledge in some appropriate formalism, to answer a particular question. The second one is the *use or simulation of the model* to get some new knowledge from it, to solve the question at hand.

In the qualitative reasoning field, both tasks have been addressed, but probably, the model building task is a more open problem, than the model simulation task which has had more technical advances. Building a model, as any creative task is more an "art" than a science, and strongly depends on the skills of the modeller.

However, many efforts and modelling techniques have been deployed in the qualitative reasoning literature.

In the next subsections, we will describe the model building task, which can be divided in the *model formulation* subtask, and the *model representation* subtask.

5.5.3.1 Qualitative Model Formulation

Model formulation has three basic components, which are being jointly merged to build a model: *the scenario, an ontology catalogue*, and a *model fragment library* implementing *a domain theory*. Let us describe each one of these components, before reviewing the methodologies for model formulation.

Scenario

The scenario is the particular mechanism, device, or more general, the system, under consideration to be modelled. It is composed by a set of objects and relations in the world. The analysis of the scenario provides the description of the system under consideration, and specifically, the initial state of the system. For instance, let us suppose that we want to analyze the dynamics of a simple device: a block-spring mechanism, like indicated in Fig. 5.33. Let us assume that the variable x will measure the position of the block, being $x = 0$ the equilibrium position. Other magnitudes of interest are:

v: velocity of the block
a: acceleration of the block
m: mass of the block, $m > 0$ and m constant

The initial state of the system must be defined. For instance, the position of the block, could be at the origin, i.e. $x(t_0) = 0$, and with an initial positive velocity, i.e. $v(t_0) = v_0$, $v_0 > 0$.

Fig. 5.33 A block-spring mechanism

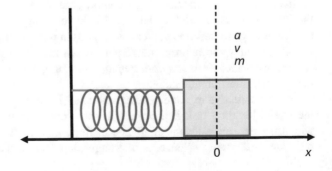

Ontologies

An *ontology* is a knowledge representation formalism intended to represent the important entities and their interrelationships in a domain. These entities are a set of concepts, their attributes, and their relationships. One relevant advantage of ontologies is that they facilitate the reuse of knowledge in a concrete domain by reusing the concepts and their related information across different software systems.

Ontologies, are important formalisms in qualitative reasoning to automatize the building process of a model, reusing different parts of existent ontologies.

For instance, an ontology describing different physical objects, like springs, blocks, tanks, etc., could be partially integrated into the formulation of the model, because, the important properties of a spring and a mass, would be already defined there, like:

k: the spring constant, $k > 0$
F: the force acting on the block, according to Newton's law, $F = m * a$
F': the spring force, according to Hooke's ideal spring law, $F' = -k * x$

This ontology is depicted in Fig. 5.34. One of the most common relations among concepts is the "*subtype of*" relation which expresses the taxonomical relation among objects. However, other relations can be defined, like "*part of*", etc. Of course, usually, more than one ontology is needed for modelling the system. Therefore, in the qualitative reasoning community research, they are developing a *catalogue of ontologies*, describing their properties and interrelationships and specifying conditions under which each ontology is appropriate. Thus, in one model formulation, several pieces of different ontologies could be used.

Probably, the most commonly used ontologies are the device ontology (de Kleer & Brown, 1984) and the process ontology (Forbus, 1984).

Component/Model Fragment Library

To complete the automation of the models, it is necessary to have available a *library of model fragments*, which can be combined to constitute, jointly with some parts of some ontologies, and with the scenario description, the model formulation. In other words, the model fragment library is the knowledge base of the modeller. The models can automatically be instantiated with objects from the given scenario, and they have explicit conditions for activation.

A *model fragment* describes part of the structure and behaviour of the system in a general way. It can be imagined as rules, which can be represented as conditions or consequences.

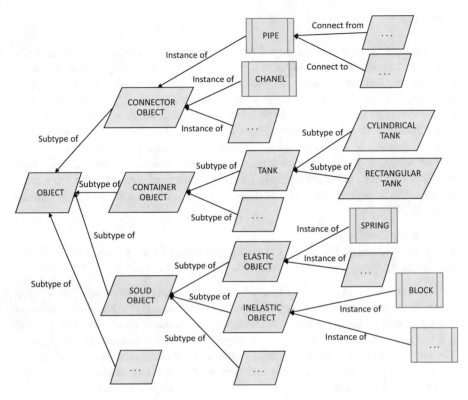

Fig. 5.34 An ontology for objects

According to QPT (Forbus, 1984), there are two basic kinds of *model fragments*:

- *Static Fragments* or *Individual Views*: The model fragments provide the structure of the system and the proportionalities or indirect influences between some of the objects in the system. An individual view consists of four components:

  - A *list of individuals*, which are the objects that must exist before it is applicable.
  - *Quantity conditions*, statements about inequalities between quantities of the individuals, and statements about whether or not certain other individual views or processes hold.
  - *Preconditions*, which are still further conditions that must be true for the view to hold.
  - A *set of relations*, statements that are true whenever the view is true.

- *Dynamic Fragments* or *Process Fragments*: The model fragments which contain at least one direct influence. They describe dynamic interactions among objects and causes of change. They act through time to change the variables of objects in a concrete situation. Examples of processes include fluid flow, boiling, motion, etc. A process fragment is composed by:

- The same components of individual views: *individuals, preconditions, quantity conditions, relations.*
- A *set of influences* imposed by the process on the variables/parameters of the individuals.

Model Formulation

Model formulation task consists in the creation of the simplest adequate model of a system for a given task. Still, many models are formulated in a hand-crafted way, but there exist several approaches to automatically obtain the formulation of the model.

The first and simple way of a model formulation is to instantiate every possible model fragment from a *domain theory*, usually implemented as a *library of model fragments*, taking into account a representation of the particular scenario under consideration. Of course, this approach makes only sense when the domain theory is very focussed, and not a broad one.

The alternative to use all the model fragments is to impose some kind of search in the space of modelling assumptions since they control which aspects of the domain theory will be instantiated. One of the simplest and very general approaches is the model formulation algorithm proposed by (Falkenhainer & Forbus, 1991). This approach instantiates all potentially relevant model fragments and uses an assumption-based truth maintenance system to find all legal combinations of modelling assumptions needed to form a model that could answer a given query. The criterion used was to minimize the number of modelling assumptions. This approach has two major drawbacks: full instantiation can be very expensive, and the number of consistent combinations of model fragments tends to be exponential.

Other approaches developed are the Nayak's algorithm (Nayak, 1994), which under at least one set of constraints, the model formulation can be carried out in polynomial time. And the work of (Rickel & Porter, 1994) which proposed to treat model formulation as a best-first search for a set of model fragments providing the simplest complete causal chain. The idea is that a model must include a complete causal chain connecting a possible changed parameter to the other parameters of interest, in generating explanations for a "what-if" simulation.

In general, model fragmentation is not a single-step task, but many times is an iterative process.

Compositional Modelling

The compositional modelling methodology (Falkenhainer & Forbus, 1991), which has become standard in qualitative physics, works in the following way.

Models are created from domain theories, which describe the kinds of entities and phenomena that can occur in a physical domain. A domain theory consists of a set of model fragments, each describing a particular aspect of the domain.

Creating a model is accomplished by instantiating an appropriate subset of model fragments, given some initial specification of the system (e.g. the propositional equivalent of a blueprint) and information about the task to be performed. Reasoning about appropriateness involves the use of modelling assumptions.

Modelling assumptions are the control knowledge used to reason about the validity or appropriateness of using model fragments. Modelling assumptions are used to express the relevance of model fragments. Logical constraints between modelling assumptions comprise an important component of a domain theory.

5.5.3.2   Qualitative Model Representation

Once the model has been formulated as the result of the integration of selected model fragments, portions of ontologies, and the initial scenario description, it must be represented as a qualitative model. Main tasks are the *representation of the outlined variables or magnitudes* from de formulation of the model, as qualitative values, and the *representation of the mathematical relationships* among those variables.

The different approaches and theories of qualitative reasoning have employed several representation formalisms, either for the variables or for the relationships. Below, we will review the main formalisms used in the literature.

Representing Continuous Magnitudes as Qualitative Values

Formalisms proposed usually have different degrees of resolution. Here, we will describe the most commonly used: status abstraction, sign algebra, quantity space, qualitative values, interval representation, and finite algebras.

Status Abstraction

It is a very low-resolution representation of a continuous quantity. The idea is just distinguishing between two basic qualitative values: one indicating normality and the other one abnormality (Abbott, 1988). Usually, these two values can be: {"normal", "abnormal"}, or {"working", "not working"} or {"open", "not-open"}, etc.

Sign Algebra

A further step is the sign algebra, which proposes to represent each continuous variable with three values $\{-, +, 0\}$, according to whether the sign of the variable is negative, positive, or zero. This representation can be also applied to the derivative of the variables, which are other variables expressible with the sign algebra. Sign algebra is quite powerful, as some theorems of differential calculus can be applied to it (de Kleer & Brown, 1984). This means that sign algebra is able to handle reasoning about system dynamics.

Usually, the sign algebra is represented with this formalism, for a variable x:

$$[x]_0 = \text{sign}(x) = \begin{cases} + & \text{if } x > 0 \\ 0 & \text{if } = 0 \\ - & \text{if } x < 0 \end{cases} \tag{5.3}$$

$$[x]_a = \text{sign}(x - a) = \begin{cases} + & \text{if } x > a \\ 0 & \text{if } = a \\ - & \text{if } x < a \end{cases} \tag{5.4}$$

And the derivative of a variable can be expressed as:

$$[\dot{x}] = \left[\frac{dx}{dt}\right] = \text{sign}\left(\frac{dx}{dt}\right) \tag{5.5}$$

Basic arithmetic operations (+, *) can be defined.

Quantity Space

Another representation method is the *quantity space* formalism (Forbus, 1984), which can extend the resolution of the qualitative values from the three values of sign algebra to the needed ones depending on the importance of the qualitative values. Therefore, a *quantity space* is a set of ordered qualitative important values. Usually, it includes the *zero* value (0), and depending on the variable range, the values $+\infty$ or/and $-\infty$. For instance, the temperature of the water can be expressed in reference to its freezing point (*FrP*) and its boiling point (*BoP*).

$$\textit{Quantity Space (TempWater)}: \ < -\infty, FrP, 0, BoP, +\infty >$$

The comparison important values can be points derived from the observation of general properties of a system and a specific situation, which are called *limit points* and *landmark values*. Landmark values are constant points of comparison introduced during the simulation to provide additional resolution (Kuipers, 1986).

Qualitative Values

Kuipers (1986) joins both the qualitative value of one variable and the corresponding one of its derivative in a pair of qualitative values that express the magnitude (*qmag*) and the direction of change of the variable (*qdir*). This representation is more concise than representing both as independent variables.

$$QV(f(t)) = <qmag, qdir>$$ (5.6)

where $qmag \in$ Quantity Space($f(t)$), thus:

$$qmag = \begin{cases} l_j & \text{if } f(t) = l_j, \text{ a landmark value} \\ (l_j, l_{j+1}) & \text{if } f(t) \in (l_j, l_{j+1}) \end{cases}$$ (5.7)

$$qdir = \begin{cases} inc & \text{if } f'(t) > 0 \\ std & \text{if } f'(t) = 0 \\ dec & \text{if } f'(t) < 0 \end{cases}$$ (5.8)

The interpretation of the landmarks by Kuipers englobes both the limit points and the landmark values as defined by Forbus.

Interval Representation

The interval representation is a well understood and widely used one in the qualitative reasoning community. It is variable–resolution representation for numerical values. The idea is to view a quantity space as the union of a set of intervals. The intervals are formed by each pair of consecutive important points (limit points and landmark values).

Therefore, if a variable x has the following quantity space:

Quantity space($x(t)$): $l_1 < l_2 < \cdots < l_{n-1} < l_n$.

Then, it generates a set of intervals: $\{[-\infty, l_1], [l_1, l_2], \ldots, [l_{n-1}, l_n], [l_n, +\infty]\}$ to which the variable can belong.

Finite Algebras

The use of finite algebras was originated with the observation that humans commonly use a finite set of ordered labels to describe qualitatively the value of a variable.

For instance, the set of labels to describe qualitatively the value of a variable *spring-length* could be: {very small, small, normal, large, very large}.

Some researchers have developed a finite algebra to operate with variables represented this way, and some operators to operate with this kind of qualitative values, have been defined.

Representing Mathematical Relationships

The mathematical relationships among the different variables of a given formulated model have been represented in different manners by the different approaches to qualitative reasoning. Probably, the most used and influential are the *confluences representation* (de Kleer & Brown, 1984), the *influences* (Forbus, 1984), and *the monotonic functions and other mathematical functions* (Kuipers, 1986). These relationships can be viewed as *qualitative constraints* that must be satisfied by the qualitative variables involved.

Confluences

Confluences are a representation of a QDE. A *confluence* joints multiple competing tendencies or influences of several variables, i.e., hence the name of *confluence* (de Kleer & Brown, 1984).

They are equations in the extended sign algebra $S' = \{-, 0, +, ?\}$ or qualitative expressions evaluating to signs, such as the combinations of some qualitative operators like $[x]_0$, $[x]_*$, $[\dot{x}] = [dx/dt]$.

For instance, considering the well-known Newton's law $F = m * a$, where the mass m is positive and constant, i.e. $\lfloor m \rfloor_0 = +$ and $[\dot{m}] = 0$. This equation can be transformed to a confluence, and some qualitative conclusions can be derived:

- $[F]_0 = [m * a]_0 = [m]_0 * [a]_0 = [a]_0$, thus, the force and the acceleration are on the same direction.
- $[\dot{F}] = \left[\frac{dF}{dt}\right] = \left[\frac{d(m*a)}{dt}\right] = [m]_0 * [\dot{a}] + [a]_0 * [\dot{m}] = [\dot{a}]$, hence a variation of the force produces a variation of the acceleration in the same direction, or vice versa.
- $[F]_* = [m * a]_* = [m]_0 * [a]_* + [a]_0 * [m]_* = [a]_*$, therefore, the perturbations to force and acceleration are in the same direction, given that there is no perturbation to mass, because is constant ($[m]_* = 0$).

Influences

Influences are the mechanism proposed in QPT (Forbus, 1984) to describe the relationships between variables. Influences can be of two types, *indirect influences* or *direct influences*. Next, there is a description of both, following the notation used in (Kuipers, 1994):

- *Indirect Influences.* An indirect influence $Q^+(y, x_i) \equiv P^+(y, x_i)$ express a functional relationship f between two variables, x_i and y, in such a way that:

$$Q^+(y, x_i) \equiv P^+(y, x_i) \equiv \ y = f(\ldots, x_i, \ldots) \text{ and } \frac{\partial f}{\partial x_i} > 0 \qquad (5.9)$$

They were also named as "Qualitative Proportionality relations". Hence the initials Q or P. The original notation from Forbus for indirect notations was: $y \propto_{Q^+} x_i$

- *Direct Influences.* A direct influence $I^+(y, x_i)$ also represent a direct relationship between two variables, x_i and y, but the influence affects the derivative of y (i.e.

the rate of change of a quantity). Therefore, only changes the magnitude of y when integrated over time. Multiple direct influences are combined additively:

$$I^+(y, x_i) \equiv dy/dt = \text{sum}(\ldots, x_i, \ldots) \tag{5.10}$$

Mathematical Functions

Kuipers (1984, 1994) propose to represent the relationship among qualitative variables by *qualitative constraints*. The set of *constraints* involve the qualitative variables and is represented by *mathematical functions expressing different relations* (monotonic relationship, algebraic, derivative, etc.). Each variable must appear in some constraint. The basic different constraints can be:

- *Monotonic Functions*. They express a monotonically increasing (M^+) or decreasing (M^-) relationship between two variables x_i and y:

$$M^+(x_i, y), \quad y = f(x_i) \quad \text{and} \quad f \in M^+ \tag{5.11}$$

- *Additive Functions*: They express an additive relationship between three variables:

$$x(t) + y(t) = z(t) \tag{5.12}$$

- *Multiplicative Functions*: They express a multiplicative relationship among three variables:

$$x(t) * y(t) = z(t) \tag{5.13}$$

- *Changing Sign Functions*: They express an opposite relationship in the magnitude of two variables, x and y:

$$y(t) = -x(t) \tag{5.14}$$

- *Derivative Functions*: They express a derivative relationship between two variables, x and y:

$$\frac{d(x(t))}{dt} = y(t) \tag{5.15}$$

- *Constant Functions*: They express a constant derivative value for a variable x:

$$\frac{d(x(t))}{dt} = 0 \tag{5.16}$$

5.5.4 Qualitative Model Simulation

The second main task in a qualitative system is the qualitative model execution, i.e. the *qualitative model simulation*. The simulation of a model computes the dynamic temporal evolution of a system, from an initial state or states.

Although the existing approaches and frameworks to qualitative simulation are different, they share the same important concepts: the *qualitative state* of the system, the temporal component (*time variable*), and the *qualitative behaviour* of the system. Let us review, the definitions of these concepts:

- *Qualitative State of the System*: The qualitative state of a system (QS) in a determined instant of time is the aggregation of the qualitative values (QV), at this concrete instant time, of all the variables describing the system. More formally, we can define:

$$QS(System(t), t_i) = \langle QV(v_1(t), t_i), QV(v_2(t), t_i), \ldots, QV(v_n(t), t_i) \rangle \quad (5.17)$$

where,

$$System(t) = \{v_1(t), v_2(t), \ldots, v_n(t)\} \quad (5.18)$$

- *Time*: The time variable is a continuous measure, but taking into account that the qualitative values of the variables describing a system have only a finite number of values (i.e. those determined by the quantity space), they partition the time in qualitative time intervals. The set of all important points (limit point or landmark value) of the quantity space of a qualitative variable, generate a discretization of time for that variable. Each variable generates a time discretization. All these time discretizations are consistently merged in the generation of the successors of one state at a given time point.

  Therefore, given a merged temporal discretization:

$$t_0 < t_1 < \cdots < t_f$$

  The temporal evolution of the system can be examined through the qualitative state of the system at the corresponding qualitative points of time:

$$QS(System(t), t_i), i \in 0, \ldots, f$$

  This way, a qualitative state is potentially representing an infinite number of quantitative states.

- *Qualitative Behaviour of the System*: The qualitative behaviour of a system (QB) is composed and described by the temporal sequence of the qualitative states of the system during a determined time period. Formally, qualitative behaviour of a system is:

$$QB(\text{System}(t), (t_0, t_f)) =$$
$$\langle \text{QS}(\text{System}(t), t_0), \text{QS}(\text{System}(t), t_1), \ldots, \text{QS}(\text{System}(t), t_f) \rangle \tag{5.19}$$

Usually, the different qualitative simulation frameworks in the literature generate the qualitative behaviours of a system, in two ways:

- *A Behaviour Tree*: A behaviour tree is a tree formed by qualitative states, which each one is the successor of his father state (node). Therefore, each possible path from the root to leave is a possible qualitative behaviour.
- *An Envisionment*: An envisionment is a state-transition graph of the possible transitions among the qualitative states of the system. The qualitative behaviours are implicitly described by the state-transition paths. Total envisionments and attainable envisionments can be distinguished. In a *total envisionment*, all possible states of the system and all possible transitions among them are generated in the graph. An *attainable envisionment* is the subset of states in the total envisionment reachable from a given initial state or states.

See the bottom of Fig. 5.32, which illustrates a behaviour tree and an envisionment.

Some approaches define the concept of *history*, which is a local evolution of one object through the time of the simulation. The union of all the histories of the different objects form the qualitative behaviour of the system.

5.5.5   Main Qualitative Reasoning Frameworks

The origins of qualitative reasoning (qualitative physics) are the de Kleer's research work on how qualitative and quantitative knowledge interacted in solving simple mechanics problems (de Kleer, 1977). After that, several researchers developed some approaches to the qualitative reasoning main issues: qualitative descriptions of the world (*qualitative model*), and qualitative reasoning techniques (*qualitative simulation*). Main qualitative reasoning approaches and frameworks developed in the research community are the following ones:

- *Qualitative Physics with Confluences* (de Kleer & Brown, 1984): Their approach is derived from qualitative physics. The qualitative variables are expressed with qualitative values corresponding to the quantity space defined for each variable. Thus, $[\cdot]_Q$ is used to indicate the qualitative value of the expression within the brackets with respect to quantity space Q. *Confluences* are a representation for a QDE. A *confluence* joints multiple competing tendencies or influences of several variables, i.e. hence, the name of *confluence*. They are equations in the extended sign algebra $S' = \{-, 0, +, ?\}$ or qualitative expressions evaluating to signs, such as the combinations of some qualitative operators like $[x]_0$, $[x]_*$, $[\dot{x}] = [dx/dt]$. For instance, a confluence expressing the qualitative relationship between the pressure across the valve (P), the area available for flow (A), and the flow through the valve (Q) would be $[dP/dt] + [dA/dt] = [dQ/dt]$. Also, some *qualitative calculus* to operate with the confluences and the qualitative variables is defined. To create the qualitative models, the device topology and some component libraries are integrated, too. The qualitative model reasoning provides behaviour predictions and causal explanations. Qualitative behaviours are represented by an *envisionment*. The simulation is frequently done under the assumption that the system is always at a point of stable equilibrium or a very near state, i.e. a *quasi-equilibrium assumption*. This environment was implemented in the ENVISION program (de Kleer & Brown, 1984).

- *Qualitative Process Theory (QPT)* (Forbus, 1984): This theory was originated as a theory for human common sense reasoning about physical systems, focussing on the generation of *causal explanations*. It can be considered as one of the first methods in AI for integrating automatically influence models from a substantial *library of model fragments*. Model fragments can be automatically instantiated with objects from a given scenario, having explicit conditions for their activation. This library can include two types of model fragments: *individual views* and *processes*. Individual views describe subsets of the given scenario from particular perspectives. Processes describe interactions among objects and causes of change. The qualitative representation of the variables is described with *quantity spaces*. A quantity space is a set of ordered qualitative important values. Usually, it includes the *zero* value (0), and depending on the variable range, the values $+\infty$ or/and $-\infty$. The *qualitative relationship* among variables is expressed by means of *influences*. Influences can be direct or indirect. Direct influences affect the rate of change of a quantity $(I^+(y, x_i), I^-(y, x_i))$. Indirect influences express a functional relationship between two variables $(P^+(y, x_i), P^-(y, x_i))$. They were also named as qualitative proportionality relations. In addition, there are *correspondences* between qualitative values of different variables. Finally, to express constraints on the corresponding magnitudes or derivatives of two variables, *inequalities* (using the operators $<, \leq, =, >, \geq$) are used. Therefore, the qualitative models are composed of the quantity spaces, continuous quantities, a set of influences, a set of correspondences, and a set of inequalities gathered from the model fragments. The behavioural predictions are expressed as an envisionment. This approach was implemented in the GIZMO software (Forbus, 1984) and updated by the QPE software (Forbus, 1990, 1992).

- *Technical Constraint Propagation* (Williams, 1984, 1986): In this approach, in addition to the qualitative representation of continuous magnitudes, the qualitative relationships are expressed as a set of Temporal Constraint Propagation (TCP). These sets of TCP are derived from a component–connection description of a physical scenario (mainly physical circuits), in a similar way to confluences. The TCP simulation model is based on the mathematics of continuous functions. The qualitative states and the variable time are represented in different ways than the other approaches. This environment was deployed in the MINIMA system (Williams, 1988).

- *Qualitative Simulation (QSIM) Framework* (Kuipers, 1986, 1994): This framework is more focussed on the qualitative representation of the QDE model, and in the qualitative simulation tasks, rather than in the model building. Model building is assumed to be done in a hand-crafted way by the observation of the given scenario and using the background knowledge of the modeller. The qualitative model, i.e. the QDE, is formed by a *set of qualitative variables*, a *set of quantity spaces*, a *set of constraints* and a *set of transitions*. The *qualitative variables* are "reasonable functions"[9] of time, and are represented as a pair of values describing the qualitative magnitude from a quantity space (*qmag*), and the direction of change (*qdir*) of the variable: $QV(v(t)) = <qmag, qdir>$, where $qmag \in$ QuantitySpace($v(t)$), and $qdir \in \{inc, dec, std\}$. Each variable has its own *quantity space*. The *set of constraints* involve the qualitative variables and can be of different types (monotonic relationship, algebraic, derivative, etc.). Each variable must appear in some constraint. The *set of transitions* are the rules defining the boundary of the domain of applicability of the ODE model. The *qualitative simulation* process, starting with the QDE and the description of an initial state or states, generates the possible behaviours consistent with the QDE constraints, using a constraint satisfaction strategy with pre-filtering tasks. The result of the simulation is a *behaviour tree*. This QSIM framework was deployed in (Kuipers, 1986, 1994).

5.5.6 An Example of a Qualitative Reasoning Model

To illustrate the qualitative reasoning model techniques, we will analyze the well-known scenario in physics of a U-tube. A U-tube system is a simple two-tank fluid-flow system. The system consists of two tanks, naming them A (the one on the left) and B (the one on the right), connected by a flow channel or pipe, as depicted in Fig. 5.35. Each one has a maximum capacity for containing an amount of fluid (water in our case), A-MAX and B-MAX, respectively. The momentum of the water flow in the pipe is not considered.

We will illustrate the basic steps using a qualitative reasoning approach: qualitative model formulation, qualitative model representation, and qualitative simulation. The example will be solved by taking the well-known QPT approach (Forbus, 1984).

---

[9]A reasonable function over $[a, b]$, as defined in Kuipers (1994, Chap. 3, page 40), is a function satisfying: it is continuous on $[a, b]$, differentiable in (a, b), has only finitely many critical points in any bounded interval, and the one-sided limits $\lim_{t \to a^+} f'(t)$ and $\lim_{t \to b^-} f'(t)$ exist in $\Re^*$, and $f'(a)$ and $f'(b)$ are equal to these limits.

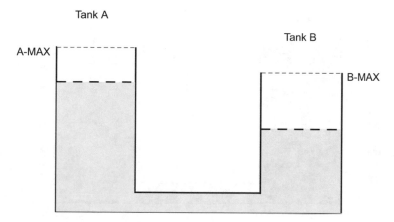

Fig. 5.35 Description of a U-tube system

5.5.6.1   Qualitative Model Formulation

Following the QPT procedures, a scenario analysis must be done, in addition to search for reusable ontologies, or parts of ontologies, in an ontology catalogue, and finally, search for instantiating model fragments from an existing library of model fragments.

Joining all the pieces, the model will be formulated.

Regarding the *scenario analysis*, it must provide the description of the system under consideration, and specifically, the initial state of the system. From the observation of the system, it is clear that each tank stores a *certain amount of water*. The amount of water in each tank produces *a certain pressure* at the bottom of each tank. There will be some *flow* from one tank to the other. Each tank has a maximum amount of water (A-MAX, B-MAX).

In addition, let us suppose that the initial state of the system we are interested and we will analyze its behaviour through the simulation will be that the tank A is full and the tank B is empty. This information coming from the scenario analysis is synthesized in Fig. 5.36.

From the object ontology described in Fig. 5.34, we need to model several objects present in our scenario (two cylindrical tanks and a pipe), and from some substance's ontology, we would need the information about water, as a liquid substance, which is contained in the tanks. Thus, these ontology components will be used in our qualitative model, as depicted in Fig. 5.37.

Therefore, we will need these pieces from these ontologies. Furthermore, from a model fragment library, we will use both *static model fragments* and *dynamic model fragments*. *Static model fragments* describe that a cylindrical tank, which is a tank and a container object, contains a liquid, which is water. Also, water, as a liquid has two properties, like the amount of water, and the pressure that water generates at the bottom of the tank. We will have one of these static model fragments for each tank in our scenario.

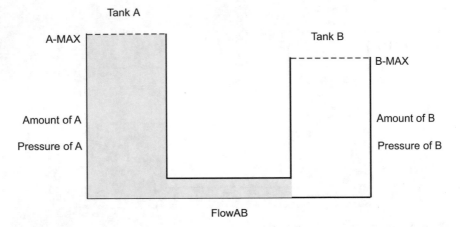

Fig. 5.36 Scenario analysis for the U-tube system

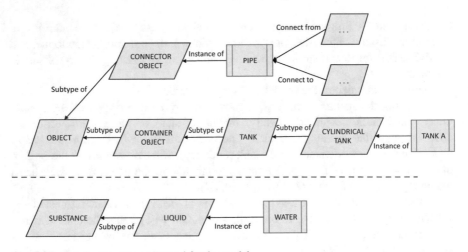

Fig. 5.37 Ontology components used for the model

In Fig. 5.38, there is the static model fragment for one tank. On the left, there is the complete static model, and o the right there is a simpler model, with only the instances of the objects.

The *dynamic model fragment* will need to represent the dynamic evolution of our process, which in this example is the flow of water from one tank to the other, through a pipe connecting both tanks. Therefore, the resulting dynamic process modelling the evolution of the system, integrated the two static model fragments (one for each tank), with the dynamic elements of the system. It is depicted in Fig. 5.39. The two static models integrated are wrapped with a dashed rectangle.

Once the model has been formulated, the qualitative representation both for the variables and for the relationships among them must be built.

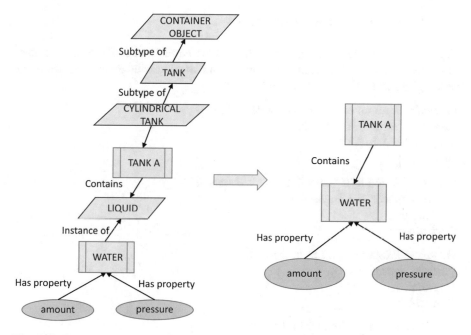

Fig. 5.38 Static model fragment for one tank

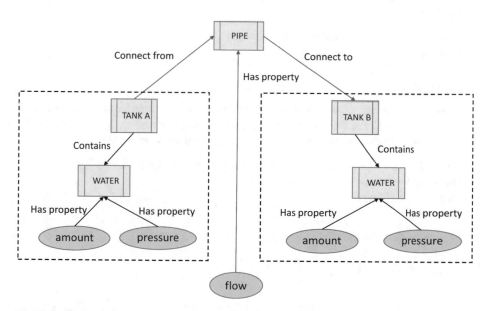

Fig. 5.39 Dynamic model fragment for the given scenario

5.5.6.2   Qualitative Model Representation

Thus, here in our scenario, several entities must be modelled. The main variables outlined both from the scenario analysis and the knowledge coming from the ontologies and model fragments are:

Variable: Amount of water in the tank A (*amtA*)
Quantity Space: $0 < $ A-MAX $ < \infty$
Magnitude: A-MAX
Derivative: −

Variable: Amount of water in the tank B (*amtB*)
Quantity Space: $0 < $ B-MAX $ < \infty$
Magnitude: 0
Derivative: +

Variable: Pressure of water in the tank A (*pressA*)
Quantity Space: $0 < \infty$
Magnitude: $(0, \infty)$
Derivative: −

Variable: Pressure of water in the tank B (*pressB*)
Quantity Space: $0 < \infty$
Magnitude: 0
Derivative: +

Variable: Flow from tank A to B (*flowAB*)
Quantity Space: $-\infty < 0 < \infty$
Magnitude: $(0, \infty)$
Derivative: −

Also, the difference of pressure between both tanks is important and, as discussed below, related to the flow from one tank to the other (*flowAB*). In addition, as the U-tube is a closed system, the amount of water is always constant.

Variable: The pressure difference between both tanks (*dpressAB*)
Quantity Space: $-\infty < 0 < \infty$
Magnitude: $(0, \infty)$
Derivative: −

Variable: Total amount of water (*total*)
Quantity Space: $0 < \infty$
Magnitude: $(0, \infty)$
Derivative: 0

The qualitative relationships between the variables must be make explicit. As in QPT, the relationships are described with influences, let us review the different influences observed in the system:

Indirect Influences or Proportionalities

There is a positive indirect influence or proportionality between *amtA* and *pressA*, because as more amount of water is in the tank A, more pressure receives the bottom of the tank A:

$$P^+(pressA, amtA)$$

There is a positive indirect influence or proportionality between *amtB* and *pressB*, because as more amount of water is in the tank B, more pressure receives the bottom of the tank B:

$$P^+(pressB, amtB)$$

There is a positive indirect influence or proportionality between *pressA* and *flowAB*, because as more pressure suffers the tank A, more flow from A to B will be generated:

$$P^+(flowAB, pressA)$$

There is a negative indirect influence or proportionality between *pressB* and *flowAB*, because as more pressure suffers the tank B, less flow from A to B will be generated:

$$P^-(flowAB, pressB)$$

Direct Influences

There is a positive direct influence between (*amtB*) and the flow from A to B (*flowAB*), because the increase rate of the amount of water in tank B, it is the flow from A to B:

$$I^+(amtB, flowAB)$$

There is a negative direct influence between (*amtA*) and the flow from A to B (*flowAB*), because the decrease rate of the amount of water in tank A, it is the flow from A to B:

$$I^-(amtA, flowAB)$$

Correspondences

In addition, some correspondences between some qualitative values of one variable and another one can be obtained:

When the *amtA* is zero then the *pressA* is 0
When the *amtA* is ∞ then the *pressA* is ∞
When the *amtB* is zero then the *pressB* is 0
When the *amtB* is ∞ then the *pressB* is ∞
When the *dpressAB* is 0 then the *flowAB* is 0
When the *dpressAB* is ∞ then the *flowAB* is ∞
When the *dpressAB* is −∞ then the *flowAB* is −∞

Inequalities

Finally, some inequalities and equalities can be deduced from the model like the following ones:

amtA + *amtB* = *total*
pressA - *pressB* = *dpressAB*
amtA ≥ 0
amtB ≥ 0
A-MAX > *B-MAX*
d(amtA)/dt < 0
d(amtB)/dt > 0

With all this new information, a final qualitative model can be represented adding all the qualitative variable information as well as the influence and inequality relations. The qualitative model is depicted in Fig. 5.40.

5.5.6.3   Qualitative Model Simulation

The qualitative model simulation will exploit the qualitative knowledge of the model to get some useful insight about the scenario given an initial state or states. In the QPT approach, the qualitative behaviour of the system, described from the qualitative values of each variable, is implicitly computed when all the transitions among possible states of the system are computed. The final *transition graph* is an envisionment of all the possible transitions among the qualitative states of the system.

Therefore, the main tasks are finding all the possible states and finding all consistent transitions among states. They are interrelated tasks, because getting some possible transitions, leads to know a new possible state, and so on.

The *qualitative states*, can be determined from the observation of the scenario, from the model fragments, and examining model elements that may change leading to the ending of one qualitative state, and thus, generating a new one. The selection

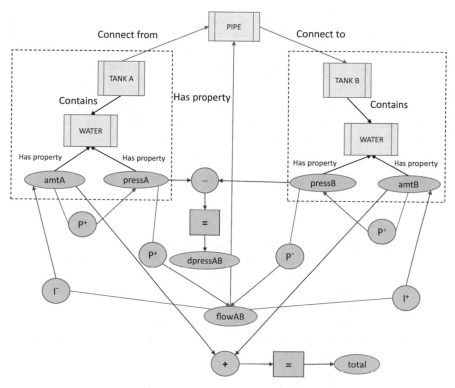

Fig. 5.40 Final qualitative model for the U-tube system

and application of model fragments, means checking for some conditions and applying the consequences.

The *qualitative transitions* mainly are obtained by using the dynamic parts of the model, *solving the influences* of the model, and applying *inequality reasoning*. Inequality reasoning may generate new inequality relations. From the set of influences, set of inequalities and the observation from external agents of the model, the different transitions are obtained.

Starting from an initial state (S_1) with the tank A full of water (*amtA* = A-MAX > 0) and the tank B empty (*amtB* = 0), is clear that the *pressA* > *pressB*, being *pressB* = 0 and *pressA* ≠ 0. Therefore, as *dpressAB* ≠ 0, then the *flowAB* ≠ 0. These means that the water will start to flow from tank A to tank B. This new situation is a qualitatively different state (S_2) from previous S_1. Now, in S_2, the tank A has some amount of water, but sure less than A-MAX (0 < *amtA* < A-MAX), and tank B is being filled with water. Thus, tank B has some amount of water (0 < *amtB* < B-MAX). As the *pressA* > *pressB*, it means *dpressAB* > 0, and hence, *flowAB* > 0.

At this stage, two interesting and qualitatively different points can be reached. Either *an equilibrium state* is reached (*pressA* = *pressB*), and then, *dpressAB* = 0, and hence, *flowAB* = 0, meaning that no more water flows from one tank to the other, or the water in tank B has reached B-MAX and water continues flowing from tank A to tank B, because still *pressA* > *pressB*, provoking the water overflows in tank B. Thus, three possible new states can be reached, i.e. three possible transitions will be possible from S_2:

- The *equilibrium state* is reached with *amtB* < B-MAX (new state S_3).
- The *equilibrium state* is reached just with *amtB* = B-MAX (new state S_4).
- The *tank B overflows*, because the equilibrium state should be reached with more water than the available capacity of the tank B (new state S_5).

Once tank B overflows (S_5), then, the equality given as a closed system:

$$amtA + amtB = total \qquad\qquad (5.20)$$

is not anymore true, and then, the system has out passed the applicability region of the model (i.e. the QDE). Then, a new qualitative state from a probably different qualitative model (QDE) should be reached. This phenomenon is modelled with *transition rules* in the QSIM framework (Kuipers, 1994) which allows to resume the simulation in another operating region of the system.

In our example, the new operating region would be related to the new scenario after the tank B overflows. Now, the amount of water in our U-tube system has diminished to a new constant value *total'*, such that:

$$amtA + amtB = total' \quad \text{and} \quad 0 < total' < total \qquad (5.21)$$

being *total − total'* the quantity overflowed in tank B. In usual conditions, tank B will burst.

Of course, in a symmetric analysis, if we start with the tank A empty and the tank B full (new state S_6), the symmetric states to S_2, S_3, S_4, and S_5 can be generated (S_7, S_8, S_9, and S_{10}).

In addition to all these states, there is another possible state (S_{11}) when both tanks are empty. Of course, if we were in this state, there is no possible transition generated from it, because there is no possible flow of water between both tanks to change the qualitative state. In addition, there is no possibility to reach this state from any other one, because no transition will empty both tanks. Therefore, this new possible state S_{11} will be an isolated state in the state transition graph.

The possible qualitative states obtained from the model fragments and discussed above are the following 11:

S_1:   Tank A is full of water (*amtA* = A-MAX), Tank B is empty (*amtB* = 0).
S_2:   Tank A has some amount of water (*amtA* < A-MAX), Tank B has some amount of water less than B-MAX (*amtB* < B-MAX). The flow between A and B is positive (*dpressAB* > 0 and *flowAB* > 0).
S_3:   Tank A has some amount of water (*amtA* < A-MAX), Tank B has some amount of water less than B-MAX (*amtB* < B-MAX), but the flow between A and B has

stopped ($dpressAB = 0$ and $flowAB = 0$). Therefore, the system has reached the *equilibrium state*.

S_4:   Tank A has some amount of water ($amtA <$ A-MAX), Tank B has an amount of water equal to the maximum ($amtB =$ B-MAX), but the flow between A and B has stopped (dpressAB = 0 and flowAB = 0). Therefore, the system has reached the *equilibrium state*.

S_5:   Tank A has some amount of water ($amtA <$ A-MAX), Tank B has an amount of water equal to the maximum ($amtB =$ B-MAX), but the flow between A and B is positive ($dpressAB > 0$ and *flowAB* > 0). Therefore, the *tank B overflows*.

S_6:   Tank B is full of water ($amtB =$ B-MAX), Tank A is empty ($amtA = 0$).

S_7:   Tank B has some amount of water ($amtB <$ B-MAX), Tank A has some amount of water less than A-MAX ($amtA <$ A-MAX). The flow between A and B is negative ($dpressAB < 0$ and *flowAB* < 0).

S_8:   Tank B has some amount of water ($amtB <$ B-MAX), Tank A has some amount of water, less than A-MAX ($amtA <$ A-MAX), but the flow between B and A has stopped ($dpressAB = 0$ and *flowAB* $= 0$). Therefore, the system has reached the *equilibrium state*.

S_9:   Tank B has some amount of water ($amtB <$ B-MAX), Tank A has an amount of water equal to the maximum ($amtA =$ A-MAX), but the flow between B and A has stopped ($dpressAB = 0$ and *flowAB* $= 0$). Therefore, the system has reached the *equilibrium state*.

S_{10}: Tank B has some amount of water ($amtB <$ B-MAX), Tank A has an amount of water equal to the maximum ($amtA =$ A-MAX), but the flow between A and B is negative ($dpressAB < 0$ and *flowAB* < 0). Water flows from B to A. Therefore, the *tank A overflows*.

S_{11}: Both Tank A ($amtA = 0$) and Tank B are empty ($amtB = 0$).

The possible transitions among states outlined above are the following:

$S_1 \rightarrow S_2$
$S_2 \rightarrow S_3$
$S_2 \rightarrow S_4$
$S_2 \rightarrow S_5$
$S_6 \rightarrow S_7$
$S_7 \rightarrow S_8$
$S_7 \rightarrow S_9$
$S_7 \rightarrow S_{10}$

S_{11} is not reachable from any other state.

Once all possible states are computed and all possible consistent transitions are obtained, the *transition graph* or *envisionment* is built. The qualitative behaviours are implicit in the possible paths. The transition graph is depicted in Fig. 5.41.

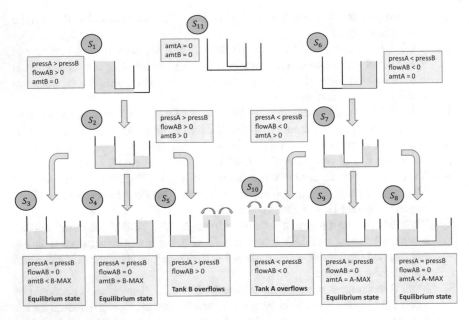

Fig. 5.41   State transition graph for the U-tube

References

Abbott, K. (1988). Robust operative diagnosis as problem solving in a hypothesis space. In *Proceedings of AAAI-88* (pp. 369–374).

Abreu, R., Außerlechner, S., Hofer, B., & Wotawa, F. (2015). Testing for distinguishing repair candidates in spreadsheets—the Mussco approach. In *ICTSS. lecture notes in computer science* (Vol. 9447, pp. 124–140). Springer.

Allen, J. (1987). *Natural language understanding* (p. 1987). Benjamin Cummings.

Bratman, M. (1987). *Intention, plans, and practical reason*. Harvard University Press.

Bredeweg, B., & Salles, P. (2009). Qualitative models of ecological systems – Editorial introduction. *Ecological Informatics, 4*(5–6), 261–262.

Bresciani, P., Perini, A., Giorgini, P., Giunchiglia, F., & Mylopoulos, J. (2004). Tropos: An agent-oriented software development methodology. *Autonomous Agents and Multi-Agent Systems, 8,* 203–236.

Brooks, R. A. (1991). Intelligence without representation. *Artificial Intelligence, 47,* 139–159.

Buchanan, B. G., & Duda, R. O. (1983). Principles of rule-based expert systems. In M. C. Yovits (Ed.), *Advances in computers* (Vol. 22, pp. 163–216). Academic Press.

Caire, G., Wim Coulier, W., Garijo, F., Gomez, J., Pavon, J., Leal, F., Chainho, P., Kearney, P., Stark, J., Evans, R., & Massonet, P. (2002). Agent oriented analysis using message/UML. In M. J. Wooldridge, G. Weiß, & P. Ciancarini (Eds.), *Agent-oriented software engineering II (AOSE 2001)* (Lecture notes in computer science) (Vol. 2222). Springer.

Cascio, F., Console, L., Guagliumi, M., Osella, M., Panati, A., Sottano, S., & Dupre, D. T. (1999). Generating on-board diagnostics of dynamic automotive systems based on qualitative models. *AI Communications, 12*(1/2).

Cioaca, E., Linnebank, F., Bredeweg, B., & Salles, P. (2009). A qualitative reasoning model of algal bloom in the Danube delta biosphere reserve (ddbr). *Ecological Informatics, 4*(5–6), 282–298.

Console, L., Dupre, D. T., & Torasso, P. (1991). On the relationship between abduction and deduction. *Journal of Logic and Computation, 1*(5), 661–690.

Console, L., Friedrich, G., & Dupre, D. T. (1993). Model-based diagnosis meets error diagnosis in logic programs. In *Proceedings 13th international joint conference on artificial intelligence* (pp. 1494–1499).

Console, L., & Torasso, P. (1990). Integrating models of correct behavior into abductive diagnosis. In *Proceedings of the European conference on artificial intelligence (ECAI)* (pp. 160–166). Pitman Publishing.

Conte, R., & Gilbert, N. (1995). Computer simulation for social theory. In N. Gilbert & R. Conte (Eds.), *Artificial societies: The computer simulation of social life* (pp. 1–15). UCL Press.

Davis, R. (1980). Meta-rules: Reasoning about control. *Artificial Intelligence, 15*(3), 179–222.

de Kleer, J. (1977). Multiple representations of knowledge in a mechanics problem solver. In *Proceedings of IJCAI-77* (pp. 299–304).

de Kleer, J. (1986). An assumption-based TMS. *Artificial Intelligence, 28*, 127–162.

de Kleer, J., & Brown, J. S. (1984). A qualitative physics based on confluences. *Artificial Intelligence, 24*, 7–83.

de Kleer, J., & Williams, B. C. (1987). Diagnosing multiple faults. *Artificial Intelligence, 32*(1), 97–130.

DeLoach, S. A. (2001). Analysis and design using MaSE and agentTool. In *Proc. of 12th Midwest Artificial Intelligence and Cognitive Science Conference (MAICS 2001)*. Miami University, Oxford.

Falkenhainer, B., & Forbus, K. (1991). Compositional modeling: Finding the right model for the job. *Artificial Intelligence, 51*, 95–143.

Ferguson, I. A. (1992). Towards an architecture for adaptive, rational, mobile agents. In E. Werner & Y. Demazeau (Eds.), *Decentralized AI 3 – Proceedings of the third European Workshop on modeling autonomous agents and multi agent worlds (MAAMAW-91)* (pp. 249–262). Elsevier.

Finin, T., Weber, J., Wiederhold, G., Genesereth, M., Fritson, R., McKay, D., McGuire, J., Pelavin, R., Shapiro, S., & Beck, C. (1993). Specification of the KQML agent communication language. In *DARPA knowledge sharing initiative external interfaces working group*.

FIPA (1999). Specification part 2 – Agent communication language. The text refers to the specification dated 16 April 1999.

Forbus, K. (1984). Qualitative process theory. *Artificial Intelligence, 24*, 85–168

Forbus, K. (1990). The qualitative process engine. In D. Weld & J. de Kleer (Eds.), *Readings in qualitative reasoning about physical systems*. Morgan Kaufman.

Forbus, K. (1992). Pushing the edge of the (QP) envelope. In B. Faltings & P. Struss (Eds.), *Recent advances in qualitative physics*. MIT Press.

Forbus, K. (1996). Qualitative reasoning. In *The handbook of computer science*. CRC.

Forgy, C. L. (1982). Rete: A fast algorithm for the many pattern/many object pattern match problem. *Artificial Intelligence, 19*, 17–37.

Friedrich, G., Gottlob, G., & Nejdl, W. (1990a). Hypothesis classification, abductive diagnosis and therapy. In *First international workshop on principles of diagnosis* (July 1990), also appeared in *Proceedings of the international workshop on expert systems in engineering, lecture notes in artificial intelligence* (Vol. 462, September 1990). Springer.

Friedrich, G., Gottlob, G., & Nejdl, W. (1990b). Hypothesis classification, abductive diagnosis and therapy. In *Proceedings of the international workshop on expert systems in engineering* (Lecture notes in artificial intelligence, 462). Springer.

Friedrich, G., Stumptner, M., & Wotawa, F. (1999). Model-based diagnosis of hardware designs. *Artificial Intelligence, 111*(2), 3–39.

Genesereth, M. R., & Fikes, R. E. (1992). Knowledge interchange format, version 3.0 reference manual. In *Technical report Logic-92-1*. Computer Science Department, Stanford University.

Georgeff, M. P., & Lansky, A. L. (1987). Reactive reasoning and planning. In *Proceedings of the sixth national conference on artificial intelligence (AAAI-87)* (pp. 677–682).

Gonzalez, A., & Dankel, D. (1993). *The engineering of knowledge-based systems. Theory and practice*. Prentice Hall.

Gray, C. S., Roxane Koitz, S. P., & Wotawa, F. (2015). An abductive diagnosis and modeling concept for wind power plants. In *9th IFAC symposium on fault detection, supervision and safety of technical processes*.

Greiner, R., Smith, B. A., & Wilkerson, R. W. (1989). A correction to the algorithm in Reiter's theory of diagnosis. *Artificial Intelligence, 41*(1), 79–88.

Hayes-Roth, F., Waterman, D. A., & Lenat, D. B. (Eds.). (1983). *Building expert systems*. Addison-Wesley Pub.

Heller, U., & Struss, P. (2001). Transformation of qualitative dynamic models – Application in hydroecology. In L. Hotz, P. Struss, & T. Guckenbiehl (Eds.), *Intelligent diagnosis in industrial applications* (pp. 95–106). Shaker Verlag.

Heller, U., & Struss, P. (2002). Consistency-based problem solving for environmental decision support. *Computer-Aided Civil and Infrastructure Engineering, 17*, 79–92. ISSN 1093-9687.

Hofer, B., Hofler, A., & Wotawa, F. (2017a). Combining models for improved fault localization in spreadsheets. *IEEE Transactions on Reliability, 66*(1), 38–53.

Hofer, B., Nica, I., & Wotawa, F. (2017b). AI for localizing faults in spreadsheets. In *ICTSS. Lecture notes in computer science* (Vol. 10533, pp. 71–87). Springer.

Huhns, M. N., & Stephens, L. M. (1999). Multi-agent systems and societies of agents. In G. Weiss (Ed.), *Multi-agent systems* (pp. 79–120). MIT Press.

Iglesias, C. A., Garijo, M., & Gonzalez, J. C. (1998). A survey of agent-oriented methodologies. In *Proceedings of the workshop on agent theories, architectures and languages*.

Jackson, P. (1999). *Introduction to expert systems* (3rd ed.). Addison-Wesley.

Jennings, N. R., Faratin, P., Lomuscio, A. R., Parsons, S., Sierra, C., & Wooldridge, M. (2001). Automated negotiation: Prospects, methods and challenges. *International Journal of Group Decision and Negotiation, 10*, 199–215.

Kansou, K., Nuttle, T., Farnsworth, K., & Bredeweg, B. (2013). How plants changed the world: Using qualitative reasoning to explain plant macroevolution's effect on the long-term carbon cycle. *Ecological Informatics, 17*, 117–142. https://doi.org/10.1016/j.ecoinf.2013.02.004

Kleene, S. C. (1967). *Mathematical logic*. John Wiley & Sons, Republication by Dover Publication. in 2002.

Koitz, R., & Wotawa, F. (2015). On the feasibility of abductive diagnosis for practical applications. In *9th IFAC symposium on fault detection, supervision and safety of technical processes*.

Kuipers, B. (1984). Common sense reasoning about causality: Deriving behaviour from structure. *Artificial Intelligence, 24*, 169–204.

Kuipers, B. (1986). Qualitative simulation. *Artificial Intelligence, 29*, 289–338.

Kuipers, B. (1994). *Qualitative reasoning*. The MIT Press.

Luck, M., McBurney, P., & Preist, C. (2003). *Agent technology: Enabling next generation computing. A roadmap for agent based computing*. Agentlink II, Agentlink Community. ISBN: 0854-327886.

Malik, A., Struss, P., & Sachenbacher, M. (1996). Case studies in model-based diagnosis and fault analysis of car-subsystems. In *Proceedings of the European conference on artificial intelligence (ECAI)*.

Mangina, E. (2002). *Review of software products for Multi-Agent Systems*. Applied Intelligence (UK) Ltd for AgentLink: http://www.AgentLink.org/

Mateis, C., Stumptner, M., & Wotawa, F. (2000). Modeling java programs for diagnosis. In *Proceedings of the European conference on artificial intelligence (ECAI)*.

Mayer, W., & Stumptner, M. (2004). Debugging program loops using approximate modeling. In *Proceedings of the ECAI'04* (pp. 843–847).

Mayer, W., Stumptner, M., Wieland, D., & Wotawa, F. (2002). Can AI help to improve debugging substantially? Debugging experiences with value-based models. In *Proceedings of the European conference on artificial intelligence (ECAI)* (pp. 417–421). IOS Press.

McCarthy, J. (1979). Ascribing mental qualities to machines. Technical report, Computer Science Dept., Stanford University, Stanford, CA 94305, USA.

Minoux, M. (1988). LTUR: A simplified linear-time unit resolution algorithm for horn formulae and computer implementation. *Information Processing Letters, 29,* 1–12.

Nayak, P. (1994). Causal approximations. *Artificial Intelligence, 70,* 277–334.

Padgham, L., & Winikoff, M. (2004). *Developing intelligent agent systems: A practical guide.* Wiley.

Peischl, B., Riaz, N., & Wotawa, F. (2012). Automated debugging of verilog designs. *International Journal of Software Engineering and Knowledge Engineering, 22*(5), 695.

Peischl, B., & Wotawa, F. (2006). Automated source-level error localization in hardware designs. *IEEE Design and Test of Computers, 23*(1), 8–19.

Pell, B., Bernard, D., Chien, S., Gat, E., Muscettola, N., Nayak, P., Wagner, M., & Williams, B. (1996). A remote-agent prototype for spacecraft autonomy. In *Proceedings of the SPIE conference on optical science, engineering, and instrumentation, volume on space sciencecraft control and tracking in the new millennium.* Society of Professional Image Engineers.

Picardi, C., Bray, R., Cascio, F., Console, L., Dague, P., Dressler, O., Millet, D., Rehfus, B., Struss, P., & Vallee, C. (2002). IDD: Integrating diagnosis in the design of automotive systems. In *Proceedings of the European conference on artificial intelligence (ECAI)* (pp. 628–632). IOS Press.

Rao, A., & Georgeff, M. (1995). BDI agents: From theory to practice. In V. Lesser (Ed.), *Proceedings of the 1st international conference on multi-agent systems (ICMAS'95).* The MIT Press.

Reiter, R. (1987). A theory of diagnosis from first principles. *Artificial Intelligence, 32*(1), 57–95.

Rickel, J., & Porter, B. (1994). Automated modeling for answering prediction questions: Selecting the time scale and system boundary. In *Proceedings of AAAI-94* (pp. 1191–1198).

Sànchez-Marrè, M., Gibert, K., Sojda, R. S., Steyer, J. P., Struss, P., Rodriguez-Roda, I., Comas, J., Brilhante, V., & Roehl, E. A. (2008). Chapter 8: Intelligent environmental decision support systems. In A. Jakeman, A. Rizzoli, A. Voinov, & S. Chen (Eds.), *State of the art and futures in environmental modelling and software* (Vol. 3, pp. 119–144). Elsevier. ISBN: 978-0-08-056886-7.

Shortliffe, E. H., & Buchanan, B. G. (1975). A model of inexact reasoning in medicine. *Mathematical Biosciences, 23*(3–4), 351–379.

Smullyan, R. M. (1968). *First order logic.* Springer.

Steinbauer, G., & Wotawa, F. (2013). Model-based reasoning for self-adaptive systems - theory and practice. In *Assurances for self-adaptive systems, lecture notes in computer science* (Vol. 7740, pp. 187–213). Springer.

Struss, P. (2008). Artificial-intelligence-based modeling for environmental applications and decision support. In *Proceedings of international congress on environmental modelling and software society.* ISBN: 978-84-7653-074-0.

Struss, P. (2009). Towards model integration and model-based decision support for environmental applications. In R. S. Anderssen, R. Braddock, & L. Newham (Eds.), *Proceedings of the 18th world IMACS congress and MODSIM09 international congress on modelling and simulation. Modelling and simulation society of Australia and New Zealand and international association for mathematics and computers in simulation.* ISBN: 978-0-9758400-7-8.

Struss, P. (2011). A conceptualization and general architecture of intelligent decision support systems. In F. Chan, D. Marinova, & R. Anderssen (Eds.), *Proceedings of the 19th international congress on modelling and simulation. Modelling and simulation society of Australia and New Zealand* (pp. 2282–2288) ISBN: 978-0-9872143-1-7.

Struss, P. (2012). Model-based environmental decision support. In *Proceedings of the 3rd international conference on computational sustainability.* http://www.computational-sustain ability.org/compsust12/

Struss, P., Bendati, M., Lersch, E., Roque, W., & Salles, P. (2003). Design of a model-based decision support system for water treatment. In *Proceedings of the 18th international joint conference on artificial intelligence (IJCAI 03)- environmental decision support systems workshop* (pp. 50–59).

Struss, P., & Heller, U. (1999). Model-based support for water treatment. In R. Milne (Ed.), *Qualitative and model based reasoning for complex systems and their control, workshop KRR-4 at the 16th international joint conference on artificial intelligence* (pp. 84–90).

Struss, P., Steinbruch, F., & Woiwode, C. (2016). Structuring the domain knowledge for model-based decision support to water management in a peri-urban region in India. In *Proceedings of the 29th international workshop on qualitative reasoning (QR16)*.

Stumptner, M., & Wotawa, F. (1999). Debugging functional programs. In *Proceedings 16th international joint conference on artificial intelligence* (pp. 1074–1079).

Sudeikat, J., Braubach, L., Pokahr, A., & Lamersdorf, W. (2004). Evaluation of agent-oriented software methodologies - examination of the gap between modeling and platform. In *Workshop on agent-oriented software engineering*. AOSE.

Tansley, D. S. W., & Hayball, C. C. (1993). *Knowledge-based systems analysis and design*. Prentice-Hall.

Thrun, S., Montemerlo, M., Dahlkamp, H., Stavens, D., Aron, A., Diebel, J., Fong, P., Gale, J., Halpenny, M., Hoffmann, G., Lau, K., Oakley, C., Palatucci, M., Pratt, V., Stang, P., Strohband, S., Dupont, C., Jendrossek, L.-E., Koelen, C., . . . Mahoney, P. (2007). The robot that won the DARPA grand challenge. In M. Buehler, K. Iagnemma, & S. Singh (Eds.), *The 2005 DARPA grand challenge* (pp. 1–43). Springer-Verlag.

Thomas, C. R., Jr., & Seibel, F. (1999). Adaptive cargo routing at southwest airlines. In *Proceedings of embracing complexity conference* (pp. 73–80).

von Martial, F. (1990). Interactions among autonomous planning agents. In Y. Demazeau & J.-P. Müller (Eds.), *Decentralized AI – Proceedings of the first European workshop on Modelling Autonomous Agents in a Multi-Agent World (MAAMAW-89)* (pp. 105–120). Elsevier Science Publishers B. V.

Weiss, G. (1999). Prologue: Multi-agent systems and distributed artificial intelligence. In G. Weiss (Ed.), *Multi-agent systems: A modern approach to distributed artificial intelligence* (pp. 1–23). MIT Press.

Williams, B. C. (1984). Qualitative analysis of MOS circuits. *Artificial Intelligence, 24,* 281–346.

Williams, B. C. (1986). Doing time. Putting qualitative reasoning on firmer ground. In *Proceedings of 5th national conference on artificial intelligence* (pp. 105–112). Morgan Kaufmann.

Williams, B. C. (1988). MINIMA: A symbolic approach to qualitative algebraic reasoning. In *Proceedings of 7th national conference on artificial intelligence* (pp. 264–269). Morgan Kaufmann.

Wooldridge, M. (2009). *An introduction to multi agent systems* (2nd ed.). Wiley.

Wooldridge, M., & Jennings, N. R. (1995). Intelligent agents: Theory and practice. *Knowledge Engineering Review, 10*(2), 115–152.

Wooldridge, M., Jennings, N. R., & Kinny, D. (2000). The Gaia methodology for agent-oriented analysis and design. *Autonomous Agents and Multi-Agent Systems, 3*(3), 285–312.

Wooldridge, M. (2002). An introduction to multiagent systems (1st ed.). Wiley.

Wotawa, F. (2011). On the use of abduction as an alternative to decision trees in environmental decision support systems. *IJAEIS, 2*(1), 63–82. https://doi.org/10.4018/jaeis.2011010104

Wotawa, F. (2014). Failure mode and effect analysis for abductive diagnosis. In *Proceedings of the international workshop on defeasible and ampliative reasoning (DARe-14)*.

Wotawa, F., Nica, M., & Moraru, I. (2012). Automated debugging based on a constraint model of the program and a test case. *The Journal of Logic and Algebraic Programming, 81*(4), 390–407.

Wotawa, F., Rodriguez-Roda, I., & Comas, J. (2010). Environmental decision support systems based on models and model-based reasoning. *Environmental Engineering and Management Journal, 9*(2), 189–195.

Wotawa, F., Rodriquez-Roda, I., & Comas, J. (2009). Abductive reasoning in environmental decision support systems. In *Proceedings of the workshop on artificial intelligence applications in environmental protection (AIAEP)*.

Further Reading

Buchanan, B. G., & Shortliffe, E. H. (1984). *Rule based expert systems: The MYCIN experiments of the Stanford heuristic programming project.* Addison-Wesley.

Faltings, B., & Struss, P. (Eds.). (1992). *Recent advances in qualitative physics.* MIT Press.

Grosz, B. J., Jones, K. S., & Webber, B. L. (1986). *Readings in natural language processing.* Morgan Kaufmann.

Weiss, G. (Ed.). (1999). *Multi-agent systems: A modern approach to distributed artificial intelligence.* The MIT Press.

Weld, D. S., & De Kleer, J. (1990). *Readings in qualitative reasoning about physical systems.* Morgan Kaufmann.

Winograd, T. (1983). *Language as a cognitive process: Syntax.* Addison-Wesley.

Chapter 6
Data-Driven Intelligent Decision Support Systems

6.1 Introduction

As explained in Chap. 4, the *data-driven IDSSs* use data as the raw material to extract relevant patterns and information for the decision-making process. The quality of the IDSSs is strictly related to the quality of the data. In fact, the data are explored to obtain some inductive model, which will be used as the core element for giving support to the decision maker. Thus, at the end, the knowledge is also compiled in a model, but with the particularity that this model has been obtained from the exploration of historical data available in a concrete domain. The data-driven IDSSs belong to the category of *Data Model Exploration DSSs*, because they explore the data, not directly, like the *Direct Data Exploration DSSs* analyzed in Sect. 3.3.2.1, but through the use of a previously induced model from the data.

Furthermore, as detailed in Sect. 3.3.2.2, there are two kinds of Data Model Exploration DSSs: *Data Confirmative Model DSSs* and *Data Explorative Model DSSs*. *Data Confirmative Model DSSs* are usually statistical and numerical models like Linear Regression models, Logistic Regression models, and Markovian models. We will not enter into details on this kind of models as they are not methods from Artificial Intelligence field.

We will focus on *Data Explorative Model DSSs*, because all Data-driven IDSSs fall in this category. These methods capture the knowledge and information from data through an explorative model. All of them use a bottom-up approach, i.e. starting from the data get a data model, which will be used for supporting decisions in the IDSS. These methods obtain the model using techniques of *inductive machine learning*, which can be also grouped as belonging to the Data Mining field.

The *Data Explorative model IDSS methods* can be classified in two major groups, unsupervised models and supervised models. Unsupervised models can be further classified in Descriptive models and Associative models. On the other hand, supervised models can be further divided in Discriminant Models and Predictive models. In addition, there is another kind of models, which are Optimization Models,

© Springer Nature Switzerland AG 2022
M. Sànchez-Marrè, *Intelligent Decision Support Systems*,
https://doi.org/10.1007/978-3-030-87790-3_6

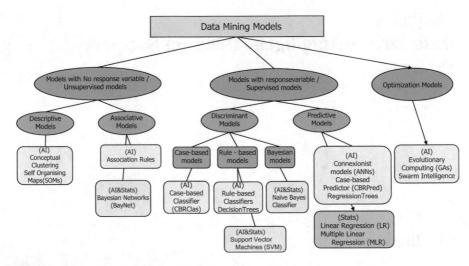

Fig. 6.1 Conceptual classification of intelligent data mining models

which can optimize the discriminant or predictive models. This taxonomy of models, with the most used families of models are depicted in Fig. 6.1.

- **Unsupervised Models:** These models aim to get some knowledge from the data (*cognitive process of obtaining knowledge patterns*), searching for relationships among elements. There is no a target or response variable being the object of interest:

  - Descriptive Models: These models search relationships among data instances or examples. Some authors in the literature name them as profiling models.
  - Associative Models: These models search relationships among random variables.

- **Supervised Models:** These models aim to get some model able to produce or predict the values of a target or response variable of interest (*recognition process of previous knowledge patterns*).

  - Discriminant Models: The variable of interest is a *qualitative variable*, which can take one of a finite set of different qualitative values (class labels). The models will be able to *discriminate* to which qualitative/class label belongs a new unseen instance, for this qualitative variable. In the literature, these models are also referred to as classification models.
  - Predictive Models: The variable of interest is a *quantitative variable*, which can take a numerical value, integer or real, among an infinite set of values. The models will be able to *predict* the quantitative value of a new unseen instance for this quantitative variable. In the literature, these models are also referred as regression models.

- Optimization Models: These models aim to optimize other models. They can be named as *second-order data mining models*, because take as input data which are previous mined models.

6.2 Data Mining, Knowledge Discovery, and Data Science

The terms of Data Mining, Knowledge Discovery, and Data Science are sometimes misunderstood and some people use them interchangeably, when they are not referring to the same process. Let us clarify each one of these terms.

Knowledge Discovery from Databases (KDD) was a term first used by Fayyad et al. (1996). Knowledge Discovery in Databases was defined as the non-trivial process of identifying valid, novel, potentially useful, and ultimately understandable patterns in data. KDD is a multi-step process, and just one of these steps is the *Data Mining* step. The KDD process is illustrated in Fig. 6.2.

The KDD process is composed by the following steps: data selection, pre-processing, data transformation, data mining, and post-processing.

Data selection is the first step where the data available must be selected according to the issue of the problem at hand. Some target data and the corresponding relevant

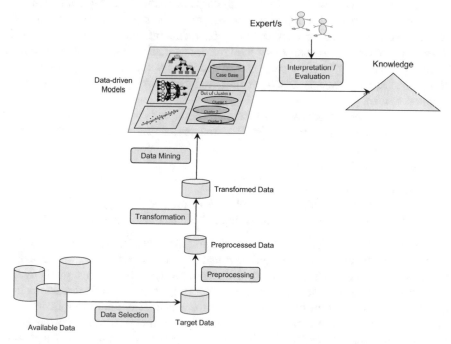

Fig. 6.2 The Knowledge Discovery process in Databases. (Adapted from Fayyad et al. (1996))

data must be collected from a more general database or databases to form the target dataset to be explored and mined to obtain the hidden knowledge patterns.

Pre-processing is the next step which consists in several subtasks to prepare the dataset previously to be analyzed with data mining techniques. As it will be described in Sect. 6.3, there are several sub-steps in the pre-processing task like meta-data analysis, visualization, descriptive statistical analysis, filtering, outlier's detection, error's detection, missing data management, etc.

Data transformation is the last step for modifying the data structure with some transformation schemes, like discretization of a numerical variable or generation of a new variable as a ratio of two previously existent variables in the dataset.

Data Mining is a step in the KDD process consisting of particular data mining algorithms that, under some acceptable computational efficiency limitations, produces a particular set of patterns, i.e. an induced model from a set of examples or data (Fayyad et al., 1996). In the Sect. 6.4, the most common data mining methods will be described.

Finally, the *post-processing* step is the last one in the chain. Once the knowledge patterns (models) have been induced from the data, they must be interpreted and validated with the help of experts on the application domain, to be sure that it can be used for further new data. The validation task is so important than usually is considered as an independent task from the post-processing step.

The post-processing tasks are described in the Sect. 6.5. Model validation is analyzed for each kind of method along the Sect. 6.4.

Last years a new term *Data Science* has emerged in the literature, but it is not the same concept than data mining. It is a broader term. *Data Science* comprises the set of principles, techniques, processes to transform data into useful patterns of knowledge to get, through the analysis of the knowledge patterns, actionable predictions which will be used to give support for decision-making in an organization or company. Data Science is based on data-analytic thinking principles plus some planned actions.

Data is a strategic asset for organizations. The issue is whether data can improve their performance and how to do it. Performance can be increased through data-analytical thinking.

Data-analytic thinking aims at:

- Extracting useful knowledge
- Identifying relevant data for some object/s of interest
- Obtaining inductive models which should be scalable to other data

An *actionable prediction* is a piece of knowledge regarding future states of a company which can be directly translated into useful actions to improve the performance of a company. Thus, the analysis is more prescriptive than just predictive.

Data Science is a *prescriptive data analysis* to generate proactive reactions in a company or organization to improve its performance. It is multidisciplinary, and include machine learning methods, visualization techniques, statistical analysis, pattern recognition strategies, decision theory, expert participation, databases, and data management.

	Name	Age	Weight	Height	Cholest-Lev	Syst-Press	Diast-Press	...	...
E_1	Mary	38	61	1.67	very-low	120	80	...	...
E_2	Mike	54	80	1.70	low	132	87	...	...
E_3	Anna	48	62	1.60	?	125	78	...	...
...	...	...	...	...	...	...	...	...	...
E_N	John	62	?	1.78	high	137	89	...	...

Fig. 6.3 An example of a medical dataset describing the information of several patients

For instance, the goal in Data Science is not just to make a good prediction whether a customer will remain loyal or will churn. The possible reactions will be analyzed and the best reaction, according to the profile of the customer, will be suggested to the decision-maker, like whether it is worth to intensify an advertising strategy for this customer or not.

6.2.1   Terminology in Data Mining

In inductive Machine Learning (ML), and hence, in Data Mining, all the data analytical process starts from a set of data. This dataset usually is organized conceptually as a two-dimensional data matrix (X of dimensions $N \times D$). The matrix is formed by N rows and D columns. Each row of the matrix corresponds to an *instance* or *example* or *observation* or *individual*,[1] like the data of a patient, the data of a customer, the data of a product, etc. Each column of the matrix refers to a characteristic of the instances or examples. These characteristics are referred as an *attribute* or a *feature* or a *dimension* or a *variable*,[2] like the age, the name, the weight or the cholesterol level of a patient, or the name, the address, the nationality, the quantity of money spent of a customer, and so on. Each cell of the matrix is the corresponding *value* (x_{ij}) of an attribute (A_j or X_j) for a given example (E_i or X_i). For instance, the value *54* is the corresponding value of *Age* attribute for the example E_2 of the dataset depicted in Fig. 6.3.

As it is shown in Fig. 6.3, the type of the attribute values can be different. Usual types are: string, numeric or quantitative (real or integer), qualitative or nominal (ordered or non-ordered). In the patient database of Fig. 6.3, the attribute *Name* has a

---

[1] All these terms referring to a row of the data matrix will be used as synonyms.

[2] All these terms referring to a column of the data matrix will be used as synonyms.

Table 6.1 Usual terminology for a SQL database in both Artificial Intelligence and Statistics field

	Artificial Intelligence	Statistics	Relational database
Dataset	Data matrix	Data	Table
Row	Instance/example	Observation/individual	Record
Column	Attribute/feature	Variable	Field
Cell	Data	Value	Value

string type, the attributes *Age* and *Weight* have a numeric (integer) type, the attribute Height has a numeric (real) type, the attribute *Cholest-Lev* has a qualitative ordered type, and the possible values are the ordered set of modalities or labels: {very-low, low, normal, high, very-high}. Finally, both *Syst-Press* and *Diast-Press* are attributes with a numeric (integer) type.

The corresponding value of the example E_3 for the *Cholest-Lev* attribute and the one of the example E_N for the *Weight* attribute is the symbol "?". This symbol is usually employed to represent an *unknown* or *missing* value. Other symbols used to represent a missing value are "null", "nil", "none", or a numerical code value which is impossible to be true for a numerical attribute like "99999" or "−99999" or similar codes. In the Sect. 6.3, were the pre-processing tasks will be detailed, the management of the missing values will be analyzed.

See the Table 6.1 for the usual terminology of a classic dataset, which can be named as SQL database,[3] both in Artificial Intelligence, Statistics, or Database field.

Since the late 1960s, the concept of NoSQL database ("Non-SQL" or "Non-relational") database existed, but they were not labelled as "NoSQL" until a growing popularity in the early twenty-first century (Leavitt, 2010). NoSQL databases provide a mechanism for storage and retrieval of data that is modelled in a different way than the tabular relations used in relational databases.

NoSQL databases are increasingly used in big data (see Sect. 6.6) and real-time web applications. NoSQL systems are also called "Not only SQL" by several people to emphasize that they may support SQL-like query languages, or coexist in addition to SQL databases in different data technologies (*polyglot persistence architecture*).

Motivations for this approach include: simplicity of design, simpler "horizontal" scaling to clusters of machines (which is a problem for relational databases) and finer control over availability. Main data structures used by NoSQL databases are, for instance, key-value structure, wide column structure, a graph structure, or a document collection structure, are different from those used by default in relational databases, making some operations faster in NoSQL.

In all the chapter, we will assume that the datasets are expressed in a two-dimensional data matrix form equivalent to a table of a relational SQL database, which is the most typical use.

Usually, all the KDD and Data Science steps can be supported by some Data Mining tools, which support all the pre-processing steps, the data mining step of

---

[3] SQL database refers to a database which is composed by a set of tabular relations in relational databases, which can be queried by SQL statements (see Sect. 3.3.2.1).

inducing the data-driven models, and the post-processing task and final validation of the models. Commonly used data mining tools are: RapidMiner, Orange, WEKA, KNIME as free data mining tools at least in some version, and IBM SPSS Modeler, Oracle Data Mining (ODM) and SAS Data Mining, as commercial tools. And the most used programming languages plus some machine learning package environments are R and Python. Some of them will be described in Chap. 8.

6.3   Pre-Processing Techniques

The pre-processing step (Pyle, 1999) is an important step in the whole KDD and Data Science concept, and specifically, is very important for the next step of data mining. It is estimated from commercial applications of KDD that the pre-processing step spends the 80% of the time in a KDD project. This means that only a 20% is devoted to the data mining and post-processing steps. Hence, the importance of the pre-processing step is outstanding.

The main goal of the pre-processing step is to collect, analyze, check, depurate, and prepare the data available. Main sub-steps of the pre-processing process (Gibert et al., 2016) are:

1. Data fusion and merge
2. Meta-data definition and analysis
3. Data filtering
4. Special variables management
5. Visualization and descriptive statistical analysis
6. Transformation and creation of new variables
7. Outlier detection and management
8. Error detection and management
9. Missing data management
10. Data reduction
11. Feature relevance determination

In the next sub-sections, we will analyze in detail each one of these steps.

6.3.1   Data Fusion and Merge

The first step in the pre-processing task is to get together different data that can come from different sources. Nowadays, some data can be originated from smart sensors capturing online *signals* which are essential data from a process. Therefore, after some *signal processing* techniques, useful signal features can be aggregated as being part of the data matrix.

In addition, in the multimedia era, useful information can be obtained from *images*. *Image processing* and feature extraction are necessary processes to extract information from images in the format of features to be added to the data matrix.

A lot of data is generated from the web. *Web mining* is an emergent field to process and extract information from *text and documents* in the web. Therefore, important features can be extracted from documents and text in the web and added to the data matrix.

Of course, most of the features in the data matrix will be extracted from available *databases*. Commonly, the information in a company or an organization is distributed into several databases. Those databases must be examined and the useful features must be *selected*, *fused* and *merged* in the final dataset. Many times, information regarding the same individual like the name of a customer, the address of the customer and the amount spent by the customer must be fused from several databases to be integrated in the same observation, i.e. the same row of the data matrix.

Other data and features could come from some *observation process* like annotating some feature which is considered important by an expert, and it is not currently monitored by any sensor, *or experimental analysis* or studies, like clinical studies on some population under study from which several data are gathered. Both *observational data* and *experimental data* can be integrated and merged to get the final data matrix or dataset to be explored with data mining methods.

6.3.2   Meta-Data Definition and Analysis

In addition to the data itself, which is organized into rows and columns in the data matrix, it is very important the definition and analysis of the information associated to the data, i.e. the *meta-data* (Mierswa et al., 2006). The meta-data is the information associated to the data describing further information about the variables or the instances themselves.

Usually, meta-data can include the following information:

- The *source of information* like the data coming from an online sensor, from an offline chemical laboratory analysis or from a subjective expert evaluation.
- The *type of the variables* like numerical or quantitative, nominal or categorical or qualitative, and in the latter case, whether it is ordered or non-ordered, or a string, etc. For the qualitative variables the set of possible qualitative values or labels or categories or modalities must be enumerated: $\{m_1, m_2, \ldots, m_Q\}$.
- The *frequency of sampling of the variable*, which can be for instance each second, each minute, each 10 min, daily, weekly, etc.
- The *unit of measurement of the variable* like m^3, m^3/s, m, cm, mg, mg/l, g, g/l, kg, kg/l, etc., which is very important to be known to avoid problems with different units of measure for the same variable due to different origin of the data.

- The *possible range of values of a numerical variable*, like for example belonging to the interval [0, 1000] because the magnitude is positive, and higher values than the upper bound of 1000 are incorrect or suspiciously erroneous.
- The *set of possible values of a qualitative variable*, like {solid, liquid, gas} as the possible labels of a substance.
- The *order between the values of an ordered qualitative variable*, like small < medium < large, for the three labels of the size of a building.
- The *degree of relevance of one variable*, i.e. the importance which is thought to have one variable in the dataset, usually expressed through a real positive numeric value or *weight* like 0.8, commonly represented in a scale belonging to the real interval [0, 1] and usually summing up to 1 for all the weights corresponding to all the variables. This information is used by several data mining methods which need to compute similarities among the different instances of the dataset.
- The *degree of importance of one instance*, i.e. the level of importance of that example in the whole database, which usually is expressed by a real positive numeric value or weight like 0.02, commonly represented in a scale belonging to the interval [0, 1] and usually summing up to 1 for all the weights corresponding to all the instances. This information is used by several data reduction and data mining methods which want to focus on the most important instances, i.e. the most difficult to be correctly discriminated or predicted by a discriminant or predictor method.
- *The degree of inexactness or uncertainty associated to a variable*, due to the real-world inexactness nature of many data sources like sensors, instrumentation devices, subjective appreciation of an expert, etc.

The metadata should be taken into account in all the data science process to avoid unexpected results and errors, and must be described in the data mining tool used.

6.3.3 Data Filtering

Data filtering is a task aiming at removing some unwanted components from raw data matrix measurements. Once data has been fused and merged to get a unique data matrix, usually a *filtering* task should be done to select the sample/s to be extracted from the original data matrix.

These means that the examples or instances to be used in the final analyzed data matrix must be *selected*. Sometimes the scope of the analysis must be restricted to some subdomain, like just only using data from last 10 years, or using only data coming from meteorological stations in European countries for a continental analysis. Some instances can be removed if they do not belong to the domain targeted in the analysis.

In addition, it is common to have some *duplicated instances* in the dataset that must be removed. Sometimes, some *inconsistency between similar or even identical instances* could exist and must be solved.

On the other hand, very probably not all the variables should be included in the final analyzed data matrix, because depending on the goals of the analysis and the decisions to be supported with it, there are some clearly irrelevant variables. Furthermore, some redundant variables can be easily detectable. For instance, after the merging process, could be the case that in a customer dataset, we have both the postal code and the name of the district/quartier in a city. Both variables are expressing very similar information about the costumer and the data scientist should avoid the duplicity of this information from the beginning.

The help of the experts at this point is very valuable, because an *expert-based selection of variables* can be done, according to the goals and the decisions involved in the problem at hand.

6.3.4   Special Variables Management

There are several kinds of special variables that commonly can appear in a dataset. These variables must be managed in a proper way, such that the data mining methods could be applied in the right way. Most common special variables are: *compositional variables* and *multi-valued variables*.

6.3.4.1   Compositional Variables

Compositional variables (Pawlowsky-Glahn & Buccianti, 2011) are a set of s strictly positive real variables among the whole set of D variables in the dataset:

$$\{X_{.1}, X_{.2}, \ldots, X_{.s}\}, \quad \text{such that}$$

$$\sum_{j=1}^{s} X_j = T, \quad T > 0 \tag{6.1}$$

T is a *total constant sum* generally representing the sum of percentages ($T = 100$) or the sum of proportions ($T = 1$) or even the sum of quantities ($T, T > 0$).

These variables are used to represent the composition or distribution of different possible values of another variable. For example, consider a dataset with the information of each *town of a country*. Each instance or observation will have the information about the population of the town, the altitude over the see, the geographical coordinates, the province or county to which it belongs, and so on. Furthermore, let us suppose that each town has the information about the percentage of population belonging to each possible profession, like lawyers = 2%, doctors = 5%, biologists = 3%, computer-scientists = 8%, philosophers = 1%, and so on. The sum of all percentages of all professions must be 100. Thus, the sum of all percentages of the different professions in each town is constant and equal to 100. In this example $T = 100$.

Another example can be in a dataset where the instances represent the data of *healthy food menus*, which are a list of ingredients and its corresponding quantities, like 100 g of chicken, 50 g of potatoes, 80 g of pumpkin, and 30 g of pear. In addition to the quantities of the ingredients of the menu, we could have several variables indicating the kcal provided by the menu according to the main basic food–family components like (fruit, vegetables, grains, protein, dairy, fats, added-sugars) presents in each menu. The final sum of kcal of each healthy menu should be of 2300 kcal for a healthy menu recommended by a nutritionist. Thus, the sum of the kcal of each food–family component (fruit, vegetables, etc.) is constant and equal to 2300 kcal. In this case, $T = 2300$, and the sum of the kcal of all food–family for each menu is equal to T. Of course, in addition to this constraint of not surpassing the bound of 2300 kcal for the whole menu, other constraints regarding the upper limit of kcal for each kind of food–family exists to consider the menu as being healthy.

The problem is that all these s variables have a linear relationship among them, as expressed by the above equation. Thus, *they are not independent*, and this fact can imply some negative consequences in the application of data mining methods. For instance, there are some data mining methods that assume independence among the variables.

The solution is to make a transformation of the originally s variables regarding some *reference* or *pivot* variable, and obtain some $s - 1$ new variables referred to the pivot variable. A usual transformation is the *log-ratio transformation*, especially to ensure normality in the data. This transformation proceed as follows:

1. Select a *reference* or *pivot* variable $X_{.r}$, such that

$$X_{.r} \in \{X_{.1}, X_{.2}, \ldots, X_{.s}\}$$

2. Build $s - 1$ new variables $X'_{.j}$, such that

$$X'_{.j} = \log\left(\frac{X_{.j}}{X_{.r}}\right), \quad \forall j \in \{1, \ldots, s\} \text{ such that } j \neq r \quad (6.2)$$

3. Use the new $s - 1$ variables $X'_{.j}$ instead of the original s variables $X_{.j}$ in the dataset.

6.3.4.2   Multi-valued Variables

Multi-valued qualitative variables are qualitative variables which can have more than one value simultaneously for the same instance.

This is not the usual case when each variable has a unique value for each instance, like the variable *colour of the eyes* of one person, when the possible values could be one of {brown, blue, green, purple, ...}, but we are assuming that nobody could have the eyes with more than one colour, which very probably is true for most of people. In those situations, it can be stated that the possible *qualitative values are mutually exclusive*. It is not possible to have more than one value.

Not with standing, *multi-valued qualitative variables* can have several values at the same time. In this situation, the *qualitative values are not mutually exclusive*, and each instance can have more than one value. For example, if we think in a variable *colours of the flag* of one country, we will obtain that a country like USA has three qualitative values for that variable {red, white, blue}, the same three values as France or Great Britain. Italy would have the three values {green, white, red}, while Germany would have {red, yellow, black} and Poland will have just two colours {white, red} and Catalonia will have {red, yellow} as the possible values for the variable.

Major problem is that most of data mining algorithms need that each variable had just one value and cannot support a multi-valued qualitative variable. The solutions aim at making some transformation to get the same information of that multi-valued variable in one or more than one new mono-valued variables. The common solutions are:

1. *Create a new binary variable for each one of the possible qualitative values or modalities.* A binary variable is a qualitative variable with just two possible qualitative values or modalities {true, false}. Binary variables are also named as *binominal variables* (i.e. nominal variables with only two possible values). Then, substitute the original multi-valued variable by the new binary variables created in the dataset. This way, a multi-valued variable in a concrete instance will have a *true* value for the new binary variables corresponding to the modalities it has, and a *false* value for the remaining binary variables corresponding to the other modalities. Thus, the procedure is a s follows:

   (a) Given a qualitative multi-valued variable X_j with the possible set of modalities $\{m_1, m_2, \ldots, m_Q\}$

   (b) Create Q binary variables: $\left\{ X_{.j_1}, \ldots, X_{.j_Q} \right\}$

   (c) Substitute X_j by the set of new binary variables $\left\{ X_{.j_1}, \ldots, X_{.j_Q} \right\}$ in the dataset. For each instance, the following process must be done: for each modality (m_q) of the multi-valued variable (X_j) in that instance, the corresponding binary variable $(X_{.j_q})$ must be set to *true*, and all the remaining binary variables corresponding to the modalities that do not appear in the instance to *false*.

2. *Recode the original multi-valued qualitative variable into a mono-valued qualitative variable.* The new value of the mono-valued variable for an instance will be the new modality which is the combination of all the original modalities that had the instance for the multi-valued variable. Therefore, the procedure is as follows:

(a) Given a multi-valued qualitative variable X_j with the possible set of modalities $S_{mod} = \{m_1, m_2, \ldots, m_Q\}$
(b) Create a new qualitative mono-valued variable X'_j with the possible set of modalities being the corresponding ones to the *power set*[4] of the original set of modalities:

$$P(S_{mod}) = \{m_\varnothing, m_1, \ldots, m_Q, m_{12}, \ldots, m_{1Q}, m_{23}, \ldots, m_{2Q}, \ldots \\ m_{Q-2,Q-1}, m_{Q-2,Q}, m_{123}, \ldots, m_{12Q}, \ldots, m_{12\ldots Q}\} \tag{6.3}$$

3. Recode all the instances, changing the set of modalities of the original variable by the corresponding unique modality from $P(S_{mod})$.

The first approach described above has the disadvantage that it is adding Q binary variables, which represent the possible modalities, with the information of the original variable. Hence, it is giving more importance to this variable because is artificially increasing the dimensionality of the data matrix, when perhaps it is not the case, and probably could introduce some bias in the data analysis.

Therefore, it is preferred the second approach of recoding the multi-valued qualitative variable into a mono-valued qualitative one. However, this approach can have the drawback related to the possible high number of modalities of the recoded variable. The power set has a large cardinality, i.e. the number of possible subsets of the original set of modalities. The cardinality expressed by this operator | |, of the power set is:

$$|P(S_{mod})| = 2^{|S_{mod}|} \tag{6.4}$$

This number could be very large, but in practice, most of the possible combinations of the modalities will not be used because that combination does not appear in the dataset.

6.3.5   Visualization and Descriptive Statistical Analysis

Visualization of data is a very powerful tool to inspect and detect possible errors or abnormalities in the data. In most of the data mining tools, there are some available graphics which can be used to analyze the data. Main graphics are bar charts, histograms, scatter plots, time series plots, distributional plots, and scatter plots with coloured points (class-labelling scatter plot), and so on.

---

[4]The *power set* of a set S is the set of all the possible subsets of S, including the empty set ($\varnothing$) and the S itself.

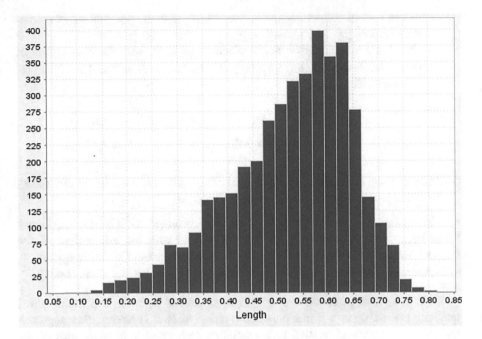

Fig. 6.4 A histogram of a *length* numeric variable

Graphical charts can help to detect possible erroneous values or possible outlier values, as well as to detect the distributional patterns of the variables involved in the dataset like unimodal distributions, normal distributions, multimodal distributions, symmetric distributions, linearity relations among two variables, etc.

In Fig. 6.4, there is an example of a histogram of a numeric variable.

In Fig. 6.5, there is a bar chart of a categorical or qualitative variable *iris-type*, which contains the possible type values of an iris flower (iris setosa, iris versicolor, iris virginica). In this dataset, a very famous one in ML (Dua & Graff, 2019), there are 50 instances of each one of the types of iris flower, as depicted in Fig. 6.5.

In Fig. 6.6, there is an example of a scatter plot with coloured points (class-labelled scatter plot) of the three class labels of an iris flower (iris setosa, iris versicolor, iris virginica). The black point with the coloured circle represents each one of the *prototypes* or *centroids* of each one of the types of the iris flower.

In addition to using graphical tools, it is very interesting to make a simple descriptive statistical analysis of the different variables of the dataset. Commonly, a descriptive analysis includes the computation of several values and statistical measures to evaluate different aspects like the size, the range, centrality, or dispersion of the distribution.

For *all kinds of variables,* the following information can be computed:

- $N(X)$: The number of observations of the variable that should be the same for all variables.
- N - missing - values(X) or Perc - missing - values(X): The number or percentage of missing values in the corresponding variable X.

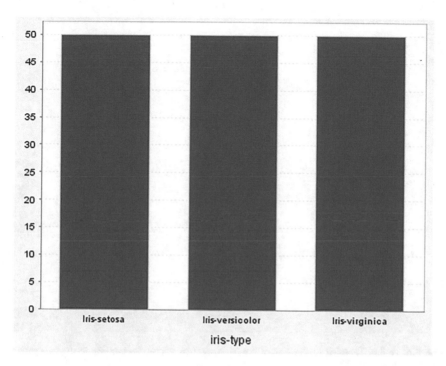

Fig. 6.5 A bar chart of *iris-type* categorical variable

For *ordered variables*, i.e. numerical or qualitative ordered, these measures can be computed:

- Median(X): The median of a variable X is the value that separates the higher part from the lower part of the variable values according to an order relation. It is the point in the middle (50% of values are lower and 50% of values are higher) on an odd number of values, or the arithmetic mean of the two central values for numeric variables and both the two central values for qualitative ordered variables, in the case of an even number of values. It can be also named as the $Q_2(X)$, i.e. the second quartile value.
- $Q_1(X)$: The first quartile value is the value that separates the 25% of lower values, and the 75% of higher values of the variable according to the order relation.
- $Q_3(X)$: The third quartile value is the value that separates the 75% of lower values, and the 25% of higher values of the variable according to the order relation.

In Fig. 6.7, the position of the three quartiles for a uniformly distributed variable is depicted. Usually, the variables are not uniformly distributed, and the quartile values are no equidistant.

The following values and statistical measures for each *numeric variable (X)*, from the available sampling data, i.e. the dataset, can be computed:

- min(X): The minimum value of the variable X.
- max(X): The maximum value of the variable X.

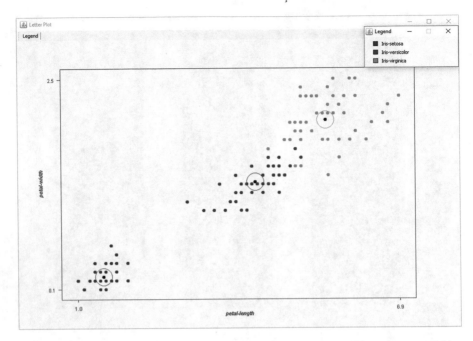

Fig. 6.6 A class-labelled scatter plot of the *iris-type* class variable according to the two variables petal-width and petal-length

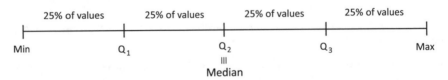

Fig. 6.7 The three quartile values of a uniformly distributed variable along the range of possible values

- Range(X): The range of the variable X, computed as:

$$\text{Range}(X) = \max(X) - \min(X) \tag{6.5}$$

- Mean(X): The *arithmetic mean* or *average* or *mean* value of a numerical variable X, computed as:

$$\bar{x} = \frac{\sum\limits_{i=1}^{n} x_i}{n} \tag{6.6}$$

- Variance(X): Variance of a numerical variable, computed as:

$$s_x^2 = \frac{\sum\limits_{i=1}^{n} (x_i - \bar{x})^2}{n} = \frac{\sum\limits_{i=1}^{n} x_i^2}{n} - \bar{x}^2 \tag{6.7}$$

- Standard deviation(X): Standard deviation of a numerical variable X, computed as:

$$s_x = \sqrt{\frac{\sum_{i=1}^{n}(x_i - \bar{x})^2}{n}} \tag{6.8}$$

And for each *pair of numeric variables* (X, Y), the following statistical measures can be computed:

- cov(X): the covariance between two variables X and Y, computed as:

$$\text{cov}(X, Y) = \frac{\sum_{i=1}^{n}(x_i - \bar{x})(y_i - \bar{y})}{n} = \frac{\sum_{i=1}^{n} x_i y_i}{n} - \bar{x}\,\bar{y} \tag{6.9}$$

- Corr(X): the correlation between two variables X and Y, computed as:

$$\text{Corr}(X, Y) = r_{xy} = \frac{\text{cov}(X, Y)}{s_x s_y}, \quad \text{where} \quad -1 \leq r_{xy} \leq 1 \tag{6.10}$$

And the following statistical measures for each *qualitative variable* (X) of the dataset:

- Mode(X): The mode value of a nominal or qualitative variable with Q possible modalities, i.e. the most frequent qualitative modality observed (m_q) in the sample, computed as:

$$\text{Mode}(X) = m_q \quad \text{s.t.} \quad q = \arg\max_{j \in Q} \text{freq}(m_j) \tag{6.11}$$

- Frequency table(X): The frequency table of a qualitative variable X, is table accounting for the frequency of all the possible modalities (m_j) of the variable

6.3.6 Transformation and Creation of Variables

Sometimes, previously to the data mining process there is the need for some transformation of a variable for different reasons like to make the data homogeneous or comparable or to get rankings or to satisfy some technical requirements. Furthermore, in some occasion, the creation of a new variable is a necessary option for summarizing information from several variables into a new one. Next, both kinds of variable operations will be analyzed.

6.3.6.1   Transformation of Existing Variables

The transformation of one variable can be due to different reasons. Main transformations can be enumerated as the following ones:

- *Transformation to get homogeneous data*. Sometimes can happen that as a result of a merging process of different databases into a final common data matrix, the measure units of some variable could be slightly different like measuring a concentration of *suspended solids in the water* measured in mg/l or in g/l. Of course, in order that the data analysis make sense all the data must be expressed on the same measure unit. In this case, the corresponding variable must be transformed either by multiplying by 10^3 or by 10^{-3} the values of the variable measured in the different unit.

- *Transformations to get comparable data*. Some variables can express some quantity in absolute values like the *population of a district in a city*, or the *sales amount of one employer*. These *absolute values* give a quantity that cannot be comparable because very probably the quantity should be referred to some kind of reference value to make sense. Thus, many times, some quantity values must become *relative to a reference value*. For example, the *population of a district in a city* should be divided by *the extension of the district* because larger districts probably will have more population. This way, we have transformed the population of a district into the *density of population of a district*, which now is comparable among all the districts in a city. The same happens when we want to compare two employers regarding their *sales amount*. If we do not know the time when that number of sales was obtained, then, the two amounts are not comparable because they could have been obtained in a different period of time. Thus, the *sales amount of one employer* should be divided by *the period of time it was obtained*. This way, the sales amount of one employer have been transformed in the *sales amount per unit of time of one employer*, which now is comparable for all employers of a company.

- *Transformations to get rankings*. Sometimes it is more important to know the *ordering* among the observations or instances than the exact distance between two observations. For example, suppose a variable expressing *the weight of the patients*, and in a clinical experiment the physician is more interested in knowing which are the patients with higher weights and the ones with lower weights than knowing the weight differences between patients. The procedure to get this ranking of the observations is to sort the observations according to the variable of interest, and then enumerating each observation with consecutive integer values: 1, 2, 3, ..., *n*. This enumerated list is the transformed variable, which shows the *relative position* or *rank of the observation* within this list. This new variable is a *qualitative ordered variable* or *ordinal variable*. In our example, the lower values or ranks are the lighter people, and the higher ranks are the heavier people.

- *Transformations to meet technical requirements.* Some data mining methods need particular technical constraints that must be satisfied. Most common transformations are the following:

  - *Discretization* of numerical variables. Some data mining methods require that numerical variables be discretized in a set of qualitative values or labels or modalities. This discretization process can be done not only in advance in the pre-processing step but also internally by the data mining method itself. For example, as it will be shown in Sect. 6.4.2, a decision tree needs a finite number of branches for each node. Thus, as the branches represent the possible values of an attribute, these must be discrete or qualitative values. If an attribute is numeric like *the height of a person*, then it must be discretized, for example in three modalities like *low*, *normal*, and *high*. Discretization of a numeric variable means that the whole range of a numeric variable (i.e. values between the min and the max value) must be divided into some *intervals* or *bins*. There are some common techniques for discretization like:

    Same-width Bins: The range of the numeric variable (max−min) is divided in a determined number of bins of equal size. Usually, this number is three or five, even though that it could be whatever number greater than one. From psychological studies is recommended that the number of bins were an odd number.

    Same-frequency Bins: The range of the numeric variable is divided in a determined number of bins, but each bin has the same number of observations in it. Usually, the bins have different length.

    User-determined Bins: The user specifies the cut-points, and hence, the number and length of the different bins.

  - *Standardization* of a numerical variable. Both standardization and normalization transformations aim to avoid variables measured in different magnitudes in the same dataset, which could bias the analysis of the data, and make not possible fair comparisons among the variables. It is especially important for many data mining methods which use distances and similarity computation. All the variables should be measured in the same interval of values. In the standardization transformation, the variable is centered and scaled to have a null mean and a deviation equal to 1.

    Each variable X_j will be transformed to a new variable X'_j:

$$X'_j = \frac{X_j - \overline{X_j}}{s_{X_j}} \tag{6.12}$$

    Now, the new variable value will be positive if the original value was above the mean, and negative if the original value was below the mean.

  - *Normalization* of a numerical variable. The normalization transformation changes the original range of one variable into the normalized interval [0,1]. It is also named as the *min–max normalization*.

Each variable X_j will be transformed to a new variable X'_j:

$$X'_j = \frac{X_j - \min\left(X_j\right)}{\max\left(X_j\right) - \min\left(X_j\right)} \tag{6.13}$$

This way, all the variables now are in the range [0, 1] and they can be compared and aggregated.

6.3.6.2   Creation of New Variables

Commonly, in a dataset, some new variables can be created as the combination of other variables in the dataset. Usually, these new variables are created to outline some interesting relationships among variables, which will be used for the decision support, once the data mining models have been induced from the data. Most common creation of new variables are the *aggregation variables* and *ratio variables*:

- *Aggregation variables*. Sometimes it is interesting the creation of a new variable that is the aggregation or total value corresponding to the sum of several related variables. For instance, one could be interested in having a new variable representing *the total quantity of a crop*, as a sum of the different quantities of the different cultivated products in a dataset of agriculture. The new aggregated variable X_{agg} is computed as:

$$X_{agg} = \sum_{j \in \text{var to be agg}} X_j \tag{6.14}$$

- *Relational variables*. Other common situation is to create a new variable which is comparing at least two variables in a given relationship. Often, this relationship is expressed between two or more variables in the form of a ratio. For example, in a dataset where we have the biometric characteristics of people, and we have the *weight* and *height* of people, probably could be very interesting for a physician, to create a new variable *BMI* (*Body Mass Index*) which express a relationship between the *weight*, expressed in kg, and the *height* of a person, expressed in m:

$$BMI = \frac{Weight}{Height^2} \tag{6.15}$$

6.3.7   Outlier Detection and Management

An *outlier* is a numeric value that it is somehow different and distant from the others (Barnett & Lewis, 1994). Outliers apparently deviates from the other observations from the same sample. They present very extreme values in one or more variables.

Commonly, an outlier can be originated from an erroneous data coming from a sensor which is wrongly operating, or just could happen that the value is an erroneous value which could be wrongly stored in the data matrix.

The *detection* and consideration of a possible value as an outlier is a difficult task, because it is not easy to know whether a value is intrinsically extreme, or it is really a wrong value. In addition, it could happen that an observation just taking into account the value of one variable does not seem an outlier, because the value seems a normal one according to that variable, but taking into account another variable or a set of variables, the combination of all values could be impossible or very strange.

Basically, there are two basic ways for the detection of a possible observation as an outlier:

- If there are some expert available that can analyse the data, and according the *expert judgement*, the value/s of an observation is clearly out of the normal range of value/s, then the observation can be considered as an outlier.
- The other method is trying to use some *mathematical definition* of what could be an outlier, based on the notion that its value or values should be deviated from the "normal" distribution values of the variable or variables. In the literature, there are some criteria which can be used. One definition is the Tukey's fences (Tukey, 1977), which John W. Tukey proposed based on the *Inter Quartile Range measure*.

The *Inter Quartile Range (IQR)* of one variable is defined as follows:

$$\text{IQR}(X) = |Q_3(X) - Q_1(X)| \qquad (6.16)$$

The lower and upper fences, i.e. bounds, are:

$$\text{Lower bound}(X) = Q_1(X) - k * \text{IQR}(X), \quad k > 0 \qquad (6.17)$$

$$\text{Upper bound}(X) = Q_3(X) + k * \text{IQR}(X), \quad k > 0 \qquad (6.18)$$

Tuckey proposed that any observation $(X_{i.})$ having a value (x_{ij}) outside this range can be considered as an *outlier*:

$$x_{ij} \notin [Q_1(X_j) - k * \text{IQR}(X_j), Q_3(X_j) + k * \text{IQR}(X_j)]$$

Specifically, he proposed the following bounds using $k = 1.5$ and $k = 3$:

$$x_{i,j} \text{ is a mild outlier} \Leftrightarrow$$
$$x_{i,j} \notin [Q_1(X_j) - 1.5 * \text{IQR}(X_j), Q_3(X_j) + 1.5 * \text{IQR}(X_j)]$$

Fig. 6.8 A box plot for a
variable *storage*

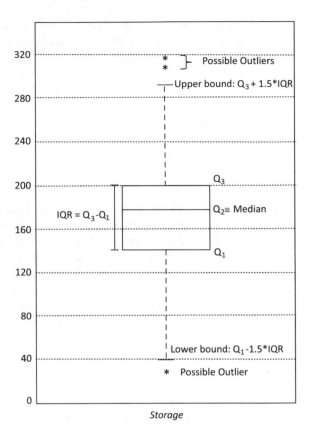

$x_{i,j}$ is a strong outlier $\Leftrightarrow$

$$x_{i,j} \notin \left[Q_1\left(X_j\right) - 3 * \mathrm{IQR}\left(X_j\right), Q_3\left(X_j\right) + 3 * \mathrm{IQR}\left(X_j\right)\right]$$

The *box plot chart* is a very useful chart for detecting possible outliers in a variable. It was introduced by John W. Tuckey in 1969. It is a method for graphically depicting groups of numerical data through their quartiles. Box plots may also have lines extending vertically from the boxes, named as *whiskers*, indicating variability outside the upper and lower quartiles. There are some variations. Ones in addition to the quartiles display the minimum and the maximum values, and others display the lowest and highest value within the *lower bound* and the *upper bound*, with $k = 1.5$. Furthermore, the values outside the lower and upper bounds, i.e. candidate values to be considered as *outliers* are depicted. Figure 6.8 depicts an example of a box-plot chart, with the main quartile values and lower an upper bound marked.

Once one outlier has been detected, the possible management actions for an outlier are the following three:

- *Remove* the whole observation containing the outlier value or values from the data matrix, especially if the outlier is an influent observation which strongly determines the results of a certain analysis.
- *Correct* the outlier, which is a wrong value, by the original value if possible.
- If it is not possible to correct the value, then the remaining option is to *substitute* the outlier value by a missing value.

The confusion of one extreme value as it were an outlier, and supressing it to avoid its perturbation on the distribution of the data is a danger. A known history happened to British scientists with some remote satellite ozone data readings from the South Pole. In 1979, these data were automatically supressed because they were extremely low, having been considered as outliers. Some years later in 1985, they reported the hole in the ozone layer which was first disregarded as it was based on some data coming from ground instruments, and the remote satellite data did not show anything unusual. However, after a careful analysis of the satellite data, they discovered that the low data values had been supressed automatically by the software. Coming back to the data from 1979, they discovered that the ozone hole could have been discovered some years before!

The management of possible outliers is a task to be done in a very careful manner.

6.3.8 Error Detection and Management

Erroneous data can be originated from several reasons. Some variables with information coming from a sensor could be erroneous due to some temporary failure in the sensor which could have generated anomalous data. In other cases, some data can be wrongly transferred from one source to another, or erroneously introduced into the dataset by a human.

The *detection of errors* in data is not easy. In the case that some expert in the domain is available, then the expert can assess whether some value can be erroneous. For example, an expert could know some kind of *relationship between two variables*, and according to that knowledge, a possible value for one variable could seem impossible according to the value of the other variable. For instance, an *environmental expert* can know that the value of a variable expressing the COD[5] *(Chemical Oxygen Demand)* of wastewater must be higher than the corresponding variable expressing the BOD[6] *(Biochemical/Biological Oxygen Demand)* of wastewater, because the biological oxidation needs less dissolved oxygen than the chemical oxidation process. These situations are very difficult to discover unless you have an expert on the data.

---

[5] *COD* denotes the amount of oxygen required for the chemical oxidation of organic material in wastewater in a specified time, i.e. it measures the level of water pollution.

[6] *BOD* denotes the amount of oxygen required for the biochemical oxidation of organic material in wastewater in a specified time, i.e., it measures the level of water pollution.

In other situations, the detection of an error can be easier, like in the typical case than the value of a variable is clearly out of the usual range of values of that variable. For example, if a variable is describing *the volume of the boot of a car*, it must be a positive value, usually less than 1000 litres. Or a variable expressing a *percentage* must be in the range [0, 100] or a *proportion* in the range [0, 1].

The management of errors in data can be solved using some of the following strategies:

- *Correct* the erroneous value, by recovering the original data value and retrieving the correct data if possible.
- *Remove* the whole observation containing the erroneous value or values from the data matrix. This probably only make sense if the data could not have been corrected, and this removal does not affect the quality of the dataset. If the number of observations is not high, it is not the preferred action.
- If it is not possible to correct the value and its removal is not a good option, then the remaining option is to *substitute* the erroneous data value with a missing value.

6.3.9   Missing Data Management

In real life, most of the datasets have some data values missed. These missed or unknown values are represented with different codes in different datasets. The symbol "?" is probably the most used to represent an *unknown* or *missing* value. Other symbols used to represent a missing value are "null", "nil", "none", "*", "NA" (not available), a blank space, or a numerical code value which is quite impossible to be true for a numerical attribute like "99999" or "−99999" or similar codes.

Some authors (Little & Rubin, 2014) distinguish between two basic kinds of missing values: *random missing values* and *non-random missing values*.

- *Random missing values* are values missed, which were randomly generated, and did not follow any concrete pattern. They appear in the data matrix by chance. In these cases, it is right to assume that they follow the same data distribution than the observed (non-missed) data. Therefore, the observed data can provide useful information to recover them. This category is the more usual like the missing values originated by a temporary faulting sensor or a manual skipping of a value in transferring data into a computer.
- *Non-random missing values* are generated by some identifiable causes. Usually, they come from specific subpopulations of data, with a different distribution of the data than the observed data. In these cases, the observed data cannot provide with useful information to manage them. Usually, the data can be missed because it was *consciously hidden* or *explicitly not provided* in order not to be known by anybody, like for instance *the salary of the executive managers* of one company, which they do not want to be known by the data scientist who is analysing the data of the company. Another motivation can be because the characteristic

measured in one variable *only makes sense in some subpopulation* of the observations. Then, the observations not belonging to that subpopulation will have a missing value for that variable. In fact, many times these kinds of missing values are representing a non-applicability situation, probably caused by a wrong coding of the information in the dataset. For example, the variable *type of transport* describing which transport means uses a citizen. For those people which do not use any transport mean, this variable can have been coded as missing values.

Missing values can be detected easily if there are already marked in the data matrix with a specific symbol like the ones detailed before. If they are not marked, a scan of the whole data matrix must be done to identify empty values in the matrix, and code them with the specific symbol used to denote a missing value.

Other missing values could have been originated as a consequence of the management of some erroneous data or some outlier management, which can generate new missing values.

In general, there are some techniques which can be used both for random missing values and non-random missing values. These techniques are *removal techniques*:

- *Removal of the observations* which have some missing value, from the dataset, i.e. removing rows from the data matrix. This means that some examples will be lost with the consequent loss of information. This strategy could be a good solution when there are observations with many missing values in the variables, and on the other hand, there are many observations in the dataset. This way, the losing of those removed observations will not harm the data mining methods.
- *Removal of the variables* which have some missing value, from the dataset, i.e. removing columns from the data matrix. This means that some variables or attributes will be lost with the consequent loss of information. This strategy could be a good solution when there are variables with a high percentage of missing values in the observations, and on the other hand, the number of variables in the dataset is high. This way, the losing of those removed variables will not harm the data mining methods.

For *random missing values*, as these values are assumed to be from the same population and distribution, the best strategies try to get valuable information from the observed data. These techniques are named as *substitution* or *imputation techniques*. They aim at substituting the missing values for a value computed through some heuristic function or procedure. They are also named as imputation strategies, which estimates a value for the missing value. There are several *imputation or substitution strategies*:

- A commonly used strategy is to substitute the missing value in a variable by *the mean or the mode value* of the observed values of the variable, depending on whether it is a numeric variable or a qualitative one. This way, it is expected that the imputed value, as a centric value of the variable will not imply a bias in the distribution of that variable.
- When it is known that there are some *temporal/spatial relationships* among one observation and the previous and following observations in adjacent time/space

units, it makes sense to substitute the missing value in a variable by *the mean or the mode value* of the observed values of the variable, but just using those *within a temporal/spatial window of length l around the missing value*. For example, the previous and the next value can be used (window of length $l = 1$), or the two previous and the two next values (window of length $l = 2$).

- Other common strategies use a *similarity-based heuristic function* to estimate the missing value. A commonly used strategy uses the most similar observation or observations in the data matrix, with known values on the required variable to impute the missing value. This strategy is commonly known *as k-Nearest Neighbour technique (k-NN)*. The imputation technique could use the *mean/mode strategy* of all the k most similar observations (neighbours) or a *weighted mean/mode* of all the k most similar observations (neighbours). The weights are directly proportional to the similarity degree of each neighbour.

For *non-random missing values*, in addition to general removal strategies, there are just a pair of general strategies depending on the kind of missing value:

- For the hidden or explicitly not provided missing values, there is just the possibility that the original value can be *recovered from the source* data or an expert on the data.
- For missing values belonging to a different subpopulation, and that the corresponding variable is qualitative, there is in addition to the previous recovery strategy, the option to *add a new "unknown-value" modality*. This new modality will encode the missing values, and thus, the variable and these modality values can be managed with the same procedures of any qualitative variable.

6.3.10   Data Reduction

The *data reduction* step aims at reducing the size of the data matrix either removing some *redundant instances*, i.e. rows of the matrix or removing some *irrelevant features* or *variables*, i.e. columns of the matrix. The process of removing instances from the data matrix is known as the *instance selection* step, and the process of removing the irrelevant features is known as the *feature selection* step.

Both *instance selection* techniques and *feature selection* techniques have the aim to reduce the size of the dataset, previously to the use of the data mining techniques.

Notwithstanding, each time some item of information, either an instance (row) or a feature (column) of a dataset is removed, there is a *loss of information*. Of course, the instance selection methods or the feature selection methods are designed to keep as much as possible the same information but reducing the number of instances or features in the dataset. The data scientist must be aware about this fact, and be very cautious before removing any kind of information from the dataset.

Instance selection reduces the number of instances, which will be used in the following data mining process. *Feature selection* reduces the number of used variables for the data mining process, which is usually named as *dimensionality*

Fig. 6.9 A general chart of a *filter* method

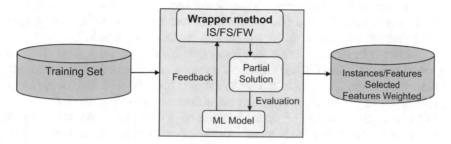

Fig. 6.10 A general chart of a *wrapper* method

reduction. The advantages of instance selection techniques and feature selection techniques are twofold: the size of the dataset is reduced, and the computational time for afterwards data mining techniques is reduced.

6.3.10.1 Filters and Wrapper Methods

The different techniques and methods used both in data reduction, i.e. instance selection and feature selection, or in feature relevance determination, i.e. feature weighting can be generally classified as belonging to two types, according to whether they are using some *bias information* or not: *filter methods* and *wrapper methods*.

Filter methods attempt to make the selection (data, features) or assess the relevance of features (feature weighting) only taking into account the available data from the training set, without any other kind of information. They use some rational idea of how to make the task and implements the idea in a corresponding algorithm to find the solution (reduced set of instances, reduced set of features, or the feature weights). See Fig. 6.9 for the general scheme of a *filter* method.

Wrapper methods (Kohavi & John, 1998) make the selection (data, features) or assess the relevance of the attributes through an *iterative process* where they also receive *feedback*, usually from a Machine Learning (ML) model (a classifier or a numerical predictor) induced from the training set. At each iteration, the current partial solution, i.e. the feature weights' (FW) vector or the instances selected (IS), or the features selected (FS) are modified according to the results of the ML method. The iterative process runs until no improvement on the ML method can be obtained, and the final solution obtained is the one giving the best performance for the ML method. See Fig. 6.10 for the general scheme of a *wrapper* method.

6.3.10.2   Instance Selection

Sometimes, redundancy of instances can produce several problems like a very huge size of the dataset, or imbalanced datasets, which can disturb the data mining process, and then, some pre-processing tasks must be required. The aim of the *instance selection* techniques is to reduce the number of examples of the dataset with a minimum loss of information (predictive accuracy, clustering accuracy, etc.). The aim is at retaining the relevant instances and removing the redundant ones, which probably are not useful for a clustering or classification task.

The aim of instance selection techniques is to obtain a subset S, $S \subseteq T$, of the original training dataset T, such that S does not contain redundant or superfluous instances, and the accuracy in the models mined (usually predictive or clustering models) is nearly equivalent when using S as the training set instead of T.

One first classification of instance selection techniques is according to the *processing way* of how S is computed. The methods could be grouped in:

- *Incremental Methods*: These instance selection methods start with an initial empty new set of instances ($S = \varnothing$) and advance incorporating only the *relevant instances* until the end of the selection process.
- *Decremental Methods*: These instance selection methods start with a new set equal to the original training set ($S = T$), and advance removing the *redundant instances* until getting the final set S along the selection process.
- *Mixed Methods*: Some instance selection methods can use a mixed incremental or decremental way of the selection process. In some steps of the algorithm, they advance incrementally adding instances or decrementally removing instances, and at some next step, they advance in the opposite way.

According to the bias dimension, the instance selection methods can be divided into *filter methods* and *wrapper methods*

- *Filter Instance Selection Methods*: These instance selection methods use several selection criteria, but none is based on the feedback of a mined model from the data (classifier, clustering, etc.). They can be classified into *size-based or percentage-based methods* and the *semantic-based methods*:

  - *Size-based* or *percentage-based methods*: They use the size or some percentage of instances to make the selection process. Sometimes they are called also *exhaustive methods*. Main techniques are:

    Random Methods: They take randomly a determined *size s* or *percentage p* of the whole dataset, or just the size/percentage of instances that the system is able to process, when the dataset is very huge. These techniques are known as *sampling methods* in the statistics field. For example, just taking a sampling of the 80% of the original dataset ($p = 80$).
    Supervised Random Methods: They take randomly a determined percentage p of the original dataset, but the same percentage of instances belonging to each class group ($p_{C_1}, \ldots, p_{C_K}$) is maintained in the reduced dataset.

Of course, these techniques only make sense when the original dataset is *supervised*, and the variable of interest is a qualitative variable acting as a class variable. These techniques are known as *stratified sampling methods* in the statistics field. For example, given that the percentages of class distribution in a dataset were (30, 40, 30) for the three classes, then in the reduced dataset, the same proportions of the distribution of all three classes are maintained.

– *Semantic-Based Methods*: the other filter methods are based on semantic criteria, based on some properties about the instances or the clusters found in the data or in the relevance of the instances. Thus, they need a *supervised* data matrix, with a variable being a class variable. Instances are defined to be *border instances* or *inner instances*. An instance i_k is a *border instance* for the class C_j if $i_k \in C_j$ and i_k is the nearest neighbour of an instance from another class C_l, $C_l \neq C_j$. On the contrary, i_k is an *inner instance* for the class C_j, if it is not a border instance. Some filter methods are based on selecting border instances. Among the *semantic-based filter methods* in the literature, we can distinguish the following categories:

Border-Instance Methods: There are several methods selecting only some border instances like the following ones:

The POP (Pattern by Ordered Projections) method (Riquelme et al., 2003) discards inner instances and selects some border instances.
The POC-NN (Pair Opposite Class-NearestNeighbor) method (Raicharoen & Lursinsap, 2005) also select border instances.

Cluster-Prototype Methods: These methods use the information provided by the clusters and the prototypes of the clusters. Some authors (Caises et al., 2009; Spillmann et al., 2006) have proposed the idea of using clustering for the selection process, taking the centres (or nearby) of the clusters as the selected instances. Some of the methods are:

The GCM (Generalized-Modified Chang Algorithm) method (Mollineda et al., 2002)
The NSB (Nearest Sub-class Classifier) method (Venmann & Reinders, 2005)
The CLU (Clustering) instance selection method (Lumini & Nanni, 2006)
The OSC (Object Selection by Clustering) (Olvera-López et al., 2007)
A filter method based on k-d Trees (Friedman et al. 1997) was proposed in (Narayan et al. 2006), where a binary tree is constructed, to obtain better clusterized data.

Weighting-Instance-Based Methods: Other authors suggested the assignment of weights to the instances and selecting only the relevant instances, i.e. the high weighted instances. Some methods are:

The WP (Weighting Prototypes) method (Paredes & Vidal, 2000)
The PSR (Prototype Selection by Relevance) method (Olvera-López et al., 2008)

- *Wrapper Instance Selection Methods*: In these instance selection methods, the selection criterion is based on the accuracy obtained by a model mined with the considering instances. Usually, those instances not contributing to the accuracy of the model are the candidates to be discarded from the training set. Usually, they need a supervised data matrix with a target qualitative variable, i.e. the class variable. As the filter methods, they can be classified in *size-based or percentage-based methods* and the *semantic-based methods*:

  - *Size-based or percentage-based methods*: they use the size or some percentage of instances to make the selection process. One of the most commonly used methods is:

    Windowing Method: it is a special wrapper method, which uses a percentage of instances to make the initial selection process followed by feedback guidance. It takes a random percentage of the data, induce a classifier/predictive model with them, and the model is applied to the remaining data. Those data that have not been correctly classified/predicted by the model are added to the initial set and the process is repeated. It has been estimated that it could save up to 20% of the data.

  - *Semantic-Based Methods*: The other type of wrapper methods are based on semantic criteria, based on some properties about the instances and the clusters found in the data. Methods can be grouped into the following types:

    k-NN-based methods: These methods use the *k*-NN classifier (see Sect. 6.4.2.1) for selection purposes. There are many methods and variants. We can enumerate the following ones:

    The Condensed Nearest Neighbour (CNN) (Hart, 1968), which is an *incremental method*, starts assigning randomly one instance from each class to S. Afterwards, each instance from T is classified using S as training set: if an instance p is misclassified then p is included in S.
    The Selective Nearest Neighbour rule (SNN) method (Ritter et al., 1975).
    The Generalized Condensed Nearest Neighbour rule (GCNN) (Chien-Hsing et al., 2006)
    Other methods focussed on discarding noisy instances in the training set are the Edited Nearest Neighbour method (Wilson, 1972) and some variations like the *all k-NN* method (Tomek, 1976) and the *Multiedit* method (Devijver & Kittler, 1980).
    The Instance-Based methods (IB2 and IB3) were proposed by Aha et al. (1991), and they are incremental methods.

Associate-Based Methods: They use the concept of *associate instances*. The *associates* of an instance *i* are those instances such that *i* is one of their *k* nearest neighbours. Some of the proposed methods are:

A *decremental method named as* Decremental Reduction Optimization Procedure (DROP1) proposed by Wilson & Martínez, (2000) discards an instance *p* from *T* if the associates of *p* in *S* are correctly classified without *p*.

Several Decremental Reduction Optimization Procedure variations of DROP1 method, like DROP2, ..., DROP5 methods (Wilson & Martínez, 2000).

The Iterative Case Filtering algorithm (ICF) proposed by Brighton & Mellish, (2002).

Optimization-Based Methods: These methods used different optimization techniques to get the set of reduced instances. Most outstanding approaches are:

Evolutionary algorithms have been proposed to be used for instance selection in the works of Cano et al. (2003), Bezdek and Kuncheva (2001), Kuncheva (1997).

A *memetic algorithm*, combining evolutionary algorithms and local search in the evolutionary process, is proposed by García et al. (2008). The Clonal Selection Algorithm (CSA) was proposed by Garain (2008) and is based on the *artificial immune system approach*.

Some methods proposed the *tabu search* for the selection of the instances (Zhang & Sun, 2002; Cerverón & Ferri, 2001).

6.3.10.3   Feature Selection

Datasets may contain irrelevant or redundant variables. As previously stated, the quality of discovered knowledge is usually dependent on the quality of the data that they operate on. Also, the success of some machine learning schemes, in their attempts to construct models of data, hinges on the reliable identification of a small set of highly predictive variables. The inclusion of irrelevant, redundant, and noisy variables in the model building process phase can result in poor predictive performance and increased computation.

Feature selectors are algorithms that attempt to identify and remove as much irrelevant and redundant information as possible prior to learning or knowledge discovery. Feature selection can result in enhanced performance, a reduced hypothesis search space, and, in some cases, reduced storage requirement.

Automated techniques for identifying and removing unhelpful or redundant variables can be divided into one of the three kinds of methods (Hall, 1999):

- *Filter Feature Selection Methods*: These techniques directly examine the relevance of candidate variables, independently of the machine learning model, usually a predictive method. Usually, we can distinguish two types of filter methods and other hybrid methods (Roffo et al., 2015; Lazar et al., 2012; Davidson & Jalan, 2010; Bekkerman et al., 2003; Caruana & de Sa, 2003; Koller & Sahami, 1996):

  - *Feature Ranking Methods*: They rank the features by a metric and eliminates all features that do not achieve an adequate score. Usually, these methods do not consider potential interactions among the features.
  - *Subset Selection Methods*: They search the set of possible features to get the best subset. Some of them can take into account the interaction among features (Huh, 2006).
  - *Mixed Approaches*: They can sequentially interleave some feature ranking operations with some subset selection operations, to try to capture all the benefits from both approaches.

- *Wrapper Feature Selection Methods*: These techniques search the best combination of features in terms of the ML model performance and get the feedback to guide the search. They utilize the ML model of interest as a black box to score subsets of features according to their predictive power (Phuong et al., 2005). They can be used in a supervised data matrix, with a target variable. Four subtypes can be distinguished:

  - *Brute-Force Methods*: They explore all the possible subsets of combinations of features and get the optimal one (Kohavi & John, 1997; Narendra & Fukunaga, 1977).
  - *Forward Methods*: Starting with an empty set of features, they add one feature at each step until getting the optimal subset of features (Reunanen, 2003; Pudil et al., 1994).
  - *Backward Methods*: Starting from a set containing all the features, they discard one feature at each step until getting the optimal set of features (Nakariyakul & Casasent, 2009; Pudil et al., 1999; Stearns, 1976).
  - *Random Methods*: They can try several subsets of features in a random way, trying to avoid to be trapped in local optima (i.e. anytime algorithms) (Sun et al., 2005; Alexandridis et al., 2005; Liu & Motoda, 1998; Yang & Honavar, 1998; Jouan-Rimbaud et al., 1995; Puch et al., 1993).

- *Embedded Methods*: They are an intermediate type of methods, which perform a feature selection task in the process of training of the ML technique and they are usually specific and included within the given ML techniques (Duval et al., 2009), such as in decision trees and artificial neural networks techniques (Mitchell, 1982).

For a survey of common feature selection techniques, see (Chandrashekar & Sahin, 2014; Molina et al., 2002).

In addition, feature selection can be performed by applying a cutting threshold over the results of a feature weighting process.

Usually, analyzing the feature subset selection provides better results than analysing the complete set of variables (Hall & Smith, 1998).

6.3.11   Feature Relevance

6.3.11.1   Relevance Detection

An important problem in the pre-processing step is to find out which are the *relevant variables* or *features* to be taken into account. When experts are available in a particular domain or application, these tasks could be easier, because the experts could give their advice on which are the relevant variables and which are the irrelevant ones. However, in general, when there is no expertise available, some automatic methods to determine the degree of relevance for each variable should be used. These methods are known as *feature weighting* methods.

6.3.11.2   Feature Weighting

Feature weighting techniques (Núñez, 2004; Aha, 1998; Wettschereck et al., 1997) can provide a ranking of the variables or attributes or features according to their degree of relevance for the dataset. The degree of relevance of a variable (X_j) is expressed by means of a weight (w_j), which is usually a real value belonging to the interval [0,1]. They assign high weights to variables that are identified as relevant, and at the same time, low weights to those that are irrelevant or redundant. A weight with a value 1 means a maximum degree of relevance (completely relevant). whereas a value 0 means a minimum value of relevance (completely irrelevant). This way, it is possible to decide what the important variables in a dataset are. These techniques are useful if the importance of the variables for the dataset can be taken into account in the analysis. For instance, Feature weight assignment is frequently used to denote the relevance of attributes in *similarity computations* in clustering techniques or in some inductive classification or predictive methods, allowing the similarity measures to emphasize the variables according to their relevance.

Feature selection can be considered as a specialization of *feature weighting*, where all the weights get binary values: 0 or 1. If the weight is 0, that means the variable is out of the set of selected variables, and if the weight is 1, that means the variable is in the set of selected variables.

One of the problems in the feature weighting methods is how to decide when a set of weights is better than the other one. They must be evaluated in terms of the performance of a task:

- In *supervised domains*, the task could be a classification task, measuring the accuracy of the label predictions for unlabelled instances.
- *Unsupervised weighting methods* assign weights to variables without any knowledge about class labels, so this task is presumed more difficult (Dash & Liu, 1999; Howe & Cardie, 2000). In fact, they use alternative measures like significant changes in similarity computations to evaluate the goodness of a set of weights.

The *feature weighting* algorithms can be divided into *filter algorithms* and *wrapper algorithms* depending on whether they used feedback knowledge by a ML model or not during the process of obtaining the weights for the features.

Feature weighting algorithms can also be distinguished by their *generality*. While most algorithms learn settings for a single set of weights that are employed globally (i.e. over the entire instance space), other algorithms assume weights differ among local regions of the instance space.

- The assumption made by *global weighting* methods that attribute relevance is invariant over the instance space is constraining, and sometimes, inappropriate. Other algorithms assume that the relevance of the attribute is not necessarily the same in the whole domain.
- Two types of *local weighting* schemes are popular. The first assigns a different weight to each qualitative value of the attribute. Although this allows feature relevance to vary over the values of the feature, it still constrains weights to be identical for all instances with the same qualitative feature value. The second local weighting scheme removes these constraints by allowing feature weights to vary as a function of the instance and its belonging to a class.

Many researchers have focussed their research on *supervised feature weighting* and several works can be found in the literature:

- Most common *filter global supervised methods* in the literature are:

  – The *Mutual Information technique* (MI) proposed in (Cover & Thomas, 1991) assigns low weights to attributes providing little information for classification, and high weights to those ones providing more reliable info. The weights are computed as follows:

$$w_d = \sum_{v \in V_d} \sum_{i \in C} p(c_i, x_d = v) * \log_2 \left(\frac{p(c_i, x_d = v)}{p(c_i) * p(x_d = v)} \right) \tag{6.19}$$

  where,

  $p(c_i)$ is the probability of an instance to have the class label c_i
  $p(x_d = v)$ is the probability of an instance to have the value v for the attribute A_d
  $p(c_i, x_d = v)$ is the probability of an instance to have the class label c_i and to have the value v for the attribute A_d.

- The *Information Gain* (IG) method proposed by (Daelemens & van den Bosch, 1992) assign weights using the Information Gain measure as the weight for each attribute, as follows:

$$
\begin{aligned}
w_d = I(X, C) - I(X, A_d) \\
= - \sum_{i \in C} p(c_i) \log_2(p(c_i)) \\
- \sum_{v \in V_d} p(x_d = v) * \left(- \sum_{i \in C} p(c_i | x_d = v) * \log_2(p(c_i | x_d = v)) \right)
\end{aligned}
\tag{6.20}
$$

- The *projection of attributes* method (PRO) proposed by (Güvenir & Akkus, 1997), that is based on the idea that instances of the *same class* in the training set should be contained in the *same region when all instances are sorted by a given attribute*. The projection of instances of the same class are grouped together. The weights are computed as follows:

$$
w_d = \frac{\sum_{n=1}^{N} \alpha(d, n)}{N}
\tag{6.21}
$$

$$
\text{where } \alpha(d, n) = \begin{cases} 1 & \text{if } C_{n,d} = C_{(n+1),d} \\ 0 & \text{otherwise} \end{cases}
$$

- The *RELIEF-F* method (Kononenko, 1994; Kira & Rendell, 1992) which uses an online sequential search. It estimates the importance of attributes according to how well their values distinguish among instances that are near each other. The procedure is as follows:

```
algorithm Relief

input: T: the training data set (N instances * D attributes)
output: w, the vector of weights

begin
   w⁽⁰⁾ ← [0.5, ..., 0.5]ᵀ
   for each i = 1, N do
      Compute hⁱ = nearest instance to tⁱ of the same class
         // i.e., it is called the nearest hit
      Compute mⁱ = nearest instance to tⁱ of a different class
         // i.e., it is called the nearest miss
      for each d = 1, D do
         w_d^(i) = w_d^(i-1) - dissim_d(t_d^i, h_d^i)/(2*N) + dissim_d(t_d^i, m_d^i)/(2*N)
      endfor
   endfor
end
```

- The *Class Value Distribution* method (CVD) (Núñez et al., 2002), which estimates how far is the attribute from the "*ideal*" attribute. The *ideal* attribute would be the one that for each possible range value would discriminate completely each possible value of the class variable. To do that, the correlation matrix between the class and value distribution for each attribute is analyzed. The heuristic value and the corresponding weight values are computed as follows:

$$H_d = \frac{1}{Q} \sum_{j=1}^{Q} \left(\frac{q_{\max j}}{q_{+j}} * \frac{q_{\max j}}{q_{\max,+}} \right) \tag{6.22}$$

$$w_d = \frac{H_d - \frac{1}{|V(A_d)|*Q}}{1 - \frac{1}{|V(A_d)|*Q}} \tag{6.23}$$

where,

Q is the number of class labels
$q_{+,j}$ is the total number of instances belonging to class label j
$q_{\max, j}$ is the maximum value of the column j
$q_{\max, +}$ is the sum of the row *max* of the correlation matrix
$|V(A_d)|$ is the number of different values of the feature A_d

- The QM2 method of Mohri and Tanaka (1994).
- The Cross-Category Feature importance (CCF) method (Creecy et al., 1992).

• On the other hand, a few *filters local supervised weighting methods* have been proposed in the literature such as:

- The value difference metric (VDM) by Stanfill and Waltz (1986) introduced one of the first similarity algorithms assigning local weights to the attributes according to its values. It assigns higher weights to features whose distribution of values across classes is highly skewed. The weights are computed as follows:

$$w_d(x_d) = \sqrt{\sum_{q=1}^{Q} \left(\frac{N(x_d, C_q)}{N(x_d)} \right)^2} \tag{6.24}$$

where,

x_d is a value of the attribute d,
$N(x_d)$ is the number of instances that have value x_d in attribute d
$N(x_d, c_q)$ is the frequency that the value x_d is classified in the class c_q for the attribute d.

– The Entropy-Based Local weighting method (EBL) (Núñez et al., 2003). This
  method uses the entropy to assign the local weights. From the correlation
  matrix, we can obtain the entropy from each value (range) as follows:

$$H_{di} = -\frac{q_{i+}}{q_{++}} \sum_{j=1}^{Q} \frac{q_{ij}}{q_{i+}} \log\left(\frac{q_{ij}}{q_{i+}}\right) \tag{6.25}$$

This entropy H_{di} belonging to value (range) i from attribute d will be the
basis to calculate the weight for the value j following this simple idea. If the
value (or range) has a maximum possible entropy ("totally random"), then the
weight must be 0. On the other hand, if the value (or range) has a minimum
possible entropy ("perfectly classified") then the weight must be 1. The
minimum possible entropy is 0 when all the instances with this value (range)
belong to the same class. The maximum possible entropy occurs when the
instances with this value (range) are equally distributed in all classes and can
be calculated as follows:

$$\max\left(H_d^i\right) = -\frac{q_{i+}}{q_{++}} \log\left(\frac{1}{Q}\right) \tag{6.26}$$

and then, scaling the values to the interval [0, 1]:

$$w_d^i = 1 - \left(\frac{H_d^i}{\max\left(H_d^i\right)}\right) \tag{6.27}$$

Q is the number of class labels
q_{ij} is the total number of instances showing feature value v_i belonging to class j
q_{i+} is the total number of instances showing feature value v_i
q_{++} is the total number of instances

• Some *wrapper supervised methods* have been designed like:

– The *diet algorithm* (DA) (Kohavi et al., 1997) proposes a discrete weight
  assignment among a finite group of possible values. This set of possible values
  is determined by: $\{0, \frac{1}{v}, \frac{2}{v}, \ldots, \frac{v-1}{v}, 1\}$. An initial weight is assigned to each
  one of the attributes. The algorithm performs a best-first search, guided by
  tenfold cross validation in a weight space where each feature has a weight in
  $\{0, \frac{1}{v}, \frac{2}{v}, \ldots, \frac{v-1}{v}, 1\}$. DA explores all variants that increase or decrease each
  weight by $\frac{1}{v}$ unless the maximum or minimum value has been reached. The
  best-performing weight set is used as the next starting point. Finally, authors
  adopt a halting criterion that stops search when it encounters five consecutive
  nodes with no children having scores of more than 0.1% better than their
  parent.

– Some approaches are based on the use of Genetic Algorithms (Golobardes et al., 2000; Ishii & Wang, 1998). The individuals of the population encode the weights of the attributes, and the fitness function gives the feedback for optimizing the weights.

However, less work has been done in the field on *unsupervised feature weighting*, which is clearly the required one when facing a new unknown database, which we want to mine to discover new knowledge, and usually no reference classification is available. In the literature, main works solving the problem are:

- The work done by Shiu et al. (2001) on the application of the *Gradient Descent technique* (GD) to optimize a function related to the similarity computation.

- Derived from a feature selection approach based on unsupervised entropy-based method (Dash & Liu, 1999), Núñez and Sànchez-Marrè (2004) proposed two new unsupervised feature weighting methods (UEB-1 and UEB-2). UEB-2 is an iterative version of UEB-1. For the entire data set of N instances, the entropy measure is given as:

$$P = -\sum_{i=1}^{N} \sum_{j=1}^{N} S_{ij} * \log_2\left(S_{ij}\right) + \left(1 - S_{ij}\right) * \log_2\left(1 - S_{ij}\right) \tag{6.28}$$

where S_{ij} is the similarity value between the instance i and the instance j normalized to [0,1].

The UEB-1 method computes the entropy of data by removing a feature. For D features this is repeated D times. Features are ranked in descending order of relevance by finding the descending order of the entropy after removing each of the D features one at a time. The algorithm proceed as follows:

---

algorithm UEB-1

input: T: the training data set (N instances * D attributes)
output: w, the vector of weights

begin
  P = Entropy value for all D features
  for i = 1 to D
    P_i = CalcEnt(i)   // P_i is the entropy when attribute i is discarded
  endfor
  $minP = \min\limits_{i=1,\ldots,D} P_i$
  $maxP = \max\limits_{i=1,\ldots,D} P_i$
  for i = 1 to D
    $w_i = \frac{P_i - minP}{maxP - minP}$
  endfor
end

Empiric works (Wettschereck et al., 1997) and theoretical ones (Langley & Iba, 1993), suggest that the learning complexity is exponential regarding the number of irrelevant variables. Therefore, the failures in the data mining process could be related to a similarity model, and in particular, with an incorrect weight assignment methodology. A comprehensive review of feature weighting algorithms can be found in Núñez (2004).

6.4   Data Mining Methods

In this section, the main data mining techniques from the machine learning field will be analyzed. The data-driven IDSSs are formed by inductive machine learning methods. Inductive machine learning methods, explore a data matrix of examples and induce a model. That model, once validated, can be used for understanding the different data patterns like clustering techniques, or can be used for making predictions of one variable for new examples in the future.

There are general important characteristics related to the models to be considered in inductive machine learning:

- *Quantity of Data*: Of course, as many examples we have available to induce a ML method, the obtained model should be better, because the sample of the general population is greater, and probably, the models induced would be more representative of what can happen in the general population. For instance, if there is a dataset of 100 examples representing the citizens of a city, which has 3.5 million people, the patterns and information what can be discovered from that sample are very limited. On the other hand, if we had a sample dataset of 100,000 citizens, very probably, many more patterns can be discovered, and the models induced could be more representative of what could happen for any citizen in the city. However, in general more quantity of data does not directly imply more quality of the data.
- *Quality of Data*: In Computer Science, there is a very famous saying:

"*Garbage in, garbage out*"

   Which means that in any general algorithm/program if the input data is not good or precise, the output generated by the algorithm will be also not very good or precise. This general principle, in ML becomes the following statement:

"*Poor quality data, bad induced models*"

   Which means that if the quality and representativeness of the data is not good, then, the induce model/s will not be good models, because they are not capturing all the patterns that could happen, and just some ones.

   The *representativeness of the data* is a crucial aspect. If in a dataset we have many examples, but all of them are very similar among them, we are probably missing the information that other different examples can supply to an induced model. For instance, take into account the different distribution of the data in the

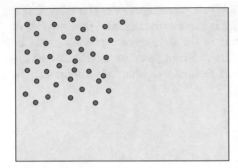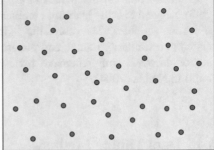

Fig. 6.11 Different representativeness of data due to a different data sample distribution

example depicted in Fig. 6.11. Of course, the data distribution on the right, is more representative than the one on the left, with the same number of observations.

It is clear that the data observations on the left-hand side are not representative of the whole space of possible observations, which here is represented in a two-dimension space. Any model induced from this data distribution could be good for making predictions or understanding patterns of similar observations, but surely will not make good predictions or could understand patterns from other data observations located very far in the observations space.

In supervised domains with a qualitative target variable, i.e. *the class variable*, sometimes could happen the problem of *imbalanced class distribution*. This means that one or more classes are underrepresented in the sample dataset. The percentage of examples belonging to that class is very low, compared to other majority classes which are overrepresented. This situation is very common in medicine, or in industrial processes, where the class or classes considered "normal" are usually, and fortunately, the most common (i.e. the class with patients not having a disease, the class corresponding to a normal operational state of a process, etc.). This is an open challenge in supervised machine learning.

As a summary it is worth mentioning two statements regarding models and modelling by two outstanding researchers in the process control community and statistical modelling community:

- Prof. Gustaf Olsson, from the University of Lund, said in an invited talk at the fourth International Congress on Environmental Modelling and Software (iEMSs 2008) in Barcelona: "Do not fall in love with models"
- Prof. George Box stated in his book entitled *"Empirical model-building and response surfaces"* (Box & Draper, 1987) and in a previous journal article: "All models are wrong, but some are useful"

Models are an abstract and simplified representations of reality. Specifically, inductive models are abstractions obtained from data. Therefore, the goodness and representativeness of the models is directly related to the goodness and representativeness of the data.

6.4.1 Unsupervised Models

As explained before, the unsupervised models try to discover the hidden relationships between the examples, i.e. *descriptive models*, or between the variables, i.e. *associative models*. In this section, both kinds of methods will be analyzed and the main techniques used will be described.

6.4.1.1 Descriptive Models

There are problems that require discovering the underlying hidden concepts in a dataset or describing the observations/instances by means of obtaining *groups* or *clusters* of instances sharing some similarities. A *clustering technique* is a Machine Learning task that partitions a dataset and groups together the most similar instances. It separates a set of instances into a number of groups so that instances in the same group, called *a cluster*, are more similar to each other than to those in other groups.

For example, in Fig. 6.12, it is shown a possible characterization of the observations in four groups or clusters. Observations of the same cluster are depicted in the same colour. Therefore, four clusters can be distinguished in the data. One cluster has its observations depicted in green, another in yellow, another in blue, and the last one in red. If this clustering process is meaningful, it will provide a *characterization* or *profile* of the observations belonging to the same cluster. The rationale is that two observations of the same cluster are more similar between them than to other observations from other clusters. Therefore, with a clustering technique, the main features that characterize a group can be extracted. This information is very valuable to describe and know the different *segmentation of the observations*, like the customers of a company, etc.

Fig. 6.12 A possible four clusters of observations of a dataset

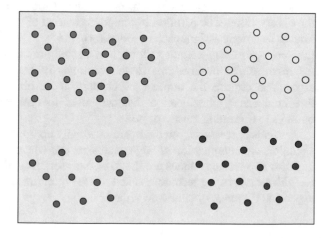

Clustering techniques are unsupervised learning techniques. Once the clusters are properly interpreted and validated, it is common to assign a different label to each cluster, creating this way a new qualitative variable in the dataset.

In addition to clustering techniques, in the literature, there are also the Self-Organizing Feature Maps or Self-Organizing Maps (SOMs) (Kohonen, 1989). SOMs provides both a data visualization technique and grouping functionality. It helps to understand high-dimensional data by reducing the dimensions of data to a map. It also represents a clustering concept by grouping similar data together.

According to the literature (Jain & Dubes, 1988), *clustering techniques* can be subdivided into *exclusive clustering* and *non-exclusive or overlapping clustering*:

- *Exclusive clustering techniques* assign a unique cluster label to each observation. Thus, one observation can only belong to a group or cluster. These kinds of techniques are the most commonly used in inductive machine learning.
- *Non-exclusive or overlapping techniques* can assign more than one label to an observation. The degree of membership to each one of the possible overlapping clusters is determined by a fuzzy or a probabilistic measure. The corresponding techniques are usually named *fuzzy clustering techniques* or *probabilistic clustering techniques*.

Exclusive clustering techniques are the most commonly used in data science because they assign just one cluster to each instance, and this way, do not make more difficult the conceptualization problem. These techniques can be divided in two major categories: *partitional clustering techniques* and *hierarchical clustering techniques* by the type of structure imposed on the data. Next, these techniques are described.

Partitional Clustering Techniques

A *partitional clustering* technique generates a single partition of the data in an attempt to discover groups present in the data. It tries to obtain a good partition of the observations. The partition is composed of a set of groups or clusters. Thus, this kind of technique assigns each observation to the "best" cluster. This "best" cluster is the one optimizing certain criteria like the minimization of the square sum of distances of the observations to the centroids of the clusters, etc. Either these algorithms require the number of clusters to be obtained, namely k, or some threshold value (classification distance) used to decide whether an observation belongs to a forming cluster or not.

Partitional clustering methods are especially appropriate for the efficient representation and compression of large databases, and when just one partition is needed.

Most popular techniques are the *K-means clustering algorithm* and the *Nearest-Neighbour clustering* technique. Next, these algorithms and a variation of *K-means* algorithm (*G-means*) will be described.

K-Means Clustering

One of the most popular partitional clustering algorithms is the *k-means clustering* algorithm (MacQueen, 1967). Starting with a randomly initial partition centroid of *k* clusters, it explores the idea of changing the current partition to another one decreasing the sum of squares of distances of the observations to the centroids of the clusters. It converges, possibly to a local minimum, but in general, can converge fast in a few iterations. *K*-means algorithm is intended to work well if the clusters to be discovered are *convex sets*.[7] It has a main parameter *k*, which is the number of desired clusters. In its original formulation, it is oriented to be used with only *numeric variables* and using the *(squared) Euclidian distance*. Other distances can be used but could prevent the algorithm to converge to a solution.

The general scheme of the algorithm is as follows:

---

algorithm *k*-means

input: *X*: the training data set (*N* instances * *D* attributes): $\{X_{1.}, ..., X_{N.}\}$;
     k: the required number of clusters
output: *C*: the set of cluster centroids: $\{C_1, ..., C_k\}$
     Class: a new attribute for each instance: $\{Class(X_{1.}), ..., Class(X_{N.})\}$
begin
    // Assign randomly k observations as the centres of the k clusters
    $C_j^{j=1,...,k} \leftarrow X_{i.}$ *randomly selected from* $1 \leq i \leq N$
  while any observation changes its cluster membership **do**
    // Assign each observation to its closest cluster centre
    $Class(X_{i.}) \leftarrow \min_{j} d(X_{i.}, C_j)$
    // Compute new cluster centres as the centroids of the clusters
    $C_j \leftarrow centroid(\{X_{i.} \mid Class(X_{i.}) = C_j\})$
  endwhile
end

G-Means Clustering

Sometimes it is hard to know in advance how many clusters can be identified in a dataset or simply it is not desired to force the algorithm to output a specific number of clusters. Gaussian-means (*G*-means) (Hamerly & Elkan, 2003) algorithm was designed to solve this issue. *G*-means use a special technique for running *k*-means multiple times while adding centroids in a hierarchical way. *G*-means has the advantage of being relatively resilient to covariance in clusters and has no need to compute a global covariance. The *G*-means algorithm starts with a small number of

---

[7] A *convex set* is a set satisfying the property that for each pair of points belonging to the set, all the points in the segment connecting both points also belong to the set. Intuitively, a convex set is more or less a "round-shaped" set.

k-means centres and grows the number of centres. Each iteration of the algorithm splits into two those centres whose data appear not to come from a Gaussian distribution. Between each round of splitting, k-*means* is run on the entire dataset and all the centres to refine the current solution.

The test used is based on the Anderson-Darling statistic (Anderson & Darling, 1954). This one-dimensional test has been shown empirically to be the most powerful normality test that is based on the empirical cumulative distribution function (ECDF).

The general scheme of the algorithm is as follows:

---

algorithm G-means

input: X: the training data set (N instances * D attributes): $\{X_1., ..., X_N.\}$;
        α: is the confidence level for the normality statistic test
output: C: the set of cluster centroids: $\{C_1, ..., C_k\}$
           Cl : a new attribute for each instance: $\{Cl_1 = \text{Class}(X_1.), ..., Cl_N = \text{Class}(X_N.)\}$
begin
    // C is the initial set of centres with just one centre C_1 as average of all instances
    $C_1 \leftarrow \{\bar{X}_{.1}, ..., \bar{X}_{.D}\}$
    $C \leftarrow \{C_1\}$
    repeat
        $C \leftarrow k - means(X, |C|)$
        for each $C_j \in C$ **do**
        // stat-test to check whether $\{X_i. \mid Class(X_i.) = Cl_j\}$ follows a Gaussian distr.
        Gaussian-distr $\leftarrow$ Anderson-Darling-stat-test(X, C_j, α)
        if Gaussian-distr **then** keep C_j
                    else replace C_j with two centres
      endif
      endfor
    until no more centres are added
end

Nearest-Neighbour Clustering

A natural way to define clusters is by utilizing the property of nearest neighbours: an observation should usually be put in the same cluster as its nearest neighbour.

A nearest neighbour of a given instance is the instance least distant from the given one. Two observations should be considered similar if they share neighbours.

One of the most used clustering algorithms which is based on the nearest neighbour idea is due to (Lu & King Sun, 1978), where the user specifies a threshold, *thr*, on the nearest-neighbour distance. If new observations are at a less distance from its nearest neighbour than *thr*, then they are assigned to the same cluster than its nearest neighbour.

The general scheme of the algorithm is as follows:

---

algorithm Nearest-Neighbour clustering

input: X: the training data set (N instances * D attributes): {$X_1., ..., X_N.$};
        thr. is the threshold in the distance value for belonging to a cluster
output: C: the set of cluster centroids: {$C_1, ..., C_k$}
        Cl : a new attribute for each instance: {Cl_1 = Class($X_1.$), ..., Cl_N = Class($X_N.$)}
begin
    // Let be k, the number of current clusters created
    $k \leftarrow 1$
    $Class(X_1.) \leftarrow Cl_1$
    for each $X_i., \ i = 2, ..., N$ **do**
        // Find the NN of $X_i.$ among the observations already assigned to clusters
        $NN_i \leftarrow \min_{1 \leq j < i} d(X_i., X_j.)$
        $m \leftarrow Class(NN_i)$
        if $d(X_i., NN_i) \leq thr$ **then** $Class(X_i.) \leftarrow Cl_m$ // assign $X_i.$ to the cluster m
                            else $k \leftarrow k + 1$ // create a new cluster
                                 $Class(X_i.) \leftarrow Cl_k$ // assign $X_i.$ to the new cluster k
    endif
    endfor
end

The distance function d used in the *Nearest-Neighbour* algorithm and, in general, in any other clustering algorithm described in Sect. 6.4.1.1 for the computation of the dissimilarity between two observations or centroids can be the *Euclidian* distance, *Manhattan* distance, *Mahalanobis* distance, etc., when all the variables are numeric. For non-ordered categorical variables, the *Hamming* distance can be used, for the ordered categorical variables a *normalized ordinal* distance can be used, and for textual variables, the *Levenshtein* distance (Levenshtein, 1966) is commonly used. In the case that the dataset has heterogeneous variables, then a heterogeneous distance measure must be used like *Gower* distance or coefficient (Gower, 1971), *L'Eixample* distance (Sànchez-Marrè et al., 1998), etc.

Hierarchical Clustering Techniques

A hierarchical clustering process is a nested sequence of partitions. Hierarchical clustering (Jain & Dubes, 1988) is a general family of clustering algorithms that build nested clusters by merging or splitting them successively. This hierarchy of clusters is represented as a tree, (named as *dendrogram*). A *dendrogram* is a special type of tree structure that provides a convenient picture of hierarchical clustering. A *dendrogram* consists of a rooted binary tree, where the nodes represent clusters. Lines connecting nodes represent clusters which are nested into one another. The y-axis shows the distance/dissimilarity value of each two clusters merged, and the x-axis has the different observations/instances to be clustered. Cutting horizontally a *dendrogram* creates a partition (set of clusters), and determines the number of clusters obtained.

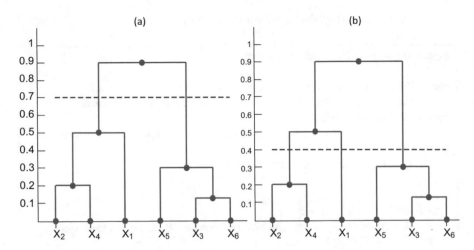

Fig. 6.13 An example of a *dendogram* from a hierarchical clustering process with six observations and two possible cutting points

Figure 6.13 provides an example of *dendogram* with six observations. The upper picture (a) corresponds to the case of cutting the *dendrogram* at the distance 0.7. This cut generates two clusters: C_1, C_2 where $C_1 = \{X_1, X_2, X_4\}$ and $C_2 = \{X_3, X_5, X_6\}$.

On the other hand, the lower picture (b) corresponds to the case of cutting the *dendrogram* at the distance 0.4. This cut generates three clusters; C_1, C_2, C_3 where $C_1 = \{X_2, X_4\}$, $C_2 = \{X_1\}$ and $C_3 = \{X_3, X_5, X_6\}$.

The best cutting point is usually the one where the clusters merged/split has the greatest distance/dissimilarity gap, which means that at that level, those clusters merged/split is different, and not as similar as other possible clusters. In the example of Fig. 6.13, the best cut seems to be at the 0.7 cutting point where the greatest gap happens among all the possible clusters merge/split. Therefore, these two clusters are supposed to be the best partition.

However, from a more mathematical point of view, there are some criteria to validate structurally the goodness of a set of clusters (partition). These criteria are known as Cluster Validity Indexes (CVIs). From them, as it will be described later in this section, the *Caliński-Harabasz index* (Caliński & Harabasz, 1974) seems to be one giving certain security that the set of clusters given by a cut point are a good choice or not.

The root of the tree is the unique cluster that gathers all the observations. The leaves are the clusters with only one observation. Hierarchical clustering techniques are useful when more than one partition is needed and/or when taxonomies are required like in Medical, Biological or Social Sciences. Anyway, *dendrograms* are impractical with just a few 100 observations.

Techniques for hierarchical clustering can be divided into two basic paradigms: *agglomerative* (bottom–up) and *divisive* (top–down) approaches. All the agglomerative and some divisive methods (viewed in a bottom–up direction) satisfy a *monotonicity property*: the dissimilarity between merged clusters is an increasing monotonic function regarding the level of the merger. Therefore, the *dendogram* can be plotted so that the height of each node is proportional to the value of the intergroup dissimilarity between its two children.

Agglomerative/Ascendant Techniques

An agglomerative or ascendant hierarchical clustering initially places each observation/instance in its own cluster, and gradually merges these atomic clusters into larger clusters until all observations are in a single cluster, like in the *dendogram* represented in Fig. 6.13. The pair chosen at each step for merging consists of the two clusters with the smallest intergroup dissimilarity (distance). There are several methods implementing this principle. Their difference relies in how they compute the distances/dissimilarities between the clusters and/or the observations.

The general algorithmic scheme for hierarchical agglomerative/ascendant clustering technique (Johnson, 1967) can be described as follows:

---

algorithm Hierarchical agglomerative clustering

input: N: the number of observations
$\quad\quad\quad$ X: the training data set (N instances * D attributes): $\{X_{1.}, ..., X_{N.}\}$
output: a *dendogram*(X,N) according to the DissCriterion

begin
$\quad CL \leftarrow \emptyset$
$\quad$ // Assign each observation to a cluster
$\quad$ **for each** $X_{i.}$, $i = 1, ..., N$ **do**
$\quad\quad CL_i \leftarrow \{X_{i.}\}$
$\quad\quad CL \leftarrow CL \cup CL_i$
$\quad$ **endfor**
$\quad$ // Compute the dissimilarities between the clusters
$\quad$ **for each** CL_i, $i = 1, ..., N$ **do**
$\quad\quad$ **for each** CL_j, $j > i$ **do**
$\quad\quad\quad DissimMatrix(i,j) = DissCriterion(CL_i, CL_j)$ $\quad\quad\quad$ // Differential Step
$\quad\quad$ **endfor**
$\quad$ **endfor**
$\quad$ // while not all observations are clustered into a single cluster of size N
$\quad$ **while** $|CL| > 1$ **do**
$\quad\quad$ // Find the closest pair of clusters.
$\quad\quad < i, j > \leftarrow \arg\min_{k,l} DissimMatrix(k,l)$
$\quad\quad$ // Merge them into a single cluster
$\quad\quad CL_{ij} \leftarrow CL_i \cup CL_j$
$\quad\quad CL \leftarrow CL - \{CL_i, CL_j\} \cup CL_{ij}$
$\quad\quad$ // Plot the merged clusters in the dendogram
$\quad\quad PlotMerge(CL_i, CL_j, DissimMatrix(i,j))$
$\quad\quad$ // Compute dissimilarities between the new cluster and the old clusters
$\quad\quad$ **for each** CL_k in CL, $k \neq ij$ **do**
$\quad\quad\quad DissimMatrix(ij,k) = DissCriterion(CL_{ij}, CL_k)$ $\quad\quad\quad$ // Differential Step
$\quad\quad$ **endfor**
$\quad$ **endwhile**
end

The different variations of the hierarchical ascendant clustering algorithms rely on the step of how the similarity criterion (*DissCriterion* in the algorithm) between two clusters is used. Main used similarity criteria in the literature are known as *single-linkage clustering, complete-linkage clustering, average-linkage clustering, centroid-linkage clustering, and Ward's method*:

- *Single-linkage clustering* (also called the *connectedness* or *minimum method*) considers the dissimilarity between one cluster and another cluster to be equal to the *shortest distance* from any member of one cluster to any member of the other cluster.

$$\text{DissCriterion}(\text{CL}_i, \text{CL}_j) = \min\left\{d(X_{k.}, X_{l.}) | X_{k.} \in \text{CL}_i, X_{l.} \in \text{CL}_j\right\} \quad (6.29)$$

- *Complete-linkage clustering* (also called the *diameter* or *maximum method*), considers the dissimilarity between one cluster and another cluster to be equal to the *greatest distance* from any member of one cluster to any member of the other cluster.

$$\text{DissCriterion}(\text{CL}_i, \text{CL}_j) = \max\left\{d(X_{k.}, X_{l.}) | X_{k.} \in \text{CL}_i, X_{l.} \in \text{CL}_j\right\} \quad (6.30)$$

- *Average-linkage clustering*, considers the dissimilarity between one cluster and another cluster to be equal to the *average distance* from any member of one cluster to any member of the other cluster.

$$\text{DissCriterion}(\text{CL}_i, \text{CL}_j) = \frac{1}{|\text{CL}_i| \cdot |\text{CL}_j|} \sum_{X_{k.} \in \text{CL}_i} \sum_{X_{l.} \in \text{CL}_j} d(X_{k.}, X_{l.}) \quad (6.31)$$

- *Centroid-linkage clustering*, considers the dissimilarity between one cluster and another cluster to be equal to the *distance* between *the centroids* of each cluster.

$$\text{DissCriterion}(\text{CL}_i, \text{CL}_j) = d\left(C_{\text{CL}_i}, C_{\text{CL}_j}\right) \quad (6.32)$$

- *Ward's method* (Ward, 1963) (also called the *minimum variance method*), which merges in a new cluster (t), the pair of clusters (q, r) minimizing the change in the square-error of the entire set of clusters ($\Delta E_{qr}^2 = e_t^2 - e_q^2 - e_r^2$) when the two clusters are merged.

Let be $C_{\text{CL}_k} = C^k = \left[C_1^k, \ldots, C_D^k\right]$ the centroid of a cluster k, which has n_k observations, where $C_j^k = \frac{1}{n_k} \sum_{i=1}^{n_k} x_{ij}^k$ is each one of the j components of the centroid.

The square-error for a cluster k is $e_k^2 = \sum_{i=1}^{n_k} \sum_{j=1}^{D} \left(x_{ij}^k - C_j^k\right)^2$, and the square-error for the whole set of K clusters is $E_K^2 = \sum_{k=1}^{K} e_k^2$, Ward's method merges the pair of clusters minimizing ΔE_{qr}^2

$$\text{DissCriterion}\left(\text{CL}_q, \text{CL}_r\right) = \Delta E_{qr}^2 = e_t^2 - e_q^2 - e_r^2$$

$$= \frac{n_q n_r}{n_q + n_r} \sum_{j=1}^{D} \left(C_j^q - C_j^r\right)^2 \qquad (6.33)$$

The general complexity for agglomerative clustering is $O(n^3)$ or $O(n^2 \log (n))$[8] (Rokach & Maimon, 2005) with the use of a heap data structure, but for some special cases, optimal efficient agglomerative methods of complexity $O(n^2)$ are known: SLINK (Sibson, 1973) for single-linkage clustering and CLINK (Defays, 1977) for complete-linkage clustering.

Divisive/Descendent Techniques

Divisive or *descendent hierarchical clustering* reverses the process of an agglomerative procedure by starting with all observations or instances in one cluster and recursively divide one of the existing clusters into two children clusters at each iteration in a top–down procedure. The split is chosen to produce two new clusters with the largest intergroup dissimilarity (distance).

In the general case, *divisive clustering techniques* have a complexity of $O(2^n)$ (Everitt, 2011). For that reason, divisive methods are not very popular, and this approach has not been studied as extensively as agglomerative methods in the clustering literature. The existent algorithms propose some *heuristic* in order not to generate all possible splitting combinations.

One of the first divisive algorithms in the literature was proposed in (MacNaughton-Smith et al., 1965). It begins by placing all observations in a single cluster G. It then chooses that observation whose average dissimilarity from all the other observations is largest. This observation forms the first member of a second cluster H. Then, it moves to the new cluster H the observations in G whose average distance from those in G is greater than the average distance to the ones in the new cluster H. The result is a split of the original cluster into two children clusters, the observations transferred to H, and those remaining in G. These two clusters represent the second level of the hierarchy. Each successive level is produced by applying this splitting procedure to one of the clusters at the previous level.

Other divisive clustering algorithm was published as the DIANA (DIvisive ANAlysis Clustering) algorithm (Kaufman & Rousseeuw, 1990). DIANA follows

---

[8] $O(f(n))$ is an asymptotic notation, named as Big O notation, used in Computer Science to measure the *execution time*, also named as *time complexity or computational complexity*, of an algorithm needed to process the amount of data n. As the main interest is the behaviour of the algorithm when the values of n are high ($n \to \infty$), the function $f(n)$ is chosen to be as simple as possible, omitting constant factors and lower order terms, which are negligible when $n \to \infty$ (asymptotic behaviour). For instance, if the execution time of an algorithm is found to be $T(n) = 3n^3 - 2n^2 + 5$, it will be said that $T(n) = O(n^3)$. See (Cormen et al., 2009) for a broader insight.

the same strategy proposed by MacNaughton-Smith et al., but chooses the cluster with the largest diameter (i.e. the one maximizing the distance among its member observations). A possible alternative could be to choose the one with the largest average dissimilarity among its member observations.

An obvious alternate choice is *k-means* clustering with $k = 2$, (Steinbach et al., 2000) but any other clustering algorithm producing at least two clusters can be used, provided that the splitting sequence possesses the *monotonicity property* required for *dendogram* representation.

Validation of Descriptive Models

In general, after a data-driven model has been induced, several *post-processing* techniques must be carried out, as it was outlined in Sect. 6.2, and it will be explained in Sect. 6.5. One of them, which is very important, is the *validation of the model*, which will be generally described along Sect. 6.4. According to the nature of the data-driven technique, there are specific validation techniques that must be done.

Once a clustering technique has been applied, the resulting set of clusters must be validated in order to ensure that the clusters are structurally well-formed, and to get the underlying meaning of the clusters. Usually, the actual partition of the data, i.e. the set of clusters, is unknown and, therefore, the results from a clustering process cannot be compared with a *reference partition* by computing misclassification rates, as in the case of supervised learning.

Note that a major problem with clustering techniques is that they always find a set of clusters, even actually there were no meaningful clusters at all in the data!

Structural Validation of Clusters

Cluster structural validation in the clustering research field is an open problem. In the literature, most of the used techniques for evaluating the clustering results are based on the numerical indexes, which evaluate the validity of the resulting partition from different points of view, known as *Cluster Validity Indexes* (CVIs). A wide number of CVIs can be found in the literature and some surveys comparing several CVIs (Halkidi et al., 2001) are available. However, there are currently no clear guidelines for deciding which is the most suitable index for a given dataset (Brun et al., 2007). In fact, there is not an agreement among those indexes, but it seems clear that each one can give some information about a different property of the partition like *homogeneity, compactness of clusters, variability*, etc. All these CVIs refer to structural properties of the partition, which are context-independent, and the evaluation based on them is mainly made in terms of the clusters' topology.

Most common CVIs in the literature are: Entropy index, Maximum Cluster Diameter index (Δ), Widest Gap index (wg), Average Within-Cluster Distance index (W), Within Cluster Sum of squares index (WSS), Average Between-Cluster Distance index (B), Minimum Cluster Separation index(δ), Separation index $(Sindex)$, Dunn index (D), Dunn-like index, Calińksi-Harabasz index (CH),

Normalized Hubert Gamma Coefficient ($\Gamma^\wedge$), Silhouettes index, Baker and Hubert index (*BH*), Within Between Ratio index (*WBR*), *C*-index, Davies-Bouldin index (*DB*)

In the work (Sevilla-Villanueva et al., 2016), it was outlined that indexes evaluate a reduced set of characteristics of a partition. Thus, all indexes can be grouped around four basic concepts:

- Indexes measuring *compactness* of clusters: *Diameter* (Δ), *wg, W, WSS*
- Indexes measuring *separation* between clusters: *B, Separation* (δ), *Sindex*
- Indexes measuring *relationships* between compactness and separation of the clusters: *CH, Silhouettes, $\Gamma^\wedge$, BH, WBR, C-Index, DB, and also D, Dunn-like*
- Indexes measuring *chaos or dispersion* in the clusters: *Entropy*

Therefore, it would be a good strategy to select one index from a different family to evaluate the different properties of the clustering result. For example, these four CVIs could be used:

- *Maximum Cluster Diameter* (Δ) is the maximum distance between any two observations or instances that belongs to the same cluster (Hennig & Liao, 2010). As lower is the value, the better is the partition *P* (set of clusters):

$$\Delta = \max_{CL_k \in P} \Delta_{CL_k} \tag{6.34}$$

$$\Delta_{CL_k} = \max_{X_{i.}, X_{j.} \in CL_k} d\left(X_{i.}, X_{j.}\right) \tag{6.35}$$

- *Minimum Cluster Separation* (δ) is the minimum distance between any two observations or instances that do not belong to the same cluster. It defines the lower separation among all the clusters. As high is the value, the better is the partition *P*:

$$\delta = \min_{CL_i, CL_j \in P} \partial_{CL_i, CL_j} \tag{6.36}$$

$$\partial_{CL_i, CL_j} = \min_{X_{k.} \in CL_i, X_{l.} \in CL_j} d(X_{k.}, X_{l.}) \tag{6.37}$$

- *Calińksi-Harabasz Index* (*CH*) is a cluster validation index based on measuring a compromise value between the *cluster separation* (between-cluster distances) and the *cluster compactness* (within-cluster distances) (Caliński & Harabasz, 1974). As high is the ratio, better is the partition $P = \{CL_1, \ldots, CL_K\}$:

$$CH = \frac{BGSS/K - 1}{WGSS/N - K} = \frac{BGSS}{WGSS} \cdot \frac{N - K}{K - 1} \tag{6.38}$$

where the Within Group Sum of Squares (*WGSS*), which is the sum for each cluster of the square of the distances of each observation ($X_{i.}$) in the cluster to the centroid of the cluster (C_{CL_k}), and the Between Group Sum of Squares (*BGSS*), which is the sum for each cluster of the square of the distance of the centroid of the cluster (C_{CL_k}) to the centroid of all the observations (C_X) weighted by the number of elements of the cluster ($|\mathrm{CL}_k|$) are defined as follows:

$$\mathrm{WGSS} = \sum_{\mathrm{CL}_k \in P} \sum_{X_{i.} \in \mathrm{CL}_k} d(X_{i.}, C_{\mathrm{CL}_k})^2 \tag{6.39}$$

$$\mathrm{BGSS} = \sum_{\mathrm{CL}_k \in P} |\mathrm{CL}_k| \cdot d(C_{\mathrm{CL}_k}, C_X)^2 \tag{6.40}$$

- *Entropy Index* measures the entropy or chaos associated with the partition *P* (Meilă, 2007). The entropy is a non-negative magnitude, which reflects the dispersion of the observations according to the clusters. It has a value 0 when there is no uncertainty, i.e. just one cluster. As lower the entropy value, better is the partition *P*:

$$\mathrm{Entropy} = -\sum_{\mathrm{CL}_k \in P} \frac{|\mathrm{CL}_k|}{N} \cdot \log\left(\frac{|\mathrm{CL}_k|}{N}\right) \tag{6.41}$$

Qualitative Validation of Clusters

In addition to the structural validation of the clusters, it is very important to make a *qualitative validation* of the clusters. Usually, the experts make this kind of validation. This validation process consists to carefully look at the composition of the obtained clusters, analyze them, and try to get an *interpretation* of each one of the clusters.

This interpretation process can be done through some *data summarization* techniques, like the *computation of the cluster centroids*. A cluster centroid is a prototype showing the most frequent characteristics of the observations belonging to that cluster. This information is very important to illustrate how the general profile of the observations belonging to a cluster is. A *centroid* is a virtual observation, which is the geometrical centre of the set of observations. It has the same number and type of components than the observations and has the average value of the numerical variables, the mode of qualitative variables, etc.

The centroid can be computed as follows:

$$C_{\mathrm{CL}_k} = C^k = \left[C_{.1}^k, \ldots, C_{.D}^k\right] = \left[\frac{1}{|\mathrm{CL}_k|} \sum_{X_{i.} \in \mathrm{CL}_k} X_{i1}, \ldots, \frac{1}{|\mathrm{CL}_k|} \sum_{X_{i.} \in \mathrm{CL}_k} X_{iD}\right] \tag{6.42}$$

for numerical variables, or

$$C_{\mathrm{CL}_k} = C^k = \left[C^k_{.1}, \ldots, C^k_{.D}\right] = \left[\mathrm{Mode}\left(X^k_{.1}\right), \ldots, \mathrm{Mode}\left(X^k_{.D}\right)\right] \qquad (6.43)$$

for categorical variables. In the case of mixed variables, each component of the centroid is computed according to its nature.

It provides a very useful information on the prototypical kind of observations of a cluster (low values of variable $X_{.1}$, high values of variable $X_{.2}$, etc.).

In addition, several *graphical visualizations of data* can help to the interpretation of the clusters (histograms, tables, bivariate scatter plots, class-labelled scatter plots, etc.). All these graphics can help to identify the characteristics of each one of the clusters.

Furthermore, graphics with a conditional distribution of the different variables according to the value of the cluster label provide a very useful information to find out the characteristic range of values of some variables in concrete clusters. Usually, this type of graphics is known as a *panel graph*. A *panel graph* is a kind of two-dimensional matrix, where in the rows there are located the different clusters obtained, and the columns are the different variables of the dataset. At each cell there is a bar plot or histogram, depending on the type of variable, showing the distribution of values for that variable among the elements of the corresponding cluster. This way, it is easy to check which variables have high values or low values for a given cluster or to check in which cluster there are the highest values or the lowest ones corresponding to one variable. Therefore, a characterization of the different clusters can be obtained from a visual inspection.

In Fig. 6.14, there is an example of a panel graph. In this panel graph, we can see K clusters, D variables, and its corresponding distributions. From the picture is easy to describe that the cluster CL_1 is formed by observations having high values of the variable $X_{.1}$, mostly showing the second modality of the variable $X_{.2}$, and so on. Furthermore, the cluster CL_2 shows a predominance of the first modality of variable $X_{.2}$, has low values of variable $X_{.D}$, etc.

An Example of a Descriptive Model

Let us suppose the following scenario. A *supermarket chain* has several stores in a country. At each store, different products can be sold like meat, fish, fruits and vegetables, cleaning products, and bakery. Furthermore, the stores are located in different municipalities and provinces. In addition, they are of different sizes. Ones are considered small (less than 100 m^2), medium (between 100 and 500 m^2), large (between 500 and 1000 m^2) and huge (more than 1000 m^2). The owner of the *supermarket chain* is wondering whether stores have a different selling behaviour or not. If that would be true, the owner would be interested in a characterization of the different selling behaviours. This way, the owner can focus efforts on the stores with less volume of sales, and detect which is the problem making them sell less than other stores.

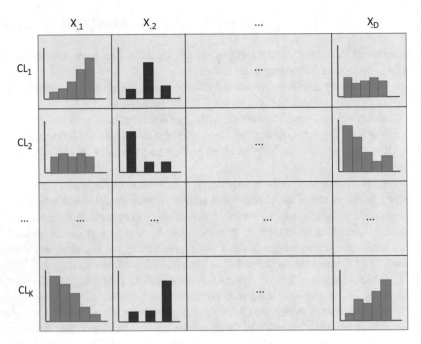

Fig. 6.14 A panel graph of a set of clusters

The company has a dataset with the data of its 174 stores over the country. The data for each store (instance) available are an *ID of the store*, the *size of the store* (small, medium, large, huge), the *shopping surface of the store*, the *municipality* where it is located, the *city type* where is the store (village, town, city, metropolitan) and the *province* where the municipality is located. Furthermore, regarding the sales at each store, it was available the average values for a period of a month of the different sales. There was the *volume of fish product sales*, the *volume of meat product sales,* the *volume of fruit-vegetable product sales,* the *volume of bakery product sales,* the *volume of cleaning product sales,* the *volume of drinking product sales* and the *total volume of sales*. Therefore, the dataset has 13 variables.

It seems that this scenario can be solved using a clustering technique which will discover which are the most similar stores among them, and more different to the others. Once the groups have been discovered, the different clusters must be validated both structurally and qualitatively.

We could apply several clustering techniques. We select the *k-means* technique, and we tried for several values of *k* (2, 3, 5, 7, and). For each partition obtained, we made a structural validation with three CVIs, which can be seen in Table 6.2.

From the Table 6.2, it seems that the best partition would be with *k* = 5. Therefore, five clusters seem to be the best partition attending to structural considerations. Once the five clusters were identified, we can try to characterize those clusters with an interpretation of the common values of the variables within them. To that end, we visualized a panel graph, and from it, we generated the characterization of the five clusters according to the variables' values, as shown in Table 6.3. The values in parenthesis on the first column are the number of instances of each cluster.

Table 6.2 The results of the Cluster Validation Indexes for each one of the possible partitions

	Maximum cluster diameter (Δ)	Minimum cluster separation (δ)	Caliński-Harabasz index (CH)
$k = 2$	13.85	3.07	37.92
$k = 3$	12.42	3.54	41.38
$k = 5$	**11.91**	**4.17**	**47.56**
$k = 7$	12.30	3.64	42.15

Table 6.3 Characterization of the five clusters

	Total sales	Meat sales	Fish sales	Drinks sales	Fr-Veg sales	Bakery sales	Size	City type
Cl 1 (16)	Low	Low	Low	Low	Low	Low	Small	Village
Cl 2 (21)	Low	Low	Low	High	Low	High	Small	Town
Cl 3 (36)	Med	Med	Med	High	Med	Med	Med	Town, city
Cl 4 (48)	High	High	Med	High	High	Low	Large	City
Cl 5 (53)	High	High	High	High	High	High	Huge	Metropolitan

There were five variables not shown in the table which have no impact on the characterization under the criteria of the manager of the company.

The manager of the company interpreted the clusters and the characterization obtained from the data made a lot of sense to the manager. The small stores in small villages were clustered together in cluster 1. Of course, they had the lowest sales. Cluster 2 is similar to cluster 1, but the drink and bakery sales are high. Probably, this can be interpreted, that the economic power of the people probably is a little bit higher, and also as most stores are in a larger town, this can increase sales. Cluster 3 is the most interesting for the manger, because it is where the company has more to gain. The sales are at a medium level, the stores are also a medium one, and are located in both towns and cities. This means that in this type of stores is where the company should focus its marketing campaigns trying to increase sales. The Cluster 4 has a high sales level, the stores are large and they are mostly located in cities: the interesting knowledge pattern here is to discover why the bakery sales are low, when all the other sales are high. Perhaps the quality of the bakery product is not good, or as we are in big cities, people prefer to buy bakery products in a specialized bakery shop? Finally, the Cluster 5 groups the huge stores which are located in metropolitan cities, and the sales level are high in all the products, because probably is where many people can go to the stores.

6.4.1.2   Associative Models

In the same way that descriptive models try to find relationships among the instances of a database, there are *associative models*, which aims to find some relationship among the variables in a dataset. There are problems that require finding meaningful relationships among variables in large datasets across thousands of values, e.g. discovering which products are bought together by customers (i.e. *market basket analysis*), finding interesting web usage patterns, or detecting software intrusion. These problems can be solved using associative models. Among the associative models, the most commonly used methods are *Association Rules* techniques, *Belief or Bayesian Network* models, *Qualitative Reasoning* models, and other statistical methods like Principal Component Analysis (PCA), etc. Qualitative reasoning techniques are model-driven techniques and have been described in Chap. 5. Association rules techniques have been used in real scenarios for its easiness of interpretation by experts. For that reason, they are described in the next subsection.

Association Rules

The main goal of the *association rules technique* is to obtain a set of association rules which express the correlation among attributes, from a database of item transactions. These techniques were originated in the field of Knowledge Discovery in large databases. Thus, accordingly, the common terminology talks about transactions, databases, and items in the transactions, because these techniques were first applied to the market basket analysis domain and the *transactions* were composed of the different *items* bought by a customer.

Given a database consisting of a set of *transactions, $T = \{t_1, t_2, \ldots, t_N\}$,* and given $I = \{i_1, i_2, \ldots, i_D\}$, be a set of D attributes called *items*.

Each transaction in T has a unique transaction ID and contains a subset of the items in I:

$$t_1 : i_2, i_3, i_4, i_6, i_9$$
$$t_2 : i_1, i_2, i_4, i_7, i_8, i_9$$
$$t_3 : i_2, i_4, i_5, i_6$$
$$t_4 : i_1, i_3, i_4, i_8, i_9, i_{10}$$
$$\ldots$$
$$t_n : i_3, i_4, i_6, i_9$$

The issue is to obtain common *patterns of co-occurrence* of the same *items* along with the database. Of course, in order that the co-occurrences found in the database have some interest, the database should have enough number of transactions in order that the co-occurrence appear a sufficient number of times. This minimum number of times required for a co-occurrence is named as the *minimum support* (*minsup*) of the rule expressing the co-occurrence.

For instance, the following *common patterns* can be obtained from the previous database:

$$i_2, i_4$$

$$i_4, i_9$$

$$i_2, i_4, i_9$$

$$i_3, i_4, i_9$$

$$i_3, i_4, i_6, i_9$$

$$\ldots\ldots$$

From a *common pattern* several *association rules* can be generated. An *association rule* is defined as an implication of the form:

$$X \rightarrow Y$$

where $X, Y \subseteq I$ and $X \cap Y = \varnothing$

Every rule is composed by two different *sets of items*, also known as *itemsets*, X and Y. X is called *the antecedent* or left-hand side (LHS) of the rule and Y is called *the consequent* or right-hand side (RHS) of the rule. The nomenclature is the same than the usual inference rules used in rule-based reasoning (see Chap. 5, Sect. 5.3.2). For instance:

$$i_2 \rightarrow i_4 \qquad\qquad i_4 \rightarrow i_2$$

$$i_4 \rightarrow i_9 \qquad\qquad i_9 \rightarrow i_4$$

$$i_2 \rightarrow i_4 \wedge i_9 \qquad i_4 \rightarrow i_2 \wedge i_9 \qquad i_9 \rightarrow i_2 \wedge i_4$$

$$i_2 \wedge i_4 \rightarrow i_9 \qquad i_2 \wedge i_9 \rightarrow i_4 \qquad i_4 \wedge i_9 \rightarrow i_2$$

$$i_3 \wedge i_4 \rightarrow i_9 \qquad i_3 \wedge i_9 \rightarrow i_4 \qquad i_4 \wedge i_9 \rightarrow i_3$$

$$i_3 \wedge i_4 \wedge i_6 \rightarrow i_9 \quad i_3 \wedge i_4 \wedge i_9 \rightarrow i_6 \quad i_3 \wedge i_6 \wedge i_9 \rightarrow i_4$$

$$i_3 \wedge i_4 \wedge i_6 \rightarrow i_9 \quad i_3 \wedge i_4 \rightarrow i_6 \wedge i_9 \quad i_3 \wedge i_6 \rightarrow i_4 \wedge i_9 \quad \cdots$$

$$\cdots$$

In the general case of application of association rules, an *item* is an attribute-value pair, and the term *itemset* is the combination of several items. *Frequent itemsets* are the itemsets which their items appear together, at least a minimum specified number of times, i.e. a minimum support *(minsup)*. Next, the main concepts related to association rules are defined:

- *Support of an Itemset* (Agrawal et al., 1993): The support value of X with respect to T is defined as the number of transactions (instances) in the database, which contains the itemset X. Also exists a relative definition according to the size of the set $|T|$.

$$\mathrm{supp}(X) = |\{t \in T | X \subseteq t\}| \qquad\qquad (6.44)$$

$$\text{supp}(X) = \frac{|\{t \in T | X \subseteq t\}|}{|T|} \tag{6.45}$$

- *Support of a rule* (Agrawal et al., 1993). The support value of a rule, $X \rightarrow Y$, with respect to T is defined as the percentage of all transactions (instances) in the database, which contains the itemset X and the itemset Y.

$$\text{supp}(X \rightarrow Y) = \frac{\text{supp}(X \cup Y)}{|T|} \tag{6.46}$$

- *Coverage of a rule.* Coverage is sometimes called antecedent support or LHS support. It measures how often a rule, $X \rightarrow Y$, is applicable in a database.

$$\text{coverage}(X \rightarrow Y) = \text{supp}(X) \tag{6.47}$$

- *Confidence/Strength of a rule* (Agrawal et al., 1993). The confidence value of a rule, $X \rightarrow Y$, with respect to a set of transactions T, is the proportion of the transactions that contains X, which also contains Y.

$$\text{conf}(X \rightarrow Y) = \frac{\text{supp}(X \cup Y)}{\text{supp}(X)} \tag{6.48}$$

- *Leverage/Piatetsky-Shapiro Measure (PS) of a rule* (Piatetsky-Shapiro, 1991). Leverage value of a rule, $X \rightarrow Y$, measures the difference between the probability of the rule and the expected probability if the items were statistically independent.

$$\text{leverage}(X \rightarrow Y) = \text{supp}(X \rightarrow Y) - \text{supp}(X) * \text{supp}(Y) \tag{6.49}$$

  It ranges from $[-1, +1]$ indicating 0 the independence condition.

- *Lift/Interest of a rule* (Brin et al., 1997). Lift value of a rule, $X \rightarrow Y$, measures how many times more often X and Y occur together than expected if they were statistically independent.

$$\text{lift}(X \rightarrow Y) = \frac{\text{supp}(X \cup Y)}{\text{supp}(X) * \text{supp}(Y)} = \frac{\text{conf}(X \rightarrow Y)}{\text{supp}(Y)} \tag{6.50}$$

  It ranges from $[0, +\infty]$, where a lift value of 1 indicates independence between X and Y, and higher values indicates a co-occurrence pattern.

Let us suppose that we had the following dataset "the *golf playing problem*" (Quinlan, 1988), described in Table 6.4 with two items outlined in boldface.

There are four attributes that can influence on the decision to *play golf* (yes or no). The *weather out* (sunny, overcast, rainy), the *temp* (hot, mild, cool), the *hum* (normal, high) and the *wind condition* (false or true).

Table 6.4 The golf playing problem dataset with {Temp = *cool*} and {Hum = *normal*} itemsets outlined

Out	Temp	Hum	Wind	Play
sunny	hot	high	false	no
sunny	hot	high	true	no
overcast	hot	high	false	yes
rainy	mild	high	false	yes
rainy	**cool**	**normal**	false	yes
rainy	**cool**	**normal**	true	no
overcast	**cool**	**normal**	true	yes
sunny	mild	high	false	no
sunny	**cool**	**normal**	false	yes
rainy	mild	**normal**	false	yes
sunny	mild	**normal**	true	yes
overcast	mild	high	true	yes
overcast	hot	**normal**	false	yes
rainy	mild	high	true	no

Examples of *Items*:

- Temp = *cool*
- Out = *sunny*
- Hum = *high*
- Wind = *false*
- Play = *yes*

Examples of *Itemsets*:

- {Temp = *cool*}
- {Temp = *cool*, Hum = *normal*}
- {Out = *sunny*, Wind = *true*, Play = *no*}

Example of rules:

- Temp = *cool* → Hum = *normal*

  - *supp*({Temp = *cool*}) = 4

because the *cool* value for attribute Temp, i.e. the itemset {Temp = cool}, appears four times in the dataset

  - *supp*({Hum = *normal*, Temp = *cool*}) = 4

because the combination of value *normal* for the attribute Hum and the value *cool* for attribute Temp, i.e. the itemset {Hum = *normal*, Temp = *cool*}, appears four times in the dataset

  - *conf*(Temp = *cool* → Hum = *normal*) =

$$= \frac{\text{supp}(\{\text{Humidity} = \text{normal, Temperature} = \text{cool}\})}{\text{supp}(\text{Temperature} = \text{cool})} = \frac{4}{4} = 100\%$$

- Hum $= normal \rightarrow$ Temp $= cool$

  - $supp(\{\text{Hum} = normal\}) = 7$

  because the combination of value *normal* for the attribute Hum, i.e. the itemset {Hum $= normal$}, appears seven times in the dataset

  - $conf(\text{Hum} = normal \rightarrow \text{Temp} = cool) =$

  $$= \frac{\text{supp}(\{\text{Hum} = \text{normal}, \text{Temp} = \text{cool}\})}{\text{supp}(\text{Hum} = \text{normal})} = \frac{4}{7} = 57.14\%$$

The rules we are interested in are those ones with *minimum support* and with *high confidence*. The different methods to induce the association rules are interested in rules with a *minimum support* (*minsup*) to outline a repetitive co-occurrence pattern, and with *high confidence*, meaning that the rules are highly accurate (both antecedent and consequent of the rule are satisfied). Also, high values of *lift* are desirable to indicate a co-occurrence pattern strength.

The general procedure to generate *association rules*, with algorithms using a *minsup* and based on generating the itemsets from a dataset of transactions can be summarized in these three steps:

1. Define the *minsup*, and the *minconf* and eventually the number of rules desired
2. Compute the *itemsets* with *supp(itemset)* ≥ *minsup*
3. For each *itemset* generated in the previous step, generate the *candidate rules* from it, checking that they have the specified *minimum confidence* (*conf (rule)* ≥ *minconf*). Commonly, they generate the rules starting first with one *itemset* in the consequent, and progress with two *itemsets*, etc.

Association Rule Methods

The first and well-known method in the literature is the *Apriori algorithm* proposed by Agrawal and Srikant in (1994).

Apriori algorithm (Agrawal & Srikant, 1994) was one of the earliest association rules methods. In fact, the *Apriori* algorithm computes just the large itemsets which their support is higher than the minimum support (*minsup*) threshold.

It uses a breadth-first search strategy to generate the itemsets: starting from large 1-itemsets, it computes afterwards large 2-itemsets, then large 3-itemsets, and so on until the maximum number of attributes available.

It uses a *candidate generation* function, which *filters* impossible large k-itemsets candidates, because they have subsets of large $k-1$-itemsets, which do not have a minimum support. It uses a hash-table to store the *itemsets*, and a lexicographical ordering for generating and storing the *itemsets* in the hash-table.

The algorithm is as follows:

---

algorithm Apriori

input T: the list of transactions (the dataset)
      minsup: the minimum support of the rules to be obtained
output $L_1 \cup L_2 \cup ... \cup L_k$

begin
  $k \leftarrow 1$
  $L_1 \leftarrow$ {large 1-itemsets};
  while $L_{k-1} \neq \varnothing$ **do**
    $k \leftarrow k + 1$
    $C_k \leftarrow$ Candidate-generation (L_{k-1})
    // New *apriori* candidates generated by extending L_{k-1} candidates
    for each $t \in T$ **do**
    $C_t \leftarrow \{c \in C_k \mid C_k \subseteq t\}$;   // Candidates contained in t
      for each $c \in C_t$ **do**
        c.count++;
      endfor
    endfor
    $L_k \leftarrow \{c \mid c \in C_k \wedge$ c.count $\geq minsup\}$
  endwhile
  return $\bigcup_k L_k$
end

function Candidate-generation (L_{k-1})
begin
  $C_k \leftarrow \varnothing$
  forall a, b $\in L_{k-1}$ such that a = $\{l_1, ..., l_{k-2}, l_{k-1}\}$ **and**
                                  b = $\{l_1, ..., l_{k-2}, l'_{k-1}\}$ **and** $l_{k-1} < l'_{k-1}$ **do**
    // join $k-1$ large itemsets with a common prefix and one item different
    // in lexicographic order to not repeat itemsets
    c $\leftarrow \{l_1, ..., l_{k-2}, l_{k-1}, l'_{k-1}\}$ // c is the join of a and b
    $C_k \leftarrow C_k \cup \{c\}$
  endfor
  foreach c such that $\exists s \mid s \subseteq c \mid s \mid = k-1$ **and** s $\notin L_{k-1}$ **do**
    $C_k \leftarrow C_k - \{c\}$ // apply filter property step
  endforeach
  return C_k
end

After the *Apriori* algorithm, the candidate rules must be generated trying all the possible combinations of the items in the antecedent or the consequent of the rule. The rules are filtered, and just only the ones with a *confidence* value higher than the minimum confidence (*minconf*) bound are shown.

Other well-known methods in the literature are the following:

- *Eclat* (Equivalence CLAss Transformation) (Zaki, 2000; Zaki et al., 1997) is a depth-first search algorithm using the set intersection. It uses a *vertical tid-list* database format where it associates with each itemset, a list of transactions in which it occurs. All frequent itemsets can be enumerated via simple tid-list intersections. In addition, a lattice-theoretic approach to decompose the original search space (lattice) into smaller pieces (sub-lattices) which can be processed independently in the main-memory is used. *Eclat* uses a prefix-based equivalence relation for the decomposition of the lattice and a bottom—up strategy for enumerating the frequent itemsets within each sub-lattice. *Eclat* requires only a few database scans, minimizing the input/output time costs, and it is suitable for both sequential as well as parallel execution with locality-enhancing properties. The association rules are generated after the *Eclat* method, using the same procedure as *Apriori* and other methods.
- *FP-growth* (Frequent Pattern growth) (Han et al., 2004, 2000) proposed a novel frequent-pattern tree structure (FP-tree), which is an extended prefix-tree structure for storing compressed, crucial information about frequent patterns, and develop an efficient FP-tree-based mining method, FP-growth, for mining the complete set of frequent patterns by pattern fragment growth. Efficiency is achieved with a large database, which is compressed into a condensed, smaller data structure, FP-tree which avoids costly, repeated database scans. The FP-tree-based mining adopts a pattern-fragment growth method to avoid the costly generation of a large number of candidates sets. Moreover, a partitioning-based, divide-and-conquer method is used to decompose the mining task into a set of smaller tasks for mining confined patterns in conditional databases, which dramatically reduces the search space.
- *Filtered-top-k Association discovery* (Webb, 2011) is an association technique that focusses on finding the most useful associations for the user's specific application. It tries to overcome the problem of not finding relatively infrequent associations (lower support) but even very interesting associations that the most frequent association mining paradigm would not discover. The user specifies three parameters: a measure of how potentially interesting an association is, filters for discarding inappropriate associations, and the number of associations to be discovered, k. Any of the numerous measures of an association's worth existing in the literature (lift, leverage, etc.) may be used. Filters can be imposed such as a requirement that associations be non-redundant (Bastide et al., 2000; Zaki, 2004), productive (Webb, 2006) or pass statistical evaluation (Webb, 2007). The system finds the k associations that optimize the specified measure within the constraints of the user-specified filters. This solves directly the problems of controlling the number of associations discovered and of focussing the results on associations that are likely to be interesting. It is often possible to derive a very efficient search by using k together with the objective function and filters to constrain the search (Hämäläinen, 2010; Pietracaprina et al., 2010). The result is that association mining can be performed efficiently, focussing on associations that are likely to be interesting to the user, without any need for a minimum support constraint.

Validation of an Association Rule Model

The validation of an association rule model can be done at two levels: assessing the rules themselves, and the general assessment of the associative model.

The assessment of the association rules can be made by checking some properties for each one of the rules. Usually, the *support* and the *confidence* of the rules. The *support* is guaranteed to be higher than the *minsup*, because the algorithms generating the rules, only generate rules with a support higher than the threshold *minsup*. As much as the support of a rule, more probabilities than the correlation outlined in the rule would happen in new data, and hence, better is the model we have.

The *confidence* of a rule shows the precision of the rule. As higher is the confidence of a rule, higher is the probability that the correlations expressed in the rule will be true for new data. Therefore, better will be the model.

A general validation of the associative model could be done if we had some other new instances, where the correlations predicted by the association rules could be checked. This new data, usually named as *validation or test set*, is a general validation methodology used in inductive machine learning. The idea is to generate a model with a portion of the data set, i.e. the *training dataset*, and verify the quality of the model with *validation or test dataset*. This validation scheme is generally used in the discriminant and predictive models and will be explained in later sections in this chapter.

This verification means to check whether the correlations asserted by the association rules are satisfied by the new data. For instance, if we had the following association rule:

$$A_2 = m_2^2 \wedge A_3 = m_1^3 \rightarrow A_1 = m_3^1$$

We must check whether in the new data this correlation happens, and how many times it happens. This correlation states that when the Attribute A_2 has the modality m_2 of the ones possible for A_2 and the A_3 the modality m_1 of the ones possible for A_3, then the attribute A_1 has the modality m_3 of the ones possible for A_1.

If the correlations stated by the association rules model are found to be happening in the new data, we can ensure to a certain degree that the correlations are valid, for any data coming from the same data distribution, and then general consequences can be derived, which can help in decision-making problems.

An Example of an Association Rule Model

Let us take the example of the *golf playing problem*, which has been used before to illustrate some of the concepts in association rules models. The dataset has 14 instances and five attributes (*Out, Temp, Hum, Wind,* and *Play*) with the possible modalities already described. The dataset is described in Table 6.5.

We want to analyze this data and extract some possible correlations among the attributes by means of association rules. We will illustrate how the *Apriori* algorithm

Table 6.5 The golf playing problem dataset

Out	Temp	Hum	Wind	Play
sunny	hot	high	false	no
sunny	hot	high	true	no
overcast	hot	high	false	yes
rainy	mild	high	false	yes
rainy	cool	normal	false	yes
rainy	cool	normal	true	no
overcast	cool	normal	true	yes
sunny	mild	high	false	no
sunny	cool	normal	false	yes
rainy	mild	normal	false	yes
sunny	mild	normal	true	yes
overcast	mild	high	true	yes
overcast	hot	normal	false	yes
rainy	mild	high	true	no

works. Let us assume that we set the minimum support for the rules to be 2, i.e. *minsup* = 2.

The *Apriori* algorithm will proceed to compute the large or frequent itemsets of length 1. This means that it must count how many times each possible combination of attribute and possible modality appears in the dataset. The ones appearing at most *minsup* times will be the large itemsets of length 1 (L_1 = {large one − itemsets}). These large 1-itemsets are shown in the first column of Table 6.6 with its corresponding support in the second column. There were 12 large or frequent one-itemsets.

After the L_1 has been computed, the algorithm expands the large one-itemsets with other large one-itemsets. This candidate generation first makes the *join* of large one-itemsets and afterwards, in the second step, uses the *filtering property* and reduce the number of candidates itemsets to be verified.

In our dataset, the large two-itemsets L_2 are computed expanding each one-itemset from L_1 {{out = sunny}, {out = overcast}, ..., {wind = false}} with all the possible other candidates from L_1 to form {{out = sunny, temp = cool}, {{out = sunny, temp = mild}, {out = sunny, temp = hot}, {out = sunny, hum = normal}, {out = sunny, hum = high}, ..., {wind = false, play = yes}, {wind = false, play = no}}.

And for each combination, it must be checked whether they have a minimum support of *minsup*. Therefore:

- {out = sunny, temp = cool} only appears one time, and will not belong to L_2
- {out = sunny, temp = mild} appears two times, and will belong to L_2
- {out = sunny, temp = cool} appears two times, and will belong to L_2
- {out = sunny, hum = normal} appears two times, and will belong to L_2
- {out = sunny, hum = high} appears three times, and will belong to L_2

  ...

Table 6.6 Computation of the large/frequent itemsets in the golf playing problem dataset

One-item sets (12)	sup	Two-item sets (47)	sup	Three-item sets (39)	sup	Four-item sets (6)	sup
out = sunny	5	out = sunny temp = mild	2	out = sunny temp = hot hum = high	2	out = sunny temp = hot hum = high play = no	2
out = overcast	4	out = sunny temp = hot	2	out = sunny temp = hot play = no	2	out = sunny hum = high wind = false play = no	2
out = rainy	5	out = sunny hum = normal	2	out = sunny hum = normal play = yes	2	out = overcast temp = hot wind = false play = yes	2
temp = cool	4	out = sunny hum = high	3	out = sunny hum = high wind = false	2	out = rainy temp = mild wind = false play = yes	2
temp = mild	6	out = sunny wind = true	2	out = sunny hum = high play = no	3	out = rainy hum = normal wind = false play = yes	2
temp = hot	4	out = sunny wind = false	3	out = sunny wind = false play = no	2	temp = cool hum = normal wind = false play = yes	2
hum = normal	7	out = sunny play = yes	2	out = overcast temp = hot wind = false	2		
hum = high	7	out = sunny play = no	3	out = overcast temp = hot play = yes	2		
wind = true	6	out = overcast temp = hot	2	out = overcast hum = normal play = yes	2		
wind = false	8	out = overcast hum = normal	2	out = overcast hum = high play = yes	2		
play = yes	9	out = overcast hum = high	2	out = overcast wind = true play = yes	2		
play = no	5	out = overcast wind = true	2	out = overcast wind = false play = yes	2		
		. . .		. . .			

And so on all the combinations will be examined to decide whether they pass to L_2 or not. The large or frequent two-itemsets are depicted in the third column of the Table 6.6, and its corresponding support values in the fourth column. There were 47 large two-itemsets.

The process is repeated now to expand the large two-itemsets L_2 into the large three-itemsets L_3 by combining each large two-itemset from L_2 with the other two-itemsets that can form three-itemsets. This means, that the candidates to be merged should have in common one same item, but not the other one. For instance, {out = sunny, temp = mild} cannot be merged with {out = sunny, temp = hot}, but yes with {out = sunny, hum = normal}. In addition, it can be also merged with {out = sunny, hum = high}. The process continues generating all the possible candidates to be in L_3. For each candidate, it must be checked whether it has a minimum support of *minsup*. Thus:

- {out = sunny, temp = mild, hum = normal} only appears one time, and will not belong to L_3
- {out = sunny, temp = mild, hum = high} only appears one time, and will not belong to L_3
- {out = sunny, temp = hot, hum = normal} never appears and will not belong to L_3
- {out = sunny, temp = hot, hum = high} appears two times, and will belong to L_3
  ...

And all the combinations will be examined to decide whether they pass to L_3 or not. The large or frequent three-itemsets are depicted in the fifth column of the Table 6.6, and its corresponding support values in the sixth column. There were 39 large three-itemsets.

The algorithm continues to expand the large three-itemsets L_3 into the large four-itemsets L_4 by combining each large three-itemset from L_3 with the other three-itemsets that can form four-itemsets. to generate the large four-itemsets. This means, that the candidates to be merged should have in common two same items, but not the other one. For instance, {out = sunny, temp = hot, hum = high} can be merged with {out = sunny, temp = hot, play = no}, but not with {out = sunny, hum = normal}. The process continues generating all the possible candidates to be in L_4. For each candidate, it must be checked whether it has a minimum support of *minsup*. Therefore:

- {out = sunny, temp = hot, hum = high, play = no} appears two times, and will belong to L_4
- {out = sunny, temp = hot, hum = high, wind = false} only appears one time, and will not belong to L_4
- {out = sunny, hum = high, windy = false, play = no} appears two times, and will belong to L_4.
  ...

And all the combinations will be examined to decide whether they pass to L_4 or not. The large or frequent four-itemsets are depicted in the seventh column of the

Table 6.6, and its corresponding support values in the eight columns. There were six large four-itemsets.

The algorithm will continue to try to generate large five-itemsets, but as five is the maximum number of attributes in the dataset, it is only possible to find large five-itemsets if there were repeated instances. As it is not the case in our dataset, there are no large five-itemsets, and the *Apriori* algorithm ends.

Now is the turn to generate the association rules from the large itemsets. For example, let us take the large three-itemset:

$$l_3 = \{\text{out} = sunny, \ \text{humidity} = high, \ \text{play} = no\}$$
$$\text{supp}(l_3) = 3$$

We will generate the rules starting first with one *itemset* in the consequent, and progress with two *itemsets*, etc. Seven rules can be generated, just from this itemset, which are listed in decreasing order of confidence:

Rules generated	Confidence
out = *sunny* $\wedge$ humidity = *high* $\rightarrow$ play = *no*	3/3 = 100%
out = *sunny* $\wedge$ play = *no* $\rightarrow$ humidity = *high*	3/3 = 100%
humidity = *high* $\wedge$ play = *no* $\rightarrow$ out = *sunny*	3/4 = 75%
out = *sunny* $\rightarrow$ humidity = *high* $\wedge$ play = *no*	3/5 = 60%
play = *no* $\rightarrow$ out = *sunny* $\wedge$ humidity = *high*	3/5 = 60%
humidity = *high* $\rightarrow$ out = *sunny* $\wedge$ play = *no*	3/7 = 42.86%
$\varnothing \rightarrow$ out = *sunny* $\wedge$ humidity = *high* $\wedge$ play = *no*	3/14 = 21.43%

The first two rules are 100% certain in the data set. The others have less precision. Note that the third one, even it has the same large one-itemsets than the two previous ones are not 100% precise. The number of rules that can be generated, in general, from just one itemset composed of n items is described by:

$$|\text{Rules}(n\text{-itemset})| = \sum_{k=1}^{n} C_{n,k} \qquad (6.51)$$

where,

$$C_{n,k} = \binom{n}{k} = \frac{n(n-1)\dots(n-k+1)}{k(k-1)\dots 1} = \frac{n!}{k!(n-k)!} \qquad (6.52)$$

which are the number of possible different *combinations* of k elements taken from a set of n elements, when the order of the elements does not matter, and the elements cannot be repeated.

In the example database, with 100% confidence and support ≥ 2, there are 58 rules. Thus, the total number of association rules that can be generated could be very large. This number can be controlled with the parameter of *minimum support* and the *confidence* of the rules.

6.4.2   Supervised Models

As it was explained at the beginning of this chapter, the supervised models are characterized because the main goal is to *estimate a variable or attribute* from the values of the other variables or attributes.

If the variable of interest to be estimated is a qualitative or categorical variable, we talk about *discriminant models*. If the variable of interest to be estimated is a numeric variable, we talk about *predictive models*. However, as we outlined at the beginning of the chapter, in the literature the discriminant models are also named as *classifier models*, and the predictive models are also named as *regression models*.

There are some general challenges to be taken into account when mining an inductive model from a dataset in supervised machine learning:

- *Overfitting of a Model*: The problem of overfitting appears when a well-performing model has been obtained from the data, but the model shows a *poor generalization* or *scalability* to new data.

  The model is *too much specialized* in the *data* used to induce the model. The problem is that the model has learned *specific patterns* instead of *general patterns,* which should be the ultimate goal of an inductive model. In general, it can be said that when models show overfitting, they are usually *too complex.*

  Commonly, the procedure to avoid this situation is to use a different dataset to learn/induce/train the model, and another dataset to validate/test the model, to ensure that the model will have a certain generalization or scalability skill.

  This procedure to split the original dataset into two or three datasets is a common practice in supervised machine learning techniques, as depicted in Fig. 6.15. The size of the datasets is usually about 60–80% for the training set, and about 10–20% for the validation set and the test set. If there is no test set, then the validation set could be about 20–30%.

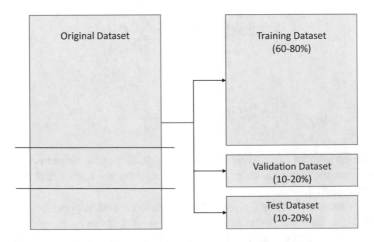

Fig. 6.15 Splitting of a dataset into training set, validation set and test set

This split of the original dataset is very useful for the validation of the supervised models, as it will be described in detail in the validation subsections of the supervised models. The basic idea is to use the *training set* to learn the model or models. Once the model has been induced, the *validation test* can be used to check that it is not overfitting, and performs accurately on the validation test.

In addition, the validation set can be used to compare different models learnt with the training set to check which one is the best. Furthermore, with the validation set, the same model with different hyper-parameters[9] can be evaluated and compared.

Finally, the *test set* can be used to test the final model selected after the validation process. If the models perform in a good way on the test data the confidence in a good fitting increase, and therefore, the overfitting problem is avoided.

- *Underfitting of a Model*: The problem of underfitting of a model means that the model is not performing well even though in the training set. Thus, the problem is that the model is intrinsically bad. Probably, important features of the domain are missing. Perhaps, there are some variables or features which are not in the dataset, and those missing ones, are very relevant for the models being induced.

To illustrate the concepts of underfitting and overfitting of a model, Fig. 6.16 shows 11 observations of two variables (X, Y), where the variable Y must be predicted from the value of variable X.

The model M0, a polynomic model of order zero $(Y = k)$ is a constant prediction, for any value of the variable X. Of course, this model is not very accurate and suffers from *underfitting*.

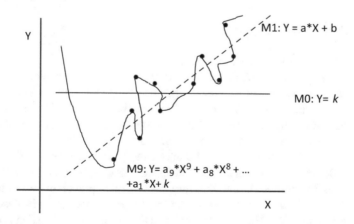

Fig. 6.16 Underfitting, overfitting and good models

---

[9] In machine learning, a *hyper-parameter* is a parameter whose value is determined before starting the learning process. On the other hand, the usual *parameters* involved in the induction of a learning model are the parameters of the model.

On the contrary, the model M9, a polynomic model of order nine $(Y = a_9 * X^9 + \cdots + a_1 * X + k)$, fits perfectly all the points of the data. This model clearly performs very well for the points on the dataset, but very probably it is an over fitted model, and it could not make good predictions for new data, because it is an *overfitting model*.

Finally, if we take into account the model M1, a polynomic model of order one, i.e. a linear regression model $(Y = a * X + b)$, it is a model that does not fits perfectly all the data points, but gives a reasonable estimation of the variable Y. It is not an overfitting model nor an underfitting model. It is just the kind of model that could have a good generalization and scalability properties for new unseen data. It is the kind of models which we aim to obtain with the inductive machine learning methods: *good models* but not perfectly fitting models.

- *Bias of a model and variance of a model*: Overfitting vs. underfitting is a trade-off between the bias of a model and the variance of a model:

  - *Bias of a Model*: The bias refers to the mistakes of the model. Thus, the bias it is related to the deviation between the predicted values and the real values. As great is the bias of a model, more inaccurate is the model.
  - *The variance of a model* refers to the variability of the results of the model for different training sets. The model has a high variance of the results when the results on different sets of training data, sampled from the same larger population are different.

Commonly, a high bias and a low variance usually matches an *underfitting* situation. On the contrary, a low bias and very high variance usually matches an *overfitting* situation.

6.4.2.1 Discriminant Models

Another common problem in Machine Learning is to obtain a *discriminant model* from a supervised dataset. Discriminant models are able to discriminate or estimate or predict the *class label*[10] of instances. The interesting case is when a new instance, not belonging to the training set is available, and the model is able to discriminate which is the corresponding label of the new instance. Discriminant models are also called classifier models or systems in the literature.

There are several kinds of discriminant methods like Decision Trees, Classification Rules, Bayesian discriminant methods, Case-Based Classifiers, Support Vector Machines, etc. In addition, in the literature, there is an approach of working with an ensemble of discriminant methods. In the next subsections, we will describe the Decision Trees, the Case-Based classifiers, and the ensemble of classifiers approach.

---

[10]The qualitative variable or attribute which is the variable of interest to be estimated is usually named in data mining as the *class variable*, and the possible qualitative modalities of this variable are named as the different *class labels* or simply *labels*.

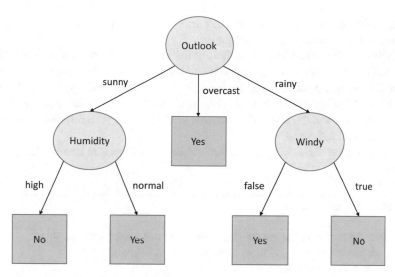

Fig. 6.17 A decision tree for the golf playing problem

Decision Trees

A *decision tree* is a hierarchical structure (a tree), which can model the decision process of deciding to which class label belongs a new example of a concrete domain. In a decision tree, the internal nodes represent qualitative attributes, or discretized numerical ones in some approaches. For each possible value of the qualitative attribute, there is a branch. The leaves of the tree have the qualitative prediction of the attribute that acts as a class label. For instance, Fig. 6.17 depicts a decision tree for the golf playing problem dataset described in Table 6.5. The leaves of the tree are depicted as squares and the internal nodes as circles.

Decision Trees[11] has some advantages over other discriminant models. The final model, the tree, is easily interpretable by an expert or end user to understand the decision process, which ends assigning a label to a new unlabelled instance. Another interesting point is that at the same time the decision tree is constructed, the attributes that have not been used in the building of the tree, are not necessary for a discrimination process. This fact probably means that those unused attributes are not very important. This way, using a decision tree has the benefit of performing an internal

---

[11] Although the basic principle is the same, i.e. to describe a decision process, there are some differences between these *induced decision trees* and the *decision tree models to describe a decision process* explained in Chap. 2. Later ones can have two kinds of internal nodes (decision nodes and random event nodes), the branches are possible alternatives for a decision, or the possible outcomes of an event, and at the leaves, there is the expected outcome for the path ending there. In an *induced decision tree* from data, all the internal nodes represent a qualitative attribute, all the branches of a node correspond to a possible qualitative value for the attribute in the node, and the leaves have a class label (qualitative value).

feature selection process as an integral part of the procedure. They can manage the presence of irrelevant predictor attributes.

There are different techniques to induce a decision tree from a supervised training dataset. All methods use a top–down recursive procedure with a greedy strategy to select the adequate attribute at each node. The strategy tries to select the most discriminant attribute at each step. The discrimination among the different classes is maximized, when the level of separation or skew among the different classes in a given node is maximized.

The general recursive algorithm for induce a decision tree can be described as follows:

---

algorithm Decision Tree

input X: the training data set (N instances * $D+1$ attributes): $\{X_{1.}, ..., X_{N.}\}$
      A: the set of the qualitative attributes: $\{A_1, ..., A_D\}$,
      Cl: is the class attribute
output a decision tree

begin
 option
    // all the examples belong to the same class
  case $\forall X_{i.} \in X,\ class(X_{i.}) = Cl_k$ **do**
    // build a tree with one node labelled with the class label Cl_k
   tree $\leftarrow$ buildTree (Cl_k)
    // not all the examples belong to the same class
  case $\neg\ (\forall X_{i.} \in X,\ class(X_{i.}) = Cl_k)$ **do**
   option
    case $A \neq \varnothing$ **do**
      $A_{discr} \leftarrow BestDiscrimAttribute(X, A, Cl)$
      // build a tree with one node labelled with the A_{discr}
     tree $\leftarrow$buildTree(A_{discr});
      // for each possible modality of A_{discr} or binary split build a subtree
     for each $v \in V(A_{discr})$ or a *binary split-point*(A_{discr}) **do**
       // build a subtree to discriminate among examples with value v for A_{discr}
       tree2 $\leftarrow$ Decision Tree($A_{discr}^{-1}(v), A - \{A_{discr}\}, Cl$)
       tree $\leftarrow$ addBranch(tree, tree2, v)
     endforeach
      // There are no remaining attributes to be used
    case $A = \varnothing$ **do**
     tree $\leftarrow$ buildTree(majorityClassLabel(X))
   endoption
  endoption
 return tree
end

The difference among the methods relies on how to estimate which is the most discriminant one, i.e. the function BestDiscrimAttribute($X, A, $Cl). Most common methods in the literature are:

- ID3 method (Quinlan, 1983, 1986) based on the *Information Gain measure.*
- CART method (Breiman et al., 1984) based on the *Impurity measure* assessment.
- C4.5 method (Quinlan, 1993) based on the *Gain Ratio measure.*

The measures used to compare different decision trees are: the compactness of the tree, the predictive accuracy of the tree, the generalization ability of the tree (scalability). In addition, some approaches propose *pruning techniques* to reduce the size of the tree and try avoiding *overfitting* problems.

Information Gain Method

One of the most well-known methods for inducing a decision tree from data is the ID3 (Quinlan, 1983, 1986) method. All the attributes must be qualitative in the original formulation of Quinlan but in practice, the numeric attributes can be discretized. On each iteration of the algorithm, it selects the best attribute according to the *Information Gain* criteria.

The *Information Gain* criteria are based on the concept of *Entropy* from information theory (Shannon, 1948). The criterion selects the attribute which maximizes the information gain. Thus, the ID3 algorithm needs to assess the information gain provided by the use of each one of the considered attributes. The *Entropy* function measures the ability of each attribute to split the instances in the possible values of the attribute in the best pure (discriminant) form. Purity means that if all instances having the same value for the attribute belongs to the same label is a better attribute than others that are mixing several instances belonging to different labels. Thus, it is measuring the *dispersion degree* of the different class labels among the possible values of an attribute.

The Information Gain function is defined as:

$$\text{Gain}(X, A_d, \text{Cl}) = \text{Info}(X, \text{Cl}) - \text{Info}(X, A_d, \text{Cl}) \qquad (6.53)$$

where,

X is the set of all instances to be discriminated at each node,
A_d is one of the D qualitative attributes to be used in the discrimination process,
Cl is the class attribute to be predicted

And the *Information function* or *Entropy* at the node before splitting is defined as:

$$\text{Info}(X, \text{Cl}) = H(X, \text{Cl}) = -\sum_{i=1}^{k} p_{x \in \text{Cl}_i} * \log_2 \left(p_{x \in \text{Cl}_i} \right) \qquad (6.54)$$

where,

X is the set of all instances to be discriminated at each node,
Cl is the class attribute to be predicted,
k is the number of different labels of the class attribute,
$P_{x \in Cl_i}$ is the probability that the instance x belongs to the class Cl_i

And the *Information function given an attribute* A_d is the amount of information needed to arrive to a perfect classification using the corresponding attribute A_d.

$$\text{Info}(X, A_d, \text{Cl}) = \sum_{j=1}^{v} P_{x \in \text{Value}_j(A_d)} * \text{Info}\left(\{x | x \in \text{Value}_j(A_d)\}, \text{Cl}\right) \qquad (6.55)$$

where,

X is the set of all instances to be discriminated at each node,
A_d is one of the D qualitative attributes to be used in the discrimination process,
$\text{Value}_j(A_d)$, is the set of instances x which have the qualitative value j for the attribute A_d
Cl is the class attribute to be predicted.

The probabilities can be estimated from the data distribution in the training set. The value of the *entropy* lies between 0 and $\log(k)$. The value is $\log(k)$ when the instances are perfectly balanced among the different classes. This corresponds to the scenario with maximum entropy. The smaller the entropy, the greater the separation in the data.

It selects the attribute which has the smallest entropy (or largest information gain) value. The set X is then split by the selected attribute to produce subsets of the data. The algorithm recursively continues on each subset, considering only attributes never selected before.

Thus, in this method:

$$\text{BestDiscrimAttribute}(X, A, \text{Cl}) = \max_{d \in D} \; \text{Gain}(X, A_d, \text{Cl})$$

$$= \min_{d \in D} \; \text{Info}(X, A_d, \text{Cl}) \qquad (6.56)$$

Gain Ratio Method

This *Information Gain measure* can be biased to select attributes with large number of possible values. In order to overcome this bias, Quinlan (Quinlan, 1993) proposed the C4.5 method that uses an extension to information gain known as *gain ratio*. It applies a kind of normalization to information gain using a *split information value*. The *split information value* represents the potential information generated by splitting the training data set X into v partitions, corresponding to the v possible values of the attribute A_d:

$$\text{SplitInfo}_{A_d}(X) = -\sum_{j=1}^{v} p_{x \in \text{Value}_j(A_d)} * \log_2\left(p_{x \in \text{Value}_j(A_d)}\right) \qquad (6.57)$$

The *Gain Ratio* is defined as follows:

$$\text{GainRatio}(X, A_d, \text{Cl}) = \frac{\text{Gain}(X, A_d, \text{Cl})}{\text{SplitInfo}_{A_d}(X)} \qquad (6.58)$$

At each node of the tree, C4.5 chooses the attribute of the data that most effectively splits its set of samples into subsets according to the Normalized Information Gain or *Gain Ratio*. The attribute with the highest Gain Ratio is chosen.

The C4.5 method also was prepared to accept numeric attributes. It has been implemented in the J4.8 method in the software WEKA. In the next evolutions of the method, Quinlan proposed the C5.0 method, where the most significant feature unique to C5.0 is a scheme for deriving rule sets.

Thus, in this method:

$$\text{BestDiscrimAttribute}(X, A, \text{Cl}) = \max_{d \in D} \text{ GainRatio}(X, A_d, \text{Cl}) \qquad (6.59)$$

Impurity Measure Method

Another well-known method is the CART method (Classification and Regression Trees) (Breiman et al., 1984), which is based on the *Impurity measure*.

For example, if $p_1, \ldots, p_k$ is the fraction of the instances belonging to the k different classes in a node N, then the *Gini-index of impurity*, Gini(X), of the current node is defined as follows:

$$\text{Gini}(X) = 1 - \sum_{i=1}^{k} p_{x \in \text{Cl}_i}^2 \qquad (6.60)$$

where X is the set of all instances to be discriminated at each node, and k is the number of different labels of the class attribute.

The value of Gini(X) lies between 0 and $1 - \frac{1}{k}$. The smaller the value of Gini(X), the greater the separation. In the cases where the classes are evenly balanced, the value is $1 - \frac{1}{k}$.

The Gini Index considers only *a binary split* for each attribute A_d, say X_1 and X_2. The possible splitting points are computed differently according to the nature of the attributes:

- *Numeric Attributes*: The midpoint between each pair of sorted adjacent values is taken as a possible split-point. The one showing a greatest reduction of the impurity measure is used.

- *Categorical Attributes*: The partitions resulting from all possible subsets of all the possible values of each attribute A_d must be examined.

  Given the possible modalities of an attribute, $V(A_d) = \{m_1, \ldots, m_v\}$, there are $\frac{2^v - 2}{2} = 2^{v-1} - 1$ possible subsets. Each subset S_{A_d} is a possible binary test of attribute A_d. The one showing a greatest reduction of the impurity measure is selected.

Once the split-point is selected at a node for a given attribute A_d, the Gini index of X given that partitioning is a weighted sum of the impurity of each partition:

$$\text{Gini}(X, A_d) = \frac{|X_1|}{|X|} * \text{Gini}(X_1) + \frac{|X_2|}{|X|} * \text{Gini}(X_2) \qquad (6.61)$$

Finally, the attribute that maximizes the reduction in impurity is chosen as the splitting attribute for a given node:

$$\Delta\text{Gini}(A_d) = \text{Gini}(X) - \text{Gini}(X, A_d) \qquad (6.62)$$

Thus, in this method:

$$\text{BestDiscrimAttribute}(X, A, \text{Cl}) = \max_{d \in D} \Delta\text{Gini}(A_d) = \min_{d \in D} \text{Gini}(X, A_d) \qquad (6.63)$$

Tree Pruning

Decision trees can also suffer from the problem of overfitting. This means that many branches of the decision tree could reflect anomalies or overspecialization in the training data due to noise or outliers.

The problem is detected when a poor accuracy value is obtained for the discrimination or classification of unseen examples, but a good accuracy was obtained with the examples in the training set.

One of the most commonly used solution to overcome the overfitting problem in decision trees is the *tree pruning* techniques. Pruning the tree means removing or just not expanding the least reliable branches of the tree.

Generally, two pruning approaches are used in the practice: *pre-pruning* and *post-pruning* techniques.

- *Pre-pruning Techniques*: The major idea is to prune the tree while building it. The rationale of these techniques is to stop the growth of the tree as soon as possible. Each time a node is considered to be split, some *goodness measure* is evaluated. If the goodness measure of the split falls below a *threshold bound*, then the node is not split.

  Major goodness measures commonly used are:

  - Statistical significance.
  - Information Gain.
  - Gini Index.

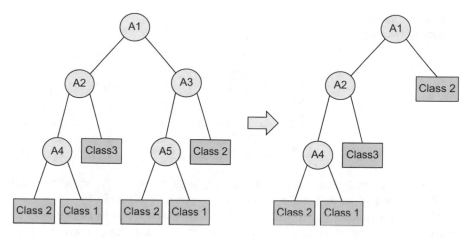

Fig. 6.18 A tree post-pruning strategy

When the tree growth stops, the current node become a leaf. The new leaf will have the most frequent class label among the remaining instances as its label.

The main drawback in *pre-pruning* techniques is the difficulty to tune the *threshold*.

- *Post-pruning Techniques*: These techniques are based on removing branches from a complete generated tree. Thus, the pruning is done after the tree has been induced. The process is iteratively done through a sequence of pruned trees. The major strategy is that being at a given node, the subtree starting at that node is pruned by replacing the subtree with a leaf. The corresponding leaf is labelled with the most frequent class label among the instances of the subtree. For instance, in the tree shown in Fig. 6.18 the subtree starting at the attribute A3, has been substituted by a leaf labelled with the majority class label of the subtree "Class2".

The most common approach used for implementing post-pruning techniques is the *cost complexity pruning* technique. Cost complexity of a tree is a function of the number of leaves and the error rate (percentage of instances misclassified by the tree).

An Example of a Decision Tree Model

Suppose that we have the dataset described in Table 6.7, which corresponds to the *human identification problem* (Quinlan, 1983). The dataset has eight examples of the description of people, and three attributes describing some characteristics of the people like the *eye-colour* (blue, brown, green), the *hair-colour* (blonde, brown), and the *height* (low, medium, tall). The qualitative variable of interest or the class attribute is an attribute named *Class*, and there are two possible labels for the class (C+ and C−). This naming of the classes in supervised machine learning when you have just two classes is very common, because means that one of the classes has the positive examples of a concept (C+), and the other one has the negative examples or counterexamples of the concept (C−) under study. In this dataset, there is no any

Table 6.7 The human identification problem dataset

	Eye colour	Hair colour	Height	Class
E1	Blue	Blonde	Tall	C+
E2	Blue	Brown	Medium	C+
E3	Brown	Brown	Medium	C−
E4	Green	Brown	Medium	C−
E5	Green	Brown	Tall	C+
E6	Brown	Brown	Low	C−
E7	Green	Blonde	Low	C−
E8	Blue	Brown	Medium	C+

positive or negative consideration of the two kinds of people described with the attributes. Simply the two classes distinguish two groups of people.

We are interested in obtaining a *decision tree* which let us to discriminate or classify new people data into the two groups. We will use the *Information Gain* approach of ID3 method.

Therefore, to build the tree we must evaluate, from the three available attributes which is the one maximizing the information gain function, and minimizing the entropy. If we apply the formulas (6.53) to (6.56) detailed in the previous section:

$$\text{Gain}(X, A_d, \text{Cl}) = \text{Info}(X, \text{Cl}) - \text{Info}(X, A_d, \text{Cl})$$

$$\text{Info}(X, \text{Cl}) = H(X, \text{Cl}) = -\sum_{i=1}^{k} p_{x \in \text{Cl}_i} * \log_2 \left(p_{x \in \text{Cl}_i} \right)$$

$$\text{Info}(X, A_d, \text{Cl}) = \sum_{j=1}^{v} p_{x \in \text{Value}_j(A_d)} * \text{Info}\left(\{x | x \in \text{Value}_j(A_d)\}, \text{Cl} \right)$$

$$\text{BestDiscrimAttribute}(X, A, \text{Cl}) = \max_{d \in D} \text{Gain}(X, A_d, \text{Cl}) = \min_{d \in D} \text{Info}(X, A_d, \text{Cl})$$

We have the following results:

$$Info(X, Cl) = H(X, Cl) = -1/2 \log_2 1/2 - 1/2 \log_2 1/2 = 1$$
$$\underset{\{E_1, E_2, E_5, E_8\}}{\quad} \underset{\{E_3, E_4, E_6, E_7\}}{\quad}$$
$$\text{C+} \qquad\qquad \text{C-}$$

$Info(X, Eye\text{-}colour, Cl) = 3/8 \, (\text{-1} \log_2 1 - 0 \log_2 0) +$ // blue
              $2/8 \, (\text{-0} \log_2 0 - 1 \log_2 1) +$ // brown
              $3/8 \, (\text{-1/3} \log_2 1/3 - 2/3 \log_2 2/3)$ // green
             $= 0.344$

$Info(X, Hair\text{-}colour, Cl) = 2/8 \, (\text{-1/2} \log_2 1/2 - 1/2 \log_2 1/2) +$ // blonde
                  $6/8 \, (\text{-3/6} \log_2 3/6 - 3/6 \log_2 3/6)$ // brown
                 $= 1$

$Info(X, Height, Cl) = 2/8 \, (\text{-1} \log_2 1 - 0 \log_2 0) +$ // tall
              $4/8 \, (\text{-1/2} \log_2 1/2 - 1/2 \log_2 1/2) +$ // medium
              $2/8 \, (\text{-0} \log_2 0 - 1 \log_2 1)$ // low
            $= 0.5$

Fig. 6.19 First attribute of
the decision tree

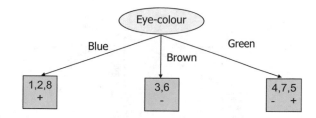

Table 6.8 The subset from
the original dataset with the
remaining examples and
attributes

	Hair Colour	*Height*	*Class*
E4	Brown	Medium	C−
E5	Brown	Tall	C+
E7	Blonde	Low	C−

Therefore, finally we have:

Gain(X, Eye-colour, Cl) = *H(X, Cl)* – *Info(X, Eye-colour, Cl)* = 1–0.344 = **0.656**
Gain(X, Hair-colour, Cl) = *H(X, Cl)* – *Info(X, Hair-colour, Cl)* = 1–1 = 0
Gain(X, Height, Cl) = *H(X, Cl)* – *Info(X, Height, Cl)* = 1–0.5 = 0.5

Hence, *Eye-colour* is the attribute maximizing the *Gain of Information*, and
minimizing the entropy. Thus, the first attribute of the decision tree will be the
attribute *Eye-colour* as depicted in Fig. 6.19.

With *eye-colour* being the first attribute in the tree, three branches appear, one for
each possible qualitative value of the attribute. The first branch (*Eye-colour = blue*) has
all its examples belonging to class C+, which means that no further discrimination
process must be done. The same happens to the second branch (*Eye-colour = brown*),
where all its examples belong to class C−, and hence, no further discrimination process
is needed. The third branch (*Eye-colour = green*) has two examples belonging to class
C− and one instance belonging to class C+. Therefore, at this node, the building of the
tree must be continued, as detailed in the algorithm. However, now we have only to
consider the set of examples to be discriminated among them: E_4, E_5 and E_7. In
addition, we have now just two remaining attributes to be used. Thus, the sub-dataset
we have to analyse for the discrimination purposes is the one described in Table 6.8.

Then, we must evaluate, for the two remaining attributes (*Hair-colour* and
Height) which is the more discriminant.

$$H(X, Cl) = -1/3 \log_2 1/3 - 2/3 \log_2 1/3 = 0.918$$

$\quad\quad\quad$ {E_5} $\quad\quad$ {E_4, E_7}
$\quad\quad\quad$ C+ $\quad\quad\quad$ C-

$Info(X, Hair\text{-}colour, Cl)$ = $1/3\ (-\ 0 \log_2 0 - 1 \log_2 1) +$ $\quad\quad\quad\quad$ // blonde
$\quad\quad\quad\quad\quad\quad\quad\quad\quad\quad$ $2/3\ (-1/2 \log_2 1/2 - 1/2 \log_2 1/2)$ $\quad\quad$ // brown
$\quad\quad\quad\quad\quad\quad\quad\quad\quad\quad$ = $2/3$
$Info(X, Height, Cl)$ = $1/3\ (-0 \log_2 0 - 1 \log_2 1) +$ $\quad\quad\quad\quad$ // medium
$\quad\quad\quad\quad\quad\quad\quad\quad\quad\quad$ $1/3\ (-1 \log_2 1 - 0 \log_2 0) +$ $\quad\quad\quad\quad$ // tall
$\quad\quad\quad\quad\quad\quad\quad\quad\quad\quad$ $1/3\ (-0 \log_2 0 - 1 \log_2 1)$ $\quad\quad\quad\quad$ //low
$\quad\quad\quad\quad\quad\quad\quad\quad$ = 0

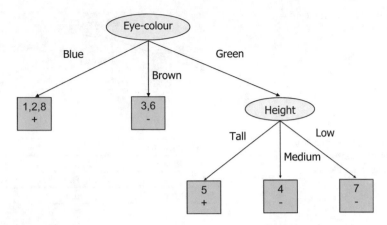

Fig. 6.20 Decision tree for the human identification problem generated with ID3

Therefore, finally:

$$\text{Gain}(X, \text{Hair-colour}, \text{Cl}) = H(X, \text{Cl}) - \text{Info}(X, \text{Hair-colour}, \text{Cl})$$
$$= 0.918 - 0.666 = 0.252$$

$$\text{Gain}(X, \text{Height}, \text{Cl}) = H(X, \text{Cl}) - \text{Info}(X, \text{Height}, \text{Cl})$$
$$= 0.918 - 0 = \mathbf{0.918}$$

Hence, *Height* is the attribute maximizing the *Gain of Information*, and minimizing the entropy. Thus, the attribute of the subtree will be the attribute *Height* as depicted in Fig. 6.20.

Then, for each possible value of the attribute *Height* (tall, medium, low) a branch appears in the tree. Now, each branch has a unique example, and therefore, there is no need to continue the discrimination process, and one leaf is created for each branch with its corresponding class label.

Finally, the decision tree has been built. Now this decision tree is a *discrimination model* which will allow that future examples could be discriminated to which class label belong.

An interesting fact that can be easily observed is that in the decision tree does not appear in any node the attribute *Hair-colour*. This means that, at least for discrimination purposes and with the dataset available, the attribute *Hair-colour* is *irrelevant*. This is an important conclusion that makes that decision trees can additionally act as a *feature selection method*, for discrimination purposes.

The decision tree is a very visual and intuitive discriminant model. Another good property is that from the tree an equivalent model formed by a *set of classification rules* can be obtained. From each path connecting the root node to a leaf, one rule can be extracted. A set of rules is probably more efficient, from a computational point of view. In our example, the discrimination or classification rules extracted from the tree are:

Eye-colour = Blue → *Class* = C+
Eye-colour = Brown → *Class* = C−
Eye-colour = Green ∧ *Height* = Tall → *Class* = C+
Eye-colour = Green ∧ *Height* = Medium → *Class* = C−
Eye-colour = Green ∧ *Height* = Low → *Class* = C−

Case-Based Discriminant Models

As it was briefly described in Chap. 4, Sect. 4.1.1.1, *Case-Based Reasoning (CBR)* is a deliberative paradigm of Artificial Intelligence to solve complex problems by analogy with previous solved problems. *Case-Based Reasoning* (Riesbeck & Schank, 1989; Kolodner, 1993; Richter & Weber, 2013) is based on the *theory of dynamic memory* proposed by Roger Schank in the late 70s and early 80s (Schank, 1982) and pioneered as a computer problem-solving technique by his Ph.D. students, especially by Janet Kolodner (1983), which states that the human memory is dynamic and change with its experiences along their lives. Humans learn new things and forget others. The acts of humans are recorded as scripts in their memory. In addition, the process of learning, understanding, reasoning, and explaining are intrinsically bind together in human memory.

This analogical reasoning is a powerful mechanism to solve problems without starting from scratch to build up a solution. Thus, the solved problems constitute the "knowledge" about the domain.

A very intuitive example of the application of CBR to daily life is the famous "*cut and paste*" strategy to write a new text or report. What we usually do to write a new report is to *search* other reports we have written in the past, *take the one most resembling* the kind of the new report, and then *modify and adapt* the text of the old report to match the goals of the new report. Finally, once *checked* that the modified report matches the goals of the new report, this new report is *stored* in our report library.

As more experienced is a CBR system better performance achieves because the experiences (cases or solved problems) are stored in the *Case Base* or *Case Library* or *Case Memory*. This way the system is *continuously learning* to solve new problems.

The main reasoning cycle in Case-Based Reasoning is depicted in Fig. 6.21. It is composed of four basic steps:

- *Retrieve* the most similar case or cases to the new case.
- *Adapt* or *Reuse* the information and knowledge in that case to solve the new case. The selected best case has to be adapted when it does not match perfectly the new case.

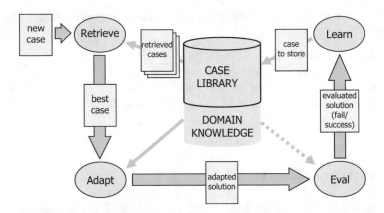

Fig. 6.21 Basic reasoning cycle of Case-Based Reasoning (CBR)

- *Evaluate* or *Revise* of the proposed solution. A CBR system usually requires some feedback to know what is going right and what is going wrong. Usually, it is performed by simulation or by asking to a human oracle.
- *Learn* or *Retain* the parts of this experience likely to be useful for future problem solving. The agent can learn both from successful solutions and from failed ones (repair).

Organization of Cases and the Case Library

The reasoning by analogy of CBR is based in collecting a lot of relevant cases or experiences in a particular domain. Storing a case means to keep a description of the experience as well as the solution provided to that experience. In CBR systems the main memory organizations, can be summarized in some general approaches:

- *Flat memories* always retrieve the set of cases best matching the input case. Moreover, adding new cases to memory is cheap. However, they have a major disadvantage: the retrieval time is very expensive since every case in the memory is matched against the current case.
- On the opposite side there are the *structured organizations*. In such kind of memories, the matching process and retrieval time are more efficient, due to the fact that only few cases are considered for similarity assessment purposes, after a prior discriminating search in a hierarchical structure. Anyway, they also have some disadvantages. Keeping a hierarchical structure in optimal conditions requires an overhead on the case library organization, and the retrieval process could miss some optimal cases searching a wrong area of the hierarchical memory. This later problem specially becomes hard in prioritized discrimination networks/trees. The structured organizations commonly used are:

- *Hierarchical/Indexed organization*: Usually, they are organized in some kind of tree or network like the following ones:

  Shared Feature Networks/Trees.
  Discriminant Networks/Trees.
  Redundant Discriminant Networks/Trees.
  k-d trees.

- *Object-Oriented Organization*: This means that the cases are organized in classes of objects and its corresponding instances.
- *Multiple Case Bases* Organization: This structure is an advanced one, which uses more than one case library to store all the cases to fit better with the different nature of cases in the domain.

- *Mixed organizations*, which use some kind of mixture between some indexation schemes and flat structures.
- *Unstructured organizations*, which are based on representing all the cases in unstructured text, i.e. in natural language.

The cases stored in the case library are real experiences, which have been captured and learned in such a way that they can be reused to solve future situations. Most commonly used *case representation schemes* are:

- *Vector of attribute-value pairs*: It is a simple but computational effective representation. Each case is a list or vector of values, where each value corresponds to one attribute characterizing the case in the descriptive part and/or characterizing the solution part.

  For example:
  Attributes of a Case:

  <type, brand, cylinders, power, fuel, colour, price, year, seats>

  Values of a Case or example or instance or observation:

  <sport, BMW, 8, 275, gasoline, blue, 35,000, 2000, 4>

- *Structures or Objects*: The case is composed of several fields or parts, like the description of the case, the solution of the case, the evaluation of the case, etc.

  A *case example* in the domain of volcanic and seismic prediction domain, could be as follows:

```
(:case-identifier      CASE-134
 :temporal-identifier   27/11/2004
 :case-situation-description ((SEISMIC-ACT Invaluable)
                   (DEFORMATIONS mean-value)
                       (GEOCHEMICAL-EVOL normal)
                       (ELECT-PHEN level-1))
 :case-diagnostics-list   (No-eruption, Seismic-pre-Alert)
 :case-solution-plan       (Alert-Emergency-Services)
 :case-solution-evaluation  correct)
```

- *Free Text in Natural Language (NL)*: Each case is represented through a description written in sentences in natural language. It is very appealing, but computationally requires intermediate representations to be effective.

  An example could be: "A red sports car with 300 horse power and 12 cylinders, with a price less than 250,000 and must be BMW".
- *Text and List of Questions and Answers* (*Conversational CBR*): A case is represented as the aggregation of an initial text given by the user, and a list of questions associated to a possible case.
- *Graphs (trees o networks)*: A case is represented as a graph, where the different relations and conditions between parts of a case can be described in a nice semantic way. Main problem happens when two cases must be compared for similarity purposes, as it is well known that the graph isomorphism problem is a NP-hard problem.

A *case* generally incorporates a set of characteristics such as:

- An *identifier* of the case, which is often a number.
- The *description* of the case, which characterize the problem or situation.
- The *diagnostic* of the case, which sometimes is used as a symbolic identifier of the case.
- The *solution* of the case, which can be simple like a diagnostic class label or a numeric value, or more complex, including a set of values or even a more complex structure like a design or a plan (a set of components), etc.
- The *derivation* of the case, i.e. from where the case has been derived/adapted.
- The *solution result*, information indicating whether the proposed case solution has been a successful one or not.
- A *utility measure* of the case in solving past cases.
- Other *relevant information* about the case.

Case Retrieval and Similarity Assessment

The task of retrieving cases in the case library is slightly more difficult than typical retrieval in databases. In database systems the recalling algorithms use an exactly matching method, whereas in a case library retrieval, because the very nature of the structure, a partial-matching strategy should be used. A retrieval method should try to maximize the similarity between the actual case and the retrieved one(s). And this task implies most of the time the use of general domain knowledge.

The retrieving process of a case (or a set of cases) from the system's memory strongly depends on the case library organization. Major case library structures are flat memories or structured ones. Flat memories have an intrinsic problem of bad performance in time, so that the retrieval time is proportional to the size of the case library. Hierarchical memories are very effective in time retrieval because only a few cases are considered for similarity assessment purposes, after a prior discriminating search in the hierarchical structure. Although sometimes could not reach optimal cases because is exploring a wrong area of the hierarchy.

The retrieval process in hierarchical case libraries usually consists of two main sub-steps:

- *Searching* the most similar cases to the new case: the goal of this stage is recalling the most promising cases—given that the subsystem has a goal and therefore the relevance of the cases depends upon that goal—based on using some direct or derived features of the new case as *indexes* into the case library.
- *Selecting* the best case(s): The best case(s) among those ones collected in the previous step are selected. Commonly, this selection is made by means of a case ranking process through a *similarity* or *distance function*. The best-retrieved case is the closest one (most similar) to the new case.

Selecting the best similar case(s), it is usually performed in most Case-based reasoning system by means of some evaluation heuristic functions or distances, possibly domain dependent. The evaluation function usually combines all the *partial matching through a dimension* or attribute of the cases, into an aggregate or *full-dimensional partial-matching* between the searched cases and the new case. Commonly, each attribute or dimension of a case has a determined importance value (weight), which is incorporated in the evaluation function. This weight could be static or dynamic depending on the Case-based reasoning system purposes. Also, the evaluation function computes an absolute match score (a numeric value), although a relative match score between the set of retrieved cases and the new case can also be computed.

Most Case-based reasoners such as REMIND (Cognitive Systems, 1992), MEDI-ATOR (Kolodner and Simpson, 1989), PERSUADER (Sycara, 1987), etc., use a generalized weighted distance function, such as:

$$d(Ci, Cj) = \sum_{k=1}^{D} w_k \cdot \text{atr\_d}\big(C_{i,k}, C_{j,k}\big) \qquad (6.64)$$

Common similarity measures are derived from Minkowski's metric:

$$d(x, y) = \left(\sum_{k=1}^{D} |x_k - y_k|^r\right)^{1/r} \qquad r \geq 1 \qquad (6.65)$$

$r = 1$, *Manhattan* or *City-block distance*
$r = 2$, *Euclidean distance*

In addition, some *normalized measures* like *Clark* and *Canberra* (Lance and Williams, 1966), or some *probabilistic measures* like the *Value Difference Metric* (*VDM*) (Stanfill & Waltz, 1986), the *SH measure* (Short & Fukunaga, 1981), or the *Minimum Risk Metric* (*MRM*) (Blanzieri & Ricci, 1999) are also used.

Others *heterogeneous measures* to manage both numeric and categorical attributes have been defined in the literature, such as the *Gower's distance* (Gower,

1971), the *Heterogeneous Valued Difference Metric* (*HVDM*) (Wilson & Martínez, 1997), *L'Eixample distance* (Sànchez-Marrè et al., 1998).

Other measures are defined and compared in the literature, such as in Leake et al. (1997), Osborne and Bridge (1998), Liao et al. (1998), Núñez et al. (2004).

Case Adaptation

When the best partial-matching case selected from the case library does not match perfectly with the new case, the old solution needs to be adapted to fit more accurately the new case solution. This reusing process can happen during the solution formulation (*adaptation*), or after some feedback has pointed out some problem in the evaluation step, which needs to be fixed (*repair*).

There are a lot of strategies that have been used in the Case-based reasoners. All these techniques can be grouped (Kolodner, 1993; Riesbeck & Schank, 1989) as *null adaptation*, *structural adaptation*, and *derivational adaptation*, although in most Case-based reasoners, several mixture kinds of adaptation methods are implemented. Here, we propose an additional kind of adaptation named as weighted average/mode adaptation, which is a variant of null adaptation:

- *Null adaptation* or *copy solution* strategy could be a right strategy in Case-based systems with very simple actions in the solution, i.e. accept/reject, a fault diagnosis, a class label, etc., such as the first adaptation method used in the PLEXUS system (Alterman, 1988). In those systems, the old solution of the most similar case is directly applied to the new case.
- *Weighted average/mode adaptation* or *copy and vote solution* strategy is suitable in the same situations than null adaptation, when the solution is composed by simple components. The difference is that here not just the most similar case is used, but the k most similar cases, and the solution given to the new case is a *weighted mean* or *mode*, depending on the nature of the solution components of the k most similar cases. Hence, it is a *copy and vote solutions* strategy.
- In *structural adaptation* methods, the adaptation process is directly applied to the solution stored in a case. The structural adaptation methods can be divided in three major techniques: *substitution* methods, *transformation* methods, and *special-purpose adaptation heuristics or critic-based* adaptation methods.

  - *Substitution methods* provide the solution of the new case with appropriate components or values computed from components or values in the retrieved solution. Most outstanding substitution techniques are: *parameter adjustment* or parameterised solutions, where the differences between the values of the retrieved case and those ones of the new case are used to guide the modification of the solution parameters in the appropriate direction. This approach has been used, for example, in HYPO (Ashley, 1990) and PERSUADER (Sycara, 1987), JUDGE (Bain, 1986). Another kind of methods, such as direct reinstantiation used in CHEF (Hammond, 1989), local search used in JULIANA (Shinn, 1988), PLEXUS and SWALE (Kass & Leake, 1988), query memory used in CYRUS (Kolodner, 1985) and JULIANA, specialised search used in SWALE, etc., can be named as *abstraction and respecialization methods*. When there is a component as an object or a value of the retrieved solution, that does not fit in the new problem, these methods look for abstractions of that component of the solution in a certain knowledge structure (concept generalization tree, etc.) that do not have the same difficulty; the last kind of substitution methods is the *case-based substitution methods*. They use the differences between the new case and the retrieved case to search again cases from the case library to eliminate these differences. These techniques have been used, for instance, in systems such as CLAVIER (Hennessy & Hinkle, 1992), JULIA (Hinrichs, 1992), CELIA (Redmond, 1992), etc.

- The *transformation methods* use either some common-sense transformation rules such as deleting a component, adding a component, adjusting values of a component, etc., as in JULIA system, or some model-guided repair transformation techniques based on a causal knowledge, such as in KRITIK (Goel & Chandrasekaran, 2014) or CASEY (Koton, 1989) systems.
- The *special-purpose adaptation* techniques or critic-based adaptation methods are based on some specific rules of repairing, called critics (Sacerdoti, 1977; Sussman, 1975), like those used in PERSUADER. Other systems such as CHEF and JULIA use some domain specific adaptation heuristics and some structure modification heuristics.

- *Derivational adaptation* methods do not operate on the original solutions, but on the method that was used to derive that solution. The goal is rerunning the same method applied to derive the old solution, to re-compute the solution for the new case. This methodology was first implemented in ARIES system, and was named as derivational replay (Carbonell, 1985). In such techniques, re-instantiation occurs when replacing a step in the derivation of the new solution, like in systems such as PRODIGY/ANALOGY (Veloso & Carbonell, 1993), JULIA (Hinrichs, 1992) or MEDIATOR (Kolodner and Simpson, 1989).

Case Evaluation

This step is one of the most important steps for a case-based reasoner. It gives the system a way to *evaluate its decisions in the real world*, allowing it to receive *feedback* that enables it to learn from success or failure.

Evaluation can be defined as the process of assessing the goodness or performance of the proposed solution for the new case derived from the solution of the best similar remembered case. The evaluation process can point out the need for additional adaptation—usually called *repair*—of the proposed solution, although this only make sense, in non-real-time world domains. Commonly, this evaluation step can be performed either by *asking to a human expert (oracle)* whether the solution is a good one or not, or by *simulating the effects of the proposed solution* in the real world such as in most planning or design domains, or by *directly getting a feedback* on the results of the proposed solution, from the real world.

Case Learning

Learning is an interesting and essential cognitive task of the Case-based systems. Mainly, there are two major kinds of learning in a Case-based system: *learning by observation* and *learning by own experience*.

Learning by observation happens when the system is provided with a set of initial cases, either by an expert or by direct observation (experimentation) of real data. Also, it can learn a new case by direct observation provided by an expert in any moment.

It is important to remark that a Case-based reasoner starts with a representative set of cases. They are like the training set of other supervised machine learning methods. To this end, the initial Case Library is usually seeded with some situations obtained by clustering procedures of historical databases. See, for example, (Sànchez-Marrè et al., 1997).

From these new discovered clusters or classes, some objects (cases) belonging to each class are selected to be included in the initial Case Library.

Learning by own experience is being done after each cycle of the Case-based reasoner. After an evaluation step appears the opportunity to increase the problem-solving capabilities of the system. So, it can learn from the new experience. If the proposed solution has been a successful one, the system can learn from this fact, in the sense that if this experience is stored in memory, when a new similar case to this one appears, it can be solved as the past one (*learning from success*). If the system has failed, it must be able to prevent itself from making the same mistake in the future (*learning from failure*). Not all the Case-based systems have both kinds of learning.

Case-Based Classifiers and Instance-Based Classifiers

Think what happens when we have a supervised dataset composed of several instances, which are described with a set of attribute values, and associated to these attribute values for each instance we have a class attribute, which is the variable of interest.

Really, in this data-driven supervised scenario, the dataset itself can be considered as a *Case Library*, where *the cases* are the instances of the dataset. The *descriptive part of the case* are the attribute values of each instance, and the *solution of the case* is the class label of each instance. Furthermore, as the solution is a simple class label a *null adaptation/copy solution* strategy or a *weighted mode adaptation/ copy and vote solutions* strategy can be used. In Figs. 6.22 and 6.23, the scheme of a null adaptation or a weighted mode adaptation strategies are depicted.

Therefore, using a *null adaptation*, what the *Case-Based classifier* is doing is assigning to the new unlabelled instance, the label of its most similar instance. This most similar instance is usually named as the *nearest neighbour*. This approach is described in Fig. 6.22, where the C_{new} is assigned to have the class label Cl_3, because this is the label of its nearest neighbour instance.

Fig. 6.22 Null adaptation or copy solution strategy for a qualitative solution variable

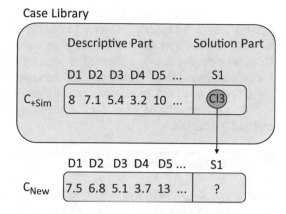

Fig. 6.23 Weighted mode adaptation or copy and vote solutions strategy for a qualitative solution variable

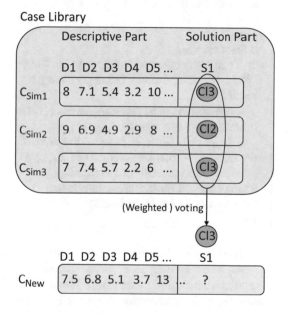

Using the *weighted mode adaptation*, what the *Case-Based classifier* does is assigning to the new unlabelled instance, the majority vote (the mode) or a weighted vote of the labels of its k most similar instances, which usually are named as *k-nearest neighbours*. This approach is described in Fig. 6.23, where the C_{new} is assigned to have the class label Cl_3, because this label is the mode value (majority vote) among its nearest neighbour instances $\{Cl_3, Cl_2, Cl_3\}$.

As it has been described earlier, the Case Library can have different organizations starting from a *flat memory*, a hierarchical organization like as a *discriminant tree or network* or a *k-d tree other hierarchical structures*, or mixed representations. When CBR uses the dataset as a *simple flat memory* scheme we talk about an *Instance-Based classifier*. On the contrary, when the organization of the Case Library is more complex, then we talk in general about a *Case-Based classifier*.

The general scheme of a *Case-Based classifier* or *Case-Based discriminant* method is as follows:

---

algorithm Case-based classifier

input *TrainingDataset*: The training dataset formed of N cases or instances
      C_{new}: case or instance to be discriminated or classified
      k: number of nearest neighbours
output *Cl*: the class label of the new case C_{new}

begin
 if first-time **then**
   // The CB is built from the dataset the first-time. This is the model learning step
   CB ← buildCaseLibrary(TrainingDataset)
 endif
   // Retrieve the most promising cases (≥ k) to be similar to C_{new} from the CB
   promCases ← retrieve(CB, C_{new}, k)
   // Assess the similarity between the promising cases and the C_{new}
   promCases ← compSim(promCases, C_{new})
   // Get the k most similar cases to C_{new}
   neighbours ← GetKmostSim(promCases, k)
   return majorityVote(neighbours)
end

When the Case Library is taken as just the set of instances as a *flat memory* and the adaptation strategy is a *null adaptation* or *weighted mode adaptation*, in the computer science and mathematics literature, people name this technique as the *k-nearest neighbour classifier* (Cover & Hart, 1967; Fix & Hodges Jr., 1951). However, what they probably are not aware is that the *k-nearest neighbour classifier*, is a particular application of Case-Based Reasoning for discrimination purposes, i.e. *a Case-Based classifier/discriminant*, with some particularities: a *flat memory*, a *feature vector representation of cases*, a *simple qualitative label as a solution*, and *null or weighted mode adaptation*. This particular application of a Case-Based classifier is named as *Instance-Based classifier*.

The general scheme of the *k-nearest neighbour classifier*, i.e. an *instance-based discriminant/classifier* method is as follows:

---

algorithm *k*-Nearest Neighbour classifier or instance-based classifier

input *TrainingDataset*: The training dataset formed of N cases or instances
     C_{new}: case or instance to be discriminated or classified
     k: number of nearest neighbours
output *Cl*: the class label of the new case C_{new}

begin
   // The dataset itself is the flat Case Library. This is the model learning step
  CB ← TrainingDataset
  neighbours ← ∅
  for each instance or case *C* ∈ *CB* **do**
   distC ← *d*(*C*, C_{new})
   if |neighbours| < *k* **then**
     // add C to neighbours in order
    orderedInsertion(C, distC, neighbours)
   else
    if distC < *d*(farthestNeighbour(neighbours), C_{new}) **then**
     remove (farthestNeighbour(neighbours), neighbours)
     orderedInsertion (C, distC, neighbours)
    endif
   endif
  endfor
  return majorityVote(neighbours)
end

Therefore, from the dataset we can induce a Case-Based Reasoning model, which is formed by the instances themselves (the cases). In CBR, as a data-driven technique, the model is not explicitly learnt from the data, like a decision tree or a set of association rules, etc. *The model is implicitly formed by all the instances.* This particular situation means that there are not algorithms for building the model of a case-based classifier/discriminant or instance-based classifier/discriminant model, except for creating a non-flat initial Case Base at the beginning. Thus, the above algorithm *does not build any model* like a decision tree algorithm or an association rule model, but *interpret* or *use the model* (i.e. the case base or dataset itself).

When in CBR happens that there is a *flat memory* as a Case Base structure, the *feature-value vector* is the case representation structure, and the solution is the class label, the CBR system is commonly named as an *Instance-Based Reasoning and Learning* system or simply an *Instance-Based Learning* system or an IBL system.

For this reason, Case-Based Reasoning in general, and particularly Instance-Based Learning is commonly known as a *lazy learning* method as opposite to most of machine learning methods which are named as *eager learning* methods. *Laziness* of a case-based classifier or instance-based classifier is referred to the fact that it does not build any model, and delay all actions until the last moment, when a new case or instance needs to be classified. This strategy is completely opposite to the *eager learning* methods, that just when they have the available data (the training set), they build the inductive model from data, using a *model builder* method. Afterwards, when it is needed to classify/discriminate a new instance, then they use a *model interpreter* method like k-nearest-neighbour algorithm or in general, a case-based classifier.

An Example of a Case-Based Classifier

Let us use a very well-known dataset, the Iris dataset (Dua & Graff, 2019), which we mentioned in Sect. 6.3.5. The dataset is formed by 150 instances and five attributes. There are 50 instances of each one of the types of iris flower (*Iris-Setosa, Iris-Versicolour, Iris-Virginica*). Each type of flower is different to the other ones regarding four basic characteristics of the Iris flower: the *petal-width*, the *petal-length*, the *sepal-width*, and the *sepal-length*. These variables are numeric ones expressed in cm. The fifth variable is the type of the Iris flower, the attribute named *Class*.

The problem under consideration with this dataset is to be able to induce a *discriminant model* which were able to estimate the class of Iris flower to which belongs a new unclassified instance. Particularly, we are now interested in use a *case-based discriminant method*.

Therefore, as described before, as a lazy method, we do not need to do anything to construct the model. The model itself will be the instances that form the dataset. As discussed previously in this chapter, and will be explained in more detail in a next subsection, the whole dataset of the 150 instances can be divided into a training set and a validation/test. Let us suppose that we randomly take the 90% of instances to form the training dataset, and the remaining 10% will be the validation test. The validation set is shown in Table 6.9. It has been formed in a *stratified* way so that we have the same number of examples (5) of each one of the classes (3) in the validation set. The Table 6.9 has seven columns. The first column is the number of example (E_1 to E_{15}) of the validation set. The second column *(ID)* is the identifier of the case (a number between 1 and 150), in the dataset. The remaining columns are the five variables. The last column corresponds to the *Class* variable, which is the variable we want to estimate or predict, based on the values of the other four variables.

Once the validation set is formed, we apply the k-nearest neighbour algorithm to each of the instances in the dataset. We tried three values of k (1, 3 and 5). Due to lack of space the results for $k = 5$ are not written, but they were the same than for the other values of k. The results, with the predicted qualitative value for the *Class* variable, compared with the actual value of the *Class* variable in Table 6.10. As all the attribute were numeric, we used the *Euclidian* distance as a dissimilarity function. We did not use any weighting scheme. Thus, all the four attributes were equally weighted in the dissimilarity computations.

As it is shown, for the 15 examples, 14 were correctly discriminated. The instance number 84 (Example E9) was the only one misclassified. It was predicted an *Iris-Virginica* label when the actual label was *Iris-Versicolor*. Therefore, the discriminant accuracy of the case-based classifier was of 93.34%, which is very good, as shown in

Table 6.9 The validation set of the Iris dataset with 15 examples or instances

	ID	Sepal-length	Sepal-width	Petal-length	Petal-width	Class
E1	7	4.6	3.4	1.4	0.3	Iris-setosa
E2	11	5.4	3.7	1.5	0.2	Iris-setosa
E3	26	5.0	3.0	1.6	0.2	Iris-setosa
E4	34	5.5	4.2	1.4	0.2	Iris-setosa
E5	48	4.6	3.2	1.4	0.2	Iris-setosa
E6	51	7.0	3.2	4.7	1.4	Iris-versicolor
E7	57	6.3	3.3	4.7	1.6	Iris-versicolor
E8	60	5.2	2.7	3.9	1.4	Iris-versicolor
E9	75	6.4	2.9	4.3	1.3	Iris-versicolor
E10	93	5.8	2.6	4.0	1.2	Iris-versicolor
E11	106	7.6	3.0	6.6	2.1	Iris-virginica
E12	113	6.8	3.0	5.5	2.1	Iris-virginica
E13	117	6.5	3.0	5.5	1.8	Iris-virginica
E14	137	6.3	3.4	5.6	2.4	Iris-virginica
E15	150	5.9	3.0	5.1	1.8	Iris-virginica

Table 6.10 Results of the case-based classifier for the validation dataset

	ID	Sepal-length	Sepal-width	Petal-length	Petal-width	Class	Predicted class ($k = 1$)	Predicted class ($k = 3$)
E1	7	4.6	3.4	1.4	0.3	Setosa	Setosa	Setosa
E2	11	5.4	3.7	1.5	0.2	Setosa	Setosa	Setosa
E3	26	5.0	3.0	1.6	0.2	Setosa	Setosa	Setosa
E4	34	5.5	4.2	1.4	0.2	Setosa	Setosa	Setosa
E5	48	4.6	3.2	1.4	0.2	Setosa	Setosa	Setosa
E6	51	7.0	3.2	4.7	1.4	Versicolor	Versicolor	Versicolor
E7	57	6.3	3.3	4.7	1.6	Versicolor	Versicolor	Versicolor
E8	60	5.2	2.7	3.9	1.4	Versicolor	Versicolor	Versicolor
E9	84	6.0	2.7	5.1	1.6	Versicolor	*Virginica*	*Virginica*
E10	93	5.8	2.6	4.0	1.2	Versicolor	Versicolor	Versicolor
E11	106	7.6	3.0	6.6	2.1	Virginica	Virginica	Virginica
E12	113	6.8	3.0	5.5	2.1	Virginica	Virginica	Virginica
E13	117	6.5	3.0	5.5	1.8	Virginica	Virginica	Virginica
E14	137	6.3	3.4	5.6	2.4	Virginica	Virginica	Virginica
E15	150	5.9	3.0	5.1	1.8	Virginica	Virginica	Virginica
Acc.							93.34%	93.34%

the last row of the Table 6.10, for all the tested values of parameter k. The algorithm computes the k-most similar instances to one given and make a majority voting among the labels of the different k nearest neighbours. For example, for the E2 (the instance 11 of the dataset), the most similar instances are the following ones: 49, 28, 37, 20, and 47. All of them has *Iris-Setosa* as a label. Therefore, the voting is unanimous, and the E2 is classified as an *Iris-Setosa*, which is true. The classifier has a success.

On the other hand, for the E9 (the instance 84 of the dataset), the most similar instances are the following ones: 134, 102, 143, 150, and 124. All of them has *Iris-Virginica* as a label. Therefore, the voting is unanimous, and the E2 is classified as an *Iris-Virginica*, but the instance has really the label *Iris-Versicolor*. Hence, here the classifier has a misclassification.

Ensemble Methods

In the literature, there are several works proposing the use of a *set of discriminant/ classifier models* to improve the accuracy of the predictions. The aim is to build a discriminant/classifier model by *combining the strengths of a collection of simpler base models*. Usually, these base models are named as *weak classifiers* or learners, and the resulting combination models of the ensemble is known as the *strong classifier* or learner.

There are several ways of implementing this approach, but there is an underlying idea common to all of them: the *diversification* concept. There are some techniques based on using diverse training sets, on using diverse level of noise in the training sets, on using diverse classifier model builder, on using diverse feature subspaces considered, on using diverse hyper-parameters of the classifier model builders, on using diverse weight of the instances, on using diverse trust or weight on the classifier models or on using diverse precision of the classifier models. Finally, the ensemble of methods is used to combine the output of each classifier. Most of them combine the predicted label of each classifier, by means of a (weighted) majority voting scheme.

A possible classification of this ensemble techniques is:

- Multi-expert combination methods (parallel classifier models), which combines the results of all classifiers models, in a parallel way:

  - Global approach (classifier models fusion), which uses the information provided by all the classifier models. Main schemes found in the literature are:

    Voting: A voting scheme, which can be weighted, is done to combine the output among different classifiers.
    Bagging: Making a resampling from the training data (*bootstrapping*), using the same classifier model, and afterwards voting.
    Stacking: A *combiner learner* (*meta-learner*), which combines the predictions of the classifiers to improve the final prediction.
    Randomizing Input Features: Using random subsets of features at each node for building decision trees (*random forests*).

  - Local approach (classifier models selection), where the output of the classifiers is used to select the best model for each instance:

    Gating: Meta selection of the best classifier/s (i.e. *best local expert/s*) to be used. It is also named as a *mixture of experts'* ensemble.

- Multistage combination methods (sequential classifier models), which combines the output of one classifier model, for the next classifier, in a sequential way:

  - *Boosting*: A sequential process where the information provided by the precedent classifier model is used to reweight the training data. This way, the next classifier model will focus on misclassified instances, because a bootstrapping from the training data is done for each new classifier model.
  - *Cascading*: An increasing complexity of classifier models is obtained in a sequential process. After each classifier model is obtained, new information about the predictions of this classifier, like probabilities to predict the different class labels, are used jointly with the training dataset to induce the next classifier model.

- Decorating methods: These approaches add artificial data, i.e. *noise data*, to the training dataset, and for each different new *decorated dataset*, a different classifier model is obtained. Usually the output of all classifiers are aggregated through a majority voting scheme.

In the next subsections, the most common approaches are detailed: *voting*, *bagging*, *random forests* and *boosting*.

Voting

This scheme is very simple. The idea is to induce several different *discriminant or classifier models* from the same dataset using different model builder techniques (MBA$_j$), and afterwards, for discriminating to which class label belongs a new unseen instance, all the different classifiers (CLM$_j$), are used to compute the estimated values of the *class label* (Pr$_j$), and finally a majority (weighted) voting scheme is used to determine the definitive *class label* predicted by the ensemble of classifiers. In this strategy, the diversity relies on the different classifier models used. As a weighting scheme, the computed accuracy of the classifiers in the past instance discrimination predictions can be used. More trust will be given to the more accurate classifier models. The general scheme is depicted in Fig. 6.24.

Bagging

The *Bagging* (*Bootstrap Aggregating*) strategy (Breiman, 1996) proposes to create ensembles by repeatedly and randomly resampling the training data. Given a training set of size n, create m samples of size n by drawing n examples from the original data, with replacement (T_m). Thus, here the diversity relies in the different samples. These are referred to as *bootstrap samples*. Each *bootstrap sample* will contain different training examples, and the rest are replicates. For each sample, the same classifier builder method (MBA) is used to induce one model for each sample. At the testing step, all models are used (CLM$_j$), and their output labels (Pr$_j$) are combined in a majority vote scheme to obtain the final class label prediction.

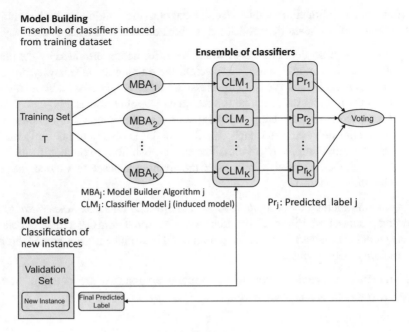

Fig. 6.24 A *voting* scheme for an ensemble of classifiers

This approach has often been shown to provide better results than single models in certain scenarios. This approach can reduce the variance of classifiers improving the *accuracy*, because of the specific random aspects of the training data. The *bagging approach* decreases the error by decreasing the variance in the results due to unstable learners whose output can change when the training data is slightly changed, like for example decision trees or other discriminant models. The bagging scheme is detailed in Fig. 6.25.

Random Forests

Random Decision Forests (Ho, 1995) or *Decision Forests* (Ho, 1998) and *Random Forests* (Ho, 1995; Breiman, 2001) are similar strategies proposed to use multiple decision trees, i.e. *a forest*, where a random subspace of features method is used in the building of the trees. This *diversity* in the features selected to grow a tree, or a subtree improves the predictive skills of single decision trees. Main difference between the *decision forests* proposed by Tin Kam Ho (1998) and the *random forests* proposed by Leo Breiman (2001) is that the former forests are grown each one with a *randomly subspace of features* which is the same for all the tree, and in the latter approach, the *random subspace of features* is selected at each node of a tree. In addition, Leo Breiman proposes that the training datasets for each tree of the forest be a *bootstrap sample* of the original dataset. This way a double *level of diversification* is obtained: diverse training sets and diverse classifier building models.

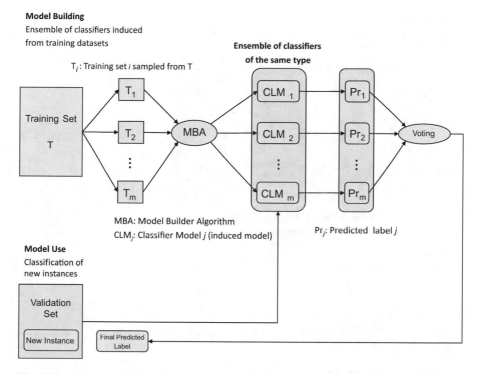

Fig. 6.25 A *bagging* scheme for an ensemble of classifiers of the same type

The *decision forest* approach is a special case of *voting scheme* because they use the same training dataset but a different model building algorithm (a different subspace of features), and the *random forest* approach is a concrete case of a bagging strategy because they use bootstrapping samples, the same model building method with the variant of using a randomly selected subspace of features at each node.

As the random forest approach offers a double diversification, we will analyze a little bit more in detail the random forests approach. In Fig. 6.26 there is the scheme of a random forest strategy.

The *random forest* approach proposes to use sets of unpruned decision trees aiming to reduce the error of the single classifiers. A number d is specified at each node, much smaller than the total number of attributes D. For example, $d = \text{int}\left(\sqrt{D}\right)$ or $d = \text{int}(\log_2 D + 1)$. At each node, d attributes are selected at random out of the D. The split used is the best split, according to the criteria used (information gain, gain ratio, Gini index of impurity, etc.), on these d attributes.

At the testing step of unclassified instances, final classification is done by majority vote across all the trees. Usually, the error rates compare favourably to *boosting* strategies like *AdaBoost* (see next subsection). It is more robust with respect to noise, and efficient on large data.

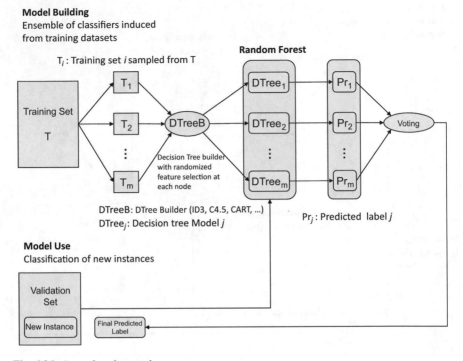

Fig. 6.26 A *random forest* scheme

In addition, at the end of building the *random forest*, it can provide an estimation of the *importance of features* in determining the classification of the class labels. This is a consequence of the analysis of the attributes more used or less frequently used in the induction of the decision trees. Those attributes which have higher frequency values are supposed to be more *relevant* than the attributes less frequently used.

Boosting

Boosting (Freund, 1995) is a common technique used in discrimination or classification tasks. Boosting is a multistage sequential combination method. The idea is to focus on successively difficult instances of the data set, in order to create models that can classify these instances more accurately, and then use the ensemble scores over all the components. A *holdout approach* (split in training and validation/test set) is used in order to determine the incorrectly classified instances of the dataset. Thus, the idea is to sequentially determine better classifiers for more difficult instances, and then combine the results in order to obtain a meta-classifier, which works well on all the dataset.

To focus on difficult instances each instance is scored with a weight, which reflects the degree of difficulty to be classified correctly. At each iteration, the dataset is resampled, and the more important instances (higher weight, great

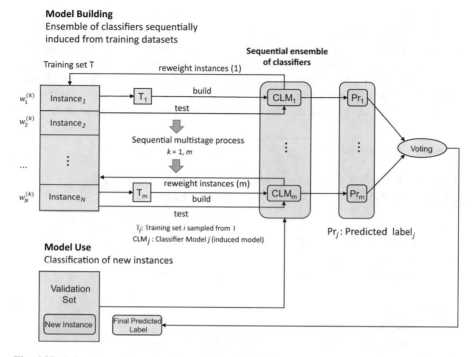

Fig. 6.27 A *boosting* scheme for an ensemble of classifiers

difficulty to be classified) have a higher probability to be selected. Then, a new classifier or hypothesis is learned and the instances are reweighted to focus the system on instances that the most recently learned classifier got wrong. In Fig. 6.27 there is the scheme of the boosting approach.

A general boosting algorithm can be expressed as follows:

```
algorithm Boosting

input X: the training data set (N instances * D+1 attributes): {X1., ..., XN.}
output EnsemClassif: the boosted ensemble of classifiers

begin
    // W is the vector of weights of the instances
    EnsemClassif ← ∅
    Set all instances to have equal uniform weights
    for each k from 1 to m do
        // Learn the next classifier CLk from the weighted examples
        CLk = buildModel(D, W)
        EnsemClassif ← EnsemClassif ∪ {CLk}
        Make a holdout evaluation of CLk with all instances
        Decrease the weights of examples CLk classifies correctly
    endforeach
    return EnsemClassif
end
```

The sequential classifiers must focus on correctly classifying the most highly weighted examples while strongly avoiding overfitting. During testing, each of the m classifiers get a weighted vote proportional to their accuracy on the training data.

One of the most used boosting approach is the *AdaBoost* (*Ada*ptive *Boost*ing) algorithm (Freund & Schapire, 1997), for building ensembles, that empirically improves generalization performance. The algorithm of *AdaBoost*, which is a particularization of the general Boosting algorithm can be described as follows:

---

algorithm AdaBoost

input X: the training dataset (N instances * $D+1$ attributes): $\{X_1., ..., X_N.\}$
output EnsemClassif: the boosted ensemble of classifiers

begin
    // W is the vector of weights of the instances
    EnsemClassif $\leftarrow \varnothing$
    for each instance $X_{i.} \in X$ **do**
    $w_i^{(0)} = \frac{1}{|X|}$
    endforeach
    for each k from 1 to m **do**
        // learn the next classifier CL_k from the weighted examples
        CL_k = buildModel(X, W)
        EnsemClassif $\leftarrow$ EnsemClassif $\cup \{CL_k\}$
        // test the instances and calculate the error, ε_t, of the model CL_k as the total sum
        // weight of the instances that it classifies incorrectly
    $\varepsilon_k = testInstances(D, W)$
    if $\varepsilon_k > 0.5$ **then** exit for **endif**
        // recompute the weights
    $\beta_k = \frac{\varepsilon_k}{1-\varepsilon_k}$
    Multiply the weights of the examples that CL_k classifies correctly by β_k
    Rescale the weights of all of the examples so the total sum weight remains 1
    endfor
    return EnsemClassif
end

When a new instance must be classified with *AdaBoost* then each classifier model (CL_k) is used to get its *class prediction* and the *class prediction* is weighted by log ($1/\beta_k$). Finally, it returns the *class label* with the highest weighted vote in total.

An Example of an Ensemble Model

To illustrate the use of an *ensemble of discriminant methods*, we will use a dataset containing information about images for breast tumour diagnose. The dataset is named "Wisconsin Diagnostic Breast Cancer" (WDBC) dataset and it is in the UCI repository of datasets (Dua & Graff, 2019). It was donated from researchers from the University of Wisconsin. The dataset is composed by 569 instances and 32 attributes. There is one attribute which is the identifier of the instance (*ID*), another is the *diagnosis of the breast tumour* (with two possible values: M meaning malignant and B meaning benign). The remaining 30 features are real-valued, and they were computed from a digitized image of a fine needle aspirate (FNA) of a breast mass. They describe characteristics for each cell nucleus present in the image and are: *the radius* (i.e. mean of distances from centre to points on the perimeter), *the texture* (standard deviation of gray-scale values), *the perimeter, the area, the smoothness* (local variation in radius lengths), *the compactness* (perimeter^2/area − 1.0), *the concavity* (severity of concave portions of the contour), *the concave points* (number of concave portions of the contour), *the symmetry*, and *the fractal dimension* ("coastline approximation"—1).

The main goal is to obtain a *discriminant* or *classifier model* which be as much accurate as possible in order to have a good diagnostic model that can be used for making the diagnosis of new breast tumours.

Of course, first the dataset was pre-processed to solve possible problems before entering into the data mining step. The dataset has no missing values.

The idea is to use some combination of discriminant models as explained in the previous subsection. We propose to use a *voting* scheme, a *bagging* scheme, and a *random forest* approach. The basic classifiers to be used will be Decision Trees and Case-Based/Instance-Based Classifiers. To set-up the experimentation to evaluate which are the best methods, first we divided the dataset into a training set (70%, 398 instances) and a validation set (30%, 170 instances). The splitting was done in a stratified way, and the different model builder algorithms were run. Afterwards, the models obtained were applied to the validation set.

The accuracy results obtained with the different methods are shown in Table 6.11. First five rows show the results of the basic discriminant/classifiers models tried (decision trees and instance-based classifiers). Last eight rows show the results of the ensemble of classifiers approaches.

Among the basic classifiers the best ones were a *decision tree using the Gini index* (DT-CART) with 95.91 of accuracy and an instance-based classifier with $k = 1$ with 93.57 of accuracy. The best results obtained a 96.49% of accuracy (in bold in table), which is really good. It was obtained by both a *random forest* ensemble of 500 *decision trees grew with the gain ratio approach,* and a *voting scheme* with two classifiers (a decision tree with Gini index approach and an instance-based classifier with $k = 1$). The second-best accuracy (95.91 in italics in table) was obtained by both a *decision tree using the Gini index of impurity* (DT-CART) and a *random forest* ensemble of 100 decision trees grew with the gain ratio approach.

From the analysis of the results of the table, it is quite clear that the ensemble approach is generally a good option as provide better results than the base classifiers, with just the exception of the *decision tree using the Gini index*, which get the second-best accuracy. All the variations of *random forests* give a very good accuracy. Thus, random forests seem to be a good approach, in general, for getting accurate models through an ensemble of classifiers approach.

Table 6.11 Comparative accuracy results from the different methods applied to WDBC dataset

	Accuracy on validation set
Decision Tree (Gain Ratio)	92.40
Decision Tree (Gini index)	*95.91*
Instance-based classifier ($k = 1$)	93.57
Instance-based classifier ($k = 3$)	91.81
Instance-based classifier ($k = 5$)	91.81
Voting (DT-GainR, DT-CART. 1-NN, 3-NN, 5-NN)	94.15
Voting (DT-CART, 1-NN)	**96.49**
Bagging (DT-GainR, 100 classifiers)	94.74
Bagging (DT-Cart, 100 classifiers)	95.32
Random Forest (100 trees with GainR)	*95.91*
Random Forest (500 trees with GainR)	**96.49**
Random Forest (100 trees with CART)	94.74
Random Forest (500 trees with CART)	95.32

Table 6.12 *Confusion matrix* of the *random forest* ensemble of 500 decision trees grew with the gain ratio approach

	Actually malignant (M)	Actually benign (B)
Predicted malignant (M)	60	2
Predicted benign (B)	4	105

Table 6.13 *Confusion matrix* of a *voting* scheme with two classifiers (a decision tree with Gini index approach and an instance-based classifier with $k = 1$)

	Actually malignant (M)	Actually benign (B)
Predicted malignant (M)	62	4
Predicted benign (B)	2	103

In addition, as it will be explained in the next subsection, another important tool to analyse the quality of the discriminant models is the so-called *confusion matrix*. The confusion matrix shows the number of instances predicted as belonging to a class label, and compare this number regarding the actual class labels of the instances. Tables 6.12 and 6.13 show the confusion matrices of the two best approaches which gave the best accuracy value of 96.49.

It can be noticed from the Table 6.12 that the random forest approach classified correctly 165 (60 + 105) instances, and only 6 (4 + 2) instances were wrongly classified.

Regarding the *voting scheme* with the two classifiers, it can be seen that the number of correctly classified is the same than with the *random forest* approach (165), but they are slightly differently distributed. The *voting scheme* classified correctly two more instances belonging to the class *Malignant (M)* than the *random forest*, and on the contrary, the *random forest* classified correctly two more instances belonging to the class *Benign (B)* than the *voting ensemble*. Thus, it seems from the experimental work that the *voting scheme* is a little bit better for detecting the malignant tumours and the *random forest* is a little bit better for detecting the benign tumours.

From a medical point of view is better to confuse a predicted malignant tumour with an actual benign tumour (*a false positive*) than confuse a predicted benign tumour with an actual malignant one (*a false negative*). Thus, taking this into account, we will select the *voting scheme* as the best one for the diagnosis of the breast tumours.

Validation of Discriminant Models

As all inductive models, *discriminant models* must be validated, in order to ensure that they can be used afterwards to make good predictions for other unseen data. The validation must be done at two levels: from a qualitative point of view, and from a quantitative point of view.

The *qualitative validation* consists in the interpretation of the model by an expert on the domain. The graphic visualization of the model is usually very important to get a meaningful insight on the knowledge patterns mined from data. For instance, a decision tree model, has a very intuitive interpretation easily made by an expert, as the different features relevant for the discrimination can be traced and the different classification paths (rules) can be interpreted, and an expert can assess whether the discrimination paths make sense or not from an expert perspective.

In addition, the *quantitative validation* of the discriminant models must be done to assess the predictive accuracy or error of the model. The generalization ability of the model must be ensured to be able to use the model for further predictions with other data. This way, we check the lack of overfitting of the model, and its ability to be scalable to new data.

There are some important concepts and methodologies which are common to both *discriminant models* and *predictive models*: the estimation of the true error and the different strategies for the validation of models and computation of the error/accuracy rates.

Estimation of the Error for Supervised Models

It is fundamental to understand that when we want to compute the *true error* of one discriminant or predictive model (M), which means the predictive error for all the possible data (x) in the domain or population (D), is impossible because the number of data belonging to the domain can be infinite.

The *true error* could be computed, if possible, as the probability for all the possible data/instances belonging to the domain (D) that the prediction of the model $M(x)$ were wrong, i.e. different from the actual value of the instance x, ($f(x)$):

$$\text{error}_D(M(x)) = P_{x \in D}[f(x) \neq M(x)] \tag{6.66}$$

As the true error cannot be computed, what can be computed is the *sample error*, which is the error for all the data in the sample (S) or training set. It can be computed as the sum of all the errors in the predictions of the value of the model ($M(x)$) and the actual value of the instance x, ($f(x)$), divided by the number of elements in the sample (N):

$$\text{error}_S(M(x)) = \frac{1}{N} \sum_{x \in S} (1 - \delta(f(x), M(x))) \tag{6.67}$$

We compute the *sample error*, which is in fact the percentage of errors shown by the model in the sample data. The function δ is a generalization of the δ of *Kronecker*[12] function. The *sample error* is an *estimator* of the *true error*.

Now that it is clear that we can compute an estimation of the true error, i.e. the sample error, we will talk just only about the error of a discriminant or classifier model or the error of a predictive model. Sometimes, we compute the error, or the opposite measure, the *accuracy of the model*. Anyway, we must be conscious that the error or accuracy are just estimations of the real magnitudes.

Quantitative Validation of Supervised Models

The main objective of the *quantitative validation* of a supervised data mining model, either a discriminant or a predictive model, is to obtain a good estimation of the true error of the model. As previously explained in the preceding subsection, we compute an estimation of the true error with the sample dataset that we have used to induce the model. Thus, we compute the sample error or just the error from the available dataset. Of course, we want good models with a small error or high accuracy.

As explained in the beginning of the Sect. 6.4.2, to be sure that the estimated accuracy or error is a really good measure, and at the same time to be sure that the models have not a *problem of overfitting*, the usual procedure is to use a different dataset to learn/induce/train the model, and another dataset to validate/test the model, to ensure that the model will have a certain *generalization* or *scalability* ability.

This procedure to split the original dataset into two or three datasets is a common practice in supervised machine learning techniques, as depicted in Fig. 6.15. The typical use of the three datasets is the following:

---

[12]The function δ *of Kronecker* was originally defined for positive integer arguments as $\delta(i,j) =$

$\begin{cases} 0 & \textit{if } i \neq j \\ 1 & \textit{if } i = j \end{cases}$. Here we generalize it for any numeric or qualitative value.

- The *training set* is used to *learn or induce the model or models*. Usually, the greatest part of the original dataset is devoted to the training set. Common size of the training dataset is about 60–80% of the original dataset.
- The *validation test* is used to *check that the model is not overfitting*, and performs accurately on the new instances in the validation test. In addition, it can be used to *compare different models* learnt with the training set to find out which one is the best. Furthermore, the same model with different hyper-parameters can be evaluated and compared. The usual size of the validation set is about 10–20% of the original dataset.
- The *test set* is used to *test the final model* selected after the validation process. If the model performs in a good way on the test data, the confidence on a good fitting increase, and therefore, the overfitting problem is discarded, and the final estimation of the error is more reliable. The common size of the test set is about 10–20% of the original dataset.

Sometimes the test set is not used, and *only a validation set and a training set* are used. If there is no test set, then the validation set could be about 30%. This is a rather common situation, in order not to lose many instances in the test set, which will be only used for testing the final model, but those instances could have been used in the validation test to ensure about the scalability property of the model or models. Hence, especially when working with not very large datasets, only the validation test is used.

Assuming that we have just a training set and a validation set, the issue is how to get the best estimation for the error or accuracy of a discriminant or predictive model. Usually, to estimate the error or accuracy rate, there are different *strategies for the validation process*, which will be described next:

- *Simple Validation or Holdout Validation*: is the basic strategy that just makes a cut on the original dataset and splits it into the *training set* and the *validation set*. The model is induced from the training set, and the instances of the validation set are used to test the model. The model is used to make the prediction values for the instances of the validation set. The error is computed from the result of comparing the predicted values and the real values of validation instances. This procedure is depicted in Fig. 6.28.
- *Cross validation or k-fold cross validation or out-of-sample validation*: this strategy splits the initial dataset into k subsets (named as folds). It is also named as *k-fold validation* or *k-fold cross validation*. Usually, the number of folds, k, is higher than 3, and it is defined trying that each subset had a minimum number of instances. *Each subset* is used as a *validation set* and the remaining $k − 1$ subsets are used as a *training set* at each iteration, to compute an estimation of the error/accuracy for each *validation set*. The *final error/accuracy rate* is computed as the *average of error/accuracy rates* obtained for each one of the k subsets. This procedure is shown in Fig. 6.29. A common value for the number of folds when the dataset is large is 10. This procedure decreases the variability in the estimation of the error/accuracy of the model.

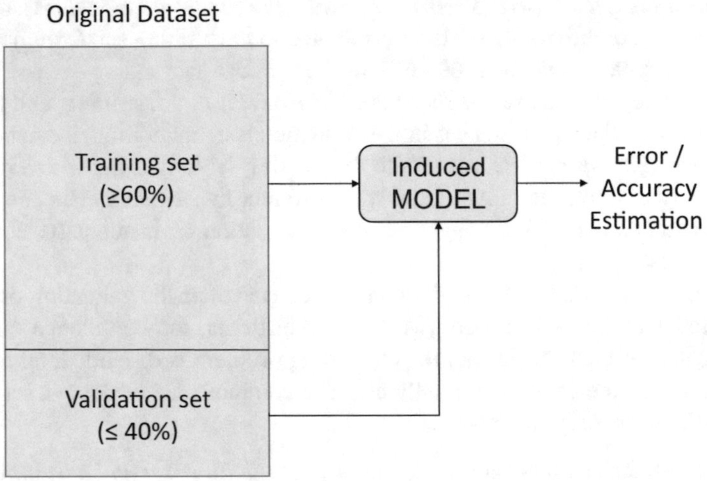

Original Dataset
Training set
(≥60%)
Validation set
(≤ 40%)
Induced
MODEL
Error /
Accuracy
Estimation

Fig. 6.28 The *simple validation* strategy

Original
Dataset
Fold1
Fold2
Foldk
Validation1
Training2
Trainingk
Training1
Validation2
Trainingk
Training1
Training2
Validationk
Induced
MODEL1
Induced
MODEL2
Induced
MODELk
Error/
Accuracy
Estimation1
Error/
Accuracy
Estimation2
Error/
Accuracy
Estimationk
Average
Final
Error/
Accuracy
Estimation

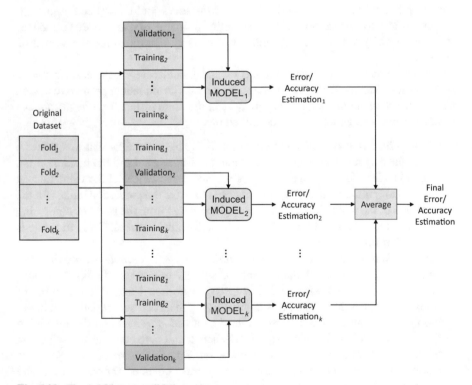

Fig. 6.29 The *k*-fold cross validation strategy

- *Leave-one-out Validation*: This strategy is an extreme application of the k-fold cross validation strategy. The initial dataset is split into k subsets, with the particularity that $k = N$, being N the number of instances in the dataset. This means that the number of instances per fold is just one instance. *Each instance* is used one time as a *validation set*, and all other instances are used as a *training set*. The *final error/accuracy rate* is computed as the *average of error/accuracy rates* obtained for each one of the instances. Usually, this strategy is adopted when the number of instances of the dataset is small.

The manner in which the *partition* of the instances is done into the training set and the validation set has two basic modalities:

- *Random Partition*: The instances belonging to the training set and the validation set are assigned completely randomly, without taking into account any consideration. Just the required number of instances for each set is satisfied. This way could happen that the distribution of the instances belonging to a class could be overrepresented or underrepresented in the set, and hence, the estimation of the error or accuracy values could not be highly significant.
- *Stratified Random Partition*: The class distribution $\{Cl_1, \ldots, Cl_M\}$ among the instances in the *training set* and in the *validation set* follows the same distribution than the original whole dataset, as much as possible. This way, the same proportion of instances of each one of the classes are maintained both in the training and the validation set, as much as possible. This strategy gets a better estimation of the error/accuracy rate than a random partition.

Validation Tools and Indicators in Discriminant Models

There are some indicators and tools that can be used for the validation of the discriminant models. Let us distinguish between the *binary discriminant or classifier models*, i.e. those ones where the *Class* variable has just two possible values, and the *n-ary discriminant or classifier model*, where the *Class* variable can have more than two values.

In *binary discriminant or classifier methods*, the major tools available for validation purposes are: the *confusion matrix*, several *measures* like the *accuracy*, the *precision*, the *recall* or *sensitivity*, the *specificity*, and the *F-measure*, the *ROC curve* and the *Gini Coefficient*. Let us analyse each one of these tools and indicators:

- *Confusion matrix*: A *confusion matrix* is a kind of table of contingency which shows the number of instances predicted as belonging to a class label (predicted C +) and actually belonging to that class (actual C+), and the number of instances predicted as belonging to a class label (predicted C+) and actually belonging to the other class (actual C−). The same frequency counting is done for the other predicted class (C−). This way, each row refers to the instances of each predicted class, and each column refers to the instances of each actual class. The confusion matrix for a binary classifier or discriminant method is a square matrix of dimension two.

Table 6.14 A Confusion matrix for a binary discriminant or classifier method

	Actual class (C+)	Actual class (C−)
Predicted class (C+)	TP	FP
		Type I error
Predicted class (C−)	FN	TN
	Type II error	

In a confusion matrix for a binary classifier, like the one depicted in Table 6.14, there are the following elements which are usually known as:

– The true positive instances (TP), which are the instances predicted as belonging to the class C+, and that actually belong to the class C+. These instances have been correctly discriminated or classified.
– The true negative instances (TN), which are the instances predicted as belonging to the class C−, and actually belonging to the class C−. These instances have been correctly discriminated or classified.
– The false positive instances (FP), which are the instances predicted as belonging to the class C+, but actually belong to the other class C−. These instances have been wrongly discriminated or classified. In Statistics, these instances constitute what are known as the *Type I error*.
– The false negative instances (FN), which are the instances predicted as belonging to the class C−, but actually belong to the other class C+. These instances have been wrongly discriminated or classified. In Statistics, these instances constitute what are known as *Type II error*.

A confusion matrix is a very visual way of detecting the erroneously classified instances and the right classified instances. In the ideal case of a perfect discriminant model with no errors on the classification, the matrix is diagonal (i.e. all the elements outside the main diagonal of the matrix are zero). These means that FP = FN = 0. In the common situation when the models are not perfect, the confusion matrix is not a diagonal matrix, and the FP and FN are the instances wrongly classified.

• *Quantitative measures to evaluate the quality of the discriminant model.* Most commonly used are the following ones:

– The *accuracy of the classifier model* (*ACC*), which is the ratio between the correctly classified instances (*TP* + *TN*) and the total number of instances (*N*), is defined as follows:

$$\text{ACC} = \frac{\text{TP} + \text{TN}}{N} = \frac{\text{TP} + \text{TN}}{\text{TP} + \text{FP} + \text{TN} + \text{FN}} \qquad (6.68)$$

where N is the total number of instances of the validation set. The *error of the classifier model* is the opposite measure to the accuracy, i.e.,

$$\text{Error} = 1 - \text{ACC} = 1 - \frac{\text{TP} + \text{TN}}{N}$$

$$= 1 - \frac{\text{TP} + \text{TN}}{\text{TP} + \text{FP} + \text{TN} + \text{FN}} = \frac{\text{FP} + \text{FN}}{\text{TP} + \text{FP} + \text{TN} + \text{FN}} \tag{6.69}$$

Sometimes, the accuracy and error rates are given as a percentage instead of a unitary rate as ACC * *100* or Error * *100*.

- The *recall* or *sensitivity* or *hit rate*, or *True Positive Rate* (*TPR*), which is the ratio between the correctly predicted positive instances (*TP*), i.e. belonging to C+, and the total number of positive instances (*TP + FN*), is defined as follows:

$$\text{Recall} = \text{Sensitivity} = \text{TPR} = \frac{\text{TP}}{\text{TP} + \text{FN}} \tag{6.70}$$

- The *specificity* or *selectivity* or *True Negative Rate* (*TNR*), which is the ratio between the correctly predicted negative instances (*TN*), i.e. belonging to C−, and the total number of negative instances (*FP + TN*), is defined as follows:

$$\text{Specificity} = \text{Selectivity} = \text{TNR} = \frac{\text{TN}}{\text{FP} + \text{TN}} \tag{6.71}$$

- The *precision* or *Positive Predictive Value* (*PPV*), which is the ratio between the correctly predicted positive instances (*TP*), i.e. belonging to C+, and the total number of predicted positive instances (*TP + FP*), is defined as follows:

$$\text{Precision} = \text{PPV} = \frac{\text{TP}}{\text{TP} + \text{FP}} \tag{6.72}$$

- The *Negative Predictive Value* (*NPV*), which is the ratio between the correctly predicted negative instances (*TN*), i.e. belonging to C−, and the total number of predicted negative instances (*FN + TN*), is defined as follows:

$$\text{NPV} = \frac{\text{TN}}{\text{FN} + \text{TN}} \tag{6.73}$$

- The *F-measure* or *F1-measure* or *F1-score*, which is the harmonic mean[13] of *precision* (*PPV*) and *recall* (*TPR*), is defined as follows:

$$F\text{-measure} = 2 \cdot \frac{\text{PPV} * \text{TPR}}{\text{PPV} + \text{TPR}} = \frac{2 * \text{TP}}{2 * \text{TP} + \text{FP} + \text{FN}} \tag{6.74}$$

---

[13]The harmonic mean of n values is defined as follows: $H(x_1, \ldots, x_n) = \frac{n}{\frac{1}{x_1} + \cdots + \frac{1}{x_n}}$

- The *ROC* (Receiver/Relative Operating Characteristic) *Curve*:[14] a ROC curve, is a graphical plot that illustrates the diagnostic ability of a binary classifier model as its discrimination threshold is varied. It is useful to compare several binary classifier models to assess its quality.

  The ROC curve is created by plotting the *recall* or *sensitivity* or *true positive rate (TPR)* in the *y*-axis against $(1 - specificity)$ or *false positive rate (FPR)* in the *x*-axis at different threshold settings. The *false positive rate* is also known as the fall-out or probability of false alarm and can be calculated as FPR = FP/(FP + TN). As both measures range from 0 to 1, the square of the plot has an area of 1.

  The *threshold settings* can be set over some probabilistic scores generated by a probabilistic classifier (Bayesian classifier, ANN classifier, etc.), which predict a class label with an associated probability or scoring value. The values higher than a fixed threshold are predicted as being of the class positive, and the values below the threshold are predicted as belonging to the negative class. Then a confusion matrix is built and the TPR and FPR values are computed. If the classifier is not probabilistic, in the sense that does not generate a score or a probability, like a Decision Tree, the points of the curve can be obtained by different validation tests, which generate a different confusion matrix, and from it, the TPR and FPR values can be computed.

  In Fig. 6.30, there is the picture of a ROC curve of four binary classifiers. The 45-degree line depicted represents the curve of a random classifier, i.e. the one that randomly makes predictions for the positive or the negative class. The curves above the 45-degree line (blue and green) are good performance classifiers, and the curve below the 45-degree line (red curve) is a bad performance classifier. However, in the case of a bad classifier, like the red curve, it can be converted in a good one, just changing its prediction output: negative by positive, and positive by negative. Thus, as much nearer is a ROC curve of a classifier to the upper left corner of the square, i.e. the point (0,1) meaning 100% of specificity (0% of $1 - specificity$) and 100% of sensitivity, better is the classifier performance.

  Related to the ROC curves, there is a measure used to compare the performance of several classifiers, which is the *Area Under Curve* (AUC) indicator. The AUC is the area lying below the ROC curve and the axis. It is value ranging from 0 to 1. As higher is the value of AUC better is the classifier. A value of AUC equal to 1 means the perfect classifier, whose ROC curve reaches the upper left corner. A value of AUC equal to 0 is the worst classifier whose ROC curve reaches the lower right corner. A random classifier has a value of AUC equal to 0.5.

- The *Gini Coefficient*: The Gini coefficient, sometimes called Gini index[15] or Gini ratio, is a measure of statistical dispersion intended to represent the income or

---

[14] The ROC curve was developed by radar engineers during World War II for detecting enemy objects in battlefields. Its use has been generalized into other scientific fields like psychology, medicine, etc. to assess the quality of discriminant methods.

[15] The *Gini coefficient*, sometimes named as *Gini index*, has nothing to do with the *Gini index of impurity* used in CART method for inducing a decision tree.

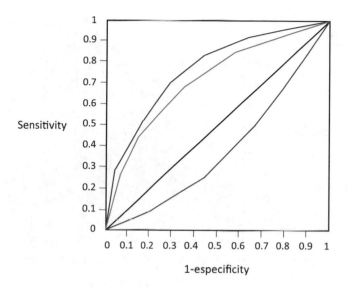

Fig. 6.30 ROC curves of several classifiers

wealth distribution of a nation's residents, and is the most commonly used measurement of inequality. It was proposed by Corrado Gini (1909, 1912).

The plot of the distribution of the population and its income in the axis has a similar, but symmetrical curve (the Lorenz curve) to the ROC curve of a classifier. The Gini index is defined as the ratio of the area below the 45-degree line and the Lorenz curve related to the whole triangular area. Exploiting this similarity of the ROC curves and the Gini coefficient, an equivalent to the Gini coefficient can be used to assess the quality of classifiers in a similar way than the AUC. Using this similarity, the Gini coefficient can be computed as follows in a ROC curve:

$$
\text{Gini-coeff} = \frac{1/2 - (1 - \text{AUC})}{1/2}
$$
$$
= 2(1/2 - (1 - \text{AUC})) = 2 * \text{AUC} - 1
$$
(6.75)

The Gini coefficient ranges from -1 to $+1$. A random classifier has a Gini coefficient equal to 0. The worst classifier has a value of -1, and the perfect classifier has a value of $+1$. However, as the classifiers are usually above the 45-degree line, the usual values for the Gini coefficient range from 0 to 1. As higher is the value of the Gini coefficient, better is the classifier.

In Fig. 6.31, there is depicted the AUC for a classifier, and the different areas involved in the computation of the Gini coefficient.

In *n-ary discriminant or classifier methods*, the major tools available for validation purposes are: the *confusion matrix*, and some *quantitative measures* like the

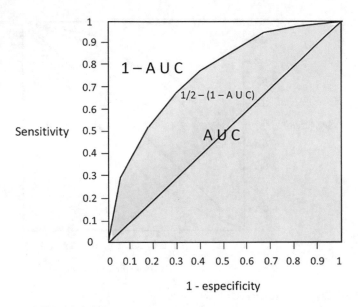

Fig. 6.31 An AUC for a classifier

Table 6.15 A Confusion matrix for a n-ary discriminant or classifier method

	Actual class (C_1)	Actual class (C_2)	...	Actual class (C_K)
Predicted class (C_1)	TC_1	FC_{12}	...	FC_{1K}
Predicted class (C_2)	FC_{21}	TC_2	...	FC_{2K}
...	...	...	...	...
Predicted class (C_K)	FC_{K1}	FC_{K2}	...	TC_K

accuracy and the *True Class Rate* (TClR). Let us analyze each one of these tools and indicators:

- *Confusion Matrix: A confusion matrix for a n-ary discriminant or classifier method is very similar to a confusion matrix for a binary classifier. Main difference is that the matrix is not a square matrix of dimension two, but a square matrix of dimension K, where K is the number of different class labels of the Class* variable. The row i has the accounting of how many instances are being predicted as belonging to class Cl_i and actually are from all the possible class labels {Cl_1, ..., Cl_K}. At each column k, there is the distribution of how many instances are actually of the class Cl_k, and are predicted as being of all the possible class labels {Cl_1, ..., Cl_K}. An element $cfm_{i, j}$ from the confusion matrix CfM accounts for the frequency of instances predicted to be from class label Cl_i and actually being from class label Cl_j.

  In Table 6.15, there is a scheme of a general confusion matrix for a n-ary classifier, where we can distinguish the following elements:

– The True Class instances (TC$_i$), which are the instances predicted as belonging
to the class C_i, and that actually belong to the class C_i. These instances have
been correctly discriminated or classified.
– The False Class instances (FC$_{ij}$), which are the instances predicted as belonging to the class C_i but actually belong to the class C_j. These instances have
been wrongly discriminated or classified.

Like in the binary classifier case, the confusion matrix of a *n*-ary classifier
method allows for a meaningful visual analysis for detecting the erroneously
classified instances and the right classified instances. In the ideal case of a perfect
discriminant model with no errors on the classification, the matrix is diagonal (the
elements FC$_{ij}$ are zero). In the common situation when the models are not perfect,
the confusion matrix is not a diagonal matrix, and all the FC$_{ij}$ are the instances
wrongly classified.

• *Quantitative measures to evaluate the quality of the discriminant model.* Most
commonly used are the following ones:

– The *accuracy of the classifier model* (ACC), which is the ratio between the
correctly classified instances (TC$_k$) and the total number of instances (*N*), is
defined as follows:

$$ACC = \frac{\sum_{k=1}^{K} TC_k}{N} = \frac{\sum_{k=1}^{K} TC_k}{\sum_{k=1}^{K} TC_k + \sum_{i=1}^{K} \sum_{\substack{j=1 \\ j \neq i}}^{K} FC_{ij}} \tag{6.76}$$

where *N* is the total number of instances of the validation set. The *error of
the classifier model* is the opposite measure to the accuracy, i.e.,

$$Error = 1 - ACC = 1 - \frac{\sum_{k=1}^{K} TC_k}{N}$$

$$= \frac{\sum_{i=1}^{K} \sum_{\substack{j=1 \\ j \neq i}}^{K} FC_{ij}}{N} = \frac{\sum_{i=1}^{K} \sum_{\substack{j=1 \\ j \neq i}}^{K} FC_{ij}}{\sum_{k=1}^{K} TC_k + \sum_{i=1}^{K} \sum_{\substack{j=1 \\ j \neq i}}^{K} FC_{ij}} \tag{6.77}$$

Sometimes, the accuracy and error rates are given as a percentage instead of
a unitary rate as *ACC * 100* or *Error * 100*.

- The *True Class Rate of Class k* (TClR$_k$), which is the ratio between the correctly predicted instances of the class C_k (TC$_k$), and the total number of instances belonging to the class C_k (TC$_k + \sum_{\substack{i=1 \\ i \neq k}}^{K} FC_{ik}$), is defined as follows:

$$TClR_k = \frac{TC_k}{TC_k + \sum_{\substack{i=1 \\ i \neq k}}^{K} FC_{ik}} \tag{6.78}$$

The *True Class Rate of Class k* (TClR$_k$) measure for each possible class or modality k allows to make an analysis of the accuracy of the classifier model for each modality or class label. The TClR$_k$ accounts for the percentage of corrected instances of the class label Cl$_k$. Thus, these measures for each modality give a clear scenario of the accuracy of the classifier model for each one of the modalities or class labels.

6.4.2.2 Predictive Models

A last common problem in Machine Learning is to obtain a *predictive model* from a supervised dataset. Predictive models are able to estimate or predict the *numeric values* of a numeric variable in the dataset. The *quantitative variable* or *attribute* which is the variable of interest to be predicted is usually named in machine learning and data mining as the *target variable*. The interesting situation is when a new instance, not belonging to the training set is available, and the model is able to predict which is the corresponding value of the new instance for the *target variable*. Predictive models are also called *regression models* in the literature. Here, we will avoid this nomenclature because it could be misleading regarding the linear regression models (either simple or multiple linear regression models).

There are several kinds of predictive methods like *Artificial Neural Networks* (*ANNs*), *Case-Based Predictors*, etc. In addition, the same idea of an ensemble of discriminant models can be applied to predictive models. In the next subsections, we will describe the *Artificial Neural Networks* (*ANNs*), the *Case-Based predictors*, the *Multiple Linear Regression* (*MLR*)[16] model, and the *ensemble of predictors* approach.

---

[16]The Multiple Linear Regression (MLR) model is a method coming from Statistics field, and not from Machine Learning field, but as it is commonly used for predicting a target variable with a linear relationship with the other variables, it will be explained in this chapter.

Artificial Neural Network Models

This approach is inspired by the biological *neural networks*, which are in the human brain. In the brain, there are a numerous set of nerve cells, which are highly interconnected. These cells are the *neurons*. Neurobiologists have estimated that the human brain is approximately formed by 10^{11} interconnected neurons, and each one is connected on average to other 10^4 neurons (Shepherd & Koch, 1990). The signals to perform different functions in the brain are transmitted through the neurons.

A neuron can be divided into the *cell corpus*, also named as *soma*, which contains the *nucleus* of the cell, and the *nervous terminations*. There are two kinds of nervous terminations: a great number of network of fibres or terminations around the soma called *dendrites*, and a unique large termination named *axon*, which connects the dendrites of other neurons. The transmission of information through the signals is done at the synaptic connection in the *synapse* points. The structure and interconnection of the neurons is depicted in Fig. 6.32.

The transmission of information starts to be performed when a neuron gathers signals from others neurons through its own numerous sets of dendrites. The neuron sends out spikes of electrical activity through the axon, which splits into thousands of branches. At the end of each branch, a structure called a *synapse* converts the activity from the axon into electrical effects that inhibit or excite activity in the connected neurons.

Signals are transmitted from one neuron to another by complex electrochemical reactions. Chemical substances released from the synapse cause a change in the electrical potential of the cell corpus. When this excitatory electrical potential is sufficiently large compared with its inhibitory input, it sends a spike of electrical activity to its axon. The spikes travelling along the axon trigger the release of

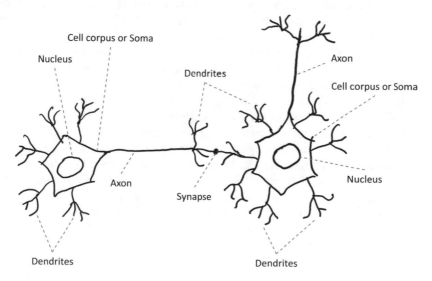

Fig. 6.32 A neural network

Fig. 6.33 An artificial neuron

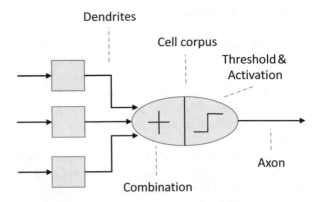

neurotransmitter substances at the synapse. The neurotransmitters cause *excitation* or *inhibition* in the dendrite of the post-synaptic neuron. The integration of the excitatory and inhibitory signals may produce spikes in the post-synaptic neuron. The contribution of the signals depends on the strength of the synaptic connection. This way the signals are propagated through the network of neurons in the brain.

Learning occurs by changing the effectiveness of the synapses so that the influence of one neuron on its interconnected neurons changes. Thus, the human brain can be considered as a highly complex, non-linear, and parallel information processing system.

The model of an *Artificial Neural Network (ANN)* mimics the biological neural networks by building an artificial system composed by the interconnection of several basic processing units, i.e. the *artificial neurons*. An artificial neuron is displayed in Fig. 6.33.

In an artificial neural network, the *input value* is the equivalent to the *dendrites* in a real neuron, the *cell corpus* or *soma* is the *artificial neuron*, the *output value* is the equivalent to the *axon*, and the *synapse* regulating the strength of the propagation are modelled with the *weights* associated to each value moving from one neuron to another. Each artificial neuron makes a combination of its input values, and afterwards, depending on whether the combined value reaches a threshold or not, a different output value is computed through an *activation function*.

A typical architecture of an ANN is composed by one *input layer*, several *intermediate layers* and one *output layer*, as depicted in Fig. 6.34.

ANNs can have several intermediate layers, usually named as *hidden layers*, between the *input layer* and the *output layer*. The *input layer* has just the input values to the network and propagate them to the next neurons in the next layer, but does not include any processing of the values. Each neuron in a *hidden layer* combines all its input values, from preceding layers, and generate an output, which will be the input for the neurons in the next layer. Finally, the neurons in the *output layer* combine all its input values, from the preceding layer, and generate an output value, which will be an output value of the network.

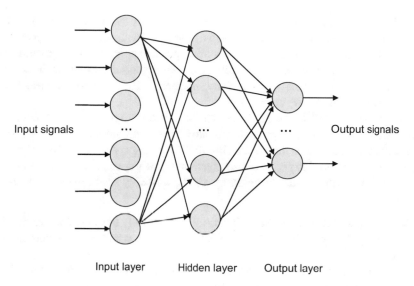

Input signals   ···   ···   ···   Output signals

Input layer   Hidden layer   Output layer

Fig. 6.34 A typical architecture of an ANN

The ANN will produce an output result, as an answer to several input information from the input layer neurons, emulating the neural networks of the brain, which propagate signals among all the neurons interconnected. The output result can be *a single numeric value* (just one single neuron in the output layer) or a *vector of numeric values* (several neurons in the output layer). The numeric output value can be a discrete one (integer value) or a continuous value (a real value). Therefore, an ANN, in addition to predict a real variable or attribute, also can be used to predict a discrete numeric value like several modalities (codified in a numeric way: 0, 1, 2, 3, . . . , K) of a class variable. These means that an *ANN can be used also as a classifier model*. Notwithstanding, as ANNs are *general approximation functions,* which is very useful in nonlinear conditions, major interest in data mining relies in using *ANNs as numerical predictors*. Therefore, they can be used as prediction models for a variable of interest or several variables of interest.

Neurons in an ANN are connected by links. Each connection has an associated score, which is named as the *weight of the connection*. Weights measure the strength or relevance of each neuron input. As it will be detailed later, an ANN during the training step, makes repeated updates of the weights, and finally learns the optimal weights that performs the best predictions of the output values.

The development of ANNs is based on the early works of some theorists like McCulloch, Pitts, Culbertson, Kleene, and Minsky, which were focussing on the functional similarity between a neuron and the on–off units of computers. They provided several brain models focussed on some logical aspects to perform specific algorithms in response to a sequence of stimulus. One of the most influential work was done by Warren McCulloch and Walter Pitts in 1943 (McCulloch & Pitts, 1943). They stated that the nervous activity and the neural events and their inter-relationships can be managed by means of propositional logic. In other words, they stated that for any logic expression satisfying certain conditions, an artificial neural net, can be found to behave in the same manner.[17] They settled the basis for the development of the *perceptron*.

There are many different ANNs which have been proposed in the literature for supervised tasks. It is not easy, and probably there is no unique possible classification, but they can be classified according to the topology of the network (number of layers), according to the type of the connections (forward, backward), etc. A possible classification, with the most important ANNs, probably incomplete, could be the following:

- *Single-Layer Perceptron* (SLP): This ANN model only has the input layer and the output layer. The basic one is the so-called *perceptron* with just one neuron in the output layer.
- *Multiple-Layer Perceptron* (MLP): They have intermediate or hidden layers between the input layer and the output layer.

  - *Static ANNs*: there is no dynamic temporal behaviour needed to be used.

    Feedforward Neural Networks (FFNN): They are ANNs where connections among the neurons do not form a cycle.

    Radial Basis Function Neural Networks (RBFNNs): They are special feedforward ANNs which has just one hidden layer, and use radial basis functions as activation function in the hidden layer. The activation of a hidden unit is determined by the distance between the input vector and a prototype vector.

    Deep Learning Neural Networks (DLNN) are feedforward networks which alternate between *convolutional layers* and *max-pooling layers*, topped by several fully or sparsely connected layers followed by a final classification layer.

    Convolutional Deep Neural Networks (CNNs): A class of deep neural networks, most commonly applied to analyzing visual imagery.

  - *Dynamic ANNs*: They capture a dynamic temporal behaviour using some kind of internal state (memory) to process the sequence of inputs or the updating of information (believes). Some of them will be analyzed in Chap. 9.

---

[17] Although this statement was demonstrated not to be completely true by Marvin Minsky and Seymour Papert (Minsky & Papert, 1969), their work was a pioneer for the deployment of ANNs. Minsky and Papert proved that the Exclusive-OR operation cannot be learnt by a basic ANN, called as *perceptron* as it will be explained next, as well as all the functions which were *not linearly separable*.

Time Delay Neural Networks (TDNNs): They are Multilayer Feedforward ANNs where the previous *k* input values, for a given *temporal window of length k*, are delayed in time until the final input is available, synchronizing all the input elements of the window.

Recurrent Neural Networks (RNNs): They use connections among the neurons that form a cycle. They allow connections to neurons in the *previous layers* or even *self-connections* in the same layer. These *recurrent connections* act as a *short-term memory* and allows the network to remember what happened in the past exhibiting a temporal dynamic behaviour. Some types are:

> *Simple Recurrent Neural Networks* (SRNNs): They are RNNs where only some neurons or units are connected backward. Main used architectures in the literature are:
>
> > *Elman Network*: An *Elman Network* is a three-layer network with the addition of a *context layer*, i.e. a set of *context units*. The first layer is the *input layer*. The second layer is a *hidden layer*, and the third one is the *output layer*. The *hidden layer* is connected to these *context units* with a fixed weight of one.
> >
> > *Jordan Network*: A *Jordan Network* is similar to an *Elman network*. It has three-layers too, and an additional *context layer*: the *input layer*, the *hidden layer*, the *output layer*, and the *context layer*. Main difference is that the *context units* are fed from the *output layer* instead of the *hidden layer*, and that they have a recurrent connection to themselves.
>
> *Fully Recurrent Neural Networks* (FRNNs): They are RNNs which connect the outputs of all neurons to the inputs of all neurons.
>
> *Long Short-Term memory Networks* (LSTMs): *is a RNN* using a convenient and appropriate gradient-based learning algorithm avoiding the *exploding* or *vanishing gradient problems*. This is achieved by an efficient gradient-based algorithm enforcing a constant error flow through *internal states of special units*.

Probabilistic Neural Networks (PNNs): are feedforward networks used in classification and pattern recognition tasks, which estimates the parent probability distribution, derive from Bayesian Networks and are organized in fours layers: input, pattern, summation and output.

Here in this chapter, we will analyze the Single-Layer Perceptron architecture, as the basis to understand how ANNs work and learn the weights from the training instances to approximate the output function, either a class label (numeric) or a numeric variable. After that, we will analyse the Multi-Layer Perceptron, and the *backpropagation algorithm* used to learn the weights.

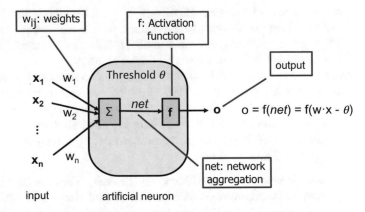

Fig. 6.35 Basic behaviour of one artificial neuron

The Perceptron, and the Basic Behaviour of an Artificial Neuron

The basic idea of how to operate an artificial neuron, was proposed by Rosenblatt (1958). His proposal is still the basis for the operation of most kind of ANNs. This type of ANN based on just one neuron at the output layer was named as a *perceptron*.

The basic idea of the perceptron is that the neuron combines the information from the input signals to a neuron using a *weighted sum of the input signals*. Afterwards, it compares the result of the combination against a *threshold value θ*. If this combination is less than the threshold, the neuron output is −1. On the other hand, if the combination is greater than or equal to the threshold, then the neuron becomes activated, and the neuron output a +1 value. This general operation is depicted in Fig. 6.35.

Thus, the aggregation or combination of the inputs to get the net value is computed as the network aggregation (net):

$$net = \sum_{i=1}^{n} w_i x_i \tag{6.79}$$

In general, the output of the neuron is the result of the application of an *activation function*, to the aggregated value of the inputs, taking into account the threshold θ:

$$o = f(net) = f(w \cdot x - \theta) \tag{6.80}$$

And concretely, it was proposed to use:

$$o(x_1, \ldots, x_n) = f(\text{net})$$

$$= f_{\text{sgn}}(w \cdot x - \theta) = \begin{cases} +1 & \text{if } \sum_{i=1}^{n} w_i x_i - \theta > 0 \\ -1 & \text{otherwise} \end{cases} \qquad (6.81)$$

This activation function is known as the *sign* or *signum function*. The *sign function* is defined as follows:

$$f_{\text{sgn}}(x) = y_{\text{sgn}}(x) = \begin{cases} +1 & \text{if } x > 0 \\ -1 & \text{if } x \leq 0 \end{cases} \qquad (6.82)$$

In fact, the original *sgn* function is defined to be 0 for the value 0. Other activation functions used in ANNs are the *step function* or *Heaviside step function*, the *sigmoid function* or *logistic function* and the *linear* or *identity function*.

The definition of these functions is as follows:

$$f_{\text{step}}(x) = y_{\text{step}}(x) = \begin{cases} +1 & \text{if } x \geq 0 \\ 0 & \text{if } x < 0 \end{cases} \qquad (6.83)$$

$$f_{\text{sigmoid}}(x) = y_{\text{sigmoid}}(x) = \sigma(x) = \frac{1}{1 + e^{-x}} \qquad (6.84)$$

$$f_{\text{linear}}(x) = y_{\text{linear}}(x) = x \qquad (6.85)$$

The *sign function* (6.82) and the *step function* (6.83) are named as *hard limiting activation functions* and are used for discrete ANN models (*ANN classifier models*), as they produce a binary output in output neurons (+1/−1 or 1/0). To use an ANN for discrimination/classification task is straightforward. If the class variable hast two possible labels (binary classifier), it can be solved with just one neuron in the output layer with the sign activation (or step) function. This way, an output value near to +1 will classify a new instance as being of one class and an output value near to −1 will classify the new instance as belonging to the other class. Of course, if the output value of the network were a value around 0, we would have a doubt about which is the label to be predicted. To solve this situation is better to have an ANN with two neurons at the output layer. This way if the vector output value were near to (+1, −1) or (1, 0) the predicted class will be one, and if the vector output were near to (−1, +1) or (0, 1) the predicted class will be the other one. With this configuration, some threshold value can be set, like for example, 0.7 to decide that the output value being higher than the threshold would be the class label predicted or deciding the final label being the one associated with the highest value between both values.

This configuration of having one neuron for each possible class label at the output layer can be extended to the case of a multi-class label problem (*n*-ary classifier).

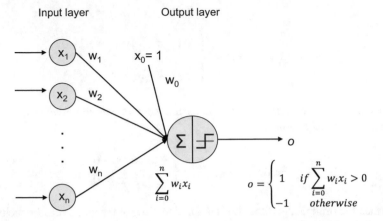

Fig. 6.36 A perceptron

The *sigmoid function* (6.84) and the *linear function* (6.85) are named as *soft limiting activation functions* and are used for continuous ANN models (*ANN predictor models*), as they produce continuous output values in the range [0, 1] and in $[-\infty, +\infty]$ respectively in output neurons.

Thus, the perceptron network has an architecture as depicted in Fig. 6.36.

The quantity of the threshold θ is represented as $-w_0$. This way, if we add a constant input $x_0 = 1$, with the weight w_0 the notation of the linear combination of the input signals can be simplified as:

$$\sum_{i=1}^{n} w_i x_i - \theta = \sum_{i=0}^{n} w_i x_i \tag{6.86}$$

The *perceptron* represents a *hyperplane decision surface* in the n-dimensional space of instances. The perceptron outputs a value $+1$ for instances lying on one side of the hyperplane and outputs a value -1 for instances lying on the other side.

The equation for this decision hyperplane is:

$$w \cdot x = \sum_{i=0}^{n} w_i x_i = 0 \tag{6.87}$$

Thus, from a geometrical point of view is clear that not all the input examples or instances can be separated by this linear hyperplane. The perceptron can only classify correctly a set of examples which are *linearly separable*.

The big question now is how the perceptron can learn the vector of weights, $(w_0, \ldots, w_n)$, to produce the desired correct classification for the training examples.

The correct classification means to outputs a value $+1$ or a value -1 for each one of the instances of the training set. There are several known algorithms to solve this learning task. Here, we will explain the two most used. They provide a procedure that converges, under some different conditions, and obtain an optimal weight vector. These algorithms are the *perceptron learning/training rule* or simply *perceptron rule* and the *delta rule*.

The *perceptron rule* was first proposed by Rosenblatt in 1960 (Rosenblatt, 1960). The rule proposes to learn the weight vector making repeated adjustments to the weights until all the examples are correctly classified. The procedure starts with *random weights*. Next, iteratively the *perceptron* is applied to *each training example*, modifying the *perceptron weights* whenever it misclassifies an example.

This process is performed again, iterating through all training examples as many times needed until the perceptron *classifies* all training examples correctly. Commonly, each iteration testing all the instances in the training set is known as an *epoch*.

Weights are modified at each step according to *the perceptron training rule*, which updates the weight w_i associated with the corresponding input $x_{d_m i}$ according to the rule:

$$w_i \leftarrow w_i + \Delta w_i \qquad (6.88)$$

where,

$$\Delta w_i \leftarrow \eta \left(t_{d_m} - o_{d_m} \right) x_{d_m i} \qquad (6.89)$$

and,

d_m is the training example X_{d_m}.
$x_{d_m i}$ is the value of the variable $X_{.i}$ for the training example X_{d_m}. or simply d_m
t_{d_m} is the target output for the current training example d_m
o_{d_m} is the output of the linear perceptron for the training example d_m
η is a small constant called the *learning rate*

The *learning rate* aims at controlling the degree to which weights are changed at each step. It is usually set to some small value (for instance, 0.1) and it is sometimes decreased as the number of epochs increases.

It can be proved that the algorithm will converge if training data is linearly separable and η is sufficiently small. However, if the *data is not linearly separable*, the *convergence is not assured*.

The algorithm of the *perceptron training rule* can be described as follows:

algorithm Perceptron Training Rule

input D: the training data set (M instances * $N+1$ attributes): $\{X_{d_1}, \ldots, X_{d_M}\}$
    t: is the target variable codified with integer modalities (the class variable)
output a weight vector $(w_0, w_1, \ldots, w_n)$

begin
  // Set the learning rate to a small value ($0 < \eta < 1$). For instance, 0.05 or 0.1
  $\eta \leftarrow 0.05$
  // Set the initial weights and threshold to small random numbers
  // For instance, in the range [-0.5, 0.5] or use other heuristics
  for each $w_i, \ i \in 0, \ldots, n$ **do**
  $w_i \leftarrow random(-0.5, 0.5)$
  endforeach
  repeat
    for each $d_m, \ m \in 1, \ldots, M$ **do**
    // Compute the output of the perceptron for each example
    $o_{d_m} = f_{step}\left(\sum_{i=0}^{n} w_i x_{d_m i}\right)$ or $f_{sgn}\left(\sum_{i=0}^{n} w_i x_{d_m i}\right)$
    // Update the weights of the perceptron applying the *perceptron rule*
    for each $w_i, \ i \in 0, \ldots, n$ **do**
    $w_i \leftarrow w_i + \eta \left(t_{d_m} - o_{d_m}\right) x_{d_m i}$
    endforeach
    endforeach
  until all examples are correctly classified
  return $(w_0, w_1, \ldots, w_n)$
end

The second algorithm, the *delta rule,* was designed to overcome the drawback of the perceptron training rule regarding the non-linearly separable examples. The delta rule converges to a best-fit approximation of the target variable.

The main idea of the delta rule is to use the *gradient descent*[18] technique to search the space of possible weight vectors to find the weights that best fit the training examples. This rule provides the *basis* for the *backpropagation algorithm*, which can learn networks with several layers and many interconnected units.

For a better understanding of the *delta rule*, let us assume that we have a perceptron with a *linear activation function*. These means that the output of the network is the same value of the linear combination of the input values of the network:

---

[18] The *gradient (descent or ascent) technique* is a commonly used general optimization method of a real-valued function, $f(x_1, \ldots, x_n)$, which iteratively moves from one point to the next taking steps proportional to the gradient of the function $\nabla f = \left[\frac{\partial f}{\partial x_1}, \ldots, \frac{\partial f}{\partial x_n}\right]$, which gives the direction of maximum ascent (positive gradient) to the maximum of the function or the maximum descent (negative gradient) to the minimum of the function.

Fig. 6.37 A linear
perceptron

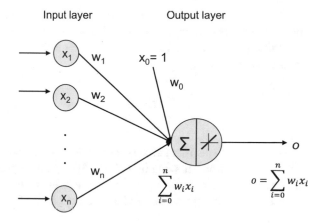

$$o(w \cdot x) = f_{\text{linear}}(w \cdot x) = w \cdot x = \sum_{i=0}^{n} w_i x_i \qquad (6.90)$$

This linear perceptron is depicted in Fig. 6.37.

To derive the *delta rule* using the gradient technique, we need a function to be optimized. In our case, we want the optimal weight vector making that the predicted values by the network (o) for the training examples were as much similar to the actual values of the target variable (t). This function assessing the similarity between both predicted and actual values will be called as the *Training Error* function.

The *Training Error* can be computed as the following squared error function:

$$E(w) = E(w_0, w_1, \ldots, w_n) = \frac{1}{2} \sum_{m=1}^{M} (t_{d_m} - o_{d_m})^2 \qquad (6.91)$$

where,

M is the total number of training examples,
d_m is the training example X_{d_m}.
t_{d_m} is the target output for the training example d_m
o_{d_m} is the output of the linear perceptron for the training example d_m

E is *a* function of the *weight vector* because the output o depends on this weight vector. The error for each training example, $(t_d - o_d)$, is squared because all the errors must be summed to get the final error of the whole training set, and some of the errors could be positive and other negative and the error magnitude can be cancelled. To avoid this situation, each error is squared. In addition, the constant term 1/2 is used, because in the process of derivation of the gradient of the error function will be algebraically very useful to cancel a constant equal to 2.

To compute the weight vector minimizing the *Training Error* using the gradient method, we must compute the partial derivatives of the function regarding all the variables (weights). This vector of derivatives is the *gradient* of E with respect to the vector $[w_0, w_1, \ldots, w_n]$.

$$\nabla E(w) = \nabla E([w_0 w_1 \ldots w_n]) = \left[\frac{\partial E}{\partial w_0}, \frac{\partial E}{\partial w_1}, \ldots, \frac{\partial E}{\partial w_n} \right] \qquad (6.92)$$

The gradient specifies the *direction* that produces the *steepest increase* in E. As we want to minimize the function E, we must use the opposite direction. Thus, using the *negative of this vector* we will obtain the *direction of steepest decrease*.

Thus, the training rule for gradient descent is the following:

$$w \leftarrow w + \Delta w \qquad (6.93)$$

where,

$$\Delta w = -\eta \nabla E(w) \qquad (6.94)$$

The same training rule can be expressed using the weights:

$$w_i \leftarrow w_i + \Delta w_i \qquad (6.95)$$

$$\Delta w_i = -\eta \frac{\partial E}{\partial w_i} \qquad (6.96)$$

Thus, finally we have:

$$w_i \leftarrow w_i - \eta \frac{\partial E}{\partial w_i} \qquad (6.97)$$

And η is small positive constant named the *learning rate* as in the perceptron training rule. The *learning rate* is used to adjust the step size in the gradient descent search.

To obtain the final updating formula for the weights we need to compute the gradient of the *training error* function for each weight ($\frac{\partial E}{\partial w_i}$). To obtain it, we must compute the partial derivative of the function regarding each weight. Thus, applying the derivative rules and algebraic manipulations we obtain:

$$\frac{\partial E}{\partial w_i} = \frac{\partial}{\partial w_i} \left(\frac{1}{2} \sum_{m=1}^{M} (t_{d_m} - o_{d_m})^2 \right) = \frac{1}{2} \frac{\partial}{\partial w_i} \left(\sum_{m=1}^{M} (t_{d_m} - o_{d_m})^2 \right)$$

$$= \frac{1}{2} \sum_{m=1}^{M} \frac{\partial}{\partial w_i} (t_{d_m} - o_{d_m})^2 = \frac{1}{2} \sum_{m=1}^{M} 2(t_{d_m} - o_{d_m}) \frac{\partial}{\partial w_i} (t_{d_m} - o_{dm})$$

$$= \sum_{m=1}^{M} (t_{d_m} - o_{d_m}) \frac{\partial}{\partial w_i} (t_{d_m} - w \cdot x_{d_m}) = \sum_{m=1}^{M} (t_{d_m} - o_{d_m}) (-x_{d_m i})$$

$$\frac{\partial E}{\partial w_i} = \sum_{m=1}^{M} (t_{d_m} - o_{d_m}) (-x_{d_m i})$$

(6.98)

where,

$x_{d_m i}$ describes the value of the variable $X_{.i}$ for the training example X_{d_m}. or simply d_m.

Finally, the weight update rule is:

$$w_i \leftarrow w_i - \eta \frac{\partial E}{\partial w_i}$$

(6.99)

$$w_i \leftarrow w_i - \eta \sum_{m=1}^{M} (t_{d_m} - o_{d_m}) (-x_{d_m i})$$

(6.100)

$$w_i \leftarrow w_i + \eta \sum_{m=1}^{M} (t_{d_m} - o_{d_m}) x_{d_m i}$$

(6.101)

The *Training Error* function due to the square polynomial form regarding the weights has a *paraboloid shape* (i.e. *a convex function*), which means that it has just only one extreme point, i.e. the minimum. This means that *independently of the starting point* (initial weight vector) and *independently whether the training examples are linearly separable or not*, and *given that the learning rate is enough small*, the *gradient descent method will converge* to the minimum error point, because the local minimum will be the global minimum.

The algorithm of the *gradient descent* which used the *delta rule* for training a linear perceptron can be described as follows:

algorithm Gradient Descent

input D: the training data set (M instances * $N+1$ attributes): $\{X_{d_1}, ..., X_{d_M}\}$
    t: is the target variable
output a weight vector $(w_0, w_1, ..., w_n)$

begin
  // set the learning rate to a small value ($0 < \eta < 1$). For instance, 0.05 or 0.1
  $\eta \leftarrow 0.05$
  // Set the initial weights and threshold to small random numbers.
  // For instance in the range [-0.5, 0.5] or use other heuristics
  for each w_i, $i \in 0, ..., n$ **do**
  $w_i \leftarrow random(-0.5, 0.5)$
  endforeach
  repeat
  $\Delta w_i \leftarrow 0$
  // Compute the update of the weights for all training examples
  for each d_m, $m \in 1, ..., M$ **do**
  // Compute the output of the linear perceptron for each example
  $o_{d_m} = \sum_{i=0}^{n} w_i x_{d_m i}$
  // Compute the update of the weights
    for each w_i, $i \in 0, ..., n$ **do**
    $\Delta w_i \leftarrow \Delta w_i + \eta \left(t_{d_m} - o_{d_m}\right) x_{d_m i}$
    endforeach
  endforeach
  // Update the weights of the linear perceptron applying the *delta rule*
  for each w_i, $i \in 0, ..., n$ **do**
  $w_i \leftarrow w_i + \Delta w_i$
  endforeach
  until termination condition is satisfied
  return $(w_0, w_1, ..., w_n)$
end

The determination of the learning rate η is very important because if it is too small, the number of iterations needed to reach the minimum could be very high. On the other hand, if it is too large, there is the possibility that the gradient descent oversteps the minimum, and sometimes some kind of zigzag oscillation around the minimum can happen which delays the approximation to the minimum. In order to solve this kind of problems, a possible solution is to reduce the value of η if the number of iterations increases very much. Another possible solution is a common modification of the gradient descent, which is named as *incremental gradient descent* or *stochastic gradient descent*.

The main idea of the incremental gradient descent is to compute the update of weights incrementally after processing each training example error, and not after summing all the training example errors to update the weights. Thus, the stochastic gradient descent algorithm can be described as follows:

---

algorithm Stochastic Gradient Descent

input D: the training data set (M instances * $N+1$ attributes): $\{X_{d_1}, ..., X_{d_M}\}$
    t: is the target variable
output a weight vector $(w_0, w_1, ..., w_n)$

begin
    // set the learning rate to a small value ($0 < \eta < 1$). For instance, 0.05 or 0.1
    $\eta \leftarrow 0.05$
    // set the initial weights and threshold to random numbers in the range [-0.5, 0.5]
    for each w_i, $i \in 0, ..., n$ **do**
    $w_i \leftarrow random(-0.5, 0.5)$
    endforeach
    repeat
        // Compute the update of the weights for all training examples
        for each d_m, $m \in 1, ..., M$ **do**
        // Compute the output of the linear perceptron for each example
        $o_{d_m} = \sum_{i=0}^{n} w_i x_{d_m i}$
        for each w_i, $i \in 0, ..., n$ **do**
            // Update the weights of the linear perceptron applying the *delta rule*
            $w_i \leftarrow w_i + \eta \left(t_{d_m} - o_{d_m}\right) x_{d_m i}$
        endforeach
        endforeach
    until termination condition is satisfied
    return $(w_0, w_1, ..., w_n)$
end

Taking a close look to the *delta rule* of the stochastic gradient and the *perceptron rule*, they seem to be identical, but actually the output value for each training example (o_{d_m}) is different. In the perceptron rule, it is the value of the sign or step activation function, $o_{d_m} = f_{\text{step}}\left(\sum_{i=0}^{n} w_i x_{d_m i}\right)$ or $f_{\text{sgn}}\left(\sum_{i=0}^{n} w_i x_{d_m i}\right)$, and in the delta rule, it is the value of the linear (identity) function, $o_{d_m} = \sum_{i=0}^{n} w_i x_{d_m i}$.

Now, that the gradient descent approach has been explained, we will describe how to extend it for a *multi-layer perceptron*, and how operates the *backpropagation algorithm* for learning the optimal weights, which is a generalization of the delta rule.

Multi-Layer Perceptron and the Backpropagation Algorithm

As it has been explained in the previous subsection, the *perceptron model* can learn only linear separable functions, and as the activation function is not differentiable, the gradient descent cannot be applied. Furthermore, the *linear perceptron*, or even a multi-layer network of linear units, only can approximate linear functions.

As many of the prediction problems involving a target variable require a non-linear approximation, the general architecture of a multi-layer perceptron network must be based on other activation functions. One of the most used is the *sigmoid function* or *logistic function*, because is a very smooth function producing values in the range of [0,1] and it is differentiable:

$$f_{\text{sigmoid}}(x) = \sigma(x) = \frac{1}{1 + e^{-x}} \tag{6.102}$$

Furthermore, the derivative of $\sigma(x)$ has a nice property:

$$\sigma'(x) = \sigma(x)(1 - \sigma(x)) \tag{6.103}$$

Other activation functions used are a variant of the sigmoid with a parameter $\lambda(\lambda > 0)$:

$$f_{\text{sigmoid}}(\lambda x) = \sigma(\lambda x) = \frac{1}{1 + e^{-\lambda x}} \tag{6.104}$$

And the hyperbolic tangent:

$$f_{\text{tanh}}(x) = \tanh(x) = \frac{e^x - e^{-x}}{e^x + e^{-x}} \tag{6.105}$$

Therefore, a *multi-layer perceptron* is a *feedforward artificial neural network* which has an input layer, at least one hidden layer, and an output layer. Usual ANNs have from one to three hidden layers, and each layer, for a commercial application can have up to 1000 neurons per layer. Usually, the architecture of an ANN is determined in a trial-and-test basis.

The neurons of the hidden layer and the ones in the output layer have as an activation function a continuous and differentiable activation function. Usually, the *sigmoid function* is used. In Fig. 6.38 there is a multi-layer perceptron architecture with the input layer, one hidden layer, and the output layer. For each connection between two neurons i and j, there is a weight (w_{ij}) associated to it.

There are several different algorithms for learning the weights in multi-layer perceptron, but the most popular and used is the *backpropagation algorithm* (Rumelhart et al., 1986). The basis of this algorithm was born from the works in the Control Theory field by Henry J. Kelley in 1960 (Kelley, 1960) and by Arthur E. Bryson in 1961 (Bryson, 1961). Bryson and Ho in 1969 (Bryson & Ho, 1969) described it as a multi-stage dynamic system optimization method, but due to its high level of computation according to the hardware available by then, it was not much used. It was rediscovered in the mid-80s due to the progress of hardware.

The *backpropagation algorithm* learns the weights for a multilayer network, given a network with a fixed set of units and interconnections. It uses a *gradient descent* to minimize the squared error between the network output values and the target values.

Let us redefine the *training error function (E)* to fit the actual architecture of the multi-layer perceptron. The error function, now has to take into account *all the output units*, and not only the unique output neuron of the perceptron. Therefore:

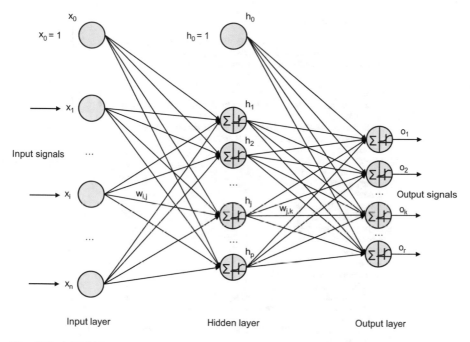

Fig. 6.38 A Multi-layer perceptron

$$E[w] = \frac{1}{2} \sum_{m=1}^{M} \sum_{k=1}^{r} \left(t_{d_m k} - o_{d_m k}\right)^2 \qquad (6.106)$$

where,

M is the total number of training examples
r is the number of output units
d_m is the training example X_{d_m}.
$t_{d_m k}$ is the target value for the training example d_m and the kth output unit
$o_{d_m k}$ is the output value of the multi-layer perceptron for the training example d_m and
the kth output unit

The *backpropagation algorithm* has two basic steps. The first one is the *propagation* through the network. A training example is given to the network input layer, the network propagates the input values, from the input layer to the output layer, passing through the hidden layers. Then, the output values are generated. The second step starts computing the error between the output values and the target values, and then it is *propagated backwards* through the network from the output layer to the input layer, passing through the hidden layers. As the error is propagated, the different weights are updated.

The derivation of the updating rules of the weights is similar to the delta rule. The backpropagation algorithm for a multi-layer feedforward network with one hidden layer, using the stochastic gradient version can be described as follows:

algorithm Backpropagation

input D: the training data set (M instances * $N+1$ attributes): $\{X_{d_1}, \dots, X_{d_M}\}$
t: is the vector of the r target variables: $\{t_{d_m 1}, \dots, t_{d_m r}\}$
output two weight vectors $[w_{ij}]_{\substack{i=0,\dots,n \\ j=1,\dots,p}}$ and $[w_{jk}]_{\substack{j=0,\dots,p \\ k=1,\dots,r}}$

begin
    // Set the learning rate to a small value ($0 < \eta < 1$). For instance, 0.05 or 0.1
    $\eta \leftarrow 0.05$
    // Set the initial weights and thresholds to small random numbers.
    // For instance, in the range [-0.5, 0.5] or other heuristics
    for each $i \in 0, \dots, n$ **do**
     for each $j \in 1, \dots, p$ **do**
     $w_{ij} \leftarrow random(-0.5, 0.5)$
     endforeach
    endforeach
    for each $j \in 0, \dots, p$ **do**
     for each $k \in 1, \dots, r$ **do**
     $w_{jk} \leftarrow random(-0.5, 0.5)$
     endforeach
    endforeach
    repeat
     for each d_m, $m \in 1, \dots, M$ **do**
        // Forward propagation step
        // Compute the output of the multi-layer network for each example
        // Compute the outputs of the neurons in the hidden layer
        for each $j \in 1, \dots, p$ **do**
        $h_{d_m j} = \sigma(\sum_{i=0}^{n} w_{ij} x_{d_m i})$ // aggregation and activation through the sigmoid
        endforeach
        // Compute the outputs of the neurons in the output layer
        foreach $k \in 1, \dots, r$ **do**
        $o_{d_m k} = \sigma(\sum_{j=0}^{p} w_{jk} h_{d_m j})$ // aggregation and activation through the sigmoid
        endforeach
        // Backward propagation step
        // Propagate the errors backward through the network
        // Compute the error for the neurons in the output layer
        for each $k \in 1, \dots, r$ **do**
        $\delta_{d_m k} = o_{d_m k}(1 - o_{d_m k})(t_{d_m k} - o_{d_m k})$ // $\delta_{d_m k}$ is the term error for output unit k
        endforeach
        // Compute the error for the neurons in the hidden layer
        for each $j \in 1, \dots, p$ **do**
        $\delta_{d_m j} = h_{d_m j}(1 - h_{d_m j}) \sum_{k=1}^{r} w_{jk} \delta_{d_m k}$ // $\delta_{d_m j}$ is the term error for hidden unit j
        endforeach
        // Update all network weights
        // Update weights between the input layer and the hidden layer
        for each $i \in 0, \dots, n$ **do**
         for each $j \in 1, \dots, p$ **do**
         $w_{ij} \leftarrow w_{ij} + \eta \, \delta_{d_m j} \, x_{d_m i}$ // α momentum can be added here
         endforeach
        endforeach
        // Update weights between the hidden layer and the output layer
        for each $j \in 0, \dots, p$ **do**
         for each $k \in 1, \dots, r$ **do**
         $w_{jk} \leftarrow w_{jk} + \eta \, \delta_{d_m k} \, h_{d_m j}$ // α momentum can be added here
         endforeach
        endforeach
     endforeach
    until termination condition is satisfied
    return $[w_{ij}]_{\substack{i=0,\dots,n \\ j=1,\dots,p}}$ and $[w_{jk}]_{\substack{j=0,\dots,p \\ k=1,\dots,r}}$
end

Both δ_{d_mk} and δ_{d_mj} denote the *error term* associated with output neuron k and hidden neuron j respectively. They play a similar role than the error quantity $(t - o)$ for the *delta training rule*. The derivation of the *weight-adapting rule* for the *backpropagation algorithm* using a stochastic gradient descent can easily be derived in the same way that the delta rule was derived. Just, the function error for each training example,

$$E_{d_m}[w] = \frac{1}{2} \sum_{k=1}^{r} (t_{d_mk} - o_{d_mk})^2 \tag{6.107}$$

must be derivated regarding all the weights $\left(\frac{\partial E_{d_m}[w]}{\partial w_{ij}} \right)$, and the expressions for both δ_{d_mk} and δ_{d_mj} can be obtained.

This is an *incremental (stochastic) gradient descent version* of the backpropagation algorithm. To use the gradient descent version, one should sum the weight-updating quantities $\delta_{d_mj} x_{d_mi}$ and $\delta_{d_mj} x_{d_mi}$ over all training examples before updating the weight values.

Most common *termination conditions* which can be used to stop the algorithm are:

- The algorithm stops after a fixed number of iterations (*epochs*).
- The algorithm stops once the error on the training examples falls below some predefined threshold.
- The algorithm stops once the error on a separate validation set of examples meets some criteria.

For the *initialization of the weights*, there are some other heuristics, like the one proposed in Haykin (1999), who proposes to set each initial weight to random numbers uniformly distributed in the following small range:

$$\left(-\frac{2.4}{F_i}, +\frac{2.4}{F_i} \right)$$

where F_i is the total number of inputs of neuron i in the network. Thus, the weight initialization is different for each neuron.

The *backpropagation algorithm* spends time making *extensive computations*. To decrease the computation time, some heuristics are used in its actual application for solving real problems. Most commonly used are:

- The use of the *hyperbolic tangent function* $f_{\tanh}(x)$, as the *activation* or *transfer function* instead of the sigmoid function usually accelerates the training process in the ANN.
- The use of some *heuristics* for *adapting the learning rate* η can commonly improve the speed of the convergence of the training (Jacobs, 1988). Some intuitive heuristics are the following two:

- If the the sum of the squared errors is increasing for *consecutive epochs*, then the *learning rate η should be decreased*. For instance, if it is higher than a predefined ratio, typically 1.04, the learning rate should be decreased, typically multiplying it by 0.7.
- If the the sum of the squared errors is *less than a previous epoch*, the *learning rate η should be increased*. For instance, typically multiplying it by 1.05.

- A very common modification of the algorithm is to add a *momentum parameter α*, to modulate the *weight-update rule* in the algorithm by making the weight update on the *n*th iteration partially dependent on the update that occurred during the (*n* − 1)th iteration, as follows:

$$\Delta w_{ij}(n) = \eta \, \delta_{d_m j} \, x_{d_m i} + \alpha \Delta w_{ij}(n-1) \tag{6.108}$$

$$\Delta w_{jk}(n) = \eta \, \delta_{d_m k} \, h_{d_m j} + \alpha \Delta w_{jk}(n-1) \tag{6.109}$$

where,

$\Delta w_{ij}(n)$ is the weight update performed during the *n*th iteration
*n*th iteration update depends on the (*n* − 1)th iteration
α is a constant, $0 \leq \alpha < 1$, called the *momentum*.

The main role of *momentum term* α relies on keeping the algorithm exploring the error function in the same direction from one iteration to the next. In addition, it can gradually increase the step size of the search in regions where the gradient is unchanging, thereby speeding convergence.

The *backpropagation algorithm* described above is useful for just a feedforward ANN with one hidden layer. However, the generalization of this algorithm for a *feedforward ANN with any number of hidden layers* is straightforward. If there are more hidden layers, the *weight-update rule/s* are the same. The only change is that more δ values must be computed. If there are more hidden layers, the corresponding δ_s for a neuron *s* in a new hidden layer *m* can be computed from the δ values at the next hidden layer *m* + 1, as follows:

$$\delta_s = h_s(1 - h_s) \sum_{t \in \text{Layer}(m+1)} w_{st} \delta_t \tag{6.110}$$

ANNs as a predictor model to approximate non-linear functions are only effective if the input values, usually real numbers, are constrained to suitable ranges, typically belonging to [0,1] or [−1, +1] ranges. The range of the output values depends on the activation function selected. Of course, if a *sigmoid function* is used, the output values are in the range [0,1], which is suitable for a classification or discrimination task but not for a general numeric prediction.

Therefore, if the input data to be used is outside these ranges, the usual procedure is to *scale the data*, usually linearly, before being fed to the ANN, and the output data must be *re-scaled to the original values*. Usually, these scaling transformations are performed automatically in the software packages implementing ANNs models.

Another solution for the output data can be to use a *linear activation function* at the output level instead of a sigmoid function.

An Example of an Artificial Neural Network Model

Let us take as an example for the use of an *ANN predictive model*, a dataset related to the *Personal Selection Process of one company*. The dataset has 200 examples, which are anonymized assessments of 200 candidates for a position in the company. There are six numeric attributes. The *target variable* to be predicted is the final scoring of the candidates in a final test of the Human Resources Dept. *The scoring of the test* ranges from 0 to 1000 points, which came from 100 questions, which can be scored with 0–10 points. The other five attributes are the assessment given to each candidate for five experts, belonging to several departments of the company.

The main task is try to obtain a model that could predict the *test scoring* of a candidate just with the assessment of the five experts of the company. Thus, we are facing a numeric prediction problem, and a *predictive model* is an adequate choice.

After a pre-processing of the dataset, it was checked that no missing values were in the dataset. In addition, no errors were detected and none outliers were suspected.

After a descriptive statistical analysis, it seemed that there was no apparent linear relationship between the response variable (*test scoring*) and the other variables. Thus, probably some kind of non-linear relationship could exist. A good model could be an ANN predictive model, which can approximate non-linear functions.

For the testing we used a simple validation scheme. The dataset was split into a training set and a validation set. The training set had 70% of the instances (140) and the validation set the remaining 30% (60 instances).

To build the ANN, we must take into account that the input layer will have five neurons (one for each predicting attribute), and the output layer just only one /the target variable). Of course, one of the drawbacks of an ANN is to determine the best architecture for the network. Usually, it is done in a trial-and-error basis. We tested several configurations, starting with one hidden layer, and we tried with one up to five nodes. We also tried two hidden layers with some nodes, but at the end, it was confirmed that the best configuration was with one hidden layer and four neurons. Of course, when we state that the input layer has five neurons for the five input variables, you must remember that a new neuron is automatically added for holding the threshold value, as well as in the hidden layer. The neurons in the hidden layer use the *sigmoid function* as activate function, and the output neuron in the output layer use the *linear function* to predict the numeric *test scoring* variable.

Furthermore, some values for the parameters of the backpropagation algorithm, *learning rate* and *momentum* were tried. The best *learning rate* value was $\eta = 0.05$ and the best momentum value was $\alpha = 0.3$. The number of epochs was tested from 300 up to 45,000.

To measure the goodness of the models, the common indicator is to measure the error between the predicted values and the actual values. The *Root Mean Squared Error* (RMSE) was computed. As it will be explained in the validation subsection of predictive models, the RMSE is one of the most common error measures. Of course, as less is the RMSE, better is the accuracy of the model.

In Table 6.16, there are the RMSE values obtained for the different configurations, as well as the different parameters of the configuration. From the first experiments it seemed that just one hidden layer is better than two hidden layers, and that the best configurations would be with 4 or 5 neurons. We varied the number of epochs to see how was the behaviour of both configurations (ANN-1-4 and ANN-1-5). With 3000 epochs, it was clear the difference. The ANN-1-4 was clearly better, because the RMSE was of 10.80 against the 28.51 of the ANN-1-5.

Since then, we tried to make some variations of the momentum and the learning rate, but we found that the best configuration was the initial one with $\eta = 0.05$ and $\alpha = 0.3$.

Table 6.16 Results for different configurations of the ANN for the *Personal Selection dataset*

	Hidden layers	Neurons per hidden layer	Learning rate (η)	Momentum (α)	Epochs	RMSE
ANN-1-1	1	1	0.05	0.3	300	41.18
ANN-1-2	1	2	0.05	0.3	300	40.96
ANN-1-3	1	3	0.05	0.3	300	38.91
ANN-1-4	1	4	0.05	0.3	300	37.87
ANN-1-5	1	5	0.05	0.3	300	37.78
ANN-1-6	1	6	0.05	0.3	300	38.58
ANN-2-5-4	2	5-4	0.05	0.3	300	40.69
ANN-2-4-3	2	4-3	0.05	0.3	300	39.24
ANN-2-3-2	2	3-2	0.05	0.3	300	41.10
ANN-1-4	1	4	0.05	0.3	600	37.66
ANN-1-5	1	5	0.05	0.3	600	37.45
ANN-1-4	1	4	0.05	0.3	1500	34.03
ANN-1-5	1	5	0.05	0.3	1500	37.64
ANN-1-4	1	4	0.05	0.3	3000	10.80
ANN-1-5	1	5	0.05	0.3	3000	28.51
ANN-1-4	1	4	0.1	0.3	3000	10.82
ANN-1-4	1	4	0.025	0.3	3000	34.57
ANN-1-4	1	4	0.1	0.2	3000	10.70
ANN-1-4	1	4	0.1	0.1	3000	10.67
ANN-1-4	1	4	0.1	0.05	3000	10.67
ANN-1-4	1	4	0.1	0	3000	10.67
ANN-1-4	1	4	0.05	0.2	3000	12.01
ANN-1-4	1	4	0.05	0.1	3000	18.91
ANN-1-4	1	4	0.05	0.05	3000	23.66

(continued)

Table 6.16   (continued)

	Hidden layers	Neurons per hidden layer	Learning rate (η)	Momentum (α)	Epochs	RMSE
ANN-1-4	1	4	0.05	0	3000	26.73
ANN-1-4	1	4	0.1	0.3	6000	9.38
ANN-1-4	1	4	0.1	0.2	6000	9.85
ANN-1-4	1	4	0.1	0.1	6000	9.66
ANN-1-4	1	4	0.1	0.05	6000	9.58
ANN-1-4	1	4	0.1	0	6000	9.51
ANN-1-4	1	4	0.1	0.4	6000	9.36
ANN-1-4	1	4	0.1	0.5	6000	9.12
ANN-1-4	1	4	0.1	0.6	6000	9.29
ANN-1-4	1	4	0.1	0.3	12,000	9.11
ANN-1-4	1	4	0.1	0.4	12,000	9.07
ANN-1-4	1	4	0.1	0.5	12,000	9.13
ANN-1-4	1	4	0.1	0.6	12,000	9.42
ANN-1-4	1	4	0.05	0.3	12,000	8.64
ANN-1-4	1	4	0.05	0.4	12,000	8.64
ANN-1-4	1	4	0.05	0.5	12,000	8.66
ANN-1-4	1	4	0.05	0.6	12,000	8.74
ANN-1-4	1	4	0.05	0.3	15,000	8.48
ANN-1-4	1	4	0.05	0.4	15,000	8.50
ANN-1-4	1	4	0.05	0.3	20,000	8.32
ANN-1-4	1	4	0.05	0.3	40,000	8.07
ANN-1-4	1	4	0.05	0.3	80,000	7.96
ANN-1-4	*1*	*4*	*0.05*	*0.3*	*100,000*	*7.94*

Regarding the RMSE, it decreased from an initial value of 37.87 with 300 epochs, to a final one 7.94 with 100,000 epochs. Of course, the final RMSE obtained by the ANN is really fantastic, because it means that making the predictions the mean error is about 8 points in the test scoring of the candidates, with just 140 training examples!

Analysing the variation of the RMSE with the best configuration (ANN-1-4, $\eta = 0.05$ and $\alpha = 0.3$) as shown in Table 6.17, it can be outlined, that the great decrease in the error is obtained when passing from 1500 epochs to 3000 epochs. In addition, the reduction of the error is somehow significant until the value of 8.64 with 12,000 epochs.

The other values obtained with a great increase in the number of epochs up to 100,000, just produce a reduction of 0.7 but multiplying the number of epochs by a factor of 8.3. Thus, probably the *best model* would be the one with 12,000 epochs and a RMSE of 8.64. The other models can have a problem of overfitting, trying to fit too much to the data, and perhaps could not generalize the predictive accuracy to new data.

Table 6.17 RMSE values for the best configuration (ANN-1-4, $\eta = 0.05$, $\alpha = 0.3$) for the *Personal Selection dataset, with different epoch values*

	Hidden layers	Neurons by hidden layer	Learning rate (η)	Momentum (α)	Epochs	RMSE
ANN-1-4	1	4	0.05	0.3	300	37.87
ANN-1-4	1	4	0.05	0.3	600	37.66
ANN-1-4	1	4	0.05	0.3	1500	34.03
ANN-1-4	1	4	0.05	0.3	*3000*	*10.80*
ANN-1-4	1	4	0.05	0.3	6000	9.30
ANN-1-4	1	4	0.05	0.3	*12,000*	*8.64*
ANN-1-4	1	4	0.05	0.3	15,000	8.48
ANN-1-4	1	4	0.05	0.3	20,000	8.32
ANN-1-4	1	4	0.05	0.3	40,000	8.07
ANN-1-4	1	4	0.05	0.3	80,000	7.96
ANN-1-4	1	4	0.05	0.3	1,000,000	7.94

Anyway, the *ANN predictive model* obtained seems to be very accurate for a difficult estimation problem, as it will be confirmed when some other predictive models will be applied, in next subsections.

Case-Based Predictive Models

As previously described, when we have a data-driven supervised dataset composed of several instances, which are described with a set of attribute values, and associated to these attribute values for each instance we have a *numeric attribute*, which is *the target variable* or variable of interest, the dataset itself can be considered as a *Case Library*, where *the cases* are the instances of the dataset. The *descriptive part of the case* are the attribute values of each instance, and the *solution of the case* is the numeric target variable of each instance. Furthermore, as the solution is a simple numeric value, a *null adaptation* or *copy solution* strategy or a *weighted mean adaptation* or *copy and average solutions* strategy can be used. In Figs. 6.39 and 6.40, the scheme of a null adaptation or a weighted mean adaptation strategy for numerical values are depicted.

Therefore, using a *null adaptation*, what the *Case-Based predictor* is doing is predicting for the new instance, the same numeric value of its most similar instance. As previously explained, this most similar instance is usually named as the *nearest neighbour*. This approach is described in Fig. 6.39, where for the C_{new} is predicted to have the numeric value of 50 for the solution part, because this is the numeric value of the solution of its nearest neighbour instance.

Using the *weighted mean adaptation*, what the *Case-Based predictor* does is predicting for the new instance, the mean or the weighted mean of the numeric values of the solutions of its k most similar instances, which usually are named as *k-nearest neighbours*. This approach is described in Fig. 6.40, where the C_{new} is

Fig. 6.39 Null adaptation or copy solution strategy for numeric solution variables

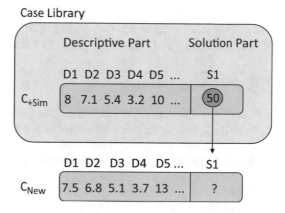

Fig. 6.40 Weighted mean adaptation or weight and copy solutions strategy for numeric solution variables

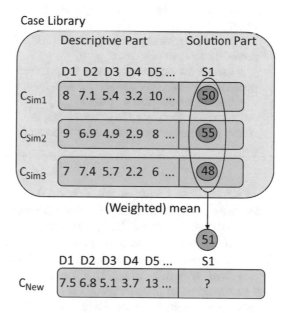

predicted to have the value 51, because this value is the mean value among its nearest neighbour instances *{50, 55, 48}*.

As it has been described before, the Case Library can have different organizations starting from a *flat memory*, a hierarchical organization like as a *discriminant tree or network* or a *k-d tree other hierarchical structures*, or mixed representations. When CBR uses the dataset as a *simple flat memory* scheme we talk about an *Instance-Based predictor*. On the contrary, when the organization of the Case Library is more complex, then we talk in general about a *Case-Based predictor*.

The general scheme of a *Case-Based predictor* method, which is the same than for a Case-Based classifier, just changing the final computation of the predicted value, which now is a numerical one, is as follows:

```
algorithm Case-based predictor

input TrainingDataset: The training dataset formed of N cases or instances
      C_new: case or instance, which solution is to be predicted
      k: number of nearest neighbours
output o: the output numeric solution value for the new case C_new

begin
 if first-time then
   // The CB is built from the dataset the first-time. This is the model learning step
   CB ← buildCaseLibrary(TrainingDataset)
 endif
   // Retrieve the most promising cases ( ≥ k) to be similar to C_new from the CB
   promCases ← retrieve(CB, C_new, k)
   // Assess the similarity between the promising cases and the C_new
   promCases ← compSim(promCases, C_new)
   // Get the k most similar cases to C_new
   neighbours ← GetKmostSim(promCases, k)
   return arithmetic-mean(solution(neighbours))
 end
```

When the Case Library is taken as just the set of instances as a *flat memory*, this technique can be named as the *k-nearest neighbour predictor*. The *k-nearest neighbour predictor*, is a particular application of Case-Based Reasoning for prediction purposes, i.e. *a Case-Based predictor*, with some particularities: a *flat memory*, a *feature vector representation of cases*, a *simple numeric value as a solution*, and *null or weighted mean adaptation*. This particular application of a Case-Based predictor can be named as an *Instance-Based predictor*.

The general scheme of the *k-nearest neighbour predictor*, i.e. an *instance-based predictor* method, which is the same than for a instance-based classifier, just changing the final computation of the predicted value, which now is a numerical one, is as follows:

```
algorithm k-Nearest Neighbour predictor or instance-based predictor

input TrainingDataset: The training dataset formed of N cases or instances
      C_new: case or instance to be predicted
      k: number of nearest neighbours
output o: the output numeric solution value for the new case C_new

begin
   // The dataset itself is the flat Case Library. This is the model learning step
   CB ← TrainingDataset
   neighbours ← ∅
   for each instance or case C ∈ CB do
     distC ← d(C, C_new)
     if |neighbours| < k then
        // add C to neighbours in order
       orderedInsertion(C, distC, neighbours)
     else
       if distC < d(farthestNeighbour(neighbours), C_new) then
         remove (farthestNeighbour(neighbours), neighbours)
         orderedInsertion (C, distC, neighbours)
       endif
     endif
   endfor
   return arithmetic-mean(solution(neighbours))
end
```

Therefore, in the same manner than with a Case-based classifier model, *the model of a Case-based predictor is implicitly formed by all the instances.* This particular situation means that there are not specific algorithms for building the model of a case-based predictor or instance-based predictor model. Thus, the above algorithm *does not build any model,* but *interpret* or *use the model* (i.e. the case base or dataset itself).

An Example of a Case-Based Predictive Model

As an example to illustrate the use of a *case-based predictor,* we will use the same dataset that we used for the *ANN predictor model:* The *Personal Selection Process of one company* dataset. The dataset has 200 examples, which are anonymized assessments of 200 candidates for a position in the company. There are six numeric attributes. The *target variable* to be predicted is the final scoring of the candidates in a final test of the Human Resources Dept.

We use the same validation scheme than before. The dataset was split into a training set and a validation set. The training set had 70% of the instances (140) and the validation set the remaining 30% (60 instances).

Table 6.18 RMSE values for
the different configurations of
the *k-nearest neighbour predictor* for the *Personal
Selection* dataset

	RMSE
k-NN (k = 1)	260.03
k-NN (k = 3)	72.61
k-NN (k = 5)	*71.56*
k-NN (k = 7)	74.01
ANN-1-4	*8.64*

Once the validation set is formed, we apply the *k*-nearest neighbour predictor algorithm to each of the validation instances in the dataset. We tried four values of *k* (1, 3, 5 and 7). The results, with the error in the predicted numeric values and the actual values for the instances in the validation set was computed. The Root Mean Squared Error (RMSE) was used, and the results for the different configurations are in Table 6.18. As all the attributes were numeric, we used the *Euclidian* distance as a dissimilarity function. We did not use any weighting scheme. Thus, all the five attributes were equally weighted in the dissimilarity computations.

As it can be seen in Table 6.18, the best RMSE corresponded to the value of $k = 5$, with an RMSE of 71.56. It is quite evident, the increase in performance changing from $k = 1$ to $k = 3$. However, even that reduction in the RMSE, the best error value was of 71.56 corresponding to a $k = 5$.

This value compared with the least error obtained with an ANN model, which was 8.64, is not good. This is due because very probably, the target variable, i.e., *the test scoring*, has a non-linear relationship with the other five variables, and the target variable is very difficult to predict.

Linear Regression Models

In general, *Linear Regression (LR) models* make predictions of one numeric target variable (Y), also named as the *response variable* or *independent variable*, from a set of numeric variables which are named as the *dependent variables* or *explicative variables* (X_i), assuming that there is a *linear relationship* among the target variable and the explicative variables.

The linear relationship between a response variable (Y) and an explicative variable (X) can be easily understood with a geometrical interpretation like the one in Fig. 6.41.

The *linear relationship* or *linear dependence* between one variable Y and one variable X, means that when the variable X increases in one unit ($x_i \rightarrow x_i + 1$), the dependent variable increases by a factor of β ($y_i \rightarrow y_i + \beta$). In two dimensions, the factor β is the slope of the line, i.e. *the regression line*, and the factor α is the point where the line cuts the *y*-axis, or in other words, the value of Y, for the value $x = 0$.

In general, in an *n-dimensional* space, when we have n dependent variables, the linear dependence means that when one of the dependent variables (X_i) increase in one unit ($x_i \rightarrow x_i + 1$) while the other dependent variables are hold constant, the dependent variable (Y) increases by a factor of β_i ($y_i \rightarrow y_i + \beta_i$).

Fig. 6.41 Linear
relationship between
response variable Y and an
explicative variable X

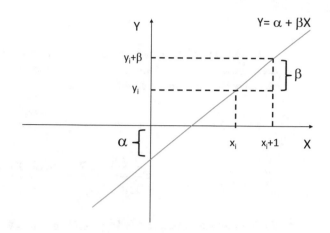

The *regression* name was formulated from observation and work of Sir Francis Galton (in late 1860s) in Biology. He was working and analyzing a dataset of height of people, and height of their children. After some experimentation, he observed the following facts:

- Tall parents had tall children, but on average not so tall, being lower than their parents.
- Low parents had low children, but on average not so low, being taller than their parents.

He concluded that there was a *regression* to the mean. Thus, this regression to the mean gave the name to these linear models, where the best prediction is based on the average value plus some variability.

There are two basic *Linear Regression Models*:

- *Simple Linear Regression Model (LR)*, which is a Linear Regression model, where there is just *one dependent variable* (X_i). The model is formulated as follows:

$$y_i = \alpha + \beta * x_i + \varepsilon_i \qquad (6.111)$$

where,

y_i is the response variable to be predicted
x_i is the dependent variable
$\widehat{y}_i = \alpha + \beta * x_i$ is the predicted value
$\varepsilon_i = y_i - \widehat{y}_i$ is the error term (noise)

Or in matrix form:

$$Y = \alpha + \beta X + \varepsilon \tag{6.112}$$

As it is a particular case of the general Multiple Linear Model, we will not detail how are derived the parameters of the model (α, β), but they can be estimated through the following formulas:

$$\alpha = \bar{y} - \beta \bar{x} = \bar{y} - \frac{\mathrm{cov}(X, Y)}{\sigma_x^2} \bar{x} \tag{6.113}$$

$$\beta = \frac{\dfrac{\sum_i x_i y_i}{n} - \bar{x}\,\bar{y}}{\dfrac{\sum_i x_i^2}{n} - \bar{x}^2} = \frac{\mathrm{cov}(X, Y)}{\sigma_x^2} \tag{6.114}$$

- *Multiple Linear Regression Model (MLR)*, which is a Linear Regression model, where the aim is to predict a numeric variable Y, i.e. the independent or response variable, from several numeric variables X_i, i.e. the dependent variables, using a linear dependency. The model is formulated as follows:

$$y_i = \beta_0 + \beta_1 * x_{i1} + \beta_2 * x_{i2} + \cdots + \beta_k * x_{ik} + \varepsilon_i \tag{6.115}$$

where,

y_i is the response variable to be predicted
$x_1, x_2, \ldots, x_k$ are the dependent variables
$\widehat{y}_i = \beta_0 + \beta_1 * x_{i1} + \beta_2 * x_{i2} + \cdots + \beta_k * x_{ik}$ is the predicted value
$\varepsilon_i = y_i - \widehat{y}_i$ is the error term (noise)

or in vector form:

$$Y = \beta_0 + \beta_1 X_1 + \beta_2 X_2 + \cdots + \beta_k X_k + \varepsilon \tag{6.116}$$

Multiple Linear Regression Models

Some examples of this situation could be the following ones detailed in the Table 6.19. For instance, one can assume that the *sales amount quantity of one*

Table 6.19 Some examples of Multiple Linear Models

Y	X_1	X_2	X_3	ε
Sales' amount	*Number of appearances in mass media per week*	*Advertising budget*	*Quality of Products*	*Number of competitor products, ...*
Manager's salary	*Experience*	*Benefit of company*	*Personal abilities*	*Academic Degree/s, ...*
Annual budget of an university	*Number of students*	*Number of academics*	*Number of administration staff*	*Research position in rankings, ...*

company regarding one new product launched recently in the market can be predicted by means of some factors (variables) like the number of appearances in mass media per week, the advertizing budget spent and the quality of the product. If we translate this into a multiple linear model means that the dependent variable will be predicted with the values of the other three explicative variables. Of course, probably we are missing some other variables that can influence on the determination of the sales amount like, the number of competitor products, etc. These missed variables in the Multiple Linear Regression Model (MLR) are the term error or noise.

The same situation happens in the other two examples described in the table.

Let us analyze the Multiple Linear Regression (MLR) Model. This statistical model is a data-driven model, which can be classified as a top–down approach and confirmative model, as stated in Chap. 3. First, the model is hypothesized, and afterwards, the data must confirm the hypothesis.

The model has several hypotheses that must be verified to be sure that the model has sense, and can be used for numerical prediction of new testing examples.

The hypotheses are the following ones:

1. Linearity:

$$E[Y|X] = \mu_{Y|X} = \beta_0 + \beta_1 X_1 + \beta_2 X_2 + \cdots + \beta_k X_k$$

$$E[\varepsilon_i] = 0$$

2. Normality:
   $\varepsilon_i \approx N(0, \sigma^2)$ $\forall i$ the residuals are normally distributed
3. Homocedasticity:
   $\mathrm{Var}(\varepsilon_i) = \sigma^2$ is constant and independent of X_i
4. Independence:

$$\mathrm{cov}(\varepsilon_i, \varepsilon_j) = 0 \quad i \neq j$$

   Perturbations (residuals) are independent among them
5. Availability of Data:
   Available number of observations (n), $n \geq k + 1$
   ($k + 1$ parameters to be estimated)
6. Linear Independence:
   X_i are linearly independent among them

To obtain the parameters of the model, we must deduce how to estimate these parameters. The parameters (β_0, β_1, ..., β_k) must be obtained as a result of the optimization process of the error function. We want the parameters that will minimize the error function. Let us analyze how to formulate the error function, and the process of solving the involved optimization problem.

The error of one observation is:

$$y_i - \widehat{y}_i$$

The error of all observations (n) is:

$$\sum_{i=1}^{n} (y_i - \widehat{y}_i)$$

As always, to avoid the problem with cancellation of positive and negative errors, usually the square of the errors is considered. Thus,

The Total error is:

$$\sum_{i=1}^{n} (y_i - \widehat{y}_i)^2$$

And, as the error depends on the parameters, the error function can be written as a function of the parameters:

$$\text{Error}(\beta_0, \beta_1, \ldots, \beta_k) = \sum_{i} (y_i - (\beta_0 + \beta_1 X_1 + \beta_2 X_2 + \ldots + \beta_k X_k))^2 \quad (6.117)$$

Thus, the optimal $\beta_0, \beta_1, \ldots, \beta_k$ which we want to find are the ones minimizing the error function: $\text{Error}(\beta_0, \beta_1, \ldots, \beta_k)$. Solving this minimization problem is known as solving the optimization problem of Least Square Minimization (LSM):

$$\min_{\beta_0, \ldots, \beta_k} \text{Error}(\beta_0, \beta_1, \ldots, \beta_k) = \min_{\beta_0, \beta_1, \ldots, \beta_k} \sum_{i} (y_i - (\beta_0 + \beta_1 X_1 + \beta_2 X_2 + \cdots + \beta_k X_k))^2$$

$$(6.118)$$

To solve the optimization problem, we can obtain the partial derivatives regarding each parameter, and after equalling them to 0, we have a $k + 1$ equation system with $k + 1$ variables:

$$\frac{\partial \text{Error}(\beta_0, \beta_1, \ldots, \beta_k)}{\partial \beta_0} = \sum_{i} 2(y_i - (\beta_0 + \beta_1 x_1 + \beta_2 x_2 + \cdots + \beta_k x_k)) * (-1) = 0$$

$$(6.119)$$

$$\frac{\partial \text{Error}(\beta_0, \beta_1, \ldots, \beta_k)}{\partial \beta_1} = \sum_i 2(y_i - (\beta_0 + \beta_1 x_1 + \beta_2 x_2 + \cdots + \beta_k x_k)) * (-x_1) = 0$$

$$(6.120)$$

$$\ldots$$

$$\frac{\partial \text{Error}(\beta_0, \beta_1, \ldots, \beta_k)}{\partial \beta_k} = \sum_i 2(y_i - (\beta_0 + \beta_1 x_1 + \beta_2 x_2 + \cdots + \beta_k x_k)) * (-x_k) = 0$$

$$(6.121)$$

Using algebraic manipulations, we get the equation system:

$$\sum_i y_i = n\beta_0 + \beta_1 \sum_i x_{1i} + \cdots + \beta_k \sum_i x_{ki}$$

$$\sum_i y_i x_{1i} = \beta_0 \sum_i x_{1i} + \beta_1 \sum_i x_{1i}^2 + \cdots + \beta_k \sum_i x_{ki} x_{1i}$$

$$\ldots \qquad \ldots \qquad \ldots \qquad \ldots$$

$$\sum_i y_i x_{ki} = \beta_0 \sum_i x_{ki} + \beta_1 \sum_i x_{1i} x_{ki} + \cdots + \beta_k \sum_i x_{ki}^2$$

Which is equivalent to the following one in matrix form:

$$
\begin{bmatrix} 1 & \cdots & 1 \\ x_{11} & \cdots & x_{1n} \\ \cdots & \cdots & \cdots \\ x_{k1} & \cdots & x_{kn} \end{bmatrix} * \begin{bmatrix} y_1 \\ y_2 \\ \cdots \\ y_n \end{bmatrix} = \begin{bmatrix} 1 & \cdots & 1 \\ x_{11} & \cdots & x_{1n} \\ \cdots & \cdots & \cdots \\ x_{k1} & \cdots & x_{kn} \end{bmatrix} * \begin{bmatrix} 1 & x_{11} & \cdots & x_{k1} \\ \cdots & \cdots & \cdots & \cdots \\ \cdots & \cdots & \cdots & \cdots \\ 1 & x_{1n} & \cdots & x_{kn} \end{bmatrix}
$$

$$(k+1) * n \qquad\qquad n * 1 \qquad\qquad (k+1) * n \qquad\qquad n * (k+1)$$

$$
* \begin{bmatrix} \beta_0 \\ \beta_1 \\ \cdots \\ \beta_k \end{bmatrix}
$$

$$(k+1) * 1$$

Being X the data matrix (x_{ij}) plus adding a first column of 1's:

$$X = \begin{bmatrix} 1 & x_{11} & \cdots & x_{k1} \\ \cdots & \cdots & \cdots & \cdots \\ \cdots & \cdots & \cdots & \cdots \\ 1 & x_{1n} & \cdots & x_{kn} \end{bmatrix}$$
$$n * (k+1)$$

Then,

$$X^T Y = X^T X \beta \tag{6.122}$$

And being $X^T X$ non-singular,[19] we can compute the estimation of the parameters:

$$\beta = \left(X^T X \right)^{-1} X^T Y \tag{6.123}$$

To understand the process of estimation of the parameters is very interesting to make a geometrical interpretation of the linear regression model.
The multiple linear model:

$$y_i = \beta_0 + \beta_1 * x_{i1} + \beta_2 * x_{i2} + \cdots + \beta_k * x_{ik} + \varepsilon_i$$

Can be expressed in vector form, this way:

$$y^T = (y_1, \ldots, y_n) \quad 1^T = (1, \ldots, 1)$$
$$x^T = (x_1, \ldots, x_n) \quad \varepsilon^T = (\varepsilon_1, \ldots, \epsilon_n)$$

$$Y = \beta_0 * 1 + \beta_1 * X_1 + \beta_2 * X_2 + \cdots + \beta_k * X_k + \varepsilon$$

$$\begin{bmatrix} y_1 \\ y_2 \\ \cdots \\ y_n \end{bmatrix} = \beta_0 \begin{bmatrix} 1 \\ 1 \\ \cdots \\ 1 \end{bmatrix} + \beta_1 \begin{bmatrix} x_{11} \\ x_{12} \\ \cdots \\ x_{1n} \end{bmatrix} + \beta_2 \begin{bmatrix} x_{21} \\ x_{22} \\ \cdots \\ x_{2n} \end{bmatrix} + \cdots + \beta_k \begin{bmatrix} x_{k1} \\ x_{k2} \\ \cdots \\ x_{kn} \end{bmatrix} + \begin{bmatrix} \varepsilon_1 \\ \varepsilon_2 \\ \cdots \\ \varepsilon_n \end{bmatrix}$$

If we take just two regressors, i.e. dependent variables, to obtain a geometrical interpretation we get the situation depicted by Fig. 6.42.

---

[19] A square matrix A is *non-singular* if A is invertible, i.e., exists A^{-1}, such that $AA^{-1} = I$, where I is the identity matrix. Equivalently, is invertible if and only if the determinant of A is non zero (det $(A) \neq 0$).

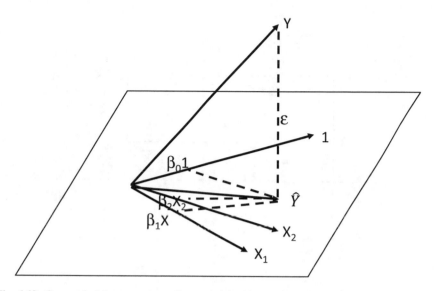

Fig. 6.42 Geometrical Interpretation a linear model with two dependent variables

$$Y = \beta_0 + \beta_1 X_1 + \beta_2 X_2 + \cdots + \beta_k X_k + \varepsilon$$

We want to minimize $\varepsilon = Y - \widehat{Y}$

Which means that ε should be orthogonal to a linear variety generated by $(1, X_1, \ldots, X_k)$:

$$\varepsilon^T * 1 = 0$$
$$\varepsilon^T * X_1 = 0$$
$$\ldots$$
$$\varepsilon^T * X_k = 0$$

which are the estimation eqs. (6.119), (6.120), until (6.121)

Validation of a Multiple Linear Regression Model

As a linear regression model is a hypothesis that must be validated. It is very important to validate the model, checking whether the hypotheses are satisfied.

The least square method always produces a regression model, whether or not there is a linear relationship between all the dependent variables (X_i) and the target variable (Y).

It must be assessed how well the linear model fits the data. Usual methods to validate the linear model are:

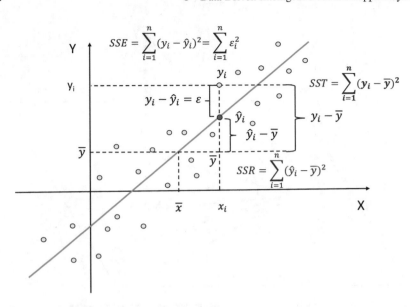

Fig. 6.43 Decomposition of the sum of squares

- *Sum of Squares of Errors (SSE)*: in the estimation of the predicted values, there is a useful decomposition of the errors $(y_i - \widehat{y}_i)$ in terms of $(\widehat{y}_i - \overline{y})$ and $(y_i - \overline{y})$. For a better understanding, this decomposition is depicted in Fig. 6.43.

In this decomposition the *Total Sum of Squares* (SST) can be decomposed as the *Sum of Squares of the Regression model* (SSR) and the *Sum of Squares of the Errors* (SSE), as follows:

$$SST = SSR + SSE \tag{6.124}$$

$$\sum_{i=1}^{n} (y_i - \overline{y})^2 = \sum_{i=1}^{n} (\widehat{y}_i - \overline{y})^2 + \sum_{i=1}^{n} (y_i - \widehat{y}_i)^2 \tag{6.125}$$

In Table 6.20, there is the description of each sum of squares, and related information.

The *sum of square of errors (SSE)* is defined as:

$$SSE = \sum_{i=1}^{n} (y_i - \widehat{y}_i)^2 = \sum_{i=1}^{n} \varepsilon_i^2 \tag{6.126}$$

As lower are the values of SSE, lower are the errors, and hence, better fit of the model.

Thus, the criterion is to have a lower value as possible for SSE.

Table 6.20 Description of the sum of squares decomposition

Source of variation	Sum of squares	Degrees of freedom	Variance
V. Explained by the regression model (VE)	$SSR = \sum_{i=1}^{n} (\widehat{y}_i - \bar{y})^2$	k	$\widehat{s}_e^2 = \frac{\sum_{i=1}^{n} (\widehat{y}_i - \bar{y})^2}{k}$
V. Not explained by the model (VNE) or V. residual (VR) or V. errors (VEr)	$SSE = \sum_{i=1}^{n} (y_i - \widehat{y}_i)^2$	$n - k - 1$	$\widehat{s}_R^2 = \frac{\sum_{i=1}^{n} (y_i - \widehat{y}_i)^2}{n-k-1}$
V. Total (VT)	$SST = \sum_{i=1}^{n} (y_i - \bar{y})^2$	$n - 1$	$\widehat{s}_y^2 = \frac{\sum_{i=1}^{n} (y_i - \bar{y})^2}{n-1}$

- *Standard Error of Estimate or Residuals* $(\widehat{s}_e/\widehat{s}_R)$: The estimation of the deviation of the residuals should be as small as possible.

  As we know that $E(\varepsilon_i) = 0$, which means that in average the errors are zero, if the deviation of the errors (σ_{ε_i}) is small, the errors tend to be closer to zero (mean error). Thus, the model fits well.

  As σ_{ε_i} is estimated by $\widehat{s}_e$, we must compute it as follows:

$$\widehat{s}_e = \widehat{s}_R = \sqrt{\frac{SSE}{n - (k+1)}} = \sqrt{\frac{\sum_{i=1}^{n} (y_i - \widehat{y}_i)^2}{n - k - 1}} \qquad (6.127)$$

  $n - (k + 1)$ are the final degrees of freedom, because we have $k + 1$ equations (estimation equations) constraining the initial n degrees of freedom.

- *Significance of the Model Test* $(F_{k,n-k-1})$: It is a *statistical test*[20] to check the linear relation between the predicted variable Y and all explicative variables (X_i). It is a test of the whole model.

  The test at *α-level* of *significance* is the following:

$$F^* = \frac{\widehat{s}_e^2}{\widehat{s}_R^2} \sim F_{(k,n-k-1;\alpha/2)} \qquad (6.128)$$

  if $F^* > F_{(k, n-k-1; \alpha/2)}$, then we accept the significance of the relation between X_i and Y. Then, effectively there is a linear relation between Y and X_i.

- *Significance of one Model Parameter Test* $(H_0 : \beta_k = 0)$: It is a statistical test to check whether there is a linear relationship between one explicative variable (X_i) and the predicted variable (Y).

  If there is no linear relationship between X_i and Y variables, it means that the corresponding coefficient in the model (β_i) must satisfy $\beta_i = 0$

---

[20] A *statistical test* is a method for verifying a statistical hypothesis. A *statistical hypothesis* is every conjecture concerning the unknown distribution F of a random variable X. A *significance test* is a test aiming to verify whether one statistical hypothesis is not false.

The *Linear Hypothesis test* is:

$$H_0 : \beta_i = 0 \quad \text{(no linear relation)}$$
$$H_1 : \beta_i \neq 0 \quad \text{(linear relation)}$$

As $\widehat{\beta}_i \sim N(\beta_i, \sigma\sqrt{q_{ii}})$

H_0 is true $\Leftrightarrow t = \frac{\widehat{\beta} - \beta}{s_{\widehat{\beta}}} \sim t_{n-k-1}$ where $s_{\widehat{\beta}} = \sigma\sqrt{q_{ii}}$, and

q_{ii} is the element i of the diagonal of the matrix $(X^T X)^{-1}$

$$t = \frac{\left(\widehat{\beta}_i - \beta_i\right)}{\widehat{s}_R \sqrt{q_{ii}}} \sim t_{n-k-1} \tag{6.129}$$

$$\text{If } \beta_i = 0 \Rightarrow t^* = \frac{\widehat{\beta}_i}{\widehat{s}_R \sqrt{q_{ii}}} \sim t_{n-k-1} \tag{6.130}$$

if $t^* > t_{n-k-1}$ then $H_0 : \beta_i = 0$ is false (there is a linear relation)
A *confidence interval* can be formulated for the coefficient β_i:
At $(100 - \alpha)$ level of confidence (95% for instance, $\alpha = 0.5$), the interval of confidence is:

$$\beta_i \in \widehat{\beta}_i \pm t_{\alpha/2; n-k-1} * \widehat{s}_R \sqrt{q_{ii}}$$

- *Coefficient of Determination (R^2): this coefficient measures the* strength of the linear relationship. It is defined as follows:

$$R^2 = \frac{SSR}{SST} = \frac{\sum_{i=1}^{n} (\widehat{y}_i - \overline{y})^2}{\sum_{i=1}^{n} (y_i - \overline{y})^2} = \frac{SST - SSE}{SST} \tag{6.131}$$

Thus,

$$R^2 = 1 - \frac{SSE}{SST} \tag{6.132}$$

It can be demonstrated that,

$$R = \frac{\text{cov}\left(Y, \widehat{Y}\right)}{s_y s_{\widehat{y}}} = r_{\widehat{yy}} \tag{6.133}$$

which is the correlation coefficient of Y and $\widehat{Y}$

$$| R^2 |= 1, 0 \le R^2 \le 1,$$

If $R^2 = 1$, it means that there is a *perfect linear relationship*, and on the other hand, if $R^2 = 0$, it means that there is *no linear relationship*.

However, the R^2 coefficient has a drawback. If there are more new variables added to the model, the value of R^2 increases making the R^2 value not fully reliable.

To solve that, a *Corrected Coefficient of Determination* $(\overline{R^2})$ is defined as:

$$\overline{R^2} = 1 - \frac{\widehat{s}_R^2}{\widehat{s}_y^2} \tag{6.134}$$

Thus,

$$\overline{R^2} = 1 - \frac{\sum_i e_i^2/(n-k-1)}{\sum_{i=1}^{n}(y_i - \bar{y})^2/(n-1)} = 1 - \frac{n-1}{n-k-1}\frac{\sum_i e_i^2}{\sum_{i=1}^{n}(y_i - \bar{y})^2}$$

and making algebraic manipulations and using the equation of

$$R^2 = 1 - \frac{\sum_i e_i^2}{\sum_{i=1}^{n}(y_i - \bar{y})^2} \tag{6.135}$$

Finally, we have:

$$\overline{R^2} = 1 - (1 - R^2)\frac{n-1}{n-k-1} \tag{6.136}$$

- *Significance of the Model Test* (revisited): with the definition of R^2 there is another way to compute the F test.
  The test is:

$$F = \frac{\widehat{s}_e^2}{\widehat{s}_R^2} = \frac{SSR/k}{SSE/n-k-1} = \frac{SSR}{SSE} * \frac{n-k-1}{k} \tag{6.137}$$

Taking into account that:

$$1 - R^2 = \frac{SSE}{SST}$$

Then,

$$\frac{\text{SSR}}{\text{SSE}} = \frac{\text{SSR/SST}}{\text{SSE/SST}} = \frac{\text{SST} - \text{SSE/SST}}{\text{SSE/SST}} = \frac{1 - (\text{SSE/SST})}{\text{SSE/SST}} = \frac{R^2}{1 - R^2}$$

Finally,

$$F = \frac{\widehat{s}_e^2}{\widehat{s}_R^2} = \frac{\text{SSR}/k}{\text{SSE}/n - k - 1} = \frac{\text{SSR}}{\text{SSE}} * \frac{n - k - 1}{k} = \frac{R^2}{1 - R^2} * \frac{n - k - 1}{k}$$

This way we have an alternative way for computing the Test:

$$F^* = \frac{\widehat{s}_e^2}{\widehat{s}_R^2} = \frac{R^2}{1 - R^2} * \frac{n - k - 1}{k} \tag{6.138}$$

The Test at α-level of significance:

$$F^* = \frac{\widehat{s}_e^2}{\widehat{s}_R^2} = \frac{R^2}{1 - R^2} * \frac{n - k - 1}{k} \sim F_{(k, n-k-1; \alpha/2)} \tag{6.139}$$

if $F^* > F_{(k, n-k-1; \alpha/2)}$, then reject H_0 (all $\beta_i = 0$), which means that there is some linear relation.

An Example of a Multiple Linear Regression Model

As an example of the use of a multiple linear regression model, we will take a dataset taken from the US Department of Commerce, Bureau of the Census (1977). This dataset contains several data for each one of the States of the USA. The dataset has 50 instances (each one accounting for one State) and eight variables describing the following statistics: *population*, which has a population estimate as of July 1, 1975; *income*, which is the per capita income by 1974; *illiteracy*, which reflects the illiterate percent of population by 1970; *murder*, which describes murder and non-negligent manslaughter rate per 100,000 population (1976); *frost*, which Accounts for the mean number of days with minimum temperature below freezing (1931–1960) in the capital or a large city of the State; and *area*, which is the land area in square miles.

Our goal will be to try to predict the *murder rate* for each state as a function of the other possible variables: *population, income, illiteracy, frost, area*.

The dataset was analyzed and pre-processed. No missing values were found. There was the intuition that some of them like the income or the illiteracy can have some linear relationship with the murder rate. Therefore, a Multiple Lineal Regression model could be a reasonable option. Thus, we were assuming the following model:

MLR1 : Murder
$$= \beta_0 + \beta_1 \text{Pop} + \beta_2 \text{Inc} + \beta_3 \text{Illit} + \beta_4 \text{Frost} + \beta_5 \text{Area} + \varepsilon \qquad (6.140)$$

We estimated the parameters and computed several indicators of the model using the R system. The results were the following:

```
Residuals:

   Min      1Q    Median     3Q      Max
-3.8542 -1.8890  -0.2493 1.5340 7.6742

Coefficients:

                Estimate   Std. Error  t value  Pr(>|t|)
(Intercept)     4.718e+00   4.131e+00   1.142   0.259611
Population      2.344e-04   8.782e-05   2.669   0.010623 *
Income         -6.284e-04   7.476e-04  -0.841   0.405151
Illiteracy      3.515e+00   9.033e-01   3.891   0.000334 ***
Frost          -3.124e-03   9.912e-03  -0.315   0.754111
Area            9.421e-06   4.721e-06   1.996   0.052182 .
---
Signif. codes:  0 '***' 0.001 '**' 0.01 '*' 0.05 '.' 0.1 ' ' 1


Residual standard error: 2.455 on 44 degrees of freedom
Multiple R-squared: 0.6029,   Adjusted R-squared: 0.5578
F-statistic: 13.36 on 5 and 44 DF,  p-value: 6.176e-08
```

The Linear regression model (MLRM1) obtained with the coefficients is:

$$\text{Murder} = 4.718 + 0.0002344 \text{Pop} - 0.0006284 \text{Inc} + 3.515 \text{Illit}$$
$$- 0.003124 \text{Frost} + 0.000009421 \text{Area} + \varepsilon \qquad (6.141)$$

From the analysis of the different validation indicators, it can be observed that the $R^2 = 0.6029$, which means that the model account for the 60.29% of the variance in murder rates across all the states.

Regarding the Residual standard error is equal to 2.455 with 44 degrees of freedom, which is quite acceptable.

The model test is related to the F-statistic, confirms that there is a linear relation between the target variable and the explicative ones, because the p-value[21] has a value of 6.176e-08, much less than 0.001 probability values.

---

[21] The *p-value* or *probability value* is defined as the probability, under the null hypothesis H_0 about the unknown distribution F of the random variable X, for the variable to be observed as a value equal to or more extreme than the actual value observed. If the *p-value* is $<\alpha$ (a level of significance like 0.05 or 0.01 or 0.005 or 0.001), then the H_0 is rejected.

   Regarding whether really the coefficients are different from 0 (a linear relation exists) for each explicative variable, the results are quite different:

- The *Illiteracy* coefficient (3.515) is significantly different from zero at the 0.001 probability level (it has a probability of 0.000334 of being 0).
- The *Population* coefficient (2.344e-04) is significantly different from zero at the 0.05 probability level (it has a probability of 0.010623 of being 0).
- The *Area coefficient* (9.421e-06) is significantly different from zero at the 0.1 probability level (it has a probability of 0.052182 of being 0).

   The other coefficients of *Frost* and *Income* are not significantly different from 0, because they have respectively the probabilities (0.754111 and 0.405151). This suggests that probably *Frost* and *Income* are not linearly related to the murder rate when the other variables are controlled.

   This suggests the idea that perhaps a linear model with less explicative variables, could make sense, so that Frost and Income seems not to be linearly dependent to the *murder rate*.

   We tried this new model:

$$\text{MLRM2} : \text{Murder} = \beta_0 + \beta_1 \text{Pop} + \beta_2 \text{Illit} + \beta_3 \text{Area} + \varepsilon \qquad (6.142)$$

   The results for this new simpler model were:

```
Residuals:

    Min        1Q    Median        3Q      Max
-4.2454   -1.9212   -0.2257    1.5764   7.3256


Coefficients:

               Estimate   Std. Error   t value   Pr(>|t|)
(Intercept)   1.228e+00    8.243e-01     1.490    0.14314
Population    2.221e-04    7.795e-05     2.850    0.00653 **
Illiteracy    4.002e+00    5.725e-01     6.990   9.42e-09 ***
Area          7.410e-06    4.067e-06     1.822    0.07497 .
---
Signif. codes:  0 '***' 0.001 '**' 0.01 '*' 0.05 '.' 0.1 ' ' 1

Residual standard error: 2.422 on 46 degrees of freedom
Multiple R-squared: 0.596,   Adjusted R-squared: 0.5696
F-statistic: 22.62 on 3 and 46 DF,  p-value: 3.811e-09
```

   In this new model, the $R^2 = 0.596$, which means that the model accounts for the 59.6% of the variance in murder rates across all the states. Compared with the previous one, it can be observed that they are practically the same (59.6% vs. 60.29%).

Taking a look to the median of the residuals, now is of −0.2257, and in the previous model it was −0.2493. Of course, they are quite similar but the median value now is lower.

Regarding the Residual standard error is equal to 2.422 on 46 degrees of freedom, which is less than the 2.455 in the previous model.

The model test is related to the F-statistic, confirms that there is a linear relation between the target variable and the explicative ones, because the p-value has a value of 3.811e-09, much less than 0.001 probability value.

Regarding whether really the coefficients are different from 0 (a linear relation exists) for each explicative variable, the results now are:

- The *Illiteracy* coefficient (4.002) is significantly different from zero at the 0.001 probability level (it has a probability of 9.42e-09 of being 0).
- The *Population* coefficient (2.221e-04) is significantly different from zero at the 0.01 probability level (it has a probability of 0.00653 of being 0).
- The *Area coefficient* (7.410e-06) is significantly different from zero at the 0.05 probability level (it has a probability of 0.07497of being 0).

The new Linear regression model (MLRM2) obtained with the new coefficients is:

$$Murder = 1.228 + 0.0002221 Pop + 3.515 Illit + 0.00000741 Area + \varepsilon \quad (6.143)$$

If we analyse the scatter plot of the residuals and the fitted values of both linear models, as depicted in Fig. 6.44, it can be seen that they are very similar.

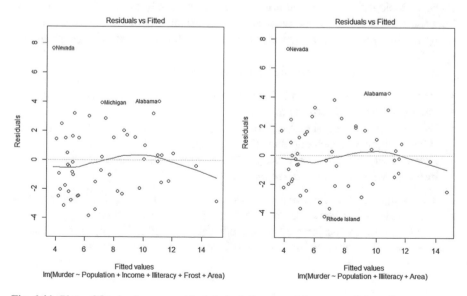

Fig. 6.44 Plots of fitted values vs. residuals in both linear models, generated from R system

The plot on the left corresponds to the first and more complex model (MLRM1) and the plot on the right corresponds to the second and simpler model (MLRM2).

The Nevada state seems to be the one with a highest residual, thus it is one of the worst predicted by the model. Apart from Nevada, all the other seems to follow a good residual pattern fitting. The case of Nevada is strange because even though it has a high *murder rate* of 11.5 (the highest one is Alabama with 15.1), it has low values of *illiteracy* and of *population*, when usually the States with high *murder rates* are correlated with a *high population* and *high illiteracy* values. Thus, clearly is following a different pattern from all other similar States. Perhaps, it could be related with Las Vegas and its business, that could have some impact on the *murder rate*. It is only a hypothesis.

Finally, as being the two models rather equivalent, buy being the second one simpler than the first one, this model (MLRM2) would be the selected. Of course, we have to be aware that it accounts just for the 60% of variance of the target variable approximately. This means that other factors or variables are the responsible for the value of the *murder rate* (the error term of the model, ε), and they must be investigated to get a better fitting model.

An Ensemble of Predictive Models

The same idea exposed in a previous subsection about the *ensemble of discriminant or predictive* methods can be extended to the *predictive models*. The idea is to *combine several predictor models* to improve the accuracy of the numeric predictions. Main difference is that now what has to be combined are different numeric quantities instead of different class labels. The usual way of the combination is to compute the *mean* or *weighted mean* of all the output values of each predictive model.

Most commonly used ensemble techniques for predictive models are:

- *Voting*: a voting scheme, which can be weighted, is done to combine the output among different predictive models.
- *Bagging*: making a resampling from the training data (*bootstrapping*), using the same predictive model, and afterwards voting.
- *Randomizing input features*: using random subsets of features at each node for building *regression trees* (*random forests*).

The *voting scheme for predictive models* is depicted in Fig. 6.45. As it can be shown, it is very similar to the voting scheme for classifier methods.

The bagging scheme for predictive models is depicted in Fig. 6.46. As it can be shown, it is very similar to the bagging scheme for classifier methods.

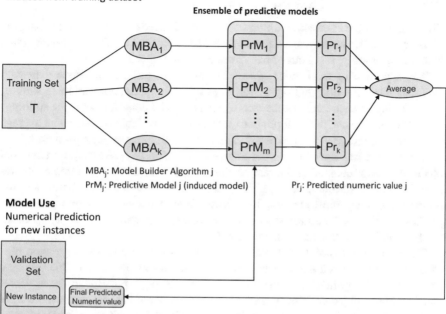

Fig. 6.45 A voting scheme for an ensemble of predictive models

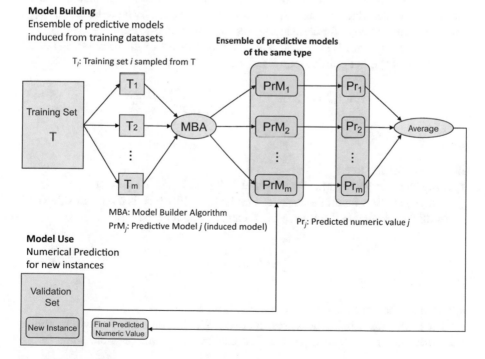

Fig. 6.46 A bagging scheme for an ensemble of predictors of the same type

Regression Trees and Predictive Random Forests

The *random forest* approach with predictive models needs that the trees forming the forest can predict numerical values at their leaves instead of discrete values (class labels). Commonly, this kind of trees are named as *regression trees*.

The approach of the *random forest* for predictive tasks is the same than with decision trees. The different regression trees are built upon a bootstrapping sample, and selecting at each node a subspace of features to be considered for the splitting.

At the testing step, the final prediction of the target variable is done by an average or weighted average of the predicted values of the different regression trees.

The use of a *regression tree* is the same than a *decision tree*. Starting from the root of the tree, after consulting the test on the node, the traversal of the tree continues into the subtree matching the splitting condition on one of the attributes, and the process continues until arriving to a leave. In the leave is where the numeric prediction is done. The numeric prediction is the aggregation or averaging of all the training instances which reached that leave.

To build a regression tree, the aim is to obtain the best possible regression tree. This means to obtain a tree that makes the predictions as best as possible. Imagine that you had a very huge tree that it had as many leaves as possible different values had for the predicted variable Y. This tree, for all the training instances would have an error value equal to 0, because each instance will traverse the tree until it reached the corresponding leave where the instance is, and then the prediction would be the actual value of Y for that instance.

The basic idea for building a good regression tree is to build a tree that minimized the error function which accounts for the difference between each predicted value $(\widehat{y}_i)$ and the actual value (y_i) for all the instances used to build the tree.

Thus, the error function, which is the sum of squares of the errors (SSE) of all predictions of the tree, can be expressed as:

$$\text{Error}(T) = \text{SSE}(T) = \sum_{\text{lf} \in \text{leaves}(T)} \sum_{i \in \text{lf}} (y_i - \widehat{y}_{\text{lf}})^2 \qquad (6.144)$$

where,

If is a leave of the tree T

y_i is the numeric value of the target variable for the instance i ($i \in \text{lf}$)

$\widehat{y}_{\text{lf}}$ is the prediction for all the instances in the leave lf, which is the mean value $(\overline{y}_{\text{lf}})$ of all the instances of that leave lf. It can be computed as:

$$\widehat{y}_{\text{lf}} = \overline{y}_{\text{lf}} = \frac{1}{n_{\text{lf}}} \sum_{i \in \text{lf}} y_i \qquad (6.145)$$

where n_{lf} is the number of instances in the leave lf,

The procedure to build the regression tree is to select the splits that will minimize the error function SSE at each node. At each node, all the possible splits for all possible explicative variables must be evaluated. Usually the splits are binary, and the candidate splits are different depending on whether the variables are categorical or numeric. See the Sect. 6.4.2.1 where these binary splits are explained for the CART method. The basic algorithm for building a regression tree can be described as follows:

---

algorithm Regression Tree

input X: the training data set (N instances * $D+1$ attributes): $\{X_{1.}, ..., X_{N.}\}$
        A: the set of attributes: $\{A_1, ..., A_D\}$,
        Y: is the numeric attribute to be predicted
        mininst: minimum number of instances required at each node
        thres: maximum SSE reduction threshold
output a regression tree

begin
 option
    // all the instances have the same value for all the explanatory attributes
   case $\forall X_{i.} \in X, \forall X_{.j} \in A, \; X_{ij} = X_{1j}$ **do**
    // return a tree with a node holding the set of instances and
    // their mean as the prediction
   tree $\leftarrow$ buildTree(node($X, \underset{\forall X_{i.} \in X}{mean}(Y_i)$))
    // not all the instances have the same value for all the explanatory attributes
   case $\neg \, (\forall X_{i.} \in X, \forall X_{.j} \in A, \; X_{ij} = X_{1j} \,)$ **do**
   split$_{max}$ $\leftarrow$ $\underset{\forall X_{.j} \, \forall bin-split(X_{.j})}{max} (SSE(T) - \sum_{i=1}^{2} SSE(T_i))$
   maxSSEReduction $\leftarrow$ $\underset{\forall X_{.j} \, \forall bin-split(X_{.j})}{max} (SSE(T) - \sum_{i=1}^{2} SSE(T_i))$
  option
    // The SSE reduction is not worthy ($<$ *thre*) or the number of instances
    // in some subtree is very low ($|T_i| <$ *mininst*)
   case (maxSSEReduction $<$ *thre*) $\vee$ ($\exists \, T_i: |T_i| <$ *mininst*) **do**
     // return a tree with a node holding the set of instances and
     // their mean as the prediction
    tree $\leftarrow$ node($X, \underset{\forall X_{i.} \in X}{mean}(Y_i)$)
     // The SSE reduction is worthy ($\geq$ *thre*) and the number of instances
     // in all subtrees is reasonable ($\forall T_i, \; |T_i| \geq$ *mininst*)
   case $\neg \, ((\text{maxSSEReduction} < \textit{thre}) \vee (\exists \, T_i: |T_i| < \textit{mininst}))$ **do**
     // for the two nodes of the best binary split build a subtree
    tree $\leftarrow$ buildTree(split-attribute(split$_{max}$))
    for each $T_i \in$ split$_{max}$ **do**
     tree2 $\leftarrow$ RegressionTree($T_i, A, Y, mininst, thres$)
     tree $\leftarrow$ addBranch(tree, tree2, split-value(split$_{max}$))
    endforeach
   endoption
  endoption
 return tree
end

Validation of Predictive Models

Predictive models must be validated as all inductive models. The scalability and generalization abilities must be tested to ensure that they can be used regularly to make good predictions for other datasets. As with all data-driven models, the validation must be done at two stages: from a qualitative point of view, and a quantitative point of view.

The *qualitative validation* consists of the interpretation of the model by an expert on the domain. The graphic visualization of the model is usually very important to get a meaningful insight into the knowledge patterns mined from data. Regarding the predictive models like Artificial Neural Networks, even though they have a clear architecture, they are usually known as a *blackbox model*, because it is not clear what is happening inside the ANN to perform good predictions. The updating of the weights is what makes the ANN to approximate the target function, but it is not easy to get knowledge patterns from it or understand why it works properly. A Case-based predictor model, if the Case Base is organized hierarchically provide an interesting structuration of the knowledge patterns (the cases). Finally, a Multiple Linear Regression model provides a nice geometric interpretation with the regression line or in general with a linear hyperplane approximating the target variable.

Furthermore, the *quantitative validation* of the predictive models must be done to assess the predictive accuracy or error of the model. The generalization ability of the model must be ensured to be able to use the model for further predictions with other data. This way, we check the lack of overfitting of the model, and its ability to be scalable to new data. The same concepts explained in Sect. 6.4.2.1 for discriminant models regarding the true error and the estimated error, the different types of validation schemes (simple, k-fold cross-validation, leave-one-out) are valid for predictive models.

The estimation of the accuracy of a predictive model is slightly different than a discriminant or classifier method. Here, it is not possible to account for the failures and successful predictions of class labels, because predictive models compute an estimation of a numeric quantity. Thus, the measurement of the accuracy is done taking into account how much is the deviation or error from the actual values of the target variable and the estimated values from the predictive model.

In the literature, there are several *Error measures* that have been proposed. Most commonly used are the following:

- *Mean Absolute Error* (*MAE*), which computes the average difference or error in absolute value for the estimated values:

$$\text{MAE}\left(\widehat{Y}, Y\right) = \frac{1}{N} \sum_{i=1}^{N} |\widehat{y}_i - y_i| \qquad (6.146)$$

being,

$\widehat{Y}_i$ the predicted values,
Y_i the actual values

- *Mean Absolute Percentage Error (MAPE)*, which computes in percentage the mean deviation in absolute value between the estimated values and the actual ones:

$$\text{MAPE}\left(\widehat{Y}, Y\right) = \frac{100}{N} * \sum_{i=1}^{N} \frac{|\widehat{y}_i - y_i|}{|y_i|} \tag{6.147}$$

or alternatively substituting the actual values y_i by the mean of all actual values $\overline{y}_i$ to overcome division by zero and some other problems (sometimes named as weighted MAPE, WMAPE):

$$\text{MAPE}\left(\widehat{Y}, Y\right) = \frac{100}{N} * \sum_{i=1}^{N} \frac{|\widehat{y}_i - y_i|}{|\overline{y}_i|} = 100 * \frac{\sum_{i=1}^{N} |\widehat{y}_i - y_i|}{\left|\sum_{i=1}^{n} y_i\right|} \tag{6.148}$$

being,

$\widehat{Y}_i$ the predicted values,
Y_i the actual values
$\overline{Y}_i$ the mean of the actual values

- *Mean Square Error (MSE)*, which computes the average square error or deviation between the estimated values and the actual values.

$$\text{MSE}\left(\widehat{Y}, Y\right) = \frac{1}{n} \sum_{i=1}^{n} (\widehat{y}_i - y_i)^2 \tag{6.149}$$

being,

$\widehat{Y}_i$ the predicted values
Y_i the actual values

- *Root Mean Square Error (RMSE)*, which is the root square of the MSE. It has the advantage over MSE, that the value of the error is in the same magnitude that the numeric values of the target variable.

$$\text{RMSE}\left(\widehat{Y}, Y\right) = \sqrt{\frac{1}{n} \sum_{i=1}^{n} (\widehat{y}_i - y_i)^2} \tag{6.150}$$

being,

$\widehat{Y}_i$ the predicted values
Y_i the actual values

- *Normalized Root Mean Square Error* (*NRMSE*), which computes the same value of RMSE, but afterwards is normalized according to the range of the target variable.

$$\text{NRMSE/NRMSD}\left(\widehat{Y}, Y\right) = \frac{1}{\left(y_i^{\max} - y_i^{\min}\right)} \sqrt{\frac{1}{n} \sum_{i=1}^{n} (\widehat{y}_i - y_i)^2} \qquad (6.151)$$

being,

$\widehat{Y}_i$ the predicted values,
Y_i the actual values

- *Corrected Coefficient of Determination* or *R-squared* (R^2), which computes how correlated are the estimated values and the actual values of the target variable.

$$R^2\left(\widehat{Y}, Y\right) = \frac{\left[\sum_{i=1}^{n}(y_i - \overline{y}_i) * \left(\widehat{y}_i - \overline{\widehat{y}}_i\right)\right]^2}{\sum_{i=1}^{n}(y_i - \overline{y}_i)^2 \sum_{i=1}^{n}\left(\widehat{y}_i - \overline{\widehat{y}}_i\right)^2} \qquad (6.152)$$

being,

$\widehat{Y}_i$ the predicted values
Y_i the actual values
$\overline{\widehat{Y}}_i$ the mean of the predicted values,
$\overline{Y}_i$ the mean of the actual values

Probably one of the most used is the Root Mean Square Error (RMSE).

6.4.3   Optimization Models

There is an artificial intelligence technique, which can act as a *second-order data mining technique*. The data mining methods we have analyzed and explained in Sect. 6.4, learn a model from data. *Genetic Algorithms* (GAs) are a general optimization technique which can be applied to a special data, i.e. *to previously mined models,* in order to mine again the new data, i.e. the models, to optimize them and give the best data mining model. Thus, they can be considered as a second-order data mining technique.

6.4.3.1   Genetic Algorithms

Evolutionary computation, and especially the most commonly used approach, *the genetic algorithm* paradigm, is a bio-inspired approach mimicking *theory of evolution* by Charles Darwin, and Weismann's theory of *natural selection* in biological populations (Holland, 1975; Goldberg, 1989). Biological populations evolve with time and only the best fitted to the environment survive. There is a competition among the individuals of the population to survive.

Genetic algorithms are stochastic methods that search for the best solution/hypothesis from the population of possible solutions/hypotheses. The function to be optimized is usually named as the *Fitness function.*

A genetic algorithm solves a problem by evolving its solution incrementally. It starts its execution with an initial *population* which is nothing but a set of possible *encoded solutions* to the problem referred as *chromosomes* or *individuals.* From the current population, a new population is evolved using several *genetic operators* in the hope of improving the solutions. The three basic genetic operators are *selection, crossover* and *mutation.*

This process is depicted in Fig. 6.47. It represents the single genetic cycle called as *generation.* After each generation, the *termination criteria* will be checked whether it is satisfied or not. If it is not satisfied, will continue with the new generation. Otherwise, the current population will be returned as the final result.

Fig. 6.47 A Genetic
Algorithm cycle

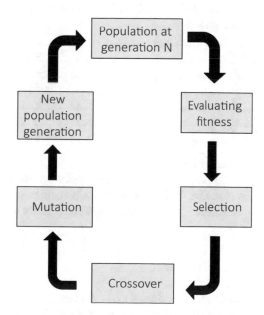

Chromosome Encoding

A *chromosome* is formed by *genes*. Each entity of a chromosome is a gene.

Let us suppose that we have the following two examples in a dataset with three attributes:

Example	Age	Height (cm)	Weight (kg)
1	22	176.78	67
2	30	182.8	71.5

There are two basic encoding techniques for chromosomes:

- *Binary Coding*

   Binary coding is the standard way of encoding. This way, as we have the minimum unit of information at a computer level, a 0 or a 1, the modifications of these genes are smoother if the coding is binary. Regarding the numerical data, commonly both integer and floating-point values are handled.

Example	Age	Height (cm)	Weight
1	10110	10110000.1001110	1000011
2	11110	10111110.1000	1000111.0101

   For categorical data: the coding is done by pooling into a list of all valid values.

   Invalid values could arise in the application of genetic operators. They must be managed.

Index	Nationality	Encoded value
0	India	000
1	Catalonia	001
2	USA	010
3	France	011
4	China	100

- *Direct Encoding*

   Direct encoding means really no encoding and handle the data as it is. For instance, if we want to use a dataset as the individuals of a population, each example from a dataset could be directly a chromosome, without any codification.

Fitness Function

The fitness function, $f(x)$, is the function that encodes the function to be optimized. Usually, the fitness values of all individuals are normalized such that all will lie between 0 and 1.

$$nf_i = \frac{f_i}{\sum_{i=1}^{n} f_i} \tag{6.153}$$

where,

f_i is the fitness value of the ith individual

nf_i is the normalized fitness of the ith individual

The fitness function could range from a simple mathematical function to a complex validation process to get an accuracy score, for the *testing of an instance set* with the model "encoded" in an individual.

Genetic Operators

The basic genetic operators are:

- *Selection Operator*: the selection operator selects individuals from current population to generate the new population. There are different strategies to select the individuals:

  - *Fitness Proportional Selection*: the probability of selection of an individual is based on its fitness. The individuals with better fitness will have more chance of getting selected as a parent. See Fig. 6.48.
  - *Tournament selection*: The best individuals according to successive tournaments among them are selected. See Fig. 6.49, where a tournament among four chromosomes is depicted.

- *Crossover Operator*: The aim of the crossover operator is to generate offspring to inject diversity in the population. The offspring have some traits of both parents, like in animal reproduction crossover of chromosomes. There are several variations of the crossover operator. Most commonly used are:

  - *Single point crossover* or *simple crossover*: there is just one point of crossover, and all the genetic material (genes) after the point will be crossed in the offspring. See Fig. 6.50.

Fig. 6.48 Fitness proportional selection

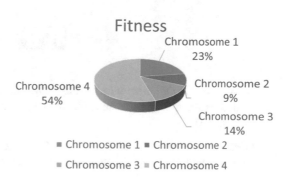

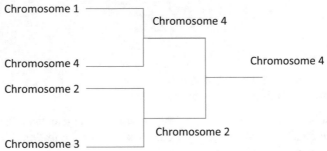

Fig. 6.49 Tournament selection

Fig. 6.50 Single point
crossover operator

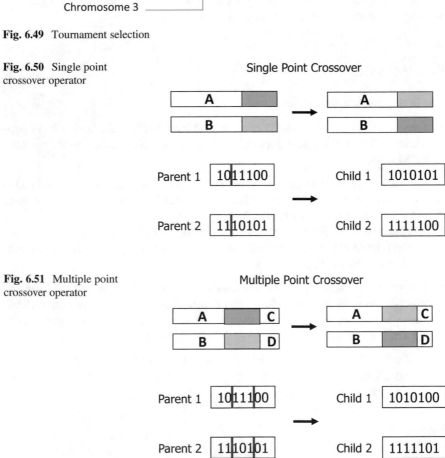

Fig. 6.51 Multiple point
crossover operator

- *Multiple Point Crossover*: There are several points of crossover, which
  delimitate the genetic material (genes) to be crossed in the offspring. See
  Fig. 6.51.

- *Mutation Operator*: The mutation operator resembles the gene mutation that
  happens with some probability in living beings. It randomly changes the treats

Fig. 6.52 Mutation
operator

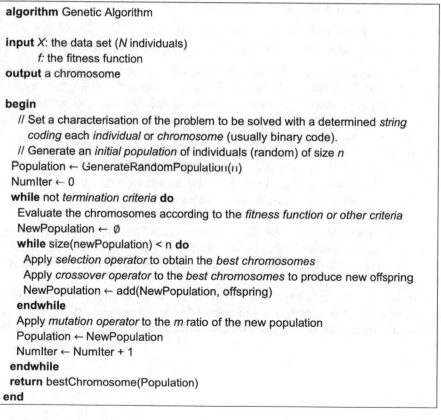

Mutation

(genetic material) of the offspring. The probability for a chromosome to get mutated depends on the mutation rate.

The mutation operator in a binary coded chromosome is just to flip the bit.

In a direct encoding of a chromosome, some specific operator must be implemented for the mutation, like an arithmetic operator for numeric values. See Fig. 6.52.

General Scheme

A *Genetic Algorithm* can be algorithmically described as follows:

```
algorithm Genetic Algorithm

input X: the data set (N individuals)
      f: the fitness function
output a chromosome

begin
   // Set a characterisation of the problem to be solved with a determined string
      coding each individual or chromosome (usually binary code).
   // Generate an initial population of individuals (random) of size n
   Population ← GenerateRandomPopulation(n)
   NumIter ← 0
   while not termination criteria do
   Evaluate the chromosomes according to the fitness function or other criteria
   NewPopulation ← ∅
   while size(newPopulation) < n do
     Apply selection operator to obtain the best chromosomes
     Apply crossover operator to the best chromosomes to produce new offspring
     NewPopulation ← add(NewPopulation, offspring)
   endwhile
   Apply mutation operator to the m ratio of the new population
   Population ← NewPopulation
   NumIter ← NumIter + 1
   endwhile
   return bestChromosome(Population)
end
```

Most common *termination criteria* are the following:

- Stop the optimization till a specified generation number.
- Stop the optimization process when fitness function does not improve.
- Run for a specified number of generations before checking for the improvement in fitness.
- Check for improvement in fitness until it reaches a specified number of generations.

Another variant of the selection operator is sometimes used, i.e. the *Elitism operator*, which means that best chromosomes will be passed to the next generation without changes.

GAs have been used for *optimization* and *search problems*. They have been used in several data mining-related problems like:

- Feature selection.
- Feature weighting.
- Rule generation.
- Classification (discrimination or categorical variable prediction).
- Prediction (numerical variable prediction).

In general Applications, GAs have been used as:

- Parameter Optimization of functions/processes.
- Computer Program generation, i.e., Genetic Programming (Koza, 1992).

We are interested in use them as optimization of *previous data mining models* we have obtained. Thus, the idea is that the chromosomes or individuals will be models. For instance, let us suppose that we want to try several ANNs models with different configurations. Thus, a chromosome could code the layers and number of neurons per layer, and the fitness function will be the accuracy of the ANN encoded in the chromosome through a cross validation of the ANN model with the dataset used.

This means each time a chromosome (an ANN model) must be evaluated for the fitness, the ANN must be trained with a training set and validated, for example, with a validation scheme using a k-fold cross-validation technique.

This way, the GA will find the optimal configuration of the ANN model.

Major advantages of GAs are that they are a general optimization methodology, usually escaping from local optima, and determining the global optimum. In addition, they are applicable to many problems, and they are an easily parallelizable technique.

As major limitations there are the problem of *crowding*, which means the overpopulation of best individuals in a great fraction of the population, which can disturb the right optimization process, and the fact that the computational cost can be high.

6.5   Post-Processing Techniques

Post-processing techniques are devoted to transform the direct results of the data mining step, i.e. the models and knowledge patterns induced from data into directly understandable, useful knowledge for later decision-making.

Major techniques could be summarized in the following tasks (Bruha & Famili, 2000):

- *Knowledge Filtering*: The knowledge induced by data-driven models should be normally filtered. For instance, if you think of a *set of association rules*, you must check all of them, because probably you have obtained some rules that are equivalent, and you probably must filter the ones that are equivalent. In addition, perhaps you want to filter the rules with less precision.
- *Knowledge Consistency Checking*. Also, we may check the new knowledge for potential conflicts with previously induced knowledge. This is a typical problem in inductive Machine Learning. As almost all the models are not incremental, i.e. if new data is available, the old models cannot be incrementally fused with the new model, and then the new models must be checked to be consistent with the previous knowledge patterns previously obtained. For instance, with a new set of rules (association or classification rules), it must be checked that there are no inconsistencies with a previously set of rules. It makes no sense that a rule specifies one conclusion in one model, and another rule in the other models specifies just the contrary conclusion.
- *Interpretation and Explanation*. The mined knowledge model could be directly used for prediction, but it would be very adequate to document, interpret and provide explanations for the knowledge discovered. This task usually is done by the experts on the domain related to the dataset being mined. Of course, there are models easily interpretable like a set of rules, or a decision tree or a regression tree. These models that can provide easy interpretations and explanations are usually named as *white-box models*, as the opposite to other models, like ANNs, which are named as *blackbox models* because is difficult to interpret those models.
- *Visualization*. Visualization of the knowledge (Cox et al., 1997) is a very useful technique to have a deeper understanding of the new discovered knowledge. It is a complementary task to the interpretation and explanation. Again, there are models which through a visualization provide the user with a lot of information like a decision tree, or a regression tree.
- *Knowledge Integration*. The traditional decision-making systems have been dependent on a single technique, strategy, or model. New sophisticated decision-supporting systems combine or refine results obtained from several models, produced usually by different methods. This process increases accuracy and the likelihood of success. In fact, this integration and interoperability of data-driven models, and even, integration with model-driven techniques is one of the open challenges in Intelligent Decision Support Systems.

- *Evaluation*. After a learning system induces concept hypotheses (models) from the training set, their evaluation (or testing) should take place. As we outlined early, as the evaluation of the models is really very important, both from a *qualitative point of view*, and from a *quantitative point of view*, evaluation of the models have been deeply explained for each type of model analyzed through this chapter. Generalization and scalability properties must be ensured before a model can be used for a regular application in the prediction.

6.6   From Data Mining to Big Data

At the beginning of the chapter different terms like data mining, knowledge discovery, and data science were analyzed and described. In the last decade, a new term has been coined: *Big Data* (Marz and Warren, 2015; Hurwitz et al., 2013).

Big Data scenario is a collection of data sets so large and complex that it becomes difficult to process using on-hand database management tools or traditional data processing applications. This trend to have larger data sets is due to several reasons. The increase of storage capacities, the increase of processing power, the increasing availability of data with the spread of internet access and open data policies from many government agencies, the additional information value derivable from the analysis of a single large set of related data.

To get you an idea of the volume level, as of 2012, limits on the size of data sets that were feasible to process in a reasonable amount of time were on the order of *exabytes*[22] of data.

Main outcomes of big data processing are that it could be more tolerant to errors than traditional database processing as the volume of data is higher. This has a direct consequence because the data mining algorithms can be more reliable if they had as an input more reliable data. Finally, they allow to discover some not so common cases, because of processing a lot more data.

Here we must state a very interesting fact: *No more data directly implies better data-driven models, unless new data are representative for other unknown patterns of knowledge*. This means that Big Data per se, does not guarantee better results, just for the volume of data.

Main features distinguishing a *Big Data scenario* are what are popularly known as the three Vs: *variety, velocity,* and *volume* as depicted in Fig. 6.53.

The main problems Big Data has to solve are not restricted only to the *volume* of data. The *Velocity* of data is another problem. The *velocity* in which data is generated and needs to be analyzed and mined is a real problem. The paradigm of data mining is evolving from batch-data analysis to real on-line streaming data which is needed to analyze. The other factor is the *Variety* of data. Nowadays, there is a lot of diversity

---

[22] An *Exabyte* (*EB*) is a unit to measure information or data. It is a multiple of a byte. It is equivalent to 10^{18} bytes or 10^3 Petabytes (PB) or 10^6 Terabytes (TB) or 10^9 Gigabytes (GB).

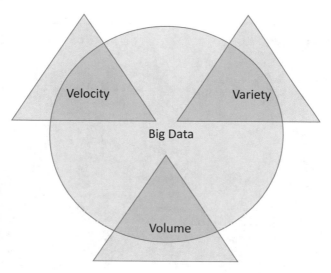

Fig. 6.53 The three Vs of Big Data

in the data at different levels. Data can be structured like in classical relational databases (SQL databases) but also in no relational databases (NoSQL databases). Also, data is varied regarding the sources they are obtained: from datasets in one company, but also through internet web servers analysing the behaviour of the customers, etc. Probably this factor, from my subjective point of view is the crux of Big Data matter: how to analyse and exploit all this diverse data which are interrelated? Finally, of course, the high *volume* of data to be analyzed creates difficulties in data management and in the processing of data mining algorithms. The union of these three factors generate an extreme complexity for managing Big Data scenarios.

Thus, managing Big Data scenarios requires different approaches regarding new techniques, new tools, and new architectures. Main challenges raised from Big Data scenarios are related to the different tasks and stages data need from its gathering until its analysis and exploiting through data-driven models. Those tasks are usually: the data gathering, data pre-processing, data storage, data querying and sharing, data transfer, and finally data analysis and visualization.

Another important factor intrinsically related to Big Data is the rising value of *social data* and its corresponding *social mining* techniques. Thus, in the data mining scenario has appeared the social mining field. This Big Data jungle is described in the picture of Fig. 6.54. Social mining is intended to cope with the *data variety* aspect of Big Data.

The mining of Social Networks has gained a lot of attention for marketing purposes, for influencing trends in public opinion, etc.

There are several issues Social Mining is focussed on such as:

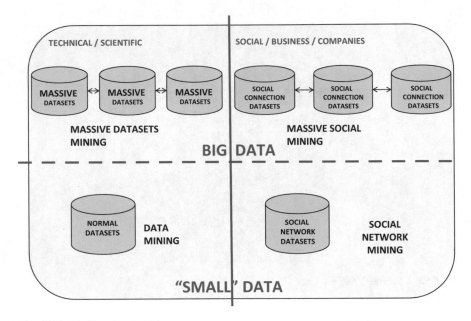

Fig. 6.54 Big Data jungle

- Analyze the context and the social connections of people.
- Analyze what you do.
- Analyze where you go.
- Analyze what you choose.
- Analyze what you buy.
- Analyze in what you spend time.
- Analyze who is influencing you.

Therefore, social mining has changed the paradigm of data mining. As it can be seen in Fig. 6.55, thereare some important differences between *scientific data mining* and *social mining* approach. In scientific data mining, the main issue is to obtain some prototypical knowledge patterns like clusters of data, similar predictive behaviour in similar examples, etc. The final goal is to set-up reasoning techniques about prototypes. It mines individuals' data (data about individuals), and obtain averages, indices, types or prototypes, clusters, etc.

However, in social mining, the main issue is the analysis of social connections, like who is liking what, who is majorly influencing in a social community, etc. The ultimate goal is to set up reasoning techniques about individuals, and not in prototypes like scientific data mining. It is very related to the recent concept in IDSS of *recommender systems* (see Chap. 9, Sect. 9.5), which gives personalized suggestions for a specific individual. *Social mining* not only mines individuals' datasets, but also the *interactions between individuals* through social networks, or their interaction over different related datasets (what have they bought, where they have gone, etc.).

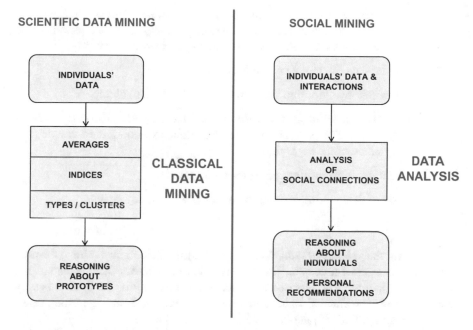

Fig. 6.55 Scientific Data mining vs. Social Mining

Social mining is one of the main launchers of the Big Data scenarios.

One of the main questions in the Big Data scenario regarding the *volume of data* is how to scale from "small data" to Big Data, at least from a technical point of view? (Bekkerman et al., 2012).

Assuming that we cope with a (distributed) high Data volume or/and a possibly slow algorithm to process the high volume of data, there are some different situations:

- High Volume Data does not fit in the computer memory.

    - The data mining algorithm performs slowly due to the volume.

        A possible solution would be to *parallelize the algorithm* and *distribute* its computation, plus in addition *slicing data*. The parallelization must be done exploiting the *intrinsic parallelism* of the algorithm. The slicing of data could be *horizontal* (by blocks or chunks of individuals or examples) or *vertical* (by blocks or chunks of attributes). Each *chunk of data* is distributed to a node of computation. The slicing of data implies that afterwards, the different patterns obtained must be recombined in a fast way.

    - The data mining algorithm performs as expected.

        A possible solution would be to *slicing data*. The slicing of data could be *horizontal* (by blocks or chunks of individuals or examples) or *vertical* (by blocks or chunks of attributes). Each *chunk of data* is distributed to a node of computation. The slicing of data implies that afterwards, the different patterns obtained must be recombined in a fast way.

- High volume data fits in the computer memory.

  - The data mining algorithm performs slowly due to the volume.

    A possible solution is to *parallelize the algorithm* and *distribute* its computation. The parallelization must be done exploiting the *intrinsic parallelism* of the algorithm.

  - The data mining algorithm performs as expected.

    Use the data mining technique/s as usual without any change.

Regarding the *velocity of data*, there is an emerging field in data mining devoted to how to process real-time data. This field is called *Data Stream Mining*, and all the examples of data must be processed online, and just examined one time. It is very related with the incremental Machine Learning techniques. Some interesting references are Gama (2010), Aggarwal (2007).

In the literature, there is a growing jungle of tools to cope with Big Data. Here is a list of current techniques available, but they increase continuously:

- NoSQL databases.

  - MongoDB, CouchDB, Cassandra, Redis, BigTable, Hbase, Hypertable, Voldemort, Riak, ZooKeeper.

- MapReduce.

  - Hadoop, Hive, Pig, Cascading, Cascalog, mrjob, Caffeine, S4, MapR, Acunu, Flume, Kafka, Azkaban, Oozie, Greenplum.

- Storage.

  - S3, Hadoop Distributed File System.

- Servers.

  - EC2, Google App Engine, Elastic, Beanstalk, Heroku.

- Processing.

  - R, Yahoo! Pipes, Mechanical Turk, Solr/Lucene, ElasticSearch, Datameer, BigSheets, Tinkerpop.

References

Aggarwal, C. (2007). *Data streams: Models and algorithms*. Springer.

Agrawal, R., Imielinski, T., & Swami, A. (1993). Mining associations between sets of items in large databases. In *Proceedings of the ACM SIGMOD International Conference on Management of Data*, Washington, DC, May 1993 (pp. 207–216).

Agrawal, R., & Srikant, R. (1994). Fast algorithms for mining association rules in large databases. In *Proceedings of the 20th International Conference on Very Large Data Bases (VLDB 94)*, Santiago, Chile, September 1994 (pp. 487–499).

Aha, D. (1998). Feature weighting for lazy learning algorithms. In H. Liu & H. Motoda (Eds.), *Feature extraction, construction and selection: A data mining perspective*. Kluwer.

Aha, D. W., Kibler, D., & Albert, M. K. (1991). Instance-based learning algorithms. *Machine Learning, 6*, 37–66.

Alexandridis, A., Patrinos, P., Sarimveis, H., & Tsekouras, G. (2005). A two-stage evolutionary algorithm for variable selection in the development of rbf neural network models. *Chemometrics and Intelligent Laboratory Systems, 75*, 149–162.

Alterman, R. (1988). Adaptive planning. *Cognitive Science, 12*, 393–422.

Anderson, T. W., & Darling, D. A. (1954). A test of goodness-of-fit. *Journal of the American Statistical Association, 49*, 765–769. https://doi.org/10.2307/2281537

Ashley, K. D. (1990). *Modelling legal argument: Reasoning with cases and hypotheticals*. The MIT Press.

Bain, W. (1986). Case-based reasoning: A computer model of subjective assessment. Ph. D. Dissertation. Dept. of Computer Science. Yale University, 1986.

Barnett, V., & Lewis, T. (1994). *Outliers in statistical data. Wiley series in probability and mathematical sciences* (3rd ed.). Wiley.

Bastide, Y., Pasquier, N., Taouil, R., Stumme, G., & Lakhal, L. (2000). Mining minimal non-redundant association rules using frequent closed itemsets. In *1st international conference on computational logic (CL 2000)* (pp. 972–986). Springer.

Bekkerman, R., Bilenko, M., & Langford, J. (2012). *Scaling up machine learning: Parallel and distributed approaches*. Cambridge University Press.

Bekkerman, R., El-Yaniv, R., Tishby, N., & Winter, Y. (2003). Distributional word clusters vs. words for text categorization. *Journal of Machine Learning Research, 3*, 1183–1208.

Bezdek, J. C., & Kuncheva, L. I. (2001). Nearest prototype classifier designs: An experimental study. *International Journal of Hybrid Intelligent Systems, 16*(12), 1445–1473.

Blanzieri, E., & Ricci, F. (1999). Probability based metrics for nearest neighbor classification and case-based reasoning. In *Proc. of 3rd international conference on case-based reasoning* (pp. 14–28). Springer.

Box, G. E. P., & Draper, N. R. (1987). *Wiley series in probability and mathematical statistics. Empirical model-building and response surfaces*. Wiley.

Breiman, L. (1996). Bagging predictors. *Machine Learning, 24*(2), 123–140.

Breiman, L. (2001). Random forests. *Machine Learning, 45*(1), 5–32.

Breiman, L., Friedman, J. H., Olshen, R. A., & Stone, C. J. (1984). *Classification and regression trees*. Wadsworth & Brooks/Cole Advanced Books & Software.

Brighton, H., & Mellish, C. (2002). Advances in instance selection for instance-based learning algorithms. *Data Mining & Knowledge Discovery, 6*(2), 153–172.

Brin, S., Motwani, R., Ullman, J. D., & Tsur, S. (1997). Dynamic itemset counting and implication rules for market basket data. In *Proceedings of the ACM SIGMOD International Conference on Management of Data (ACM SIGMOD '97)* (pp. 265–276).

Bruha, I., & Famili, A. (2000). Postprocessing in machine learning and data mining. *ACM SIGKDD Explorations Newsletter, 2*(2), 110–114.

Brun, M., Sima, C., Hua, J., Lowey, J., Carroll, B., Suh, E., & Dougherty, E. R. (2007). Model-based evaluation of clustering validation measures. *Pattern Recognition, 40*(3), 807–824.

Bryson, A. E. (1961). A gradient method for optimizing multi-stage allocation processes. In *Proceedings of the Harvard University Symposium on Digital Computers and Their Applications*.

Bryson, A. E., & Ho, Y. C. (1969). *Applied optimal control*. Blaisdell.

Caises, Y., González, A., Leyva, E., & Pérez, R. (2009). SCIS: Combining instance selection methods to increase their effectiveness over a wide range of domains. In E. Corchado & H. Yin (Eds.), *IDEAL 2009, LNCS 5788* (pp. 17–24). Burgos.

Caliński, T., & Harabasz, J. (1974). A dendrite method for cluster analysis. *Communications in Statistics, 3*(1), 1–27.

Cano, J. R., Herrera, F., & Lozano, M. (2003). Using evolutionary algorithms as instance selection for data reduction in KDD: An experimental study. *IEEE Transactions on Evolutionary Computation, 7*(6), 561–557.

Carbonell, J. G. (1985). *Derivational analogy: A theory of reconstructive problem solving and expertise acquisition*. Computer Science Dept., paper 1534. Carnegie-Mellon University.

Caruana, R., & de Sa, V. (2003). Benefitting from the variables that variable selection discards. *Journal of Machine Learning Research, 3*, 1245–1264.

Cerverón, V., & Ferri, F. J. (2001). Another move toward the minimum consistent subset: A tabu search approach to the condensed nearest neighbour rule. *IEEE Transactions on Systems Man and Cybernetics Part B, 31*(3), 408–413.

Chandrashekar, G., & Sahin, F. (2014). A survey on feature selection methods. *Computers & Electrical Engineering, 40*(1), 16–28.

Chien-Hsing, C., Bo-Han, K., & Fu, C. (2006). The generalized condensed nearest neighbor rule as a data reduction method. In *Proceedings of the 18th International Conference on Pattern Recognition* (pp. 556–559). IEEE Computer Society.

Cognitive Systems. (1992). *ReMind Developer's reference manual* (Vol. 1992).

Cormen, T. H., Leiserson, C. E., Rivest, R. L., & Stein, C. (2009). *Introduction to algorithms* (3rd ed.). MIT Press.

Cover, T. M., & Hart, P. E. (1967). Nearest neighbor pattern classification. *IEEE Transactions on Information Theory, IT-13*(1), 21–27.

Cover, T. M., & Thomas, J. (1991). *Elements of information theory*. Wiley.

Cox, K. C., Eick, S. G., Wills, G. J., & Brachman, R. J. (1997). Visual data mining: recognizing telephone calling fraud. *Data Mining and Knowledge Discovery, 1*(2), 225–231.

Creecy, R. H., Masand, B. M., Smith, S. J., & Waltz, D. L. (1992). Trading MIPS and memory for knowledge engineering. *Communications of the ACM, 35*, 48–64.

Daelemans, W., & van den Bosch, A. (1992). Generalization performance of backpropagation learning on to syllabification task. In *Proceedings of TWLT3: Connectionism natural and language processing* (pp. 27–37).

Dash, M., & Liu, H. (1999). Handling large unsupervised dates via dimensionality reduction. In *SIGMOD Data Mining and Knowledge Discovery Workshop (DMKD)*, Philadelphia, 1999.

Davidson, J. L., & Jalan, J. (2010). Feature selection for steganalysis using the Mahalonobis distance. In: *Proceedings of SPIE 7541, Media Forensics and Security II 7541*.

Defays, D. (1977). An efficient algorithm for a complete-link method. *The Computer Journal. British Computer Society, 20*(4), 364–366.

Devijver, P. A., & Kittler, J. (1980). On the edited nearest neighbour rule. In: *Proceedings of the 5th International Conference on Pattern Recognition*, Los Alamitos, CA (pp. 72–80).

Dua, D., & Graff, C. (2019). UCI machine learning repository. University of California, School of Information and Computer Science. http://archive.ics.uci.edu/ml.

Duval, B., Hao, J.-K., & Hernandez-Hernandez, J. C. (2009). A memetic algorithm for gene selection and molecular classification of a cancer. In *Proceedings of the 11th ACM Annual conference on Genetic and evolutionary computation, GECCO '09*, New York, (pp. 201–208).

Everitt, B. (2011). *Cluster analysis*. Wiley. ISBN 9780470749913.

Fayyad, U., Piatetsky-Shapiro, G., & Smyth, P. (1996). From data mining to knowledge discovery in databases. *AI Magazine, 17*(3), 37–54.

Fix, E., & Hodges Jr., J. L. (1951). *Discriminatory analysis, nonparametric discrimination*. USAF School of Aviation Medicine, Randolph Field, Tex., Project 21-49-004, Rept. 4, Contract AF41 (128)-31, February 1951.

Freund, Y. (1990). Boosting a weak learning algorithm by majority, information and computation 121, no. 2 (September 1995), 256–285; an extended abstract appeared in *Proceedings of the Third Annual Workshop on Computational Learning Theory*.

Freund, Y. (1995). Boosting a weak learning algorithm by majority. *Information and Computation, 121*(2), 256–285.

Freund, Y., & Schapire, R. (1997). A decision-theoretic generalization of online learning and application to boosting. *Journal of Computer and System Sciences, 55*(1), 119.

Friedman, J. H., Bentley, J. L., & Finkel, R. A. (1997). An algorithm for finding best matches in logarithmic expected time. *ACM Transactions on Mathematical Software, 3*(3), 209–226.

Gama, J. (2010). *Knowledge discovery from data streams*. Chapman and Hall/CRC.

Garain, U. (2008). Prototype reduction using an artificial immune model. *Pattern Analysis and Applications, 11*, 353–363.

García, S., Cano, J. R., & Herrera, F. (2008). A memetic algorithm for evolutionary prototype selection: A scaling up approach. *Pattern Recognition, 41*, 2693–2709.

Gibert, K., Sànchez-Marrè, M., & Izquierdo, J. (2016). A survey on pre-processing techniques: Relevant issues in the context of environmental data mining. *AI Communications, 29*(6), 627–663.

Gini, C. (1909). Concentration and dependency ratios (in Italian). *English translation from Rivista di Politica Economica, 87*(769–789), 1997.

Gini, C. (1912). *Variabilità e mutabilità: contributo allo studio delle distribuzioni e delle relazioni statistiche (Variability and Mutability)*. Bologna: P. Cuppini.

Goel, A., & Chandrasekaran, B. (1992). Case-based design: A task analysis. In C. Tong & D. Sriram (Eds.), *Artificial intelligences approaches to engineering design* (Vol. 2.: Innovative design). Academic Press.

Goldberg, D. E. (1989). *Genetic algorithms in search, optimization and machine learning*. Addison-Wesley.

Golobardes, E., Llora, X., & Garrell, J. M. (2000). Genetic Classifier System as a heuristic weighting method for a Case-Based Classifier System. In *Proceedings of the 3rd Catalan Conference on Artificial Intelligence (CCIA2000)*.

Gower, J. C. (1971). A general coefficient of similarity and some of its properties. *Biometrics, 27*, 857–874.

Güvenir, H. A., & Akkus, A. (1997). Weighted K nearest neighbor classification on feature projections. In S. Kuru, M. U. Caglayan, & H. L. Akin (Eds.), (Oct. 27-29, 1997) *Proceedings of the twelfth international symposium on computer and information sciences (ISCIS XII)* (pp. 44–51).

Halkidi, M., Batistakis, Y., & Vazirgiannis, M. (2001). On clustering validation techniques. *Journal of Intelligent Information Systems, 17*(2), 107–145.

Hall, M. A. (1999). *Feature selection for discrete and numeric class machine learning*. Technical Report, Department of Computer Science, University of Waikato, Working Paper 99/4.

Hall, M. A., & Smith, L. A. (1998). Practical feature subset selection for machine learning. In *Proceeding of 21st Australian Computer Science Conference* (pp. 181–191). Springer.

Hämäläinen, W. (2010). Efficient discovery of the top-k optimal dependency rules with the Fisher's exact test of significance. In *Proceedings of the 10th IEEE International Conference on Data Mining* (pp. 196–205).

Hamerly, G., & Elkan, C. (2003). *Proceedings of the 17th Annual Conference on Neural Information Processing Systems (NIPS'2003)*, December 2003 (pp. 281–288).

Hammond, K. (1989). *Case-based planning: Viewing planning as a memory task*. Academic Press.

Han, J., Pei, J., & Yin, Y. (2000). Mining frequent patterns without candidate Generation. In *Proceedings of ACM-SIGMOD International Conference on management of Data (SIGMOD'00)*, Dallas (pp. 1–12).

Han, J., Pei, J., Yin, Y., & Mao, R. (2004). Mining frequent patterns without candidate generation: A frequent-pattern tree approach. *Data Mining and Knowledge Discovery, 8*, 53–87.

Hart, P. E. (1968). The condensed nearest neighbor rule. *IEEE Transactions on Information Theory, 14*, 515–516.

Haykin, S. (1999). *Neural networks: A comprehensive foundation* (2nd ed.). Prentice-Hall.

Hennessy, D., & Hinkle, D. (1992). Applying case-based reasoning to autoclave loading. *IEEE Expert, 7*(5), 21–26.

Hennig, C., & Liao, T. F. (2010). *Comparing latent class and dissimilarity based clustering for mixed type variables with application to social stratification*. Technical report.

Ho, T. K. (1995). Random decision forests. In *Proceedings of 3rd IEEE International Conference on Document Analysis and Recognition* (Vol. 1, pp. 278–282).

Ho, T. K. (1998). The random subspace method for constructing decision forests. *IEEE Transactions on Pattern Analysis and Machine Intelligence, 20*(8), 832–844.

Holland, J. H. (1975). *Adaptation in natural and artificial systems*. University of Michigan Press. New edition, MIT Press.

Howe, N., & Cardie, C. (2000). Feature subset selection and order identification for unsupervised learning. In *Proceedings of 17th International Conference on Machine Learning*. Morgan Kaufmann.

Hinrichs, T. R. (1992). *Problem solving in open worlds: A case study in design*. Lawrence Erlbaum.

Huh, M. Y. (2006). Subset selection algorithm based on mutual information. In A. Rizzi & M. Vichi (Eds.), *Compstat 2006 - proceedings in computational statistics*. Physica-Verlag HD.

Hurwitz, J. S., Nugent, A., Halper, F., & Kaufman, M. (2013). *Big data for dummies*. John Wiley & Sons.

Ishii, N., & Wang, Y. (1998). Learning feature weights for similarity using genetic algorithms. In *Proceedings of IEEE International Joint Symposia on Intelligence and Systems* (pp. 27–33). IEEE.

Jacobs, R. A. (1988). Increased rates of convergence through learning rate adaptation. *Neural Networks, 1*, 295–307.

Jain, S. K., & Dubes, R. C. (1988). *Algorithms for clustering data*. Prentice-Hall.

Johnson, S. C. (1967). Hierarchical clustering schemes. *Psychometrika, 2*, 241–254.

Jouan-Rimbaud, D., Massart, D. L., Leardi, R., & Noord, O. E. D. (1995). Genetic algorithms as a tool for wavenumber selection in multivariate calibration. *Analytical Chemistry, 67*(23), 4295–4301.

Kass, A. M., & Leake, D. B. (1988). Case-based reasoning applied to constructing explanations. In *Proc. of DARPA workshop on case-based reasoning* (pp. 190–208).

Kaufman, L., & Rousseeuw, P. J. (1990). *Finding groups in Data - an introduction to cluster analysis. A Wiley-science publication*. Wiley.

Kelley, H. J. (1960). Gradient theory of optimal flight paths. *ARS Journal, 30*(10), 947–954.

Kira, K., & Rendell, L. (1992). A practical approach to feature selection. In *Proceedings of the 9th International Conference on Machine Learning* (pp. 249–256). Morgan Kaufmann.

Kohavi, R., & John, G. H. (1997). Wrappers for feature subset selection. *Artificial Intelligence, 97*, 273–324.

Kohavi, R., & John, G.-H. (1998). The wrapper approach. In H. Liu & H. Motoda (Eds.), *Feature selection for knowledge discovery and data mining* (pp. 33–50). Kluwer Academic.

Kohavi, R., Langley, P., & Yun, Y. (1997). The utility of feature weighting in nearest-neighbor algorithms. In *Proceedings of the European Conference on Machine Learning (ECML97)*.

Kohonen, T. (1989). *Self-organization and associative memory* (3rd edn). Springer Series in Information Sciences. Springer.

Koller, D., & Sahami, M. (1996). Towards optimal feature selection. *ICML, 96*, 284–292.

Kolodner, J. L. (1993). *Case-based reasoning*. Morgan Kaufmann.

Kolodner, J. L., & Simpson, R. L. (1989). The MEDIATOR: Analysis of an early case-based problem solver. *Cognitive Science, 13*(4), 507–549.

Kolodner, J. L. (1985). Memory for experience. In G. Bower (Ed.), *The psychology of learning and motivation* (Vol. 19). Academic Press.

Kolodner, J. L. (1983). Reconstructive memory: A computer model. *Cognitive Science, 7*(4), 281–328.

Kononenko, I. (1994). Estimating attributes: Analysis and extensions of RELIEF. In *Proceedings of European Conference on Machine Learning (ECML 1994)* (pp. 171–182). Springer.

Koton, P. (1989). *Using experience in learning and problem solving.* Ph. D. dissertation. Dept. of Computer Science.

Koza, J. R. (1992). *Genetic programming: On the programming of the computers by means of natural Selection.* MIT Press.

Kuncheva, L. I. (1997). Fitness functions in editing k-NN referent set by genetic algorithms. *Pattern Recognition, 30,* 1041–1049.

Lance, G. N., & Williams, W. T. (1966). Computer programs for hierarchical polythetic classification ("similarity analyses"). *Computer Journal, 9,* 60–64.

Langley, P., & Iba, W. (1993). Average-case analysis of a nearest neighbor algorithm. In *Proceedings of the thirteenth international joint conference on artificial intelligence* (pp. 889–894). Chambery.

Lazar, C., Taminau, J., Meganck, S., Steenhoff, D., Coletta, A., Molter, C., et al. (2012). A survey on filter techniques for feature selection in gene expression microarray analysis. *IEEE/ACM Transactions on Computational Biology and Bioinformatics, 9,* 1106.

Leake, D. B., Kinley, A., & Wilson, D. (1997). Case-based similarity assessment: Estimating adaptability from experience. In *Proc. of American Association of Artificial Intelligence (AAAI-97)* (pp. 674–679).

Leavitt, N. (2010). Will NoSQL databases live up to their promise? *IEEE Computer,* pp. 12–14.

Levenshtein, V. I. (1966). Binary codes capable of correcting deletions, insertions, and reversals. *Soviet Physics Doklady, 10*(8):707–10. English version of a previously published article in 1965 (in Russian).

Liao, T. W., Zhang, Z., & Mount, C. R. (1998). Similarity measures for retrieval in case-based reasoning systems. *Applied Artificial Intelligence, 12*(4), 267–288.

Little, R. J., & Rubin, D. B. (2014). *Statistical analysis with missing data.* John Wiley & Sons.

Liu, H., & Motoda, H. (1998). *Feature selection for knowledge discovery and data mining.* Kluwer Academic.

Lu, S.-Y., & King Sun, F. (1978). A sentence-to-sentence clustering procedure for pattern analysis. *IEEE Transactions on Systems, Man, and Cybernetics, 8*(5), 381–389.

Lumini, A., & Nanni, L. (2006). A clustering method for automatic biometric template selection. *Pattern Recognition, 39,* 495–497.

MacNaughton-Smith, P., Williams, W., Dale, M., & Mockett, L. (1965). Dissimilarity analysis: A new technique of hierarchical subdivision. *Nature, 202,* 1034–1035.

MacQueen, J. B. (1967). Some methods for classification and analysis of multivariate observations. In *Proceedings of 5th Berkeley Symposium on Mathematical Statistics and Probability* (Vol. 1, pp. 281–297). University of California Press.

Marz, N., & Warren, J. (2015). *Big Data.* Manning Publications.

McCulloch, W. S., & Pitts, W. (1943). A logical calculus of the ideas immanen in nervous activity. *Bulletin of Mathematic Biophysics, 5,* 115–137.

Meilă, M. (2007). Comparing clusterings? An information-based distance. *Journal of Multivariate Analysis, 98*(5), 873–895.

Mierswa, I., Wurst, M., Klinkenberg, R., Scholz, M., & Euler, T. (2006). Yale: Rapid prototyping for complex data mining tasks. In *Proceedings of the 12th ACM SIGKDD International Conference on Knowledge Discovery and Data Mining* (pp. 935–940). ACM.

Minsky, M. L., & Papert, S. A. (1969). *Perceptrons.* The MIT Press.

Mitchell, T. M. (1982). Generalization as search. *Artificial Intelligence, 18*(2), 203–226.

Mohri, T., & Tanaka, H. (1994). An optimal weighting criterion of case indexing for both numeric and symbolic attributes. In *Workshop on Case-Based Reasoning.* AAAI Press.

Molina, L., Belanche, L., & Nebot, A. (2002). Feature selection algorithms: A survey and experimental evaluation. In *ICDM 2002: Proceedings of the IEEE international conference on Data Mining* (pp. 306–313).

Mollineda, R. A., Ferri, F. J., & Vidal, E. (2002). An efficient prototype merging strategy for the condensed 1-NN rule through class-conditional hierarchical clustering. *Pattern Recognition, 35*, 2771–2782.

Nakariyakul, S., & Casasent, D. P. (2009). An improvement on floating search algorithms for feature subset selection. *Pattern Recognition, 42*, 1932–1940.

Narayan, B. L., Murthy, C. A., & Pal, S. K. (2006). Maxdiff kd-trees for data condensation. *Pattern Recognition Letters, 27*, 187–200.

Narendra, P., & Fukunaga, K. (1977). A branch and bound algorithm for feature subset selection. *IEEE Transactions on Computers, 26*(9), 917–922.

Núñez, H., Sànchez-Marrè, M., Cortés, U., Comas, J., Rodríguez-Roda, I., & Poch, M. (2002). Feature weighting techniques for prediction tasks in environmental processes. In *Proc. of 3rd ECAI'2002 workshop on binding environmental sciences and artificial intelligence (BESAI'2002).* pp. 4:1-4:9.

Núñez, H. (2004). *Feature weighting in plain case-based reasoning*. Ph.D. Thesis, Doctoral Program on Artificial Intelligence, Universitat Politècnica de Catalunya.

Núñez, H., & Sànchez-Marrè, M. (2004). Instance-based learning techniques of unsupervised feature weighting do not perform so badly! In *Proceedings of 16th European Conference on Artificial Intelligence (ECAI 2004)* (pp. 102–106). IOS Press.

Núñez, H., Sànchez-Marrè, M., & Cortés, U. (2003). Improving similarity assessment with entropy-based local weighting. In *Proceedings of the 5th International Conference on Case-Based Reasoning (ICCBR2003). Lecture Notes in Artificial Intelligence, (LNAI-2689)* (pp. 377–391). Springer.

Núñez, H., Sànchez-Marrè, M., Cortés, U., Comas, J., Martínez, M., Rodríguez-Roda, I., & Poch, M. (2004). A comparative study on the use of similarity measures in case-based reasoning to improve the classification of environmental system situations. *Environmental Modelling & Software, 19*(9), 809–819.

Olvera-López, J. A., Carrasco-Ochoa, J. A., & Martínez-Trinidad, J. F. (2007). Object selection based on clustering and border objects. In *Computer recognition systems 2, ASC 45*, Wroclaw, Poland (pp. 27–34).

Olvera-López, J. A., Carrasco-Ochoa, J. A., & Martínez-Trinidad, J. F. (2008). Prototype selection via prototype relevance. In *Proceedings of CIARP 2008, LNCS5197*, Habana, Cuba (pp. 153–160).

Osborne, H. R., & Bridge, D. G. (1998). A case base similarity framework. In *Proc. of 4th European Workshop on Case-Based Reasoning (EWCBR'98)* (pp. 309–323).

Paredes, R., & Vidal, E. (2000). Weighting prototypes. A new editing approach. In *Proceedings of the International Conference on Pattern Recognition ICPR* (Vol. 2, pp. 25–28).

Pawlowsky-Glahn, V., & Buccianti, A. (2011). *Compositional data analysis: Theory and applications*. Wiley.

Phuong, T. M., Lin, Z., & Altman, R. B. (2005). Choosing SNPs using feature selection. In *Proceedings/IEEE Computational Systems Bioinformatics Conference, CSB* (pp. 301–309).

Piatetsky-Shapiro, G.. Discovery, analysis, and presentation of strong rules. Knowledge discovery in databases, MIT Press pp. 229–248, 1991.

Pietracaprina, A., Riondato, M., Upfal, E., & Vandin, F. (2010). Mining top-k frequent itemsets through progressive sampling. *Data Mining and Knowledge Discovery, 21*(2), 310–326.

Puch, W., Goodman, E., Pei, M., Chia-Shun, L., Hovland, P., & Enbody, R. (1993). Further research on feature selection and classification using genetic algorithm. In *International conference on genetic algorithm* (pp. 557–564).

Pudil, P., Novovicova, J., & Kittler, J. (1994). Floating search methods in feature selection. *Pattern Recognition Letters, 15*, 1119–1125.

Pudil, P., Novovicova, J., Kittler, J., & Paclik, P. (1999). Adaptive floating search methods in feature selection. *Pattern Recognition Letters, 20*, 1157–1163.

Pyle, D. (1999). *Data preparation for data mining. The Morgan Kaufmann series in data management systems*. Morgan Kaufmann.

Quinlan, J. R. (1983). Learning efficient classification procedures and their application to chess end games. In R. S. Michalski, J. G. Carbonell, & T. M. Mitchell (Eds.), *Machine learning: An artificial intelligence approach* (pp. 463–482). Tioga/Morgan Kaufmann.

Quinlan, J. R. (1986). Induction of decision trees. In *Machine learning* (Vol. 1, pp. 81–106). Kluwer Academic.

Quinlan, J. R. (1988). Induction, knowledge and expert systems. In J. S. Gero & R. Stanton (Eds.), *Artificial intelligence developments and applications* (pp. 253–271). Elsevier.

Quinlan, J. R. (1993). *C4.5: Programs for machine learning*. Morgan Kaufmann.

Raicharoen, T., & Lursinsap, C. (2005). A divide-and-conquer approach to the pairwise opposite class-nearest neighbour (POC-NN) algorithm. *Pattern Recognition Letters, 26*(10), 1554–1567.

Redmond, M. A. (1992). Learning by observing and understanding expert problem solving. Georgia Institute of Technology. College of Computing. Technical report GIT-CC-92/43, 1992.

Reunanen, J. (2003). Overfitting in making comparisons between variable selection methods. *Journal of Machine Learning Research, 3*, 1371–1382.

Richter, M. M., & Weber, R. O. (2013). *Case-based reasoning: A textbook*. Springer.

Riesbeck, C. K., & Schank, R. C. (1989). *Inside case-based reasoning*. Lawrence Erlbaum Associates Publishers.

Ritter, G. L., Woodruff, H. B., Lowry, S. R., & Isenhour, T. L. (1975). An algorithm for a selective nearest neighbor decision rule. *IEEE Transactions on Information Theory, 21*(6), 665–669.

Riquelme, J. C., Aguilar-Ruíz, J. S., & Toro, M. (2003). Finding representative patterns with ordered projections. *Pattern Recognition, 36*, 1009–1018.

Roffo, G., Melzi, S., & Cristani, M. (2015). Infinite feature selection. In *International Conference on Computer Vision*. http://www.cv-foundation.org.

Rokach, L., & Maimon, O. (2005). Clustering methods. In *Data mining and knowledge discovery handbook* (pp. 321–352). Springer.

Rosenblatt, F. (1958). The perceptron: A probabilistic model for information storage and organization in the brain. *Psychological Review, 65*, 386–408.

Rosenblatt, F. (1960). Perceptron simulation experiments. *Proceedings of the Institute of Radio Engineers, 48*, 301–309.

Rumelhart, D. E., Hinton, G. E., & Williams, R. J. (1986). Learning representations by back-propagating errors. *Nature (London), 323*, 533–536.

Sacerdoti, E. D. (1977). *A structure for plans and behavior*. North-Holland.

Sànchez-Marrè, M., Cortés, U., Roda, I. R., Poch, M., & Lafuente, J. (1997). Learning and adaptation in WWTP through case-based reasoning. *Microcomputers in Civil Engineering, 12*(4), 251–266.

Sànchez-Marrè, M., Cortés, U., R-Roda, I., & Poch, M. (1998). L'Eixample distance: A new similarity measure for case retrieval. In *Proceedings of 1st Catalan Conference on Artificial Intelligence (CCIA'98)*. ACIA Bulletin 14–15 (pp. 246–253).

Schank, R. C. (1982). *Dynamic memory: A theory of learning in computers and people*. Cambridge University Press.

Sevilla-Villanueva, B., Gibert, K., & Sànchez-Marrè, M. (2016). Using CVI for understanding class topology in unsupervised scenarios. In *Proceedings of 17th Conference of the Spanish Association for Artificial Intelligence (CAEPIA 2016)*. Lecture Notes in Artificial Intelligence (Vol. 9868, pp. 135–149). Springer.

Shannon, C. E. (1948). A mathematical theory of communication. *Bell System Technical Journal, 27*(379–423), 623–656.

Shepherd, G. M., & Koch, C. (1990). Introduction to synaptic circuits. In G. M. Shepherd (Ed.), *The synaptic organisation of the brain* (pp. 3–31). Oxford University Press.

Shinn, H. S. (1988). Abstractional analogy: A model of analogical reasoning. In *Proc. of DARPA workshop on case-based reasoning* (pp. 370–387).

Shiu, S. C. K., Yeung, D. S., Sun, C. H., & Wang, X. Z. (2001). Transferring case knowledge to adaptation knowledge: An approach for case-base maintenance. *Computational Intelligence, 17*(2), 295–314.

Short, R. D., & Fukunaga, K. (1981). The optimal distance measure for nearest neighbour classification. *IEEE Transactions on Information Theory, 27*, 622–627.

Sibson, R. (1973). SLINK: An optimally efficient algorithm for the single-link cluster method. *The Computer Journal. British Computer Society, 16*(1), 30–34. https://doi.org/10.1093/comjnl/16.1.30

Spillmann, B., Neuhaus, M., Bunke, H. P., Ekalska, E., & Duin, R. P. W. (2006). Transforming strings to vector spaces using prototype selection. In: D.-Y. Yeung, et al. (Eds.), *SSPR&SPR 2006, LNCS 4109*, Hong-Kong (pp. 287–296).

Stanfill, C., & Waltz, D. (1986). Toward memory-based reasoning. *Communications of the ACM, 29*(12), 1212–1228.

Stearns, S. (1976). On selecting features for pattern classifiers. In: *Proceedings of the 3rd International Conference on Pattern Recognition* (pp. 71–75).

Steinbach, M., Karypis, G., & Kumar, V. (2000). A comparison of document clustering techniques. In *Proceedings of KDD Workshop on Text Mining* (Vol. 400, No. 1, pp. 525–526).

Sussman, G. J. (1975). *A computer model of skill acquisition*. American Elsevier.

Sun, Y., Babbs, C., & Delp, E. (2005). A comparison of feature selection methods for the detection of breast cancers in mammograms: Adaptive sequential floating search vs. genetic algorithm. *Conference Proceedings: Annual International Conference of the IEEE Engineering in Medicine and Biology Society, 6*, 6532–6535.

Sycara, K. (1987). Finding creative solutions in adversarial impasses. In *Proc. of 9th annual conference of the cognitive science society*.

Tomek, I. (1976). An experiment with the edited nearest-neighbor rule. *IEEE Transactions on Systems, Man, and Cybernetics, 6-6*, 448–452.

Tukey, J. W. (1977). *Exploratory data analysis*. Addison-Wesley.

U.S. Department of Commerce, Bureau of the Census. (1977). *Statistical abstract of the United States, and County and City Data Book*.

Veloso, M. M., & Carbonell, J. G. (1993). Derivational analogy in PRODIGY: Automating case acquisition, storage and utilization. *Machine Learning, 10*(3), 249–278.

Venmann, C. J., & Reinders, M. J. T. (2005). The nearest sub-class classifier: A compromise between the nearest mean and nearest neighbor classifier. *IEEE Transactions on Pattern Analysis and Machine Intelligence, 27*(9), 1417–1429.

Ward, J. H. (1963). Hierarchical grouping to optimize an objective function. *Journal of the American Statistical Association, 58*(301), 236–244.

Webb, G. I. (2006). Discovering significant rules. In *Proceedings of the 12th ACM SIGKDD International Conference on Knowledge Discovery and Data Mining (KDD-2006)* (pp. 434–443).

Webb, G. I. (2007). Discovering significant patterns. *Machine Learning* (pp. 1–33).

Webb, G. I. (2011). Filtered-top-k association discovery. *Data Mining and Knowledge Discovery, 1*(3), 183–192.

Wettschereck, D., Aha, D. W., & Mohri, T. (1997). A review and empirical evaluation of feature weighting methods for a class of lazy learning algorithms. *Artificial Intelligence Review*, Special Issue on lazy learning Algorithms.

Wilson, D. L. (1972). Asymptotic properties of nearest neighbor rules using edited data. *IEEE Transactions on Systems, Man, and Cybernetics, 2*, 408–421.

Wilson, D. R., & Martínez, T. R. (1997). Improved heterogeneous distance functions. *Journal of Artificial Intelligence Research, 6*, 1–34.

Wilson, D. R., & Martínez, T. R. (2000). Reduction techniques for instance-based learning algorithms. *Machine Learning, 38*, 257–286.

Yang, J., & Honavar, V. (1998). Feature subset selection using a genetic algorithm. *IEEE Intelligent Systems, 13*(2), 44–49.

Zaki, M. J. (2000). Scalable algorithms for association mining. *IEEE Transactions on Knowledge and Data Engineering, 12*(3), 372–390.

Zaki, M. J. (2004). Mining non-redundant association rules. *Data Mining and Knowledge Discovery, 9*, 223–248.

Zaki, M. J., Parthasarathy, S., Ogihara, M., & Li, W. (1997). Parallel algorithms for discovery of association rules. *Data Mining and Knowledge Discovery, 1*, 343–373.

Zhang, H., & Sun, G. (2002). Optimal reference subset selection for nearest neighbor classification by tabu search. *Pattern Recognition, 35*, 1481–1490.

Further Reading

Bekkerman, R., Bilenko, M., & Langford, J. (Eds.). (2011). *Scaling up machine learning: Parallel and distributed approaches* (1st ed.). Cambridge University Press.

Curwin, J., & Slater, R. (2001). Quantitative methods for business decisions. *Thomson Business Press, 2001.*

Gama, J. (2010). *Knowledge discovery from Data streams*. Chapman and Hall/CRC.

Hurwitz, J. S., Nugent, A., Halper, F., & Kaufman, M. (2013). *Big Data for dummies. For dummies.* Wiley.

Leskovec, J., Rajaraman, A., & Ullman, J. D. (2014). *Mining of massive datasets* (2nd ed.). Cambridge University Press.

Marz, N., & Warren, J. (2015). *Big data*. Manning Publications.

Shannon, C. E. (1948). A mathematical theory of communication. *Bell System Technical Journal, 27*, 379–423, 623–656.

Chapter 7
The Use of Intelligent Models in Decision Support

Miquel Sànchez-Marrè
Contributor: Franz Wotawa

Along with the chapter, several case studies will be analyzed to illustrate the application of intelligent methods in the deployment of IDSSs. We will follow the methodologies for the deployment of IDSSs proposed in previous chapters of this book. Most of the case studies are based on true stories about some companies and organizations. The names of the companies and other details have been changed to preserve confidentiality issues when required, and sometimes adapted to better illustrate fundamental aspects of the deployment of IDSSs.

7.1 Using Model-Driven Methods in IDSS

In this section, several case studies will be described to illustrate the use of some *model-driven methods* in the development of IDSSs.

7.1.1 The Use of Agent-Based Simulation Models

This case study is based on the development and results of a real research project in an MSc. Thesis by Thania Rendón-Sallard (Rendón-Sallard, 2009; Rendón-Sallard & Sànchez-Marrè, 2007) undertaken in the Knowledge Engineering and Machine

---

The original version of this chapter was revised. The correction to this chapter is available at https://doi.org/10.1007/978-3-030-87790-3_11

Prof. Franz Wotawa has written the Sect. 7.1.3 about The Use of Model-Based Reasoning Techniques.

F. Wotawa
Institute for Software Technology, Technische Universität Graz, Graz, Austria
e-mail: wotawa@ist.tugraz.at

© Springer Nature Switzerland AG 2022, Corrected Publication 2023
M. Sànchez-Marrè, *Intelligent Decision Support Systems*,
https://doi.org/10.1007/978-3-030-87790-3_7

Learning Group (KEMLG), from Universitat Politècnica de Catalunya (UPC),[1] regarding the deployment of a Multi-Agent System framework to support the decision-making in complex real-world domains. This research was done in cooperation with the Chemical and Environmental Engineering Laboratory (LEQUIA), from the University of Girona (UdG),[2] and the public entity "Consorci Besòs Tordera" (CBT).[3] Later partners were previously collaborating in a research project entitled Besòs, for the development of an Environmental DSS (EDSS) for the management of the hydraulic infrastructure to preserve the water quality in the *Besòs* Catchment (Devesa, 2006; Devesa et al., 2005). In this previous project, the DSS was built up as the integration of three simulation systems for each one of the subsystems: InfoWorks© CS[4] (Collection System) for simulating the sewer systems, GPS-X™[5] for simulating the WWTPs, and InfoWorks© RS[6] (River Simulation) for the water river quality analysis.

Case Study: River Basin Management System

The Scenario

As described in (Rendón-Sallard, 2009), the complexity of environmental problems makes the management of environmental systems especially difficult to be undertaken by traditional software systems. Particularly, *river catchment or basin systems* are very difficult to manage in order to achieve good quality and quantity of water at the river.

River catchments are important social, economic, and environmental units. They sustain ecosystems, which are the main source of water for households, agriculture, and industry. Therefore, the protection of all surface waters and ground waters must be assured in their quality and quantity. The best way to fulfil these requirements is with a *management system at catchment scale* that integrates all the water systems involved: the sewer systems, the Wastewater Treatment Plants (WWTPs) and the river) (Devesa, 2006).

(continued)

---

[1] The Knowledge Engineering and Machine Learning Group (KEMLG, https://kemlg.upc.edu), is a research group belonging to the Research Centre on Intelligent Data Science and Artificial Intelligence (IDEAI-UPC, https//ideai.upc.edu) devoted to the analysis, design, implementation and application of Artificial Intelligence systems to support real-world complex domains.

[2] The Chemical and Environmental Engineering Laboratory (LEQUIA, http://www.lequia.udg.edu/), is a research group of the University of Girona (UdG) devoted to the development of eco-innovative environmental solutions. LEQUIA is a part of the UdG's Institute of the Environment (IMA).

[3] The "Consorci Besòs Tordera" (CBT, http://www.besos-tordera.cat/) is a public consortium entity which joins 64 municipalities and 4 other public bodies for the integrated water management in the catchments of Besòs and Tordera rivers, in Catalonia.

[4] InfoWorks© CS was a software for sewer systems simulation, superseded by InfoWorks© ICM SE software developed by Innovvyze® (https://www.innovyze.com).

[5] GPS-X™ is a dynamic wastewater treatment plant simulator software developed by Hydromantis (https://www.hydromantis.com/GPSX.html).

[6] InfoWorks© RS was a software for river system simulation, integrated in InfoWorks© ICM software developed by Innovvyze® (https://www.innovyze.com).

The management of river basins involves many interactions between physical, chemical, and biological processes. Therefore, a river basin management system is a very complex system. Common problematic features found in the river basin domain are intrinsic instability, uncertainty and imprecision of data or approximate knowledge and vagueness, huge quantity of data, heterogeneity and different time scales, just to outline some of them. See (Poch et al., 2004) or (Cortés et al., 2000) for a broader analysis.

The main goal of the research was to build an IDSS to help in the decision-making for the river basin management, with the ultimate goal to have a good water quality in the river, as the receiving environment.

Problem Analysis

The case study was focussed on a region located in the Besòs river basin. The Besòs basin is located on the North East of the Mediterranean coast of the Iberian Peninsula. The catchment area is one of the most populated catchments in Catalonia, having more than two million people connected. The scope of the study area is around the final reaches of the Congost River. The river sustains, in an area of 70 km², the discharges of four towns which are connected to two Waste Water Treatment Plants (WWTPs) (Devesa, 2006). For the development of an IDSS prototype two sanitation systems were taken into account, La Garriga and Granollers, with their respective sewage systems and Waste Water Treatment Plants, and one section of the Congost river, an effluent of the Besòs river, as a receptor environment for their waste water.

Hydraulic infrastructures for sanitation have traditionally been managed separately, taking into account only the characteristics of the water at the entry and exit points of each installation.

The current tendency is to treat the hydrographic basin as a single area of operations within which hydraulic infrastructures have to be managed in an integrated manner, bearing in mind the condition of the receptor environment.

The water system has three key elements which are depicted in Fig. 7.1: sewer systems, WWTPs, and the river.

The main elements of the environmental system are the following ones:

- *Sewer System.* There are two sewer systems, one that drains the area of the town La Garriga and another one that drains the area of Granollers and some small surrounding villages.
- *WWTPs.* There are two WWTPs, one for each sewer system. Both plants have a biological treatment. The average flows are 6000 m³/d for La Garriga (WWTP1) and 26,000 m³/d for Granollers (WWTP2).
- *River.* The studied reach of the Congost River has a length of 17 km. The Congost is a typical Mediterranean river with seasonal flow variations. Before the two WWTP, the average flow is about 0.5 m³/s, but can easily reach a punctual maximum flow of 200 m³/s.

(continued)

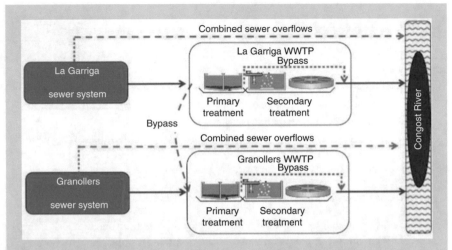

Fig. 7.1   Water System scenario. (Extracted from (Rendón-Sallard, 2009))

- *Industry.* There are 24 industries connected to Granollers system and 4 connected to La Garriga. These industries have a very high rate of potential pollution.
- *Sewer channel.* Joins the two WWTPs, allowing to bypass the flow from the La Garriga-WWTP to the Granollers-WWTP when required.

Other considered elements are rain control stations, river water quality control stations, flow retention, and storage tanks. An essential element of this system is the sewer channel joining both WWTPs, which allows to bypass the flow from the La Garriga-WWTP to the Granollers-WWTP.

Decisions involved

As described above the management of a river basin is complex because there are several subsystems integrated into it with direct interrelationships. The daily management of this system involves several decisions to be made like the following ones:

- How to manage critical episodes that can happen related to overloading situations, storm situations, and so on?
- How to minimize the discharge of poorly treated wastewater, in some situations when not all the inflow can be treated within the WWTPs?
- How to maximize the use of the installation's treatment capacity, in order to get a maximum water quality outflowing the WWTPs?
- How to minimize the economic costs of new investments and daily management, like for example where to build new storage tanks and which capacity they should have?
- How to maintain a minimum flow in the river guaranteeing an acceptable ecological state?

(continued)

Requirements

Hence, the requirements of the IDSS should be to help in the decision-making processes providing a system able to:

- Model the information about the complex domain.
- Evaluating the consequences of critical processes in the decision-making.
- Supervise the processes taking place at the domain.
- To be able to simulate and predict the evolution of the system.

And more concretely, the IDSS system should be able to provide end users with valuable information and knowledge on whether or how to perform several tasks, like the following ones:

- By-passing the water flow from La Garriga WWTP to Granollers WWTP.
- The management of the storage tanks (open/close hatches).
- The control of the sewer systems.
- The monitoring of the river basin system.

Data availability

Before designing an IDSS always is interesting to check whether there are data available or not to get some data-driven model. In our case, there was some data available about the water flow and the water quality in each subsystem (sewer systems, WWTPs and the river system). However, there was not enough data for inducing good data-driven models.

Expert or First-principles knowledge availability

There was a general understanding among the stakeholders that there was valuable knowledge from experts who manage day by day these kinds of situations that could be used. Thus the major models envisioned were *expert-based models* describing how to cope with different situations.

Moreover, the possibility of having some *first-principles knowledge* coming from some theoretical principles should be analyzed too. In this sense, some mathematical water quality models can be used to simulate the behaviours of the different components of the river basin system.

A Possible Solution

After the analysis of the scenario and the concrete problem to be solved, i.e. the river basin management, it became clear that a good solution could be the use of *multi-agent systems* (MAS) for the *simulation* of this scenario, and also the use of *rule-based reasoning* for exploiting the *expert-based models* that can be obtained from the experts on the system. These models could lead to the agent-based simulation system.

MAS is able to cope with the intricacy (*e.g,* uncertainty, approximate knowledge) related to the decision-making processes of complex real-world domains by integrating several agents that model real situations and work collaboratively for achieving the system's goals.

(continued)

Type of IDSS

Therefore, the IDSS will be mainly based on an *agent-based simulation system*, which will use *rule-based reasoning* as a latent reasoning mechanism.

Kind of tasks involved (Analysis, Synthesis, Prognosis)

The main task of the IDSS will be a *prognosis task* because the simulation of future consequences of possible scenarios will be the main goal of the IDSS. This simulation of the system will allow the stakeholders of the river basin management to make the most accurate decisions on how to act to tackle different possible situations.

Model selection

The simulation process will be performed with an *agent-based simulation model*. Following the methodology proposed in Sect. 5.2.2.1, after analyzing the system and identifying the main entities, we proceed to identify the different types of agents.

Next, the type of agents defined for the Agent-based simulation model is described:

- *Sewer Agents*: La Garriga Sewer system, Granollers Sewer system. These agents are responsible for the management of the sewer systems. They are aware of the rainfall, the runoff produced by industrial discharges or rainfall incidence, and the level of the water flow in the sewer systems.
- *WWTP Agents*: Data Gathering, Diagnosis, Decision support, Plans and actions, Connectors. They receive information from the sewer system agents and the storage tanks to start working on the water flow. The agents perform various processes like data gathering, diagnosis of the water state using a rule-based system, formulating an action plan, user-validation of the plan, etc.
- *River agent*: Data gathering (data required: meteorological, physical, kinetic, water quality). It collects valuable data in order to monitor the state of the river.
- *Storage Tanks Agents*: Industrial parks, Rainfall. There are two types: the industrial parks control the flow from the industrial area and the rain retention can manage the rain flow.
- *Manager Agent*. It is responsible for the coordination between all the elements of the system. This agent interacts and supports the user in the decision-making. It starts and terminates all other agents and handles the communication between them.
- *Environment Agent*: This agent creates the Graphical User Interface (GUI), which has two main components: the "WWTP Input" and "Simulation" tabs to communicate with the end-user.

(continued)

- *User Agent*: This agent initiates the FIPA Request Protocol (RP), and gathers the data provided by the end user to start the simulation process, and to obtain the final simulation results.
- *Rule-based reasoning agent*: this agent implements a rule-based reasoning engine, which is able to make inferences from a knowledge base. In our case, the Drools[7] engine was used.

IDSS Functional architecture

The agent's graphic representation and the dependences among them are depicted in Fig. 7.2.

Implementation and software tools

For the design and development of a prototype for our case study, the Prometheus methodology (Padgham & Winikoff, 2004) and the agent platform Jadex (Braubach et al., 2004) were selected respectively. The prototype features the main elements of the hydraulic infrastructure and aims to manage the environmental system as a single area, integrating the two sanitation systems (La Garriga and Granollers) with their respective sewage systems and WWTP's, as well as the Congost river as the receptor for their waste water. Other elements are rain control stations, river water quality control stations, flow retention, and storage tanks. There is also a sewer channel that connects both WWTPs, allowing to bypass the flow from the La Garriga-WWTP to the Granollers-WWTP.

Another step in the deployment of the agent-based simulation is the assignment of roles to the agents. They are depicted in Fig. 7.3. The *roles* defined for the agents are depicted as blue rectangles, like for instance "By-passing Control" or "Storage Tank management". The *goals* are depicted as yellow ovals, for example "To minimize the economic costs" or "To manage critical episodes". The *perceptions* are depicted as red-pink multi-polygonal form, like "StorageTankMeasurement" or "RainfallDetected". Finally, the *actions* associated with the roles are depicted as thick green edges, like "WaterDischarge" or "WaterRetention".

The system overview diagram shown in Fig. 7.4 depicts the main types of agents of the MAS along with their *perceptions* (red-pink multi-polygonal forms) and *actions* (depicted as thick green edges). In addition, the major *data repositories* (depicted as yellow disk storages) are described as well as some *messages* (depicted as violet envelopes) that are sent between agents.

The next step is the identification of suitable and important *scenarios* to be simulated that can afford important consequences for the management of the river basin. A benchmarking strategy was used. The benchmark is a simulation

(continued)

---

[7] https://www.drools.org/

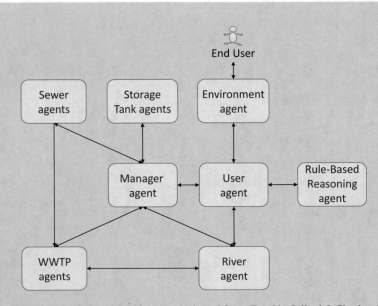

Fig. 7.2 River Basin MAS Architecture. (Adapted from (Rendón-Sallard & Sànchez-Marrè, 2007))

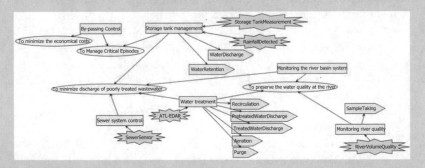

Fig. 7.3 Roles of the main agents in the River Basin MAS system. (Extracted from (Rendón-Sallard & Sànchez-Marrè, 2007))

environment defining a plant layout, a simulation model, influent loads, test procedures, and evaluation criteria. See http://www.benchmarkwwtp.org/ for further details. For each of these items, compromises were pursued to combine plainness with realism and accepted standards.

(continued)

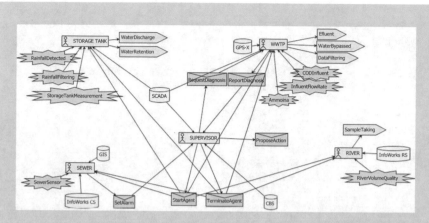

Fig. 7.4 River Basin MAS Overview. (Extracted from (Rendón-Sallard & Sànchez-Marrè, 2007))

Three possible scenarios were defined as preliminary ones for the evaluation of the system. The focus was on the behaviour of a given WWTP defined in the benchmark:

- *Dry weather:* acts as the reference scenario; one can notice in the profiles the daily and weekly typical variations.
- *Rain weather:* (dry weather + long rain period), the scenario's data contains 1 week of dry weather and a long rain event during the second week.
- *Storm weather:* (dry weather + 2 storm events), the scenario's data contains 1 week of dry weather and two storms during the second week.

The scenario's data structure for the simulated plant has these components:

- *Time:* measured in days.
- *Flow:* measured in cubic meters per day (m³/day).
- Concentrations of the following pollutants in milligram per litre (mg/L). The pollutants were defined by (Metcalf et al., 2004):

  - *COD:* (Chemical Oxygen Demand). The COD test is used to measure the oxygen equivalent of the organic material in wastewater that can be oxidized chemically using dichromate in an acid solution, at high temperatures.
  - *BOD5:* (Biochemical Oxygen Demand). It is a measure of the amount of oxygen required to stabilize a waste biologically.
  - *TSSe:* (Total Suspended Solids). Suspended solids can lead to the development of sludge deposits and anaerobic conditions when untreated wastewater is discharged to the aquatic environment.
  - *TKN:* (Total Kjeldahl Nitrogen). It is a measure of organic and ammonia nitrogen.

(continued)

In order to obtain the Final Concentration (*FConc*) for each pollutant the system uses the simple following equation for each pollutant:

$$FConc = (IFlow * IConc + WWTPFlow * WWTPConc)/FFlow \quad (7.1)$$

where,

FConc, is the Final Concentration.
IConc, is the Initial Concentration.
WWTPConc, is the WWTP Concentration.
WWTPFlow, is the WWTP Flow.
IFlow, is the Initial Flow.
FFlow, is the Final Flow, which can be computed as FFlow = IFlow + WWTPFlow.

The WWTP performance is obtained using the actuation rules defined by the experts, and also taking into account the results of previous simulations with the previous hydraulic simulation system built-up. Assuming that the WWTP operates at an optimum level when it receives an average load of about 7000 kg of COD per day. We can assume that a lower load of pollutants reduces the WWTP performance, perhaps slightly above the average load the performance is maintained or increased somewhat, but ultimately it also ends up reducing the performance due to overload.

The process of agent's communication and interaction in the MAS System is the following:

1. The manager agent starts the Sewer and Tank Storage agents, the WWTP agent, the Rule-Based Reasoning (RBR) agent, the River agent, the Environment agent, and the User agent.
2. The Rule-based reasoning agent publishes its service in the DF (Directory), the GUI shows the system's views and options, and the rest of the agents initialize their belief bases.
3. The human user through the Environment agent selects a scenario, an execution mode and clicks the start button; this broadcasts an internal message that triggers the simulation.
4. First the User agent requests the initial values for the flow and the pollutant concentration to the River agent (via the RP Protocol).
5. The River agent uses an expression to query its belief base and sends the result back to the User agent.
6. Then the User agent searches for services in the DF, finds the rule-based reasoning service and sends a request (via the RP Protocol) to the RBR agent.
7. The receiver side of the RP Protocol, the RBR agent, receives and accepts the request.

(continued)

8. The RBR agent creates the fact and knowledge bases, and then it uses the Drools engine to perform the deduction. When it finishes it sends the result back to the User agent.

9. The User agent calculates the final concentration for the pollutant and generates and stores the data series for constructing the output charts through the Environment agent. In addition, it is always aware of possible alarm messages sent by the Sewer Agents regarding the corresponding situation of any sewer system. In case that it would be necessary to actuate on some storage tanks it will send the corresponding messages to the StorageTank agents in order to discharge water or to retain water in the tanks, as part of the corresponding action plan.

10. The Environment agent through the GUI shows the results in the output section.

Results and Evaluation

Finally, once the MAS river basin system was defined and modelled, and after the definition of the scenarios, the different simulations of scenarios were run to extract the important facts and conclusions to give support to the management of the river basin according to the different scenarios.

In order to help evaluate and fully understand the meaning of data from the simulated scenarios, the experts' opinion was taken into account.

On the basis of the results produced, experts were able to interpret the data and perform an analysis of the results from each simulated scenario. This analysis showed that relevant and useful knowledge can be obtained about the river water quality and that the knowledge can be used to support the final user (e.g. WWTP manager) in his/her decision-making.

The simulation results of the different scenarios were analyzed by the experts. As an example, the *dry weather* and *stormy weather* scenarios analysis are presented here to illustrate the evaluation of the simulation results.

Dry Weather Scenario Analysis

The dry weather scenario acts as the reference scenario. We can note the daily and weekly typical variations in the profiles.

As regard the output charts for the *dry scenario* shown in Figs. 7.5 (Load of COD at WWTP effluent) and 7.6 (COD at the river), experts concluded that each peak, in which the WWTP has failed to eliminate the COD pollutant (See Fig. 7.5), coincides with a peak on the chart of the river's final state. This shows the great influence that plant spills have on the river. This influence is due to several reasons:

- The river has very little dilution capacity because its flow is very low due to dry weather.
- Almost all the river flow after the WWTP spill comes from the spill flow itself. This is typical behaviour of Mediterranean rivers in dry weather where there is very little rain or, no rain at all.

(continued)

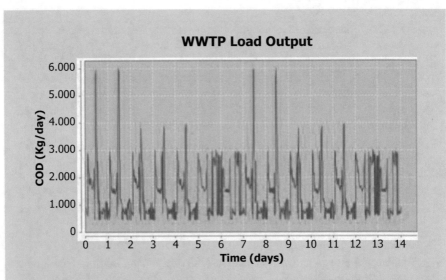

Fig. 7.5 Load of COD (Flow * COD concentration) at the WWTP effluent for the *dry weather scenario*. (Extracted from (Rendón-Sallard, 2009))

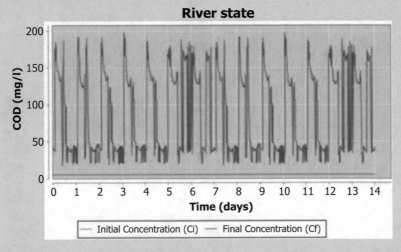

Fig. 7.6 COD concentration in the river for the *dry weather scenario*. (Extracted from (Rendón-Sallard, 2009))

(continued)

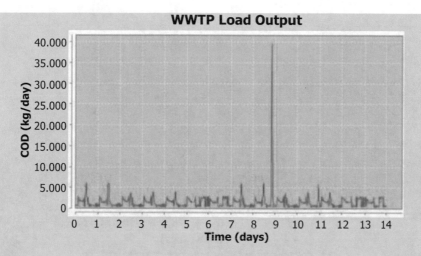

Fig. 7.7 Load of COD (Flow * COD concentration) at the WWTP effluent for the *storm weather scenario*. (Extracted from (Rendón-Sallard, 2009))

An environmentalist's analysis could perhaps start by investigating whether or not the river state meets the legislative requirements, and what these values mean for the river life.

Storm Weather Scenario Analysis

The input data file contains 1 week of dry weather followed by two storms during the second week. Figures 7.7 and 7.8 show the Load of COD at the WWTP effluent and the COD concentration at the river for the *storm weather scenario*.

The basic hypothesis is that pollutants accumulate during dry weather periods before they are transferred into and along with the system. The accumulation process occurs both on catchment surfaces and in sewer pipes. Pollutants accumulated on catchment surfaces are washed off during storm events.

Following a storm event, large amounts of pollutants that were previously dormant in sediment deposits spread around the drainage system (sewers, streets, etc.), and are carried to the WWTP all at the same time. If the WWTP lacks the capacity to process such large volumes of water then it has to be bypassed but, nevertheless, a very high concentration of pollutants will still enter into the WWTP.

This surge of pollutants affects the performance of the WWTP, and as such, we can note on the graph of Fig. 7.7 that the exit flows register a "high concentration peak" on day nine. On the other hand, a prolonged period of rain also increases the river's capacity to dilute pollutants and so this explains why there is just one peak after which a load of pollutants diminishes very fast.

(continued)

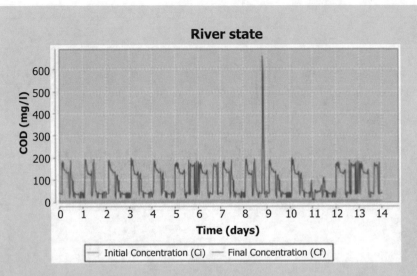

Fig. 7.8 COD concentration in the river for the *storm weather scenario*. (Extracted from (Rendón-Sallard, 2009))

We should also note that the same effect is seen in the final concentration of pollutants in the river, i.e. an extraordinary peak of pollution concentration. As a result of this peak, the river has very little dilution capacity which, indeed, is the same as would be the case during a dry period. Thus, the water discharged from the WWTP can make the final concentration of COD in the river increase substantially, which in turn can lead the levels of oxygen in the river to diminish considerably, a phenomenon known as "oxygen depletion".

Accuracy and efficiency

The accuracy or precision of the system is intrinsically related to the quality of the inference rules, which gathered the experts' knowledge, about the possible actions to be carried out for the management of a river basin. Thus, as much accurate as the knowledge, as much better will be the efficiency and accuracy of the IDSS.

Response time

The agent-based simulation system generated the simulation results in a very short time. Therefore, the response time of the system was very fast and reasonable.

Quality of the output generated (alternatives, support given)

The simulation results for each scenario give very useful information to check which will be the consequences of some actions and decide the best actions to be done for each one of the possible scenarios. The managers of a river basin can inspect the evolution of several important variables, like COD

(continued)

at the output of the WWTPs, or the river, and have the necessary information about the pollution levels, and the corresponding action planning to manage the situations.

Business opportunities (benefit/cost)
This prototype of IDSS could be very useful for the managers of some water treatment installations, and specifically for the river basin management of many rivers in a country. The integral management of a whole river basin is an important functionality for a stakeholder. This kind of IDSS lets obtain the manager a great benefit in terms of the scope and power of analysis and planning over the river basin.

7.1.2   The Use of Expert-Based Models

This case study is based on the real case of one gardening company. It will demonstrate the practical use of expert-based models to be used in the deployment of an IDSS.

Case Study: Garden Design at KAMACO

The Scenario
The KAMACO gardening company is introducing advanced information technology and artificial intelligence techniques to its daily management. For this reason, his head of computing, who is a lover of Artificial Intelligence and Gardening, is thinking about the possibility of building an Intelligent Decision Support System (IDSS) for the *design of gardens* for his/her clients.

Problem Analysis
The design of gardens is a very creative task, but it shows many patterns of behaviour that can be generalized and used in a computer. There are a number of features that must be taken into account when designing a garden: The *climate of the area* (temperature, rain, wind, etc.); the *type of soil* that can be very heavy and resistant to water (clay), light (sandy), with a certain amount of salt (saline), etc. The *depth, drainage, and soil pH* are also important characteristics; The *orography of the plot*: If there are high areas or depressed areas, flat areas, areas with strong or soft slopes, etc.; *Water for irrigation*: if there is too much or too little water, and the type of water (saline, limestone, etc.); *Plants to be included*: trees, shrubs, hedges, climbing plants, lawns, upholsterer plants, aromatic plants, fruit trees, etc.; *Elements of the garden*, such as furniture, stairs, rocks, constructions, retaining walls, rockeries, pool, etc.; *Garden uses* that can involve various areas of use such as sporting uses, family reunions, relaxation and rest, of a functional nature, etc.; The *maintenance of*

(continued)

the garden that varies according to the vegetal species; The *budget* that the client wants to spend.

In the short life of the company, it has accumulated some knowledge and information on the design of gardens. In fact, basically there is a set of main types of gardens with certain characteristics: *Mediterranean garden* with predominance of pines, shrubs, roses, soft grass, aromatic plants, such as rosemary, thyme, lavender, savoury, etc.; the *Tropical garden* were bamboos, palm trees, lawns, and leafy vegetation predominate; *Alpine garden* with cold-resistant lawns, fir trees, plants resistant to water and cold; the *Arabian garden* with plants resistant to the drought, autochthonous plants, a fountain and/or a pond; the *Classic garden* with hedges, lawns, terraces with colourful flowers, ponds, climbing plants; the *Japanese garden* with hedges, sand and white gravel, ponds, garden lanterns, boxwood, wooden furniture, rocks, chrysanthemums, bamboos; and a few more types of the garden (*English* or *Landscape garden*, *French garden*, *Romantic garden*, *Medieval garden*, etc.).

In addition, they usually identify the different areas aimed at different uses in the garden.

We want to build an IDSS that, given the customers' preferences, the budget available, but taking into account the characteristics of the available plot and the surroundings, obtain the best possible design of the garden.

Decisions involved

Analyzing the problem of the KAMACO Company, it seems very reasonable to think that the design of the garden for a given customer, taking into account the characteristics of the plot, and the customer's preferences, can be decomposed into two main decisions:

- The *type of the garden*: this decision must select the best type of garden (Mediterranean, Tropical, Alpine, Arabian, Classic, Japanese, etc.), taking into account both the main characteristics of the plot where the garden must be located and the climatic and environ conditions. Perhaps some preferences on the type of garden form the customer can be considered too.
- The *composition of the garden*: once the type of garden has been selected, the spatial composition of the garden must be decided. This means to decide which plants will be planted, and how will be disposed of in the garden. Probably, different zones in the garden must be configured and some other elements like furniture, stairs, rocks, etc., must be allocated in the garden.

Requirements

The IDSS should have the following functional requirements:

- The IDSS must allow that the customer's preferences, the available budget, the climatic and environmental conditions and the characteristics of the

(continued)

available terrain, i.e. the plot, for the garden could be entered into the system.
- The IDDS for the KAMACO Company must be able to generate the most suitable garden design for a given customer and a given set of preferences and constraints. The system can generate more than one possible solution, i.e. type of garden and its composition, ordered by degree of plausibility.

Data availability

The KAMACO Company is a relatively new company, which was launched just a few years ago. Therefore, they do not have an extensive record of garden designs previously done. This means that there is no enough data to be explored and exploited to get some data-driven model.

What it could be done is to keep the small number of garden designs to make some testing of the IDSS built.

Expert or First-principles knowledge availability

The company had a lot of expert knowledge which they use to build up the gardens for its customers. Thus, the best way to use this knowledge was to develop several *expert-based models*, coded as inference rules.

The possibility of having some first-principles knowledge coming from some theoretical principles was not contemplated in this problem, because there were no important physical or chemical processes leading the problem-solving task.

A Possible Solution

After analyzing the scenario and the concrete problem with the two consecutive decisions involved, a clear idea of the IDSS prototype appeared. The IDSS could be composed of two subsystems, each one addressing the support for each one of the decisions.

Type of IDSS

This IDSS must belong to the type of static IDSS. In this problem, there is not a changing process or system that must be continuously controlled or supervised. On the contrary, the system is a static system that given a set of constraints (user preferences, ground restrictions, climatic conditions, available budget, etc.) must design a garden for a given customer. Therefore, is a static process, and the IDSS must be executed just once to get the support information for solving the concrete problem.

Kind of tasks involved (Analysis, Synthesis, Prognosis)

This problem can be divided into two sub-problems: the selection of the type of the garden and the generation of the composition of the garden.

The *selection of the garden* is a clear *analytical task*. It must be selected a possible type of garden among a prefixed set of garden types like the following ones:

(continued)

- Mediterranean garden.
- Tropical garden.
- Alpine garden.
- Arabian garden.
- Classic garden.
- Japanese garden.
- English/Landscape garden.
- French garden.
- Romantic garden.
- Medieval garden.
- ....

The *generation of the composition of the garden* is a *synthetic task*. The composition of the garden or each one of the possible zones of the garden must be configured (plants, elements, etc.), taking into account the user preferences, the environmental conditions, budget constraints, etc. For instance, a final composition of a garden could be like the following one:

Type-of-garden: Alpine.
Composition-of-the-garden:

Number-of-zones: 3.

Zone: 1.
Geolocation point: <Latitude-1, longitude-1>.
Surface: 30 m^2.
Location in the plot: West.
Composition: one apple-tree, one prunus-pisardii, cold-resistant-lawn.
Cost: 200 €.

Zone: 2.
Geolocation point: <Latitude-2, longitude-2>.
Surface: 30 m^2.
Location in the plot: East.
Composition: one fir-tree-abies-nordmanniana
        one fir-tree-picea-globosa, cold-resistant-lawn.
Cost: 280 €.

Zone: 3.
Geolocation point: <Latitude-3, longitude-3>.
Surface: 90 m^2.
Location in the plot: Middle.
Composition: three rose plant, cold-resistant-lawn.
Cost: 540 €.

(continued)

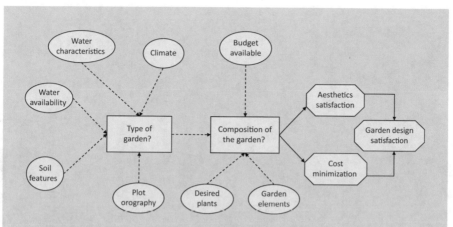

Fig. 7.9 Diagram influence modelling both decisions involved

Total-cost-garden: 1200 €.

The two decisions and their corresponding problem-solving tasks can be depicted in the influence diagram of Fig. 7.9.

Model selection

The *analytical task* of *selecting the type of garden* can be performed using an *expert-based model*. This analytical task can be implemented as a *heuristic classification problem-solving method* (Clancey, 1985).

A *heuristic classification* process is a non-hierarchical association between data and solutions (possibly class labels) that usually requires intermediate inference processes and possibly involving concepts from another taxonomy.

The intermediate processes are usually the following three: *data abstraction* process, *heuristic association,* and *solution refinement.* See (Jackson, 1999) for further description.

In our case, the *heuristic classification* process is depicted in Fig. 7.10. The concrete values for some features describing the piece of terrain must be abstracted to get a general qualitative data that can be matched with general types of garden. Furthermore, from a general type of garden, a particular one will be refined using additional expert knowledge and data from the desired garden. The knowledge required for each one of the three processes will be coded with inference rules and stored in knowledge base (KB) modules.

The *synthetic task* of the *composition of the garden* will be performed also by an *expert-based model*. This time, the task is completely different from the preceding one. In the selection of the type of garden, the number of possible solutions is usually a finite number and they are already known in advance, and the goal is to determine which one best matches the available data from the

(continued)

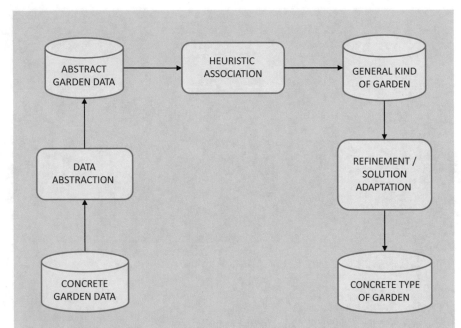

Fig. 7.10 Type of garden selection through a heuristic classification process

problem. In the composition of the garden generation, the solution does not exist and must be synthesized/created. The problem-solving technique commonly used is named as the *constructive problem-solving technique* (Jackson, 1999), which uses some *operators* to construct the solution. A solution is a complex object formed by several components. In addition, the *constraints* of the solution must be represented. For each type of garden, a different KB module will be created containing the different operators and constraints expressed in form of rules:

- Solution: is composed of a set of zones and an associated cost.

    - Solution: (list (zones), cost).

- Zone: is a set of plants, a list of elements in the zone, and its cost.

    - Zone: (list (plants (name, location), list (elements (name, location)), cost).

- Operators.

    - addPlant (zone, plant, costPlant, location),
    - removePlant (zone, plant),
    - interchangePlants (zone, plant1, location1, plant2, location2),
    - addElement (zone, element-type, element, location),

(continued)

- removeElement (zone, element),
- ...

- Constraints.

  - Total cost ≤ budget available.
  - If type-garden is alpine-garden then no warm-plant can be included.
  - If type-garden is Arabian-garden then no cold-plant can be included.
  - ...

This way, an *expert-based model* can be used (model-driven intelligent component) and exploited through a rule-based reasoning mechanism. This model will be composed of several KB modules.

Model Integration
The different models used should be integrated in a sequential way, because the two decisions involved are sequential. First, the type of the garden must be determined, and afterwards the composition of the garden must be generated.

IDSS Functional architecture
The functional architecture of the IDSS system is depicted in Fig. 7.11.

Implementation and software tools
The implementation of the IDSS basically relies on the implementation of a rule-based reasoning mechanism. As we told in a previous case study, there are several inference engines that can be used (Drools, CLIPS,[8] etc.). Independently of the final reasoning engine, we must define the fact base and the knowledge base.

Fact Base

Garden-type: (Mediterranean, tropical, alpine, Japanese, Arabian, English, classic, French, medieval, . . .),
Precipitation-average: numeric value in mm/month.
Mean-temperature: numeric value in °C.
Wind-speed: numeric average value in m/s.
Soil-pH: numeric value.
Garden-class: (warm-garden, mild-garden, cold-garden).
Precipitation-average-qual: (low, normal, high).
Mean-temperature-qual: (low, normal, high, very-high).
Wind-speed-qual: (low, moderate, high).
Soil-pH-qual: (acid, neutral, alkaline).
Available-budget: numeric value in Euros.
. . . .

(continued)

---

[8]http://www.clipsrules.net/

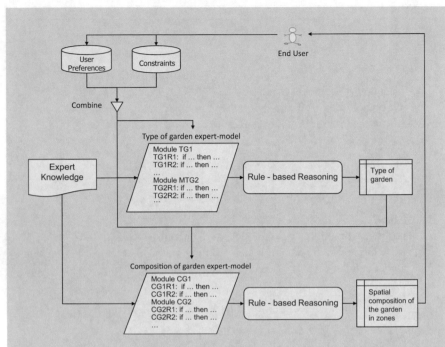

Fig. 7.11 Functional Architecture of the KAMACO Company IDSS

Plant: (rose, fir-tree, lawn, shrubs, pine, rosemary, lavender, thyme, savoury, Bamboo, palm tree, . . .).

Element: (table-furniture, chair-furniture, rocks, stairs, pool, hedge, . . .).

Location: (East, West, North, South, Northeast, Northwest, Southeast, Southwest).

Solution: (list (zones), cost).

Zone: (list (plants (name, location), list (elements (name, location), cost))).

. . . .

Knowledge Base

The knowledge base will be structured in several modules. One will be implementing the necessary knowledge for the selection of the *garden type* (heuristic classification process), and the others for implementing the necessary knowledge for the generation of the *composition of the garden* (constructive problem solving). In the middle, the *determination of the number of zones* in the plot seems to be a good strategy before entering to derive the composition of the garden. The scheme of the modular KB is depicted in Fig. 7.12.

(continued)

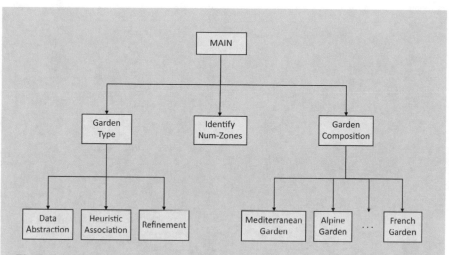

Fig. 7.12 Knowledge Base modular structure

Some examples of the rules that will be on each of the modules are described next in Fig. 7.13.

Results and Evaluation

After the IDSS was deployed, the system was assessed. For testing, some data collected in the short time that the company was operating was used to check whether the gardens previously designed by the company matched or not the garden design suggested by the IDSS.

Accuracy and efficiency

The IDSS system showed good accuracy with several garden scenarios which were tested. The types of the possible garden were correctly recommended, discarding the very unfeasible ones. The composition of the gardens was also considered satisfactory by the experts of the company. Even though, they thought that the expert model, i.e. the inference rules, could be improved to get a better performance of the IDSS.

Response time

The response time of the IDSS was really good because it just took a few seconds to generate the possible designs of the garden.

Quality of the output generated (alternatives, support given)

As told above, the quality of the generated compositions of the gardens and their compositions were very good in the tested case studies. The IDSS system proposed several feasible alternatives for a garden, like an Alpine garden or an English garden. In addition, the composition of the gardens assessed regarding the distribution of the zones taking into account the available surface and other features were successful according to experts' criteria.

(continued)

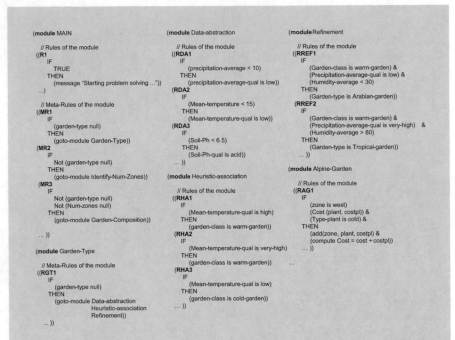

```
(module MAIN                          (module Data-abstraction              (module Refinement

   // Rules of the module                // Rules of the module                // Rules of the module
   ((R1                                  ((RDA1                                ((RREF1
      IF                                    IF                                    IF
         TRUE                                  (precipitation-average < 10)          (Garden-class is warm-garden) &
      THEN                                  THEN                                     (Precipitation-average-qual is low) &
         (message "Starting problem solving …"))   (precipitation-average-qual is low))   (Humidity-average < 30)
   …)                                    (RDA2                                 THEN
                                            IF                                    (Garden-type is Arabian-garden))
   // Meta-Rules of the module               (Mean-temperature < 15)          (RREF2
   ((MR1                                  THEN                                    IF
      IF                                     (Mean-temperature-qual is low))       (Garden-class is warm-garden) &
         (garden-type null)             (RDA3                                     (Precipitation-average-qual is very-high)   &
      THEN                                  IF                                    (Humidity-average > 80))
         (goto-module Garden-Type))        (Soil-Ph < 6.5)                 THEN
   (MR2                                   THEN                                    (Garden-type is Tropical-garden))
      IF                                     (Soil-Ph-qual is acid))         … ))
         Not (garden-type null)         … ))
      THEN                                                                 (module Alpine-Garden
         (goto-module Identify-Num-Zones))   (module Heuristic-association
   (MR3                                                                       // Rules of the module
      IF                                    // Rules of the module             ((RAG1
         Not (garden-type null)          ((RHA1                                  IF
         Not (Num-zones null)              IF                                       (zone is west)
      THEN                                    (Mean-temperature-qual is high)       (Cost (plant, costpl)) &
         (goto-module Garden-Composition))  THEN                                   (Type-plant is cold) &
                                              (garden-class is warm-garden))   THEN
   … ))                                   (RHA2                                     (add(zone, plant, costpl) &
                                            IF                                       (compute Cost = cost + costpl))
(module Garden-Type                          (Mean-temperature-qual is very-high)  … ))
                                          THEN
   // Meta-Rules of the module               (garden-class is warm-garden))
   ((RGT1                                 (RHA3
      IF                                    IF
         (garden-type null)                   (Mean-temperature-qual is low)
      THEN                                  THEN
         (goto-module Data-abstraction        (garden-class is cold-garden))
                    Heuristic-association  … ))
                    Refinement))
   … ))
```

Fig. 7.13 Implementation of the modular Knowledge Base for the garden design problem

Business opportunities (benefit/cost)

The KAMACO Company evaluated the IDSS very positively for many reasons. The system had a good performance and efficiency and greatly simplified their task. They reduced the time spent on the design task, and hence, they saved money. Furthermore, they were confident that with the addition of a GIS component, the general design of the gardens could improve a lot, achieving a graphical interface for presenting to their customers, the final appearance of the garden, with the different zones, its plants, and elements, to get a more informed approval from their customers.

7.1.3   The Use of Model-Based Reasoning Techniques

This case study will illustrate the use of model-based reasoning techniques in the domain of fault localization in combinatorial circuits. This is a very common use case in companies that design and ensemble digital combinatorial circuits and must verify the behaviour of the circuits, prior to their integration in other circuits.

Case Study: Combinatorial Circuit Fault Localization

The Scenario

In the following, we outline the use of model-based reasoning for *automatically localizing faults in combinatorial circuits*. Combinatorial circuits are digital circuits comprising the basic Boolean gates like *AND-gates*, *OR-gates*, or *Inverters*. The functionality of such a circuit origin from the structure of the circuit, i.e. the gates and their interconnections. For example, let us have a look at a simple combinatorial circuit implementing a Boolean *full adder*. A full adder is used to add two binary numbers, i.e. 0 and 1, using a special carry bit for considering that the sum of two binaries may be larger than 1. In addition, such a basic full adder also allows a carry bit as input in order to combine full adders to deal with binaries having more than 1 bit.

Let us have a deeper look into the functionality of a full adder. If we want to add 1 and 0 with an input carry bit of 0, the result should be obviously 1 with no carry, i.e. the output carry bit is 0. If we add 1 to 1 and not changing the input carry bit, the full adder returns 0 as output and sets the output carry bit to 1. Hence, a full adder provides the following basic functionality:

$$a + b + c^I = c^O s \tag{7.2}$$

where,

a and b serve as inputs,
c^I is the input carry bit,
s the result,
and c^O the output carry bit.

The binary $c^O s$ stands for a two digits output where the s is the rightmost element. Table 7.1, which is called a truth table, specifies the input–output behaviour of a full adder, i.e. its functionality.

Using basic Boolean gates, we are able to implement a full adder that covers the discussed behaviour. See Fig. 7.14 for a schematic diagram of a full adder outlining the components and connections comprising the circuit.

This circuit comprises two *XOR-gates* $\mathbf{X_1}$ and $\mathbf{X_2}$, two *AND-gates* $\mathbf{A_1}$ and $\mathbf{A_2}$ and one *OR-gate* $\mathbf{O}$. All these basic gates have two Boolean inputs and one Boolean output. An *XOR-gate* is returning a 1 only if exactly one input is 1, and 0 otherwise. An *AND-gate* returns 1 if and only if both inputs are 1, and 0 otherwise. And an *OR-gate* delivers 1 if at least one input is 1, and 0 otherwise. It can be easily shown that the above circuit implements a *full adder*. We only need to take the input depicted in the truth table of a *full adder* and compute the outputs using the schematics of the circuit. In this case we would obtain the output that is specified in the truth table.

(continued)

Table 7.1 Behaviour of a
Boolean *full adder* circuit

a	b	c^I	c^O	s
0	0	0	0	0
0	0	1	0	1
0	1	0	0	1
0	1	1	1	0
1	0	0	0	1
1	0	1	1	0
1	1	0	1	0
1	1	1	1	1

Once we implement the *full adder* as given in the schematics using hardware components, e.g. integrated circuits (ICs), we obtain a correct circuit that returns the expected result for any arbitrary given input. However, due to external issues like aging or overheating a gate might break, and the circuit could not be working as expected anymore. In such a case, we are interested in identifying the root cause behind it. For example, let us consider the case where we set $a = 1$, $b = 1$, $c^I = 0$, and obtain $s = 0$ and $c^O = 0$ as result. Obviously, the sum s is correct but c^O is not. We would expect accordingly to the truth table of the *full adder* circuit c^O to be 1. Hence, we want to obtain the reason behind the deviation between the expected and the observed behaviour. In particular, we want to *identify the gates* in the circuit that are responsible for the observed *faulty behaviour*.

In general, such a problem is a *diagnosis problem*, which can be summarized for the case of combinatorial circuits as follows:

- *Given*: A combinatorial circuit comprising gates and their interconnections, and observations covering at least the inputs and one output of the circuit that deviate from the expected output.
- *Find*: The gates that when assumed to be faulty explain the deviation, i.e. a diagnosis.

The question of how to obtain a diagnosis from the given information is called the *diagnosis problem*. Note that in the context of this case study, we restrict the more general diagnosis problem where we are interested in root causes to the case of identifying gates of the circuit that have to be faulty to explain the deviation. In the rest of this section, we discuss the diagnosis problem in more detail and show why *model-based reasoning* methods discussed in Chap. 5, Sect. 5.4 provides the right means for solving this problem.

Problem Analysis

The introduced diagnosis problem relies on the given information, i.e. the structure of a combinatorial circuit and its gates with a known behaviour, and observations, for identifying the gates that are responsible for observed

(continued)

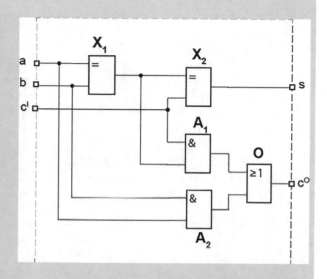

Fig. 7.14 A schematics of a Boolean *full adder* circuit

misbehaviour. The decision behind the diagnosis problem is to classify certain gates as being faulty or correct. In addition, we assume that a fault is permanent, i.e. when using the same input values, we always obtain the observed output values. Hence, we require that a faulty gate remains faulty over time unless we replace it with a spare part. In addition, we require that we know the structure of the circuit and the behaviour of its comprising gates.

These requirements assure that the underlying diagnosis problem is deterministic and we do not have to consider time as a parameter. Note also that we do not require a faulty gate to behave—for all its inputs—differently from its correct behaviour. Neither do we expect that a faulty gate always provides only one fixed output. Such a deterministic faulty behaviour corresponds to specific faults like stuck at zero, where a digital gate always provides a 0 at the output regardless of its inputs. However, in the context of this case study, we assume that we do not know the behaviour of a faulty gate. Therefore, a faulty gate might deliver different output values depending on the current inputs.

For particular combinatorial circuits, it is possible to observe their behaviour over time and to store it for further use, e.g. for obtaining data storing correct and also incorrect behaviour. However, since digital circuits and gates usually have a very high mean time between failures (MTBF) of at least several years, it might be the case that most of the data only represent correct behaviour. As a consequence, using observations as the basis for applying machine learning for diagnosis directly seem to be rather unlikely. Of course, MTBF depends on various factors like environmental conditions, e.g. the operating temperature, and design parameters, e.g. the value of the voltage supply. Hence, it would be–in principle–possible to obtain observational data

(continued)

from controlled experiments. However, there is always the need to identify the diagnosis manually to augment the observations.

Alternatively, someone might use the simulation for obtaining data that can be used later on for applying machine learning to solve the diagnosis problem. Such simulation would include introducing faults in the system and comparing the output of the system with its expected value, which can also be obtained using simulation based on the assumption that all component, e.g. gates, work as expected. In this case, someone has to provide a very large number of simulations. In order to deal with all possible diagnoses, we would need to obtain an exponential number of cases. If we have n components in a system, we have to simulate all 2^n potential diagnoses, which is of course infeasible. This is necessary in order to obtain data for all different diagnoses, i.e. a subset of the set of all components. In addition, we also have to make use of various input combinations. Hence, using data as the basis of diagnosis, in this case, seems to be not a feasible choice except for very small circuits comprising only a few components.

What is available to provide a solution to the combinatorial circuit diagnosis problem is the structure of the circuits and behavioural models of their components, i.e. the gates. Hence, we can make use of this information to provide an *intelligent decision support system* that computes diagnoses given a *model of the system*, and observations.

In summary, we have the following information available for the diagnosis of combinatorial circuits: the decision involved, the requirements, the data available, and the expert/first-principles models.

Decisions involved

Main decisions involved in the problem-solving task are to *identify gates of a combinatorial circuit to be faulty in order to explain deviations* between the observed and the expected behaviour.

Requirements

Main requirements are the availability of the system structure and the behaviour of its components, as well as the observations. In addition, we assume faults to be permanent. We do not require to know the behaviour of components in case of a fault.

Data availability

Data to be used to apply machine learning for diagnosis are hardly available. This is due to the fact of larger mean time between failures (MTBF), which as a consequence would lead to less data capturing the faulty behaviour. In addition, such data would need to capture an exponential number of faulty scenarios, which is hardly feasible.

(continued)

Expert or First-principles knowledge availability

In this case, instead of data, what are available are some *models of the system*, i.e. its structure and the behaviour of its subcomponents. Hence, we have *first-principles knowledge* available to solve the diagnosis problem for arbitrary combinatorial circuits.

A Possible Solution

In the following, we outline a possible solution to the *diagnosis problem for the combinatorial circuit*, where we discuss the basic principles behind modelling and implementation of a troubleshooting system relying on the principles of *model-based reasoning*. In particular, we make use of *model-based diagnosis* where we only assume to know the correct behaviour of system components.

In order to come up with a model for diagnosis, we distinguish two parts:

- The *model fragments* representing the behaviour of components, i.e. in our application domain the basic gates, and.
- The *system model*, i.e. the structure of the Boolean circuit we want to diagnose.

For each of these two parts, we first discuss a formal model and afterwards a concrete implementation based on the input language of the MINION constraint solver.[9]

A *constraint solver* takes more or less equations over variables as input and delivers variable assignments as output that are solutions for the given set of equations. If no solution can be found by the constraint solver the equations are contradictory and some of them have to be either deleted or changed. In *model-based diagnosis* we make use of constraint solving to check for satisfiability of solutions, i.e. diagnoses.

Let us start with formalizing the behaviour of the *basic components*, i.e. the Boolean gates used to compose digital circuits. For the *full adder* example, we used the *gates*: *AND*, *OR*, and *XOR*. In Fig. 7.15 we summarize the graphical representations of the gates as well as their behaviour. For this purpose, we use the truth tables of the gates and added also the one for inverter gates implementing the *NOT* function. Note that using these gates, we are able to construct all different kinds of combinatorial circuits. We are also able to represent other gates like *NAND-gates* or *NOR-gates*, which are combinations of an *AND-gate* respective *OR-gate* connected with an *inverter* at the output. It is also worth noting that for the gates depicted in Fig. 7.15, we define their behaviour referencing to their ports *in*, in_1, in_2, and *out*, where *in*, in_1, and in_2 are inputs and *out* is their output.

(continued)

---

[9]https://constraintmodelling.org/minion/

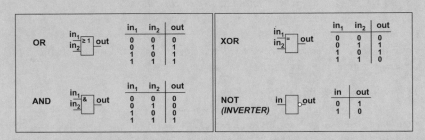

Fig. 7.15 Boolean gates and their correct behaviour

To use knowledge about correct behaviour for diagnosis, we have to formalize it. In *model-based diagnosis* this is done using a logical implication of the form:

$$\neg Ab(C) \rightarrow behaviour\,(in_1, \ldots, in_k, out_1, \ldots, out_m)$$

where,

The predicate *Ab* stands for abnormal (and therefore, $\neg Ab$ for correct),
And *behaviour* is a relation specifying which input/output tuples are correct.

In the case of the Boolean gates, we have only one to two inputs and one output. The behavioural relation is specified using the respective truth tables from Fig. 7.15. For example, the behaviour of the *XOR-gate*, *XOR*, it would be represented in logical form as follows:

$$\neg Ab(XOR) \rightarrow behaviour_{XOR}(in_1, in_2, out)$$

with the facts $behaviour_{XOR}(0,0,0)$, $behaviour_{XOR}(0,1,1)$, $behaviour_{XOR}(1,0,1)$, $behaviour_{XOR}(1,1,0)$.

Such logical rules and facts can be used together with a theorem prover directly to check whether a certain behaviour is contradicting the expected one.

In order to use such a formal representation together with the MINION *constraint solver*, we have to represent the rules using the available input language. There are many different ways of presenting such models in MINION. One would be to implement the truth tables directly. Unfortunately, in this case, MINION would not be as efficient as it could. Hence, we represent this knowledge using other *constraints* offered by the MINION solver:

(continued)

Table 7.2 Right gate behaviour in MINION

	Right behaviour
OR	max([in_1,in_2], out_I)
AND	min([in_1,in_2], out_I)
XOR	reify(diseq(in_1,in_2), out_I)
NOT	diseq(in, out_I)

- In particular, there is a *minimum constraint* that is evaluated to be true if the output of the constraint is equivalent to the minimum value at the input variables. Such a constraint can be used to represent an *AND-gate* that only delivers 1 at the output in cases where all inputs are 1.
- Similarly, the *maximum constraint* can be used to represent the behaviour of the *OR-gate*.
- For the *XOR-gate* and also the *NOT-gate*, we can make use of the *not equal constraint* (named *diseq* in MINION). In the case of the *XOR-gate* we need the *reify constraint*, which evaluates to true if the evaluation of the constraint, which is the first parameter, is equivalent to the second parameter. An *XOR-gate* only delivers 1 if both inputs are not equivalent.

In the following Table 7.2, we summarized the MINION representation of the right behaviour of the different gates.

Note that in this MINION representation, we do not represent the $\neg$ *Ab* predicate. This can be easily done via adding another constraint that relates the output *out_I* used in the MINION representation with the output *out* of the gate (but only when being correct). We can use the following constraint for this purpose:

$$\text{reifyimply}(\text{eq}(\text{out\_I}, \text{out}), !\text{Ab\_C})$$

There we assume that gate C should be correct in case the outputs are the same. In MINION the '!' is used to present the *negation*, and *reifyimply* is an *implication* that is only applied in case the first constraint parameter evaluates to true.

The *second part of the model* captures the structure of a *combinatorial circuit*. For each connection except for known inputs and outputs of the circuit, we introduce a unique variable. For each component, we make use of these variables and also the input and outputs to specify the behaviour of the gates using the model fragments. Let us explain this modelling step using the *full adder* example.

In the *full adder* example, we additionally introduce variables x, y, z that represent the *output connection* of gate X_1, A_1, and A_2, respectively. In this circuit, we also have the variables a, b, and c^I representing the inputs, and

(continued)

s and $\mathbf{c^O}$ representing the outputs. For the formal logical model, we instantiate the model fragments using these variables. For $\mathbf{X_1}$, $\mathbf{X_2}$ we obtain:

$$\neg Ab(\mathbf{X_1}) \rightarrow behaviour_{XOR}(\mathbf{a}, \mathbf{b}, \mathbf{x})$$
$$\neg Ab(\mathbf{X_2}) \rightarrow behaviour_{XOR}(\mathbf{x}, \mathbf{c^I}, \mathbf{s})$$

Because both gates are *XOR-gates*, we can rely on the same behaviour. For the other gates we obtain the following rules:

$$\neg Ab(\mathbf{A_1}) \rightarrow behaviour_{AND}(\mathbf{c^I}, \mathbf{x}, \mathbf{y})$$
$$\neg Ab(\mathbf{A_2}) \rightarrow behaviour_{AND}(\mathbf{b}, \mathbf{a}, \mathbf{z})$$
$$\neg Ab(\mathbf{O}) \rightarrow behaviour_{OR}(\mathbf{y}, \mathbf{z}, \mathbf{c^O})$$

In the corresponding MINION representation, we have first to define the *variables*. Instead of subscript or superscript we use a '_' before the subscript or superscript. *Variables* are defined in MINION using the *type* followed by the *variable name*. Hence, for the full-adder circuit we make use of the following MINION code:

```
**VARIABLES**
BOOL A
BOOL B
BOOL C_I
BOOL X
BOOL Y
BOOL Z
BOOL S
BOOL C_O
BOOL X_I
BOOL Y_I
BOOL Z_I
BOOL S_I
BOOL C_O_I
BOOL Ab_X_1
BOOL Ab_X_2
BOOL Ab_A_1
BOOL Ab_A_2
BOOL Ab_O
```

The following MINION constraints model the full-adder:

```
**CONSTRAINTS**
reify(diseq(A, B), X_I)
reifyimply(eq(X_I, X), !Ab_X_1)
```

(continued)

```
reify(diseq(X, C_I), S_I)
reifyimply(eq(S_I, S), !Ab_X_2)
min([C_I, X], Y_I)
reifyimply(eq(Y_I, Y), !Ab_A_1)
min([B, A], Z_I)
reifyimply(eq(Z_I, Z), !Ab_A_2)
max([Y, Z], C_O_I)
reifyimply(eq(C_O_I, C_O), !Ab_O)
```

Using these constraints together with the ones capturing the observations we are able to compute diagnoses using MINION. Observations can be easily represented in MINION using the *equality constraint eq*. Let us assume that we have the following observations: $a = 1$, $b = 1$, $c^I = 0$, $s = 0$ and $c^O = 0$. The MINION constraints for these observations are:

```
eq(A, 1)
eq(B, 1)
eq(C_I, 0)
eq(S, 0)
eq(C_O, 0)
```

Using these constraints, and storing them together with some additional code in the file *full_adder.minion*, and calling the constraint solver using the following command line call, we obtain 55 solutions, where each solution is a setting of the variables that correspond to the health state of gates, i.e. Ab_X_1, Ab_X_2, Ab_A_1, Ab_A_2, Ab_O:

./minion-findallsols full_adder.minion

The reason behind is that MINION searches for really all diagnoses when using the parameter -findallsols. Included are all minimal diagnoses and all supersets and also duplicates. If we are only interested in diagnoses of size 1, i.e. *single fault diagnoses*, we have to add the following two constraints to the *full_adder.minion* file:

sumgeq([Ab_X_1, Ab_X_2, Ab_A_1, Ab_A_2, Ab_O], 1),
sumleq([Ab_X_1, Ab_X_2, Ab_A_1, Ab_A_2, Ab_O], 1).

These constraints restrict the number of faulty components to 1 because they construct the sum of the variable values and compare them with 1. One constraint is used to check for *greater or equal* and the other for *less or equal*. When calling MINION again, we obtain only the two solutions:

Ab_O = 1 and
Ab_A_2 = 1.

(continued)

Stating that either the *OR-gate* **O** or the second *AND-gate* **A₂** are faulty. Both solutions are reasonable because they explain the wrong output c^O. A diagnosis **A₁** alone would be wrong because **A₁** is already delivering a 0 at its output and thus cannot explain why there is a 0 at c^O and not a 1.

What we have seen so far is how to obtain a *model* that can be used by a constraint solver directly in order to compute single fault diagnoses. If someone is interested in *all minimal diagnoses* we have to implement an algorithm that calls the constraint solver more often. We start with searching for all single faults, using the *sumgeq* and *sumleq* constraints. Afterwards, we search again but this time for double faults using the same constraints but extending the search space. Instead of the 1 value at the rightmost position, we take a 2. In addition, we add constraints stating that we are no longer interested in the obtained *single fault diagnoses*. This process continues until we have computed all diagnoses of a given size. For more information, we refer the interested reader to (Nica & Wotawa, 2012).

Next, we summarize the findings obtained when discussing a possible solution for the troubleshooting problem under consideration.

Type of IDSS
This IDSS would be a dynamic one because the IDSS continuously must monitor the combinatorial circuit to detect a new faulty output and must diagnose the faulty components of the combinatorial circuit for the problem. The IDSS built is a diagnosis system based on *model-based reasoning* methods.

Kind of tasks involved (Analysis, Synthesis, Prognosis)
This IDSS involves two main tasks. The first one is the *analysis* of the application domain in order to *identify* the structure of systems including comprising components and their correct behaviour. The second one is the *synthesis of a model* from the model fragments coming from the analysis phase. Using the particular model together with observations it searches for faulty components explaining failures.

Model selection
Both tasks are implemented through *model-based reasoning* methods. The whole modelling approach is based on a circuit structure model and model fragments. The underlying type of the model fragments, e.g., models based on Boolean algebra or equations, depending on the application domain, i.e. first-principles models. It might be necessary to focus on abstract models like qualitative representations of a quantitative value space in order to simplify the model fragments.

Model Integration
The integration of the different models is based on the composition of the system model using the model fragments describing the components' behaviour using the known system structure.

(continued)

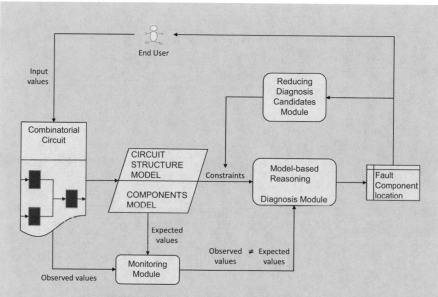

Fig. 7.16 Functional architecture of the IDSS of the combinatorial circuit fault localization

IDSS Functional architecture

The underlying IDSS comprises a *monitoring* module for obtaining observations and a *diagnosis* module for computing diagnosis candidates,
i.e. system components that are responsible for the deviation between the
observed and the expected behaviour. The architecture may also include a
module that allows for further providing probing information for *reducing the
number of diagnosis candidates*.

The architecture of the IDSS is depicted in Fig. 7.16.

Implementation and software tools

The fault localization of the components of the circuit is done by analyzing
the models and its translation in a *set of constraints* that must be solved to find
the possible faulty components. A *constraint solver software tool* like MIN-
ION together with additional algorithms or a theorem prover, SAT solver, or
SMT solver together with an implementation of a hitting-set algorithm can be
used. See (Greiner et al., 1989) for further information.

Results and Evaluation

The presented modelling approach for troubleshooting is applicable for all
combinatorial circuits. For a particular circuit, we only need to generate its
corresponding model considering the circuit's structure and the model

(continued)

fragments that describe the behaviour of the basic digital gates. The remaining question is whether the approach is good enough to be used in practice. We are going to answer this question considering qualitative and quantitative aspects. Let us start with a discussion of the pros and cons of the approach ignoring quantitative values like response time, accuracy, or efficiency.

The positive aspects of *model-based reasoning* include:

- The *model-based diagnosis* approach can be fully automated including model generation (based on model fragments) and diagnosis computation.
- The generated diagnoses are based on models. Hence, if the models are correctly representing a system, the diagnoses themselves have to be correct.
- The diagnosis approach does not have real requirements on side of the observations. The number of observations of course influences the number of generated diagnoses. However, there is no need to apply changes neither on side of the model nor on the underlying computational mechanisms in case of changes in the observations.

Of course, there are also some drawbacks, we have to mention:

- We need a *model of the system*. Most of the effort needed to come up with an IDSS relying on *model-based reasoning* goes into *modelling*. In technical domains, there are often at least partial models available. However, making more abstract models that capture important parts of the system required to diagnose the system is often time-consuming.
- *Model-based diagnosis* is at least NP-complete[10] in general. Hence, computing all minimal diagnoses is exponential in the number of system components. In practice, this is not so much problematic, because usually, someone is more interested in single or double-faults only, which can be obtained in polynomial time.

Let us discuss the last aspect, i.e. the run-time required for computing diagnosis, in more detail. For this purpose, we present the diagnosis run-time using the MINION constraint solver reported in (Nica et al., 2013). In this paper, the authors compared the run-time of different diagnosis algorithms using the ISCAS85 combinatorial benchmark examples.[11]

In Table 7.3, we depict statistical information of the ISCAS85 benchmark circuits used together with the corresponding data of their constraint

(continued)

---

[10] In Computational Complexity Theory, *NP-complete* problems are a set of problems to each of which any other NP-problem can be reduced in polynomial time and whose solution may still be verified in polynomial time. See (Cormen et al., 2009; Hopcroft & Ullman, 1979).

[11] http://www.cbl.ncsu.edu:16080/benchmarks/ISCAS85/

Table 7.3 Statistical information of the ISCAS85 benchmark circuits

Circuit	#Inputs	#Ouputs	#Gates	Function	#V	#Co
c432	36	7	160	27-ch. Interrupt controller	197	205
c499	41	32	202	32-bit SEC circuit	244	277
c880	60	26	383	8-bit ALU	444	471
c1355	41	32	546	32-bit SEC circuit	588	621
c1908	33	25	880	16-bit SEC/DED circuit	914	940
c2670	233	140	1193	12-bit ALU and controller	1427	1492
c3540	50	22	1559	8-bit ALU	1720	1743
c5315	178	123	2307	9-bit ALU	2486	2610
c6288	32	32	2406	16-bit multiplier	2449	2482
c7752	207	108	3512	32-bit adder/comparator	3720	3828

representation, i.e. the number of variables (#V) and constraints (#Co). We see that the benchmarks range from small circuits comprising 160 gates to larger ones having more than 3500 gates.

In (Nica et al., 2013), the authors used an Apple MacPro4,1 (early 2009) computer featuring an Intel Xeon W3520 quad-core processor, 16 GB of RAM, and running version of OS X 10.8 for carrying out the experiments. In addition, they made use of different test sets where they injected faults in the circuits so that they obtained single faults, double faults, and triple faults for the test sets TS1, TS2, and TS3, respectively. When carrying out the evaluation they also introduced a time limit of 200 s. In Table 7.4, we see the empirical results obtained when using the MINION constraint solver. There we depict the minimum, maximum, average, and median run-time for the different test sets.

The empty cells indicate that not all problem instances could be solved within the given 100 s time limit. In particular, for the single fault case all instances could be solved, for the double fault case 68%, and for triple faults only 27%. Nevertheless, in the case of searching for all single faults the maximum time needed for diagnosis never exceeded about 3.1 s. Diagnosis computation is fast enough even for larger circuits comprising more than 3500 gates. Therefore, we can conclude that *model-based reasoning* applied to troubleshooting digital circuits is fast enough for practical applications.

The question of accuracy for troubleshooting digital circuits cannot be easily answered. Because diagnosis in the case of Boolean circuits is based on models that correctly represent the real system, all the computed diagnoses are valid explanations. However, of course, in one particular situation, there is only one diagnosis. Hence, when defining accuracy as the ratio between the number of real diagnoses, which is 1, and all computed diagnoses, it ranges

(continued)

Table 7.4 Run-time in seconds for computing single-, up to double-, and up to triple-fault diagnoses for TS1, TS2, and TS3 respectively (top to bottom)

	Circuit	MIN	MAX	AVG	MED
TS1	c432	0.042	0.086	0.051	0.047
	c499	0.047	0.098	0.066	0.054
	c880	0.049	0.099	0.072	0.075
	c1355	0.103	0.677	0.218	0.106
	c1908	0.187	0.227	0.209	0.211
	c2670	0.081	0.345	0.237	0.231
	c3540	0.230	0.604	0.475	0.495
	c5315	0.138	1.341	0.843	0.860
	c6288	0.294	1.260	0.953	1.166
	c7752	1.735	3.081	2.116	1.906
TS2	c432	0.064	0.297	0.197	0.215
	c499	0.326	0.367	0.344	0.340
	c880	0.179	3.539	2.016	2.687
	c1355	10.27	12.10	10.57	10.39
	c1908	0.884	131.5	44.07	42.42
	c2670	0.277			52.22
	c3540	37.94			
	c5315	0.335			
	c6288	16.50			
	c7752	0.176			
TS3	c432	0.320	6.315	1.553	0.710
	c499	12.19	14.56	13.13	12.88
	c880	0.365			8.907
	c1355				
	c1908				
	c2670				
	c3540				
	c5315				
	c6288				
	c7752	2.062			

between 1 and 1 divided by the number of system components depending on the structure of the system.

When combining diagnosis computation with other functionalities like probing, diagnosis becomes a process starting with computing all valid diagnoses, identifying the most probable place in the circuit for probing, making additional measurements, and repeat this process until we identify one single diagnosis. *Model-based reasoning* provides means for establishing such a process, which would lead again to an accuracy of finally 100%. In addition,

(continued)

because of relying on optimal measurements, the whole approach can also be considered optimally efficient.

It is worth noting, that in cases the model is not completely correctly representing the system's behaviour, the accuracy and efficiency of *model-based reasoning* might be degraded. This can be the case when using abstract system models based on qualitative reasoning. In such cases, we obtain spurious diagnosis candidates leading to too many required measurements.

In practice, the *model-based reasoning* approach should be used in all cases where models are available or can be easily obtained. This is due to the fact that accuracy, efficiency as well as the quality of generated output in terms of providing all alternative solutions and support for probing is very high and algorithms are available. In some cases, e.g. when systems are huge, the approach might be less applicable. However, here we often can make use of hierarchical approaches and abstract models covering the most important functionality of systems. A drawback of model-based reasoning is modelling itself, which can be time consuming and requiring a lot of effort. This limits the direct applicability of *model-based reasoning* in practice.

Accuracy and efficiency

Model-based reasoning has high accuracy and can be used to efficiently solve the fault localization problem depending on the underlying models. Thus, the IDSS is highly accurate.

Response time

The response time for smaller up to larger systems comprising up to several thousands of components is acceptable for computing single fault diagnosis. For obtaining double faults, triple faults, or all minimal diagnoses the response time might be too high for practical applications.

Quality of the output generated (alternatives, support given)

The quality of the diagnosis output is high. All computed diagnoses explain the observed misbehaviour. The approach is flexible and does not have any restrictions regarding the observations. *Model-based reasoning* allows for computing all alternatives and also provides support for the whole diagnosis process in order to reduce the number of diagnosis candidates. Using probing based on entropies allows for asking the user about new measurements in an optimal fashion for reducing the number of required diagnosis steps.

Business opportunities (benefit/cost)

Because of flexibility and also easy adaptively *model-based reasoning* offers a lot of business opportunities whenever an IDSS performing diagnosis should be developed. The only drawback is the modelling issue that usually is the largest driver for costs. In cases where models are already available, this seems not to be an issue. Applying *model-based reasoning* requires modelling

(continued)

skills as well as good knowledge of formal representations, logics, and mathematics. Application domains include diagnosis of plants and other technical systems, software, and also self-adaptive systems like autonomous vehicles where flexibility and adaptively is required.

7.1.4   The Use of Qualitative Reasoning Models

This case study will illustrate the use of special *qualitative reasoning* models, which are *Causal Loop Diagrams* (CLDs). CLDs provide very nice qualitative models which represent the cause-effect relationships between variables. These models allow us to analyze the consequences of the increase or decrease of different variables involved in a concrete scenario, performing a qualitative simulation of the global system.

Case Study: Sustainable Action Planning in INTELLICITY

The Scenario

Nowadays, all responsible cities in the world are worried about designing and implementing sustainable plans to improve the *sustainability* in the daily life of their citizens. *Sustainability* issues involve three connected dimensions: the *social dimension*, the *environmental dimension,* and the *economic dimension*. In the social dimension there are some goals like improving the quality of life, the education level of society, and to get equal opportunities for the citizens. The environmental dimension is oriented to improve environmental protection, resource management, and preservation of the habitat reducing the contamination levels of air, water and land. The economic dimension is worried about a smart growth of society, to reduce the cost of living in a city, and fostering cost savings in general.

Thus, in the sustainable city named INTELLICITY, its major and the corresponding local municipality government is wondering which would be the best possible actions to be taken in their municipality to improve the sustainability degree of the city, and the general satisfaction of their citizens.

Problem Analysis

Planning the *best sustainable action plans* to be taken in a city is not an easy task. There are a lot of features involved and interrelated. Furthermore, there are several social agents in a city who want to influence the decisions that a local municipal government must make. The citizens, the political parties, the huge companies, the government members, etc.

It is very difficult to assess which are the best alternatives or options that will improve as much as possible all the sustainable society dimensions in a

(continued)

city. It is difficult for many reasons, but especially because *the consequences and effects* of one decision are not easily forecasted, and can have more consequences that are not envisioned.

Therefore, the *cause–effect relationships* are not easy to be made explicit, especially because the cause-effect chains can be long. For instance, even though the municipality government thinks that *introducing or increasing a car sustainability tax* could be a good measure, they are not sure about the possible negative consequences of this decision.

They want to build an IDSS that can help them to decide which the best possible actions are to be made to increase the degree of sustainability of IN TELLICITY, and at the same time, to increase the satisfaction degree of the citizens.

Decisions involved

Main decisions are to select which are the best possible actions to be done in the city of INTELLICITY given the constraints in the budget, and other considerations that the municipality has regarding the sustainable management of the city.

There are several possible actions that can be done like:

- Invest or not in renewable energies.
- Invest or not in water drinking and treatment systems.
- Invest or not in waste and recycling management.
- Invest or not in education.
- Introduce/increase or not a car sustainable tax.

However, it is not easy for the municipality to decide which the best options are, because it is not clear which are the cause-effect relationships among possible actions to be taken and other variables relevant for the city management.

Requirements

The IDSS to be deployed must be able to solve the following functional requirements:

- The user of the IDSS must be able to input to the system possible candidate sustainable actions.
- The IDSS must be able to assess how other variables will evolve, taking into account several cause–effect relationships originated by the initial sustainable actions.

Data availability

As sustainable actions have not been regularly implemented in the history of INTELLICITY, there is no available dataset with related information which could be used to provide data-driven methods to be integrated into the IDSS system.

(continued)

Expert or First-principles knowledge availability

There was not a first-principles knowledge or theory available, because the sustainability field is not a classic subject with well-established theories and principles which could be applied.

Instead, within the municipality of INTELLICITY, there was a reduced set of enthusiastic Sustainability Master's students who recently graduated, who had acquired a lot of expert knowledge in their Master studies, and were recently hired by the municipality. They were keen to analyze and study the possible consequences of some actions under the sustainability view. They had a lot of cause-effect knowledge that could be used to build up some model incorporating these cause–effect relationships between some relevant variables.

A Possible Solution

After the analysis of the scenario and the concrete problem to be solved, it became to be clear that an IDSS using some kind of *qualitative model* would be a good solution. In a *qualitative model*, it would be easier to model the cause-effect relationships between two variables *A* and *B*, because many times there is no quantitative relation between the involved variables, but just a *qualitative relationship* like if the variable *A* increases, then the variable *B* will increase too.

Type of IDSS

This IDSS must be static IDSS because in this problem there is not a changing system or process, which must be continuously monitored to make suitable decisions in a regular period. On the other hand, the system is a static problem were given a set of possible sustainable actions to be taken into account, it must help the end user to select the best possible action plan according to the possible good or bad consequences of candidate actions.

This IDSS will be used, from time to time, by the INTELLICITY municipality, in order to select the best actions to be undertaken in the city.

Kind of tasks involved (Analysis, Synthesis, Prognosis)

This IDSS mainly involves a basic task which is the *prognosis of the different possible sustainable actions*, taking into account the cause-effect relationships between the involved relevant variables in a qualitative model created with the expert knowledge available in the domain.

Model selection

To implement the prognosis task, the Sustainability Technicians hired by the municipality proposed to use a special kind of qualitative model, which is known as the *Causal Loop Diagrams (CLDs)* technique. The Causal Loop Diagram concept was first proposed by Jay Forester (1961) and further elaborated by other researchers such as Meadows et al. (1972), Rosnay (1979), Richardson and Pugh (1981), Senge (1990), and Sterman (2000).

(continued)

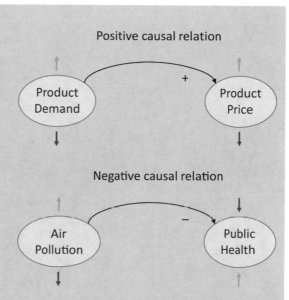

Fig. 7.17 Representation of positive and negative causal relations in a CLD

A *Causal Loop Diagram (CLD)* is a directed graph formed by a set of nodes or vertices and a set of edges connecting the nodes in a directed way. The *nodes* represent the *relevant variables* describing the problem at hand. The *edges* linking two nodes represent the *relation between two variables*.

The relationships between two variables can be of two types:

- *Positive relations*: a *positive causal relation* means that the two variables change in the *same direction* given that all the other variables remain stable. Thus, if the variable on the origin of the link increases, the connected variable at the end of the link also increases. And vice versa, if the variable on the origin of the link decreases, the connected variable at the end of the link also decreases. As illustrated in Fig. 7.17, the positive causal relations are marked with a *positive sign* in the graph, like the causal relation between the variable "Product Demand" and the variable "Product Price". If the demand for a product increases (represented by a green arrow), then the price of the product also increases (green arrow). On the contrary, if the demand for a product decreases (represented by a red arrow), the price of the product also decreases (red arrow).
- *Negative Relations*: A *negative causal relation* means that the two variables change in the *opposite direction* given that all the other variables remain stable. Thus, if the variable on the origin of the link increases, the connected variable at the end of the link decreases. And vice versa, if the variable on the origin of the link decreases, the connected variable at the end of the link increases. As illustrated in Fig. 7.17, the negative causal relations are

(continued)

marked with a *negative sign* in the graph, like the causal relation between the variable "Air Pollution" and the variable "Public Health". If the air pollution level increases (represented by a green arrow), then public health decreases (red arrow). On the contrary, if the air pollution level decreases (red arrow), the public health increases (green arrow).

Another important feature in the CLDs is the *closed feedback loops* that can appear in the diagram. There are two kinds of loops:

- *Reinforcing Feedback Loops*: A reinforcing feedback loop or simply a reinforcing loop is a cycle in which the effect of a variation in any variable propagates through the loop and returns to the variable reinforcing the initial variation in the same direction. Therefore, if a variable increases in a reinforcing loop, the effect through the cycle will return an increase to the same variable and vice versa. These loops reinforce the trend of one variable and usually are represented with an *R* letter (i.e. *reinforcing*) in green, in the diagrams as depicted in Fig. 7.18. If the "Plant Photosynthesis" increases then the "Plant Growth" will increase, and vice versa. When one variable has a variation and gets a reinforcement variation, this new variation will produce a new variation as well, and the system will be trapped in a cycle of circular chain reactions if the loop is not broken. They are related to *exponential increases or decreases* of variable magnitudes.

Fig. 7.18 Representation of a reinforcing loop and a balancing loop in a CLD

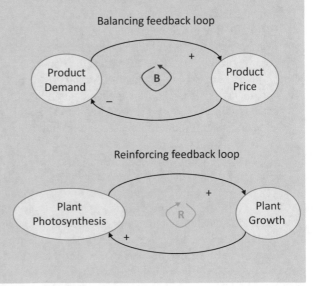

(continued)

- *Balancing Feedback Loops*: A balancing feedback loop or simply a balancing loop is the cycle in which the effect of a variation in any variable propagates through the loop and returns to the variable an opposite variation to the initial one. Thus, if a variable increases in a balancing loop the effect through the cycle will return a decrease to the same variable and vice versa. These loops *balance the trend of variables* and usually are represented with a *B* letter (i.e. *balancing*) in red, in the diagrams as depicted in Fig. 7.18. If the "Product Demand" increases then the "Product Price" increases, but then if the "Product Price" increases, then the "Product Demand" will start to decrease, until some final equilibrium point is found, as the demand-offer law states. Therefore, balancing loops are associated with subsystems reaching a *steady or equilibrium state*.

In the example of Fig. 7.18 the loops only have two nodes, but in general, the feedback loops can have any number of nodes. To determine whether a causal loop is a *reinforcing* one or a *balancing* one, it is needed just to start at some node in the loop and consider some trend or variation, and follow the loop until arriving at the initial node. Then, if the final result trend is aligned with the same direction (increase or decrease) than the initial node, the loop is a *reinforcing feedback loop*. On the other hand, if the final trend is opposite to the trend of the initial node, then the loop is a *balancing feedback node*.

From a practical point of view, the determination of the type of feedback loop is very easy: if the number of negative causal relations in the loop is zero or an even number, the loop is a *reinforcing loop*. On the other hand, if the number of negative causal relations is an odd number, then the loop is a *balancing loop*.

The identification of closed loops in a CLD is very important to identify possible *dynamic behaviour patterns of the system*. Sometimes, to indicate that there is some *delay* in a cause–effect relationship, which is marked with a double short line across the causal link, the whole system might have some fluctuations.

Another interesting concept in a CLD is the *propagation of a trend or variation along a path*. This *cause–effect analysis along a path* is based on computing what will be the effect caused by a variation in the initial node of the path to the last node at the end of the path, following the *cause-effect relationships* around the trajectory of all intermediate nodes in the path. The final sign of the trend (increasing/decreasing) can be computed at the ending node of any path, starting with the sign of the trend at the initial node, and computing step by step the sign of the trend at each intermediate node.

Given a path described as a sequence of connected nodes $\{n_0, n_1, \ldots, n_f\}$, then the final sign of the trend along a path can be computed as follows:

(continued)

$$\text{Sign}_{\text{Path}}(n_0, n_1, \ldots, n_f) = \text{Sign}_{\text{Trend}}(n_0) * \prod_{j=0}^{f-1} \text{Sign}_{\text{Relation}}(n_j, n_{j+1}) \quad (7.3)$$

where,

$$\text{Sign}_{\text{Trend}}(n) = \begin{cases} +1 & \text{if the trend at node } n \text{ is increasing} \\ -1 & \text{if the trend at node } n \text{ is decreasing} \end{cases} \quad (7.4)$$

$$\text{Sign}_{\text{Relation}}(n_j, n_{j+1}) = \begin{cases} +1 & \text{if the relation between } n_j \text{ and } n_{j+1} \text{ is positive} \\ -1 & \text{if the relation between } n_j \text{ and } n_{j+1} \text{ is negative} \end{cases}$$
$$(7.5)$$

This propagation of the trend along a path also can be used to compute the final *sign of a loop* to decide whether the loop is a reinforcing loop or a balancing loop:

$$\text{Sign}_{\text{Loop}}(n_0, n_0) = \text{Sign}_{\text{Path}}(n_0, n_1, \ldots, n_{f-1}, n_0) \quad (7.6)$$

Therefore, you just need to propagate the trend along the circular path of the loop, starting and ending at the same node. If the final value is +1, the loop is a *reinforcing loop*, and if the result is −1, the loop is a *balancing loop*.

For the deployment of the CLD model the technicians started to enumerate some possible actions and several variables which can be affected by causality relations. This way they obtain the following variables:

- Invest or not in renewable energies.
- Invest or not in water drinking and treatment systems.
- Invest or not in waste and recycling management.
- Invest or not in education.
- Introduce/increase or not a car sustainable tax.
- Budget available.
- Health cost.
- Public health.
- Water diseases.
- Drinking water quality.
- City cleanliness.
- Renewable energies use.
- Conventional Energies use.
- Energy cost.
- Private car transport.
- Sustainable public transport.
- Greenhouse gas emissions.

(continued)

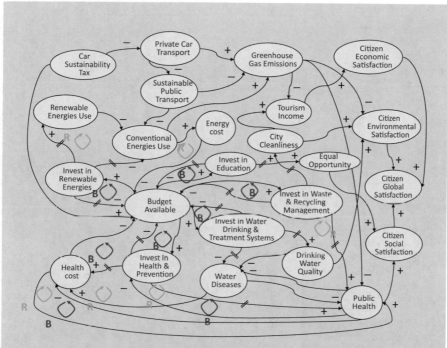

Fig. 7.19 Causal Loop Diagram model for the INTELLICITY scenario

- Tourism income.
- Equal opportunity.
- Citizen economic satisfaction.
- Citizen environmental satisfaction.
- Citizen social satisfaction.
- Citizen global satisfaction.

After that, they analyzed the cause-effect relationships between all the variables, and finally obtained the CLD model depicted in Fig. 7.19. For instance, analyzing the variable "Invest in renewable energies", they thought that it had a clear effect on the variable "Renewable energies use". If the investment in renewable energies increases, then the renewable energies used by the municipality will also increase, i.e. there is a positive cause–effect relation. Furthermore, if the "Renewable energies use" variable increases, then the variable "Conventional energies use" would decrease because there is more use on renewable energies in the municipality. Thus, there is a negative cause-effect relation, as depicted in Fig. 7.19. The relationship between "Invest in Renewable energies" is marked as delayed, because the effect of the investments usually is not immediate, and could it be delayed.

(continued)

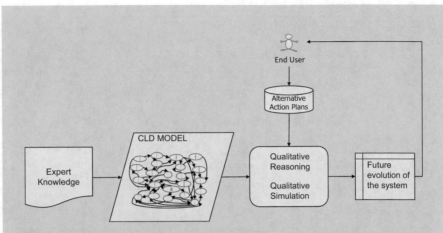

Fig. 7.20 Functional Architecture of the INTELLICITY IDSS

This way they were constructing several *cause–effect chains* as depicted in the final model in Fig. 7.19. As shown in the diagram of the model, the different possible investment actions, finally have some consequences on several variables reflecting the citizen social, environmental, economic, or global satisfaction. Moreover, the most important reinforcing loops and balancing loops in the CLD are depicted in green or red loop-arrows with the corresponding (R or B) letter within it.

Model Integration
In this IDSS, there is a basic qualitative model, the *CLD model*, which will be deployed into the final architecture. Thus, no other model will be integrated with it.

IDSS Functional architecture
The IDSS for INTELLICITY is mainly composed of the CLD model which is exploited by the end user to get the evolution of the system for a given possible action or action plan. The architecture is depicted in Fig. 7.20.

Implementation and software tools
As the municipality had no available tool for implementing and running a CLD model, main efforts in the implementation of the IDSS relay on implementing those three functionalities for the CLD model:

- *Building of the CLD model* by the experts. This means the implementation of a graphical editor to build a CLD. The interface allowed to create of nodes and cause-effect relations between nodes. In addition, the feedback loops could be automatically detected and marked in the CLD.

(continued)

- *Qualitative simulation of the evolution of the CLD model.* Given information about the trend or variation of one variable (cause), the system was able to propagate the cause-effect relationships and compute the final trends (effects) of several variables along several cause–effect paths.
- *Reverse propagation on a path in the CLD model.* Given a concrete trend in one variable (effect), the system could automatically analyze, which are the necessary trends/variations (causes) in the related variables through the reverse propagation of the cause-effect paths.

These functionalities can be implemented in a general programming language like Python, Java, etc.

Results and Evaluation

After the IDSS system was implemented, it had to be evaluated to ensure that the results generated by the system were the expected ones. As the main model used was an association model implemented with a qualitative model, i.e. a CLD, the assessment of the system was based on the evaluation of the cause-effect paths (associations) and the final evolution of the model for different initial scenarios. Those initial scenarios were the different possible sustainable action alternatives for the municipality of INTELLICTY.

The different possible actions were given as input to the IDSS, and the results were assessed by the municipality experts on sustainability and management tasks in the city. Those actions were to invest or not in several sustainable actions like:

- Invest or not in renewable energies.
- Invest or not in water drinking and treatment systems.
- Invest or not in waste and recycling management.
- Invest or not in education.
- Introduce/increase or not a car sustainable tax.

They assessed the different associations found by the system, like:

- If the municipality makes an increase in "Invest in renewable energies", and following the CLD model, it produce an increase in the "Renewable energies use" by the municipality (positive relation), as well as a decrease in the "Budget available" (negative relation). At the same time, the decrease in the "Budget available" will generate a decrease in the "Invest in renewable energy" making a *balancing feedback loop*. On its way, the increase in "Renewable energies use" would produce a decrease in the "Conventional energies use" (negative relation) by the municipality. Next, this decrease in the "Conventional energies use" will produce a decrease in the "Greenhouse gas emissions", and also a decrease in the "Energy cost" (both positive relations), due to the higher costs of conventional energies. This

(continued)

savings in "Energy cost" available will produce an increase in the "Budget available" will be produced (negative relation).

This increase in the "Budget available" could balance the decrease in "Budget available" generated by the initial increase in the "Invest in renewable energies" (negative relation).

Furthermore, the decrease in "Greenhouse gas emissions" will generate an increase in the "Citizen environmental satisfaction", an increase in "Tourism income", and an increase in "Public health" (negative relations). On its way, the increase in "Tourism income" will generate an increase in "Citizen economic satisfaction". The increase in "Public health" will produce a decrease in the "Health cost" (negative relation) of the municipality, which in turn, will increase the "Budget available" (negative relation). This increase in "Public health" will generate an increase in the "Citizen social satisfaction", and a decrease in "water diseases". The corresponding decrease in "Water diseases" will generate a decrease in "Health cost" too, which in turn, will increase the "Budget available" (negative relation).

Finally, "Citizen economic satisfaction", "Citizen environmental satisfaction" and "Citizen social satisfaction" will increase the "Citizen global satisfaction" (positive relations).

Summarizing, an increase in "Invest in renewable energies" will generate an increase in "Citizen global satisfaction" and also an increase in the "Budget available". Thus, this action seems to be a very good one.

See the corresponding cause-effect paths marked with dashed-blue links in Fig. 7.21.

- If the municipality increases or creates a "Car sustainable tax", then the "Budget available" will increase (positive relation). Following another causal–effect path, it will decrease the "Private car transport" (negative relation), as many people will decide not to use the car. This will generate an increase in the "Sustainable public transport" (negative relation). Next the "Conventional energies use" will decrease (negative relation, which in turn will decrease the "Energy cost" (direct relation), which in turn will increase the "Budget available" (negative relation).

On the other hand, the decrease in "Private car Transport", the increase of "Sustainable Public transport" and the decrease in "Conventional energies use" will generate a decrease in the "Greenhouse gas emissions". The decrease in "Greenhouse gas emissions" will produce an increase in the "Tourism income" in the city (negative relation), an increase in "Public health" and an increase in "Citizen Environmental satisfaction". The "Public health" increase will increase the "Citizen social satisfaction", and the "Tourism income" increase will generate an increase in the "Citizen economic satisfaction". Finally, all kinds of citizen satisfaction will increase the "Citizen global satisfaction" variable.

(continued)

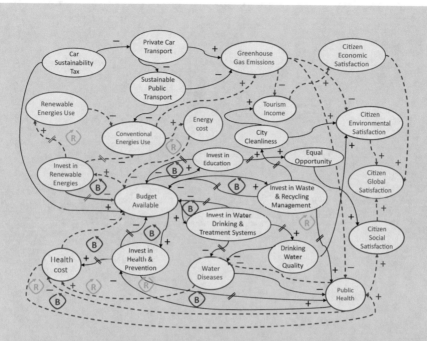

Fig. 7.21 Causal-effect paths activated by an increase in "Invest in renewable energies" variable

Summarizing an increase in or a creation of "Car sustainable tax" will increase the "Budget available", will diminish the "Greenhouse gas emissions", and will increase the "Citizen global satisfaction".

See the corresponding cause-effect paths marked with dashed-blue links in Fig. 7.22.

Accuracy and efficiency

The experts' evaluations were very satisfactory. They said that the cause-effect paths were meaningful and that the final effects of some variations in the value of some variables were computed in a very accurate way.

On the other hand, the CLD model implemented was very efficient because the computations of final path variations were very fast. The other secondary and derived paths also must be explored. However, the computational cost is not very high, because the ramification factor in the exploration is a very low number.

Response time

The response time of the IDSS for computing the final trends of the different cause-effect was very short. In fact, the main path exploration is

(continued)

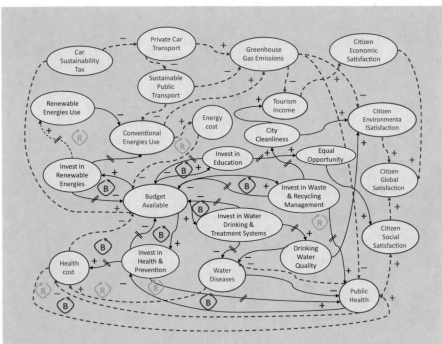

Fig. 7.22 Causal-effect paths activated by an increase/creation in/of "Car sustainable tax" variable

proportional to the length of the path (*linear time*). The other secondary paths usually are just a few and very short, and thus, the final response time of the systems was very good.

Quality of the output generated (alternatives, support given)

As detailed above, the quality of the IDSS was very accurate and valuable, because the system provided a very valuable information to support the decision-making process in the municipality. For a given possible sustainable action, i.e. a variation in some variable, the system computed the possible trends in several related variables through a cause–effect chains of reasoning. As this can be done for several possible alternatives, all of them could be assessed in terms of their possible consequences in the municipality. This way the major and the government of INTELLICITY can make the final decision having a lot of useful information.

In addition, the reverse path propagation functionality was very nice to support the municipality government to decide which should be the variations needed to obtain a given desired trend in a concrete variable of interest, like "Citizen social satisfaction", etc.

(continued)

Business opportunities (benefit/cost)

The government of the municipality of INTELLICITY evaluated the IDSS very positively, and they thought that the system produced a great value or benefit for the municipality. The support provided by the system can help very much the municipality in performing the best possible sustainable actions to increase the global satisfaction of citizens in the municipality, and at the same time, to increase the sustainable level of INTELLICITY. Furthermore, some costs can be reduced by not making wrong actions.

They also thought that this IDSS could be reused, with minor modifications, by other similar municipalities, and therefore, some extra business opportunities for INTELLICITY could appear with the licensing or similar economic agreement with other municipalities.

7.2   Using Data-Driven Methods in IDSS

In this section, several case studies will be described to illustrate the use of some *data-driven methods* in the development of IDSSs.

7.2.1   The Use of Descriptive Models

This case study will show the use of descriptive models in the deployment of IDSSs. Concretely, this case will illustrate the use of *clustering* techniques to help identifying similar observations in a Crime Analysis scenario within a Police Department in any country or state of the world. Clustering techniques are the most used descriptive models in Data Science to characterize homogeneous groups of observations in a dataset.

Case Study: Crime Analysis, Control, and Prevention

The Scenario

One of the main tasks in a Police Department in any country of the world is to analyze and identify how the crimes happen and are distributed within its jurisdictional territory to crime control and crime prevention. This case study will be based on real data coming from the Maryland State in the USA. Therefore, the data and the information used are real. In order to describe a full case study, we will imagine ourselves as we were the new Information Technology (IT) director at the Maryland State Police (MDSP), and we want to use Information Technologies, and especially Artificial Intelligence

(continued)

techniques, to use all available data and knowledge at MDSP to fight against crime.[12] Therefore, the goal of the IT director is to develop an IDSS to help the MDSP in crime control and prevention.

Problem Analysis

Fighting against crime is a difficult task due to many reasons. There are different kinds of crimes (violent crimes, property crimes, etc.), and usually, depending on the extension of the state or country, the territory extension to be controlled can be huge. Furthermore, the crimes are not happening in a uniform way within the jurisdiction of a state or country. A *crime* occurs when someone breaks the law by an overt act, omission, or neglect that can result in punishment. A person who has violated a law, or has breached a rule, is said to have committed a *criminal offense*. In general, the following types of *criminal offenses* are commonly considered in the USA, according to Charles Montaldo (2019)[13]:

- Violent Crimes: A violent crime occurs when someone harms, attempts to harm, threatens to harm, or even conspires to harm someone else. Violent crimes are offenses that involve force or threat of force, such as rape, robbery, or homicide.

  Some crimes can be both property crimes and violent at the same time, for example, carjacking someone's vehicle at gunpoint or robbing a convenience store with a handgun. Most common violent crimes are the following:

  - Homicide: A homicide is any killing of a human being by another human being. Homicides can be justifiable, excusable, or criminal.
  - Murder: Usually classified as first-degree or second-degree, murder is a criminal homicide to the wilful taking of another person's life.
  - Rape: Rape occurs when someone forces sexual contact with another person without their consent.
  - Sexual Assault: Although definitions vary by state, generally it occurs when a person or persons commit a sexual act without the consent of the victim.

(continued)

---

[12] Of course, by no means we want to impersonate or pretend to be the Maryland State Police. We act as we supposedly were them, just as a literary license to get a coherent situation for the case study.

[13] See https://www.thoughtco.com/common-criminal-offenses-970823, https://www.thoughtco.com/what-is-a-crime-970836

- Robbery: Robbery involves stealing from another person by the use of physical force or by putting the victim in fear of death or injury.
- Assault: Criminal assault is defined as an intentional act that results in a person becoming fearful of imminent bodily harm.

  Aggravated Assault: Aggravated assault is causing or attempting to cause serious bodily harm to another or using a deadly weapon during a crime.

- Property Crimes: A property crime is committed when someone damages, destroys or steals someone else's property, such as stealing a car or vandalizing a building. Usual property crimes are the following:

  - Break and Enter (B and E): Breaking and entering, which is usually committed by forcing entry into a building without the intention of committing a crime.
  - Burglary: Any entry into a building or structure at any time of day or night, without permission and with the intent to commit a crime inside.
  - Larceny Theft: It is unlawful taking and carrying away of the property of another person, with the intent to permanently deprive them of its use.
  - Motor/Vehicle Theft: is the criminal act of stealing or attempting to steal a motor vehicle.

Any police department needs to analyze and identify how the crimes are being committed, in what geographical zones there are more violent crimes or where are more property crimes, etc. All this knowledge about the crimes structure will enable any police department to have a clear understanding about the crime environment and will let them organize better to control and prevent crimes.

Our IT director wants to build an IDSS that lets them make better crime control and prevention action plans.

Decisions involved

Analyzing and thinking about the crime control and prevention in a police department, like the MDSP, it becomes clear that a helpful IDSS should give support to some decisions which must be taken periodically to control the crimes in the territory, like the following:

- How to assign the police human resources to the different geographical regions in the territory like cities, metropolitan areas, counties, etc.? How many policemen units must be assigned to the different territories to better control crimes?
- According to the most common crimes committed in the jurisdiction under the control of the police department, which should be the best formation courses on specialized crimes to prepare police specialists to fight against

(continued)

the most common crimes in the territory to control and to prevent crime commitment?

Requirements
The aimed IDSS should meet the following functional requirements:

- The IDSS must allow that the corresponding end user, i.e. a police staff, can load into the system a dataset with statistical information regarding the different crimes committed and where they happened in their jurisdictional territory during a concrete period of time. The period could be whatever: 1 year, several years, a decade, etc.
- Optionally, some geographical information about the territory (maps) can be loaded to let some geographical information visualization complement the basic IDSS functionalities (a GIS subsystem).
- Furthermore, the IDSS could support the functionality of entering the different reporting data along time to create the dataset from the different crimes by the department.
- The IDSS must be able to process all the input data, and automatically discover which are the territories where more crimes are committed, detect which territories have a similar behaviour regarding the crime commission, and what are the types of crime more frequent, what are the less frequent, and so on.

Data availability
In this case study, the police department can use a dataset that is periodically built up with the reporting data obtained from all the territory.

This dataset was downloaded from the Maryland Open data portal.[14] All data includes observations from 1975–2016 by each county but also including data from Baltimore City about different types of crimes. The data are provided by the Maryland Statistical Analysis Center (MSAC), within the Governor's Office of Crime Control and Prevention (GOCCP). MSAC, in turn, receives these data from the Maryland State Police's annual Uniform Crime Reports. We are very grateful to all of them (MSAC, GOCCP, and MDSP) for gathering and compiling all the data and for allowing everybody, including us, to use these data. This data transparency allows us to illustrate this case study. We strongly appreciate and require this open data policy in public administrations.

The dataset has the Maryland Crime information by each county (Allegany County, Anne Arundel County, Baltimore County, Calvert County, ..., Wicomico County, and Worcester County). There are 23 counties plus Baltimore City, which is accounted separately. Furthermore, there are data from the

(continued)

---

[14] https://opendata.maryland.gov/

year 1975 until the year 2016. Therefore, 42 years for each county. Thus, the dataset has 1008 rows (42 * 24) and 38 variables. The variables are the following:

- JURISDICTION: The corresponding county or Baltimore City.
- YEAR: the year (between 1975–206).
- POPULATION: The population of the county in that year.
- MURDER: tTe total number of murder crimes in the corresponding county and year.
- RAPE: The total number of rape crimes in the corresponding county and year.
- ROBBERY: The total number of robbery crimes in the corresponding county and year.
- AGG-ASSAULT: The total number of aggravated assault crimes in the corresponding county and year.
- B and E: The total number of breaking and entering episodes in the corresponding county and year.
- LARCENY THEFT: The total number of larceny thefts in the corresponding county and year.
- MOTOR VEHICLE THEFT: The total number of motor vehicle thefts in the corresponding county and year.
- GRAND TOTAL: the total number of crimes in the corresponding county and year.
- PERCENT CHANGE: The percent change of total crimes regarding the previous year for the same county.
- VIOLENT CRIME TOTAL: The total number of violent crimes in the corresponding county and year.
- VIOLENT CRIME PERCENT: The percent of violent crime regarding the total crimes in the corresponding county and year.
- VIOLENT CRIME PERCENT CHANGE: The percent change of violent crimes regarding the previous year for the same county.
- PROPERTY CRIME TOTAL: The total number of property crimes in the corresponding county and year.
- PROPERTY CRIME PERCENT: The percent of property crime regarding the total crime in the corresponding county and year.
- PROPERTY CRIME PERCENT CHANGE: The percent change of property crimes regarding the previous year for the same county.
- OVERALL CRIME RATE PER 100,000 PEOPLE: The total number of crimes per 100,000 people in the corresponding county and year.
- OVERALL PERCENT CHANGE PER 100,000 PEOPLE: The percent change of total crimes per 100,000 people regarding the previous year for the same county.

(continued)

- VIOLENT CRIME RATE PER 100,000 PEOPLE: The total number of violent crimes per 100,000 people in the corresponding county and year.
- VIOLENT CRIME RATE PERCENT CHANGE PER 100,000 PEOPLE: The percent change of violent crimes per 100,000 people regarding the previous year for the same county.
- PROPERTY CRIME RATE PER 100,000 PEOPLE: The total number of property crimes per 100,000 people in the corresponding county and year.
- PROPERTY CRIME RATE PERCENT CHANGE PER 100,000 PEO-PLE: The percent change of property crimes per 100,000 people regarding the previous year for the same county.
- MURDER PER 100,000 PEOPLE: The total number of murder crimes per 100,000 people in the corresponding county and year.
- RAPE PER 100,000 PEOPLE: The total number of rape crimes per 100,000 people in the corresponding county and year.
- ROBBERY PER 100,000 PEOPLE: The total number of robbery crimes per 100,000 people in the corresponding county and year.
- AGG-ASSAULT PER 100,000 PEOPLE: The total number of aggravated assault crimes per 100,000 people in the corresponding county and year.
- B & E PER 100,000 PEOPLE: The total number of breaking and entering episodes per 100,000 people in the corresponding county and year.
- LARCENY THEFT PER 100,000 PEOPLE: The total number of larceny thefts per 100,000 people in the corresponding county and year.
- MOTOR/VEHICLE THEFT PER 100,000 PEOPLE: The total number of motor vehicle thefts per 100,000 people in the corresponding county and year.
- MURDER RATE PERCENT CHANGE PER 100,000 PEOPLE: The percent change of murder crimes per 100,000 people regarding the previous year for the same county.
- RAPE RATE PERCENT CHANGE PER 100,000 PEOPLE: The percent change of rape crimes per 100,000 people regarding the previous year for the same county.
- ROBBERY RATE PERCENT CHANGE PER 100,000 PEOPLE: The percent change of robbery crimes per 100,000 people regarding the previous year for the same county.
- AGG-ASSAULT RATE PERCENT CHANGE PER 100,000 PEOPLE: The percent change of aggravated assault crimes per 100,000 people regarding the previous year for the same county.
- B and E RATE PERCENT CHANGE PER 100,000 PEOPLE: The percent change of breaking and entering episodes per 100,000 people regarding the previous year for the same county.

(continued)

- LARCENY THEFT RATE PERCENT CHANGE PER 100,000 PEOPLE: The percent change of larceny thefts per 100,000 people regarding the previous year for the same county.
- MOTOR/VEHICLE THEFT RATE PERCENT CHANGE PER 100,000 PEOPLE: The percent change of motor vehicle thefts per 100,000 people regarding the previous year for the same county.

Expert or First-principles knowledge availability

Although there was not a well-established first-principles knowledge or theory about crime control and prevention, there were a lot of experimented policemen, which owned very valuable knowledge, i.e. *expert-based knowledge*, about their daily activity against crime.

This expert knowledge can be used to validate the recommendations and suggestions of the IDSS in the evaluation step.

A Possible Solution

As the outcome of the analysis of the case scenario, it became apparent that a *cognitive descriptive process* should be done to discover the similarities between the different territories, i.e. counties, in the state regarding the different crimes. In addition, the discovery of the distribution of the different crime ratios along the whole jurisdiction can afford very interesting information to support the decisions of the police department head.

Therefore, the IDSS should use some *descriptive model*, which will mine the data to discover hidden knowledge patterns. The IDSS prototype will have two basic components: a *grouping component* to obtain homogeneous groups of counties, and a *profiling/interpretation component* to characterize each one of the groups obtained regarding the different values of the corresponding variables.

Type of IDSS

This IDSS will be a static IDSS because there is no need to manage and control a dynamic system in constant evolution. On the contrary, at some time, the IDSS system will be run to analyze a gathered dataset from the police crime reports and statistics collected in the state during a given period, which probably will be 1 year or several ones. From the dataset, several homogeneous groups will be discovered, interpreted, and labelled (profiling task). Finally, all these *data-driven knowledge patterns* will be used to plan the guidelines for the crime control and prevention strategies of the police state.

Kind of tasks involved (Analysis, Synthesis, Prognosis)

The main task in this IDSS is *analytical*. The dataset must be deeply analyzed to get the most valuable cognitive information from it. *Descriptive techniques* and models can be used to identify the most similar observations (territories/counties in concrete years) and characterize what are the common

(continued)

features of these similar counties regarding crime volumes and crime distribution.

Model selection

Among the descriptive models for mining knowledge patterns from data, the adequate in this scenario are the *clustering techniques. Clustering techniques* can obtain, from an unsupervised dataset, some groups of observations that are more similar to observations belonging to the same group and more different from observations belonging to the other groups. This way different counties with similar criminal behaviour along some concrete year could be identified. Furthermore, the distribution of crimes in the different groups is also a valuable information to design strategies for crime control and crime prevention in the state.

Model Integration

As it can be outlined in Fig. 7.23 showing the IDSS architecture, the IDSS has two main components which are executed sequentially. In the first step, a clustering model will be used to obtain the groups or clusters of similar behaviours, i.e. similar counties regarding the crime distribution.

After having obtained the groups, in the second step, these clusters must be validated, structurally and semantically. The qualitative/semantic validation can be done by the expert/s in the MDSP, like the MDSP head for instance. The expert/s must identify the main features of each cluster and recognize that the clusters make sense and are meaningful. This step is named as a *profiling step*, and usually ends giving a semantically appealing label to each one of the groups discovered in the precedent clustering step.[15]

IDSS Functional architecture

The functional architecture of the IDSS for the MDSP case study is depicted in Fig. 7.23.

Implementation and software tools

To implement the IDSS, both the clustering task and the validation and profiling task must be done. The primary source of knowledge is the available dataset. To apply a data-driven model like clustering, the first step is the pre-processing task.

The original dataset had 38 variables and 1008 observations (42 years * 24 counties). The observations are the crime statistics of one county in a concrete year. There was data from 1975 until 2016. There were 23 counties

(continued)

---

[15]Note that there are some people that consider that the *clustering* process includes both the *grouping process* itself (obtaining the partition, *i.e*, the set of clusters) and the *profiling process* (characterization and labelling of the clusters).

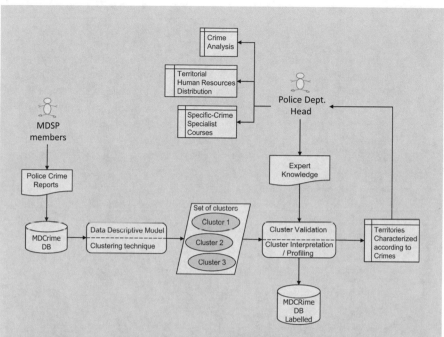

Fig. 7.23 Functional Architecture of the IDSS for the MDSP case study

plus Baltimore city-data. The dataset was analyzed to detect possible missing values, possible errors in the data, possible outliers, redundant variables, and so on.

It was detected that there were some variables having missing values. Those variables were the variables expressing the change in the crime statistics on that county (observation) regarding the previous year. The missing values were on the first year of the data (1975) because obviously, it was not possible to compute the change with previous information. Thus, the missing values were on those variables for the year 1975 for all the 24 counties. As all the variables indicating this change were recorded to analyze the trends or temporal behaviour on the counties along time, we decided that for the required analysis, the temporal component was not the main issue. Thus, 13 variables expressing a change related to the previous year were discarded. From the original 38 variables, 25 variables were kept. The dataset for all variables were examined regarding possible erroneous values, but none was found.

Furthermore, in this dataset, there were several variables measuring related crime data information, like for example the "Grand Total" which is the sum of "Violent Crime Total" plus "Property Crime Total". In a similar way, the variables expressing "Murder", "Rape", "Robbery", "Agg-Assault", "B and

(continued)

E", "Larceny Theft", and "Motor Vehicle Theft" were accounting for the total number of crimes in the county. These numbers are not comparable, because the population in the counties is different. Therefore, the number of crimes must be related to the population. As in the dataset, there was the same number of crimes per 100,000 people, just these variables were used. This means that another seven variables were removed. From the 18 variables remaining, there were some that were correlated with other ones like for instance "Violent Crime Percent" and "Property Crime Percent", because both always sum 100 (100%), etc. In addition, there were the variable "Population" of the county that it was already taken into account in other variables. Finally, there were two variables like "Jurisdiction" and "Year" which are clearly *identifier variables* that identify the observation information (the county and the year), but are not information about the crimes themselves. These two variables can be used as *illustrative variables* as we will explain later.

Finally, the following 10 variables were kept, as the only ones not correlated with others, and being comparable among the different counties:

- Violent crime percent.
- Violent crime rate per 100,000 people.
- Property crime rate per 100,000 people.
- Murder per 100,000 people.
- Rape per 100,000 people.
- Robbery per 100,000 people.
- Agg-assault per 100,000 people.
- B & E per 100,000 people.
- Larceny theft per 100,000 people.
- Motor/vehicle theft per 100,000 people.

As explained above the two followings will be used as *illustrative variables*, once the clusters were formed:

- Jurisdiction.
- Year.

Once the definitive feature selection process was ended, a feature weighting technique was applied, in order to obtain the degree of relevance of the variables, which has great influence in the similarity computation during the clustering process. The UEB1 method (Núñez & Sànchez-Marrè, 2004) being able to cope with unsupervised datasets was used.

The weights obtained for all the ten variables are described in Table 7.5.

From the weights is clear that almost all the variables used are really relevant, with the exception of the variable "Violent Crime percent", which is correlated with "Violent Crime rate per 100,000 people" and "Property

(continued)

Table 7.5 Feature weights for the ten variables

Attribute/variable	Weight
Violent crime percent	0.0
Violent crime rate per 100,000 people	9.9
Property crime rate per 100,000 people	9.5
Murder per 100,000 people	8.8
Rape per 100,000 people	6.0
Robbery per 100,000 people	9.9
Agg-assault per 100,000 people	8.8
B & E per 100,000 people	7.8
Larceny theft per 100,000 people	8.8
Motor/vehicle theft per 100,000 people	10.0

Crime rate per 100,000 people" variables. Notwithstanding, we decided to maintain it for descriptive and illustrative criteria. Furthermore, the variable "Rape per 100,000 people" had a middle weight (6.0) regarding all other variables. It can be noticed that really the number of rapes per 100,000 people, is not varying so much in all the counties compared with other crimes, even though it has some variation ranging from 0.0 to 99.8.

Once the final dataset was obtained and the weights for each remaining variable were determined, some descriptive statistical analysis was done for the 10 variables. The probability distribution of the variables showed some skewness to the left, meaning that the lower values were more probable than the high ones in all the crimes, which is a great new for society!

Afterwards, the *clustering process* was done. It was decided to use the *k*-means method. It was tried with several values of the parameter k, i.e. the number of clusters. It was tried for $k = 3$, $k = 4$, and $k = 5$.

From some analysis and structural validation using CVIs, it seemed that the best partition was obtained with $k = 3$. The obtained partition was the following:

- Class 1 formed by 332 instances.
- Class 2 formed by 616 instances.
- Class 3 formed by 60 instances.

To validate from a qualitative point of view the clusters, an interpretation of the different groups was needed. To help with that the different *prototypes* or *centroids* for each cluster were computed. The resulting centroids are described in Table 7.6.

From the centroids, it can be outlined that the higher values of "Violent crimes per 100,000 people" and "Property crimes per 100,000 people" are in *class 3*. Those values, 1578.59 and 7091.56 are the average crime rate per 100,000 inhabitants in the counties belonging to class 3 (marked in bold in

(continued)

Table 7.6 Description of the centroids of the three clusters

	Class 1 (332 instances)	Class 2 (616 instances)	Class 3 (60 instances)
VIOLENT CRIME PERCENT	12.3587	12.061	17.66
VIOLENT CRIME RATE PER 100,000 PEOPLE	623.8173	*327.9275*	**1578.5896**
PROPERTY CRIME RATE PER 100,000 PEOPLE	4411.7954	*2400.4568*	**7091.5596**
MURDER PER 100,000 PEOPLE	7.2277	3.1577	21.62
RAPE PER 100,000 PEOPLE	31.7234	21.2410	61.47
ROBBERY PER 100,000 PEOPLE	161.6794	52.5089	690.5549
AGG-ASSAULT PER 100,000 PEOPLE	423.1873	251.0176	804.9568
B & E PER 100,000 PEOPLE	1060.2961	622.8991	1762.7787
LARCENY THEFT PER 100,000 PEOPLE	2964.161	1649.8429	4496.3994
MOTOR/VEHICLE THEFT PER 100,000 PEOPLE	387.3437	127.7206	832.3851

Table 7.7 Profiling of the three clusters

	Number of instances	Size of the cluster	Label of the cluster
Class 1	332	Medium	Middle crime ratios
Class 2	616	Large	Lowest crime ratios
Class 3	60	Small	Highest crime ratios

table). Furthermore, the lower values in the same two variables, 327.93 and 2400.46 (marked in italics in table) correspond to the counties belonging to the *class 2*. Finally, the counties belonging to the class 1, have middle values, 623.82 and 4411.80, of crime ratios for violent and property crimes.

In addition, taking a look at the other average crime ratios for the different categories (murder, rape, robbery, aggravated assault, breaking and entering, larceny-theft, and motor vehicle theft), the trend pattern is the same: class 3 has the highest values, class 2 the lowest, and class 1 the middle ones.

From this first interpretation of the clusters, we can make a first interpretation and profiling of the three clusters summarized in Table 7.7.

From Table 7.7 can be checked that the majority class is the class 2, with 616 instances meaning that fortunately the "Lowest crime ratios" is the most common situation and that class 3, corresponding to the "Highest crime ratios" is, fortunately, the minority class, with just 60 observations (counties in a concrete year).

(continued)

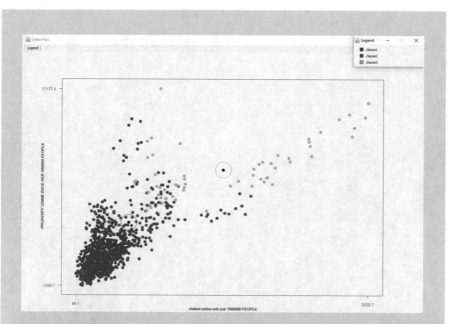

Fig. 7.24 Scatter plot with the class label information regarding the variables "Violent Crime per 100,000 people" versus the "Property Crime per 100,000 people"

In Fig. 7.24, a scatter plot with the variables "Violent Crime per 100,000 people" versus the "Property Crime per 100,000 people" gives a nice picture of the different clusters obtained. Each observation point is depicted with a different colour corresponding to the different clusters. Thus, class 1 observations are depicted in red and correspond to the "Middle crime ratios" counties. Class 2 observations are depicted in blue and correspond to the "Lowest crime ratios" counties. Finally, class 3 observations are depicted in green and correspond to the "Highest crime ratios" counties.

All three clusters are characterized by the volume of crime ratios and the number of observations belonging to each cluster. We wonder whether they share some other common characteristics. One way to obtain that information is to use *illustrative variables*. *Illustrative variables* are those ones that have not been used in the clustering process to obtain the clusters. However, they afford interesting information for identifying which are the observations in the clusters. Especially, the identifier variables like the "Jurisdiction" and the "Year" can be very useful.

The results of using the illustrative variables and see how they behave within the cluster can be seen in Table 7.8.

(continued)

Table 7.8 Illustrative variables in the centroids of the three clusters

	Class 1 (332 instances)	Class 2 (616 instances)	Class 3 (60 instances)
Jurisdiction	Queen Anne's county = 1, Baltimore City = 13, Calvert County = 1, Cecil County = 17, Worcester County = 32, Somerset County = 8, Dorchester County = 36, Harford County = 6, Talbot County = 9, Wicomico County = 32, Allegany County = 1, Caroline County = 3, Howard County = 23, Frederick County = 1, Prince George's county = 31, Baltimore County = 29, St. Mary's county = 3, Montgomery County = 23, Charles County = 28, Anne Arundel county = 35	Kent County = 42, Queen Anne's county = 41, Calvert County = 41, Washington County = 42, Cecil County = 25, Worcester County = 1, Somerset County = 34, Dorchester County = 6, Harford County = 36, Talbot County = 33, Wicomico County = 2, Allegany County = 41, Caroline County = 39, Carroll County = 42, Howard County = 19, Frederick County = 41, Prince George's county = 3, Baltimore County = 7, St. Mary's county = 39, Montgomery County = 19, Garrett County = 42, Charles County = 14, Anne Arundel county = 7	Prince George's county = 8, Baltimore City = 29, Worcester County = 9, Baltimore County = 6, Wicomico County = 8
Year	1/1/1988 = 8, 1/1/1987 = 10, 1/1/1986 = 9, 1/1/1985 = 8, 1/1/1984 = 7, 1/1/2009 = 7, 1/1/1983 = 9, 1/1/2008 = 8, 1/1/1982 = 9, 1/1/2007 = 8, 1/1/1981 = 13, 1/1/2006 = 8,	1/1/1988 = 13, 1/1/1987 = 12, 1/1/1986 = 14, 1/1/1985 = 15, 1/1/1984 = 16, 1/1/2009 = 17, 1/1/1983 = 14, 1/1/2008 = 16, 1/1/1982 = 13, 1/1/2007 = 16, 1/1/1981 = 9, 1/1/2006 = 16,	1/1/2000 = 1, 1/1/1979 = 3, 1/1/1978 = 2, 1/1/1977 = 1, 1/1/1976 = 1, 1/1/1975 = 2, 1/1/2013 = 1, 1/1/1989 = 2, 1/1/1988 = 3, 1/1/1987 = 2, 1/1/1986 = 1, 1/1/1985 = 1,

(continued)

1/1/1980 = 10, 1/1/2005 = 8, 1/1/2004 = 8, 1/1/2003 = 8, 1/1/2002 = 7, 1/1/2001 = 7, 1/1/2000 = 7, 1/1/1979 = 11, 1/1/1978 = 8, 1/1/1977 = 11, 1/1/1976 = 12, 1/1/1975 = 10, 1/1/1999 = 8, 1/1/1998 = 8, 1/1/1997 = 9, 1/1/1996 = 9, 1/1/1995 = 8, 1/1/1994 = 9, 1/1/1993 = 7, 1/1/1992 = 8, 1/1/1991 = 8, 1/1/2016 = 2, 1/1/1990 = 8, 1/1/2015 = 3, 1/1/2014 = 4, 1/1/2013 = 3, 1/1/2012 = 6, 1/1/2011 = 6, 1/1/2010 = 6, 1/1/1989 = 9	1/1/1980 = 12, 1/1/2005 = 16, 1/1/2004 = 16, 1/1/2003 = 16, 1/1/2002 = 16, 1/1/2001 = 16, 1/1/2000 = 16, 1/1/1979 = 10, 1/1/1978 = 14, 1/1/1977 = 12, 1/1/1976 = 11, 1/1/1975 = 12, 1/1/1999 = 15, 1/1/1998 = 15, 1/1/1997 = 14, 1/1/1996 = 11, 1/1/1995 = 13, 1/1/1994 = 12, 1/1/1993 = 13, 1/1/1992 = 11, 1/1/1991 = 11, 1/1/2016 = 22, 1/1/1990 = 13, 1/1/2015 = 21, 1/1/2014 = 20, 1/1/2013 = 20, 1/1/2012 = 18, 1/1/2011 = 18, 1/1/2010 = 18, 1/1/1989 = 13	1/1/1984 = 1, 1/1/1983 = 1, 1/1/1982 = 2, 1/1/1981 = 2, 1/1/1980 = 2, 1/1/1999 = 1, 1/1/1998 = 1, 1/1/1997 = 1, 1/1/1996 = 4, 1/1/1995 = 3, 1/1/1994 = 3, 1/1/1993 = 4, 1/1/1992 = 5, 1/1/1991 = 5, 1/1/1990 = 3, 1/1/2002 = 1, 1/1/2001 = 1

(continued)

Analyzing the jurisdiction distribution in each cluster, the following statements can be formulated:

- The counties with the highest crime ratios are the Baltimore City, which 39 times (i.e. 39 years) got high crime ratios. Other ones that some years got also high crime ratios are the Worcester county (9 years), the Prince George's county (8 years), the Wicomico county (8 years), and the Baltimore county (6 years). This means that those counties and, especially the city of Baltimore, are the territories where the Police Dept. should pay more attention and dedicate more human resources to prevent the crimes.
- The counties with the lowest crime ratios are the Kent county, Washington county, Carroll county and Garrett county, which all the 42 years had the lowest crime ratios. Also Queen Anne's county, Calvert county, Allegany county, Frederick county had the lowest values in 41 years. In addition, Caroline county (39 years), St. Mary's county (39 years), Harford county (36 years), Somerset county (34 years), Talbot county (33 years) and the Cecil county (25 years) are the counties with lowest crime ratios. Therefore, probably they could need less human resources from the Police dept. to control the crime.
- The counties with middle crime ratios are the Dorchester county (36 years), the Anne Arundel county (35 years), the Worcester county and Wicomico county (32 years), the Prince George's county (31 years), the Baltimore county (29 years), the Charles county (28 years), and the Howard county and Montgomery county (23 years).

Taking into account the Year variable, it can be stated the following:

- The counties having the highest crime ratios (i.e. counties belonging to the class 3) were reached in the years from 1975 to 2002, with a mean frequency about 2 counties. Also in 2013, just one county reached the highest values. This means that starting from 2003 to 2016 there were no counties with those highest values. Therefore, it seems that the crime ratios were decreasing starting from 2003. This is an interesting fact to be taken into account.
- The number of counties with the lowest crime ratio has been on average 15 each year from 1975 until 2009. Starting from 2010 until 2016 the number of counties started to increase from 18 until 22. Thus, these means that the number of counties were the crime ratios are within the lowest is increasing since 2010.
- The number of counties with middle crime ratios on average are 7 from the year 1975 until 2009. Starting from 2010 until 2016, the number of counties started to decrease from 6 until 2. Therefore, the number of counties having middle crime ratios decreased since 2010.

(continued)

Summarizing, all these facts related to the number of counties having low and middle crime ratios point out that the crime ratios have been decreasing since last years (2010–2016), which is a very positive fact.

All this interpretation and profiling of the clusters will help the Police dept. head to make the corresponding decisions about how to structure the human resources on the territory, and to know in what region there are more crimes.

Furthermore, knowing which the most common crimes are is very interesting to prepare the policemen to prevent and control these crimes. In this sense, to get valuable information about the different crime ratios per each county, an average computation for each county averaging all the different years can be computed. The table showing this information is described in Table 7.9. The highest crime ratios are marked in bold in the table.

Thus, the next patterns can be extracted from the table:

- *Murder crimes* on average are higher in Baltimore City, Prince George's county, Dorchester county, Somerset county, and Wicomico county.
- *Rape crimes* on average are higher in Baltimore City, Worcester County, Prince George's county, Wicomico county, and Somerset county.
- *Robbery crimes* on average are higher in Baltimore City, Prince George's county, Baltimore county, Wicomico county, and Anne Arundel county.
- *Aggravated assault crimes* on average are higher in Baltimore City, Baltimore county, Worcester county, Wicomico county, and Dorchester county.
- *Breaking and entering crimes* on average are higher in Baltimore City, Worcester county, Prince George's county, Wicomico county, and Baltimore county.
- *Larceny thefts* on average are higher in Baltimore City, Worcester county, Prince George's county, Wicomico county, and Baltimore county.
- *Motor vehicle thefts* on average are higher in Prince George's county, Baltimore City, Baltimore county, Anne Arundel county, and Howard county.

As conclusions, it can be stated:

- Baltimore City is always the worst jurisdiction for *all the crimes but motor vehicle thefts*.
- The five jurisdictions were the *crime rates are the highest* are Baltimore City, Prince George's county, Wicomico county, Baltimore county, and Worcester county.
- Other conflictive jurisdictions in *some types of crime* are: Dorchester county and Somerset county (*murder*), Somerset county (*rape*), Anne Arundel (*robbery*), Dorchester county (*aggravated assault*), Anne Arundel county and Howard county (*motor vehicle thefts*).

(continued)

Table 7.9 Average values along all years of crime ratios per 100,000 inhabitants by county

County	Murder	Rape	Robbery	Agg. Assault	B & E	Larceny theft	Motor vehicle theft
Allegany	2.7	22.6	34.8	247.0	639.3	2101.9	97.7
Anne Arundel	3.6	23.4	**126.7**	320.5	876.9	2723.1	**323.0**
Baltimore city	**36.9**	**59.9**	**990.3**	**950.9**	**1714.6**	**4180.3**	**984.3**
Baltimore	4.1	28.1	**217.0**	**546.6**	987.2	3018.4	**471.9**
Calvert	3.8	18.9	23.1	282.3	599.1	1403.7	102.0
Caroline	4.6	28.8	45.3	322.2	766.3	1594.1	131.4
Carroll	1.4	19.4	28.7	167.3	477.0	1406.1	95.5
Cecil	3.7	23.6	63.2	398.3	907.3	2018.1	228.5
Charles	4.8	24.2	96.3	369.9	691.6	2370.3	288.9
Dorchester	**6.8**	31.2	90.2	**480.8**	912.7	2527.8	159.3
Frederick	2.5	19.1	65.7	314.4	527.4	1735.4	122.5
Garrett	2.6	15.9	10.0	179.9	590.7	1172.6	92.3
Harford	2.8	21.3	73.4	226.8	662.4	1783.7	163.2
Howard	2.6	20.5	83.3	188.9	755.1	2577.6	**314.8**
Kent	2.1	21.0	45.0	262.0	702.7	1409.0	101.1
Montgomery	2.4	20.7	109.3	112.9	622.6	2415.4	305.4
Prince George's	**12.1**	**43.6**	**415.7**	404.8	**1230.8**	**3238.9**	**1083.6**
Queen Anne's	3.0	21.8	33.5	242.8	674.6	1507.5	109.5
Somerset	**6.8**	**35.0**	53.8	344.0	949.9	1616.3	109.6
St. Mary's	3.8	22.9	44.0	279.0	717.0	1718.3	113.9
Talbot	5.0	26.4	65.6	272.0	701.6	2109.7	115.0
Washington	3.2	16.2	70.3	220.5	617.4	1633.5	149.2
Wicomico	**6.8**	**39.7**	**157.9**	**513.2**	**1208.8**	**3069.8**	210.3
Worcester	6.3	**46.2**	91.8	**529.7**	**1502.6**	**4720.6**	251.3

All these patterns extracted are a very valuable information that the IDSS offers to support the final decisions of the Police Dept. head, in order to analyze, control, and prevent crimes in the state. This information helps to manage the human resources, where to focus the attention of the policemen, where the most conflictive jurisdictions are, and where the highest crime ratios for specific types of crime are.

Results and Evaluation

The results from the IDSS are the obtaining of some clusters and their characterization and profiling. All these information let make an accurate characterization of the different territories or jurisdictions of the state or country according to the crimes.

(continued)

With that output, the Police Dept. Head with her/his corresponding support team can make a very precise crime analysis in the territory, as well as planning the human resources allocation to each one of the jurisdictions, and also to plan which crimes are the most worrying that should be controlled and prevented with more specialists in those crimes.

Furthermore, automatic preliminary guidelines and plans for crime control and prevention can be generated automatically from all the clustering characterization.

The evaluation of the utility of IDSS output should primarily be done for the IT director and by the experts in the Police State Dept. regarding the clusters and profiles obtained.

In our case, assuming our role of IT director, the clustering techniques and its profiling were really accurate and offered valuable knowledge patterns hidden in the data, which will be very useful for the Police Dept. Head and her/his team. This knowledge could be used by the Police Dept. Head to plan the crime control and prevention in the jurisdiction or territory under her/his supervision.

Accuracy and efficiency

From the technical point of view, the clusters obtained were validated both structurally and qualitatively, with some CVIs computation and with qualitative validation and interpretation of the meaning of the different clusters.

From an end user point of view, the accuracy should be assessed by a Police Dept. Head. As in our case study we have not this person available, we cannot verify this, but comparing the information extracted and the data used, it seems to be accurate.

Response time

Regarding the efficiency, the IDSS is really efficient because the computation of the clusters and the next interpretations and centroid computations are polynomial in time. In addition, there is no need that the IDSS showed a real-time response because the results do not require it.

Quality of the output generated (alternatives, support given)

The knowledge patterns obtained from the profiling of the clusters are of great value and accuracy to support the decision-making process in the Police Dept. The characterization of the clusters provided a great picture about the crime distribution in the territory. All this information is very important to plan the crime control and prevention for the Police Dept.

Business opportunities (benefit/cost)

Of course, this IDSS prototype can be used and generalized to any other Police Dept., provided that the required data is available. The data-driven methods used, and in particular, the descriptive methods can be used in any other similar problem.

(continued)

From a business point of view, this IDSS, after some extravalidation and extension of functionalities could be used for other Police departments. No economic compensation should be obtained as the prototype was developed using open data coming from public administration. It should be granted for free to other Police departments.

7.2.2   The Use of Associative Models

In this case study, the use of associated models will be illustrated by means of *association rules* model. The domain of application is a common problem in business: the market basket analysis problem.

Case Study: Market Basket Analysis for GIFTCO

The Scenario

All retailing companies are very interested in analyzing the *shopping behaviour* of their customers. In this case study we will focus on a real company of gift shopping by internet. We will use real data from a real online retailing company from UK. We will denominate the company as GIFTCO. The aim of the general manager of GIFTCO is developing an IDSS for helping them to analyze the shopping behaviour of their customers and investigate possible product relationship patterns through the customers' transactions. This analysis is usually denominated as *market basket analysis*.

Problem Analysis

Market basket analysis (Kamakura, 2012) encompasses a broad set of analytics techniques aimed at uncovering the associations and connections between specific items, discovering customer's behaviours and relations between items. The underneath idea is that if a customer buys a certain group of items, is more/less likely to buy another group of items. For example, it is known that when a customer buys beer, in most of the cases, buys chips as well. These behaviours produced in the purchases is what the general manager of GIFTCO was interested in. The manager was interested in analyzing which items are purchased together in order to create new strategies that improved the benefits of the company and customer's experience.

Decisions involved

Analyzing the market basket problem, we think that three types of decisions can be supported by the IDSS in this scenario:

(continued)

- Decisions on what can be personally recommended to a customer. This methodology of the *creation of personalized recommendations* (Portugal et al., 2018) is well known nowadays. During the explosion of e-commerce, personalized recommendations have appeared as a part of the marketing process. It consists in suggesting items to a customer based on his/her preferences. There are several ways to do it, as it will be explained in Chap. 9. The basis for these recommendations rely on knowing what products are bought together by which kind of customers.
- Decisions on how *should be created discounts and promotions*. Based in customer's behaviour, special sales can be offered. For example, if the client knows which items are often purchased together, he/she can create new offers for his/her customers.
- Decisions on how to *spatially distribute the items in physical stores* (Lee & Pace, 2005). Due the increasing number of products that nowadays exist, physical space in stores has started to be a problem. Increasingly, stores invest money and time trying to find which distribution of items can lead them to obtain more profit. Knowing in advance which items are commonly purchased together, the distribution of the store can be changed to obtain more benefits.

In all these decisions there is a key feature which must be known: what products are bought together? And if known, by who are bought these products jointly?

Requirements

The IDSS imagined by the general manager of GIFTCO should meet the following requirements:

- The end-user, i.e. the general manager of the GIFTCO Company, must be able to load into the IDSS a dataset containing retailing transactions of the different products bought by the different customers of the company during a period.
- The IDSS system must be able to discover the different relationships among the different products, i.e. what products are most frequently bought together by their customers?

Data availability

In this case study, the available data comes from the UCI ML Repository (Dua & Graff, 2019), and it was collected and donated by Dr. Daqing Chen, who analyzed this dataset in the work (Chen et al., 2012)[16]

(continued)

---

[16] We are very grateful to them and to UCI ML repository.

This dataset is a transnational dataset that contains all the transactions occurring between 01/12/2010 and 09/12/2011 for a UK-based and registered non-store online retail, named here as GIFTCO. The company mainly sells *unique all-occasion gifts*. Many customers of the company are wholesalers.

The dataset was formed by 541,909 instances and 8 attributes. Each instance represents one product sold in a certain customer order (invoice). Thus, different products of the same order are in different rows forming different observations, but sharing the same number of order (invoice). The eight attributes of the dataset were:

- *Invoice-No*: The invoice number, which is a six-digit number uniquely assigned to each transaction. If this code starts with letter 'c', it indicates a cancellation.
- *Stock-Code*: The product (item) code, which is a five-alphanumeric code uniquely assigned to each distinct product.
- *Description*: The product (item) name, which is a nominal value.
- *Quantity*: The quantities of each product (item) per transaction, which is a numeric value.
- *Invoice-Date*: The invoice/order date and time. It contains the day and time when each transaction was generated.
- *Unit-Price*: The unit price, which is a number expressing the product price per unit in sterling pounds.
- *Customer-ID*: The customer number, which is five-digit number uniquely assigned to each customer.
- *Country*: The country name, which is a nominal value where each customer resides.

Due to some implementation issues, instead of using the original whole dataset with 541,909 instances, we used a sampled dataset from the original of 10,000 instances.

Expert or First-principles knowledge availability

The knowledge acquired by the general manager of GIFTCO and her/his team could be very useful, for the validation of the models obtained through the data-driven methods used here.

However, there was not a first-principles knowledge or theory about the market basket analysis.

A Possible Solution

After the analysis of the given scenario in the GIFTCO Company, and having in mind that the main goal was to find out which products were bought together by the customers, an *associative method* could be used.

The goal of the IDSS is to detect the relationships and associations among several attributes. One of the most common used methods are the *association rules*, which find these associations among the attributes represented as rules (see Sect. 6.4.1.2).

(continued)

The IDSS system will have two basic steps. The first one will be the *preparation of the dataset*. The original dataset had the information of each transaction spread among several observations. The information must be arranged in such a way that the association rules techniques can be used. The second step is the *generation of the association rules* by means of some techniques.

Type of IDSS

The IDSS will belong to the type of static IDSSs. In this case study, there is no need to control and manage a system in a continuous way. The IDSS will be used periodically, when new transactions information from the GIFTCO sales has been accumulated, and it is reasonable to analyze again the behaviour of its customers to check whether their purchase behaviour has changed or not. The knowledge patterns extracted in form of association rules will be generated from the dataset and will be analyzed for the general manager of GIFTCO, in order to make the corresponding decision about marketing, discounts, etc.

Kind of tasks involved (Analysis, Synthesis, Prognosis)

Clearly the main task in this IDSS is *analytical*. The system must analyse the transactions dataset to obtain the correlations and relationships among the different products bought by the customers.

Furthermore, if the IDSS must generate some recommendations for a given customer, which can be an additional requirement/functionality of the system, then a *synthetic* task must be done.

Model selection

From available associative models for extracting correlations patterns among variables, the most suitable seems to be the *association rules* technique. Association rules describe the correlation between the antecedent of the rule and the consequent of the rule. In association rules, both the antecedent and the consequent can be formed by more than one variable. For instance, prod1 $= beer \land$ prod2 $= chips \rightarrow$ prod3 $= cheese$.

Association rules provide very useful information for detecting products that regularly are bought together by customers. Of course, regularity is a key point. As much frequent are the repeated purchase patterns, much credibility can be given to those patterns.

Model Integration

The architecture of the IDSS is shown in Fig. 7.25. It outlines that the system can be divided into three subsystems. The first component is the *data preparation component*, which starting from the original transaction dataset generates an aggregated and grouped dataset ready for using an associative model. Next, the *associative model* is used and the association rules are generated. The final component is the *completion of the association rules* with the identifiers of the products.

(continued)

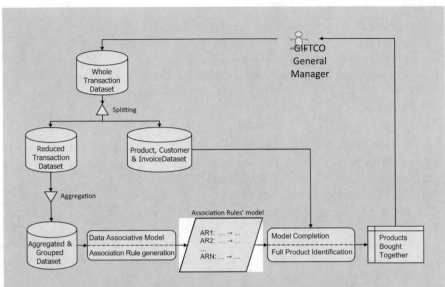

Fig. 7.25 Functional Architecture of the IDSS for the GIFTCO case study

IDSS Functional architecture

The functional architecture of the IDSS for the GIFTCO Company is depicted in Fig. 7.25.

Implementation and software tools

The three components of the IDSS system were implemented. The first component was the *data preparation*, in order that the associative models could be applied. The original dataset had the information of each transaction spread among several observations. Thus, each observation of the dataset had information about:

1. <Invoice/order code>
2. <Product code>
3. <Product description>
4. <Quantity product>
5. <Invoice date and time>
6. <Product unit price>
7. <Customer identifier>
8. <Customer country>.

Obviously, there are many redundant information about the product and the customer, which is not necessary for the analysis of the product sale

(continued)

correlations. Thus, the *original dataset* was split into two datasets. The first one was a *reduced transaction dataset* with just two variables:

1. <Invoice/order code>
2. <Product code>.

The second dataset contained the *secondary information about products, invoices, and customers*. It had information about each product like the description, the quantity, and the unit price and the information about the customer (code and country) and the time and date of the transaction:

1. <Product description>
2. <Quantity product>
3. <Invoice date and time>
4. <Product unit price>
5. <Customer identifier>
6. <Customer country>.

The *reduced transaction dataset* has the same invoice/order code repeated for all products bought at the same invoice/order. Thus, all the products belonging to the same invoice/order must be aggregated and grouped to get a final transaction dataset, which can be used for the association rule method. This dataset has the following structure for each observation (transaction):

1. <Invoice/order code>
2. <Product code 1>
3. <Product code 2>
4. ...
5. <Product code N>

The second component was the *mining of the association rules* from the aggregated and grouped transaction dataset. This process was done using the FP-growth technique. As explained in Sect. 6.4.1.2, the generation of association rules depends on two main parameters: the minimum support (*minsup*) of the frequent itemsets and the minimum confidence (*conf*) of the rules. Depending on these values more or less frequent itemsets, and hence, more or less association rules can be obtained. After some initial trials, we obtained the following results using the *FP-growth* algorithm summarized in Table 7.10.

It was clear that values of *minsup* less than 0.04 (4%) give a huge number of frequent itemsets and association rules, which makes impossible to analyze the obtained rules. We decide that a reasonable parameter would be *minsup* = 0.04 (4%) give a reasonable result of 24 association rules. The 4% of support means that in 400 (0.04 * 10000) transactions that pattern happened, which seems that it could be a reasonable pattern to be taken into account by the manager of the company.

(continued)

Table 7.10 Results of the association rule generation according to *minsup* and *conf* parameters

FP-growth configuration	Number of frequent itemsets	Number of association rules
minsup = 0.2, conf = 0.8	0	0
minsup = 0.2, conf = 0.7	0	0
minsup = 0.1, conf = 0.7	6	0
minsup = 0.05, conf = 0.7	42	3
minsup = 0.04, conf = 0.7	**90**	**24**
minsup = 0.03, conf = 0.7	862	12,832
minsup = 0.025, conf = 0.7	27,119	8,343,635

The list of rules obtained was the following:

Association Rules (*minsup* = 0.04, *conf* = 0.7).

R1: [37370] → [21071] (confidence: 0.704).
R2: [22865, 22867] → [22866] (confidence: 0.704).
R3: [84029E] → [84029G] (confidence: 0.714).
R4: [22866] → [22865] (confidence: 0.729).
R5: [84029G] → [84029E] (confidence: 0.732).
R6: [84029E, 84029G] → [85123A] (confidence: 0.733).
R7: [37370] → [84029E] (confidence: 0.741).
R8: [37370] → [84029E, 84029G] (confidence: 0.741).
R9: [22866, 22867] → [22865] (confidence: 0.760).
R10: [37370] → [84029G] (confidence: 0.778).
R11: [22632, 22865] → [22866] (confidence: 0.793).
R12: [82482] → [82494L] (confidence: 0.826).
R13: [22834] → [22865] (confidence: 0.864).
R14: [21071] → [37370] (confidence: 0.864).
R15: [82482] → [85123A] (confidence: 0.870).
R16: [22745] → [22748] (confidence: 0.870).
R17: [22748] → [22745] (confidence: 0.870).
R18: [85123A, 84029G] → [84029E] (confidence: 0.880).
R19: [22632, 22866] → [22865] (confidence: 0.920).
R20: [71053] → [85123A] (confidence: 0.950).
R21: [71053] → [84029G] (confidence: 0.950).
R22: [84029G, 37370] → [84029E] (confidence: 0.952).
R23: [85123A, 84029E] → [84029G] (confidence: 0.957).
R24: [84029E, 37370] → [84029G] (confidence: 1.000).

Of course, this list is no meaningful as it is because in the rules appear the product codes, and we want that the names or description of the product appeared. To obtain that, the third component of the IDSS makes the

(continued)

transformation using the product, invoice and customer dataset where the information of the description of the product can be extracted.

Therefore, the list of association rules can be transformed into the following one, ordered by decreasing confidence value, which is really meaningful for the manager of GIFTCO:

Association Rules ($minsup = 0.04$, $conf = 0.7$)

R1: [RETRO COFFEE MUGS ASSORTED] → [VINTAGE BILLBOARD DRINK ME MUG] (confidence: 0.704).

R2: [HAND WARMER OWL DESIGN, HAND WARMER BIRD DESIGN] → [HAND WARMER SCOTTY DOG DESIGN] (confidence: 0.704).

R3: [RED WOOLLY HOTTIE WHITE HEART] → [KNITTED UNION FLAG HOT WATER BOTTLE] (confidence: 0.714).

R4: [HAND WARMER SCOTTY DOG DESIGN] → [HAND WARMER OWL DESIGN] (confidence: 0.729).

R5: [KNITTED UNION FLAG HOT WATER BOTTLE] → [RED WOOLLY HOTTIE WHITE HEART] (confidence: 0.732).

R6: [RED WOOLLY HOTTIE WHITE HEART, KNITTED UNION FLAG HOT WATER BOTTLE] → [WHITE HANGING HEART T-LIGHT HOLDER] (confidence: 0.733).

R7: [RETRO COFFEE MUGS ASSORTED] → [RED WOOLLY HOTTIE WHITE HEART] (confidence: 0.741).

R8: [RETRO COFFEE MUGS ASSORTED] → [RED WOOLLY HOTTIE WHITE HEART, KNITTED UNION FLAG HOT WATER BOTTLE] (confidence: 0.741).

R9: [HAND WARMER SCOTTY DOG DESIGN, HAND WARMER BIRD DESIGN] → [HAND WARMER OWL DESIGN] (confidence: 0.760).

R10: [RETRO COFFEE MUGS ASSORTED] → [KNITTED UNION FLAG HOT WATER BOTTLE] (confidence: 0.778).

R11: [HAND WARMER RED RETROSPOT, HAND WARMER OWL DESIGN] → [HAND WARMER SCOTTY DOG DESIGN] (confidence: 0.793).

R12: [WOODEN PICTURE FRAME WHITE FINISH] → [WOODEN FRAME ANTIQUE WHITE] (confidence: 0.826).

(continued)

R13: [HAND WARMER BABUSHKA DESIGN] → [HAND WARMER OWL DESIGN] (confidence: 0.864).

R14: [VINTAGE BILLBOARD DRINK ME MUG] → [RETRO COFFEE MUGS ASSORTED] (confidence: 0.864).

R15: [WOODEN PICTURE FRAME WHITE FINISH] → [WHITE HANGING HEART T-LIGHT HOLDER] (confidence: 0.870).

R16: [POPPY'S PLAYHOUSE BEDROOM] → [POPPY'S PLAYHOUSE KITCHEN] (confidence: 0.870).

R17: [POPPY'S PLAYHOUSE KITCHEN] → [POPPY'S PLAYHOUSE BEDROOM] (confidence: 0.870).

R18: [WHITE HANGING HEART T-LIGHT HOLDER, KNITTED UNION FLAG HOT WATER BOTTLE] → [RED WOOLLY HOTTIE WHITE HEART] (confidence: 0.880).

R19: [HAND WARMER RED RETROSPOT, HAND WARMER SCOTTY DOG DESIGN] → [HAND WARMER OWL DESIGN] (confidence: 0.920).

R20: [WHITE METAL LANTERN] → [WHITE HANGING HEART T-LIGHT HOLDER] (confidence: 0.950).

R21: [WHITE METAL LANTERN] → [KNITTED UNION FLAG HOT WATER BOTTLE] (confidence: 0.950).

R22: [KNITTED UNION FLAG HOT WATER BOTTLE, RETRO COFFEE MUGS ASSORTED] → [RED WOOLLY HOTTIE WHITE HEART] (confidence: 0.952).

R23: [WHITE HANGING HEART T-LIGHT HOLDER, RED WOOLLY HOTTIE WHITE HEART] → [KNITTED UNION FLAG HOT WATER BOTTLE] (confidence: 0.957).

R24: [RED WOOLLY HOTTIE WHITE HEART, RETRO COFFEE MUGS ASSORTED] → [KNITTED UNION FLAG HOT WATER BOTTLE] (confidence: 1.000).

Analyzing the list of association rules obtained, they reveal some interesting knowledge patterns. For instance, tasking the rule R24, which is the one with confidence = 1, meaning that always it can be applied (the antecedent is true) then the consequent is also true. In more practical words: if a customer buys the "RED WOOLLY HOTTIE WHITE HEART" and the "RETRO COFFEE MUGS ASSORTED", also buys the "KNITTED UNION FLAG HOT WATER BOTTLE", which makes sense for the GIFTCO manager

(continued)

because all three items are somehow related and especially complementary for making a gift or gifts to the same person or to several people.

In addition, R22 sets up a correlation among the same three items but in a different order. If a customer buys the "KNITTED UNION FLAG HOT WATER BOTTLE" and the "RETRO COFFEE MUGS ASSORTED", also buys the "RED WOOLLY HOTTIE WHITE HEART", which again makes sense, and set up a clear correlation between these three kinds of objects, which do not belong to the same type of items, but could be displayed in usual packs in the website of the GIFTCO company or even could be distributed physically in the nearby in a store. Furthermore, it can provide some useful recommendations for people who have bought some of them, and the others can be suggested as an interesting joint purchase.

Rules R20 and R21 show that if a customer has bought the item "WHITE METAL LANTERN" is very probable that the item "WHITE HANGING HEART T-LIGHT HOLDER or the "KNITTED UNION FLAG HOT WATER BOTTLE" would be bought with very high confidence (confidence: 0.950).

From rules R16 and R17 it can be outlined that whenever the item "POPPY'S PLAYHOUSE BEDROOM" is purchased, the item "POPPY'S P LAYHOUSE KITCHEN" is also purchased, and the inverse is also true, with high confidence (confidence: 0.870). These clearly mean that usually both items are bought together by customers, which is very reasonable because both objects are objects to be kept in a poppy's playhouse. Clearly, a good discount for both items bought together could be a very interesting business option for the GIFTCO Company.

Rule R12 shows that wooden frames like "WOODEN PICTURE FRAME WHITE FINISH" and "WOODEN FRAME ANTIQUE WHITE" are bought together by most customers with high confidence as well (confidence: 0.826).

Rules R2, R4, R9, R11, and R19 outline that most hand warmers items are bought together, which is very reasonable, and again, give business opportunities for the manager of GIFTCO.

Rules R14 and R1 show that with high confidence (confidence: 0.864), if the "VINTAGE BILLBOARD DRINK ME MUG" is bought then the "RETRO COFFEE MUGS ASSORTED" item is also purchased. The reverse correlation is also true with slightly lower confidence of 0.704. Anyway, both rules are pointing that both gifts are bought frequently together, which is very understandable because both belong to the category of mugs. Again this information is very useful for the manager of GIFTCO.

Results and Evaluation

The results from the IDSS built up were the *set of association rules* obtained applying the association rules techniques which derived these rules

(continued)

from a transaction database of historic purchases in the GIFTCO online retailing company.

These rules outlined several interesting *purchase patterns* gathered from the customers' behaviour data. These patterns are very useful for the GIFTCO Company manager to know which products are purchased together. This valuable information can be used in a smart way for offering adequate discounts to products usually bought together, for recommending a joint purchase of these products on the website of the company or even to physically arrange them near locations in the shelves of the stores to try to maximize the sales of the company.

The manager of the company evaluated the IDSS with very enthusiastic positive feedback. The output of the IDSS gives highly valuable information to support several decisions that a manager has to make related to the sales of the company.

Accuracy and efficiency

The validity of the purchase patterns was relatively accurate regarding the transactions dataset available from the customers. Of course, as in any data-driven model, its accuracy highly depends on the quality of the data. If the data are representative of the purchase behaviour of all customers, then the association rules are meaningful. On the contrary, if they are not, then the association rules will be not representative of the customers' behaviour, and hence, no reliable conclusions can be extracted.

In our case study and assuming that the data were representative, the obtained purchase patterns were accurate, as the associations were confirmed by the sales manager team.

Response time

The response time of the IDSS is highly correlated to the association rule algorithm used to mine the frequent itemsets. In our case, as the FP-growth method was used, and it is a very efficient algorithm to compute them, the general efficiency of the IDSS was very satisfactory.

Furthermore, there was no need of a real-time response from the IDSS, because the management strategies are not requiring a real-time computation.

Quality of the output generated (Alternatives, Support Given)

The association rules generated from the available dataset are very valuable because they outline common sales patterns of the customers of the company. These frequent purchase patterns can be used for making the corresponding decisions in the marketing and discount campaigns.

In addition, they can be used to create recommendations for given users based on her/his past preferences and to her/his similarity to other users and/or to other similar products liked by the user.

(continued)

Furthermore, they can be used to spatially distribute the items either in physical stores or in virtual website page views to maximize the sales volume.

Business opportunities (benefit/cost)

This deployed IDSS is general enough to be used by other companies. Probably, some pre-processing components should be minimally updated, according to the transaction dataset structure.

However, the core of the IDSS, the association rules generation, can be completely reused without any change. This way, the IDSS has an added value of generality, which can be used by other companies to cope with the same decisions in the sales department. Hence, there are good economic perspectives by the GIFTCO for getting some benefit from the IDSS if it is sold under any kind of license.

7.2.3   The Use of Discriminant Models

This case study will illustrate the use of *discriminant models* in the development of IDSSs. *Decision trees* will be used as an example of discriminant models which can predict a qualitative variable. Decision trees are one of the most commonly used discriminant methods in Data Science due to their high interpretability by the end user of the model deployed, i.e. the tree. This case study is based on the Manufacturing Industry and focussed on one of its major problems: the quality control process.

Case Study: SECOM Manufacturing Industry Quality Control

The Scenario

Manufacturing industry is great and fertile terrain for the application of Artificial Intelligence techniques. In this sense, it has been coined the term *Industry 4.0*, as one of the targets for research and innovation projects in the industry in the next years around the world. The term refers to the trend in the fully *automation* and *digitalization* of industries. The broad use of Information and Communication Technologies (ICT) like wireless connectivity, sensors, data gathering and analysis, internet of things, cloud computing, and artificial intelligence techniques characterize these efforts in the industry.

The production of articles in a large scale through machinery keeps space for improvement in several issues in the manufacturing process. *Maintenance* and *predictive maintenance* of the machines involved in the manufacturing process to reduce faults in the production is one of them. The *tuning of the several machines* involved in the manufacture of a concrete product is another issue because usually, this tuning of the different parameters of the machines spends high costs and time in a trial-and-test approach until the right product is

(continued)

fabricated satisfying the corresponding quality parameters. *Quality control* of the manufactured products is another interesting goal in manufacturing industry. It consists of the inspection of the final products to check whether they satisfy or not some minimum quality parameters.

Summarizing, the manufacturing industry shows great potential to apply Intelligent Decision Support Systems. In this case study, we will focus on a particular semi-conductor manufacturing process in a manufacturing industry. We will use real data coming from a real Semi-Conductor Manufacturing of one piece. We will name this manufacturing industry shortly as SECOM Company.

Problem Analysis

This case study will be focussed on the *quality control goal* of manufacturing industry. *Quality control* in manufacturing industry is still usually done in a human-based inspection process of the final products. Of course, many times this process can be automated, and even the final quality assessment of a manufactured piece could be estimated from past historical data. Therefore, it seems that data-driven techniques for the prediction of the final quality of the products can be used.

Usually the final quality control result, which is a final decision regarding each product in the manufacturing chain, assigns a "pass" or "fail" label to each fabricated item. This label is assigned after checking several quality parameters which depend on the type of piece being manufactured. For instance, in a steel manufacturing process, a metallic cylindrical piece probably must be checked regarding the inner diameter of the piece, the thickness of the cylinder, the outer diameter of the piece, and other parameters. Main problem is that many times the different quality parameters can depend on a large number of other parameters collected in the manufacturing process. Usually, all this examination of many parameters is done by experts of the quality control department of the manufacturing industry.

Thus, in the SECOM manufacturing company we will formulate the hypothesis that having a lot of information of signals, variables, and parameters from all the manufacturing processes of a piece, it could be possible to predict whether the piece will satisfy the quality control inspection getting a "pass" label or will not satisfy the quality control inspection getting a "fail" label.

Therefore, the SECOM problem is to have an IDSS that lets staff from Quality Control Dept. to be able to obtain predictions about the quality of each new piece being manufactured. Furthermore, a derived problem is how to identify what is the relevant information related to quality control from the huge quantity of variables in the manufacturing process.

(continued)

Decisions involved

Main decisions in the quality control problem, which have to be supported by the IDSS, are the following two:

- What are the actually relevant information to be taken into account to decide whether a manufactured product satisfies the quality requirements or not?
- Given a manufactured product, should it receive the "pass" label or the "fail" label, according to its satisfaction or not of the quality requirements?

The later decision should be made continuously for each fabricated product in the company. The former one should be made just one time unless new information from the manufacturing process were available.

Requirements

The IDSS designed for the SECOM manufacturing company should meet the following requirements:

- The users of the IDSS, the staff from the Quality Control Dept. of SECOM Company, must be able to load a dataset containing as much as possible information gathered from the whole manufacturing process of one type of piece. For each type of piece, a specific dataset can be loaded.
- The IDSS should be able to make predictions for the quality of a given piece from some values and information gathered during the fabrication process. This prediction will give very valuable support to the Quality Control Dept. staff for deciding whether the fabricated piece pass or not the quality control.

Data availability

For this case study, there was available data coming from the UCI ML Repository (Dua & Graff, 2019), which was collected and donated by Michael McCann, and Adrian Johnston.[17]

This SECOM dataset has data about a *semi-conductor manufacturing process*. Manufacturing processes are under continuous surveillance through the monitoring of signals and variables collected from sensors at several process measurement points. In our case, the semi-conductor manufacturing process data were collected into two datasets.

One dataset contained 590 measures of several variables of the manufacturing process of one unit. The names of the features of the manufacturing process were unknown, which made the problem even more difficult because the *meta-data* associated to the features was missed. The available features were all numeric and were named as F1, F2, ..., F590. There were 1567

(continued)

---

[17] We are very grateful to them and to UCI ML repository.

examples of manufactured pieces in the manufacturing line. An important characteristic of this dataset was the high number of missing values which were 41,951 missing values. This means 4.5% of all the data set.

The other dataset contained the time stamp, date and time of the fabrication process, and the quality control label, which was codified as -1 for the "pass" good pieces and as 1 for the "fail" bad pieces. Thus it was a data matrix of 1567×2. The fabrication dates ranged from July 2008 until October 2008. There were 104 "fail" labels and 1463 "pass" labels. Therefore, the "fail" class was clearly under-represented, which probably will make it very difficult to predict the "fail" labels on the dataset. This problem when you have a minority class regarding the other ones is usually named as an *unbalanced classification problem*. It will have some important consequences as it will be shown afterwards.

Expert or First-principles knowledge availability

There was not available any first-principles knowledge or theory about the semi-conductor manufacturing process and its associated quality parameters.

However, the knowledge and experience acquired by the staff of the Quality Control Dept. (QCD) of the SECOM manufacturing company is a plus for the company. The years of experience and the acquired knowledge let them be able to assess the quality of the manufactured pieces in a safe and trustable way. This knowledge can be used within the IDSS to validate that the predictive models/s obtained from the historical data are really reliable or not.

A Possible Solution

From the analysis of the scenario within the SECOM manufacturing company, it was rather clear that to get a predictive model for the quality of the fabricated pieces, the "pass/fail" label, a *discriminant method* would be a reasonable solution.

Among the discriminant or classifiers methods, one of the most used are the decision trees. *Decision trees* are very useful because their high interpretability. The experts in the application can easily understand which are the features being used in the discrimination process, and how the discrimination rules are formed.

The main goal of the IDSS is to be able to make a good prediction about whether the quality of one piece is accepted or not, given the different measures and features obtained during the manufacturing process.

This IDSS will have three steps. The first one will be the *preparation of the dataset*. The information on the SECOM case study is spread into two datasets, which must be joined to form the final dataset to be analyzed. The second step is the *pre-processing of the dataset*. As described previously there were many missing values, which need to be properly managed. Probably, there will be also some errors or/and outliers that must be managed. In

(continued)

addition, very probably a feature selection process must be undertaken because the 590 descriptive features are really a very high number of features, and probably a dimensionality reduction will be necessary to get the most relevant features in the dataset. The final step will be the *induction of the decision tree* for the prediction of the quality for each manufactured piece.

Type of IDSS

The IDSS developed will be a dynamic one, because continuously it must give the corresponding support information, in the form of a predicted quality label "pass" or "fail" to the QCD staff, for a given semi-conductor manufactured piece. The predicted labels can be used as valuable support information for making the final decision about the quality of the manufactured semi-conductor piece by the QCD staff.

At irregular points of time, new datasets with new information about the manufacture process can be uploaded into the IDSS to be analyzed and possibly to generate different predictive models taking into account the new data, with possible new features of the manufacturing process.

Kind of tasks involved (Analysis, Synthesis, Prognosis)

The main task in this IDSS is an *analytical task*. The system must analyze the data to induce a predictive model for the qualitative variable "Quality Control", from the set of measured features and variables from the manufacturing process.

Notwithstanding, a lot of effort must be devoted to the pre-processing step, and also to the feature selection step to reduce the huge dimensionality of the data.

Model selection

From the possible *discriminant models*, *decision trees* will be used. Other approaches could be case-based discriminator models or nearest-neighbour models, ensemble methods, etc. Decision trees have the advantage that the model itself has a very nice graphical interpretation which is meaningful for an end user regarding the explanation about the discrimination process for labelling new instances.

Model Integration

The architecture of the IDSS is shown in Fig. 7.26. The picture outlines that the system can be divided into four subsystems. The first component is the *data preparation*, where the two original datasets are joined to produce the needed dataset with all the required information. The second component is the *pre-processing* step where the dataset is filtered and cleaned from errors, managing the outliers and missing values. In the third component, some *feature selection* and *feature weighting* techniques are applied to reduce the high dimensionality of the dataset. Finally, the *induction of the decision tree/s* is the last step of the system. The decision tree model will be used for the prediction of the quality assessment of each piece.

(continued)

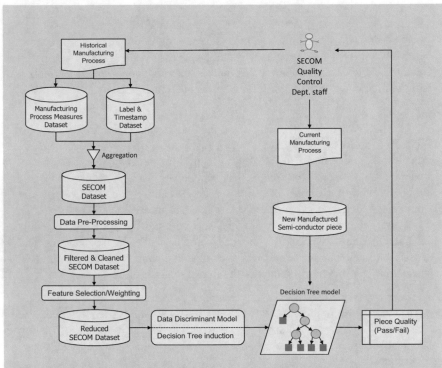

Fig. 7.26 Functional Architecture of the IDSS for the SECOM case study

IDSS Functional architecture

The architecture of the IDSS deployed for the SECOM manufacturing case study is depicted in Fig. 7.26, where the four steps of the system can be shown.

Implementation and software tools

The four major components of the IDSS were implemented. The first subsystem was the *data preparation*. First, it was necessary to join the two available datasets into a single one with all the required information: The dataset with the 590 features from the manufacturing process of all the pieces during the period of July 2008 to October 2008, and the dataset containing the other two features: the quality control final label ("pass/fail") and the timestamp ("Date and Time") of the manufactured piece. Once joined the dataset, the codification of −1 and 1 for the "Quality Control" label was changed to "pass" and "fail".

(continued)

The second great component was the *pre-processing* step. After a first statistical descriptive analysis and graphical exploration of the distribution of the values, the following facts were observed:

- There were several features that had *invariant values*, and hence, they were completely irrelevant, like the features F6, F14, F43, ..., F539.
- There were several features with higher percentages of *missing values* (higher than 45%), like the features F73, F74, ... F582.
- There were several features with some *outlier values*, which probably were erroneous values, like the features F92, F94, ..., F575.
- There were some features with *missing values*, which should be handled in some way.

The invariant-value features were the following ones: F6, F14, F43, F50, F53, F70, F75, F98, F142, F150, F179, F180, F187, F190, F191, F192, F193, F194, F195, F207, F210, F227, F230, F231, F232, F233, F234, F235, F236, F237, F238, F241, F242, F243, F244, F250, F257, F258, F259, F260, F261, F262, F263, F264, F265, F266, F267, F277, F285, F314, F315, F316, F323, F326, F327, F328, F329, F330, F331, F343, F348, F365, F370, F371, F372, F373, F374, F375, F376, F379, F380, F381, F382, F388, F395, F396, F397, F398, F399, F400, F401, F402, F403, F404, F405, F415, F423, F450, F451, F452, F459, F462, F463, F464, F465, F466, F467, F479, F482, F499, F502, F503, F504, F505, F506, F507, F508, F509, F510, F513, F514, F515, F516, F522, F529, F530, F531, F532, F533, F534, F535, F536, F537, F538, F539.

All these 125 features were removed from the dataset. This way, the original 592 features were reduced to 467 features.

The features with percentages of *missing values* higher than 45%, were removed from the dataset. Those features were the following ones: F73, F74, F86, F110, F111, F112, F113, F158, F159, F221, F245, F246, F247, F248, F293, F294, F346, F347, F359, F383, F384, F385, F386, F493, F517, F518, F519, F520, F579, F580, F581 and F582.

All these 32 features were removed from the dataset. This way, the already reduced 467 features were further reduced to 435 features.

In addition, some *suspicious outlier values* were recognized in several features as very probably being error values. Therefore, these erroneous values were converted into *missing values*.

Finally, all the remaining *missing values* in the features were substituted by the average value of each one of the features, aiming not to disturb the distribution probability of the known values. Then, the dataset was prepared to be further processed.

The third component was the *dimensionality reduction* step, where some feature selection or/and feature weighting were applied. This step, in fact, is

(continued)

usually included as a pre-processing step, but its nature it is a little bit different from the simple data cleaning and data preparation.

First, we applied some feature selection process to explore whether the set of 433 attributes available could be reduced losing the minimum quantity of information. Of course, we take out from the feature selection process both the class feature "Quality Control", which is the variable to be predicted, and the "Date and Time" variable, which is an identifier of the manufactured piece. We tried the following techniques:

- *Backward Elimination* process, which is a *decremental feature selection* technique that starts with all the features in the final subset, and assess which are the ones that contribute less to the performance of the candidate subset when inducing a classifier model.

  Surprisingly, the result of this technique just dropped off one feature {F54}. This way, the subset of selected features will be formed by all the features minus this one, i.e. a total of 432 features.

  This probably means that it will be difficult to predict the "Quality Control" target variable, very possibly due to the fact that we have an imbalanced dataset. The imbalance is because we have just 104 examples (6.6%) of the "fail" class label against the 1463 examples (93.4%) of the "fail" class, which is by the way very important to forecast. Therefore, the majority class label "pass" will be clearly easier to predict than the "fail" label which will be more difficult to be predicted. Thus, in general, the prediction of both kinds of labels will be difficult, and this is what is pointing to the backward elimination process, where just only one feature could be removed.

- *Forward Selection* process, which computes the best subset of features starting from an empty subset in an *incremental* way, adding the ones that contribute more to the performance when inducing a classifier model.

  The resulting subset of features obtained was just formed by two features: {F127, F406}. This means that adding more features did not improve the average accuracy. Thus, with just these two features the "pass" label could be easily predicted, but if just a new feature were added, the other class "fail" could not be well predicted, because the space search of features is not stable, and the one by one adding strategy does not work, because some correlation among attributes exists.

These results mean that the *feature selection methods* are suitable when the class distribution is balanced, but not when it is unbalanced. In the latter situation, just a few features are selected or deselected due to the unbalanced problem and to the possible correlation among several attributes.

(continued)

Table 7.11 One confusion matrix of a CART decision tree from a cross-validation process having used a *backward elimination* feature selection process

	True "pass"	True "fail"	Predicted rate
Predicted "pass"	1368	88	93.96%
Predicted "fail"	95	16	14.41%
Class rate	93.51%	15.38%	

Table 7.12 One confusion matrix of a CART decision tree after a cross-validation process having used a *forward selection* feature selection process

	True "pass"	True "fail"	Predicted rate
Predicted "pass"	1452	96	93.80%
Predicted "fail"	11	8	42.11%
Class rate	99.25%	7.69%	

We checked the classification accuracy of both feature selection methods, using a cross-validation scheme, with tenfolds, to induce a decision tree (the CART method was used with no pruning techniques) and to check the accuracy of the decision tree. The confusion matrices obtained were the following ones described in Tables 7.11 and 7.12.

The general accuracy of the *decision tree* having used the backward elimination strategy (i.e., keeping 432 features) was:

$$Acc_{Back\,Elim} = 88.32 \pm 2.38\%$$

Although it seems a good accuracy value of 88.32% (11.68% of error), it can be easily shown in Table 7.11 that this average is misleading. The "pass" class accuracy is 93.51%, whereas the "fail" class accuracy is just of 15.38%. This really means that the "fail" class label is very difficult to predict because it is an under-represented class label in the dataset. Usually, in those unbalanced datasets, a new measure is defined, which is not so misleading as the global accuracy or error. This measure is the so-called *Balanced Error Rate (BER)*, which is defined as the average of the *False Positive Rate (FPR)* and the *False Negative Rate (FNR)*:

$$BER = \frac{FPR + FNR}{2} \tag{7.7}$$

Thus, in this case:

(continued)

$$\text{BER}_{\text{Back Elim}} = \frac{95/(1368 + 95) + 88/(88 + 16)}{2} = \frac{95/1463 + 88/104}{2}$$

$$= \frac{0.065 + 0.846}{2} = \frac{0.911}{2} = 0.4555$$

This *BER* measure (45.55% of averaged error) is by far more meaningful than the 11.68% of global error and states that the error on the "fail" class label is 84.6%, very high compared against the error rate of the "pass" class label which is just of 6.5%. Therefore, the *BER* measure is the criteria to be used once comparing different classifiers when there is an unbalanced distribution of the classes.

The general accuracy of the *decision tree* having used the forward selection strategy (i.e. keeping only two features) was:

$$\text{Acc}_{\text{Forw Sel}} = 93.17 \pm 1.08\%$$

Again, even though it seems a good accuracy value of 93.17% (6.83% of error), it can be easily shown in Table 7.12 that this average is misleading. The "pass" class accuracy is 99.25%, whereas the "fail" class accuracy is just of 7.69%.

In this case, the *BER* measure is:

$$\text{BER}_{\text{ForwSel}} = \frac{11/(1452 + 11) + 96/(96 + 8)}{2} = \frac{11/1463 + 96/104}{2}$$

$$= \frac{0.0075 + 0.923}{2} = \frac{0.9305}{2} = 0.4653$$

This *BER* measure (46.53% of averaged error) is again more meaningful than the 6.83% of global error and states that the error on the "fail" class label is 92.3%, very high compared against the error rate of the "pass" class label which is only of 0.75%.

To complete the *dimensionality reduction process*, several well-known *feature weighting methods* were used: Information Gain (IG) method, Information Gain Ratio (IGR) method, Deviation (DEV) method, RELIEF method, Principal Component Analysis (PCA) method, and Class Value Distribution (CVD) method.

Several trials were done considering different thresholds on the weights to get more or less features selected, and different decision tree induction methods, as well as the use of pruning techniques. In Table 7.13, there is a summary of several parameter values and the final accuracy and *BER* results obtained in the cross-validation process.

(continued)

Table 7.13 Average accuracy and *BER* values for the decision trees of the cross-validation process having used different feature weighting techniques and different decision tree induction methods

Decision tree method	Pruning	Feature weight	Threshold/# features	Global accuracy	Balanced error rate (BER)
ID3 (Inf. Gain)	Yes	IG	(≥ 0.23) / 33	90.11 ± 1.57	*45.04*
ID3 (Inf. Gain)	Yes	IGR	33	89.60 ± 2.22	50.23
ID3 (Inf. Gain)	Yes	DEV	33	89.47 ± 1.61	**48.06**
ID3 (Inf. Gain)	Yes	RELIEF	33	90.43 ± 1.93	**44.87**
ID3 (Inf. Gain)	Yes	PCA	33	88.32 ± 2.01	**47.34**
ID3 (Inf. Gain)	Yes	CVD	33	90.81 ± 1.36	**46.90**
ID3 (Inf. Gain)	Yes	–	All (433)	88.83 ± 1.79	**48.45**
C4.5 (gain ratio)	Yes	IG	(≥ 0.23) / 33	92.92 ± 1.01	48.45
C4.5 (gain ratio)	Yes	IGR	33	93.49 ± 0.26	*47.25*
C4.5 (gain ratio)	Yes	DEV	33	93.30 ± 0.33	50.03
C4.5 (gain ratio)	Yes	RELIEF	33	92.85 ± 0.71	49.83
C4.5 (gain ratio)	Yes	PCA	33	93.24 ± 0.44	50.07
C4.5 (gain ratio)	Yes	CVD	33	93.11 ± 0.27	49.69
C4.5 (gain ratio)	Yes	–	All (433)	92.85 ± 1.08	48.48
CART (impurity)	Yes	IG	(≥ 0.23) / 33	91.32 ± 2.01	*43.95*
CART (impurity)	Yes	IGR	33	91.96 ± 1.04	**44.50**
CART (impurity)	Yes	DEV	33	92.72 ± 0.76	49.89
CART (impurity)	Yes	RELIEF	33	91.13 ± 2.07	45.83
CART (impurity)	Yes	PCA	33	92.41 ± 1.35	50.06
CART (impurity)	Yes	CVD	33	92.34 ± 0.80	48.31
CART (impurity)	Yes	–	All (433)	90.36 ± 1.02	48.92

From the analysis of Table 7.13, it can be stated that the feature weighting techniques getting the best *BER* values were, in descending order: IG, IGR, RELIEF, CVD, PCA, DEV, and the worst configuration was taking all the features. This means that really from the 433 features, just taking the most 33 relevant, the prediction improves. This also means that from the 433 features there are many of them that are redundant regarding the other ones, and they can be removed without affecting the prediction ability of the trees.

In addition, it could be noted that the best approaches (IG, IGR) are obtained using the CART method for inducing the tree, and the other ones using the ID3 scheme. Therefore, the C4.5 method gets the worst results. The best result for each feature weighting method is marked in bold in the table. Furthermore, the best result for each decision tree induction method is marked in italics in the table.

(continued)

Table 7.14 Average accuracy and *BER* values for the decision trees of the cross-validation process with no pruning option and using ID3 and CART decision tree induction methods

Decision Tree method	Pruning	Feature weight	Threshold/# features	Global accuracy	Balanced error rate (BER)
ID3 (Inf. Gain)	No	IG	($\geq$ 0.23) / 33	89.09 ± 1.64	*43.36*
ID3 (Inf. Gain)	No	IGR	33	86.92 ± 2.56	48.98
ID3 (Inf. Gain)	No	RELIEF	33	88.39 ± 2.13	44.62
ID3 (Inf. Gain)	No	CVD	33	86.79 ± 3.35	47.27
ID3 (Inf. Gain)	No	–	All (433)	87.62 ± 1.83	**47.27**
CART (impurity)	No	IG	($\geq$ 0.23) / 33	89.47 ± 1.76	*41.81*
CART (impurity)	No	IGR	33	89.02 ± 1.90	**42.05**
CART (impurity)	No	RELIEF	33	87.30 ± 2.41	46.10
CART (impurity)	No	CVD	33	87.88 ± 1.64	45.34
CART (impurity)	No	–	All (433)	87.56 ± 1.67	47.75
ID3 (Inf. Gain)	No	IG	40	89.15 ± 2.34	46.45
ID3 (Inf. Gain)	No	IGR	40	88.83 ± 2.33	45.28
ID3 (Inf. Gain)	No	RELIEF	40	87.94 ± 2.34	*44.42*
ID3 (Inf. Gain)	No	CVD	40	87.30 ± 2.15	49.23
ID3 (Inf. Gain)	No	–	All (433)	87.62 ± 1.83	47.27
CART (impurity)	No	IG	(> 0.22) / 40	89.73 ± 1.63	*42.12*
CART (impurity)	No	IGR	40	88.90 ± 1.42	42.57
CART (impurity)	No	RELIEF	40	88.71 ± 2.20	**43.12**
CART (impurity)	No	CVD	40	87.11 ± 1.98	46.20
CART (impurity)	No	–	All (433)	87.56 ± 1.67	47.75
CART (impurity)	No	IG	($\geq$ 0.21) / 50	89.53 ± 2.41	*42.22*
CART (impurity)	No	IGR	50	88.44 ± 2.27	43.25
CART (impurity)	No	RELIEF	50	88.26 ± 1.87	43.80
CART (impurity)	No	CVD	50	87.81 ± 1.97	**44.93**
CART (impurity)	No	–	All (433)	87.56 ± 1.67	47.75

In order to try more combinations, the C4.5 method was not used, and only the IG, IGR, RELIEF and CVD methods were tested. Furthermore, to get more accuracy, the pruning option was not set in the induction of the trees to compare the results with the pruning option. Of course, with no pruning, the classifier model could be overfitting to the data, but given the difficulty of the dataset, it was not a negligible choice. In addition, different thresholds (number of features) were evaluated for all the weighting schemes: 33, 40, and 50 features were used.

In Table 7.14 there are the results obtained in this new round of experiments.

(continued)

Table 7.15 Average accuracy and *BER* values of the cross-validation process with no pruning option, using CART decision tree induction method from both the original and the reduced matrix

Decision Tree method	Pruning	Feature weight	Threshold/# features	Global accuracy	Balanced error rate (BER)
CART (impurity)	No	IG	(≥ 0.23)/33	89.47 ± 1.76	*41.81*
CART (impurity)	No	IGR	33	89.02 ± 1.90	**42.05**
CART (impurity)	No	RELIEF	33	87.30 ± 2.41	46.10
CART (impurity)	No	CVD	33	87.88 ± 1.64	**45.34**
CART (impurity)	No	–	All (433)	87.56 ± 1.67	47.75
CART (impurity)	No	IG	(> 0.22) / 40	89.73 ± 1.63	*42.12*
CART (impurity)	No	IGR	40	88.90 ± 1.42	42.57
CART (impurity)	No	RELIEF	40	88.71 ± 2.20	**43.12**
CART (impurity)	No	CVD	40	87.11 ± 1.98	46.20
CART (impurity)	No	–	All (433)	87.56 ± 1.67	47.75
CART (impurity)	No	IG	(≥ 0.23) / 33	90.17 ± 1.92	***39.20***
CART (impurity)	No	IGR	33	89.28 ± 1.76	44.60
CART (impurity)	No	RELIEF	33	88.26 ± 1.72	43.81
CART (impurity)	No	CVD	33	87.11 ± 3.19	48.88
CART (impurity)	No	–	All (590)	88.26 ± 2.21	**46.93**
CART (impurity)	No	IG	(> 0.22) / 40	89.02 ± 1.72	*43.39*
CART (impurity)	No	IGR	40	88.06 ± 1.58	43.01
CART (impurity)	No	RELIEF	40	86.79 ± 1.39	47.27
CART (impurity)	No	CVD	40	86.53 ± 2.26	48.44
CART (impurity)	No	–	All (590)	88.26 ± 2.21	**46.93**

In Table 7.14 the best value of the *BER* for each feature weighting method is marked in bold, and the best value of the *BER* for each configuration of the decision tree induction method is marked in italics. Table 7.14 outlines that the best/lowest *BER* value, 41.81%, is obtained with the IG method, selecting the features with weights ≥0.23, i.e. 33 features, using the CART method to induce the tree with no pruning option.

Finally, to test how reliable was the reduced matrix of 433 features compared with the original matrix of 590 features, the accuracy and *BER* values of CART decision trees were computed using the original dataset, with its 590 original features, and the same feature weighting techniques, with the two thresholds selecting both 33 or 40 features. The results from both the original matrix with the 590 features and the reduced matrix with the 433 features are in Table 7.15.

The table confirms that the IG weighting approach is the one giving the best *BER* value for all the configurations. Comparing the original dataset and the reduced one, it can be seen that using the no feature weighting approach, the

(continued)

Table 7.16 One confusion matrix of a CART decision tree after a cross-validation process having used the IG method with the first 33 features, from the original dataset of 590 features

	True "pass"	True "fail"	Predicted rate
Predicted "pass"	1385	76	94.80%
Predicted "fail"	78	28	26.42%
Class rate	94.67%	26.92%	

Table 7.17 One confusion matrix of a CART decision tree after a cross-validation process having used the IG method with the first 33 features, from the reduced dataset of 433 features

	True "pass"	True "fail"	Predicted rate
Predicted "pass"	1379	81	94.45%
Predicted "fail"	84	23	21.50%
Class rate	94.26%	22.12%	

BER values are 46.93% for the original dataset and 47.75% for the reduced dataset. Therefore, there is a minimum reduction of 0.82% but getting the gain of working with a reduced dataset with 157 features less than in the original dataset.

Of course, what is even more interesting is that using some feature weighting techniques like IG or IGR taking just into account 33 or 40 features instead of all of them, the error of the prediction with the decision trees is reduced from 47.75% and 46.93% to 41.81% and 39.20%. Thus, using much less information the prediction error of the discriminant models of the decision trees improve!

Two confusion matrices from these configurations using the top 33 features obtained according to IG methods for the two datasets, the original and the reduced one, are described in Tables 7.16 and 7.17. The general *Accuracy* and *BER* values are:

$$\mathrm{ACC}_{\mathrm{IG,33,orig}} = 90.17 \pm 1.92\%$$

$$\mathrm{BER}_{\mathrm{IG,33,orig}} = 39.20\%$$

$$\mathrm{ACC}_{\mathrm{IG,33,red}} = 89.47 \pm 1.76\%$$

$$\mathrm{BER}_{\mathrm{IG,33,red}} = 41.81\%$$

The best configurations were obtained with the IG method and selecting just 33 features. The most relevant features according to the IG method were the following ones described in Table 7.18 with their corresponding weights, computed both from the reduced dataset (433 features) and from the original dataset (590 features).

(continued)

Table 7.18 The top 33 features with higher weights (weight ≥ 0.23 in the reduced dataset, and weight ≥ 0.25 in the original dataset) according to IG method and its corresponding weights with decreasing order

Feature (433)	IG weight	Feature (590)	IG weight
F60	1.0	F60	1.0
F104	0.6657	F104	0.6769
F349	0.4839	F349	0.4985
F65	0.4825	F65	0.4971
F131	0.4748	F131	0.4895
F66	0.4633	F478	0.4718
F478	0.4474	F66	0.4658
F34	0.4382	F34	0.4536
F206	0.4311	F206	0.4535
F211	0.4103	F211	0.4263
F153	0.3804	F342	0.3983
F511	0.3754	F511	0.3920
F342	0.3737	F153	0.3854
F288	0.3716	F288	0.3769
F426	0.3639	F426	0.3752
F291	0.3274	F248	0.3738
F130	0.3083	F520	0.3738
F563	0.2815	F291	0.3432
F127	0.2746	F113	0.3349
F29	0.2723	F130	0.3262
F64	0.2720	F563	0.2998
F281	0.2630	F127	0.2931
F39	0.2618	F29	0.2909
F427	0.2618	F281	0.2817
F442	0.2618	F39	0.2805
F122	0.2570	F427	0.2805
F128	0.2552	F442	0.2805
F332	0.2489	F122	0.2758
F80	0.2458	F128	0.2741
F133	0.2349	F332	0.2679
F430	0.2346	F64	0.2655
F429	0.2331	F80	0.2648
F156	0.2304	F133	0.2542

From Table 7.18, it can be outlined that the set of most important attributes is almost the same both when computed from the original dataset and the reduced dataset. This means that with the reduced dataset we are almost not losing information!

From the whole 33 features, 30 of them are in both lists. Thus, just only these features {F430, F429, F156} are in the list computed from the reduced dataset and those ones {F113, F248, F520} are in the list computed from the original dataset. These latter ones could have not been obtained in the list

(continued)

computed from the reduced dataset, because these features have a percentage of missing values of 45.6%, and were removed in the reduced dataset. Furthermore, the first 15 features are the same in both lists, with 11 in the same positions and 4 swapped. The remaining ones share the relative order in most of the cases minus the different features.

Summarizing, the results and conclusions obtained are the following:

- The reduced dataset with 433 features having being pre-processed removing invariant attributes, removing attributes with missing percentages higher than 45%, and having managed the outliers and errors found, and replacing the missing values by the average value of the attributes has a minimal loss of information regarding the complete and original dataset of 590 attributes.
- The application of feature weighting techniques even gets an improved accuracy or Balanced Error Rate (*BER*) reduction from values of 46.93% or 47.75% to values of 39.20% or 41.81%, like the IG method, which was the one giving a greater reduction among the tested techniques. Therefore, passing from the whole dataset with 590 or 433 features to a dataset with just 33 attributes improves the error, which is very nice.
- Taking just these 33 attributes, the *BER* obtained by a Decision tree using CART method with no pruning option with a cross-validation process gives the values of 39.20% or 41.81%. This means that the fold tested in each iteration (156 instances) of the cross-validation process can be predicted with an approximate 60% of accuracy, having used just the remaining ninefolds (1411) to train the decision tree. Therefore, the conclusion is that we can expect a 60% of the accuracy of the quality control prediction. Of course, this is not a very high value, but the problem, as pointed previously is the unbalanced distribution of the two-class labels. The "fail" class label is clearly under-represented. Thus, if more data were available as new instances of "fail" quality control pieces, then the model could be improved.
- Independently of whether more instances were available or not, if we have to build the best decision tree model, we should use all the instances available for training. If we do that with no pruning option, the decision trees obtained are able to *predict correctly all the instances*. Therefore, the models can predict with an accuracy of 100%. Main possible problem, as commented previously, can be that those models surely are overfitting the data, and perhaps will not be generalizable to other data with the same accuracy. However, if we should produce the best models, those would be the ones trained with all the available data. The obtained trees with the two following configurations are described in the following Figs. 7.27 and 7.28.

(continued)

```
F60 > 8.080
|  F334 > 7.207
|  |  F118 > 58.245: FAIL {PASS=0, FAIL=10}
|  |  F118 d 58.245
|  |  |  F16 > 401.393: PASS {PASS=6, FAIL=0}
|  |  |  F16 d 401.393: FAIL {PASS=0, FAIL=1}
|  F334 d 7.207
|  |  F249 > 0.010
|  |  |  F542 > 11.854
|  |  |  |  F394 > 0.132
|  |  |  |  |  F229 > 0.020
|  |  |  |  |  |  F10 > 0.028: PASS {PASS=1, FAIL=0}
|  |  |  |  |  |  F10 d 0.028: FAIL {PASS=0, FAIL=4}
|  |  |  |  |  F229 d 0.020: PASS {PASS=12, FAIL=0}
|  |  |  |  F394 d 0.132: FAIL {PASS=0, FAIL=6}
|  |  |  F542 d 11.854
|  |  |  |  F3 > 2161.978
|  |  |  |  |  F39 > 84.937
|  |  |  |  |  |  F65 > 7.572
|  |  |  |  |  |  |  F268 > 0.021
|  |  |  |  |  |  |  |  F23 > 664.875
|  |  |  |  |  |  |  |  |  F23 > 3484.250: FAIL {PASS=0, FAIL=1}
|  |  |  |  |  |  |  |  |  F23 d 3484.250
|  |  |  |  |  |  |  |  |  |  F52 > 42.677
|  |  |  |  |  |  |  |  |  |  |  F60 > 8.119
|  |  |  |  |  |  |  |  |  |  |  |  F72 > 213.582: FAIL {PASS=0, FAIL=1}
|  |  |  |  |  |  |  |  |  |  |  |  F72 d 213.582
|  |  |  |  |  |  |  |  |  |  |  |  |  F89 > 1637.419
|  |  |  |  |  |  |  |  |  |  |  |  |  |  F587 > -0.002
|  |  |  |  |  |  |  |  |  |  |  |  |  |  |  F32 > 4.703
|  |  |  |  |  |  |  |  |  |  |  |  |  |  |  |  F1 > 3014.535: PASS {PASS=1, FAIL=0}
|  |  |  |  |  |  |  |  |  |  |  |  |  |  |  |  F1 d 3014.535: FAIL {PASS=0, FAIL=1}
|  |  |  |  |  |  |  |  |  |  |  |  |  |  |  F32 d 4.703
|  |  |  |  |  |  |  |  |  |  |  |  |  |  |  |  F68 > 0.770: PASS {PASS=168, FAIL=0}
|  |  |  |  |  |  |  |  |  |  |  |  |  |  |  |  F68 d 0.770
|  |  |  |  |  |  |  |  |  |  |  |  |  |  |  |  |  F1 > 3073.720: FAIL {PASS=0, FAIL=1}
|  |  |  |  |  |  |  |  |  |  |  |  |  |  |  |  |  F1 d 3073.720: PASS {PASS=1, FAIL=0}
|  |  |  |  |  |  |  |  |  |  |  |  |  |  F587 d 0.002: FAIL {PASS=0, FAIL=1}
|  |  |  |  |  |  |  |  |  |  |  |  |  F89 d 1637.419: FAIL {PASS=0, FAIL=1}
|  |  |  |  |  |  |  |  |  |  |  |  F60 d 8.119: FAIL {PASS=0, FAIL=1}
|  |  |  |  |  |  |  |  |  |  F52 d 42.677: FAIL {PASS=0, FAIL=1}
|  |  |  |  |  |  |  |  |  F23 d 664.875: FAIL {PASS=0, FAIL=1}
|  |  |  |  |  |  |  |  F268 d 0.021
|  |  |  |  |  |  |  |  |  F1 > 2997.320: PASS {PASS=1, FAIL=0}
|  |  |  |  |  |  |  |  |  F1 d 2997.320: FAIL {PASS=0, FAIL=2}
|  |  |  |  |  |  |  F65 d 7.572: FAIL {PASS=0, FAIL=2}
|  |  |  |  |  |  F39 d 84.937: FAIL {PASS=0, FAIL=2}
|  |  |  |  F3 d 2161.978: FAIL {PASS=0, FAIL=2}
|  |  F249 d 0.010
|  |  |  F549 > 78.612: PASS {PASS=6, FAIL=0}
|  |  |  F549 d 78.612
|  |  |  |  F12 > 0.947: FAIL {PASS=0, FAIL=9}
|  |  |  |  F12 d 0.947: PASS {PASS=1, FAIL=0}
F60 d 8.080
|  F65 > 31.648
|  |  F426 > 8.265
|  |  |  F16 > 405.673: FAIL {PASS=0, FAIL=7}
|  |  |  F16 d 405.673: PASS {PASS=3, FAIL=0}
|  |  F426 d 8.265
|  |  |  F93 > 0.003
```

Fig. 7.27 CART tree with no pruning from reduced dataset with all 433 features

```
|   |   |   F3 > 2195.050: PASS {PASS=2, FAIL=0}
|   |   |   F3 d 2195.050: FAIL {PASS=0, FAIL=4}
|   |   F93 d 0.003
|   |   |   F152 > 22.078: FAIL {PASS=0, FAIL=1}
|   |   |   F152 d 22.078: PASS {PASS=31, FAIL=0}
| F65 d 31.648
|   | F154 > 0.004
|   |   F430 > 10.403
|   |   |   F91 > 8895.610: PASS {PASS=5, FAIL=0}
|   |   |   F91 d 8895.610
|   |   |   |   F1 > 3224.820: PASS {PASS=1, FAIL=0}
|   |   |   |   F1 d 3224.820: FAIL {PASS=0, FAIL=5}
|   |   F430 d 10.403
|   |   |   F122 > 16.025
|   |   |   |   F121 > 6.292: PASS {PASS=8, FAIL=0}
|   |   |   |   F121 d 6.292
|   |   |   |   |   F1 > 2981.980: FAIL {PASS=0, FAIL=6}
|   |   |   |   |   F1 d 2981.980: PASS {PASS=2, FAIL=0}
|   |   |   F122 d 16.025
|   |   |   |   F575 > 1.622
|   |   |   |   |   F17 > 10.794
|   |   |   |   |   |   F35 > 50.433: FAIL {PASS=0, FAIL=4}
|   |   |   |   |   |   F35 d 50.433: PASS {PASS=9, FAIL=0}
|   |   |   |   |   F17 d 10.794
|   |   |   |   |   |   F31 > 0.285: FAIL {PASS=0, FAIL=1}
|   |   |   |   |   |   F31 d 0.285
|   |   |   |   |   |   |   F136 > 613.500: FAIL {PASS=0, FAIL=1}
|   |   |   |   |   |   |   F136 d 613.500
|   |   |   |   |   |   |   |   F226 > 3244.250: FAIL {PASS=0, FAIL=1}
|   |   |   |   |   |   |   |   F226 d 3244.250
|   |   |   |   |   |   |   |   |   F333 > 6.203
|   |   |   |   |   |   |   |   |   |   F2 > 2527.625: PASS {PASS=3, FAIL=0}
|   |   |   |   |   |   |   |   |   |   F2 d 2527.625: FAIL {PASS=0, FAIL=2}
|   |   |   |   |   |   |   |   |   F333 d 6.203
|   |   |   |   |   |   |   |   |   |   F3 > 2310.683
|   |   |   |   |   |   |   |   |   |   |   F1 > 3028.915: PASS {PASS=1, FAIL=0}
|   |   |   |   |   |   |   |   |   |   |   F1 d 3028.915: FAIL {PASS=0, FAIL=1}
|   |   |   |   |   |   |   |   |   |   F3 d 2310.683
|   |   |   |   |   |   |   |   |   |   |   F72 > 232.512
|   |   |   |   |   |   |   |   |   |   |   |   F1 > 3126.510: FAIL {PASS=0, FAIL=1}
|   |   |   |   |   |   |   |   |   |   |   |   F1 d 3126.510: PASS {PASS=1, FAIL=0}
|   |   |   |   |   |   |   |   |   |   |   F72 d 232.512
|   |   |   |   |   |   |   |   |   |   |   |   F89 > 1955.150
|   |   |   |   |   |   |   |   |   |   |   |   |   F1 > 2939.120: FAIL {PASS=0, FAIL=1}
|   |   |   |   |   |   |   |   |   |   |   |   |   F1 d 2939.120: PASS {PASS=1, FAIL=0}
|   |   |   |   |   |   |   |   |   |   |   |   F89 d 1955.150
|   |   |   |   |   |   |   |   |   |   |   |   |   F332 > 0.025
|   |   |   |   |   |   |   |   |   |   |   |   |   |   F357 > 2.642
|   |   |   |   |   |   |   |   |   |   |   |   |   |   |   F1 > 3008.410: PASS {PASS=1, FAIL=0}
|   |   |   |   |   |   |   |   |   |   |   |   |   |   |   F1 d 3008.410: FAIL {PASS=0, FAIL=1}
|   |   |   |   |   |   |   |   |   |   |   |   |   |   F357 d 2.642
|   |   |   |   |   |   |   |   |   |   |   |   |   |   |   F436 > 0.301
|   |   |   |   |   |   |   |   |   |   |   |   |   |   |   |   F524 > 0.028
|   |   |   |   |   |   |   |   |   |   |   |   |   |   |   |   |   F562 > 93.945
|   |   |   |   |   |   |   |   |   |   |   |   |   |   |   |   |   |   F1 > 3013.955: FAIL {PASS=0, FAIL=1}
|   |   |   |   |   |   |   |   |   |   |   |   |   |   |   |   |   |   F1 d 3013.955: PASS {PASS=1, FAIL=0}
|   |   |   |   |   |   |   |   |   |   |   |   |   |   |   |   |   F562 d 93.945
|   |   |   |   |   |   |   |   |   |   |   |   |   |   |   |   |   |   F62 > 10.921
|   |   |   |   |   |   |   |   |   |   |   |   |   |   |   |   |   |   |   F17 > 10.052: FAIL {PASS=0, FAIL=2}
|   |   |   |   |   |   |   |   |   |   |   |   |   |   |   |   |   |   |   F17 d 10.052: PASS {PASS=8, FAIL=0}
```

Fig. 7.27 (continued)

```
| | | | | | | | | | | | | | | | | F62 d 10.921
| | | | | | | | | | | | | | | | | F575 > 111.135
| | | | | | | | | | | | | | | | | | F1 > 3094.890: FAIL {PASS=0, FAIL=1}
| | | | | | | | | | | | | | | | | | F1 d 3094.890: PASS {PASS=1, FAIL=0}
| | | | | | | | | | | | | | | | | F575 d 111.135
| | | | | | | | | | | | | | | | | | F41 > 84.770
| | | | | | | | | | | | | | | | | | | F1 > 3093.635: FAIL {PASS=0, FAIL=1}
| | | | | | | | | | | | | | | | | | | F1 d 3093.635: PASS {PASS=2, FAIL=0}
| | | | | | | | | | | | | | | | | | F41 d 84.770
| | | | | | | | | | | | | | | | | | | F78 > 0.090
| | | | | | | | | | | | | | | | | | | | F3 > 2190.028: FAIL {PASS=0, FAIL=1}
| | | | | | | | | | | | | | | | | | | | F3 d 2190.028: PASS {PASS=2, FAIL=0}
| | | | | | | | | | | | | | | | | | | F78 d 0.090
| | | | | | | | | | | | | | | | | | | | F101 > -0.001
| | | | | | | | | | | | | | | | | | | | | F157 > 0.015
| | | | | | | | | | | | | | | | | | | | | | F186 > 2.570
| | | | | | | | | | | | | | | | | | | | | | | F212 > 0.012
| | | | | | | | | | | | | | | | | | | | | | | | F489 > 985.866
| | | | | | | | | | | | | | | | | | | | | | | | | F3 > 2236.961: FAIL {PASS=0, FAIL=1}
| | | | | | | | | | | | | | | | | | | | | | | | | F3 d 2236.961: PASS {PASS=3, FAIL=0}
| | | | | | | | | | | | | | | | | | | | | | | | F489 d 985.866
| | | | | | | | | | | | | | | | | | | | | | | | | F302 > 0.365: PASS {PASS=1157, FAIL=0}
| | | | | | | | | | | | | | | | | | | | | | | | | F302 d 0.365
| | | | | | | | | | | | | | | | | | | | | | | | | | F1 > 3061.920: FAIL {PASS=0, FAIL=1}
| | | | | | | | | | | | | | | | | | | | | | | | | | F1 d 3061.920: PASS {PASS=4, FAIL=0}
| | | | | | | | | | | | | | | | | | | | | | | F212 d 0.012
| | | | | | | | | | | | | | | | | | | | | | | | F1 > 3008.660: PASS {PASS=3, FAIL=0}
| | | | | | | | | | | | | | | | | | | | | | | | F1 d 3008.660: FAIL {PASS=0, FAIL=1}
| | | | | | | | | | | | | | | | | | | | | | F186 d 2.570
| | | | | | | | | | | | | | | | | | | | | | | F5 > 1.847: FAIL {PASS=0, FAIL=1}
| | | | | | | | | | | | | | | | | | | | | | | F5 d 1.847: PASS {PASS=3, FAIL=0}
| | | | | | | | | | | | | | | | | | | | | F157 d 0.015
| | | | | | | | | | | | | | | | | | | | | | F1 > 2943.695: PASS {PASS=3, FAIL=0}
| | | | | | | | | | | | | | | | | | | | | | F1 d 2943.695: FAIL {PASS=0, FAIL=1}
| | | | | | | | | | | | | | | | | | | | F101 d -0.001
| | | | | | | | | | | | | | | | | | | | | F2 > 2554.895: FAIL {PASS=0, FAIL=1}
| | | | | | | | | | | | | | | | | | | | | F2 d 2554.895: PASS {PASS=3, FAIL=0}
| | | | | | | | | | | | | | | | | | F524 d 0.028
| | | | | | | | | | | | | | | | | | | F1 > 2988.955: PASS {PASS=1, FAIL=0}
| | | | | | | | | | | | | | | | | | | F1 d 2988.955: FAIL {PASS=0, FAIL=1}
| | | | | | | | | | | | | | | | | F436 d 0.301
| | | | | | | | | | | | | | | | | | F1 > 3055.645: FAIL {PASS=0, FAIL=1}
| | | | | | | | | | | | | | | | | | F1 d 3055.645: PASS {PASS=1, FAIL=0}
| | | | | | | | | | | | | | | | F332 d 0.025
| | | | | | | | | | | | | | | | | F1 > 2967.065: PASS {PASS=1, FAIL=0}
| | | | | | | | | | | | | | | | | F1 d 2967.065: FAIL {PASS=0, FAIL=1}
| | | | | | F575 d 1.622
| | | | | | | F1 > 3141.250: FAIL {PASS=0, FAIL=3}
| | | | | | | F1 d 3141.250: PASS {PASS=4, FAIL=0}
| | | | F154 d 0.004
| | | | F4 > 1347.238: FAIL {PASS=0, FAIL=3}
| | | | F4 d 1347.238: PASS {PASS=1, FAIL=0}
```

Fig. 7.27 (continued)

```
F60 > 8.080
|  F342 > 4.120
|  |  F60 > 15.015
|  |  |  F34 > 9.230: FAIL {PASS=0, FAIL=3}
|  |  |  F34 d 9.230
|  |  |  |  F153 > 0.309: PASS {PASS=14, FAIL=0}
|  |  |  |  F153 d 0.309: FAIL {PASS=0, FAIL=2}
|  |  F60 d 15.015: FAIL {PASS=0, FAIL=9}
|  F342 d 4.120
|  |  F39 > 84.937
|  |  |  F563 > 247.583
|  |  |  |  F511 > 34.255
|  |  |  |  |  F133 > 2.277
|  |  |  |  |  |  F64 > 30.070: FAIL {PASS=0, FAIL=1}
|  |  |  |  |  |  F64 d 30.070
|  |  |  |  |  |  |  F153 > 5.106: FAIL {PASS=0, FAIL=1}
|  |  |  |  |  |  |  F153 d 5.106
|  |  |  |  |  |  |  |  F206 > 13.195
|  |  |  |  |  |  |  |  |  F291 > 0.155
|  |  |  |  |  |  |  |  |  |  F29 > 65.750: FAIL {PASS=0, FAIL=3}
|  |  |  |  |  |  |  |  |  |  F29 d 65.750: PASS {PASS=1, FAIL=0}
|  |  |  |  |  |  |  |  |  F291 d 0.155: PASS {PASS=6, FAIL=0}
|  |  |  |  |  |  |  |  F206 d 13.195
|  |  |  |  |  |  |  |  |  F128 > 0.531
|  |  |  |  |  |  |  |  |  |  F130 > 0.378
|  |  |  |  |  |  |  |  |  |  |  F34 > 8.769: PASS {PASS=13, FAIL=0}
|  |  |  |  |  |  |  |  |  |  |  F34 d 8.769
|  |  |  |  |  |  |  |  |  |  |  |  F342 > 2.851: PASS {PASS=2, FAIL=0}
|  |  |  |  |  |  |  |  |  |  |  |  F342 d 2.851: FAIL {PASS=0, FAIL=3}
|  |  |  |  |  |  |  |  |  |  F130 d 0.378: PASS {PASS=108, FAIL=0}
|  |  |  |  |  |  |  |  |  F128 d 0.531
|  |  |  |  |  |  |  |  |  |  F281 > 0.015
|  |  |  |  |  |  |  |  |  |  |  F127 > 2.609: PASS {PASS=28, FAIL=0}
|  |  |  |  |  |  |  |  |  |  |  F127 d 2.609: FAIL {PASS=0, FAIL=1}
|  |  |  |  |  |  |  |  |  |  F281 d 0.015
|  |  |  |  |  |  |  |  |  |  |  F29 > 68.933: FAIL {PASS=0, FAIL=3}
|  |  |  |  |  |  |  |  |  |  |  F29 d 68.933
|  |  |  |  |  |  |  |  |  |  |  |  F206 > 6.790: PASS {PASS=9, FAIL=0}
|  |  |  |  |  |  |  |  |  |  |  |  F206 d 6.790: FAIL {PASS=0, FAIL=2}
|  |  |  |  F133 d 2.277
|  |  |  |  |  F29 > 70.972: PASS {PASS=2, FAIL=0}
|  |  |  |  |  F29 d 70.972
|  |  |  |  |  |  F29 > 64.306: FAIL {PASS=0, FAIL=3}
|  |  |  |  |  |  F29 d 64.306: PASS {PASS=1, FAIL=0}
|  |  |  F511 d 34.255
|  |  |  |  F427 > 1.464
|  |  |  |  |  F65 > 25.947: PASS {PASS=1, FAIL=0}
|  |  |  |  |  F65 d 25.947: FAIL {PASS=0, FAIL=6}
|  |  |  |  F427 d 1.464: PASS {PASS=6, FAIL=0}
|  |  |  F563 d 247.583
|  |  |  |  F130 > 0.449: PASS {PASS=3, FAIL=0}
|  |  |  |  F130 d 0.449
|  |  |  |  |  F127 > 2.737: FAIL {PASS=0, FAIL=5}
|  |  |  |  |  F127 d 2.737
|  |  |  |  |  |  F332 > 0.077: PASS {PASS=3, FAIL=0}
|  |  |  |  |  |  F332 d 0.077: FAIL {PASS=0, FAIL=2}
|  |  F39 d 84.937: FAIL {PASS=0, FAIL=3}
F60 d 8.080
|  F65 > 31.648
|  |  F426 > 8.265
```

Fig. 7.28 CART tree with no pruning from reduced dataset with IG top 3 features

```
| | | F104 > -0.007
| | | | F29 > 63.139: PASS {PASS=3, FAIL=0}
| | | | F29 d 63.139: FAIL {PASS=0, FAIL=1}
| | | F104 d -0.007: FAIL {PASS=0, FAIL=6}
| | F426 d 8.265
| | | F29 > 68.583
| | | | F104 > -0.007
| | | | | F29 > 70.656: PASS {PASS=1, FAIL=0}
| | | | | F29 d 70.656: FAIL {PASS=0, FAIL=4}
| | | | F104 d -0.007: PASS {PASS=4, FAIL=0}
| | | F29 d 68.583
| | | | F288 > 0.211
| | | | | F34 > 8.794: PASS {PASS=3, FAIL=0}
| | | | | F34 d 8.794: FAIL {PASS=0, FAIL=1}
| | | | F288 d 0.211: PASS {PASS=25, FAIL=0}
| F65 d 31.648
| | F427 > 0.394
| | | F430 > 10.403
| | | | F66 > 26.346: PASS {PASS=4, FAIL=0}
| | | | F66 d 26.346
| | | | | F104 > -0.010: FAIL {PASS=0, FAIL=5}
| | | | F104 d -0.010: PASS {PASS=2, FAIL=0}
| | | F430 d 10.403
| | | | F122 > 16.025
| | | | | F211 > 0.121: FAIL {PASS=0, FAIL=4}
| | | | | F211 d 0.121
| | | | | | F442 > 0.572: PASS {PASS=10, FAIL=0}
| | | | | | F442 d 0.572: FAIL {PASS=0, FAIL=2}
| | | | F122 d 16.025
| | | | | F104 > 0.004
| | | | | | F29 > 69.189. FAIL {PASS=0, FAIL=1}
| | | | | | F29 d 69.189: PASS {PASS=1, FAIL=0}
| | | | | F104 d 0.004
| | | | | | F153 > 5.791
| | | | | | | F29 > 67.533: PASS {PASS=1, FAIL=0}
| | | | | | | F29 d 67.533: FAIL {PASS=0, FAIL=1}
| | | | | | F153 d 5.791
| | | | | | | F332 > 0.025
| | | | | | | | F130 > 0.970
| | | | | | | | | F39 > 87.359: FAIL {PASS=0, FAIL=2}
| | | | | | | | | F39 d 87.359
| | | | | | | | | | F332 > 0.103
| | | | | | | | | | | F29 > 69.339: FAIL {PASS=0, FAIL=3}
| | | | | | | | | | | F29 d 69.339: PASS {PASS=2, FAIL=0}
| | | | | | | | | | F332 d 0.103
| | | | | | | | | | | F211 > 0.055: PASS {PASS=49, FAIL=0}
| | | | | | | | | | | F211 d 0.055
| | | | | | | | | | | | F29 > 68.122: FAIL {PASS=0, FAIL=1}
| | | | | | | | | | | | F29 d 68.122: PASS {PASS=1, FAIL=0}
| | | | | | | | F130 d 0.970
| | | | | | | | | F427 > 0.453
| | | | | | | | | | F342 > 5.916
| | | | | | | | | | | F29 > 67.244: PASS {PASS=4, FAIL=0}
| | | | | | | | | | | F29 d 67.244: FAIL {PASS=0, FAIL=1}
| | | | | | | | | | F342 d 5.916
| | | | | | | | | | | F60 > -5.163
| | | | | | | | | | | | F65 > 18.483
| | | | | | | | | | | | | F153 > 0.188
| | | | | | | | | | | | | | F478 > 10.874
| | | | | | | | | | | | | | | F29 > 72.311: FAIL {PASS=0, FAIL=1}
```

Fig. 7.28 (continued)

```
| | | | | | | | | | | | | | | F29 d 72.311: PASS {PASS=13, FAIL=0}
| | | | | | | | | | | | | | F478 d 10.874
| | | | | | | | | | | | | | | F427 > 0.542: PASS {PASS=611, FAIL=0}
| | | | | | | | | | | | | | | F427 d 0.542
| | | | | | | | | | | | | | | | F427 > 0.542: FAIL {PASS=0, FAIL=1}
| | | | | | | | | | | | | | | | F427 d 0.542: PASS {PASS=51, FAIL=0}
| | | | | | | | | | | | | F153 d 0.188
| | | | | | | | | | | | | | F34 > 8.999: FAIL {PASS=0, FAIL=1}
| | | | | | | | | | | | | | F34 d 8.999: PASS {PASS=6, FAIL=0}
| | | | | | | | | | | | F65 d 18.483
| | | | | | | | | | | | | F60 > 1.525
| | | | | | | | | | | | | | F104 > -0.015
| | | | | | | | | | | | | | | F64 > 16.276
| | | | | | | | | | | | | | | | F39 > 86.358
| | | | | | | | | | | | | | | | | F332 > 0.104
| | | | | | | | | | | | | | | | | | F29 > 69.794: PASS {PASS=1, FAIL=0}
| | | | | | | | | | | | | | | | | | F29 d 69.794: FAIL {PASS=0, FAIL=1}
| | | | | | | | | | | | | | | | | F332 d 0.104
| | | | | | | | | | | | | | | | | | F122 > 15.605: PASS {PASS=44, FAIL=0}
| | | | | | | | | | | | | | | | | | F122 d 15.605
| | | | | | | | | | | | | | | | | | | F60 > 4.788: PASS {PASS=2, FAIL=0}
| | | | | | | | | | | | | | | | | | | F60 d 4.788: FAIL {PASS=0, FAIL=1}
| | | | | | | | | | | | | | | | F39 d 86.358
| | | | | | | | | | | | | | | | | F39 > 86.129: FAIL {PASS=0, FAIL=3}
| | | | | | | | | | | | | | | | | F39 d 86.129: PASS {PASS=1, FAIL=0}
| | | | | | | | | | | | | | | F64 d 16.276
| | | | | | | | | | | | | | | | F127 > 3.883
| | | | | | | | | | | | | | | | | F29 > 69.072: PASS {PASS=2, FAIL=0}
| | | | | | | | | | | | | | | | | F29 d 69.072: FAIL {PASS=0, FAIL=1}
| | | | | | | | | | | | | | | | F127 d 3.883
| | | | | | | | | | | | | | | | | F153 > 1.209
| | | | | | | | | | | | | | | | | | F60 > 2.596: PASS {PASS=5, FAIL=0}
| | | | | | | | | | | | | | | | | | F60 d 2.596: FAIL {PASS=0, FAIL=1}
| | | | | | | | | | | | | | | | | F153 d 1.209
| | | | | | | | | | | | | | | | | | F29 > 65.883: PASS {PASS=301, FAIL=0}
| | | | | | | | | | | | | | | | | | F29 d 65.883
| | | | | | | | | | | | | | | | | | | F29 > 65.756: FAIL {PASS=0, FAIL=1}
| | | | | | | | | | | | | | | | | | | F29 d 65.756: PASS {PASS=8, FAIL=0}
| | | | | | | | | | | | | | F104 d -0.015
| | | | | | | | | | | | | | | F29 > 70.783: PASS {PASS=1, FAIL=0}
| | | | | | | | | | | | | | | F29 d 70.783: FAIL {PASS=0, FAIL=1}
| | | | | | | | | | | | | F60 d 1.525: FAIL {PASS=0, FAIL=1}
| | | | | | | | | | | | F60 d -5.163
| | | | | | | | | | | | | F29 > 75.650: FAIL {PASS=0, FAIL=1}
| | | | | | | | | | | | | F29 d 75.650
| | | | | | | | | | | | | | F60 > -5.165: FAIL {PASS=0, FAIL=1}
| | | | | | | | | | | | | | F60 d -5.165
| | | | | | | | | | | | | | | F430 > 1.829
| | | | | | | | | | | | | | | | F563 > 248.848: PASS {PASS=74, FAIL=0}
| | | | | | | | | | | | | | | | F563 d 248.848
| | | | | | | | | | | | | | | | | F29 > 71.611: FAIL {PASS=0, FAIL=1}
| | | | | | | | | | | | | | | | | F29 d 71.611: PASS {PASS=3, FAIL=0}
| | | | | | | | | | | | | | | F430 d 1.829
| | | | | | | | | | | | | | | | F332 > 0.096: FAIL {PASS=0, FAIL=2}
| | | | | | | | | | | | | | | | F332 d 0.096: PASS {PASS=10, FAIL=0}
| | | | | | | | | F427 d 0.453
| | | | | | | | | | F130 > 0.142
| | | | | | | | | | | F34 > 9.324: PASS {PASS=1, FAIL=0}
| | | | | | | | | | | F34 d 9.324: FAIL {PASS=0, FAIL=2}
| | | | | | | | | | F130 d 0.142
```

Fig. 7.28 (continued)

```
| | | | | | | | | |  F80 > -0.025: PASS {PASS=20, FAIL=0}
| | | | | | | | | |  F80 d -0.025: FAIL {PASS=0, FAIL=1}
| | | | | | | |  F332 d 0.025
| | | | | | | |  F29 > 65.883: PASS {PASS=1, FAIL=0}
| | | | | | | |  F29 d 65.883: FAIL {PASS=0, FAIL=1}
| |  F427 d 0.394
| | |  F29 > 65: FAIL {PASS=0, FAIL=3}
| | |  F29 d 65: PASS {PASS=1, FAIL=0}
```

Fig. 7.28 (continued)

From the analysis of both trees, it can be extracted the conclusion that they are very similar in depth. The tree potentially using all the 433 features from Fig. 7.27 has indeed used just 56. This means that these 56 features are the more relevant ones for discrimination purposes according to the impurity measure used in the induction of the tree. This really outlines the fact that probably the 433 features are not needed to make a good prediction model.

On the other hand, the other tree depicted in Fig. 7.28, which can use just the 33 variables previously selected by the IG method as the most relevant, has the same accuracy and almost equal size. From all 33 features it has used only 29 features. This means that there are 4 features out of all 33 which could not be very relevant for the discrimination purpose. Then, it would be preferable to use this tree with just the 33 (29) more relevant features as the simplest model to predict the quality control of the manufactured piece.

Anyway, as stated before, this model would be the best one to predict new manufacturing piece quality, if we have no more data, but as outlined previously the small number of "fail" instances is a problem that should be solved to get a better scalable data-driven model to be integrated into the IDSS.

Results and Evaluation

The IDSS constructed can answer to the two decisions required by the QCD of the SECOM Company: what are the more relevant information (features) for assessing the quality of the manufactured pieces, and the automatic assessment of the quality of the pieces.

Both decisions can be supported by the different components implemented into the system: the feature selection/weighting component and the induction of the predictive model through a decision tree.

Accuracy and efficiency

The obtained discriminant models, i.e. decision trees, were validated from a quantitative point of view as described in the previous section. With cross-validation processes could be estimated that the generalization power of the models was about 60% of balanced accuracy. Notwithstanding, using all the data available, the induced trees had a 100% accuracy on these data.

(continued)

Of course, as pointed several times in the case study analysis, it should be mandatory to obtain more data, especially from the "fail" manufactured pieces, which will allow the improvement of the predictive component of the system. This way, the predictive model/s will be more scalable than the current one.

In general, the accuracy and the efficiency of the IDSS must be assessed by the QCD of the SECOM Company testing the system predictions against the quality assessment given by its experts in the manufacturing process.

Response time
Regarding the efficiency in a time of the IDSS, it can be stated that the predictive component is really fast because the traversal of a decision tree is linear regarding the number of levels. In the induced decision trees, the depth of the trees are about 20. What is a more expensive process is the iterative process carrying out the pre-processing of the dataset. Thus, if new datasets or new data were available, the pre-processing component can spend a lot of time.

However, with the experience gained with this first pre-processing of the original data, the whole process can be fully automated to make it really fast.

Quality of the output generated (Alternatives, Support Given)
Both the list of relevant features obtained from the IDSS, and the discrimination model to predict the quality of the manufactured pieces was of great value to the SECOM Company.

This knowledge was extremely useful for the company in its daily manufacturing process regarding the quality assessment.

Business opportunities (benefit/cost)
The IDSS system built up for the SECOM company can be generalized to be used for other manufacturing companies both in the same semiconductor manufacturing domain or in other ones, following the same architecture.

Anyway, for a real business opportunity, the system should be validated with more data, and especially, to be more robust, the number of "fail" instances should be increased, to improve the reliability and scalability of the IDSS constructed.

7.2.4 The Use of Predictive Models

In this case study, the use of predictive models in the development of IDSSs will be illustrated. Several methods like Artificial Neural Networks (ANNs), a case-based predictor (k-nearest neighbour or instance-based predictor) and also a multiple linear regression model will be used. This case study is based on a common Sales Management problem: the product pricing problem.

Case Study: Sales Planning in COATEX

The Scenario

Sales Management is an important business area, which is focussed on the deployment and application of several *sales strategies* for the management of sales operations in any firm. One relevant task is the *sales planning* which includes the definition of sales targets, product pricing, sales forecasting and demand management.

The *product pricing* subtask is an actually very important area for the business of any firm. Depending on the product price selected, different sales volume, and consequently, different benefit will be obtained by the company.

In this case study we will focus the particular problem of a new textile industry specialized in the production of coats. This company will be named as COATEX. They produce different types of coats. Although COATEX is relatively new, they have some data from its past experience in the product pricing task.

Problem Analysis

Product pricing is an important problem for sales management in COATEX. They need to set a particular price for a new product, a new coat in their case, aiming at maximizing the sales amount. Setting a price is not an easy task, because to determine the best optimal price, the number of possible units sold, and hence, the sales benefit must be estimated.

Furthermore, an accurate estimation of sales forecasting depends on many factors. The *number of competitor products* to the new one has its influence, but also the *prices of the competitor products*, and especially the price of the main competitor product has a clear influence.

Other factors that can influence the number of units sold could be related to the *marketing strategies* undertaken like whether they were done through several media channels (*multi-channel*) like internet, radio, television, e-mailing, press, etc. or just were done using one channel (*mono-channel*).

The *selling format* could also have some impact. Some formats can be through the internet, in a producer to consumer mode, producer to retailer, etc. In addition, the *application of discount* or not to the new product could have some influence.

Furthermore, the *payments options* like by card or metallic could have some influence on the sales amount.

And finally, the *type of product* probably could be a factor to be taken into account for any company in this estimation process of the sales.

COATEX is a new company, and its Chief Technology Officer (CTO) strongly believes that new technologies like artificial intelligence can be applied to support the sales planning tasks, and especially to the product

(continued)

pricing problem. Therefore, what the CTO is expecting is an IDSS that helped them in the product price setting task of the company.

Decisions Involved

In this sales planning problem, the main decisions to be supported by the IDSS are the following:

- What is the *best price* to be set for a new product of COATEX? Of course, the best price for the company means the price maximizing the sales level.
- Additionally, other decisions that could be supported are what are the *best marketing strategies* for the new product?
- Or what are the *best-selling formats* for the new product?

Last two decisions usually could be supported once the first main decision has been made. Once the best price is determined, it could be characterized and correlated with the best marketing strategy and with the best-selling format.

Requirements

The required IDSS for the COATEX company should meet the following requirements regarding its functionality:

- The users of the IDSS, i.e. the staff from the Sales Department and especially the CTO, must be able to upload into the IDSS, a dataset with historical information about other products including the price set, the number of units sold, the benefit obtained, the competitor's information and all gathered information related to the coat sales done in the past. These datasets could be loaded periodically to update the existing information in the system.
- The main requirement of the IDSS will be that it can make a good prediction of the "optimal" price to be set for each new coat they produce, and maximize the sales amount. This optimal price estimation will be done very probably through the estimation of the number of units sold of the new coat, i.e. estimating the sales amount.

Data availability

In this case study, the COATEX company had available some historical data they gathered in the short period that the company has been launched until now. This dataset was collected from the different coats they have produced and sold in the market during recent years.

The dataset contained data about 44 coats produced by COATEX during the last years. They have gathered 5 variables from its sales. In this small dataset were no missing values. The information gathered for each new coat was:

(continued)

- "Prod-ID": the identifier of the coat product.
- "Unit-Price": the unitary price of the coat set.
- "Comp-Prods": the number of competitor products of the coat.
- "Main-Comp-Price": the price of the main competitor product of the coat.
- "Units-Sold": the number of units sold of the coat.

In addition, they have collected information related to the marketing strategies, the selling formats and the possible discounts. Nevertheless, COATEX has used in these years the same marketing strategy (internet), the same selling format (internet), and no discounts were applied. Therefore, this information is formed by invariable values which are completely irrelevant for the prediction of the sales amount.

Expert or First-principles knowledge availability

In principle, there was not any first-principles knowledge or theory regarding how is the correlation between a determined price of a textile product, a coat in this case study, and the volume of sales.

However, the staff from the Sales Department have started to accumulate some *rules of thumb* that they currently apply to make their estimation of the sales amount. This knowledge can be used to validate, from an "expert" point of view the estimated quantity of product sold or the sales benefit obtained from the *predictive models* used in the IDSS. Of course, this kind of knowledge is not yet very reliable because the company was launched just a few years ago, and this expertise is not yet well established.

A Possible Solution

As a result of the analysis in the COATEX company scenario, it was clear that the IDSS must help them to assess different possible price settings, and for each of them use some *predictive model* to estimate the number of units to be sold or the sales benefit.

This way, the staff from the Sales Department could try to evaluate some candidate prices, which they probably hypothesize are the most reliable ones, and check the corresponding estimated units sold, or equivalently the sales amount to make the final decision about what is the "optimal" price.

There were several *numerical predictor methods* that could be used. As the prediction problem seemed to be very difficult, it seemed a good idea to select more than one to try to model the possible correlation among the numeric variables in different ways. Thus, an *Artificial Neural Network* (*ANN*) model was selected to try to model possible non-linear relationships. A *Multiple Linear Regression* (*MLR*) method was selected to try to model possible linear relationships, and finally, a *Case-Based Predictor* (*CBPred*) was also selected because it can be used in many situations. Afterwards, all the models will be compared to evaluate which is the best model.

(continued)

The IDSS will have three steps, but the first one, which is the *pre-processing step* will be very simple because the dataset is very small, i.e. 44 instances and five variables, and there are no missing values. The important step will be the *mining of the dataset* to induce the different predictive models and to make a comparison of the different predictive models to obtain the best model, which will be the one used for making the sales forecasting. The last step will be the *prediction step*, where the forecasting of the sales volume is made.

Type of IDSS
The nature of the IDSS developed for the COATEX case study will be a *dynamic one* because it must continuously give support to the Sales Department. It will provide the estimation of the number of units sold, for a given unit price of the new coat. This information will be used by the Sales Department to make the final decision about which is the best price setting for the new coat.

Of course, if COATEX has available more information in new datasets, those can be uploaded to the IDSS to be analyzed, and if needed, to generate updated predictive models. In the case that COATEX started another textile product manufacturing different from coats, like sweaters or shirts, then specific datasets with information about the new textile products could be uploaded into the system. Then, new predictive models should be induced to provide the system with the corresponding estimation models for the new products.

Taking into account the possible bad results of a wrong price setting for the business of the company, it can be argued that the IDSS will be a *hard-constrained dynamic IDSS*, because the support provided by it to the end users must be very reliable in order not to compromise the business of COATEX.

Kind of tasks involved (Analysis, Synthesis, Prognosis)
Main tasks in the COATEX IDSS will be an *analytical task* and a *prognosis task*. The *analytical task* will be the data analysis to be done for afterwards inducing the predictive models.

Anyway, the key task of the IDSS will be the *prognosis task* of forecasting the sales with some predictive models. Through this prognosis task, the staff from the Sales Department will be able to decide which is the optimal price of a new textile item produced.

Model selection
As mentioned before, there were many different numerical predictor methods to be used, but as explained before three kind of models were selected according to their diverse modelling properties:

(continued)

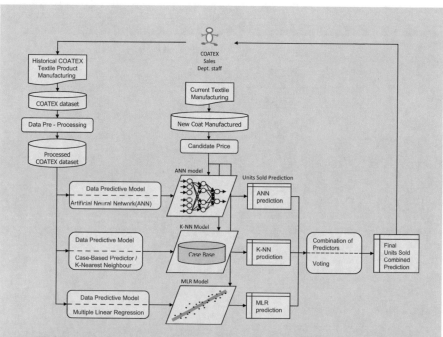

Fig. 7.29   Functional Architecture of the COATEX IDSS case study

- An *Artificial Neural Network* (*ANN*) method to try to model possible non-linear relationships.
- A *Multiple Linear Regression* (*MLR*) method was selected to try to model possible linear relationships.
- A *Case-Based Predictor* (*CBPred*) method, particularly a *k-Nearest Neighbour predictor* trying to exploit its general applicability.

Model Integration

The architecture of the COATEX IDSS is depicted in Fig. 7.29. From the figure, it can be outlined that the system has three basic components. The first component is the *data pre-processing*, which is not very intensive.

The second component is the *data mining component* where the different models are induced. As it will be explained later in the implementation section, it was decided to integrate all three predictors in an *ensemble of predictors* through a *voting scheme*, which showed the best performance as described afterwards.

Finally, the third component is the *prediction component* where for a given new coat and its main characteristics, the system assesses different possible price settings and estimates the number of units sold at that price to support the final decision about the final price that the Sales Department has to determine.

(continued)

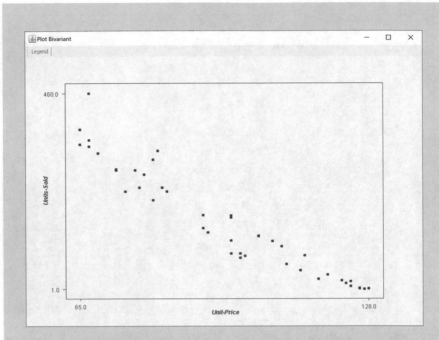

Fig. 7.30   Scatter plot between variables "Unit-Price" and "Units-Sold"

IDSS Functional architecture

The functional architecture of the IDSS deployed for the COATEX case study is depicted in Fig. 7.29. It shows the three main components.

All three components of the IDSS were implemented. The first component was the *data pre-processing*. In this case, this task was not very hard because the dataset was small and there were no missing values. A descriptive statistical analysis was done to check for possible outliers and errors, but they were not found.

The most important outcome from this statistical analysis was the information revealed by the scatter plots done with the target variable "Units-Sold" against the other three variables. As it could be expected, in the plot of the target variable, "Units-sold", against the "Unit-Price" variable it could be appreciated *an inverse linear relationship*. Therefore, as the unit price of the product increases, the number of units sold decreased in a linear way approximately. This correlation can be appreciated in Fig. 7.30.

This fact suggests that a possible predictive model could be a *Multiple Linear Regression* (*MLR*) model because at least between these two variables a linear relationship exists. However, from the other scatter plots, it could be outlined that no linear relationship existed with the other variables. Thus,

(continued)

probably other predictive models than the linear regression model should be checked to capture the other relationships.

The second main component was the *data mining step* where several models were induced from the dataset. The first predictive model was the MLR model. Thus, the target variable was tried to be predicted from the other three explanatory variables. The Root Mean Square Error of the model after a simple validation process (70% for training and 30% for validation) was:

$$RMSE_{MLR} = 29.056$$

This value is pretty good and confirms that there was a high negative correlation between the target variable, "Units-Sold", and the "Unit-Price" variable. Note the negative coefficient $\beta_{UnitPrice} = -6.178$ of the "Unit-Price" variable indicating the negative linear correlation. The obtained model was the following one described in the equation:

$$UnitsSold = 686.826 - 6.178 * UnitPrice - 1.466 * CompProds + 0.887 \\ * MainCompPrice$$

In order to try to capture the possible influence or correlation of the other two variables, which in principle is not linear, other methods were tried. Another model, which was thought that could have some good results was a Case-Based Predictor method, and particularly, a k-Nearest Neighbour (k-NN) predictor.

k-Nearest Neighbour predictors are of wide application and are useful to make good local estimations. However, the power of its model relies on the whole set of instances in the dataset. Therefore, in this case, study, the dataset just has 44 instances which is really a very low number. Thus, probably the results will not be very good. Anyway, we tried it, because in the near future, when more historical data will be available, then probably the k-NN predictor could start improving.

The RMSE results of the k-NN predictor with several values of k after a simple validation process (70% for training and 30% for validation) are depicted in Table 7.19.

Therefore, from the Table 7.19, the best error value for the k-NN predictor is obtained with $k = 2$. Thus:

$$RMSE_{kNN} = 30.494$$

This value is quite similar to the RMSE obtained by the MLR method, which is quite good taking into account that only 44 instances were available. This means that with a great number of instances probably the error could be reduced.

(continued)

Table 7.19 Root Mean Square Error values for different configurations of the k-NN predictor

Predictor model	Root mean square error (RMSE)
k-Nearest Neighbour ($k = 1$)	36.237
k-Nearest Neighbour ($k = 2$)	30.494
k-Nearest Neighbour ($k = 3$)	35.709
k-Nearest Neighbour ($k = 4$)	35.534
k-Nearest Neighbour ($k = 5$)	33.397
k-Nearest Neighbour ($k = 6$)	38.650
k-Nearest Neighbour ($k = 7$)	45.968
k-Nearest Neighbour ($k = 8$)	50.610

Finally, a third model was tried. In order to capture possible non-linear relationships among the explicative variables and the target variable, an *Artificial Neural Network (ANN) model* was tested.

The same scheme of simple validation was used. The 70% for the training set (31 instances) and the remaining 30% (13 instances) for validation. The input layer will have three neurons, one for each predictor variable, and the output layer just only one, i.e. the *target variable*. Remember that from the five variables, one is an identifier of the instance ("Prod-ID"). Thus, from the remaining four variables, one is the target variable to be predicted ("Units-Sold"), and the other three will be the ones to be inputted to the ANN.

To determine the best architecture for the network several trial-and-error iterations were done. We tested several configurations, starting with two hidden layers, and we tried with one up to three nodes. You must remember that a new neuron is automatically added for holding the threshold value at all layers minus the output layer. The neurons in the hidden layer use the *sigmoid function* as activate function, and the output neuron in the output layer uses the *linear function* to predict the numeric *"Units-Sold"* variable.

Furthermore, some values for the parameters of the backpropagation algorithm, *learning rate,* and *momentum* were tried. The best *learning rate* value was $\eta = 0.01$ and the best *momentum value* was $\alpha = 0.9$. The *number of epochs* was tested from 100 up to 300, but in this dataset was clear that more epochs would not give better results but worse results. To measure the goodness of the models, the RMSE was also used as in the previous models.

In Table 7.20, there are the RMSE values obtained for the different configurations, as well as the different parameters of the configuration. The testing was done with two hidden layers varying the number of nodes accordingly (*ANN-2-3-3, ANN-2-3-2, ANN-2-2-2, ANN-2-2-1, ANN-2-1-1*) and testing the values of $\eta = 0.005$ and $\eta = 0.01$ and varying the values of α ($\alpha = 0.9, \alpha = 0.6, \alpha = 0.3.6, \alpha = 0.2$). We varied the number of epochs to

(continued)

Table 7.20 Results with two hidden layer configurations of the ANN for the COATEX dataset

	Hidden layers	Neurons per hidden layer	Learning rate (η)	Momentum (α)	Epochs	RMSE
ANN-2-3-3	2	3–3	0.005	0.9	100	129.414
ANN-2-3-3	2	3–3	0.005	0.9	200	129.398
ANN-2-3-3	2	3–3	0.005	0.9	300	128.977
ANN-2-3-3	2	3–3	0.01	0.9	100	128.673
ANN-2-3-3	2	3–3	0.01	0.9	200	26.842
ANN-2-3-3	2	3–3	0.01	0.9	300	28.042
ANN-2-3-3	2	3–3	0.01	0.6	200	128.417
ANN-2-3-3	2	3–3	0.01	0.3	200	128.794
ANN-2-3-3	2	3–3	0.01	0.2	200	128.889
ANN-2-3-2	**2**	**3–2**	**0.01**	**0.9**	**200**	**26.811**
ANN-2-3-2	2	3–2	0.01	0.6	200	130.257
ANN-2-3-2	2	3–2	0.01	0.3	200	129.655
ANN-2-3-2	2	3–2	0.01	0.2	200	129.573
ANN-2-2-2	2	2–2	0.01	0.9	200	108.740
ANN-2-2-1	2	2–1	0.01	0.9	200	130.923
ANN-2-1-1	2	1–1	0.01	0.9	200	128.618

see how was the behaviour of the configurations. We tried 100, 200, and 300 epochs and it was outlined that the best results were obtained with 200 epochs. This behaviour was confirmed afterwards with other configurations. Thus, it can be concluded that not necessarily in an ANN model, more epochs mean more accuracy or less error. Clearly, the best configuration having two layers got an RMSE of 26.811. It was *ANN-2-3-2*, $\eta = 0.01$, $\alpha = 0.9$, and 200 *epochs*, which is marked in bold in Table 7.20.

Furthermore, some trials were done to check whether the RMSE could be decreased using a configuration with one hidden layer, which is depicted in Table 7.21. The hidden layer was tried with one, two, or three nodes, and checking several values of the *momentum* α (0.9, 0.6, 0.3, 0.2, and 0.1). The learning rate was maintained as $\eta = 0.01$ because it seemed the best value from precedent trails, and the epochs were maintained in 200. The best option for each different layer configuration is marked in italics, and the best one among all is marked in bold in Table 7.21.

Other configurations were tried to test several parameter values, but all gave worse results. Two of them, in the last two rows, are shown because they will be used in the final combination of models. Finally, the best configuration was the *ANN-1-1* with $\eta = 0.01$, $\alpha = 0.2$ and 200 *epochs* with:

(continued)

Table 7.21 Results with one hidden layer configuration of the ANN for the COATEX dataset

	Hidden layers	Neurons per hidden layer	Learning rate (η)	Momentum (α)	Epochs	RMSE
ANN-1-3	1	3	0.01	0.9	200	29.233
ANN-1-3	1	3	0.01	0.6	200	27.876
ANN-1-3	1	3	0.01	0.3	200	26.275
ANN-1-3	1	3	0.01	0.2	200	*25.035*
ANN-1-3	1	3	0.01	0.1	200	25.496
ANN-1-2	1	2	0.01	0.9	200	27.628
ANN-1-2	1	2	0.01	0.6	200	26.994
ANN-1-2	1	2	0.01	0.3	200	24.609
ANN-1-2	1	2	0.01	0.2	200	*24.014*
ANN-1-2	1	2	0.01	0.1	200	28.014
ANN-1-1	1	1	0.01	0.9	200	28.661
ANN-1-1	1	1	0.01	0.6	200	26.825
ANN-1-1	1	1	0.01	0.3	200	24.049
ANN-1-1	**1**	**1**	**0.01**	**0.2**	**200**	**22.997**
ANN-1-1	1	1	0.01	0.1	200	24.052
ANN-1-1	1	1	0.05	0.2	200	29.049
ANN-1-1	1	1	0.1	0.2	200	28.321

$$RMSE_{ANN} = 22.997$$

Although the final RMSE obtained by the *ANN model* was good, a *combination of predictors* was tried to check whether this RMSE could be decreased. Thus, a voting scheme was adopted to combine the predictions of the *ANN model*, the *k-NN model,* and the *MLR model*.

The combination of usually weaker predictors results in a stronger predictor. The results obtained varying the configuration of the *ANN model* but maintaining the best *k-NN model* and the *MLR model* are depicted in Table 7.22.

Therefore, from the analysis of Table 7.22, it can be concluded that the best combination, marked in bold in the table, is not the one using the best ANN model. This happens sometimes due to the cancellation of errors among the different weak predictors composing the ensemble.

Thus, the best error obtained with the voting model, which is, of course, the best among all the models is:

$$RMSE_{Voting} = 18.238$$

(continued)

Table 7.22 Results of the *Voting prediction* for several configurations for the COATEX dataset

	RMSE
Voting (ANN-1-3(0.01,0.2200), 2-NN, MLR)	19.784
Voting (ANN-1-2(0.01,0.2200), 2-NN, MLR)	19.690
Voting (ANN-1-1(0.01,0.2200), 2-NN, MLR)	19.450
Voting (ANN-1-1(0.05,0.2200), 2-NN, MLR)	**18.238**
Voting (ANN-1-1(0.1,0.2200), 2-NN, MLR)	18.530

Table 7.23 Results of the *leave-one-out validation* for the best models for the COATEX dataset

	RMSE
ANN-1-1 (0.01, 0.2, 200)	24.335 ± 21.858
k-NN (k = 2)	30.053 ± 23.197
Multiple linear regression (MLR)	25.562 ± 21.512
Voting (ANN-1-1 (0.05,0.2200), 2-NN, MLR)	21.519 ± 18.603
Voting (ANN-1-1 (0.1,0.2200), 2-NN, MLR)	**21.092 ± 18.564**
Voting (ANN-1-1 (0.01,0.2200), 2-NN, MLR)	22.253 ± 19.942

As explained, the validation scheme used was a *simple validation* strategy. In order to confirm the results of simple validation, a cross-validation process was done with as many folds as instances were available, i.e. using the *leave-one-out validation* strategy. This means that each instance is predicted using all the others as a training set. This scheme is recommended when the size of the dataset is small, as in our case study.

The results of the leave-one-out validation are described for the best configurations of each model in Table 7.23. These results are coherent with the results of the simple validation process. The best model with the cross-validation process is a voting scheme using the *ANN-1-1 (0.1, 0.2, 200)* which was the second best one with the simple validation, but with an almost equal RMSE error.

Therefore, the best predictive model for the estimation of the number of units sold of a new coat is a *voting scheme*, using the three base models, with an *ANN* configuration of *learning rate* of 0.1 or 0.05, a *momentum* of 0.2, and 200 *epochs*.

Results and Evaluation

The IDSS developed for the COATEX company can give the necessary support for the Sales Department staff of the company. They must make the decision about what is the best price for a new coat in the company.

The predictive component of the IDSS can be used to try several candidate prices and obtain the forecast of how many units of the product will be sold.

(continued)

Based on this, the benefit of the company can be estimated. This way they will be able to make the right decision and select the best price according to the sales planning of the company.

Accuracy and efficiency

The accuracy of the IDSS relies in the accuracy of the predictive component. The predictive component is formed by the different predictive models. AS it has been explained before the best predictive model, which was a combination of models can obtain in average an *accuracy* of 79–80%, which is really good, taken into account that the *data-driven predictive models* were obtained just from a dataset of 44 examples.

This means that when new data be available, the IDSS can process the new data and obtain new models probably more accurate. In addition, the *efficiency* of the system is very high, because the predictive component, which is the most complex part, is using very fast predictive models.

On the other hand, the IDSS was evaluated by the Sales Department staff and their CTO, because they have some expertise acquired in the few years that COATEX has been running. Their assessment of the IDDS was really good and they evaluated the IDSS tool as very promising and useful.

Response time

Although this IDSS can be classified as belonging to the *hard-constrained dynamic* type of IDSS, the hard limitations are not in the response time, which is very fast for the IDSS, but on the quality of the sales forecasting of the system.

Those estimations must be accurate enough, in order not to harm the business of the company. From the assessment of the COATEX staff, it seemed that the IDSS provided consistent estimations.

Quality of the output generated (Alternatives, Support Given)

The numerical estimations produced by the data-driven predictive models included in the IDSS were evaluated satisfactorily. It was outlined by the Sales Department staff that they could try several alternatives, i.e. prices, and the system output the new units sold estimation for each possible price.

This *prognostic capacity* was really very important by the Sales Department staff, and the quality of the forecasts was really good.

In addition, the system capabilities could be enlarged to give support to the other decisions enumerated before, related to the selection of the *best marketing strategies* and the *best-selling formats*. After this information be sufficiently recorded in the new datasets, it can be used to check the forecasting scenarios not only changing the price but changing one, two, or both of those values. This way, in addition to the best price, the best marketing strategy and the best-selling formats could be obtained.

(continued)

Business opportunities (benefit/cost)

The generalization of this IDSS for the COATEX company can be easily scaled to other textile products from the same company, provided that data from other sold products were available.

Furthermore, the system can be generalized to other textile industries, and even, to other companies of a different nature. The only requirement is to have data available from past product processes, the number of units sold, information about the competitors, etc.

Of course, before a real business opportunity, the system must be validated with more data. Notwithstanding, the business value of the system to be licensed to other companies is really high.

References

Braubach, L., Pokahr, A., & Lamersdorf, W. (2004). Jadex: A short overview. In *Main Conference Net.ObjectDays*. AgentExpo.

Chen, D., Sain, S. L., & Guo, K. (2012). Data mining for the online retail industry: A case study of RFM model-based customer segmentation using data mining. *Journal of Database Marketing and Customer Strategy Management, 19*(3), 197–208.

Clancey, W. J. (1985). Heuristic classification. *Artificial Intelligence, 27*, 289–350.

Cortés, U., Sànchez-Marrè, M., Ceccaroni, L., R-Roda, I., & Poch, M. (2000). Artificial intelligence and environmental decision support systems. *Applied Intelligence, 13*(1), 77–91.

Devesa, F. (2006). *Development of an Environmental Decision Support System for the Management of Hydraulic Infrastructures, with the Objective of Guaranteeing Water Quality in the Besòs Basin.* (In Catalan). PhD. Thesis, University of Girona.

Devesa, F., De Letter, P., Poch, M., Rubén, C., Freixó, A., & Arráez, J. (2005). Development of an EDSS for the management of the hydraulic infrastructure to preserve the water quality in the Besòs catchment. In *Proceedings of EU-LAT workshop on e-environment* (pp. 31–45).

Dua, D., & Graff, C. (2019). *UCI machine learning repository*. University of California, School of Information and Computer Science. http://archive.ics.uci.edu/ml.

Forrester, J. (1961). *Industrial dynamics*. MIT Press.

Greiner, R., Smith, B. A., & Wilkerson, R. W. (1989). A correction to the algorithm in Reiter's theory of diagnosis. *Artificial Intelligence, 41*(1), 79–88.

Kamakura, W. A. (2012). Sequential market basket analysis. *Marketing Letters, 23*(3), 505–516.

Lee, M.-L., & Pace, R. K. (2005). Spatial distribution of retail sales. *The Journal of Real Estate Finance and Economics, 31*(1), 53–69.

Meadows, D. H., Meadows, D. L., Randers, J., Behrens, I. I. I., & William, W. (1972). *The limits to growth; a report for the Club of Rome's project on the predicament of mankind.* Universe Books. ISBN 0876631650.

Metcalf, L., Eddy, H. P., & Tchobanoglous, G. (2004). *Wastewater engineering: Treatment, disposal and reuse* (Fourth ed.). McGraw Hill.

Montaldo, C. (2019). https://www.thoughtco.com/what-is-a-crime-970836, https://www.thoughtco.com/common-criminal-offenses-970823.

Nica, I., & Wotawa, F. (2012). ConDiag – Computing minimal diagnoses using a constraint solver. In *Proceedings of the International workshop on principles of diagnosis (DX)*.

Nica, I., Pill, I., Quaritsch, T. and Franz Wotawa (2013). The route to success - A performance comparison of diagnosis algorithms. Proceedings of 23rd International Joint Conference on Artificial Intelligence (IJCAI).

Núñez, H., & Sànchez-Marrè, M. (2004). Instance-based learning techniques of unsupervised feature weighting do not perform so badly! In *Proc. of 16th European conference on artificial intelligence (ECAI 2004)* (pp. 102–106). IOS Press.

Padgham, L., & Winikoff, M. (2004). *Developing intelligent agent systems: A practical guide.* Wiley.

Poch, M., Comas, J., Rodríguez-Roda, I., Sànchez-Marrè, M., & Cortés, U. (2004). Designing and building real environmental decision support systems. *Environmental Modelling and Software, 19*(9), 857–873.

Portugal, I., Alencar, P., & Cowan, D. (2018). The use of machine learning algorithms in recommender systems: A systematic review. *Expert Systems with Applications, 97*, 205–227.

Rendón-Sallard, T. (2009). *A multi-agent system framework for simulating scenarios in complex real world problems.* MSc. Thesis in artificial intelligence, Universitat Politècnica de Catalunya.

Rendón-Sallard, T., & Sànchez-Marrè, M. (2007). *Multi-agent prototype for simulating scenarios for decision-making in river basin systems.* Research report LSI-07-6-R. Dept. of computer science. Universitat Politècnica de Catalunya, January.

Richardson, P. G., & Pugh, A., III. (1981). *Introduction to system dynamics modelling.* Productivity Press.

Rosnay, J. (1979). *The Macroscope: A new world scientific system.* Harper & Row.

Senge, P. (1990). *The fifth discipline, the art and practice of the learning organisation.* Century Business.

Sterman, J. D. (2000). *Business dynamics: Systems thinking and modelling for a complex world.* McGraw-Hill Education; HAR/CDR edition.

Further Reading

Cormen, T. H., Leiserson, C. E., Rivest, R. L., & Stein, C. (2009). *Introduction to algorithms* (3rd ed.). MIT Press.

Hopcroft, J. E., & Ullman, J. D. (1979). *Introduction to automata theory, languages and computation.* Addison-Wesley.

Jackson, P. (1999). *Introduction to expert systems* (2nd ed.). Addison Wesley.

Witten, I. H., Frank, E., Hall, M. H., & Pal, C. J. (2016). *Data mining. Practical machine learning tools and techniques* (4th ed.). Morgan Kaufman.

Part III
Development and Application of IDSS

Chapter 8
Tools for IDSS Development

8.1 Introduction

Until now the reader has acquired knowledge about the different models, methods, and techniques (in Chaps. 5 and 6) that can be integrated in an IDSS for helping in the decision-making process in a particular domain. In addition, in Chap. 7, several case studies have been analyzed in order that users get a thorough insight into the process of development of IDSSs.

The final step is to implement the whole IDSS in a computer. We must reveal to the reader, who probably has guessed it, that there is no general tool for deploying an IDSS, as far as we know. This is probably a consequence of not existing a general agreed framework or architecture for the development of IDSS and to the complexity of IDSS, which many times must integrate and interoperate several models and reasoning methods. People build IDSSs in its own way, in an ad hoc manner. However, not all are bad news! As it will be explained in this chapter, there are some useful software tools or software platforms that can be used to make easier the implementation step of an IDSS.

Mainly there are two solutions to implement an IDSS. The first one is to implement all the models, methods, and techniques in a programming language (Java™, C++, etc.) and write all the necessary code to implement the IDSS from scratch. Alternatively, there is the option to use some software tools that provide already implemented data-driven models or model-based techniques. This way you do not have to write all the code, but to ensemble the different calls to modules, packages, functions of the software tools, and write the remaining code to build up your IDSS.

In the next sections of this chapter, we will review the most used available tools for the development of IDSS. First, the tools for data-driven models will be analyzed. Afterwards, the tools for the model-driven techniques. Finally, we will review some general tools which are programming languages that integrate a lot of

© Springer Nature Switzerland AG 2022
M. Sànchez-Marrè, *Intelligent Decision Support Systems*,
https://doi.org/10.1007/978-3-030-87790-3_8

data-driven models and some model-driven techniques, as well as other packages or modules to implement many useful facilities.

8.2   Tools for Data-Driven Methods

In the literature and internet, there are available many Data Mining platforms, suites, or software tools that can help to implement data-driven methods. We will briefly review the most used ones, but this analysis is not exhaustive. A good reference for additional information on Data Mining platforms and software is KDnuggets™ (https://www.kdnuggets.com/), which is a website for AI, Analytics, Big Data, Data Mining, Data Science, and Machine Learning and is edited by Gregory Piatetsky-Shapiro and Matthew Mayo, and launched since 1997. Another interesting source website is The Data Mine (https://the-data-mine.com/), which is a collaborative platform established since 1994.

Here, we will analyze some free software tools, or at least having some free licensed product versions like *Weka*, *RapidMiner Studio*, *KNIME Analytics Platform*, *KEEL*, and *Orange*. In addition, we will analyze two of the most used commercial licensed tools like *IBM SPSS® Modeller* and *SAS® Enterprise Miner™*. The information about the software tools has been extracted from descriptive information in the corresponding websites of the developers.

8.2.1   Weka

Weka which stands for Waikato environment for knowledge analysis (Hall et al., 2009) is an open-source software containing a collection of visualization tools and algorithms for data analysis and predictive modelling, together with graphical user interfaces for easy access to this functionality. Main characteristics are:

- Tool: Weka
- Developers: a research group at the Computer Science Department in the University of Waikato, New Zealand (Witten et al., 2016). Most of Weka 3 code was written by Eibe Frank, Mark Hall, Peter Reutemann, Len Trigg, and significant contributions of other authors.
- Website: http://www.cs.waikato.ac.nz/ml/weka
- Date since it is available: 1999
- Latest stable version (by July 2021): Weka 3.8.5
- Implementation Language: Java™
- Type of license: open-source software

Weka is tried and tested open-source machine learning software that can be accessed through a graphical user interface, standard terminal applications, or a Java™ API. It is widely used for teaching, research, and industrial applications, contains a plethora of built-in tools for standard machine learning tasks, and additionally gives transparent access to well-known toolboxes such as Scikit-learn, R, and Deeplearning4j.

Weka supports several standard DM tasks: data pre-processing, clustering, classification, regression, visualization, and feature selection. Weka provides access to SQL databases using Java™ Database Connectivity and can process the result returned by a database query.

Weka contains several components. The main user interface is managed by the component named *Explorer*, but essentially the same functionality can be accessed through the component-based *Knowledge Flow interface* and from the command line. There is also the component *Experimenter*, which allows the systematic comparison of the predictive performance of Weka's machine learning algorithms on a collection of datasets rather than a single one. Weka is a general-purpose package, freely available on the Internet and it became rather famous in the Artificial Intelligence community.

8.2.2 RapidMiner Studio

RapidMiner comprises a set of platforms for data science. It is one of the most used Data Mining tools. It is available in several products: a stand-alone application for data analysis (RapidMiner Studio), a AutoML tool built for anyone (RapidMiner Go), a product enabling writing code in a managed notebook environment (RapidMiner Notebooks), and an enterprise-wide platform for collaboration, decision automation, deployment and control (RapidMiner AI Hub, formerly RapidMiner Server).

Here we will focus on RapidMiner Studio, which is a comprehensive data science platform with visual workflow design and full automation. Main features are:

- Tool: RapidMiner Studio.
- Developer: © RapidMiner GmbH 2001–2021.
- Website: http://rapidminer.com/
- Date since it is available: 2001
- Latest stable version (by July 2021): RapidMiner Studio 9.9.2.
- Implementation Language: Java™.
- Type of license: Free, Professional, Enterprise. Also, an educational license program exists.

A lot of applications of RapidMiner have been constructed. Its main outstanding functionalities are: GUI interface, Data Integration, Analytical ETL, Data Analysis, and Reporting in one single suite; Powerful but intuitive graphical user interface for the design of analysis processes; Repositories for process, data and metadata handling; Only solution with metadata transformation: forget trial and error and inspect results already during design time; Only solution which supports on-the-fly error recognition and quick fixes; Complete and flexible: Many formats of data loading, data transformation, data modelling, and data visualization methods.

8.2.3   KNIME Analytics Platform

KNIME (Konstanz Information Miner) is a user-friendly and comprehensive data integration, processing, analysis, and exploration platform. In early 2004, at the Chair for Bioinformatics and Information Mining at the Department of Computer and Information Science of the University of Konstanz (Germany), a team of developers from a Silicon Valley software company specializing in pharmaceutical applications started working on a new open-source platform as a collaboration and research tool. When the first version of KNIME was released in 2006, several pharmaceutical companies began using it and, soon thereafter, software vendors started building KNIME-based tools.

KNIME has two platforms: the KNIME Analytics platform, which is the free, open-source software for creating data science, and the KNIME Server, which is the commercial solution for productionizing data science.

Here on, we will focus on the KNIME Analytics platform. KNIME Analytics Platform is a modern data analytics platform that allows performing sophisticated statistics and data mining on data to analyze trends and predict potential results. Its visual workbench combines data access, data transformation, initial investigation, powerful predictive analytics and visualization. KNIME Analytics Platform also provides the ability to develop reports based on information or automate the application of new insight back into production systems. KNIME Analytics Platform is open-source and available under GPL license. It can be extended to include professional support and large enterprise functionality, providing the best of both worlds. Main characteristics are:

- Tool: KNIME Analytics platform
- Developer: © KNIME AG, Zurich, Switzerland
- Website: https://www.knime.org/
- Date since it is available: 2006
- Latest stable version (by July 2021): KNIME Analytics platform 3.7.2
- Implementation Language: Java™, and based on Eclipse platform
- Type of license: Free, open-source software

The KNIME Analytics Platform base version already incorporates hundreds of processing nodes for data I/O, pre-processing and cleansing, modelling, analysis, and data mining as well as various interactive views, such as scatter plots, parallel coordinates. and others. It integrates all analysis modules of the well-known WEKA data mining environment and additional plugins allow R-scripts[1] to be run, offering access to a vast library of statistical routines.

---

[1] A script is a sequence of instructions expressed in some programming language or commands expressed in some operating system.

KNIME Analytics Platform is based on the Eclipse[2] platform, which permits easy building and delivery of integrated tools. Through its modular API, it is easily extensible. When desired, custom nodes and types can be implemented in the KNIME Analytics Platform within hours thus extending the KNIME Analytics Platform to comprehend and provide first-tier support for highly domain-specific data. This modularity and extensibility permit the commercial platform KNIME Server to be employed in commercial production environments as well as teaching and research prototyping settings.

8.2.4 KEEL

KEEL (Knowledge Extraction based on Evolutionary Learning) is an open-source (GPLv3) Java™ software tool that can be used for a large number of different knowledge data discovery tasks (Triguero et al., 2017).

Main characteristics of the tool are:

- Tool: KEEL
- Developers: Five research groups from Spanish Universities (Univ. of Granada, Univ. of Oviedo, Univ. of Jaén, Univ. of Córdoba and Univ. Ramon Llull)
- Website: http://www.keel.es/
- Date since it is available: 2006
- Latest stable version (by July 2021): KEEL 3.0
- Implementation Language: Java™
- Type of license: open-source software (GPLv3)

KEEL provides a simple GUI based on the data flow to design experiments with different datasets and computational intelligence algorithms (paying special attention to evolutionary algorithms) in order to assess the behaviour of the algorithms.

It contains a wide variety of classical knowledge extraction algorithms, pre-processing techniques (training set selection, feature selection, discretization, imputation methods for missing values, among others), computational intelligence-based learning algorithms, hybrid models, statistical methodologies for contrasting experiments, and so forth.

It allows to perform a complete analysis of new computational intelligence proposals in comparison to existing ones. Moreover, KEEL has been designed with a twofold goal: research and education. For a detailed description, see the section "Description" on the left menu.

---

[2]Eclipse (https://www.eclipse.org/) is an open-source multi-language software Integrated Development Environment (IDE), extendible with many plug-ins. It is provided by the Eclipse Foundation, a not-for-profit organization.

8.2.5   Orange

Orange (Demsar et al., 2013) Data Mining fruitful and fun is a machine learning and data mining suite for data analysis through Python scripting and visual programming. Orange was conceived in the late 1990s and is one of the earlier Python's software packages. It focusses on simplicity, interactivity through scripting, and component-based design. Main features are:

- Tool: Orange.
- Developer: Bioinformatics Lab at University of Ljubljana, Slovenia, in collaboration with the open-source community.
- Website: https://orange.biolab.si/
- Date since it is available: 2011
- Latest stable version (by July 2021): Orange3 3.29.3
- Implementation Language: Python
- Type of license: Free software

Orange library is a hierarchically organized toolbox of data mining components. The low-level procedures at the bottom of the hierarchy, like data filtering, probability assessment, and feature scoring, are assembled into higher-level algorithms, such as classification tree learning. This allows developers to easily add new functionality at any level and fuse it with the existing code.

The tool is designed to simplify the assembly of data analysis workflows and the crafting of data mining approaches from a combination of existing components.

As it is a component-based software, the components of orange are called widgets. These widgets range from data visualization and pre-processing to an evaluation of algorithms and predictive modelling. Widgets offer major functionalities like: showing data table and allowing to select features, reading the data, training predictors, and to compare learning algorithms, visualizing data elements, etc.

Additionally, Orange brings a more interactive and fun visual interface to the dull analytic tools. Data coming to Orange gets quickly formatted to the desired pattern and it can be easily moved where needed by simply moving/flipping the widgets. Orange allows users to make smarter decisions in a short time by quickly comparing and analyzing the data.

8.2.6   IBM SPSS® Modeler

IBM® SPSS® Modeler (formerly Clementine) was one of the first commercial tools oriented to Data Mining. Later absorbed by the firm SPSS, which also commercializes a very popular and widely used statistical package (SPSS). Afterwards, it was acquired by IBM which commercializes it under the name IBM® SPSS® Modeler.[3] Main characteristics are:

- Tool: IBM® SPSS® Modeler.
- Developer: IBM® (formerly SPSS® and Clementine).
- Website: https://www.ibm.com/products/spss-modeler
- Date since it is available: June 1994 (Clementine 1.0)
- Latest stable version (by July 2021): IBM® SPSS® Modeler 18.3
- Implementation Language: scripts are written in Jython, a Java™-based implementation of Python
- Type of license: commercial

Clementine was designed to support the CRISP-DM, the de facto standard data mining methodology. It provides a visual interactive workflow interface supporting the data mining process and has an open architecture for integrating with other systems and all SPSS predictive analytics. It includes facilities for database access, text, survey and web data preparation, model management, automatic version control, user authentication, etc.

IBM® SPSS® Modeler is a visual, drag-and-drop tool that speeds operational tasks for data scientists and data analysts, accelerating time to value. It enables users to consolidate all types of data sets from dispersed data sources across the organization and build predictive models— all without the requirement of writing code. SPSS® Modeler offers multiple machine learning techniques–including classification, segmentation and association algorithms including out-of-the-box algorithms that leverage Python and Spark. And users can now employ languages such as R and Python to extend modelling capabilities.

8.2.7   SAS® Enterprise Miner™

In 1966, vast amounts of agricultural data were being collected through USDA[4] grants – but no computerized statistics program existed to analyze the findings. A consortium of eight universities came together under a grant from the National Institutes of Health (NIH) in USA to solve that problem. The resulting program, the Statistical Analysis System, gave SAS® both the basis for its name and its corporate beginnings. The project was led by North Carolina State University (NCSU).

In 1976, James Goodnight, Anthony James Barr, Jane Helwig, and John Sall left NCSU and formed SAS Institute Inc., a private company devoted to the maintenance and further development of SAS.

---

[3]This software is probably replacing the previous Data Mining tool from IBM, Intelligent Miner, which is not supported by IBM anymore.

[4]USDA stands for United States Department of Agriculture.

SAS® Enterprise Miner™ (Hall et al., 2014) streamlines the entire data mining process from data access to model deployment by supporting all necessary tasks within a single, integrated solution, all while providing the flexibility for efficient workgroup collaborations. Main characteristics are:

- Tool: SAS® Enterprise Miner™
- Developer: SAS® Institute Inc.
- Website: https://www.sas.com/en_us/software/enterprise-miner.html
- Date since it is available: 1999
- Latest stable version (by July 2021): SAS® Enterprise Miner™ 15.2.
- Implementation Language: Java™, C, and formerly the original SAS® in PL/I.
- Type of license: commercial

SAS® Enterprise Miner™ provides tools for graphical programming, avoiding manual coding, which makes it easy to develop complex data mining processes. It was designed for business users and provides several tools to help with pre-processing data (descriptive analysis, advanced statistical graphics) together with advanced predictive modelling tools and algorithms, including decision trees, artificial neural nets, auto-neural nets, memory-based reasoning, linear and logistic regression, clustering, association rules, time series, rule induction, market basket analysis, weblog analysis among others, as well as the facility for direct connection with data warehouses. It also offers tools for comparing the results of different modelling techniques.

It is integrated with other tools from the wider SAS suite of analysis software. SAS® Enterprise Miner™ is delivered as a distributed client-server architecture with a Java™ based client allowing parallel processing and grid computing, and it is especially well suited for large organizations.

The user interface has an easy to use data-flow Graphical User Interface (GUI) similar to that pioneered in Clementine or the Weka's Knowledge Flow interface. You can also integrate code written in the SAS language and write code in the R language.

8.3 Tools for Model-Driven Techniques

Depending on the kind of model-driven techniques there are not so many tools for model-driven techniques as tools exist for data-driven methods. In this section, some of the available tools or software platforms will be analyzed. The different kinds of model-driven techniques in IDSS will be reviewed in each one of the following subsections: *agent-based simulation tools*, *expert-based model tools*, *model-based reasoning techniques,* and *qualitative reasoning techniques*.

8.3.1 Agent-Based Simulation Tools

In the literature, there are many agent-based simulation tools. Two types of agent-based simulation tools are used: ones that are implemented in a general *multi-agent system tool*, and others implemented in a particular *agent-based simulation tool*.

Regarding the *general multi-agent system tools*, among the most commonly used are *Prometheus* (Padgham and Winikoff, 2004) for the design of the system, and *Jadex* (Braubach et al., 2004) for the implementation. In (Rendón-Sallard and Sànchez-Marrè, 2006) there is a review on several multi-agent system tools.

There are many *agent-based simulation tools*. In (Abar et al., 2017) there is a wide review of agent-based simulation tools. Taking into account some properties like the generality of the tool, the type of license, the degree of scalability, and the complexity of its development some of them can be outlined.

Among the general-purpose tools, being free open-source tools, having a medium-large scalability, and being of moderate complexity, we can enumerate: *Ascape*, *FLAME*, and *Janus*. It is also remarkable *Netlogo*, for its easy level of development, large scalability, easiness for education and research, even though it is focused on 2D/3D simulations of natural and social sciences.

Regarding the proprietary licensed tools being of general-purpose, and being of moderate difficulty and having also a medium to large scalability, *Anylogic* can be enumerated.

In next sub-sections we will briefly analyze main characteristics of these tools. Most part of the information has been extracted from their corresponding websites and publications.

8.3.1.1 Prometheus

Prometheus (Padgham and Winikoff, 2004) is a mature and well-documented methodology. It supports BDI-concepts and additionally provides a CASE-Tool, the Prometheus Design Tool (PDT), for drawing, and using the notation.

The Prometheus methodology consists of three phases that can be done simultaneously: system specification, architectural design and detailed design. The three phases are:

- The first phase, system specification focuses on identifying the goals and basic functionalities of the system, along with inputs (percepts) and outputs (actions).
- The architectural design phase uses the outputs from the previous phase to determine which agent types the system will contain and how they will interact.
- The detailed design phase looks at the internals of each agent and how it will accomplish its tasks within the overall system.

Prometheus is supported by the Prometheus Design Tool (PDT), which provides forms to enter design entities. It performs cross-checking to help ensure consistency and generates a design document along with overview diagrams.

Main characteristics of the PDT tool are:

- Tool: Prometheus Design Tool (PDT).
- Developer: PDT is developed and supported by the RMIT Intelligent Agents Group, RMIT University, Australia.
- Website: https://sites.google.com/site/rmitagents/software/prometheusPDT
- Date since it is available: Old PDT version 0.1.1 (2002 approximately)
- Latest stable version (by July 2021): Eclipse-based PDT v0.5.1 released in 2014. An extended version (called TDF) is under development.
- Implementation Language: Java™, Eclipse-based.
- Type of license: free, Eclipse plugin

8.3.1.2   Jadex

Jadex (Braubach et al., 2004) is a Java™-based framework that allows the creation of goal-oriented agents and provides a set of development tools to simplify the creation and testing of agents.

The Jadex BDI reasoning engine allows the development of rational agents using mental notions. It allows for programming intelligent software agents in XML and Java™ and can be deployed on different kinds of middleware such as JADE. In contrast to all other available BDI engines, Jadex fully supports the two-step practical reasoning process: *goal deliberation* and *means-end reasoning*. This means that Jadex allows the construction of agents with explicit representation of mental attitudes (beliefs, goals, and plans) and that automatically deliberate about their goals and subsequently pursue them by applying appropriate plans (Bellifemine et al., 2007).

The Jadex BDI reasoning engine enables the construction of complex real-world applications by exploiting the ideas of intentional systems (Dennett 1987; McCarthy, 1978).

Practical reasoning is handled via *goal deliberation* and *means-end reasoning*. The *goal deliberation* is mainly state-based and has the purpose of selecting the current non-conflicting set of goals.

The *means-end reasoning* has incoming messages, internal events and new goals as input, and then it dispatches these events to plans selected from the plan library for further processing. Plans execution may access and modify the belief base, send messages to other agents, create new goals, and cause internal events.

Main characteristics of the tool are:

- Tool: Jadex.
- Developers: Originally by a research group at the University of Hamburg mainly formed by Prof. Lars Braubach and Dr. Alexander Pokahr. Currently, under a software company called © Actoron GmbH (https://www.actoron.com).
- Website: https://www.activecomponents.org/
- Date since it is available: 2004
- Latest stable version (by July 2021): 4.0.241 released on June 2021.
- Implementation Language: Java™
- Type of license: open-source software under GPL-License Version 3.

Jadex uses a hybrid approach for defining and programming agents. The structural part comprises the agent's static design composed of beliefs, goals, plans, and the agent's initial state. All these aspects are specified in the Jadex XML language following an XML-Schema defining the BDI meta-model. The behavioural part of the BDI agent is encoded in Jadex plans using plain Java™.

8.3.1.3 Ascape

Ascape (Inchiosa and Parker, 2002; Parker, 2001; Parker, 1999) is an innovative tool for developing and exploring general-purpose agent-based models. It is designed to be flexible and powerful, but also approachable, easy to use and expressive. Models can be developed in Ascape using far less code than in other tools. Ascape models are easier to explore, and profound changes to the models can be made with minimal code changes. Ascape offers a broad array of modelling and visualization tools.

A high-level framework supports the complex model design, while end user tools make it possible for non-programmers to explore many aspects of model dynamics. Ascape is written entirely in Java™, and should run on any Java™-enabled platform.

Main characteristics of the tool are:

- Tool: Ascape.
- Developers: Leaded by Miles T. Parker at The Brookings Institution, was developed also with Mario Inchiosa, Josh Miller, Oliver Mannion, and other contributors (see website).
- Website: http://ascape.sourceforge.net/
- Date since it is available: Ascape version 1.1–2000.
- Latest stable version (by July 2021): Ascape 5.6.1 released on April 2011.
- Implementation Language: Java™
- Type of license: Open-source, BSD,[5] Free.

---

[5] BSD licenses are a family of permissive free software licenses, imposing minimal restrictions on the use and distribution of covered software. BSD takes its name from an original Berkeley Software Distribution of a Unix-like operating system.

8.3.1.4   FLAME

FLAME (Flexible Large-scale Agent Modelling Framework) (Holcombe et al., 2006) is a flexible and generic agent-based modelling platform that can be used to develop models and simulations for complex system applications in many areas such as economics, biology and social sciences, to name a few. It generates a complete agent-based application that can be compiled and deployed on many computing systems ranging from laptops to distributed high-performance super computing environments.

Models are created based upon a model of computation called (extended finite) state machines. The behaviour model is based upon state machines which are composed of a number of states with transition functions between those states. There is a single start state and by traversing states using the transition functions the machine executes the functions until it reaches an end state. This happens to each agent/machine as one-time step or iteration is completed. A time step or iteration of the model is when each agent goes from their start state to an end state.

By defining agent-based models in this way the FLAME framework can automatically generate simulation programs that can run models efficiently on High-Performance Computers (HPCs).

Main characteristics of the tool are:

- Tool: FLAME.
- Developer: © 2013 Software Engineering Group, Scientific Computing Dept., SFTC (https://www.scd.stfc.ac.uk/Pages/Software-Engineering-Group.aspx).
- Website: http://flame.ac.uk/
- Date since it is available: 2006
- Latest stable version (by July 2021): FLAME Xparser version 0.17.1 released on December 2016, FLAME Libmboard version 0.3.1 released on November 2016.
- Implementation Language: C
- Type of license: open-source, GNU Lesser General Public license, free.

Source code for FLAME (Xparser and libmboard) is now available on the FLAME-HPC GitHub (FLAME-HPC). Xparser is the program that parses a model file and produces source code for a simulation program. Libmboard is the communication library used by simulation programs.

8.3.1.5   Janus

Janus (Cossentino et al., 2007) is an open-source multi-agent platform fully implemented in SARL (implemented with Java™ for its versions 1 and 2). Janus enables developers to quickly create web, enterprise and desktop multi-agent-based applications. It provides a comprehensive set of features to develop, run, display and monitor multi-agent-based applications.

Janus-based applications can be distributed across a network. Janus could be used as an agent-oriented platform, an organizational platform, and/or an holonic platform. It also natively manages the concept of recursive agents and holons.

Main characteristics of the tool are:

- Tool: Janus.
- Developer: a research group in agent-based modelling and simulation at Univ. Bourgogne Franche-Comté, UTBM, France, mainly formed by Stéphane Galland, Nicolas Gaud, Alexandre Lombard and Sebastián Rodríguez.
- Website: http://www.sarl.io/runtime/janus/
- Date since it is available: 2007–2008.
- Latest stable version (by July 2021): Janus 3.0.12.0.
- Implementation Language: SARL Agent-Oriented Programming Language.
- Type of license: open-source, GPLv3 for non-commercial use or JIUL/JCRL usage/redistribution commercial license, free.

Janus platform was initially published during the 2007–2008 period. Since 2014, Janus is fully re-implemented to support the SARL Agent-Oriented Programming Language.

8.3.1.6   Netlogo

NetLogo (Wilensky and Rand, 2015; Wilensky, 1999) is a multi-agent programmable modelling environment. It was authored by Uri Wilensky in 1999 and has been in continuous development ever since at the Centre for Connected Learning and Computer-Based Modelling.

Netlogo is used by many hundreds of thousands of students, teachers, and researchers worldwide. It also powers HubNet participatory simulations. NetLogo is a programmable modelling environment for simulating natural and social phenomena.

NetLogo is particularly well suited for modelling complex systems developing over time. Modellers can give instructions to hundreds or thousands of "agents" all operating independently. This makes it possible to explore the connection between the micro-level behaviour of individuals and the macro-level patterns that emerge from their interaction.

Main characteristics of the tool are:

- Tool: Netlogo.
- Developer: Uri Wilensky at Center for Connected Learning and Computer-Based Modelling (CCL), University of Northwestern, since 1999.
- Website: http://ccl.northwestern.edu/netlogo/
- Date since it is available: Netlogo v1.0 (April 2002).
- Latest stable version (by July 2021): Netlogo v6.2.0 released on September 2019.
- Implementation Language: Scala, Java™.
- Type of license: open source, GPL, free.

NetLogo has extensive documentation and tutorials. It also comes with the Models Library, a large collection of pre-written simulations that can be used and modified. These simulations address content areas in the natural and social sciences including biology and medicine, physics and chemistry, mathematics and computer science, and economics and social psychology. Several model-based inquiry curricula using NetLogo are available and more are under development.

8.3.1.7   Anylogic

AnyLogic (Borshchev et al., 2002) is a professional software for building industrial strength agent-based simulation models. Moreover, agent-based simulation models can be easily combined with discrete event or system dynamics elements, for complete, no compromise, modelling. This can be seen, for instance, with warehouses that behave on a supply chain as agents but are modelled internally using discrete event modelling.

AnyLogic develops models using all three modern simulation methods: Discrete Event, Agent-Based and System Dynamics. The three methods can be used in any combination, with one software, to simulate business systems of any complexity. In AnyLogic, you can use various visual modelling languages: process flowcharts, state charts, action charts, and stock and flow diagrams.

AnyLogic was the first tool to introduce multimethod simulation modelling, and still remains the only software that has that capability.

Main characteristics of the tool are:

- Tool: AnyLogic.
- Developer: © The AnyLogic Company (former XJ Technologies).
- Website: https://www.anylogic.com/
- Date since it is available: 2000.
- Latest stable version (by July 2021): AnyLogic 8.7.5 released on June 2021.
- Implementation Language: Java™
- Type of license: Closed source, proprietary, free personal learning edition available.

AnyLogic models enable analysts, engineers, and managers to gain deeper insights and optimize complex systems and processes across a wide range of industries.

8.3.2   Expert-Based Model Tools

As described in Chap. 5, most common *expert-based models* represent the knowledge from experts using inference rules. Therefore, the expert-based model tools are mainly rule-based system tools.

These systems are general inference engines, also known as *expert system shells*, that can be applied to any domain, once the knowledge base and the fact base have been defined. In the literature, the most common used rule-based reasoning systems are *CLIPS*, *Drools* among the free tools, and *Jess*® among the commercial tools. Next, a review of all these tools is done.

8.3.2.1 CLIPS

The C Language Integrated Production System (CLIPS) (Giarratano and Riley, 2004) is an expert system tool originally developed by the Software Technology Branch (STB) at NASA/Lyndon B. Johnson Space Center from 1985 to 1996.

Since its first release in 1986, CLIPS has undergone continual refinement and improvement. Gary Riley, while working at NASA's Johnson Space Center, was responsible for the design and development of the rule-based components of CLIPS. Since 1996, Gary Riley left NASA and he has continued to independently develop and maintain a public domain version of CLIPS.

CLIPS is a rule-based programming language useful for creating expert systems and other programs where a heuristic solution is easier to implement and maintain than an algorithmic solution. Written in C for portability, CLIPS can be installed and used on a wide variety of platforms.

Like many other expert system languages, CLIPS has a LISP-like syntax which uses parentheses as delimiters. Although CLIPS is not written in LISP, the style of LISP has influenced the development of CLIPS.

CLIPS is designed to facilitate the development of software to model human knowledge or expertise. There are three ways to represent knowledge in CLIPS:

- *Rules*, which are primarily intended for heuristic knowledge based on the experience.
- *Deffunctions* and *generic functions*, which are primarily intended for procedural knowledge.
- *Object-oriented programming*, also primarily intended for procedural knowledge. The five generally accepted features of object-oriented programming are supported: classes, message-handlers, abstraction, encapsulation, inheritance, and polymorphism. Rules may pattern match on objects and facts.

Users can develop software using only rules, only objects, or a mixture of objects and rules.

Main characteristics of the tool are:

- Tool: CLIPS.
- Developer: NASA's Johnson Space Center (1985–1996). Since 1996, Gary Riley left NASA and he has continued to independently develop and maintain a public domain version of CLIPS.
- Website: http://www.clipsrules.net/
- Date since it is available: CLIPS 3.0 summer 1986.
- Latest stable version (by July 2021): CLIPS 6.4 released on April 2021.

- Implementation Language: C
- Type of license: public domain software.

CLIPS is a forward-chaining rule-based programming language written in C that also provides procedural and object-oriented programming facilities.

Descendants of the CLIPS tool include Jess (rule-based portion of CLIPS rewritten in Java™ which later grew up in a different direction) and FuzzyCLIPS, which adds the concept of relevancy into the tool and language.

8.3.2.2   Drools

Drools (Proctor, 2007) is a business rule management system (BRMS) with a forward and backward chaining inference-based rules engine, also known as a production rule system, using an enhanced implementation of the Rete algorithm.

Drools provide a core Business Rules Engine (BRE), a web authoring and rules management application (Drools Workbench), full runtime support for Decision Model and Notation (DMN) models at Conformance level 3 and an Eclipse IDE plugin for core development.

Drools support the Java™ Rules Engine API (Java™ Specification Request 94) standard for its business rule engine and enterprise framework for the construction, maintenance, and enforcement of business policies in an organization, application, or service.

Main characteristics of the tool are:

- Tool: Drools.
- Developer: © Red Hat Inc. (https://www.redhat.com/).
- Website: https://www.drools.org/
- Date since it is available: Drools 3.0.6.GA (August 2007) is the first documented at website.
- Latest stable version (by July 2021): Drools 7.56.0. Final released on June 2021.
- Implementation Language: Java™
- Type of license: open-source software, under the Apache License 2.0.

8.3.2.3   Jess

Jess® (Friedman-Hill, 1997; Friedman-Hill, 2003) is a rule engine and scripting environment written entirely in Oracle's® Java™ language by Ernest Friedman-Hill at Sandia National Laboratories in Livermore, CA, the USA. Jess® was originally conceived as a Java clone of CLIPS, but nowadays is a superset of CLIPS with many differences between them.

Using Jess® allows to build Java™ software that has the capacity to reason using knowledge supplied in the form of declarative rules. Jess® is an expert system shell.

Jess® is small, light, and one of the fastest rule engines available. Its powerful scripting language gives access to all of Java™ APIs. Jess® includes a full-featured development environment based on the Eclipse platform.

Jess® uses an enhanced version of the Rete algorithm to process rules. Rete is a very efficient mechanism for solving the difficult many-to-many matching problem (see Chap. 5, Sect. 5.3.3.1). Jess® has many unique features including backwards chaining and working memory queries, and of course, Jess® can directly manipulate and reason about Java™ objects. Jess® is also a powerful Java™ scripting environment, from which can create Java™ objects, call Java™ methods, and implement Java™ interfaces without compiling any Java™ code.

Main characteristics of the tool are:

- Tool: Jess® the Rule Engine for the Java™ Platform.
- Developer: Ernest Friedman-Hill, at Sandia National Laboratories (https://www.sandia.gov/) in Livermore, CA, the USA.
- Website[6]: https://jess.sandia.gov/, http://www.jessrules.com/
- Date since it is available: late 1995.
- Latest stable version (by July 2020): Jess® 7.1p2 released on November 2008. Currently, Jess® 8 is a test version.
- Implementation Language: Java™
- Type of license: proprietary software, no cost for academic use (research-based academic license) and can be licensed for commercial use.

8.3.3 Model-Based Reasoning Tools

There are some model-based reasoning tools. Most of the existing ones are devoted to the *consistency-based diagnosis problem* (see Chap. 5, Sect. 5.4.3). In the literature, most tools are general-purpose *constraint solvers* like *MINION, Gecode,* or *Choco solver*. These tools will be reviewed in the next subsections. See the Constraint Solving website[7] for an extensive list of constraint solvers.

There are other useful tools like *Conjure, MiniZinc,* or *Savile Row* tools that can be used before a constraint solver. In addition, there are some solvers focussed on a particular application domain like software verification and software analysis (*Z3*), 3D Solver for Chemically Reacting flows (*Open3DCFD*), or for solving Navier–Stokes equations (*ADFC, OpenFlower*).

Conjure[8] is an automated constraint modelling tool, which accepts problems specified in the *Essence*[9] language (Frisch et al., 2008) and outputs constraint models in the language *Essence*, suitable for input to tools, like *Savile Row* for instance,

---

[6]Both official websites for Jess® had time-out connection problems by July 2021.

[7]http://www.constraintsolving.com/solvers is maintained by Martine Ceberio and her research students from the Constraint Research and Reading Group (CR²G), at the University of Texas at El Paso.

[8]https://conjure.readthedocs.io/en/latest/.

[9]*Essence* is an abstract constraint language for specifying combinatorial problems.

which then tailors these models to the input requirements of particular constraint solvers.

Savile Row[10] takes as input constraint models in the solver-independent modelling language *Essence*. These models may be produced automatically by some tool like *Conjure*, or written directly by the user. *Savile Row* translates a given model into input suitable for a particular constraint solver, such as the constraint solver MINION.

MiniZinc[11] is a free and open-source constraint modelling language. *MiniZinc* can be used to model constraint satisfaction and optimization problems in a high-level, solver-independent way, taking advantage of a large library of pre-defined constraints. The produced model is then compiled into FlatZinc, a solver input language that is understood by a wide range of solvers.

8.3.3.1 Minion

MINION (Gent et al., 2006) is a relatively new constraint solver which is fast and scales well as problem size increases. MINION is a general-purpose constraint solver, with an expressive input language based on the common constraint modelling device of matrix models. Focussing on matrix models supports a lean, highly optimized implementation. This contrasts with other constraint toolkits, which, in order to provide ever more modelling and solving options, have become progressively more complex at the cost of both performance and usability.

MINION is a black box from the user point of view, deliberately providing few options. This, combined with its raw speed, makes MINION a substantial step towards Puget's `Model and Run' constraint solving paradigm.

MINION has been developed mainly by Ian Gent, Christopher Jefferson, Ian Miguel, and Peter Nightingale, at the Artificial Intelligence Research Group at the University of St Andrews, Scotland, the UK.

Main characteristics of the tool are:

- Tool: MINION.
- Developer: Artificial Intelligence Research Group at the University of St Andrews, Scotland, the UK.
- Website: https://constraintmodelling.org/minion/
- Date since it is available: MINION 0.1.0 (2006).
- Latest stable version (by July 2021): MINION 1.8 released in February 2015.
- Implementation Language: C++
- Type of license: open-source software, licensed under GNU General Public License Version 2.

---

[10] https://savilerow.cs.st-andrews.ac.uk/.

[11] https://www.minizinc.org/.

8.3.3.2   Gecode

Generic constraint development environment (Gecode) (Schulte and Tack, 2006) is an open-source C^{++} toolkit for developing constraint-based systems and applications. Gecode provides a constraint solver with state-of-the-art performance while being modular and extensible.

Gecode is open for programming: It can be easily interfaced to other systems. It supports the programming of new constraints, branching strategies, and search engines. New variable domains can be programmed at the same level of efficiency as the variables that come predefined with Gecode.

Gecode has a comprehensive set of features: Constraints over integers, Booleans, sets, and floats (it implements more than 70 constraints from the Global Constraint Catalogue and many more on top); C^{++} modelling layer; advanced branching heuristics (accumulated failure count, activity); many search engines (parallel, interactive graphical, restarts); automatic symmetry breaking (LDSB); no-goods from restarts; *MiniZinc* support; and many more.

Gecode offers excellent performance with respect to both runtime and memory usage. It won all gold medals in all categories at the *MiniZinc* Challenges from 2008 to 2012.

Prof. Christian Schulte[12] led the development of Gecode until early 2020. Gecode is currently developed by Dr. Guido Tack, Monash University, Australia and Mikael Zayenz Lagerkvist, formerly at KTH, Sweden, and currently as an independent researcher.

Main characteristics of the tool are:

- Tool: Gecode.
- Developers: Christian Schultze, Guido Tack and Mikael Zayenz Lagerkvist.
- Website: https://www.gecode.org/
- Date since it is available: Gecode 1.0.0 (2005).
- Latest stable version (by July 2021): Gecode 6.2.0 released on April 2019.
- Implementation Language: C^{++}
- Type of license: free software under the MIT license.

8.3.3.3   Choco Solver

Choco solver (Laburthe, 2000; Fages and Prud'Homme, 2017) is an open-source Java™ library for constraint programming. The user models its problem in a declarative way by stating the set of constraints that need to be satisfied in every solution. Then, the problem is solved by alternating constraint filtering algorithms with a search mechanism.

---

[12]Prof. Christian Schulte (1967–2020), from School of Electrical Engineering and Computer Science, KTH Royal Institute of Technology, Sweden, was the designer and main developer of Gecode.

Choco solver comes with the commonly used types of variables: Integer variables with either bounded domain or enumerated one, boolean variables and set variables. Views but also arithmetical, relational, and logical expressions are supported.

Up to 100 constraints and more than 150 propagators are provided: from classic ones, such as arithmetical constraints, to must-have global constraints, such as *allDifferent* or *cumulative*, and include less common even though useful ones, such as a *tree*. One can pick some existing propagators to compose a new constraint or create its own one in a straightforward way by implementing a filtering algorithm and a satisfaction checker.

Choco solver has been carefully designed to offer a wide range of resolution configurations and good resolution performances. Backtrackable primitives and structures are based on trailing. The propagation engine deals with seven priority levels and manage either fine or coarse grain events which enable to get efficient incremental constraint propagators.

Main characteristics of the tool are:

- Tool: Choco solver.
- Developers: Charles Prud'Homme and Jean-Guillaume Fages, IMT-Atlantique, France, and some collaborators.
- Website: https://choco-solver.org/
- Date since it is available: 1999
- Latest stable version (by July 2021): Choco solver 4.10.6 released in December 2020.
- Implementation Language: Java™
- Type of license: open-source software

8.3.4 Qualitative Reasoning Tools

In the literature, there are a limited number of qualitative reasoning (modelling and simulation) software tools. Some of them are the experimental and general research qualitative modelling and simulation tools like the formers *Envision* program (De Kleer and Brown, 1984), the *Qualitative Process Engine (QPE)* (Forbus, 1990, 1992) and the *Qualitative Simulation (QSIM) framework* (Kuipers 1986, 1994). These tools were described in Chap. 5, Sect. 5.5.5. These tools were implemented in Lisp, and are available only for research, as a professional courtesy of the developers.

More recently, some tools like *Garp3* or *GQR* are available as open-source software. Furthermore, some specific qualitative software tools are oriented to a concrete target of application like in spatial reasoning (*SparQ*), in genetic regulatory networks (*GINSim*), or in children education (*VModel*).

On the other hand, there are some tools for deploying *Causal Loops Diagrams* (*CLDs*), a special type of qualitative model (see Chap. 7, Sect. 7.1.4). Main proprietary software tools are *Stella*[13] and *Vensim*.[14] Both are licensed software, but *Vensim PLE* (Personal Learning Edition) version is free for educational and personal use, and *Stella Online* is a free limited online version. *Simantics System Dynamics* is an open-source tool.

Next, *Garp3*, *GQR,* and *Simantics System Dynamics* software tools will be analyzed.

8.3.4.1 Garp3

Garp3 (Bredeweg et al., 2009) has been developed and implemented as part of the NaturNet-Redime European Research Project. Garp3 is a workbench for building, simulating, and inspecting qualitative models.

The Garp3 qualitative modelling and simulation software allows users to develop explicit conceptual (causal) representations of scientific theories and particular systems to which they apply. The simulations allow scientists to test their ideas for consistency and agreement with observations in reality.

The Build environment in Garp3 provides nine workspaces for creating model ingredients, divided into two categories: building blocks and constructs. Building blocks are used to define objects that can be reused and assembled into constructs. The model ingredients are:

- Building blocks: they can be entities, agents, assumptions, attributes, configurations, quantities, and quantity spaces.
- Constructs: they can be scenarios or model fragments.

Garp3 is based on previously developed software (Garp, Homer, and VisiGarp) and provides a seamless workbench for building, simulating, and inspecting qualitative models. Garp3 is implemented in SWI-Prolog[15] and can be downloaded online.

Main characteristics of the tool are:

- Tool: Garp3
- Developer: Qualitative Reasoning group, Informatics Institute, Faculty of Science, University of Amsterdam, The Netherlands.
- Website: https://ivi.fnwi.uva.nl/tcs/QRgroup/QRM/software/
- Date since it is available: 2005
- Latest stable version (by July 2021): Garp3 1.5.2 released on November 2012
- Implementation Language: SWI-Prolog version 6.2.0 (or later)
- Type of license: free software

---

[13] https://www.iseesystems.com/store/products/index.aspx.

[14] https://vensim.com/vensim-software/.

[15] http://www.swi-prolog.org/.

8.3.4.2   GQR

GQR (Generic Qualitative Reasoner) (Gantner et al., 2008) is a solver for binary qualitative constraint networks. GQR takes a calculus description and one or more constraint networks as input, and tries to solve the networks using the path consistency method and backtracking.

Both spatial calculi, e.g. RCC-5 and RCC-8, and temporal ones, like Allen's interval algebra, are supported. New logics can be added to the system using a simple text format; no programming is necessary.

GQR (Generic Qualitative Reasoner), developed at the University of Freiburg, is a solver for binary qualitative constraint networks. GQR takes a calculus description and one or more constraint networks as input, and tries to solve the networks using the path consistency method and (heuristic) backtracking. In contrast to specialized reasoners, it offers reasoning services for different qualitative calculi, which means that these calculi are not hard-coded into the reasoner.

Currently, GQR supports arbitrary binary constraint calculi; new calculi can be added to the system by specifications in a simple text format or in an XML file format.

Main characteristics of the tool are:

- Tool: GQR.
- Developers: Zeno Gantner, from University of Hildesheim, Germany; Matthias Westphal and Stefan Wölfl from University of Freiburg, Germany.
- Website: http://openscience.org/gqr/, http://gki.informatik.uni-freiburg.de/tools/gqr/
- Date since it is available: 2006
- Latest stable version (by July 2021): GQR 1500 released on September 2012.
- Implementation Language: mostly C^{++} and Python.
- Type of license: free software distributable under the GNU General Public License.

8.3.4.3   Simantics System Dynamics

Simantics System Dynamics (Lempinen et al., 2011) is a ready-to-use system dynamics modelling and simulation software application for understanding different organizations, markets, and other complex systems and their dynamic behaviour.

Simantics System Dynamics is used for modelling and simulating large hierarchical models with multidimensional variables. The models are created in a traditional way with stock and flow diagrams and causal loop diagrams. Simulation results and the model structure can be analyzed with different visual tools.

Import and Export features can be used, for example, to create different versions, create backups, or to transport modelling assets like models or modules to another database.

The Simantics platform has server–client architecture with a semantic ontology-based modelling database and Eclipse framework-based client software with a plugin interface.

Main characteristics of the tool are:

- Tool: Simantics System Dynamics.
- Developers: Simantics System Dynamics is developed by VTT Technical Research Centre of Finland Ltd., Finland and Semantum company, Finland.
- Website: http://sysdyn.simantics.org/
- Date since it is available: 2011
- Latest stable version (by July 2021): Simantics System Dynamics 1.35.0 released on September 2018.
- Implementation Language: Java™, OpenModelica.[16]
- Type of license: open-source software, with Eclipse Public License (EPL)

OpenModelica is used to build and simulate the models. Simantics platform has integrated OpenModelica 1.9.0 beta4 for Windows environments. For other versions and other environments, OpenModelica must be installed. In addition to OpenModelica, a development version of a purpose-built Modelica solver is embedded in the tool.

8.4   General Development Environments

As explained in the introduction section, there are no comprehensive thorough software tools to develop a whole IDSS. However, there are general tools which are programming languages or systems that integrate many data-driven models and some model-driven techniques, as well as other packages or modules to implement many useful functionalities. Using these platforms make it easier to build up an IDSS. The different functions, packages, modules can be called when required to do the corresponding tasks. This way is actually better than programming all the tasks involved in the IDSS (diagnosis, synthesis, and prognosis) from scratch in any programming language.

In other cases, probably when the IDSS is not requiring to implement all kinds of tasks, a good option could be also to use some software tools from the ones reviewed in previous Sects. 8.2 and 8.3.

The most commonly used general development environments due to their great number of available functionalities and extensive quantity of different models available are the *R environment* and the *Python programming language*. In the

---

[16]OpenModelica (https://www.openmodelica.org/) is a free and open-source environment based on the Modelica modelling language.(https://www.modelica.org/modelicalanguage) for modelling, simulating, optimizing, and analyzing complex dynamic systems intended both for industrial and academic usage.

next subsections, these two environments will be detailed. Most of the information described for these tools have been extracted from their corresponding websites. Finally, a tool named GESCONDA, which is under development by the author in his research group, will be described too.

8.4.1 R

R (Becker et al., 1988; R Core Team, 2021) is both a language and an environment for statistical computing and graphics. It was originally focussed on statistical computation, but it has been extended with many other functionalities, with some packages, and now many data-driven methods are usable and some model-driven techniques can be also used.

The initial version of R was developed by Ross Ihaka and Robert Gentleman, from University of Auckland, in 1996.

R implements a dialect of language S, which was developed by John Chambers, Rick Becker and Allan Wilks from AT&T Bell Laboratories (1976), slightly different from commercial version S-PLUS. Since the mid-1997, there has been a core group, the "R Core Team", with write access to the R source leading its development.

R is a well-supported platform, open-source, command-line driven, specialized in statistics, and related methods. There are hundreds of extra *packages* freely available, which provide all sorts of data mining, machine learning and statistical techniques. It has a large number of users, particularly in the areas of bio-informatics and social science; also, a large number of developers that continuously enlarge its functionalities providing new *packages*.

It compiles and runs on a wide variety of UNIX platforms and similar systems (including FreeBSD and Linux), Windows, and MacOS.

R is an integrated suite of software facilities for data manipulation, calculation, and graphical display. It includes:

- an effective data handling and storage facility,
- a suite of operators for calculations on arrays, in particular matrices,
- a large, coherent, integrated collection of intermediate tools for data analysis,
- graphical facilities for data analysis and display either on-screen or on hardcopy, and,
- a well-developed, simple, and effective programming language which includes conditionals, loops, user-defined recursive functions, and input and output facilities.

R can be extended (easily) via *packages*. There are about eight packages supplied with the R distribution and many more are available covering a very wide range of modern statistics. Currently, by July 2021, there are 17,893 packages available,

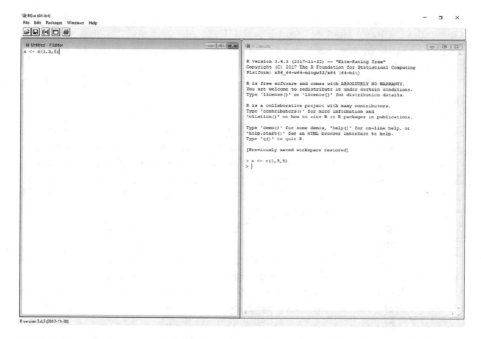

Fig. 8.1 R standard graphical user interface

which can be obtained from several CRAN[17] sites at https://cran.r-project.org/web/packages/index.html. The number of available packages continuously increases.

R allows to implement user functions to extend their functionalities through a programming language. Main characteristics of the R environment are:

- Tool: R environment.
- Developer: © The R Foundation.
- Website: https://www.r-project.org/
- Date since it is available: R-0.49 (April 1997)
- Latest stable version (by July 2021): R version 4.1.0 released on May 2021
- Implementation Language: C
- Type of license: open source, free software under GNU General Public License

In Fig. 8.1, there is a screenshot of R standard graphical user interface where the *R Console* and an *R script editor* are depicted.

There are some tools that have been deployed to make easier the use of R environment. Among them, *R-Commander, Rattle,* and *RStudio* can be mentioned.

R-Commander[18] (Fox, 2017; Fox 2005) adds to the two basic functionalities mentioned above (*R Scripts editor* and *R Console*) several *menus* already set up

---

[17]CRAN (Comprehensive R Archive Network) is a network of ftp and web servers around the world that store identical, up-to-date, versions of code, and documentation for R.

[18]https://www.rcommander.com/.

which give access to main common statistical models like Linear regression models, Generalized Linear models, Multinomial logit models, and Ordinal regression models. Other menus provide statistical distributions, graphics, or statistical analysis.

It can be obtained after the installation of the package *Rcmdr*, which is available in the CRAN repository.

Rattle[19] (R Analytic Tool to Learn Easily) (Williams 2011) offers a graphic user interface (GUI) for Data Science in R. It gives the user point-and-click access to many of the R functions related to unsupervised and supervised data-driven models.

Rattle also supports the ability to transform and score data and offers a number of data visualization tools for evaluating models. A key feature is that all user interactions through the graphical user interface are captured as an R script that can be readily executed in R independently of the Rattle interface.

Rattle is a free open-source software. The *rattle* package can be obtained from the CRAN repository.

RStudio[20] (Verzani, 2011) comprises a set of Integrated Development Environments (IDEs) for R. RStudio tools are created by RStudio® Corporation, which is a Delaware Public Benefit Corporation (PBC) and a Certified B Corporation®. From all the products, RStudio Desktop is an open-source edition of the corresponding professional product, with a GNU Affero GPL v3 license.

In Fig. 8.2, there is a screenshot of RStudio Desktop.

RStudio Desktop IDE provides many functionalities that help users to develop R programs:

- Source editor for *R script edition* which provides nice syntax highlighting, code completion, and smart indentation
- *Execution of R code directly from the source editor*, and quickly jump to function definitions
- A *console* where R commands and their results can be visualized.
- A window for *managing the environment* where all the active objects are depicted, and also shows the user *command history* and recently a *tutorial window*
- A window for *files and folders management* in the by-default workspace, another for *plot depiction*, another for *package management* which comprises package development tools and another one for *help information* supply
- Interactive *Debugging* functionality is integrated

One of the powerful and useful characteristics of R environment is the possibility of using the multiple *packages* of R code developed by people collaborating in the R open-source environment effort.

---

[19] https://rattle.togaware.com/.

[20] https://rstudio.com/.

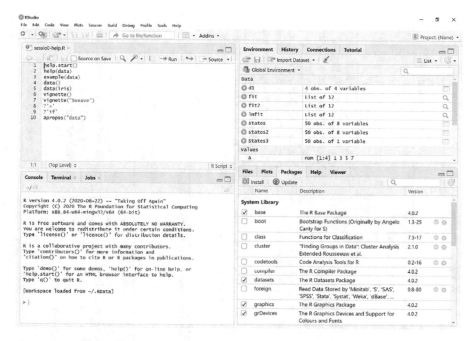

Fig. 8.2 RStudio Desktop IDE screenshot

This is very important for the development of IDSSs because there are many software packages available that implement many data-driven methods and some model-driven methods. All the functions implementing the interested methods can be used, once the corresponding packages have been installed and loaded.

Some useful readings for learning the R environment, especially related to data mining and machine learning methods are (Kabacoff, 2015; Zumel and Mount, 2014).

In the next Table 8.1, there is a list of the most important data-driven and model-driven techniques implemented in software *packages* available from the CRAN software *package* repository.[21] By no means is extensive, because the number of *packages* continuously increases, but can be a first guideline for an interested user to deploy an IDSS in the R environment.

---

[21] The version of the packages is the current one by July 2021.

Table 8.1 Software packages in R environment implementing data-driven and model-driven techniques

Kind of Method	Package	Main Functions	Description	Authors and website
Descriptive/ Clustering	*stats* 4.0.2	kmeans(), hclust(), plot. hclust()	Statistical package implementing the k-means method, and hierarchical agglomerative clustering techniques	R Core team and contributors worldwide. It is already released in the basic R distribution
	cluster 2.1.2	agnes(), plot.agnes(), diana(), plot.diana()	Many clustering-related functions like agglomerative nested hierarchical clustering (AGNES), divisive hierarchical clustering (DIANA method)	Martin Maechler, Peter Rousseeuw, Anja Struyf, and Mia Hubert https://cran.r-project.org/web/pack ages/cluster/index.html
	NbClust 3.0	NbClust()	An examination of 30 CVIs for Determining the ideal number of clusters	Malika Charrad, Nadia Ghazzali, Veronique Boiteau, and Azam Niknafs https://cran.r-project.org/web/pack ages/NbClust/index.html
Associative/ Association rules	*arules* 1.6–8	apriori(), eclat(), ruleInduction()	Mining association rules and frequent Itemsets. Apriori and Eclat algorithms are available	Michael Hahsler, Christian Buchta, Bettina Gruen, and Kurt Hornik https://cran.r-project.org/web/pack ages/arules/index.html
Associative/ Bayesian networks	*bnlearn* 4.6–1	bn.fit(), cpquery()	Bayesian network structure learning, parameter learning and inference	Marco Scutari and contributor Robert ness https://cran.r-project.org/web/pack ages/bnlearn/index.html
	BayesianNetwork 0.1.5	BayesianNetwork()	It is an interactive web application for Bayesian network modelling and analysis, powered by *bnlearn* and *networkD3* packages	Paul Govan https://cran.r-project.org/web/pack ages/BayesianNetwork/index.html
Discriminant/ Decision trees	*rpart* 4.1–15	rpart(), predict.rpart(), plot.rpart()	Recursive partitioning and regression trees. It builds decision trees using Gini index or information split criteria	Terry Therneau, Beth Atkinson and Brian Ripley (producer of the initial R translation) https://cran.r-project.org/web/pack ages/rpart/index.html

(continued)

Table 8.1 (continued)

Kind of Method	Package	Main Functions	Description	Authors and website
Discriminant/ Support vector machines (SVM)	*kernlab* 0.9–29	ksvm(), predict.ksvm()	Kernel-based machine learning lab. Several kernel functions are available	Alexandros Karatzoglou, Alex Smola, and Kurt Hornik https://cran.r-project.org/web/pack ages/kernlab/index.html
	e1071 1.7-7	svm(), predict.svm()	Misc functions of the Department of Statistics, probability theory group (formerly: E1071), TU Wien. Several kernel functions are available	David Meyer, Evgenia Dimitriadou, Kurt Hornik, Andreas Weingessel, and Friedrich Leisch https://cran.r-project.org/web/pack ages/e1071/index.html
Discriminant/ k-nearest neighbour	*class* 7.3–19	knn()	Discriminant/classifiers methods	Brian Ripley https://cran.r-project.org/web/pack ages/class/index.html
Discriminant/ Ensemble methods	*Boot* 1.3.28	Boot()	Resampling and bootstrapping methods	Angelo Canty (original S version) and Brian Ripley (translation into R) https://cran.r-project.org/web/pack ages/boot/index.html
	adabag 4.2	bagging(), predict. bagging(), boosting(), predict.boosting()	Applies multiclass AdaBoost.M1, SAMME and bagging, using decision trees as individual classifiers	Esteban Alfaro, Matías Gámez, and Noelia García https://cran.r-project.org/web/pack ages/adabag/index.html
	randomForest 4.6–14	randomForest(), predict. randomForest()	Breiman and Cutler's random forests for classification and regression	Fortran original by Leo Breiman and Adele cutler, R port by Andy Liaw and Matthew wiener https://cran.r-project.org/web/pack ages/randomForest/index.html

(continued)

Table 8.1 (continued)

Kind of Method	Package	Main Functions	Description	Authors and website
Predictive/ Artificial Neural Networks (ANNs)	*nnet* 7.3–14	nnet(), predict.nnet()	Feed-forward neural networks with a single hidden layer, and multinomial log-linear models	Brian Ripley https://cran.r-project.org/web/pack ages/nnet/index.html
	neural 1.4.2.2	mlptrain(), mlp(), rbftrain(), rbf()	RBF and MLP neural networks with graphical user interface	Adam Nagy. https://cran.r-project.org/web/pack ages/neural/index.html
Deep Learning Architectures	*deepnet* 0.2	dbn.dnn.train(), sae.dnn. train()	Implement some deep learning architectures and artificial neural network algorithms, including BP, RBM, DBN, deep autoencoder and so on	Xiao Rong https://cran.r-project.org/web/pack ages/deepnet/index.html
Predictive/ Multiple linear regression (MLR)	*stats* 4.0.2	lm(), predict.lm()	Statistical package implementing multiple linear regression models	R Core team and contributors worldwide. It is already released in the basic R distribution
Predictive/ Regression trees	*rpart* 4.1–15	rpart(), predict.rpart(), plot.rpart()	Recursive partitioning and regression trees. It builds regression trees	Terry Therneau, Beth Atkinson and Brian Ripley (producer of the initial R translation) https://cran.r-project.org/web/pack ages/rpart/index.html
Optimization/ Genetic Algorithms	*GA* 3.2.1	ga(), plot.Ga-method()	Flexible general-purpose toolbox implementing genetic algorithms (GAs) for stochastic Optimisation	Luca Scrucca https://cran.r-project.org/web/pack ages/GA/index.html
Agent-based simulation	*NetlogoR* 0.3.8	agentmatrix(), plot. agentmatrix()	Build and run spatially explicit agent-based models. It is a translation in R of the structure and functions of "NetLogo" (Wilensky, 1999)	Sarah Bauduin, Eliot J. B. McIntire and Alex M Chubaty https://cran.r-project.org/web/pack ages/NetlogoR/index.html

(continued)

Table 8.1 (continued)

Kind of Method	Package	Main Functions	Description	Authors and website
Rule-based system	*Rdrools*[a] *1.1.1*	runRulesDrl()	An interface for using the Java-based Drools rule engine. This package provides an intuitive interface to execute rules on datasets through an R interface	Ashwin Raaghav, SMS Chauhan, Naren Srinivasan, Dheekshitha PS, Zubin Dowlaty, Mayukh Bose and Arushi Khattri https://cran.r-project.org/web/packages/Rdrools/index.html
	Rdroolsjars 1.0.1	By using rJava, the Java layer Communicates with the R layer and provides the functionality of using drools in R	The "Rdroolsjars" package collects all the external jar files required for the "Rdrools" package	Ashwin Raaghav, SMS Chauhan https://cran.r-project.org/web/packages/Rdroolsjars/index.html
	frbs 3.2–0	frbs.learn(), predict.frbs()	Fuzzy rule-based Systems for Classification and Regression Tasks	Lala Septem Riza, Christoph Bergmeir, Francisco Herrera, and José Manuel Benítez https://cran.r-project.org/web/packages/frbs/index.html
Model-based reasoning	*NlcOptim 0.6*	solnl()	Solve non-linear optimization with nonlinear constraints. Linear or nonlinear equality and inequality constraints are allowed	Xianyan Chen and Xiangrong yin https://cran.r-project.org/web/packages/NlcOptim/index.html
Qualitative reasoning/ System Dynamics	*readsdr 0.2.0*	sd_simulate()	Translate models from system Dynamics software into R. it can parse "XMILE" files (*Vensim* and *Stella*) models into R objects to construct Networks	Jair Andrade https://cran.r-project.org/web/packages/readsdr/index.html

(continued)

Table 8.1 (continued)

Kind of Method	Package	Main Functions	Description	Authors and website
	dynr 0.1.16-2	dynr.cook(), dynr.model()	Dynamic modelling in R. it implements a set of computationally efficient algorithms for handling a broad class of linear and non-linear discrete- and continuous-time models	Lu Ou, Michael D. Hunter, Sy-Miin chow, Linying Ji, Meng Chen, hui-Ju hung, Jungmin lee, Yanling li, and Jonathan Park https://cran.r-project.org/web/packages/dynr/index.html

[a]Package *Rdrools* was removed from the CRAN repository. Note archived on 2020-05-06 as check problems were not corrected. Last version can be downloaded from https://www.rdocumentation.org/packages/Rdrools/versions/1.1.1 or https://cran.r-project.org/src/contrib/Archive/Rdrools/, but seems that *Rdroolsjars* package solve the problems calling directly to Drools tool through *rJava*

8.4.2   *Python*

Python (Python Software Foundation, 2021) is a general-purpose, multi-paradigm programming language, supporting object-oriented, functional, and procedural coding structures. Python is an interpreted, and interactive programming language. It can be used both for standalone applications and for scripting applications in many domains. It is comparable to Perl, Ruby, Scheme, or Java.

Python outstanding features as a programming language are:

- Combining remarkable power with easiness to use
- Simple and consistent programming language syntax and semantics
- Cross-platform availability
- Highly modular: code can be grouped into *modules* and *packages*
- Suited both for rapid prototyping and large-scale programming
- The language supports raising and catching exceptions, resulting in cleaner error handling exceptions
- Supporting object-oriented programming with multiple inheritances.
- Data types are strongly and dynamically typed
- Powerful large standard library is included for numeric processing, image manipulation, user interfaces, web scripting, etc.
- There are interfaces to many systems calls and libraries, as well as to various windowing systems.
- New built-in modules are easily written and compiled in C or C++ or other languages for higher speed.
- Usable as an extension language for applications written in other languages that need easy-to-use scripting or automation interfaces.

Python was created in the early 1990s by Guido van Rossum at Stichting Mathematisch Centrum (CWI[22]) in the Netherlands as a successor of a language called ABC. Guido van Rossum remains Python's principal author, although it includes many contributions from others.

In 1995, Guido van Rossum continued his work on Python at the Corporation for National Research Initiatives (CNRI[23]) in Reston, Virginia where he released several versions of the software.

In May 2000, Guido van Rossum and the Python core development team moved to BeOpen.com to form the BeOpen PythonLabs team. In October of the same year, the PythonLabs team moved to Digital Creations (now Zope Corporation[24]). In 2001, the Python Software Foundation (PSF[25]) was formed, a non-profit organization created specifically to own Python-related Intellectual Property. Zope Corporation is a sponsoring member of the PSF.

---

[22] https://www.cwi.nl.

[23] https://www.cnri.reston.va.us.

[24] https://www.zope.org.

[25] The Python Software Foundation (PSF, https://www.python.org/psf) is a 501(c)(3) non-profit corporation that holds the intellectual property rights behind the Python programming language.

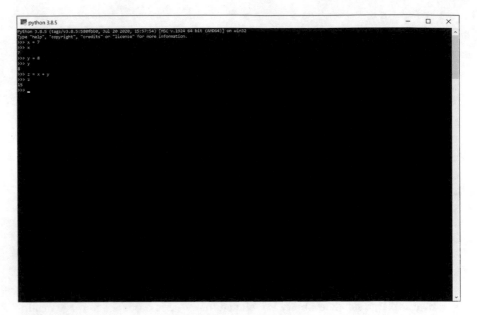

Fig. 8.3 Python basic interactive environment screenshot

Main characteristics of the tool are:

- Tool: Python programming language
- Developer: since 2001, © Python Software Foundation. Created and designed by Guido van Rossum.
- Website: https://www.python.org/
- Date since it is available: Python 1.0.1 (1994)
- Latest stable version (by July 2021): Python 3.9.6 released in June 2021.
- Implementation Language: mainly C
- Type of license: open-source software, and starting with Python versions higher than 2.1 are compatible with GNU General Public License (GPL)

Python's interactive mode makes it easy to test short snippets of code. The Python release offers the original *basic command-line mode*. This *basic interactive environment* allows to access to the Python interpreter and execute the commands. In Fig. 8.3, there is a screenshot of the basic Python interactive environment.

In addition, the Python release provides an improved environment to access to the Python interpreter. It has an Integrated Development Environment called *IDLE*. *IDLE* is the *Python's Integrated Development and Learning Environment*.

It has a shell window, i.e. the interactive interpreter, with colorizing of code input, output, and error messages. Also, it has a multi-window text editor with multiple undo, Python colourizing, smart indent, call tips, auto completion, and other features.

It can search within any window, replace within editor windows, and search through multiple files. Debugging facility with persistent breakpoints, stepping, and viewing of global and local namespaces.

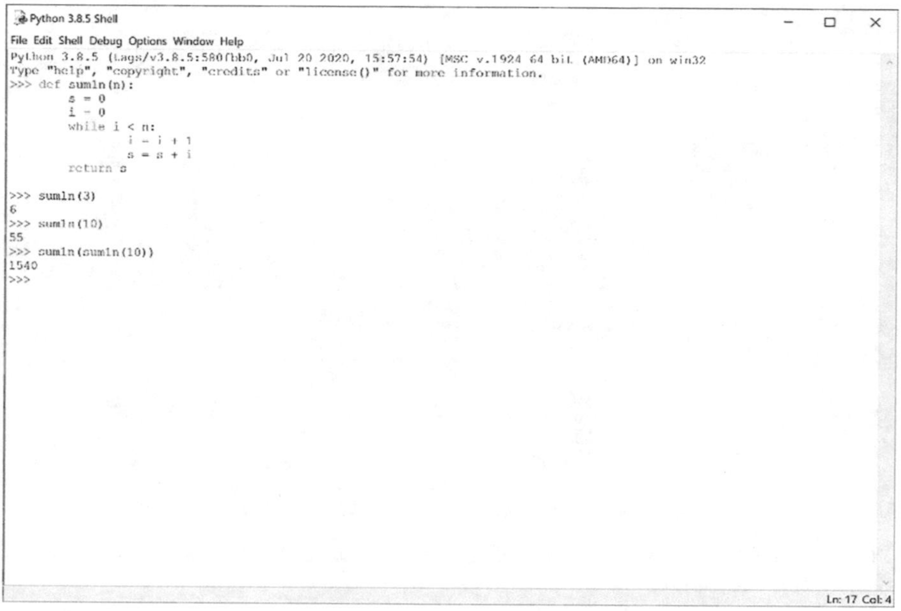

Fig. 8.4 Python IDLE shell

Finally offers configuration settings, browsers, and other dialogs. It is a simple but efficient IDE for starting to learn Python. In Fig. 8.4, there is a screenshot of the Python IDLE shell.

In addition, in 2001 appeared Interactive Python (*IPython*[26]) (Pérez and Granger, 2007), which is a command shell for interactive computing in multiple programming languages, originally developed for the Python programming language.

IPython offers introspection, rich media, shell syntax, tab completion, and history. IPython provides the following features: interactive shells (terminal and Qt-based), support for interactive data visualization and use of GUI toolkits, flexible and embeddable interpreters to load into one's own projects, tools for parallel computing, and finally, a browser-based *notebook interface* with support for code, text, mathematical expressions, inline plots, and other media.

IPython is open-source software released under the revised BSD license.

This capability of managing *notebooks*, which can integrate source code, results of the code execution and text, similarly to automatic generation of reports in R, originated that in 2014 Fernando Pérez launched a new open-source spin-off project from IPython called Project *Jupyter*.[27] IPython continued to exist as a Python shell and kernel for Jupyter, but the notebook interface and other language-agnostic parts of IPython were moved under the Jupyter name.[28]

---

[26] © The IPython Development Team (https://ipython.org/).

[27] © Project Jupyter (https://jupyter.org/).

[28] Jupyter is language agnostic and its name is a reference to core programming languages originally supported by Jupyter: Julia, Python, and R.

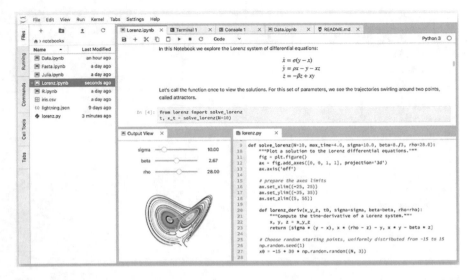

Fig. 8.5 JupyterLab web-based interactive environment (extracted from https://jupyterlab. readthedocs.io/en/latest/)

Jupyter Notebook (formerly IPython Notebooks) is a web-based interactive computational environment for creating, executing, and visualizing Jupyter notebooks.

JupyterLab is the next-generation web-based user interface for Project Jupyter. JupyterLab is a web-based interactive development environment for Jupyter notebooks, code, and data.

JupyterLab is open-source software and released under the liberal terms of the modified BSD license. In Fig. 8.5 there is a screenshot of JupyterLab interface.

Although the IDLE shell is useful and the IPython shell or JupyterLab are extremely useful, there are some even more complete IDEs in the literature, which integrate the capabilities of IDLE shell and IPython with other functionalities, such as *PyDev*, *PyCharm* (community edition), *Spyder* among the free software, and *Komodo IDE*[29] and *Wing Python IDE*[30] among the proprietary licensed software.

From the free software, *PyDev* and *PyCharm* are based on mature and well-established Eclipse IDE and IntelliJ IDEA IDE, and *Spyder* is quite similar to RStudio or MATLAB[®][31] configuration. A brief description of each one with a screenshot of each of them follows.

---

[29] Komodo   IDE   (https://www.activestate.com/products/komodo-ide/)   is   developed   by © ActiveState Software Inc.

[30] Wing Python IDE (https://wingware.com/) is developed by © Wingware.

[31] MATLAB® (https://www.mathworks.com/products/matlab.html) combines a desktop environment tuned for iterative analysis and design processes with a programming language that expresses matrix and array mathematics directly. It includes the Live Editor for creating scripts that combine code, output, and formatted text in an executable notebook. It is developed by © The MathWorks®, Inc.

Fig. 8.6 PyDev IDE interface (extracted from https://www.pydev.org)

PyDev[32] is a Python IDE for Eclipse, which may be used in Python development. It provides many features such as: Django[33] integration, code completion with auto import, type hinting, code analysis, go to definition, refactoring, a debugger, a tokens browser, an interactive console, a unit-test integration, code coverage, code analysis, and the find references functionality.

The recommended way of using PyDev is bundled in LiClipse,[34] which provides PyDev built-in as well as support for other languages.

PyDev was originally created by Aleks Totic in July 2003, but Fabio Zadrozny became the project's main developer in January 2005. In September 2005, PyDev Extensions was started as a commercial counterpart of PyDev, offering features such as code analysis and remote debugging. In July 2008, Aptana acquired PyDev, retaining Zadrozny as the project head. They open sourced PyDev Extensions in September 2009, and merged it with PyDev.

PyDev is open-source software since 2009. Current version of PyDev, by July 2021, is 8.3.0 released on April 2021.

In Fig. 8.6 there is a screenshot of PyDev.

---

[32] http://www.pydev.org/.

[33] Django (https://www.djangoproject.com/) is a high-level open-source Python Web framework that encourages rapid development and clean, pragmatic design.

[34] LiClipse (http://www.liclipse.com/), which is a fast editor supporting many languages and providing usability improvements for all Eclipse editors, is a proprietary licensed software by © Brainwy Software Ltda-Me.

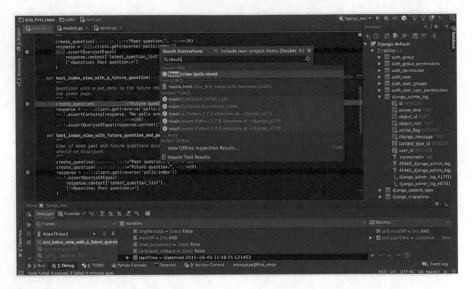

Fig. 8.7 PyCharm IDE interface (extracted from https://www.jetbrains.com/pycharm/)

PyCharm[35] (community edition) (Nguyen, 2019) is a free open-source Python IDE providing smart code completion, code inspections, on-the-fly error highlighting, and quick-fixes, along with automated code refactoring and rich navigation capabilities.

PyCharm offers a huge collection of tools out of the box: an integrated debugger and test runner; a built-in terminal and integration with major Version Control Systems.

PyCharm is developed by © JetBrains s.r.o. They offer improved versions of PyCharm community edition: PyCharm professional edition, which is proprietary licensed, and the PyCharm educational edition, which is free for educational or learning purposes. Current version, by July 2021, is PyCharm 2021.1.13 released on June 2021.

In Fig. 8.7 there is a screenshot of PyCharm interface.

Spyder[36] (Scientific PYthon Development EnviRonment) (Raybaut, 2017) is an open-source cross-platform powerful scientific environment for Python, written in Python, and designed by and for scientists, engineers, and data analysts. It offers a unique combination of the advanced editing with a multi-language editor, analysis, interactive debugging, and profiling functionality of a comprehensive development tool with the data exploration, interactive execution, deep inspection, and beautiful visualization capabilities of a scientific package.

---

[35] https://www.jetbrains.com/pycharm/.

[36] https://www.spyder-ide.org/.

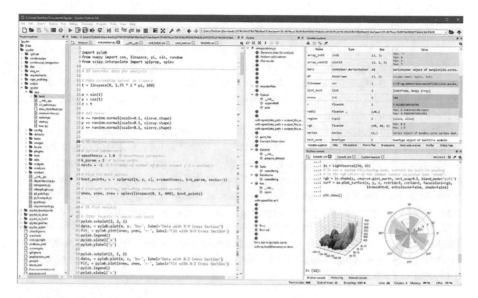

Fig. 8.8 Spyder IDE interface (extracted from https://www.spyder-ide.org/)

Furthermore, Spyder offers built-in integration with many popular scientific packages, including NumPy, SciPy, Pandas, IPython, QtConsole, Matplotlib, SymPy, and more.

Beyond its many built-in features, its abilities can be extended even further via its plugin system and API. Furthermore, Spyder can also be used as a PyQt5[37] extension library, allowing developers to build upon its functionality and embed its components, such as the interactive console, in their own PyQt software.

Spyder supports as many IPython consoles as required. They allow to run code by line, cell, or file; and render plots right inline. It provides a variable explorer, which interacts with variables and modify variables on the fly, plot a histogram or time series, edit a dataframe or Numpy array, sort a collection, or dig into nested objects.

Spyder is a free open-source software licensed under the MIT license. Initially created and developed by Pierre Raybaut in 2009, since 2012 Spyder has been maintained and continuously improved by a team of scientific Python developers and the community, mainly led by Carlos Córdoba. Current version of Spyder, by July 2021, is 5.0.5 released on June 2021.

In Fig. 8.8 there is a screenshot of the Spyder interface.

Even though the *standard library* of Python has many useful valuable libraries for many tasks, the Python open-source community and collaborators have developed

---

[37] PyQt/PyQt5 is a comprehensive set of Python bindings for Qt/Qt v5, developed by Riverbank Computing Limited (https://www.riverbankcomputing.com/). Qt is a free and open-source widget toolkit for creating graphical user interfaces as well as cross-platform applications that run on various software and hardware platforms.

many *packages/libraries* of Python code which are available at the Python Package Index (PyPI) website (https://pypi.org/).

This is very important for the development of IDSSs, because there are many software packages available that implement many data-driven methods and some model-driven methods. All the functions implementing the interested methods can be used, once the corresponding *packages* have been installed and imported into your Python environment.

Some useful readings for learning Python, especially focussed on data mining and machine learning methods are (Lutz, 2014; Lutz 2013; Ceder, 2010; Grus, 2015).

In the next Table 8.2, there is a list of the most important data analysis and manipulation, efficient numeric computation, stylish static, animated, and interactive visualizations, data-driven, and model-driven techniques available in software packages available from *Python Package Index* repository or from the own developers site.[38] The same situation happens as in the case of the R list, and this Python package list is by no means extensive, because the number of *packages/libraries* continuously increases,[39] but can be a first guideline for an interested user to deploy an IDSS in Python.

8.4.3   GESCONDA

GESCONDA (https://kemlg.upc.edu/en/projects/gesconda-1) (Sànchez-Marrè et al. 2010; Gibert et al. 2006) is an Intelligent Data Analysis System under development at the Knowledge Engineering and Machine Learning Group (KEMLG), at Universitat Politècnica de Catalunya (UPC) created and led by the author of this book. It was designed for knowledge discovery from databases. Initially focussed on environmental databases, has become a general-oriented tool. On the basis of previous experiences, it was designed as a four-level architecture connecting the user with the environmental system or process: *data filtering and pre-processing* layer, *recommendation and meta-knowledge* layer, *data mining* layer, and *knowledge management and reasoning* layer.

The *Data filtering and pre-processing layer* provide a set of tools for data cleaning. Statistical one-way and two-way are provided, missing value and outlier value management, graphical visualization, and variable transformation operators are integrated into this layer. Cleaned data will be used afterwards at the other layers to produce data mining models, which will be executed at the reasoning level.

The *recommendation and meta-knowledge layer* include two modules: The recommender and meta-knowledge module, and the feature relevance module. The former one (Gibert et al., 2010) let the user to be assisted to select the most suitable

---

[38] The version of the packages is the current one by July 2021.

[39] The number of packages/projects in the Python Package Index repository (https://pypi.org) by July 17th 2021 was of 316,394 projects.

Table 8.2 Software packages/libraries in Python implementing data analysis, data visualization, numerical computation, interactive computation, data-driven and model-driven techniques

Kind of task/method	Package/Module	Description/Classes/Methods	Authors and website
Data analysis and manipulation	pandas 1.3.0	pandas is a fast, powerful, flexible, and easy to use open-source data analysis and manipulation tool, built on python. It provides additional data structures for working with datasets in python, especially the DataFrame structure	(The Pandas Development Team, 2020) https://pandas.pydata.org/
Numeric computation	NumPy 1.21.0	NumPy is the fundamental package for numerical computation with a core well optimized in C code. It defines the numerical n-dimensional arrays and basic operations on them, as well as mathematical functions, random number generators, linear algebra routines, Fourier transforms, and more	(Oliphant, 2006) https://numpy.org/
	SciPy 1.7.0	SciPy provides many user-friendly and efficient numerical routines, such as routines for numerical integration, interpolation, optimization, linear algebra, signal processing, and statistics	(Virtanen et al., 2020) https://www.scipy.org/ scipylib/index.html
	SymPy 1.8	SymPy is a python library for symbolic mathematics. It aims to become a full-featured computer algebra system (CAS) while keeping the code as simple as possible in order to be comprehensible and easily extensible	(Meurer et al., 2017) https://www.sympy.org/ en/index.html
Graphic Visualization	Matplotlib 3.3.0	Matplotlib is a comprehensive library for creating static, animated, and interactive visualizations in python.	(Hunter, 2007) https://matplotlib.org/
	seaborn 0.11.1	seaborn is a Python data visualization library based on matplotlib and closely integrated with pandas data structures. It provides a high-level interface for drawing attractive and informative statistical graphics	(Waskom et al., 2020) https://seaborn.pydata. org/index.html
Interactive computing	IPython 7.25.0	IPython provides a rich architecture for interactive computing with: a powerful interactive shell, a kernel for Jupyter notebooks, support for interactive data visualization and use of GUI toolkits, flexible, embeddable interpreters to load into current projects, and high-performance tools for parallel computing	(Pérez and Granger, 2007) http://ipython.org/
Image Processing	scikit-image 0.18.2	scikit-image builds on scipy.Ndimage module to provide a versatile set of image processing routines in python	(van der Walt et al., 2014) https://scikit-image.org/

(continued)

Table 8.2 (continued)

Kind of task/method	Package/Module	Description/Classes/Methods	Authors and website
Machine learning	scikit-learn 0.24.2	scikit-learn is a collection of algorithms and tools for machine learning and data analysis, built on NumPy, SciPy, and matplotlib. It offers routines for data pre-processing, for dimensionality reduction, for clustering, for discriminant tasks, for numerical predictive tasks and for model validation	(Pedregosa et al., 2011) https://scikit-learn.org
Descriptive/Clustering	scikit-learn/ sklearn.cluster 0.24.2	Classes: KMeans, DBSCAN, SpectralClustering, AgglomerativeClustering and others Methods: KMeans.fit_predict(), DBSCAN.fit_predict(), SpectralClustering.fit_predict() AgglomerativeClustering.fit_predict() and other methods.	(Pedregosa et al., 2011) https://scikit-learn.org
Associative/Association rules	armine 0.2.0	armine is an association rule mining package for python 3. It provides ARM class to generate association rules from a set of transactions Methods: Armine.ARM.Learn(), armine.ARM.Print()	Priyam Singh https://pypi.org/project/armine/
	arules 0.0.0	arules is an open-source python package for association rules creation over tabular data (pandas dataframe) Methods: arules.create_association-rules(), arules.present_rules_per_consequent ()	Abir Koren https://pypi.org/project/arules/
Associative/Bayesian Networks	bnlearn 0.3.21	bnlearn is a python package for learning the graphical structure of Bayesian networks, parameter learning, inference, and sampling methods. It is built on the pgmpy package Methods: bnlearn.structure_learning.Fit(), bnlearn.parameter_learning.Fit(), bnlearn.Inference.Fit()	Erdogan Taskesen https://pypi.org/project/bnlearn/
	bayespy 0.5.22	bayesPy provides tools for Bayesian inference with python. The user constructs a model as a Bayesian network, observes data, and runs posterior inference modules: bayespy.nodes, bayespy.Inference, bayespy.plot	Jaakko Luttinen and contributors: Hannu Hartikainen, Deebul Nair, Christopher Cramer, Till Hoffmann http://bayespy.org/, https://pypi.org/project/bayespy/

(continued)

Table 8.2 (continued)

Kind of task/method	Package/Module	Description/Classes/Methods	Authors and website
Discriminant/Decision trees	*scikit-learn/ klearn.tree 0.24.2*	The CART (criterion = "Gini") method and the ID3 (criterion = "entropy") methods are implemented *Class*: DecisionTreeClassifier *Methods*: DecisionTreeClassifier.Fit(), DecisionTreeClassifier.Predict(), and other methods	(Pedregosa et al., 2011) https://scikit-learn.org
Discriminant/Support vector machines (SVM)	*scikit-learn/ sklearn.svm 0.24.2*	Both binary and multi-class classification problems are managed. Several kernel functions and SVM techniques are implemented *Classes*: SVC, nuSVC, LinearSVC *Methods*: SVC.Fit(), SVC.Predict(), nuSVC.Fit(), nuSVC.Predict(), LinearSVC. Fit(), LinearSVC.Predict(), and other methods	(Pedregosa et al., 2011) https://scikit-learn.org
Discriminant/k-nearest neighbour	*scikit-learn/ sklearn.neighbors 0.24.2*	The k-nearest neighbour discriminant method is implemented *Class*: KNeighborsClassifier *Methods*: KNeighborsClassifier.Fit(), KNeighborsClassifier.Predict(), and other methods	(Pedregosa et al., 2011) https://scikit-learn.org
Discriminant/Ensemble methods	*scikit-learn/ sklearn.Ensemble 0.24.2*	Several ensembles of classifier methods are implemented like voting, bagging, boosting, random forests, stacking, etc *Classes*: VotingClassifier, BaggingClassifier, RandomForestClassifier, AdaBoostClassifier *Methods*: VotingClassifier.Fit(), VotingClassifier.Predict(), BaggingClassifier. Fit(), BaggingClassifier.Predict(),RandomForestClassifier.Fit(), RandomForestClassifier.Predict(), AdaBoostClassifier.Fit(),AdaBoostClassifier. Predict(), and other methods	(Pedregosa et al., 2011) https://scikit-learn.org
Predictive/Artificial Neural Networks (ANNs)	*scikit-learn/ sklearn. neural_network 0.24.2*	This module implements a multi-layer perceptron (feed-forward ANNs) *Class*: MLPRegressor *Methods*: MLPRegressor.Fit(),MLPRegressor.Predict(), and other methods	(Pedregosa et al., 2011) https://scikit-learn.org
Deep Learning Architectures	*torch 1.9.0*	Tensors and dynamic neural networks in Python with strong GPU acceleration PyTorch is a python package that provides two high-level features: tensor computation (like NumPy) with strong GPU acceleration, and deep neural networks built on a tape-based autograd system	PyTorch team https://pypi.org/project/ torch/

(continued)

Table 8.2 (continued)

Kind of task/method	Package/Module	Description/Classes/Methods	Authors and website
Predictive/Multiple Linear Regression (MLR)	*scikit-learn/ sklearn. linear_model 0.24.2*	This module implements several linear models and some variants like the multiple linear regression, logistic regression, etc *Classes*: LinearRegression, LogisticRegression, and others *Methods*: LinearRegression.Fit(), LinearRegression.Predict(), LinearRegression.Coefs(), LogisticRegression.Fit(), LogisticRegression.Predict(), and other methods	(Pedregosa et al., 2011) https://scikit-learn.org
Predictive/Regression trees	*scikit-learn/ sklearn.tree 0.24.2*	Several criteria ("mse", "Friedman_mse", "mae") are implemented *Class*: DecisionTreeRegressor *Methods*: DecisionTreeRegressor.Fit(), DecisionTreeRegressor.Predict(), and other methods	(Pedregosa et al., 2011) https://scikit-learn.org
Optimization/Genetic Algorithms	*geneticalgorithm 1.0.2*	*geneticalgorithm* is an easy implementation of genetic-algorithm (GA) to solve continuous and combinatorial optimization problems with real, integer, and mixed variables in python Method: Geneticalgorithm.Run()	Ryan (Mohammad) Solgi https://pypi.org/project/ geneticalgorithm/
Agent-based simulation	*Mesa 0.8.9*	*Mesa* is an agent-based modelling (or ABM) framework in python. It allows users to quickly create agent-based models using built-in core components (such as spatial grids and agent schedulers) or customized implementations; visualize them using a browser-based interface; and analyze their results using Python's data analysis tools	Project Mesa team https://pypi.org/project/ Mesa/
	dworp 0.1.0	*dworp* is a flexible framework for building agent-based modelling simulations. *Dworp* defines basic interfaces for building simulations and provides some default components to support rapid creation of agent-based models	Cash Costello https://pypi.org/project/ dworp/
Rule-based system	*experta 1.9.4*	*experta* is a Python library for building expert systems strongly inspired by CLIPS. Experta is a spin-off project derived from *Pyknow* project (it is a *Pyknow* fork)	Maintainer: Roberto Abdelkader Martínez Pérez https://pypi.org/project/ experta/

(continued)

Table 8.2 (continued)

Kind of task/method	Package/Module	Description/Classes/Methods	Authors and website
Model-based reasoning	*python-constraint* 1.4.0	The *python-constraint* module offers solvers for constraint satisfaction problems (CSPs) over finite domains in simple and pure python. The following solvers are available: Backtracking solver, recursive backtracking solver and minimum conflicts solver. Several predefined constraint types are available. *Classes*: Problem, variable, domain, solver, constraint, and others	Gustavo Niemeyer https://pypi.org/project/python-constraint/
	gecode-python 0.27	*gecode-python* offers Gecode bindings for python. It requires both python and Gecode software (version higher than 3.6.0) installed	Denys Duchier https://pypi.org/project/gecode-python/
Qualitative reasoning/ System Dynamics	*PyRCC8*[a] *1.2.2*	PyRCC8 is an efficient qualitative spatial reasoner written in pure python. It implements the RCC8 constraint language that has been popularly adopted by the qualitative spatial reasoning and GIS communities	Michael Sioutis https://pypi.org/project/PyRCC8/
	PySD 1.8.1	PySD is a package for running system dynamics models in python. PySD translates Vensim or XMILE model files into python modules, and provides methods to modify, simulate, and observe those translated models	James P. Houghton https://pypi.org/project/pysd/
	BPTK-Py 1.2.1	The business prototyping toolkit for python (BPTK-Py) is a computational modelling framework that enables to build simulation models using system Dynamics (SD) and/or agent-based modelling (ABM) and manage simulation scenarios with ease. Next to providing the necessary SD and ABM language constructs to build models directly in python, the framework also includes a compiler for transpiling system dynamics models conforming to the XMILE standard into Python code	Transentis labs GmbH[b] https://pypi.org/project/BPTK-Py/

[a]There is a note in the Python Package Index website (PyPI) stating that "This software has been outdated and will be replaced within time"
[b]Transentis consulting (https://www.transentis.com/)

methods to be applied in front of a real situation, by considering the goals of the user, main features of the domain, and their structure. In addition, the meta-information associated with the data can be managed, both from the features and the observations. The feature relevance module provides GESCONDA with a set of implemented feature weighting algorithms, which determine the weight or relevance of each one of the features describing the data. There are several unsupervised methods and some supervised methods.

The *data mining layer* is the layer joining the modules containing several data mining models, which can be induced from the data. There are five modules developed: the clustering techniques module, the association rule module, the decision tree techniques module, the rule induction module, and the statistical modelling module.

At the *knowledge management and reasoning layer* is where the new reasoning capabilities provide GESCONDA with analytical, synthetic, and predictive skills (*model execution*). Until now GESCONDA was an Intelligent Data Mining and Discovery tool, which was able to produce several knowledge models, but no problem-solving skills were given to the user. The past integration of a rule-based reasoning and case-based reasoning (Sànchez-Marrè et al., 2010) module has led GESCONDA to become a prototype tool for the Intelligent Decision Support development, which can assist the user to all the steps of the problem-solving cycle: diagnosis, planning/solution generation, prediction, and finally decision support.

Currently, the rule-based reasoning module is being interconnected with the decision tree module and the classification rule module to be able to use models produced by both modules achieving interoperability between *model producers* and *model executors*. The clustering module can interoperate with the classifier modules providing the class variable information to them. This interoperation is among *model producers*. Recently, a new evolutionary computing reasoning module (e.g. a genetic algorithm reasoning engine) has been implemented, which can be interoperated with the case-based reasoning module (Sànchez-Marrè et al., 2014).

Central characteristics of GESCONDA tool are the integration of statistical, AI and mixed methods into a single tool for extracting knowledge contained in data. All techniques implemented in GESCONDA can share information among themselves to best co-operate for extracting knowledge. It also includes capability for explicit management of the results produced by the different methods. Portability of the software between platforms is provided by a common Java™ platform.

Main drawback is that the tool is not yet stable and mature enough to be used regularly by the research community. It was partially funded at the beginning[40] of the project, but it could not advance as planned by the developers, due to the low funding amount for human resources. The tool has been partially developed with some graduate final projects and master thesis of collaborator students.

---

[40] Partially funded through projects TIN2004-01368 and TIC2000-1011 by Spanish Research and Technology Commission CICyT, during 2001–2007.

Currently, we have started to re-implement and improve GESCONDA in Python, but is still in the first stage of development. We hope that in a few years, the Python tool could be stable and mature.

References

Abar, S., Theodoropoulos, G. K., Lemarinier, P., & O'Hare, G. M. P. (2017). Agent based modelling and simulation tools: A review of the state-of-art software. *Computer Science Review, 24*, 13–33.

Becker, R. A., Chambers, J. M., & Wilks, A. R. (1988). *The new S language*. Chapman & Hall.

Bellifemine, F., Caire, G., & Greenwood, D. (2007). *Developing multi-agent systems with JADE*. John Wiley & Sons Ltd.

Borshchev, A., Karpov, Y., & Kharitonov, V. (2002). Distributed simulation of hybrid systems with AnyLogic and HLA. *Future Generation Computer Systems, 18*(6), 829–839.

Braubach, L., Pokahr, A., & Lamersdorf, W. (2004). Jadex: A short overview. In *Main Conference Net. ObjectDays* (pp. 195–207). AgentExpo.

Bredeweg, B., Linnebank, F., Bouwer, A., & Liem, J. (2009). Garp3 - workbench for qualitative modelling and simulation. *Ecological Informatics, 4*(5–6), 263–281.

Ceder, V. L. (2010). *The quick python book*. Manning Publications Co.

Cossentino, M., Gaud, N., Galland, S., Hilaire, V., & Koukam, A. (2007). A Metamodel and Implementation Platform for Holonic Multi-Agent Systems. In *Proc. of 5th European workshop on Multi-Agent Systems (EUMAS'07). 13–14 December.*

de Kleer, J., & Brown, J. S. (1984). A qualitative physics based on confluences. *Artificial Intelligence, 24*, 7–83.

Demsar, J., Curk, T., Erjavec, A., Gorup, C., Hocevar, T., Milutinovic, M., Mozina, M., Polajnar, M., Toplak, M., Staric, A., Stajdohar, M., Umek, L., Zagar, L., Zbontar, J., Zitnik, M., & Zupan, B. (2013). Orange: Data mining toolbox in python. *Journal of Machine Learning Research, 14*, 2349–2353.

Dennett, D. C. (1987). *The intentional stance*. The MIT Press.

Fages, J. G., & Prud'Homme, C. (November 2017). Making the first solution good! In *Proc. of 29th IEEE International Conference on Tools with Artificial Intelligence (ICTAI 2017), Boston, MA, United States.*

Forbus, K. (1992). Pushing the edge of the (QP) envelope. In B. Faltings & P. Struss (Eds.), *Recent advances in qualitative physics*. MIT Press.

Forbus, K. (1990). The qualitative process engine. In D. Weld & J. de Kleer (Eds.), *Readings in qualitative reasoning about physical systems*. Morgan Kaufman.

Fox, J. (2017). *Using the R commander: A point-and-click interface for R*. Chapman & Hall/CRC Press.

Fox, J. (2005). The R commander: A basic-statistics graphical user interface to R. *Journal of Statistical Software, 19*(9), 1–42.

Friedman-Hill, E. (1997). *Jess, The Java Expert System Shell*. Sandia Report SAND98–8206

Friedman-Hill, E. (2003). *Jess in action*. Manning Publications.

Frisch, A. M., Harvey, W., Jefferson, C., Martínez-Hernández, B., & Miguel, I. (2008). Essence: A constraint language for specifying combinatorial problems. *Constraints, 13*(3), 268–306.

Gantner, Z., Westphal, M., & Wölfl, S. (2008). GQR - A Fast Reasoner for Binary Qualitative Constraint Calculi. In *Workshop Notes of the AAAI-08 Workshop on Spatial and Temporal Reasoning, Chicago, USA.*

Gent, I. P., Jefferson, C., & Miguel, I. (2006). MINION: A Fast, Scalable, Constraint Solver. In *Proc. of the 17th European Conference on Artificial Intelligence (ECAI 2006).*

Giarratano, J. C., & Riley, G. D. (2004). *Expert Systems: Principles and Programming* (4th ed.). Course Technology.

Gibert, K., Sànchez-Marrè, M., & Codina, V. (2010). Choosing the right data mining technique: Classification of methods and intelligent recommendation. In *5th International Congress on Environmental Modelling and Software (iEMSs 2010). iEMSs 2010 Proceedings* (Vol. 3, pp. 1940–1947).

Gibert, K., Sànchez-Marrè, M., & Rodríguez-Roda, I. (2006). GESCONDA: An intelligent data analysis system for knowledge discovery and management in environmental data bases. *Environmental Modelling & Software, 21*(1), 116–121.

Gregory, M. P., O'Hare dDemsar, J., Curk, T., Erjavec, A., Gorup, C., Hocevar, T., Milutinovic, M., Mozina, M., Polajnar, M., Toplak, M., Staric, A., Stajdohar, M., Umek, L., Zagar, L., Zbontar, J., Zitnik, M., & Zupan, B. (August 2013). Orange: Data mining toolbox in python. *Journal of Machine Learning Research, 14*, 2349–2353.

Grus, J. (2015). *Data Science from Scratch. First Principles with Python*. O'Reilly Media Inc..

Hall, P., Dean, J., Kabul, I. K., & Silva, J. (2014). An Overview of Machine Learning with SAS® Enterprise Miner™. In *Technical Paper SAS313–2014*. SAS® Institute Inc..

Hall, M., Frank, E., Holmes, G., Pfahringer, B., Reutemann, P., & Witten, I. H. (2009). The weka data mining software: An update. *ACM SIGKDD Explorations Newsletter, 11*(1), 10–18.

Holcombe, M., Coakley, S., & Smallwood, R. A. (2006). General Framework for agent-based modelling of complex systems. In *Proceedings of the 2006 European Conference on Complex Systems*.

Hunter, J. D. (2007). Matplotlib: A 2D graphics environment. *Computing in Science & Engineering, 9*(3), 90–95.

Inchiosa, M. E., & Parker, M. T. (May 2002). Overcoming design and development challenges in agent-based modeling using ASCAPE. *Proceeding of the National Academy of Sciences of the United States of America (PNAS), 99*(suppl. 3), 7304–7308.

Kabacoff, R. (2015). R in action. In *Data analysis and graphics with R* (2nd ed.). Manning Publications.

Kuipers, B. (1994). *Qualitative reasoning*. The MIT Press.

Kuipers, B. (1986). Qualitative simulation. *Artificial Intelligence, 29*, 289–338.

Laburthe, F. (2000). Choco: implementing a CP kernel. In *Proc. of Techniques foR Implementing Constraint programming Systems (TRICS'00)* (pp. 118–133).

Lempinen, T., Ruutu, S., Karhela, T., & Ylén, P. (July 2011). Open source system dynamics with simantics and open modelica. In *Proc. of 29th International Conference of the System Dynamics Society. Washington, DC, USA*.

Lutz, M. (2014). *Python pocket reference* (5th ed.). O'Reilly Media.

Lutz, M. (July 2013). *Learning python* (5th ed.). O'Reilly Media.

McCarthy, J. (1978). Ascribing mental qualities to machines. In *Technical report*. Stanford University AI Lab.

Meurer, A., Smith, C. P., Paprocki, M., Čertík, O., Kirpichev, S. B., Rocklin, M., Kumar, A., Ivanov, S., Moore, J. K., Singh, S., Rathnayake, T., Vig, S., Granger, B. E., Muller, R. P., Bonazzi, F., Gupta, H., Vats, S., Johansson, F., Pedregosa, F., … Scopatz, A. (2017). SymPy: Symbolic computing in python. *PeerJ Comput Science, 3*, e103. https://doi.org/10.7717/peerj-cs.103

Nguyen, Q. (2019). *Hands-on application development with PyCharm: Accelerate your python applications using practical coding techniques in PyCharm*. Packt Publishing.

Oliphant, T. E. (2006). *A guide to NumPy*. Trelgol Publishing.

Padgham L., & Winikoff M. (2004). Developing intelligent agent systems: *A practical guide*. Wiley.

Parker, M. (2001). What is ascape and why should you care? *Journal of Artificial Societies and Social Simulation, 4*(1), 5. https://www.jasss.org/4/1/5.html

Parker, M. (1999). Ascape: An agent-based modeling environment in Java. In *Proceedings of agent simulation: applications, models and tools*. University of Chicago.

Pedregosa, F., Varoquaux, G., Gramfort, A., Michel, V., Thirion, B., Grisel, O., Blondel, M., Prettenhofer, P., Weiss, R., Dubourg, V., Vanderplas, J., Passos, A., Cournapeau, D., Brucher, M., Perrot, M., & Duchesnay, É. (2011). Scikit-learn: Machine Learning in Python. *Journal of Machine Learning Research, 12*, 2825–2830.

Pérez, F., & Granger, B. E. (2007). IPython: A system for interactive scientific computing. *Computing in Science and Engineering, 9*(3), 21–29. https://doi.org/10.1109/MCSE.2007.53

Proctor, M. (2007). Relational declarative programming with JBoss drools. In *Proc. of 9th International Symposium on Symbolic and Numeric Algorithms for Scientific Computing (SYNASC 2007)* (Vol. 1, p. 5). IEEE Computer Society.

Python Software Foundation (2021). *The Python Language Reference*. Available on-line at: https://docs.python.org/3/reference/.

Raybaut, P. August, 2017. *Spyder Documentation. Release 3*. Available on-line at: http://www.academia.edu/download/54731353/spyderide.pdf. .

R Core Team. (2021). *R Language Definition*. Available on-line at: https://cran.r-project.org/doc/manuals/r-release/R-lang.pdf. Version 4.1.0.

Rendón-Sallard, T., & Sànchez-Marrè, M. (2006). *A review on multi-agent platforms and environmental decision support systems simulation tools*. In *Research Report LSI-06-3-R*. Department of Computer Science, UPC.

Sànchez-Marrè, M., Gibert, K., Vinayagam, R. K., & Sevilla, B. (June 2014). Evolutionary Computation and Case-Based Reasoning Interoperation in IEDSS through GESCONDA. In *7th International Congress on Environmental Modelling & Software (iEMSs 2014). iEMSs 2014 Proceedings* (Vol. 1, pp. 493–500).

Sànchez-Marrè, M., Gibert, K., & Sevilla, B. (2010). Evolving GESCONDA to an intelligent decision support tool. In *5th International Congress on Environmental Modelling and Software (iEMSs'2010). iEMSs 2010 Proceedings* (Vol. 3, pp. 2015–2024) ISBN 978-88-903574-1-1.

Schulte, C., & Tack, G. (2006). Views and Iterators for Generic Constraint Implementations. In *Recent Advances in Constraints (2005)* (Lecture Notes in Artificial Intelligence) (Vol. 3978, pp. 118–132). Springer-Verlag.

Triguero, I., González, S., Moyano, J. M., García, S., Alcalá-Fdez, J., Luengo, J., Fernández, A., del Jesus, M. J., Sánchez, L., & Herrera, F. (2017). KEEL 3.0: An open source software for multi-stage analysis in data mining. *International Journal of Computational Intelligence Systems, 10*, 1238–1249.

The Pandas Development Team. (February 2020). *Pandas-dev/pandas: Pandas. Zenodo*. https://doi.org/10.5281/zenodo.3509134

van der Walt, S., Johannes, L., Schönberger, J. N.-I., Boulogne, F., Warner, J. D., Yager, N., Gouillart, E., Yu, T., & The Scikit-Image Contributors. (June 2014). scikit-image: Image processing in Python. *PeerJ Life & Environment, 2*, e453. https://doi.org/10.7717/peerj.453

Verzani, J. (2011). *Getting started with RStudio*. O'Reilly Media.

Virtanen, P., Gommers, R., Oliphant, T. E., Haberland, M., Reddy, T., Cournapeau, D., Burovski, E., Peterson, P., Weckesser, W., Bright, J., van der Walt, S. J., Brett, M., Wilson, J., Millman, K. J., Mayorov, N., Nelson, A. R. J., Jones, E., Kern, R., Larson, E., . . . SciPy 1.0 Contributors. (2020). SciPy 1.0: Fundamental Algorithms for Scientific Computing in Python. *Nature Methods, 17*, 261–272.

Waskom, M., Botvinnik, O., Ostblom, J., Gelbart, M., Lukauskas, S., Hobson, P., Gemperline, D. C., Augspurger, T., Halchenko, Y., Cole, J. B., Warmenhoven, J., de Ruiter, J., Pye, C., Hoyer, S., Vanderplas, J., Villalba, S., Kunter, G., Quintero, E., Bachant, P., . . . Fitzgerald, C. (April 2020). *Brian. mwaskom/seaborn: v0.10.1. Zenodo*. https://doi.org/10.5281/zenodo.3767070

Wilensky, U., & Rand, W. (2015). *An introduction to agent-based simulation*. The MIT Press.

Wilensky, U. (1999). NetLogo. (http://ccl.northwestern.edu/netlogo/). In *Center for connected learning and computer-based modeling, Northwestern University, Evanston, IL, USA.*

Williams, G. J. (2011). *Data Mining with Rattle and R: The art of excavating data for knowledge discovery. Series Use R!* Springer.

Witten, I. H., Frank, E., Hall, M. H., & Pal, C. J. (2016). Data mining. In *Practical machine learning tools and techniques* (4th ed.). Morgan Kaufman.

Zumel, N., & Mount, J. (2014). *Practical data science with R.* Manning Publications.

Further Reading

Dennett, D. C. (1987). *The Intentional Stance.* The MIT Press.

McCarthy, J. (1978). Ascribing mental qualities to machines. In *Technical report, Stanford University.* AI Lab..

Chapter 9
Advanced IDSS Topics and Applications

9.1 Introduction

Conceptual components of an IDSS have been analyzed in Chap. 4, where a proposal for an IDSS architecture was made. In addition to those main conceptual components such as the different data-driven models, both from machine learning/artificial intelligence and from statistical/numerical methods, the model-driven techniques, the ontological components, and other complementary systems, there are some *advanced topics*, which should be taken into account in the development of an IDSS.

In this chapter, main advanced topics in IDSSs will be analyzed in Sects. 9.2, 9.3, and 9.4. Those topics are the *uncertainty management* (Walker et al., 2003; Zimmermann, 2000), the *temporal reasoning issues* (Fisher et al., 2005) and the *spatial reasoning issues* (Stock, 1997).

Furthermore, in the last decades a special kind of IDSS application strongly emerged in the literature to solve a particular problem, which is to suggest the best items to be used or consumed by a concrete user given some knowledge regarding users' preferences over the items in the domain. This advanced kind of IDSSs application is known as *Recommender Systems* (Ricci et al., 2015). These systems will be described in Sect. 9.5.

9.2 Uncertainty Management

As soon as target problems to be solved move from toy problems to real-world problems, an Intelligent Decision Support System must cope with *inexactness* in the information and the knowledge it uses.

© Springer Nature Switzerland AG 2022
M. Sànchez-Marrè, *Intelligent Decision Support Systems*,
https://doi.org/10.1007/978-3-030-87790-3_9

The term *inexactness* groups several kinds of inexactness forms:

- *Uncertainty*: This situation happens when it is not possible to verify the truth or falseness of a fact or assertion. Usually, these facts or assertions refer to one variable and its corresponding value.
- *Imprecision* or *Vagueness*: This situation is produced when it is not possible to determine a precise or concrete value for the variables involved in a fact or assertion.
- *Incomplete Information*: It happens when there is a lack of information regarding some fact or assertion. This usually means that the value of one variable is unknown.

Let us show the differences among the different forms of *inexactness* with the following examples:

- Sharon is between 55 and 65 years old.

  This statement is asserting that the variable "Sharon's age" has a value that is between 55 and 65, but it cannot be precisely determined. As it is an assertive statement, and we have no other information contradicting it, the statement is *certain* but *imprecise*.
- Sharon is young.

  This statement is asserting that the variable "Sharon's age" has a value that can be considered "young", but we do not know which is the precise value of her age. Again, this is an assertive statement, and we have no other information contradicting it, the statement is *certain* but *imprecise*. In this case, the word "young" is a clear example of a vague and subjective definition of a concept (youth).
- Miquel is 56 years old.

  This statement is asserting that the variable "Miquel's age" has a concrete value of 56. Thus, the statement is *precise*. In addition, as it is an assertive statement, and we have no other information contradicting it, the statement is *certain*.
- Tomorrow it will rain.

  This statement is stating that the value of the variable "tomorrow's weather" is rainy. Thus, it determines precisely the value of the variable. Hence, the statement is *precise*. However, as nobody can surely know the weather of tomorrow unless he/she is a seer, the statement is clearly *uncertain*.
- Is probable that tomorrow would be sunny.

  This statement is guessing that the value of the variable "tomorrow's weather" could be sunny with some probability, but could have other values like rainy or cloudy as well. Therefore, it does not determine precisely the value of the variable, and hence, the statement is *imprecise*. Again, as nobody can surely know the weather of tomorrow unless he/she is a seer, the statement is *uncertain*.
- We have evidences that the murderer is 30 years old.

  This statement is stating that the value of the variable "murderer's age" is 30. Thus, it determines precisely the value of the variable. Hence, the statement is *precise*. However, as we have just some evidence, but we are not completely sure about his/her age, the statement is *uncertain*, i.e. there is some degree of uncertainty about the "murderer's age" variable.

- We have evidence that the murderer is young.

    This statement is stating that the value of the variable "murderer's age" can be considered as "young", but we do not know which is the precise value of his/her age. Hence, the statement is *imprecise*. As in the previous example, as we have just some evidences, but we are not completely sure about his/her age, the statement is *uncertain*.

- The age of Ramon is unknown.

    This statement is stating that the value of the variable "Ramon's age" is unknown. Therefore, it is a clear situation of *incompleteness*. There is a lack of information regarding the value of the variable "Ramon's age".

The sources of *inexactness* are multiple (Zimmermann, 2000):

- *Raw data* obtained with some measuring devices like *sensors* can be *uncertain* when sensors are giving erroneous values, and *imprecise*, when the measuring devices have some limitations on the measurement precision.
- Data values obtained through *subjective estimations* from users could be *uncertain*.
- *Lack of information*, i.e. *missing* data, due to several reasons can produce *incompleteness*.
- *Ambiguities* and *bad interpretations* in the description language of information, like for instance using Natural Language, can generate *imprecision* or *vagueness*.
- *Partial* or/and *incomplete knowledge* due to the complexity of the system in some expert-based models expressing the experts' knowledge about some domain, like the *uncertainty* between the antecedent and the consequent of heuristic rules, or some wrong first-principles models, which can generate *uncertain* conclusions.

Summarizing, the sources of inexactness are both on the *data*, but also the different *models* used to build-up an IDSS (Walker et al., 2003; Walley, 1996). *Data-driven models* can suffer from inexactness due to the fact that they have been obtained through data that could have some kind of inexactness. On the other hand, *model-driven techniques* can be also suffering from inexactness due to the possible inexactness knowledge represented in the models.

Nowadays, in addition to the lack of information, i.e. *incompleteness*, due to the technical advances and increasing availability of related data and information, i.e. *big data scenario*, there is the opposite problem of abundance of information that can cause *uncertainty* too (van Asselt & Rotmans, 2002).

Notwithstanding, in the literature there is a general common practice to name all kinds of *inexactness* forms as *uncertainty*. Thus, when the term "*uncertainty* management" is used in general, it should be named as "*inexactness* management". As historically both terms have been confused in the literature, we will follow the same convention and will talk about *uncertainty* in a general way, when we should refer to *inexactness*.

The development of IDSSs coping with real-world problems usually needs to manage uncertainty. This means that the different models integrated in the IDSS must have some uncertainty model and uncertainty propagation mechanisms to manage uncertainty.

In the next sections, the main models proposed in the artificial intelligence literature to manage uncertainty will be analyzed.

9.2.1 Uncertainty Models

Uncertainty models must provide a formalism to *represent the uncertainty* associated with each one of the entities managed by the different reasoning mechanisms, like facts and inference rules in expert-based models, variables in statistical models, etc. In addition, they must provide some mechanisms to *propagate the uncertainty* values of the different entities through the corresponding reasoning mechanism like rule-based reasoning, Bayesian Network inference, etc.

Commonly, in the Artificial Intelligence community, this reasoning processes using an uncertainty model are named as *approximate reasoning* process (López de Mántaras, 1990).

In the Artificial Intelligence field, several models have been proposed to model uncertainty. As historically the first successful AI systems coping with real-world problems were rule-based reasoning systems, most uncertainty models are oriented to manage the uncertainty associated with facts and inference rules.

Most commonly *uncertainty models*, which are depicted in Fig. 9.1, can be grouped as follows:

- *Probabilistic Models*: These models use probabilities or near-probability values to model the uncertainty associated to the basic entities used in the reasoning mechanisms:

  - *Pure Probabilistic Model*: This model is a very simple one based on the conditional probabilities.
  - *Near-Probabilistic Models*: These models used some measures, based on the probabilities, but they are not probabilities.

    Certainty Factor Model: This method was proposed in MYCIN expert system (Shortliffe & Buchanan, 1975) and uses certainty factors that are not probabilities.

    Subjective Bayesian Model: This method was proposed in the PROSPEC TOR expert system (Duda et al., 1977), and uses two basic measures, level of sufficiency and level of necessity that are not probabilities.

  - *Bayesian Network Model*: The probabilistic causal network or Bayesian network model (Pearl, 1988) was initially developed in the late 1970s inspired by the work of Wright (1921).

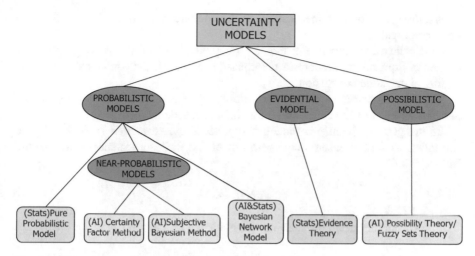

Fig. 9.1 Most common uncertainty models classification

- *Evidential Model*: This model is based on the Dempster–Shafer Evidence Theory based on Dempster (1967) work on using a lower (belief) and upper (plausibility) probability values and extended with belief functions by his student Shafer (1976) work.
- *Possibilistic Model*: This model is based on the Fuzzy Sets Theory for modelling the imprecision proposed by Lotfi Zadeh (1965), and the Possibility Theory for managing uncertainty (Zadeh, 1978), further contributed by Dubois and Prade (1988).

In the next sections, we will describe the basis of four of these methods: the Pure Probabilistic model, the Certainty Factor model, the Bayesian network model and the Possibilistic/Fuzzy Set model.

Both the pure probabilistic model and the certainty factor model are oriented to manage the uncertainty in rule-based systems. The Bayesian network model has been used in many disciplines, due to its generality. Finally, major applications of Possibilistic/Fuzzy Set Theory have been in rule-based systems, and especially in control systems.

9.2.2   Pure Probabilistic Model

Historically, it was the first method used to model reasoning with uncertainty. It was based on the probability theory.

The pure probabilistic model was developed to manage the uncertainty of a rule-based system. To model the uncertainty of this kind of system, the uncertainty originated both from the *facts*, and its propagation through the different *logic connectives* (and, or, not, implication) in the inference rules of the knowledge base must be provided.

Uncertainty can appear in a fact, in a combination of facts, in the implication operation from the antecedent to the consequent, and in the combination of paths ending at the same conclusion.

The main uncertainty propagation problem to be solved is how to assess the uncertainty level of the conclusion of a chain of reasoning steps?

In the pure probabilistic model, the *probability associated to a fact* is the formalism used to represent the *uncertainty* of that fact. For instance, if we had the following rule:

$$A \wedge B \rightarrow C$$

The uncertainty of all the facts can be modelled by the probability of them:

$$\text{uncertainty}(A) : \quad p(A)$$
$$\text{uncertainty}(B) : \quad p(B)$$
$$\text{uncertainty}(C) : \quad p(C)$$

The model must be able to provide the mechanisms for:

- Knowing the probability distribution of all its facts.
- Applying the probability operations to combine probabilities from conjunction, disjunction of facts, and negation of a fact.
- Being able to use a mechanism to propagate the probability from the antecedent to the consequent of a rule.
- Combining the probabilities of co-conclusions from different chains of reasoning which meet at the same conclusion.

The *probability distribution of all the facts* can be provided by the experts or users of the system, or alternatively can be estimated from available data.

All the probabilities must satisfy the probability axioms:

- $0 \leq p(A) \leq 1$
- $p(A) = 1$ iff[1] A is a completely true fact
- If A and B are mutually exclusive facts, then $p(A \cap B) = 0$, and hence, $p(A \cup B) = p(A) + p(B)$

The *probability propagation to combine logic connectives* is based on the basic operations on probabilities:

---

[1] *iff*, which is read as *if and only if*, is a symbol commonly used in formal mathematical or philosophical definitions which is equivalent to the logical double implication symbol $\Leftrightarrow$.

$$\text{uncertainty}(A \lor B) : \quad p(A \cup B) = p(A) + p(B) - p(A \cap B)$$

$$\text{uncertainty}(A \land B) : \quad p(A \cap B) = p(A|B) * p(B)$$

$$\text{uncertainty}(\neg A) : \quad p(\neg A) = 1 - p(A)$$

The *propagation of the probability from the antecedent to the consequent* of a rule can be modelled figuring out the implication as an assessment of the likelihood of the conclusion *conditioned* that we have observed the antecedent and we know its probability:

If we had the rule:

$$e \rightarrow h$$

were e stands for evidence, and h stands for hypothesis, and we know:

$$\text{uncertainty}(e) = p(e)$$

Once applied the rule,

$$\text{uncertainty}(h) = p(h|e)$$

The conditional probability, can be computed through the application of the Bayes theorem:

$$p(A|B) = \frac{p(B|A) * p(A)}{p(B)} \tag{9.1}$$

This way,

$$\text{uncertainty}(h) = p(h|e) = \frac{p(e|h) * p(h)}{p(e)}$$

The application of a pure probabilistic (Bayesian) model has three main drawbacks:

1. It implies a high number of a priori and *conditioned* probabilities to be computed. Let us suppose that in a rule-based system, we had:

   (a) $H = \{h_1, \ldots, h_n\}$, the set of possible hypotheses
   (b) $E = \{e_1, \ldots, e_m\}$, the set of available evidences

   Then, to set which is the most probable hypothesis set:
   $\forall e_i$, 2^n probability computations conditioned by each h_j must be computed

2. The difficulty of experts to make quantitative estimations (numerical) of probabilities
3. Probability Theory does not provide a mechanism for probability propagation between several rules concluding the same hypothesis (co-conclusion paths):

$$\left.\begin{array}{c} e_1 \rightarrow h \\ e_2 \rightarrow h \end{array}\right\} \Rightarrow \text{uncertainty}(h)?$$

In order to avoid some drawbacks some simplifying assumptions can be adopted like:

1. $\{h_i\}_{i=1,n}$ are mutually exclusive:

$$h_i \cap h_j = \varnothing \tag{9.2}$$

2. $\{h_i\}_{i=1,n}$ are collectively exhaustive:[2]

$$\bigcup_{i=1,n} h_i = \Omega \tag{9.3}$$

3. $\{e_i\}_{i=1,m}$ are conditionally independent for each h_i so that:

$$p(e_1 \cap \ldots \cap e_m | h_i) = p(e_1 | h_i) * \ldots * p(e_m | h_i) \forall h_i$$
$$\Downarrow \tag{9.4}$$

$$p(h_i | e_{i_1} \cap \ldots \cap e_{i_k}) = \frac{p(e_{i_1} \cap \ldots \cap e_{i_k} | h_i) * p(h_i)}{p(e_{i_1} \cap \ldots \cap e_{i_k})}$$
$$= \frac{\prod_{r=1}^{k} p(e_{i_r} | h_i) * p(h_i)}{\sum_{j=1}^{n} \left(\prod_{r=1}^{k} p(e_{i_r} | h_j) \right) * p(h_j)} \tag{9.5}$$

Although these assumptions simplify the computational efforts, not always they are applicable on all domains were an IDSS is built-up. Furthermore, the pure probabilistic model gives no answer to the co-conclusion's uncertainty propagation issue.

---

[2] Ω is used to refer to the whole sample space.

9.2.3 Certainty Factor Model

The Certainty Factor model was proposed in the MYCIN expert system (Shortliffe & Buchanan, 1975). It was the first system proposing a rather practical uncertainty model.

The Certainty Factor model derives from the probabilistic model, even though it is not probabilistically correct, as other near-probabilistic models.

Generally, all uncertainty models for rule-based reasoning mechanisms need to manage the combination/propagation of the uncertainty through the different logic connectives:

- Combination of evidence certainty in a disjunction in a rule:

$$e_1 \lor e_2 \to h$$

- Combination of evidence certainty in a conjunction in a rule:

$$e_1 \land e_2 \to h$$

- Propagation of certainty through a rule:

$$e_1 \to h$$

- Combination of certainty for a hypothesis obtained by different chains of reasoning:

$$e_1 \to h$$

$$e_2 \to h$$

The Certainty Factor model emerged to solve problems of the Bayesian model and for the experts' disagreement with some probabilistic axioms, like for instance:

$$p(h|e) = x \Rightarrow p(\neg h|e) = 1 - x$$

The model proposes to associate two values to each rule:

- *Measure of Belief* (*MB*): Degree to which the evidence reinforces the hypothesis
- *Measure of Disbelief* (*MD*): Degree to which the evidence counteracts the hypothesis

Each rule has these two values associated:

$$e \xrightarrow{\text{MB}(h, e), \text{MD}(h, e)} h$$

These measures are defined as follows:

$$\text{MB}(h, e) = \begin{cases} 1 & \text{if } p(h) = 1 \\ \dfrac{\max\{p(h|e), p(h)\} - p(h)}{1 - p(h)} & \text{if } p(h) \neq 1 \end{cases} \tag{9.6}$$

$$\text{MD}(h, e) = \begin{cases} 1 & \text{if } p(h) = 0 \\ \dfrac{p(h) - \min\{p(h|e), p(h)\}}{p(h)} & \text{if } p(h) \neq 0 \end{cases} \tag{9.7}$$

Main properties of these measures are the following:

- $\text{MB}(h, e) \in [0, 1]$ and $\text{MD}(h, e) \in [0, 1]$
- if $\text{MB}(h, e) > 0 \Rightarrow \text{MD}(h, e) = 0$
  The observation of the evidence increases the trust on the hypothesis
- if $\text{MD}(h, e) > 0 \Rightarrow \text{MB}(h, e) = 0$
  The observation of the evidence decreases the trust on the hypothesis

These two measures are combined in a new value, i.e. the *Certainty Factor* (CF), according to the formula:

$$\text{CF}(h, e) = \frac{\text{MB}(h, e) - \text{MD}(h, e)}{1 - \min\{\text{MB}(h, e), \text{MD}(h, e)\}} \tag{9.8}$$

Thus, each rule has associated its Certainty Factor (CF):

$$e \xrightarrow{\text{CF}(h, e)} h$$

Main properties of Certainty Factors are the following:

- $\text{CF}(h, e) \in [-1, +1]$
- if $\text{CF}(h, e) > 0$ then
  The evidence increases the trust on the hypothesis
- if $\text{CF}(h, e) < 0$ then
  The evidence decreases the trust on the hypothesis

The propagation and combination of evidences through the different logic connectives proposed are as follows:

- Conjunction of evidences: $e_1 \wedge e_2$

$$\text{CF}(e_1 \wedge e_2, e') = \min\{\text{CF}(e_1, e'), \text{CF}(e_2, e')\} \tag{9.9}$$

- Disjunction of evidences: $e_1 \vee e_2$

$$\text{CF}(e_1 \vee e_2, e') = \max\{\text{CF}(e_1, e'), \text{CF}(e_2, e')\} \tag{9.10}$$

- Propagation of evidences: $e' \to e \to h$

$$CF(h, e') = CF(h, e) * \max\{0, CF(e, e')\} \qquad (9.11)$$

being,

$CF(h, e)$: from the rule $e \to h$
$CF(e, e')$: from the evidence e

- Co-conclusions (alternative paths): $e_1 \to h$
$$e_2 \to h$$

$$CF(h, [e_1, e_2]) = \begin{cases} CF(h, e_1) + CF(h, e_2) * (1 - CF(h, e_1)) & \text{if } CF(h, e_i) > 0 \ \ i = 1, 2 \\ CF(h, e_1) + CF(h, e_2) * (1 + CF(h, e_1)) & \text{if } CF(h, e_i) < 0 \ \ i = 1, 2 \\ \dfrac{CF(h, e_1) + CF(h, e_2)}{1 - \min\{|CF(h, e_1), CF(h, e_2)|\}} & \text{otherwise} \end{cases}$$

$$(9.12)$$

9.2.3.1   An Example Using the Certainty Factor Model

Let us illustrate the use of the Certainty Factor model with the example described in Fig. 9.2. In this example, the knowledge base is formed by *three rules*. Each rule has associated its corresponding certainty factor. The certainty factors of several *facts* are known ($CF(A)$, $CF(B)$, $CF(C)$, $CF(D)$, $CF(F)$). The *goal* is the fact G. Thus, we must find out what is the certainty factor of G, which will reveal us if G is rather certain or uncertain. The *conflict resolution criterion* is the first rule in order, and the inference engine is *forward reasoning*.

Thus, in order to discover if goal G can be deduced from the knowledge base and the facts we know, and with what *level of certainty* it can be deduced, we will start a forward reasoning process.

At the first reasoning cycle, both R1 and R3 are applicable because all the certainty factors of the antecedents are known. According to the conflict resolution criterion, the first rule to be applied will be R1.

Fig. 9.2 Description of the Knowledge Base, initial Fact Base, the goal, the reasoning engine, and the conflict resolution policy in the example of the use of the Certainty Factor model

Knowledge base

R1: A ∨ B→E with CF = 0.2
R2: E ∧ F→G with CF = 0.7
R3: C ∨ D→G with CF = 0.3

Goal: G?, CF(G)?

Initial Fact Base

CF(A) = 0.1
CF(B) = 0.3
CF(C) = 0.2
CF(D) = 0.7
CF(F) = 0.6

Forward reasoning

Conflict Resolution criterion: First rule in order

R1 Application

$$CF(A \vee B) = \max \{CF(A), CF(B)\} = \max \{0.1, 0.3\} = 0.3$$
$$CF(E) = CF(A \vee B \to E) * \max \{0, CF(A \vee B)\}$$
$$= 0.2 * \max \{0, 0.3\} = 0.2 * 0.3 = 0.06$$

Thus, $CF(E) = 0.06$

Now, both R2 and R3 are applicable, because all the involved certainty factors are known. Then, following the conflict resolution criterion R2 will be applied.

R2 Application

$$CF(E \wedge F) = \min \{CF(E), CF(F)\} = \min \{0.06, 0.6\} = 0.06$$
$$\mathbf{CF(G, [R2])} = CF(E \wedge F \to G) * \max \{0, CF(E \wedge F)\}$$
$$= 0.7 * \max \{0, 0.06\} = 0.7 * 0.06 = \mathbf{0.042}$$

We got a first CF of goal G from one path through R2, but other paths are available for concluding G.

R3 Application

$$CF(C \vee D) = \max \{CF(C), CF(D)\} = \max \{0.2, 0.7\} = 0.7$$

$$\mathbf{CF(G, [R3])} = CF(C \vee D \to G) * \max \{0, CF(C \vee D)\}$$
$$= 0.3 * \max \{0, 0.7\} = 0.3 * 0.7 = \mathbf{0.21}$$

Now, a second CF of goal G has been obtained through the R3 path. Thus, both values need to be combined through the corresponding procedure.

Combination (Co-conclusion) of the Two CFs

$$\mathbf{CF(G, [R2, R3])} = CF(G, [R2]) + CF(G, [R3]) * (1 - CF(G, [R2]))$$
$$= 0.042 + 0.21 * (1 - 0.042) = 0.042 + 0.21 * 0.958 = \mathbf{0.24318}$$

Finally, the certainty factor of G, given the knowledge available and the two reasoning chains is 0.24, which means that is a little bit certain but not much. Have in

Fig. 9.3 Certainty propagation for the example in Fig. 9.2

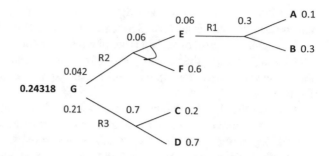

mind that a CF equal to 0 would mean that the fact is not certain but also not uncertain, meaning that probably we are in a do-not-know situation.

As a summary, in Fig. 9.3 the propagation of the certainty factors through the application of the rules in both reasoning paths can be observed.

9.2.4 Bayesian Network Model

Probabilistic causal networks or *belief networks* or *Bayesian networks* (Pearl, 1988; Neapolitan, 1990; Jensen & Nielsen, 2007) were initialled developed in the late 1970s, and were inspired by the early work by Scwall Wright (1921).

As it was already described in Chap. 4, Sect. 4.1.1.3, a *Bayesian Network* is a *probabilistic graphical model* representing a set of random variables and their conditional dependencies (cause–effect relationships) through a Directed Acyclic Graph[3] (DAG). Therefore, they are composed by a set of nodes, i.e. the random variables, and the associated conditional probabilities to each node for each possible value of the random variable, and by the set of directed edges connecting the nodes and expressing the *cause–effect dependencies* among them. If a node P is the immediate predecessor, i.e. parent, of a node S, then it is considered that P is a *direct cause* of S, or reversely, that S is a *direct effect* of P.

Actually, a *Bayesian network* is a particularization of an *influence diagram* (see Chap. 2, Sect. 2.4.2.2), where all the nodes of the diagram (random events, deterministic events, decisions, objectives) have been converted to random variables.

Bayesian Networks are a two-in-one modelling technique. On the one hand, they can be considered as an *associative data-driven model* because a Bayesian network expresses the association or dependence among several variables (an associative model) and it can be induced from available data, as it was depicted in the data mining models' chart in Fig. 6.1, in Chap. 6. The corresponding conditional

---

[3] A Directed Acyclic Graph (DAG) is a finite graph where the connections from one node to another one are directed, and that no *cycles*, i.e. circular paths starting and ending at the same node, are allowed.

probabilities of variables can be estimated from data, and the structure of the network can be also induced from data (Sangüesa & Cortés, 1997; Heckerman, 1995).

Moreover, if the set of variables are given by an expert, as well as the cause–effect relationships among them, then a *Bayesian network* can be considered as an *associative model-driven technique* because is an expert-based model, and not induced from data.

On the other hand, in addition, to be an associative model, a *Bayesian network* provides a *probabilistic Bayesian inference reasoning* model which can *manage the uncertainty* associated with the variables and provides an *inference reasoning mechanism*. Therefore, it is also an *uncertainty model*, which is based on the pure probabilistic model, and the uncertainty is modelled with a priori probabilities and *conditioned* probabilities.

The *inference reasoning mechanism* allows obtaining the new probability values of any variable within the network after some new evidences are known. When new evidences are available, the conditional probabilities must be updated according to Bayes theorem and conditional probability definition (evidence reasoning).

Summarizing, a *Bayesian network* is an *associative model* and an *uncertainty probabilistic inference reasoning model*. The uncertainty probabilistic inference reasoning model will be the focus of this section.

A classic example (Pearl & Russell, 2000) based on a former example by Murphy (1998) is the scenario where one wants to model the conditional dependences between the event of slippery when going out of your home, where you have a garden, which can be wet and other related events. Moreover, you can take into account that it could have been raining during the night or perhaps you had your sprinkler open all the night, or the current season of the year. This simple Bayesian network is depicted in Fig. 9.4. It has five random variables $\{X_1, X_2, X_3, X_4, X_5\}$. The variable $X_1 \equiv Season$, has four possible values {Summer, Winter, Spring, Fall}, but the other four variables *Rain, Sprinkler, Wet Floor,* and *Slippery* are binary variables that can have as a value the two possible values {Yes, No}.

Fig. 9.4 A classic example described in Pearl and Russell (2000) of a *Bayesian network* for the slippery causality problem

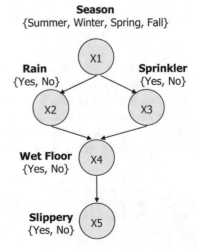

9.2.4.1 Fundamentals of *Bayesian Networks*

Main advantage of *Bayesian networks* is that they allow to manage very complex Joint Probability Distribution (JPD) concerning large number of variables, by means of conditional probabilities. The edges indicate *causality* relations between the variables (*conditional dependency*), and the absence of connection indicates conditional independence between them.

The network supports the computation of the probabilities of any subset of variables given evidence about any other subset.

Let us remind the basic *probability chain rule*:

Based on the conditioned probability,

$$p(A|B) = \frac{p(A,B)}{p(B)} \tag{9.13}$$

which implies,

$$p(A,B) = p(A|B) * p(B)$$

Then, the joint probability of n variables can be computed as follows:

$$p(X_1, X_2, \ldots, X_n) = p(X_n|X_{n-1}, \ldots, X_1) * p(X_{n-1}, \ldots, X_1)$$
$$p(X_{n-1}, \ldots, X_1) = p(X_{n-1}|X_{n-2}, \ldots, X_1) * p(X_{n-2}, \ldots, X_1)$$
$$p(X_{n-2}, \ldots, X_1) = p(X_{n-2}|X_{n-3}, \ldots, X_1) * p(X_{n-3}, \ldots, X_1)$$

$$\ldots$$

$$p(X_3, X_2, X_1) = p(X_3|X_2, X_1) * p(X_2, X_1)$$
$$p(X_2, X_1) = p(X_2|X_1) * p(X_1)$$

Thus, finally we have that:

$$p(X_1, X_2, \ldots, X_n) =$$
$$p(X_n|X_{n-1}, \ldots, X_1) * p(X_{n-1}|X_{n-2}, \ldots, X_1) * \ldots * p(X_2|X_1) * p(X_1)$$

And in a more compact formulation, the JPD can be expressed as:

$$p(X_1, X_2, \ldots, X_n) = \prod_{i=1}^{n} p(X_i|X_{i-1}, \ldots, X_1) \tag{9.14}$$

considering that $p(X_1|\varnothing) = p(X_1)$.

Dependence and Independence Relations in a *Bayesian Network*

In order to make a close analysis about the dependence and independence relations which can appear in a *Bayesian network,* let us assume the following scenario related to a *road trip* which a group of friends is planning for a near weekend. They must rent a car for the trip, and they are thinking about which could be the *Price of Renting,* the *Fuel Consumption,* and how long the *Duration of the Trip* could last, in order to know the expenses involved and the needed scheduling of the flights for being at the location where the road trip will start. In addition, they have started to think about the different features influencing on those variables. For them, it is clear that the *Type of Vehicle (TV)* has a direct influence on the *Price of Renting (PR)*. They think that the *Duration of the Trip (DT)* can be influenced both by the *Distance Travelled (DST)* and by the *Average Velocity (AV)* of the vehicle. Furthermore, they think that the *Type of Road (TR)* used in the trip can influence both on the *Distance Travelled* and on the *Average Velocity* of the vehicle. Finally, they think that the *Type of Vehicle* has some influence on the *Average Velocity* of the vehicle, the *Average Velocity* clearly influences and determines the *Fuel Consumption (FC),* and that the *Number of Friends (NF)* travelling influences the *Type of Vehicle* they will choose.

Taking into account all these influences the corresponding Bayesian network is depicted in Fig. 9.5. All the variables are discrete and can have three or four values.

Taking a close look to the network depicted in Fig. 9.5, it can be observed that there are different kind of dependence relations between neighbour nodes (variables).

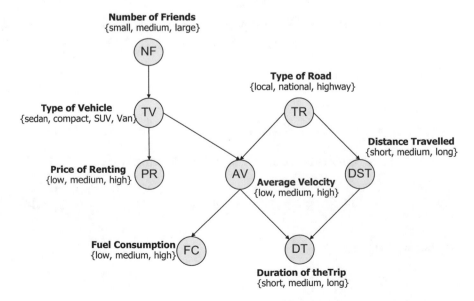

Fig. 9.5 The *Bayesian network* corresponding to the road trip example

Let us consider the subgraph formed by nodes {TR, DST, DT}. It can be represented this way: TR → DST → DT. Here we can say that the *Type of Road (TR)* directly influences on the *Distance Travelled (DST)*, and this later directly influences on the *Duration of the Trip (DT)*. In addition, if we do not know the *Distance Travelled*, the *Type of Road* has an influence on the *Duration of the Trip* variable. On the opposite, a longer or shorter *Duration of the Trip* can determine the *Type of Road* being used. However, if we know the *Distance Travelled*, the additional evidence on the *Type of Road* used does not influence on the *Duration of the Trip*. Therefore, it can be stated that a priori, *Type of Road* and *Duration of the Trip are dependent variables,* but given that the variable *Distance Travelled* is known, then *Type of Road* and *Duration of the Trip* become independent variables.

In a more formal way, it can be formulated as:

$$p(DT, TR|DST) - p(DT|DST) * p(TR|DST)$$

or equivalently:[4]

$$p(DT|TR, DST) = p(DT|DST)$$

or

$$p(TR|DT, DST) = p(TR|DST)$$

DT and TR are conditionally independent given DST.

This structure is usually named as a *causal chain* or *linear connection* subgraph. The nodes {NF, TV, PR} also form a causal chain subgraph: NF → TV → PR.

Analyzing the subgraph formed by the nodes {PR, TV, AV}, which can be represented as follows, PR ← TV → AV, there is a similar situation. A low *Price of Renting* could be due to the fact that the *Type of Vehicle* is a non-expensive one like a sedan, and hence, the probability of a low *Average Velocity* is high. However, if it is known that the *Type of Vehicle* is a SUV, whatever is the *Price of Renting* does not influence on the probability of having a low *Average Velocity*. Therefore, again it can be stated that a priori, *Price of Renting* and *Average Velocity* are *dependent variables*, but given that the *Type of Vehicle* variable is known, then *Price of Renting* and *Average Velocity* become independent variables.

In a more formal way, it can be formulated as:

---

[4]Applying the definition of conditional probability, it can be shown that the definition of conditional independence of two variables X, Y given another one Z:

$$p(X, Y|Z) = p(X|Z) * p(Y|Z)$$

and the equalities:
$p(X|Y, Z) = p(X|Z)$ and $p(Y|X, Z) = p(Y|Z)$
are all three equivalent.

$$p(\text{PR}, \text{AV}|\text{TV}) = p(\text{PR}|\text{TV}) * p(\text{AV}|\text{TV})$$

or equivalently:

$$p(\text{PR}|\text{AV}, \text{TV}) = p(\text{PR}|\text{TV})$$

or

$$p(\text{AV}|\text{PR}, \text{TV}) = p(\text{AV}|\text{TV})$$

PR and AV are conditionally independent given TV.

This structure is usually named as a *bifurcation* or *diverging connection* or *common causes* subgraph. The nodes {AV, TR, DST} also form a causal chain subgraph: AV ← TR → DST.

Finally, looking at the subgraph formed by the nodes {TV, AV, TR}, which can be represented as follows, TV → AV ← TR, shows the opposite situation. Here, the *Type of Vehicle* selected is independent of the *Type of Road*. However, if it's known the value of *Average Velocity,* for instance, a high value, then if the *Type of Vehicle* is one with low performance, this will influence in the probability that the value of the *Type of Road* be national or highway, i.e. they become dependents. Thus, the variables *Type of Vehicle* and *Type of Road* which were a priori independent turn to be dependent variables given that the variable *Average Velocity* is known.

In a more formal way, it can be formulated as:

$$p(\text{TV}, \text{TR}|\text{AV}) \neq p(\text{TV}|\text{AV}) * p(\text{TR}|\text{AV})$$
$$p(\text{TV}|\text{TR}, \text{AV}) \neq p(\text{TV}|\text{AV})$$

and

$$p(\text{TR}|\text{TV}, \text{AV}) \neq p(\text{TR}|\text{AV})$$

Thus,

$$p(\text{TV}, \text{TR}|\text{AV}) = p(\text{TV}|\text{TR}, \text{AV}) * p(\text{TR}|\text{AV})$$

TV and TR are conditionally dependent given AV.

This structure is usually named as a *concentration* or *converging connection* or *common effects* subgraph.

Analyzing the possible connections among three nodes in a DAG in such a way that the three nodes are connected, no cycles can appear, and taking into account symmetries, all the possibilities are reduced to the above *three structures*: a causal chain, a common cause, or a common effect. And as detailed above, these structures imply *conditional independence relationship* in the case of causal chains and common cause structures, and to *dependence relationship* in the case of common

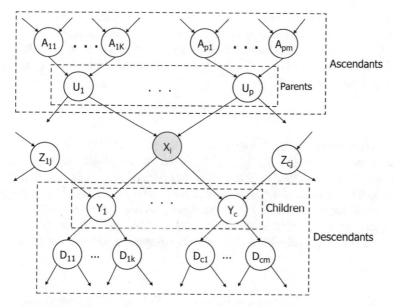

Fig. 9.6 Classification of the nodes in a *Bayesian network* regarding a concrete node X_i

effect structure. This fact is very important because leads to an important property of Bayesian networks, which is called the *Markov Condition*.

Markov Condition and Factorization of the Joint Probability Distribution

The *Markov*[5] *Condition* or *Markov Assumption* states that:

Given a node X_i in the Bayesian network, X_i is conditionally independent of its non-descendant nodes, given its parent nodes (see Fig. 9.6). Thus, formally:

$$p(X_i|\text{NonDesc}(X_i), \text{Parents}(X_i)) = p(X_i|\text{Parents}(X_i)) \qquad (9.15)$$

where:

- Parents(X_i) is the set of the immediate predecessor nodes or *parents* of the node X_i, which are directly connected to the node (nodes $U_1, \ldots, U_p$ in Fig. 9.6).
- Desc(X_i) is the set of the *descendant* nodes of the node X_i, which are the immediate *children* of the node X_i (nodes $Y_1, \ldots, Y_c$ in Fig. 9.6), or the descendant nodes of their children (nodes D_{cm} in Fig. 9.6).
- NonDesc(X_i) is the complementary set of the descendant nodes of the node X_i not belonging to Parents(X_i) regarding all the nodes of the network (*ascendant* nodes A_{pm} and *other nodes* Z_{cj} in Fig. 9.6).

---

[5]Andrey Andreyevich Markov (1856–1922) was a Russian mathematician who made important contributions on *stochastic processes*. Main subject of his research later became known as *Markov chains* and *Markov processes*.

Thus, considering the Markov condition, and the fact that all the variables in a *Bayesian network* can be numbered following a *topological ordering*[6] $(X_1, \ldots, X_n)$, in such a way that the conditional probability of each variable X_i only depends on non-descendant nodes $(X_{i-1}, \ldots, X_1)$, then each term of the probability chain rule, can be generally written as:

$$p(X_i|X_{i-1}, \ldots, X_1) = p(X_i|\text{Parents}(X_i)) \qquad (9.16)$$

The computation of the *Joint Probability Distribution* (JPD) of all the variables of the Bayesian network $\{X_1, X_2, \ldots, X_n\}$ according to the *probability chain rule*, can be done in different orderings of the set of variables of the network. In fact, for a set of n variables, there are $n!$ different permutations[7] of the variables. Although the final joint probability value (Eq. 9.14) is the same, independently of the order, given a topological ordering of the variables, and considering the Markov condition (Eq. 9.16), then the *probability chain rule* in a Bayesian network can be written as follows:

$$p(X_1, X_2, \ldots, X_n) = \prod_{i=1}^{n} p(X_i|X_{i-1}, \ldots, X_1) = \prod_{i=1}^{n} p(X_i|\text{Parents}(X_i))$$

Therefore, the *Joint Probability Distribution* $p(X_1, X_2, \ldots, X_n)$ can be *factorized* by using the *conditional independence relationships* from the *Bayesian network*:

$$p(X_1, X_2,, \ldots, X_n) = \prod_{i=1}^{n} p(X_i|\text{Parents}(X_i)) \qquad (9.17)$$

This fact is what makes *Bayesian networks* becoming actually very interesting and computationally efficient for managing the joint probability distribution among all the variables.

Without using this factorization, the *Joint Probability Distribution* (JPD) table corresponding to a *Bayesian network* with n variables, assuming they are Boolean, leads to the computation of $2^n - 1 \in O(2^n)$ probability different entries in the JPD table. Each table entry is one possible combination of values for all the n variables.

Using factorization, it is enough to compute one *Conditional Probability Distribution* (CPD) table between each node and its parents, which can be stored at each

---

[6] A *topological ordering* or *topological sort* of a directed acyclic graph (DAG) is a linear ordering of its nodes satisfying that for every directed edge $X \rightarrow Y$ from X to node Y, X comes before Y in the ordering. A DAG can have many topological orderings.

[7] The number of permutations of n different elements $\{e_1, \ldots, e_n\}$ without any repeated element is $n! = n * (n-1) * \ldots * 2 * 1$, as the first position can be occupied by n possible elements, the second one has only $(n-1)$ elements left, and the third one has only $(n-2)$ possible elements until the last position which has just one available element. The number of permutations of n elements without repeated elements is described as: $P_n = n!$

node. The maximum number of possible links in a DAG of n nodes is $n * (n - 1)/2$, because each node can be connected, at much, to $(n - 1)$ nodes, but as no cycles are allowed, just the half of all possible links can appear. Furthermore, the maximum number of parents of a node, when all the links are in the network is $n - 1$. Taking into account both facts, it can be seen that the maximum number of CPD table entries in the worst case (all possible links exist in the network) to be computed is:

$$\sum_{i=0}^{n-1} 2^i = 2^n - 1$$

Hence, in the worst case, the CPD tables would equal the same amount of computation of the JPD table. However, this value is an *upper bound* because in almost all *Bayesian networks* not all the possible links are present and consequently, the number of parents of a node is usually a small value less than $n - 1$, because the direct causes of a node are limited (for instance, 2 or three at maximum). Assuming k is the maximum number of parents of a node in a network, the CPD tables' values would be $n * 2^k \in O(n)$, which is linear regarding n, very different than $2^n - 1 \in O(2^n)$, which is exponential regarding n.

Therefore, there is a practical advantage on using the CPD tables instead of using all the JPD table.

To check that, let us analyze the example described above about the *Road Trip*, and depicted in Fig. 9.5. There are eight nodes: {NF, TV, TR, PR, AV, DST, FC, DT}. If we compute the JPD table, we will need to compute $3^7 * 4 - 1 = 8747$ probability values to complete the whole JPD table of 8748 entries. Note that here instead of having Boolean binary variables with possible values {true, false} we have seven variables which have three possible values, and one variable which has four possible values.

However, if we used the factorization provided by the CPD tables to compute the JPD, we would have:

$$p(\text{NF}, \text{TV}, \text{TR}, \text{PR}, \text{AV}, \text{DST}, \text{FC}, \text{DT}) = p(\text{DT}|\text{AV}, \text{DST}) * p(\text{FC}|\text{AV}) * p(\text{DST}|\text{TR})$$
$$* p(\text{AV}|\text{TV}, \text{TR}) * p(\text{PR}|\text{TV})$$
$$* P(\text{TR}) * p(\text{TV}|\text{NF}) * ap(\text{NF})$$

Thus, the number of probability values to be computed for the eight CPD tables would be, accordingly to each CPD table appearing in the joint probability computation:

$$p(\text{DT}|\text{AV}, \text{DST}) : 3^2 * 2 = 18$$
$$p(\text{FC}|\text{AV}) : 3 * 2 = 6$$

$$p(\text{DST}|\text{TR}) : 3 * 2 = 6$$

$$p(\text{AV}|\text{TV}, \text{TR}) : 4 * 3 * 2 = 24$$

$$p(\text{PR}|\text{TV}) : 4 * 2 = 8$$

$$P(\text{TR}) : 2$$

$$p(\text{TV}|\text{NF}) : 3 * 3 = 9$$

$$p(\text{NF}) : 2$$

Finally, the sum of all would be $18 + 6 + 6 + 24 + 8 + 2 + 9 + 2 = 75$.

Therefore, the saving in the computation cost or estimation task by an expert is really huge: 75 probability values for the 8 CPD tables against 8747 probability values of the JPD.

Actually, what a *Bayesian network* is providing is a concise and reduced representation of the JPD by using the conditional independence relations between the nodes of the network. This way, the computation of the new probability values of the variables in the network conditioned to the appearance of new evidence about the values of some variables makes easier the propagation of these evidence values and the computation of the new a posteriori probability values, as it will be described in Sect. 9.2.4.2.

Generalization of Conditional Independence Relations: D-Separation

In previous section, the main subgraph structures in a *Bayesian network* were analyzed. In order to exploit the conditional independence relations in a *Bayesian network* to design inference methods, the *general conditional independence relationships* with different set of nodes must be described. Thus, the main issue is to know whether a set of nodes X is independent of another set Y, given a set of evidence nodes E.

The concept of *direction-dependent separation* or *d-separation* (Pearl, 1988) provides the needed help to that.

Given a set of nodes X, a set of nodes Y, and a set of evidence nodes E, if every *undirected path*[8] from a node in X to a node in Y is *d-separated* by E, then X and Y are *conditionally independent* given E. D-separation feature is synonymous to *blocking*.

Therefore, it also can be stated that a set of nodes E, *d-separates* two sets of nodes X and Y, if every *undirected path* from a node in X to a node in Y is *blocked* given E.

Furthermore, a path is *blocked* given a set of nodes E, if there is a node Z_i on the path for which one of the following three conditions holds (see Fig. 9.7):

---

[8] An *undirected path* between two nodes is a path ignoring the direction of the edges.

Fig. 9.7 The three possible ways of blocking a path from set X to Y, given the set of evidences E

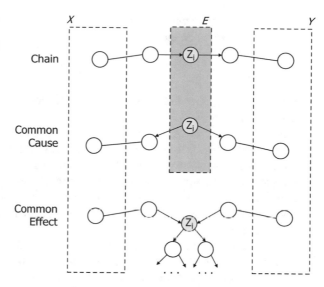

1. Z_i belongs to E and Z_i has one edge on the path leading in and one edge out (*causal chain* or *linear connection structure*)
2. Z_i belongs to E and Z_i has both edges on the path leading out (*common cause* or *diverging connection* or *bifurcation structure*)
3. Neither Z_i nor any descendant of Z_i belongs to E, and both edges on the path lead into Z (*common effect* or *converging connection* or *concentration structure*)

This d-separation concept was used by J. Pearl to derive a first inference method in a *Bayesian network* based on the propagation of the evidence through the passing of messages between a node and its parents and children.

9.2.4.2 Inference in Bayesian Networks

Once a *Bayesian network* has been obtained either from a dataset or from experts' knowledge, it can be used as an inference mechanism. The inference process usually happens when some new *event* (a set of evidence observations) is observed, and the new a posteriori probability distribution for some variables must be computed. This inference process is sometimes named as *evidence propagation* mechanism.

An *event* e is described by a set of variables, named as the *evidence variables* $E = \{E_1, E_2, \ldots, E_v\}$. The variables of interest, $X = \{X_1, X_2, \ldots, X_t\}$, are commonly referred as the *query variables*. Usually, each query is made on a single query variable, X_i. If there are more query variables, the inference process is repeated for the other query variables. The other variables of the network, which are neither evidence variables nor query variables are usually named as the *unobserved* or *hidden variables* $Y = \{Y_1, Y_2, \ldots, Y_h\}$. The complete set of all variables of the network will be named as $Vars = \{X_i \cup E \cup Y\}$.

Thus, the common query asking for a posteriori probability distribution in a *Bayesian network* can be expressed as:

$$p(X_i|e)$$

Where the variable X_i is the *query variable*, and the *event e* is formed by the *set of evidence variables*.

Let us introduce an example which will be used to show the operation of some of the methods explained. It is a variation of the classical example of the *Garden wet grass*, and its possible causes in a house. Suppose you live in a residential zone formed by clonal houses where each neighbour has her/his own garden with grass and the sprinkler system. On the morning, you can observe whether it has rained or not, whether your garden grass is wet or not, and whether the grass of your neighbour garden is wet or not.

The fact of having rained or not has a direct influence on the probability that you garden is wet or not. Also, it influences in the probability that the garden of your neighbour is wet too. Finally, the possibility that you have left your sprinkler open during the night or not influence the possibility that your garden is wet.

The variables involved are:

- *Rain (R)*: it has rained or not.
- *Sprinkler (S)*: your sprinkler has been working or not.
- *Garden (G)*: your garden is wet or not.
- *Neighbour Garden (N)*: your neighbour's garden is wet or not.

All the variables are Boolean variables with possible values being *true* or *false*.

The Fig. 9.8 shows the *Bayesian network* for this scenario with the corresponding Conditional Probability Tables (CPTs). Let us imagine that in this situation, you leave your home in the morning and see that your *Garden grass* is wet. What is more

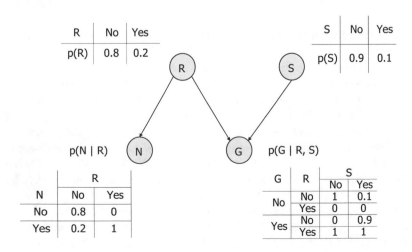

Fig. 9.8 *Bayesian network* for the garden of clonal houses scenario

probable: it has been *Raining* during the night or you have left your *Sprinkler* open during the night?

To answer these questions, the situation can be formalized this way: there is an event in which your Garden is wet, $G =$ true. To know whether it is more probable that it has been *Raining* or that you left your *Sprinkler* open during the night, the following two conditional probabilities must be computed and compared:

$$p(R \mid G = \text{true})$$

$$p(S \mid G = \text{true})$$

As it will be seen afterwards these probabilities are:

$$p(R \mid G = \text{true}) = \langle 0.7353, 0.2647 \rangle$$

$$p(S \mid G = \text{true}) = \langle 0.3382, 0.6618 \rangle$$

Therefore, it is more probable that it has been *Raining* (R) that you left the *Sprinkler* (S) open, because $0.7353 > 0.3382$.

There are two kinds of inference methods: the *exact inference methods* and the *approximate inference methods*. The exact inference methods compute the posterior probabilities in an exact[9] way through mathematical operations. The approximate inference methods compute those probabilities using different simulation techniques to obtain approximate values of those probabilities.

As it will be explained, the approximate methods are needed because depending on the size and/or topology of a network, the exact inference algorithms can be exponential in time cost.

Exact Inference Methods

In this section, the *summation out* or *marginalizing* method using the Joint Probability Distribution, the *inference by enumeration* using the Conditional Probability Tables (CPTs), the *variable elimination (VE) inference* method (Zhang & Poole, 1994, 1996), and the classic *evidence propagation method* (Pearl, 1988; Neapolitan, 1990) will be described.

Marginalization or Summation Out Method

The first approach is to use the Joint Probability Distribution (JPD) to compute the required conditional probability of the *query variable* (X_i) regarding the *evidence variables* $(E = \{E_1, \ldots, E_v\})$ forming an event e. Thus:

---

[9]Of course, having in mind the possible rounding errors produced by the computer limitations.

$$p(X_i|e) = \frac{p(X_i, e)}{p(e)} \tag{9.18}$$

Taking into account that $p(e)$ is a constant factor, i.e. it is a normalization factor for the $p(X_i|e)$ values, which does not depend on the different values of X_i, the above expression assuming that $\alpha = 1/p(e)$, can be rewritten as follows:

$$p(X_i|e) = \frac{p(X_i, e)}{p(e)} = \alpha\, p(X_i, e) \propto p(X_i, e) \tag{9.19}$$

And hence, the joint probability of X_i and the *evidence variables* (E) forming the event e, can be computed through the *summation out* over all possible combinations of values of the unobserved or hidden variables (Y):

$$p(X_i|e) = \frac{p(X_i, e)}{p(e)} = \alpha\, p(X_i, e) = \alpha \sum_{Y_k \in Y} p(X_i, e, Y) \tag{9.20}$$

Noting that $\{X_i, E, Y\}$ is the complete set of variables of the domain (*Vars*), $p(X_i, e, Y)$ is a subset of probability entries of the full JPD of all variables.

This *summation out* of the remaining variables not instanced to compute the probability of one variable through the JPD is usually named as *marginalization*, because imaging the JPD described as a rectangular table, the unconditional probability or marginal probability of one variable can be obtained summing the corresponding row and writing it at the margin of the JPD table.

In the problem of the garden described in Fig. 9.8, the JPD is shown in Table 9.1 Thus, the probabilities to be computed are:

Table 9.1 Joint probability distribution for the garden problem

R	S	G	N	p(R, S, G, N)
T	T	T	T	0.02
T	T	T	F	0
T	T	F	T	0
T	T	F	F	0
T	F	T	T	0.18
T	F	T	F	0
T	F	F	T	0
T	F	F	F	0
F	T	T	T	0.0144
F	T	T	F	0.0576
F	T	F	T	0.0016
F	T	F	F	0.0064
F	F	T	T	0
F	F	T	F	0
F	F	F	T	0.144
F	F	F	F	0.576

$$p(R = \text{true} \mid G = \text{true})$$

$$p(R = \text{false} \mid G = \text{true})$$

$$p(S = \text{true} \mid G = \text{true})$$

$$p(S = \text{false} \mid G = \text{true})$$

In order to simplify the mathematical notation, the possible values, false and true, for each Boolean variable will be described with lowercase initials of the variable name, such as $R = \text{true}$ will be written as r and $R = \text{false}$ will be written as $\bar{r}$.

Let us to compute the probabilities regarding the variable *Raining* (R) applying Eq. (9.20):

$$p(R = \text{true} \mid G = \text{true}) = p(r \mid g) \propto \sum_{S,N} p(r, g, S, N)$$

$$\propto p(r, g, s, n) + p(r, g, s, \bar{n}) + p(r, g, \bar{s}, n) + p(r, g, \bar{s}, \bar{n})$$

$$\propto 0.02 + 0 + 0.18 + 0 \propto 0.2$$

$$p(R = \text{false} \mid G = \text{true}) = p(\bar{r} \mid g) \propto \sum_{S,N} p(\bar{r}, g, S, N)$$

$$\propto p(\bar{r}, g, s, n) + p(\bar{r}, g, s, \bar{n}) + p(\bar{r}, g, \bar{s}, n) + p(\bar{r}, g, \bar{s}, \bar{n})$$

$$\propto 0.0144 + 0.0576 + 0 + 0 \propto 0.072$$

Normalizing the values:

$$p(r \mid g) = \frac{0.2}{0.2 + 0.072} = \frac{0.2}{0.272} = 0.7354$$

$$p(\bar{r} \mid g) = \frac{0.072}{0.2 + 0.072} = \frac{0.072}{0.272} = 0.2647$$

Let us to compute the probabilities regarding the variable *Sprinkler* (S):

$$p(S = \text{true} \mid G = \text{true}) = p(s \mid g) \propto \sum_{R,N} p(s, g, R, N)$$

$$\propto p(s, g, r, n) + p(s, g, r, \bar{n}) + p(s, g, \bar{r}, n) + p(s, g, \bar{r}, \bar{n})$$

$$\propto 0.02 + 0 + 0.0144 + 0.0576 \propto 0.092$$

$$p(S = \text{false} \mid G = \text{true}) = p(\bar{s} \mid g) \propto \sum_{R,N} p(\bar{s}, g, R, N)$$

$$\propto p(\bar{s}, g, r, n) + p(\bar{s}, g, r, \bar{n}) + p(\bar{s}, g, \bar{r}, n) + p(\bar{s}, g, \bar{r}, \bar{n})$$

$$\propto 0.18 + 0 + 0 + 0 \propto 0.18$$

Normalizing the values:

$$p(s \mid g) = \frac{0.092}{0.092 + 0.18} = \frac{0.092}{0.272} = 0.3382$$

$$p(\bar{s} \mid g) = \frac{0.18}{0.092 + 0.18} = \frac{0.18}{0.272} = 0.6618$$

Therefore, using the JPD table, any conditional probability can be computed summing the corresponding entries from the JPD. However, the time complexity is exponential on the number of variables of the network. On the other hand, the conditional independence relationships of the *Bayesian network* are nor exploited. In the next methods, the topology of the *Bayesian network*, the factorization of the conditional probabilities, and several mathematical properties like associative and commutative properties are applied to reduce the computational cost of the inference process.

Enumeration Method

Considering that a Bayesian network is a more concise representation of the full JPD, and recalling that the JPD can be computed through the conditional probabilities of the nodes of the network (Eq. 9.17):

$$p(X_1, X_2, , \ldots, X_n) = \prod_{i=1}^{n} p(X_i \mid \text{Parents}(X_i))$$

It follows that:

$$p(X_i \mid e) = \frac{p(X_i, e)}{p(e)} = \alpha\, p(X_i, e) = \alpha \sum_{Y_k \in Y} p(X_i, e, Y)$$

$$= \alpha \sum_{Y_k \in Y} \left(p(X_i \mid \text{Parents}(X_i)) * \prod_{E_j \in E} p(E_j \mid \text{Parents}(E_j)) * \prod_{Y_k \in Y} p(Y_k \mid \text{Parents}(Y_k)) \right)$$

$$\tag{9.21}$$

Let us consider, the above example for computing the probability of *Raining* given the evidence that your *Garden grass is wet*:

$$p(R = \text{true} \mid G = \text{true}) = p(r \mid g) = \alpha\, p(r, g) = \alpha \sum_{S, N} p(r, g, S, N)$$

Thus,

$$p(r \mid g) = \alpha \sum_{S} \sum_{N} p(r, g, S, N) = \alpha \sum_{S} \sum_{N} p(g \mid r, S) \, p(N \mid r) \, p(r) \, p(S)$$

By the moment, the time complexity of this computation in the worst case when almost all variables are hidden is $O(n2^n)$, due to the product of n conditional probabilities, for each one of the summands.

However, if we notice that in the expression we have a constant term, $p(r)$, which does not depend on the variables of the summation S, N, and can be moved outside both summations:

$$p(r \mid g) = \alpha \, p(r) \sum_{S} \sum_{N} p(g \mid r, S) \, p(N \mid r) \, p(S)$$

Furthermore, the term $p(N \mid r)$ does not depend on the variable S, and can be moved outside the summation over S (application of associative and commutative properties):

$$p(r \mid g) = \alpha \, p(r) \sum_{N} p(N \mid r) \sum_{S} p(g \mid r, S) \, p(S)$$

This way, moving factors outside the summation over the hidden variables, the number of mathematical operations is reduced, and the computational cost is reduced. The time complexity would be $O(2^n)$, which is not good but better than before, but the space required is linear on the number of variables (n).

Writing the evaluation of the above expression, we would get the tree depicted in Fig. 9.9, where the nodes marked with the symbol "+" are summation nodes and the ones not labelled are "propagation" nodes. The summation nodes have outside the corresponding summation variable. The leaves of the tree are marked with square forms, and provide the neutral element of the product, i.e. 1. In this example, the value of $p(r \mid g)$ is equal to α 0.2, the value of $p(\bar{r} \mid g)$ would be equal to α 0.072. After the normalization process, the probabilities would be $p(r \mid g) = 0.7354$ and $p(\bar{r} \mid g) = 0.2647$.

The algorithm to evaluate these expression trees is a depth-first recursive method, which starts from top node along each path until arriving to the bottom nodes (leaves), with a 1 value. It multiplies the corresponding values of the probability values associated to the edges of the path along "propagation" nodes and if the type of node is labelled as a sum, then it must sum the values provided by the recursive exploration of the subtrees arriving at it.

Although some mathematical operations have been avoided, there is still space for improving. The tree of Fig. 9.9 shows there are some subexpressions that must be evaluated twice such as $p(g \mid r, s) \, p(s)$ and $p(g \mid r, \bar{s}) \, p(\bar{s})$.

Furthermore, if we come back to the equation after moving the constant term outside the hidden variables summation:

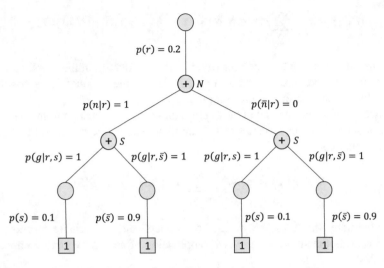

Fig. 9.9 Evaluation tree for the expression to compute $p(r \mid g)$

$$p(r \mid g) = \alpha \, p(r) \sum_S \sum_N p(g \mid r, S) \, p(N \mid r) \, p(S)$$

It could be also transformed taking into account that the term $p(g \mid r, S) \, p(S)$ does not depend on the hidden variable N and can be moved out of the summation of variable N. Thus, the probability computation could be done as:

$$p(r \mid g) = \alpha \, p(r) \sum_S p(g \mid r, S) \, p(S) \sum_N p(N \mid r)$$

And the corresponding evaluation tree is depicted in Fig. 9.10. Note that this way the number of repeated subexpression terms is just $p(n \mid r)$ and $p(\bar{n} \mid r)$, reducing the number of computations and the final time to be computed.

Observe that the *ordering* in the summation over the hidden variables has a direct influence on the time of computation. A *good heuristic* seems to be that the summation of products from the right side to the left side of the expression must be as less as possible because a large number of terms to be multiplied will generate the repetition of more subexpressions computation than if there are less number of multipliers.

Furthermore, in this particular case, observe that the last summation on the N variable, $\sum_N p(N \mid r)$ is equal to 1, because $p(n \mid r) + p(\bar{n} \mid r) = 1$ by definition. Therefore, it could be omitted, and the expression could be computed as:

$$p(r \mid g) = \alpha \, p(r) \sum_S p(g \mid r, S) \, p(S)$$

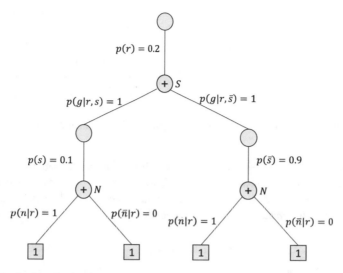

Fig. 9.10 Evaluation tree for the expression to compute $p(r \mid g)$ according to the second equation

Variable Elimination Method

The *variable elimination (VE) algorithm* (Zhang & Poole, 1994, 1996) is an improved version of the previous enumeration algorithm. The basic idea is the same: *ordering* the hidden variables in a query, in such a way that the application of commutative and associative properties moves some terms of the product of the conditional probabilities forward in the formula, out of the summation for the hidden variables decreasing the complexity of computations. In addition to that, the variable elimination algorithm *avoids repeated calculations* of some probabilities, using *factors*, which allow to keep the terms already computed to be reused later. This mechanism uses an algorithmic technique named as *dynamic programming*.[10]

The variable elimination evaluates the conditional probability expression of a query in a *Bayesian network factorized* in a list of *factors* in a *right-to-left* approach. The intermediate results are stored using *factors*, and the *hidden variables* are eliminated, one-by-one, by the summation out of the corresponding *factors*.

A *factor f* is a function over a set of variables $\{X_1, X_2, \ldots, X_p\}$ mapping each instantiation of these variables to a real non-negative number. For instance, Table 9.2 shows a factor of two variables.

As it will be described later in the *Garden grass* example, each *factor* can be represented by a *matrix* indexed by the values of its argument variables. Note that Conditional Probability Tables satisfy the definition of *factors*. In the computation of

---

[10]*Dynamic programming* is an algorithmic problem-solving technique aiming to simplify a complex problem by breaking it down into simpler sub-problems in a recursive manner, usually tabulating these intermediate sub-problems' results, and combining them to get the final solution to the original problem.

Table 9.2 Description of a
factor $f(X, Y)$ of two variables
$\{X, Y\}$

X	Y	$f(X, Y)$
0	0	2
0	1	3
1	0	4
1	1	1

conditional probabilities in the inference process three operations will be needed on *factors*:

- Product, named as *Pointwise Product*: The pointwise product of two factors $f_1(X_1, .., X_p, Y_1, \ldots, Y_q)$ and $f_2(Y_1, \ldots, Y_q, Z_1, \ldots, Z_r)$ is a new factor f_p, whose *variables* are the *union* of the variables in f_1 and f_2, and whose *elements* are computed as the *product* of the corresponding elements in the two factors. Thus:

$$f_p(X_1, .., X_p, Y_1, \ldots, Y_q, Z_1, \ldots, Z_r)$$
$$= f_1(X_1, .., X_p, Y_1, \ldots, Y_q) \times f_2(Y_1, \ldots, Y_q, Z_1, \ldots, Z_r) \qquad (9.22)$$

  The output factor of the *pointwise product* can contain more variables that any of the involved factors, and both the size of the resulting factor and the computational time cost are exponential regarding the number of variables. The output factor f_p has $2^{p + q + r}$ entries assuming that all the variables are Boolean. The *pointwise product* is a commutative and associative operation.

- *Summing Out*: Summing out a variable Y from a factor $f(X, Y, Z)$ is done by adding up the submatrices formed by fixing the variable of the summation to be eliminated to its values in turn. Thus:

$$\sum_{y_i \in Y} f(X, y_i, Z) = f(X, y_1, Z) + \ldots + f(X, y_k, Z) \qquad (9.23)$$

  Summing out is a commutative operation. It has an exponential time cost regarding the number of variables in the original factor, and it needs an exponential space cost regarding the number of variables in the resulting factor.

- *Conditioning*: Conditioning means to set-up the *evidence variables* (E_v) to its corresponding observed values (e_{v_i}) in a factor f. Thus:

$$f(X, E_v, Y) \quad \rightarrow \quad f(X, e_{v_i}, Y)$$

As mentioned already in the enumeration method, the structure of the network and the *ordering of elimination of the variables* is very important to reduce the time and space complexity of the algorithm. Generally, as larger are the factors built-up in the processing of the list of factors worse will be those complexities. Determining which is the best order of elimination for the hidden variables it is a complex problem itself. However, some heuristic criteria can be applied. Eliminating first the variable minimizing the size of the next factor to be constructed is a good choice. In other words, the heuristic would select the first variable to be eliminated as the one that involves the least possible number of the other hidden variables from the initial list of factors for its summing out.

Let us summarize the *Variable Elimination* algorithm in the following scheme:

---

algorithm Variable Elimination

input *LCPs*: the List of Conditional Probabilities from the Bayesian network
  X: the set of query variables
  E: the set of evidence or observed variables
  E_v: the set of observed values
  HiddenVars: the ordered list of hidden or unobserved variables (*Y*)

output the resulting factor which accounts for $p(X|E, Y)$

begin
 // Condition all the CPs with the observed values of the evidence variables
 // to get the initial list of factors
 Factors ← conditionProbabilities(LCPs, *E*, E_v)
 // Eliminate each hidden variable according to the ordering set in *HiddenVars*
 for each *y* **in** *HiddenVars* **do**
  // Select the factors containing *y*
  FactorsY ← selectFactors(Factors, *y*)
  // Remove the factors containing *y* from the list of factors
  Factors ← Factors − FactorsY
  // Multiply the factors containing *y* and sum out the variable *y* from the
  // resulting factor. This factor is added to the list of factors to the right
  Factors ← addFactorRight(Factors, summingOut(*y*, pointwiseProduct(FactorsY)))
 endforeach
 // Multiply the final list of factors to obtain the final resulting factor
 FactRes ← pointwiseProduct(Factors)
 // Normalize the resulting factor containing $p(X, E, Y)$ to get for all $x \in X$
 // the required $p(x|E, Y) = p(x, E, Y)/\sum_{x \in X} p(x, E, Y)$
 return normalize(FactRes)
end

To illustrate how it works, let us solve the *Garden grass wet* problem, assuming that we have the same scenario described by Fig. 9.8, and that the situation is the same. All the variables are Boolean variables with possible values being *true* or *false*. Assuming that you leave your home on the morning and see that your *Garden* is wet. What is more probable: it has been *Raining* during the night or you have left your *Sprinkler* open during the night?

Thus, we should compute $p(R = \text{true} \mid G = \text{true})$ and $p(S = \text{true} \mid G = \text{true})$ and compare what is the highest probability. Here, for simplicity just $p(R = \text{true} \mid G = \text{true})$ will be computed, but the procedure is analogous.

Let us use the variable elimination algorithm. First, the probability to be computed must be expressed as the product of conditional probabilities according to the structure of the network. Thus:

$$p(R \mid G) = \alpha\, p(R, G) = \alpha \sum_{S,N} p(R, G, S, N)$$

$$p(R|G) = \alpha \sum_{S,N} p(G|R, S)\, p(N|R)\, p(R)\, p(S)$$

The list of initial CPs must be conditioned to the observed values of the evidence variables ($G = \text{true}$, which is expressed as g), to get the initial list of factors. This way we get:

$$p(R|g) = \alpha \sum_{S,N} p(g|R, S)\, p(N|R)\, p(R)\, p(S)$$

Another important aspect is determining the *ordering* of the hidden variables. Here, we have just two variables (S, N), and hence, just two possible orderings. Applying the heuristic specified before, the variable first selected should be the one that when summing it out involves least variables of the remaining hidden variables.

In this case, variable S will involve two terms and the other variable R: $p(g \mid R, S)\, p$ (S). On the other hand, variable N will involve just one term and does not involve the other hidden variable S: $p(N \mid R)$. Therefore, the best candidate variable is N, and the ordering of variables will be: N, S.

This way the ordering of elimination leads the ordering of the summation of terms:

$$p(R|g) = \alpha \sum_{S} \sum_{N} p(g|R, S)\, p(N|R)\, p(R)\, p(S)$$

Therefore, the following factors appear in the list of initial factors:

$$\text{Factors} = \langle f_1(R, S),\ f_2(N, R),\ f_3(R),\ f_4(S) \rangle$$

Where they can be expressed in matrix form as follows:

$$f_1(R, S) = p(g|R, S) = \begin{pmatrix} p(g|r, s) & p(g|r, \bar{s}) \\ p(g|\bar{r}, s) & p(g|\bar{r}, \bar{s}) \end{pmatrix} = \begin{pmatrix} 1 & 1 \\ 0.9 & 0 \end{pmatrix}$$

$$f_2(N,R) = p(N|R) = \begin{pmatrix} p(n|r) & p(n|\bar{r}) \\ p(\bar{n}|r) & p(\bar{n}|\bar{r}) \end{pmatrix} = \begin{pmatrix} 1 & 0.2 \\ 0 & 0.8 \end{pmatrix}$$

$$f_3(R) = p(R) = \begin{pmatrix} p(r) \\ p(\bar{r}) \end{pmatrix} = \begin{pmatrix} 0.2 \\ 0.8 \end{pmatrix}$$

$$f_4(S) = p(S) = \begin{pmatrix} p(s) \\ p(\bar{s}) \end{pmatrix} = \begin{pmatrix} 0.1 \\ 0.9 \end{pmatrix}$$

And then the probability can be described as:

$$p(R|g) = \alpha \sum_S \sum_N f_1(R,S) \times f_2(N,R) \times f_3(R) \times f_4(S)$$

Then, the algorithm proceeds to *eliminate the first variable* (N) in the ordering. The factors containing the variable N must be gathered and taken out of the original list of factors. This way we have:

$$\text{Factors}_N = \{f_2(N,R)\}$$
$$\text{Factors} = \langle f_1(R,S), f_3(R), f_4(S) \rangle$$

Hence, we have,

$$p(R|g) = \alpha \sum_S f_1(R,S) \times f_3(R) \times f_4(S) \sum_N f_2(N,R)$$

Then, the pointwise product of *Factors$_N$* must be done, but this time is not necessary as there is only one factor. Afterwards, the summation out of the variable N must be done:

$$f_5(R) = \sum_N f_2(N,R) = f_2(n,R) + f_2(\bar{n},R) = \begin{pmatrix} p(n|r) \\ p(n|\bar{r}) \end{pmatrix} + \begin{pmatrix} p(\bar{n}|r) \\ p(\bar{n}|\bar{r}) \end{pmatrix}$$

$$= \begin{pmatrix} 1 \\ 0.2 \end{pmatrix} + \begin{pmatrix} 0 \\ 0.8 \end{pmatrix} = \begin{pmatrix} 1 \\ 1 \end{pmatrix}$$

$$f_5(R) = \begin{pmatrix} 1 \\ 1 \end{pmatrix}$$

Therefore, $f_5(R)$ will be added on the right to the list of factors:

$$\text{Factors} = \langle f_1(R,S), f_3(R), f_4(S), f_5(R) \rangle$$

And thus, the computation of the probability follows:

$$p(R|g) = \alpha \sum_S f_1(R,S) \times f_3(R) \times f_4(S) \times f_5(R)$$

The algorithm continues to *eliminate the second and last hidden variable* (S) in the ordering. The factors containing the variable S must be gathered and taken out of the original list of factors. This way we have:

$$\text{Factors}_S = \{f_1(R,S), f_4(S)\}$$
$$\text{Factors} = \langle f_3(R), f_5(R) \rangle$$

Hence, we have,

$$p(R|g) = \alpha\, f_3(R) \times f_5(R) \sum_S f_1(R,S) \times f_4(S)$$

Then, the *pointwise product* of *Factors$_S$* must be done, before the *summation out* of the variable S:

$$f_6(R) = \sum_S f_1(R,S) \times f_4(S) = \sum_S f_{\text{aux}}(R,S)$$

Let us compute first the *pointwise product* of $f_1(R,S) \times f_4(S)$:

$$f_{\text{aux}}(R,S) = f_1(R,S) \times f_4(S) = \begin{pmatrix} p(g|r,s) & p(g|r,\bar{s}) \\ p(g|\bar{r},s) & p(g|\bar{r},\bar{s}) \end{pmatrix} \times \begin{pmatrix} p(s) \\ p(\bar{s}) \end{pmatrix}$$

$$= \begin{pmatrix} 1 & 1 \\ 0.9 & 0 \end{pmatrix} \times \begin{pmatrix} 0.1 \\ 0.9 \end{pmatrix} = \begin{pmatrix} 1*0.1 & 1*0.9 \\ 0.9*0.1 & 0*0.9 \end{pmatrix} = \begin{pmatrix} 0.1 & 0.9 \\ 0.09 & 0 \end{pmatrix}$$

Thus,

$$f_6(R) = \sum_S f_{\text{aux}}(R,S) = f_{\text{aux}}(R,s) + f_{\text{aux}}(R,\bar{s}) = \begin{pmatrix} 0.1 \\ 0.09 \end{pmatrix} + \begin{pmatrix} 0.9 \\ 0 \end{pmatrix} = \begin{pmatrix} 1 \\ 0.09 \end{pmatrix}$$

Therefore, $f_6(R)$ will be added on the right to the list of factors:

$$\text{Factors} = \langle f_3(R), f_5(R), f_6(R) \rangle$$

And thus, the computation of the probability has eliminated all the hidden variables. Now its turn to make the *pointwise product* of the remaining factors:

$$p(R|g) = \alpha\, f_3(R) \times f_5(R) \times f_6(R)$$

$$f_3(R) \times f_5(R) \times f_6(R) = f_3(R) \times f_{56}(R) = \begin{pmatrix} 0.2 \\ 0.8 \end{pmatrix} \times \left(\begin{pmatrix} 1 \\ 1 \end{pmatrix} \times \begin{pmatrix} 1 \\ 0.09 \end{pmatrix} \right)$$

$$= \begin{pmatrix} 0.2 \\ 0.8 \end{pmatrix} \times \begin{pmatrix} 1 \\ 0.09 \end{pmatrix} = \begin{pmatrix} 0.2 \\ 0.072 \end{pmatrix}$$

Finally,

$$p(R|g) = \alpha\, f_3(R) \times f_5(R) \times f_6(R) = \alpha \begin{pmatrix} 0.2 \\ 0.072 \end{pmatrix}$$

And normalizing the probabilities:

$$p(R|g) = \begin{pmatrix} 0.2/(0.2+0.072) \\ 0.072/(0.2+0.072) \end{pmatrix} = \begin{pmatrix} 0.7353 \\ 0.2647 \end{pmatrix}$$

One interesting observation is regarding $f_5(R)$, which is the resulting factor of the summation out of variable N of the factor $f_2(N, R)$:

$$f_5(R) = \sum_N f_2(N, R) = \begin{pmatrix} 1 \\ 1 \end{pmatrix}$$

Clearly, this factor does not affect the next pointwise product, and actually it could be eliminated from the list of factors. The variable N is irrelevant for the query given (R) the evidence variable that we have (G). The reason is that when summing out the factors, we are computing the global probabilities, whose values are 1 by definition:

$$\begin{pmatrix} p(n|r) \\ p(n|\bar{r}) \end{pmatrix} + \begin{pmatrix} p(\bar{n}|r) \\ p(\bar{n}|\bar{r}) \end{pmatrix} = \begin{pmatrix} 1 \\ 1 \end{pmatrix}$$

This property can be generalized to any network. *Any leave node, which is a hidden variable for a given query and set of evidences, can be removed.*

Furthermore, another interesting filtering action is the observation that *every variable not being an ancestor of a query variable or evidence variable is irrelevant for the query.*

Therefore, a good strategy is to remove these possible irrelevant variables for a given query, before starting the variable elimination algorithm.

Evidence Propagation Through Local Message Passing Method

One of the first exact inference algorithms was developed by Pearl (Pearl, 1982, 1988; Kim & Pearl, 1983; Neapolitan, 1990; Castillo et al., 1997). This algorithm can be applied to a special kind of networks called *polytrees* or *singly connected networks*.

A *polytree* is a directed acyclic graph with a single undirected path between any two nodes, or that the equivalent undirected graph is acyclic. This means that ignoring the direction of the edges, there is no pair of nodes connected from more than one path. Figure 9.11 illustrates the concept of a polytree.

The great advantage of polytrees is that the computational cost and space cost of *exact inference algorithms* is linear with the number of nodes: $O(n)$.

In a polytree, whatever pair of nodes are connected by a unique path. Thus, it means each node divides the polytree into two unconnected polytrees: one polytree contains all its parent nodes and all other nodes it is connected through its parents, and the other polytree contains its children nodes and all other nodes it is connected through its children.

For instance, in the upper right polytree at Fig. 9.11, the node X_4 divides the network into two unconnected polytrees. The first one, $\{X_1, X_2, X_3\}$ includes the parent nodes of X_4, which are $\{X_1, X_2\}$, and the nodes accessible from X_4 through its

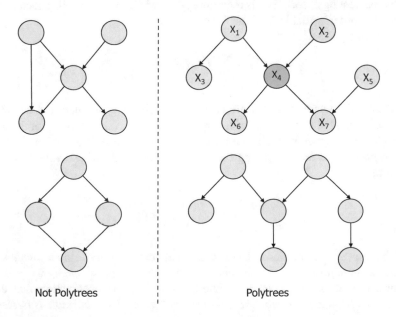

Not Polytrees   Polytrees

Fig. 9.11 Examples of DAGs being polytrees and not being polytrees

parents, $\{X_3\}$. The second polytree, $\{X_5, X_6, X_7\}$, contains the children nodes of X_4, which are $\{X_6, X_7\}$, and the nodes accessible from X_4 through its children, $\{X_5\}$.

As it is graphically shown in Fig. 9.11, the node X_4 *d-separates* these two sets: $\{X_1, X_2, X_3\}$ and $\{X_5, X_6, X_7\}$. Any path going from a node in one set and ending at another node in the other set is *blocked* by the node X_4. In other words, both sets $\{X_1, X_2, X_3\}$ and $\{X_5, X_6, X_7\}$ are conditionally independent, given the node X_4.

Propagation of evidence in this kind of networks can be efficiently performed combining the information coming from the different subnetworks through the message passing, i.e. local computations, from one subnetwork to another one.

To derive the algorithm for the propagation of messages, let us formulate the common query asking for a posteriori probability distribution in a Bayesian network:

$$p(X_i = x_i|e) = p(x_i|e)$$

Where the variable X_i is the *query variable*, and the *event e* is formed by the *set of evidence variables* (E). We write x_i instead of $X_i = x_i$, to simplify the notation. Thus,

$$p(x_i|e) = p(x_i|E) \tag{9.24}$$

The set of evidence variables E can be decomposed into two disjunctive sets, each one contained in one of the polytrees that are separated by the node X_i, in the original polytree. Hence, the set E can be divided in:

- $E_{X_i}^{|}$: is the evidence variable subset of E connected to X_i through its parents. It is the *causal support* for X_i.
- $E_{X_i}^{-}$: is the evidence variable subset of E connected to X_i through its children. It is the *evidential support* for X_i.

This way, $E = E_{X_i}^{-} \cup E_{X_i}^{+}$. See Fig. 9.12 were this separation of the evidence variables in the two subsets is depicted.

Applying the definition of conditional probability and the division of E in the two subsets, we have:

$$p(x_i|E) = p\left(x_i|E_{X_i}^{-}, E_{X_i}^{+}\right) = \frac{p\left(x_i, E_{X_i}^{-}, E_{X_i}^{+}\right)}{p\left(E_{X_i}^{-}, E_{X_i}^{+}\right)} = \frac{p\left(E_{X_i}^{-}, E_{X_i}^{+}|x_i\right) * p(x_i)}{p\left(E_{X_i}^{-}, E_{X_i}^{+}\right)} \tag{9.25}$$

Taking into account that X_i d-separates $E_{X_i}^{+}$ from $E_{X_i}^{-}$ according to the polytree structure, $E_{X_i}^{+}$ and $E_{X_i}^{-}$ are conditionally independent given X_i, i.e. $p\left(E_{X_i}^{-}, E_{X_i}^{+}|x_i\right) = p\left(E_{X_i}^{-}|x_i\right) * p(E_{X_i}^{+}|x_i)$, then:

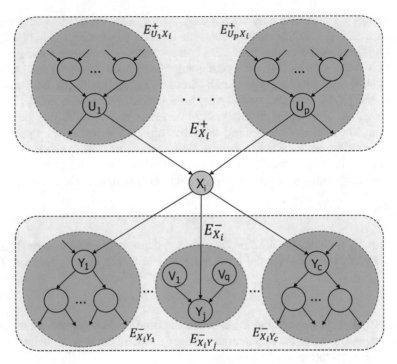

Fig. 9.12 Separation of the evidence variables (E) into the subset connected to a node X_i through its parents ($E_{X_i}^+$) and the subset connected to a node X_i through its children ($E_{X_i}^-$)

$$p(x_i|E) = \frac{p\left(E_{X_i}^-|x_i\right) * p\left(E_{X_i}^+|x_i\right) * p(x_i)}{p\left(E_{X_i}^-, E_{X_i}^+\right)}$$

$$= \frac{p\left(E_{X_i}^-|x_i\right) * p\left(x_i, E_{X_i}^+\right)}{p\left(E_{X_i}^-, E_{X_i}^+\right)} \tag{9.26}$$

$$= k * \lambda_{X_i}(x_i) * \rho_{X_i}(x_i)$$

where,

$$k = \frac{1}{p\left(E_{X_i}^-, E_{X_i}^+\right)} \tag{9.27}$$

k is a constant factor of normalization

$\lambda_{X_i}(x_i) = p\left(E_{X_i}^-|x_i\right)$ is a term capturing the evidence coming from X_i children

$\rho_{X_i}(x_i) = p\left(x_i, E_{X_i}^+\right)$ is a term capturing the evidence from X_i parents

Defining $\beta_{X_i}(x_i)$ as the *Belief* of a variable X_i:

$$\beta_{X_i}(x_i) = \lambda_{X_i}(x_i) * \rho_{X_i}(x_i) \tag{9.28}$$

Therefore, the conditional probability is the normalization of the *belief* of the variable, and can be written as:

$$p(x_i|E) = k * \beta_{X_i}(x_i) \propto \beta_{X_i}(x_i) \tag{9.29}$$

The functions $\beta_{X_i}(x_i)$, $\lambda_{X_i}(x_i)$, and $\rho_{X_i}(x_i)$ must be computed. To do that, the environ of a node X_i must be considered. Whatever node, in general, can have a set of p parents $U = \{U_1, \ldots, U_p\}$, and a set of c children $Y = \{Y_1, \ldots, Y_c\}$, as depicted in Fig. 9.12.

Again, taking into account the structure of the polytree, the set of causal evidence $E_{X_i}^+$ can be divided in a set of p disjunctive subsets, each one corresponding to each one of the parents of the node X_i: $\left\{ E_{U_1 X_i}^+, \ldots, E_{U_p X_i}^+ \right\}$. Similarly, the set of evidence $E_{X_i}^-$ can be divided in a set of c disjunctive subsets, each one corresponding to each one of the children of the node X_i: $\left\{ E_{X_i Y_1}^-, \ldots, E_{X_i Y_c}^- \right\}$.

Applying the definition of conditional probabilities, mathematical operations, and conditional independence when it holds, the final expression for the functions $\beta_{X_i}(x_i)$, $\lambda_{X_i}(x_i)$, and $\rho_{X_i}(x_i)$ can be computed. Details can be found in Castillo et al. (1997), Pearl (1988). Furthermore, the corresponding messages sent between one node and its parents and children and vice versa:

- Messages $\lambda_{X_i U_j}$: between the node X_i and its parents (U_j)
- Messages $\rho_{U_j X_i}$: between the parents of node X_i (U_j) and X_i
- Messages $\lambda_{Y_k X_i}$: between the children of node X_i (Y_k) and X_i
- Messages $\rho_{X_i Y_k}$: between the node X_i and its children (Y_k)

can be computed too. These messages sent and received by a node X_i are depicted in Fig. 9.13.

Therefore, the final formulas are:

$$\lambda_{X_i}(x_i) = \prod_{j=1}^{c} \lambda_{Y_j X_i}(x_i) \tag{9.30}$$

$$\rho_{X_i}(x_i) = \sum_{u} \left(p(x_i|u \cup e_i^+) \prod_{j=1}^{p} \rho_{U_j X_i}(u_j) \right) \tag{9.31}$$

$$\beta_{X_i}(x_i) = \lambda_{X_i}(x_i) * \rho_{X_i}(x_i) \tag{9.32}$$

Fig. 9.13 Passing of messages ρ and λ around a node X_i

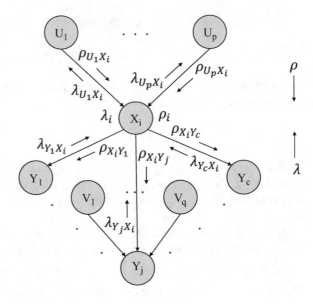

$$p(x_i|E) = k * \beta_{X_i}(x_i) \tag{9.33}$$

Thus:

$$p(x_i|E) = k * \beta_{X_i}(x_i) = k * \lambda_{X_i}(x_i) * \rho_{X_i}(x_i) \propto \lambda_{X_i}(x_i) * \rho_{X_i}(x_i) \tag{9.34}$$

$$p(x_i|E) \propto \left(\prod_{j=1}^{c} \lambda_{Y_j X_i}(x_i) \right) * \left(\sum_{u} \left(p(x_i|u \cup e_i^+) \prod_{j=1}^{p} \rho_{U_j X_i}(u_j) \right) \right) \tag{9.35}$$

And the computation of the messages is as follows:

$$\rho_{X_i Y_j}(x_i) = \rho_{X_i}(x_i) \prod_{\substack{k=1 \\ k \neq j}}^{c} \lambda_{Y_k X_i}(x_i) \tag{9.36}$$

$$\lambda_{Y_j X_i}(x_i) = \sum_{y_j} \left(\lambda_{Y_j}(y_j) \sum_{v_1, \ldots, v_q} \left(p(y_j|v, x_i) \prod_{k=1}^{q} \rho_{V_k Y_j}(v_k) \right) \right) \tag{9.37}$$

---

algorithm Local Message-Passing

input *LCPs*: the List of Conditional Probabilities from the polytree
        X: the set of query variables
        E: the set of evidence or observed variables
        e_i: the set of observed values
output $p(x_i|E)$ for each node $X_i \subset X$, $X_i \notin E$

begin
   // Initialization step
  for each $X_i \in E$ **do**
   if $x_i = e_i$ **then** $\rho_{X_i}(x_i) = 1$; $\lambda_{X_i}(x_i) = 1$
           else $\rho_{X_i}(x_i) = 0$; $\lambda_{X_i}(x_i) = 0$ **endif**
  endforeach
  for each $X_i \notin E$ **do**
   if $Parents(X_i) = \emptyset$ **then** $\rho_{X_i}(x_i) = p(x_i)$ **endif**
   if $leaf?(X_i)$ **then** $\lambda_{X_i}(x_i) = 1$ **endif**
  endforeach
   // Iterative steps
  repeat   // repeat until functions ρ_{X_i} and λ_{X_i} for each node $X_i \notin E$ are computed
   for each $X_i \notin E$ **do**
   if $\forall j \in 1..p$: $\rho_{U_j X_i}$ have been received **then**
    $\rho_{X_i}(x_i) = \sum_u \left(p(x_i|u \cup e_i^+) \prod_{j=1}^p \rho_{U_j X_i}(u_j) \right)$
   endif
   if $\forall j \in 1..c$: $\lambda_{Y_j X_i}$ have been received **then**
    $\lambda_{X_i}(x_i) = \prod_{j=1}^c \lambda_{Y_j X_i}(x_i)$
   endif
   if $\rho_{X_i}(x_i)$ has been computed **then**
    for each $Y_i \in Children(X_i)$: X_i has received λ messages from all $Y_{j,j \neq i}$ **do**
     $\rho_{X_i Y_j}(x_i) = \rho_{X_i}(x_i) \prod_{\substack{k=1 \\ k \neq j}} \lambda_{Y_k X_i}(x_i)$ // compute and send $\rho_{X_i Y_j}(x_i)$
    endforeach
   endif
   if $\lambda_{X_i}(x_i)$ has been computed **then**
    for each $U_j \in Parents(X_i)$: X_i has received ρ messages from all $U_{k,k \neq j}$ **do**
     $\lambda_{X_i U_j}(u_i) = \sum_{x_i} \left(\lambda_{X_i}(x_i) \sum_{u_{k,k \neq j}} \left(p(x_i|u_k, u_j) \prod_{k=1, k \neq j}^p \rho_{U_k X_i}(u_k) \right) \right)$ //comp & send
    endforeach
   endif
   endforeach
  until ρ_{X_i} and λ_{X_i} for each node $X_i \notin E$ are computed
  // Finalization step
  for each $X_i \notin E$ **do**
   $\beta_{X_i}(x_i) = \lambda_{X_i}(x_i) * \rho_{X_i}(x_i)$ // Compute the *Belief* of each node $X_i \notin E$
   $p(x_i|e) = \frac{\beta_{X_i}(x_i)}{\sum_{x_i} \beta_{X_i}(x_i)}$ // Normalize the Belief to obtain $p(x_i|e)$
  endforeach
  return $p(x_i|E)$ for each node $X_i \notin E$
end

The computation of $\lambda_{X_i}(x_i)$ can be done when all the messages λ from its children have arrived to the node X_i. Similarly, $\rho_{X_i}(x_i)$ can be computed when all the messages ρ from its parents have arrived to the node X_i. This means that in the propagation of the inference some particular order in the computations must be followed. The evidence propagation through local *message-passing method* can be described as the algorithm detailed above.

One advantage of the inference method through the propagation of messages among the nodes of the polytree is the possibility to *distribute the computation*. Thus, a parallel computation could be managed, for instance taking each node as a distributed autonomous computation node communicated with the other ones.

Let us illustrate, how the algorithm works with the example of the *Garden grass wet* described in Fig. 9.8. Remind that the initial scenario is that when you leave your home in the morning you notice that your *Garden grass* is wet. What is more probable: it has been *Raining* during the night or you have left your *Sprinkler* open during the night?

To answer these questions, the situation can be formalized this way: there is an event in which your *Garden grass* (G) is wet, $G =$ true. To know whether it is more probable that it has been *Raining* (R) or that you left your *Sprinkler* (S) open during the night, the following two conditional probabilities must be computed and compared:

$$p(R \mid G = \text{true})$$

$$p(S \mid G = \text{true})$$

The algorithm starts with the initialization step:

We will represent the true value with a 1 and the false with a 0 for shortening purposes.

$\rho_r = [0.8, 0.2]$
   {vector or aggregation of two components: $[\rho_r(0), \rho_r(1)]$}
$\rho_s = [0.9, 0.1]$
   {vector or aggregation of two components: $[\rho_s(0), \rho_s(1)]$}
$\lambda_n = [1, 1]$ {vector or aggregation of two components: $[\lambda_n(0), \lambda_n(1)]$}
$\lambda_g = [1, 1]$ {vector or aggregation of two components: $[\lambda_g(0), \lambda_g(1)]$}

Then, it starts the iterative steps. First, before none evidence is given, the computation of ρ and λ for all nodes is done as follows:

$$\begin{aligned} \lambda_{nr} &= [\lambda_n(0)p(n = 0|r = 0) + \lambda_n(1)p(n = 1|r = 0) \\ &\quad \lambda_n(0)p(n = 0|r = 1) + \lambda_n(1)p(n = 1|r = 1)] \\ &= [1 * 0.8 + 1 * 0.2, 1 * 0 + 1 * 1] \\ &= [1, 1] \end{aligned}$$

$$\rho_{sg} = \rho_s * 1 = [0.9, 0.1] * 1 = [0.9, 0.1]$$

$$\begin{aligned}
\lambda_{gr} = &\big[\lambda_g(0)\big(p(g=0|r=0,s=0)\rho_{sg}(0) + p(g=0|r=0,s=1)\rho_{sg}(1)\big) \\
&+ \lambda_g(1)\big(p(g=1|r=0,s=0)\rho_{sg}(0) + p(g=1|r=0,s=1)\rho_{sg}(1)\big), \\
&\quad \lambda_g(0)\big(p(g=0|r=1,s=0)\rho_{sg}(0) + p(g=0|r=1,s=1)\rho_{sg}(1)\big) \\
&+ \lambda_g(1)\big(p(g=1|r=1,s=0)\rho_{sg}(0) + p(g=1|r=1,s=1)\rho_{sg}(1)\big)\big] \\
= &[1(1*0.9 + 0.1*0.1) + 1(0*0.9 + 0.9*0.1), \\
&\quad 1(0*0.9 + 0*0.1) + 1(1*0.9 + 1*0.1)] \\
= &[1, 1]
\end{aligned}$$

$$\rho_{rg} = \rho_r * \lambda_{nr} = [0.8, 0.2] * [1, 1] = [0.8, 0.2]$$

$$\rho_{rn} = \rho_r * \lambda_{gr} = [0.8, 0.2] * [1, 1] - [0.8, 0.2]$$

$$\begin{aligned}
\lambda_{gs} = &\big[\lambda_g(0)\big(p(g=0|r=0,s=0)\rho_{rg}(0) + p(g=0|r=1,s=0)\rho_{rg}(1)\big) \\
&+ \lambda_g(1)\big(p(g=1|r=0,s=0)\rho_{rg}(0) + p(g=1|r=1,s=0)\rho_{rg}(1)\big), \\
&\quad \lambda_g(0)\big(p(g=0|r=0,s=1)\rho_{rg}(0) + p(g=0|r=1,s=1)\rho_{rg}(1)\big) \\
&+ \lambda_g(1)\big(p(g=1|r=0,s=1)\rho_{rg}(0) + p(g=1|r=1,s=1)\rho_{rg}(1)\big)\big] \\
= &[1(1*0.8 + 0*0.2) + 1(0*0.8 + 1*0.2), \\
&\quad 1(0.1*0.8 + 0*0.2) + 1(0.9*0.8 + 1*0.2)] \\
= &[1, 1]
\end{aligned}$$

$$\lambda_r = \lambda_{nr} * \lambda_{gr} = [1, 1] * [1, 1] = [1, 1]$$

$$\lambda_s = \lambda_{gs} = [1, 1]$$

$$\begin{aligned}
\rho_n = &[p(n=0|r=0)\rho_{rn}(0) + p(n=0|r=1)\rho_{rn}(1), \\
&\quad p(n=1|r=0)\rho_{rn}(0) + p(n=1|r=1)\rho_{rn}(1)] \\
= &[0.8*0.8 + 0*0.2, 0.2*0.8 + 1*0.2] \\
= &[0.64, 0.36]
\end{aligned}$$

$$\begin{aligned}
\rho_g = &[p(g=0|r=0,s=0)\rho_{rg}(0)\rho_{sg}(0) + p(g=0|r=0,s=1)\rho_{rg}(0)\rho_{sg}(1) \\
&+ p(g=0|r=1,s=0)\rho_{rg}(1)\rho_{sg}(0) + p(g=0|r=1,s=1)\rho_{rg}(1)\,\rho_{sg}(1), \\
&\quad p(g=1|r=0,s=0)\rho_{rg}(0)\rho_{sg}(0) + p(g=1|r=0,s=1)\rho_{rg}(0)\,\rho_{sg}(1) \\
&+ p(g=1|r=1,s=0)\rho_{rg}(1)\rho_{sg}(0) + p(g=1|r=1,s=1)\rho_{rg}(1)\rho_{sg}(1)] \\
= &[1*0.8*0.9 + 0.1*0.8*0.1 + 0*0.2*0.9 + 0*0.2*0.1, \\
&\quad 0*0.8*0.9 + 0.9*0.8*0.1 + 1*0.2*0.9 + 1*0.2*0.1] \\
= &[0.728, 0.272]
\end{aligned}$$

Now, assume that we know the evidence: our grass is wet $\Rightarrow g = 1$
initialization: $\lambda_g = [0,1]$ {also $\rho_g = [0, 1]$}

$$\lambda_{gr} = [\lambda_g(0)\big(p(g=0|r=0,s=0)\rho_{sg}(0) + p(g=0|r=0,s=1)\rho_{sg}(1)\big)$$
$$+ \lambda_g(1)\big(p(g=1|r=0,s=0)\rho_{sg}(0) + p(g=1|r=0,s=1)\rho_{sg}(1)\big),$$
$$\lambda_g(0)\big(p(g=0|r=1,s=0)\rho_{sg}(0) + p(g=0|r=1,s=1)\rho_{sg}(1)\big)$$
$$+ \lambda_g(1)\big(p(g=1|r=1,s=0)\rho_{sg}(0) + p(g=1|r=1,s=1)\rho_{sg}(1)\big)]$$
$$= [0(1*0.9+0.1*0.1) + 1(0*0.9+0.9*0.1),$$
$$0(0*0.9+0*0.1) + 1(1*0.9+1*0.1)]$$
$$= [0.09,1]$$

$$\lambda_r = \lambda_{gr} * \lambda_{nr} = [0.09,1] * [1,1] = [0.09,1]$$

$$\beta_r = \lambda_r * \rho_r = [0.09,1] * [0.8,0.2] = [0.072,0.2]$$

$$p(R|g=1) = k * \beta_r = k * [0.072,0.2] = [0.2647,0.7353]$$

$$\lambda_{gs} = [\lambda_g(0)\big(p(g=0|r=0,s=0)\rho_{rg}(0) + p(g=0|r=1,s=0)\rho_{rg}(1)\big)$$
$$+ \lambda_g(1)\big(p(g=1|r=0,s=0)\rho_{rg}(0) + p(g=1|r=1,s=0)\rho_{rg}(1)\big),$$
$$\lambda_g(0)\big(p(g=0|r=0,s=1)\rho_{rg}(0) + p(g=0|r=1,s=1)\rho_{rg}(1)\big)$$
$$+ \lambda_g(1)\big(p(g=1|r=0,s=1)\rho_{rg}(0) + p(g=1|r=1,s=1)\rho_{rg}(1)\big)]$$
$$= [0(1*0.8+0*0.2) + 1(0*0.8+1*0.2),$$
$$0(0.1*0.8+0*0.2) + 1(0.9*0.8+1*0.2)]$$
$$= [0.2,0.92]$$

$$\lambda_s = \lambda_{gs} = [0.2,0.92]$$

$$\beta_s = \lambda_s * \rho_s = [0.2,0.92] * [0.9,0.1] = [0.18,0.092]$$

$$p(S|g=1) = k * \beta_s = k * [0.18,0.092] = [0.6618,0.3382]$$

Therefore, as $0.7353 > 0.3382$, this means that is more probable that it has been *Raining* by night than your *Sprinkler* has been opened during the night.

Now, in addition, we have a new evidence: *Neighbour's grass* is wet too, which means $n = 1$ (and $g = 1$)

initialization: $\lambda_n = [0,1]$ {also $\rho_n = [0,1]$}

$$\lambda_{nr} = [\lambda_n(0)p(n=0|r=0) + \lambda_n(1)\,p(n=1|r=0),$$
$$\lambda_n(0)\,p(n=0|r=1) + \lambda_n(1)\,p(n=1|r=1)]$$
$$= [0*0.8 + 1*0.2, 0*0 + 1*1]$$
$$= [0.2,1]$$

$$\lambda_r = \lambda_{nr} * \lambda_{gr} = [0.2,1] * [0.09,1] = [0.018,1]$$

$$\beta_r = \lambda_r * \rho_r = [0.018,1] * [0.8,0.2] = [0.0144,0.2]$$

$$p(R|n = 1, g = 1) = k * \beta_r = k * [0.0144, 0.2] = [0.0672, \mathbf{0.9328}]$$

$$\rho_{rg} = [0.8, 0.2] * [0.2, 1] = [0.16, 0.2]$$

$$\lambda_{gs} = [\lambda_g(0)(p(g = 0|r = 0, s = 0)\rho_{rg}(0) + p(g = 0|r = 1, s = 0)\rho_{rg}(1))$$
$$+ \lambda_g(1)(p(g = 1|r = 0, s = 0)\rho_{rg}(0) + p(g = 1|r = 1, s = 0)\rho_{rg}(1)),$$
$$\lambda_g(0)(p(g = 0|r = 0, s = 1)\rho_{rg}(0) + p(g = 0|r = 1, s = 1)\rho_{rg}(1))$$
$$+ \lambda_g(1)(p(g = 1|r = 0, s = 1)\rho_{rg}(0) + p(g = 1|r = 1, s = 1)\rho_{rg}(1))]$$
$$= [0(1 * 0.16 + 0 * 0.2) + 1(0 * 0.16 + 1 * 0.2),$$
$$0(0.1 * 0.16 + 0 * 0.2) + 1(0.9 * 0.16 + 1 * 0.2)]$$
$$= [0.2, 0.344]$$

$$\lambda_s = \lambda_{gs} = [0.2, 0.344]$$

$$\beta_s = \lambda_s * \rho_s = [0.2, 0.344] * [0.9, 0.1] = [0.18, 0.344]$$

$$p(S|n = 1, g = 1) = k * \beta_s = k * [0.18, 0.344] = [0.84, 0.16]$$

Therefore, as $0.9328 > 0.16$, this implies that is *even more probable* that it has been *Raining* by night than your *Sprinkler* has been opened during the night.

Inference in Multiply-Connected Networks

The time and space complexity of *exact inference methods* in *polytrees* is linear on the size of the network, even the number of parents of a node is bounded by a constant factor. However, when the network is not a polytree, i.e. *a multiply connected network*, it becomes that both time and space complexity are exponential in the worst case, even if the number of parents is bounded by a constant factor.

Therefore, in the literature several methods have been proposed for the *multiply-connected networks*. Main approaches can be divided in:

- *Clustering Algorithms*: These methods transform the original *multiply connected network* into an equivalent polytree by clustering or merging nodes until the network does not have cycles and becomes a polytree. The first approach was formulated by Lauritzen and Spiegelhalter (1988) and latter improved by Jensen (1989). Let us imagine the network in Fig. 9.14a. This network is not a polytree because there is a cycle between all the four nodes A, B, C and D. If nodes B and C are merged into a clustered node $\{B, C\}$, then the resulting network is a polytree, as depicted in Fig. 9.14b. Of course, now a new conditional probability table must be computed for the $p(B, C|A)$. The new clustered node $\{B, C\}$ has as values: $\{TT, TF, FT, FF\}$, which are the cross product of the possible values of nodes A: $\{T, F\}$ and B: $\{T, F\}$. The new probabilities can be computed from the CPTs from the original network. This new CPT is shown in Fig. 9.14b.

Once the network has been transformed into a polytree, the exact inference algorithms like variable elimination or propagation through local messages can be

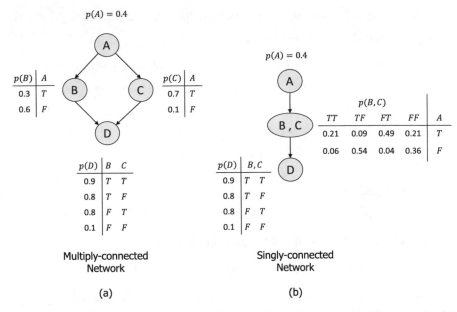

Fig. 9.14 A multiply connected Network (**a**) and its equivalent polytree obtained by clustering (**b**)

applied. At the end, the original queries over the original variable X_i in the original network can be answered through the queries over the transformed network, marginalizing the probability distribution of a joined node containing the variable X_i over all the other variables. Note that the values of variables B and C can be easily derived from the values of "variable" $\{B, C\}$.

However, if the number of joined nodes is large, and the number of variables in the junction is high, it means that the CPTs will be exponentially large in the number of variables to be joined.

Some modifications to the basic clustering algorithm were done in what was named as the *clique tree propagation algorithm* (Lauritzen & Spiegelhalter, 1988; Jensen et al., 1990; Shafer & Shenoy, 1990). The *clique tree or junction tree or join tree propagation* has a first pre-processing step that transforms a *Bayesian network* into a secondary structure called *clique tree* or *junction tree* or *join tree*. The pre-processing step consists in several sub-steps:

1. Transform the original Bayesian network, which is a DAG, into an *undirected graph*. These means that the direction of the edges is eliminated.
2. *Moralization of the resulting graph* from the previous step. The *moralization of a graph* means that an edge must be added between each pair of non-adjacent nodes having common children in the original DAG, until all nodes have their parents "married".
3. *Triangulation of the graph* resulting from the preceding step. A *triangulated graph* is a graph which all cycles of four or more nodes have a chord, which is an

edge that is not part of the cycle but connects two nodes of the cycle. This process aims at obtaining that all nodes belong to some *clique*.[11]

4. *Identification of the cliques* in the triangulated graph. Thus, identify the subset of nodes (*joined nodes* or *hypernodes*) forming subgraphs being *complete* and *maximal*.

5. *Enumeration of the cliques* obtained in the previous step: $Clq_1, \ldots Clq_n$, in such a way that the *running intersection property* would be satisfied. The *running intersection property* states:

$$\forall i \in \{1, \ldots, n\}, \exists j < i : Clq_i \cap (Clq_1 \cup \ldots \cup Clq_{i-1}) \subseteq Clq_j$$

6. *Building of a tree* where the cliques are its *hypernodes*. This is the final *clique tree*. The root node is the first clique, Clq_1 and for a given Clq_i, the candidates cliques Clq_j to be the parents of Clq_i in the tree, being $j < i$, are those ones satisfying $Clq_i \cap (Clq_1 \cup \ldots \cup Clq_{i-1}) \subseteq Clq_j$. As several candidates can appear, several clique trees can be generated.

After the clique tree is generated, then an algorithm taking into account the *intersection running property*, which computes the probabilities for each clique:

$$p(Clq_i) = p(R_i|S_i)\, p(S_i) \tag{9.38}$$

where,

$$S_i = Clq_i \cap (Clq_1 \cup \ldots \cup Clq_{i-1}) \tag{9.39}$$

$$R_i = Clq_i - S_i \tag{9.40}$$

Those probabilities contain a reduced number of variables, and this way it is not costly to compute the marginal probability distribution for an individual variable within the clique node.

• *Conditioning algorithms*: these kinds of algorithms aim at cutting the multiple paths between the nodes in the cycles, by assigning concrete values to a reduced set of variables within the different cycles (Pearl, 1986, 1988; Shachter, 1988; 1990; Peot & Shachter, 1991; Suermondt & Cooper, 1991). This set of variables selected over all the cycles are usually named as the *cutset*. This way, for each assigned set of values to the variables in the *cutset*, a polytree will be generated. Each polytree can be solved using the exact inference methods for polytrees.

This set of generated polytrees are simpler than the original network, just the contrary of the clique tree propagation algorithm, which generates a complex polytree formed of clique nodes.

---

[11] A *clique* is any subset of nodes in an undirected graph, which the subgraph formed by this subset is *complete* and *maximal*. An undirected graph is called *complete* when all pair of different nodes are adjacent. An undirected graph is called *maximal* if there is no any complete graph containing it.

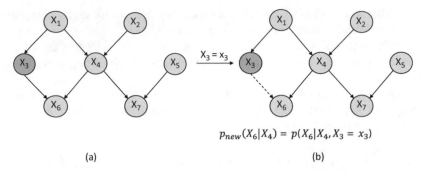

$$p_{new}(X_6|X_4) = p(X_6|X_4, X_3 = x_3)$$

(a)   (b)

Fig. 9.15 A *multiply-connected network* (**a**) and the corresponding *conditioned network* with the evidence $X_3 = x_3$ absorbed through the removed edge $X_3 \rightarrow X_6$ (**b**)

The main problem of this approach is that the number of resulting polytrees is exponential in the size of the *cutset* variables. If you have four Boolean variables in the *cutset*, then 2^4 polytrees must be solved.

Let us analyze the original network described in Fig. 9.15a. In the network, there is a cycle or loop among the nodes X_1, X_3, X_4, X_6. To cut the loop, one of the two alternative paths from traversing from X_1 to X_6 must be cut. In general, for each cycle in a network several variables are candidate to be used for cutting the alternative paths. The final *cutset* of variables is obtained choosing one variable for cutting each one of the cycles. This way, several different *cutsets* can be formed selecting different variables for each cycle. In our example, the candidate *cutsets* for the unique cycle are $\{X_1\}$, $\{X_3\}$ or $\{X_4\}$.

This process can be done using the *absorption evidence method* proposed in (Shachter, 1988, 1990), which proves that the evidence added to the node can be absorbed by changing the topology of the network. Formally, if X_i is an evidential node, all the edges from the network: $X_i \rightarrow X_j$ can be removed substituting the original conditional probability of the node X_j:

$$p\big(X_j|\text{Parents}\big(X_j\big)\big)$$

by

$$p_{new}\big(X_j|\text{Parents}\big(X_j\big) - X_i\big) = p\big(X_j|\text{Parents}\big(X_j\big) - X_i, X_i = x_i\big) \qquad (9.41)$$

Therefore, the loop can be cut by removing the edges $X_1 \rightarrow X_3$, $X_1 \rightarrow X_4$, $X_3 \rightarrow X_6$ or $X_4 \rightarrow X_6$, considering either the node X_1 or X_3 or X_4 as evidential nodes and assigning them an arbitrary value.

For instance, in our example in Fig. 9.15, if the node X_3 is transformed into an evidential node, assigning it an arbitrary value $X_3 = x_3$, then this evidence can be absorbed removing the edge $X_3 \rightarrow X_6$ as illustrated in Fig. 9.15b. To maintain the probability distribution in the new transformed network, the conditional probability table of node X_6 must be updated to:

$$p_{\text{new}}(X_6|X_4) = p(X_6|X_4, X_3 = x_3)$$

In general, following the arbitrary assignment of values to all of the *cutset* variables $C = \{C_1, \ldots, C_k\}$ several different polytrees will be created.

The conditional probability for any query variable X_i, given determined evidence variable or variables $E = e$ and non-observed variables Y, can be obtained summing up the probabilities obtained by each assignment of values $\{c_1, \ldots, c_k\}$ to the *cutset* variables $\{C_1, \ldots, C_k\}$ in each polytree:

$$p(X_i = x_i|e) = p(x_i|e) = \sum_{c_1, \ldots, c_k} p(x_i|e, c_1, \ldots, c_k) \tag{9.42}$$

And the probability at each polytree can be computed using an efficient method for a polytree as follows:

$$p(x_i|e, c_1, \ldots, c_k) = \frac{p(x_i, e, c_1, \ldots, c_k)}{p(e, c_1, \ldots, c_k)} = \sum_{y_k \in Y} \frac{p(x_i, e, c_1, \ldots, c_k, y_k)}{p(e, c_1, \ldots, c_k)}$$

$$= \frac{1}{p(e, c_1, \ldots, c_k)} \sum_{y_k \in Y} p(x_i, e, c_1, \ldots, c_k, y_k) \tag{9.43}$$

$$= \alpha \sum_{y_k \in Y} p(x_i, e, c_1, \ldots, c_k, y_k)$$

Thus:

$$p(x_i|e, c_1, \ldots, c_k) \propto \sum_{y_k \in Y} p(x_i, e, c_1, \ldots, c_k, y_k) \tag{9.44}$$

In order to decrease the number of polytrees, one should minimize the size of the *cutset*. One strategy is to select a common variable to as much as possible adjacent cycles. In the best case, if there is a common variable belonging to all cycles, selecting just this one to be in the *cutset* will work.

Approximate Inference Methods

Depending on the number of nodes and complexity of the *multiply connected network*, the *exact inference methods* could not be a reasonable approach regarding the computational time needed. To solve that, some *approximation methods*, which do not compute the exact probability values, but an approximated value to them through some stochastic sampling algorithms, usually named as *Monte Carlo methods*,[12] must be considered. These methods provide approximate values whose accuracy depends on the number of samples generated. The most commonly used methods are *direct sampling methods* and *Markov Chain Monte Carlo sampling methods*.

Direct Sampling Methods

These methods generate the random samples from scratch for a determined large number of times. The methods differ in how they manage the consistency with the evidences. Major approaches are:

- *Prior Sampling or Logic Sampling Method* (Henrion, 1988): Main idea of the method is generating random samples or events from the given probability distribution in the network, and estimate the specific query probability by counting the frequencies of interested variables in the whole set of events randomly generated. The generated events have no evidence associated with them.

  Each round of the simulation starts by sampling each variable in the network following the *topological ordering*. Sampling is done by randomly choosing a value according to the probability distribution. First, values are sampled for all the root nodes of the network. Afterwards, following the *topological ordering* of the network, the other variable values are sampled but its probability distribution is conditioned on the values already assigned to the variable's parents (*priors*).

  At the end of the simulation run, a randomly sampled event formed by all the variables in the network is obtained. Once all the runs are finished, the ratio between the number of events were the *query variables* and the *evidence variables* are *true*, divided by the number of events where the *evidence variables* are *true*, must be computed. This ratio is the estimation of the query probability of interest.

  The estimated approximate value will be better the larger the number of simulations runs.

  Let us consider, the example network of Fig. 9.14a, which have the four variables A, B, C, and D, with the given probability distribution. Assume that the topological ordering is: $[A, B, C, D]$. The first step is sampling for the root variables. We have just variable A. Thus:

  Sampling from $p(A) = \langle 0.4, 0.6 \rangle$, means that the 40% of the times we would get A = true, and the 60% of the time we would get A = false. Let us suppose that the sampling results in a value of false. The sampled event is by now: $[A = false, B = ?, C = ?, D = ?]$.

---

[12] *Monte Carlo simulation methods* are a kind of computational algorithms that rely on repeated random sampling to estimate some numerical result. They can be used to solve any problem having a probabilistic interpretation. The name refers to the grand casino in the Principality of Monaco at Monte Carlo, which is an icon of repeated random bets, i.e. gambling. The term "Monte Carlo" was first introduced in 1947 by Nicholas Metropolis.

Following the ordering, B variable must be sampled from $p(B|A = \text{false}) = \langle 0.6, 0.4 \rangle$. Let us suppose that the value sampled is *true*. Thus, $B = \text{true}$, and the sampled event by now is: $[A = \text{false}, B = \text{true}, C = ?, D = ?]$.

Now is the turn of variable C according to the ordering. The C variable must be sampled from probability $p(C|A = \text{false}) = \langle 0.1, 0.9 \rangle$. Let us suppose that the sampling gets a false value, $C = \text{false}$. The sampled event by now is: $[A = \text{false}, B = \text{true}, C = \text{false}, D = ?]$.

The final step is the sampling for the last variable D. The D variable must be sampled from probability $p(D|B = \text{true}, C = \text{false}) = \langle 0.8, 0.2 \rangle$. Let us suppose that the sampling gets a true value, $D = \text{true}$. The finally sampled event is: $[A = \text{false}, B = \text{true}, C = \text{false}, D = \text{true}]$.

This event sampling will be repeated many times getting different values for the four variables. Let us suppose that the query probability would be:

$$p(B = \text{true}|D = \text{true}) = \frac{p(B = \text{true}, D = \text{true})}{p(D = \text{true})}$$

Then, this probability would be estimated by the ratio between the number of events were $B = \text{true}$ and $D = \text{true}$ and the number of events were $D = \text{true}$:

$$\widehat{p}(B = \text{true}|D = \text{true}) = \frac{\#[A = ?, B = \text{true}, C = ?, D = \text{true}]}{\#[A = ?, B = ?, C = ?, D = \text{true}]}$$

Where $\widehat{p}$ means the estimation of the probability p ($\widehat{p} \approx p$), and $\#[\ldots]$ expresses the number of times that the event $[\ldots]$ holds.

Main problem of this method is that when the combination of values for the evidence variables is rare, many sampled events will not be useful for the estimation because they will not hold the required values. Thus, really a huge number of random events would be needed to have a minimum number of events with the required values.

- *Rejection Sampling Method*: is a sampling method aiming to *reject* the sampled events *not being consistent* with the evidence variables. This way, the method just takes into account the useful events sampled to estimate the query probability:

$$p(X_i = x_i | E_1 = e_1, \ldots, E_v = e_v, Y) \tag{9.45}$$

were the evidence values are satisfied by the sampled events.

The method proceeds as the *Prior Sampling Method*: It generates a sampled event as explained above, but if *it does not match the evidence values* $\{E_1 = e_1, \ldots, E_v = e_v\}$ *it is rejected*. This way, all the generated events satisfy the evidence values. Thus, finally to estimate the query probability, it is only needed to count how many of them satisfy $X_i = x_i$. Imagine that the number of sampled events is n, and the number of those ones satisfying $X_i = x_i$ is r, and that the variable X_i is a Boolean variable. Then:

$$p(X_i = x_i | E_1 = e_1, \ldots, E_v = e_v, Y) \approx \left\langle \frac{r}{n}, \frac{n-r}{n} \right\rangle \qquad (9.46)$$

- *Likelihood Weighting Method* (Fung & Chang, 1989; Shachter & Peot, 1989): This method also tries to overcome the problem of obtaining enough useful samples when the combination of the evidence variables is rare. Furthermore, it avoids the inefficient generation of many inconsistent samples with the evidence variables as it is done by the *rejection sampling method*.

  Main idea is that during the sample generation as described in the *logic sampling method*, whenever an evidence variable is reached, instead of randomly choosing a value according to the conditional probabilities, the given value for the evidence value is fixed. However, the conditional probabilities are used to weight how likely is that value. This way, only the non-evidence variables are sampled. Each event is *weighted by the likelihood* of the event according to the evidence, as measured by the product of the conditional probabilities for each evidence variable, given its parents.

  The final estimation after n simulation runs will be computed adding up the evidence for each value of the query variable weighted by the likelihood score. Thus, if the weight w of a given sampled event se_j is:

$$w_{se_j}(X, E, Y) = \prod_{i=1}^{v} p(E_i = e_i | \text{Parents}(E_i = e_i)) \qquad (9.47)$$

  Then, the probability of the query variable X_i assuming that is a Boolean variable will be estimated as:

$$p(X_i = x_i | E_1 = e_1, \ldots, E_v = e_v) \approx \text{Normalize}\left\langle \sum_{j=1}^{r} w_{se_j}, \sum_{j=1}^{n-r} w_{se_j} \right\rangle \qquad (9.48)$$

  Where the number of simulations runs satisfying $X_i = x_i$ is r and the number of simulations runs not satisfying $X_i = x_i$ is $n - r$.

Let us consider again the above example from the *Bayesian network* of Fig. 9.14a, which have the four variables A, B, C, and D, with the given probability distribution and the topological ordering: $[A, B, C, D]$.

Let us assume that we are interested in computing the following probability:

$$p(B | C = \text{false}, D = \text{true})$$

Thus, we have two evidence variables $\{C, D\}$, which will not be sampled, but its probability value will be weighted. The weight of the first sampled event (se_1) is initialized to the neuter element of the product, i.e. 1:

$$w_{se_1}(B, \{C, D\}, A) = w_{se_1} = 1$$

The first step is sampling for the root variable A. Suppose that the sampling from $p(A) = \langle 0.4, 0.6 \rangle$ results in a value of false. The sampled event se_1 is by now: $[A = \text{false}, B = ?, C = \text{false}, D = \text{true}]$.

Following the ordering, B variable must be sampled from $p(B|A = \text{false}) = \langle 0.6, 0.4 \rangle$. Let us suppose that the value sampled is true. Thus, $B = \text{true}$, and the sampled event se_1 is: $[A = \text{false}, B = \text{true}, C = \text{false}, D = \text{true}]$.

The third variable according to the order is the variable C. Variable C is an evidence variable with value false. Therefore, the weight for this evidence value given its parents must be computed and accumulated to the general weight of the sampled event se_1;

$$w_{se_1} \leftarrow w_{se_1} * p(C = \text{false}|A = \text{false}) = 1 * 0.9 = 0.9$$

The sampled event is: $[A = \text{false}, B = \text{true}, C = \text{false}, D = \text{true}]$.

The final step is for the last variable D. Variable D is an evidence variable with value true. Therefore, the weight for this evidence value given its parents must be computed and accumulated to the general weight of the sampled event se_1;

$$w_{se_1} \leftarrow w_{se_1} * p(D = \text{true}|B = \text{true}, C = \text{false})$$

$$w_{se_1} \leftarrow 0.9 * 0.8 = 0.72$$

Thus, the first sampled event se_1 event is: $[A = \text{false}, B = \text{true}, C = \text{false}, D = \text{true}]$ and its computed weight is $w_{se_1} = 0.72$.

Finally, this weight is accumulated for the estimation of $p(B = \text{true}|C = \text{false}, D = \text{true})$.

This process will be repeated n times and both accumulated weights, $\left\langle \sum_{j=1}^{r} w_{se_j}, \sum_{j=1}^{n-r} w_{se_j} \right\rangle$ will be normalized to obtain the associated probabilities for both $p(B = \text{true}|C = \text{false}, D = \text{true})$ and $p(B = \text{false}|C = \text{false}, D = \text{true})$.

Usually, *the likelihood weighting method* converges faster than *prior sampling method*, and this way, is preferable for large networks. In addition, it is more efficient than *rejection sampling method*, because the samples not being consistent with evidence variables are not generated. However, it can suffer from a decreasing performance in the estimations as the number of evidence variables increases, especially if they are located at the end of the variable ordering, because then, they have a very low influence with the generated samples.

Furthermore, all sampling techniques can take a long time to generate accurate estimations for unlikely events.

Markov Chain Monte Carlo Sampling Methods

The Markov Chain Monte Carlo (MCMC) methods generate the random samples by making a *random change* to the preceding generated sample. Each sample can be thought as a *state* described with the corresponding values of all variables. The transition from the *current state* to the *next state* is done by making random changes to the *current state*. Most known methods are the Gibbs sampling method, the Metropolis-Hastings algorithm and other methods.

- *Gibbs Sampling Method* (Geman & Geman, 1984): This method sets up an arbitrary initial state for the non-evidential variables (X, Y) and with the corresponding fixed observed values for the evidential variables (E) evolves generating the next state by randomly sampling a value for one of the non-evidential variables (X, Y). The non-evidential variable sampled at each step can be randomly selected or simply repeatedly cycling over all of them. The sampling for each non-evidential variable (NE) is done conditioned on the current variables in the Markov blanket[13] of NE. Taking into account that a variable in a Bayesian network is independent of all other variables given its Markov blanket, then:

$$p(\text{NE}|\{X, Y\} - \text{NE}, E) = p(\text{NE}|\text{Mb}(\text{NE})) \tag{9.49}$$

And it can be shown that:

$$p(\text{NE}|\text{Mb}(\text{NE})) = \alpha * p(\text{NE}|\text{parents}(\text{NE}))$$
$$* \prod_{Y_c \in \text{Children}(\text{NE})} p(Y_c|\text{parents}(Y_c)) \tag{9.50}$$

Therefore, the method randomly traverses the state space by turning over one non-evidential variable (NE) at each time while the evidential variables (E) are maintained fixed with its observed values.

At the end, the method accounts how many traversed states, without taking into account the initial state, had each one of the possible values for the query variable (X_i). Then after a normalizing procedure the estimated probabilities for each possible value of the query variable (X_i) are obtained.

Let us consider again the above example from the Bayesian network of Fig. 9.14a, and assume that we are interested in computing the same probability than before:

$$p(B|C = \text{false}, D = \text{true})$$

First, the *evidential variables* $\{C, D\}$ are set to its observed values to obtain the initial state and the *non-evidential variables* are randomly initialized. Let us suppose that $A = \text{true}$ and $B = \text{false}$. Thus, the *initial random state* is:

---

[13] The *Markov blanket* (Mb) of a variable X_i in a *Bayesian network* is the set of nodes formed by its parents, its children and its children's parents. Therefore, $\text{Mb}(X_i) = \{\text{Parents}(X_i), \text{Children}(X_i), \text{Parents}(\text{Children}(X_i))\}$.

$$s_0 \equiv [A = \text{true}, B = \text{false}, C = \text{false}, D = \text{true}]$$

From now on, the *non-evidential variables* $\{A, B\}$ are repeatedly sampled following a random or cycling order. For instance:

Variable B is sampled according to the conditional probability:

$$p(B|A = \text{true}, C = \text{false}, D = \text{true}) = p(B|\text{Mb}(B))$$
$$= \alpha * p(B|A = \text{true})$$
$$* p(D = \text{true}|B, C = \text{false})$$

For $B = \text{true}$ and $\{A = \text{true}, C = \text{false}, D = \text{true}\}$, which are represented as $(b, a, \bar{c}, d)$, we have:

$$p(b|a, \bar{c}, d) = \alpha * p(b|a) * p(d|b, \bar{c}) = \alpha * 0.3 * 0.8 = \alpha * 0.24$$

Reciprocally for $B = \text{false}$, we have:

$$p(\bar{b}|a, \bar{c}, d) = \alpha * p(\bar{b}|a) * p(d|\bar{b}, \bar{c}) = \alpha * 0.7 * 0.1 = \alpha * 0.07$$

Thus, after the normalization process:

$$p(B|a, \bar{c}, d) = \left\langle \frac{0.24}{0.24 + 0.07}, \frac{0.07}{0.24 + 0.07} \right\rangle = \langle 0.77, 0.23 \rangle$$

Note that the conditional probabilities can be obtained from the CPTs of the *Bayesian network*. Assume that the sampling from B according to this distribution results in a *true* value. This way the *new current state* will be:

$$s_1 \equiv [A = \text{true}, B = \text{true}, C = \text{false}, D = \text{true}]$$

Now variable A will be sampled according to the conditional probability:

$$p(A|B = \text{true}, C = \text{false}, D = \text{true}) = p(A|\text{Mb}(A))$$
$$= \alpha * p(A|\emptyset) * p(B = \text{true}|A) * (C = \text{false}|A)$$

For $A = \text{true}$ and $\{B = \text{true}, C = \text{false}, D = \text{true}\}$, which are represented as $(a, b, \bar{c}, d)$, we have:

$$p(a|b, \bar{c}, d) = \alpha * p(a) * p(b|a) * p(\bar{c}|a) = \alpha * 0.4 * 0.3 * 0.3 = \alpha * 0.036$$

Reciprocally for $A = \text{false}$, we have:

$$p(\bar{a}|b, \bar{c}, d) = \alpha * p(\bar{a}) * p(b|\bar{a}) * p(\bar{c}|\bar{a}) = \alpha * 0.6 * 0.6 * 0.9 = \alpha * 0.324$$

Thus, after the normalization process:

$$p(A|b, \bar{c}, d) = \left\langle \frac{0.036}{0.036 + 0.324}, \frac{0.324}{0.036 + 0.324} \right\rangle = \langle 0.1, 0.9 \rangle$$

Assume that the sampling from A, according to this distribution results in a false value. This way the new current state will be:

$$s_2 \equiv [A = \text{false}, B = \text{true}, C = \text{false}, D = \text{true}]$$

And this process will be continued for a determined number n of times. Supposing this process traversed 56 states were the *query variable* B was true and 34 states were the *query variable* B was false, then the estimated probabilities would be:

$$\left\langle \frac{56}{56 + 34}, \frac{34}{56 + 34} \right\rangle = \langle 0.62, 0.38 \rangle$$

- *Metropolis-Hastings Algorithm*: The Metropolis-Hastings method (Hastings, 1970) is a Markov Chain Monte Carlo algorithm aiming at generating samples x for a probability distribution $p(x)$, which is usually difficult to be sampled. This probability distribution $p(x)$ is usually named as the *target distribution*. The interesting fact is that the algorithm simulates a *Markov Chain*, which asymptotically reaches a *unique stationary distribution*,[14] $\pi(x)$, which matches the target probability $p(x)$. Thus:

$$\pi(x) \approx p(x) \tag{9.51}$$

This means, that with a *long number* of transitions in the Markov Chain simulation, the sampled states from the Markov Chain are equivalently distributed to the samples from the target distribution $p(x)$.

Another important feature is the need for a *transition kernel*, $q(x'|x)$, which express the probability distribution of a state x' given the state x. In many applications of the method, especially when the states are described with numerical random variables, the *transition kernel* is described adding a normal perturbation (pert $\approx N$ $(0, 1)$) to the current state random variables. This way, the new state is $x' = x + N$ $(0, 1)$ and the kernel is $q(x'|x) \approx N(x, 1)$. This transition kernel is usually known as a *random walk transition kernel*.

---

[14] A *stationary distribution* π of a *Markov Chain*, is a probability distribution which remains unchanged by the repeated transition operations from one state to the next one. Thus, if the transition matrix storing the transition probabilities between states is P, the stationary distribution satisfies:

$$\pi = \pi P.$$

As the *transition kernel* provides the proposal value for the candidate new state of the Markov Chain, it is usually named as the *proposal distribution*.

The method iterates two basic steps until the end. First, a candidate new state x' is proposed according to the proposal distribution $q(x'|x)$ given the current state x, and secondly, it probabilistically accepts or rejects the candidate state x' according to an *acceptance probability*, which is defined as:

$$\alpha(x'|x) = \min\left(1, \frac{\pi(x') * q(x|x')}{\pi(x) * q(x'|x)}\right) \tag{9.52}$$

This acceptance probability $\alpha(x'|x)$ is introduced into the transition probability by modifying the proposal distribution with an acceptance of the transition. It can be derived from the condition of *detailed balance*[15] of the Markov Chain which sufficiently guarantees the *stationary distribution*:

$$p(x'|x) * \pi(x) = p(x|x') * \pi(x') \tag{9.53}$$

Which can be described as:

$$\frac{p(x'|x)}{p(x|x')} = \frac{\pi(x')}{\pi(x)} \tag{9.54}$$

Then, the idea is to decompose the *transition probability* into the *proposal distribution* and an *acceptance probability* of the transition:

$$p(x'|x) = q(x'|x) * \alpha(x'|x) \tag{9.55}$$

From this expression and substituting into the detailed balance equation, we can get:

$$\frac{q(x'|x) * \alpha(x'|x)}{q(x|x') * \alpha(x|x')} = \frac{\pi(x')}{\pi(x)}$$

And it follows:

$$\frac{\alpha(x'|x)}{\alpha(x|x')} = \frac{\pi(x') * q(x|x')}{\pi(x) * q(x'|x)}$$

And hence:

---

[15] The *detailed balance* means that for every pair of states x, x' the probability of being in state x and transitioning to state x' must be equal to the probability of being in state x' and transitioning to state x.

$$\alpha(x'|x) = \frac{\pi(x') * q(x|x')}{\pi(x) * q(x'|x)} * \alpha(x|x') \tag{9.56}$$

From the above relation, it follows that the ratio is essential for determining whether the transition from x to x' should be done or not.

The Metropolis-Hastings proposal for the acceptance probability satisfies the previous equation.

The algorithm for a *discrete target probability distribution* can be summarized as follows:

---

algorithm Metropolis-Hastings

input: $\pi(x)$: the *target probability* distribution
         $q(x'|x)$: the *transition kernel* or *proposal distribution*

output the estimated $\hat{p}(x) \equiv \hat{\pi}(x)$

begin
   // Initialization: set the *accumFreq* to 0 and t to 0
   // Initialization: set the current state to a random initial state
   $accumFreq \leftarrow 0$
   $t \leftarrow 0$
   $x_t \leftarrow$ randomState()
   // Iteration to generate T new states from the initial one x_0
   for each t **in** $0, \dots, T-1$ **do**
     // Sample a new state candidate x' from the proposal distribution
     $x' \leftarrow$ sampleFrom($q(x'|x_t)$)
     // Compute the acceptance probability for the transition: $x_t \rightarrow x'$
     $\alpha(x'|x_t) = min\left(1, \frac{\pi(x') * q(x_t|x')}{\pi(x_t) * q(x'|x_t)}\right)$
     // Make the transition or not (accept or reject)
     $u \leftarrow$ unifRand(0,1)   // generate a uniform random number $u \in [0,1]$
     if $u \leq \alpha(x'|x_t)$ **then** $x_{t+1} \leftarrow x'$ // accept the transition
                 else $x_{t+1} \leftarrow x_t$ // reject the transition
   endif
     // accumulate the frequency of interested states (all or just some)
     $accumFreq \leftarrow accumFreq + Freq(x_{t+1})$
   endforeach
     // compute the estimation of $\pi(x)$ through the T generated states
   $\hat{\pi}(x) \leftarrow accumFreq/T$
   return $\hat{\pi}(x)$
end

In the particular application of the *Metropolis-Hastings algorithm* for estimating the probabilities in a *Bayesian network*, we will have:

$$\pi(x) = p(x_i|E) \tag{9.57}$$

Which is the query conditional probability to be approximated, and the *transition kernel probability* must be defined:

$$q(x'|x)$$

Note that if the *proposed transition* selects one non-evidential variable each time, this method would be the *Gibbs sampling method*. The *proposed transition*, could select for sampling either more than one non-evidential variable, or all the non-evidential variables, or even all the variables randomly. This selection can have a strong influence to the convergence of the *Markov Chain* to the stationary distribution probability.

On the other hand, if in the *Metropolis-Hastings algorithm*, the *proposed transition distribution* is a *symmetric function* on its arguments, i.e. it satisfies the following:

$$\forall x, x' \quad q(x'|x) = q(x|x') \tag{9.58}$$

Then, the *acceptance probability* is:

$$\alpha(x'|x) = \min\left(1, \frac{\pi(x')}{\pi(x)}\right) \tag{9.59}$$

This particular case of the *Metropolis-Hastings algorithm* when the transition kernel $q(x'|x)$ is a symmetric function, was presented in Metropolis et al. (1953), and due to this fact, it is usually referred as the *Metropolis algorithm*.

Other Methods

Other methods used in the literature are the *variational methods* and the *loopy belief propagation method*:

- *Variational Methods*: This family of approximation methods are based on proposing a reduced version of the original problem that is simpler to work with, but resembling the original problem as much as possible. The simplified problem is described by some *variational parameters* λ. These parameters are adjusted to minimize a distance function d between the original and the simplified problem. Thus, usually the parameters are determined by numerically solving the following system of equations through an iterative approximation:

$$\frac{\partial d}{\partial \lambda} = 0$$

Commonly, strict lower and upper bounds can be obtained. These methods have been applied in other fields like statistical physics. A particular commonly used method is the *mean-field approximation method*, which exploits the law of large numbers to approximate large sums of random variables by their means. Thus, the original variables are assumed to be independent. The method decouples all the nodes, and introduces the variational parameter for each node, and iteratively update these parameters to minimize the *Kullback – Leibler Divergence* (*KL Divergence*) between the approximate and true probability distributions.

The foundations for applying variational methods to *Bayesian networks* were developed by Saul et al. (1996), and they obtained accurate lower-bound approximations for sigmoid belief networks. This methodology was extended in a later work (Jaakkola & Jordan, 1996) obtaining both lower and upper bounds. A nice survey on the literature of variational methods was written in the book (Wainwright & Jordan, 2008).

- *Loopy Belief Propagation Methods*: These methods are based on the application of the original *direct message-passing propagation method* for polytrees proposed by Pearl (1982). Even the method is only guaranteed to be correct for polytrees, Pearl (1988) already suggested that it could be applied to general *Bayesian networks* containing loops, i.e. undirected cycles. Yair Weiss (2000) and other authors proved how a loopy propagation method works and when the approximation is correct. Intuitively, consider that in certain cases like in a single loop, events are double counted equally, and hence, they cancel to give the right answer. Furthermore, Murphy et al. (1999) presented empirical studies about the performance of the loopy belief propagation approach.

9.2.5   Fuzzy Set Theory/Possibilistic Model

Fuzzy Sets Theory was born for modelling the concept of *imprecision* or *vagueness*. It was proposed by Lotfi Zadeh (1965). At the same time, other researchers have worked in similar concepts like the work on extension of the notion of classical sets by Dieter Klaua (1965), and Salii (1965) who defined a more general structure called an *L-relation* on the context of abstract algebra.

Later on, the *Possibility Theory* was introduced for managing uncertainty in fuzzy systems (Zadeh, 1978).

In the exact words of Lotfi Zadeh (1965): "Essentially, such a framework provides a natural way of dealing with problems in which the source of imprecision is the absence of sharply defined criteria of class membership rather than the presence of random variables".

In this section, the fundamentals of fuzzy sets will be explained, as well as the basis of the possibility theory developed upon fuzzy sets. Furthermore, the practical representation of fuzzy sets will be introduced, and the computation of the fuzzy connectives within the scope of the fuzzy logic concept will be described. Finally, the fuzzy inference process in a rule-based system will be analyzed. The fuzzy set theory has been applied to many real applications, but the fuzzy inference process has been deeply analyzed in rule-based reasoning systems.

9.2.5.1 Fundamentals on Fuzzy Sets

Fuzzy sets were proposed to model the *imprecision* or *vagueness* of natural language expressions like the contained in propositions such as:

The temperature is *high*
Turn the steering wheel *a little bit on the right*

What is a *high* temperature? Is it 27 °C a *high* temperature? Or is it 30 °C a *high* temperature?

What exactly means *turn the steering wheel a little bit on the right*? Perhaps an angle of 5°? Or an angle of 10°?

Surely the concept of *high temperature* is not precisely defined, and in addition can be subjective. The same happens with the action *turn the steering wheel a little bit on the right*.

Furthermore, fuzzy sets were proposed to model the degree of possibility described in expressions like:

It is *very possible* that if she appears yellowish, then she has a kind of hepatitis.

In classical set theory, the ordinary sets or *crisp sets*,[16] usually are defined by enumerating the elements that form part of it or by describing a property that the elements satisfy. For instance:

$$B = \{a, b, c, d, e\}$$

where B is the set formed by the first fifth lowercase letters of the alphabet, or

$$A = \{x \in \mathbb{R}^+ \mid 15 \le x \le 20\}$$

where A is formed by all the positive real values belonging to the closed interval [15, 20].

In any case, there exists a membership function, the so-called *characteristic function* of a set A, that indicates whether a given element x belongs or not to the set:

---

[16] Usual ordinary sets are named as *crisp sets* in the context of fuzzy set theory.

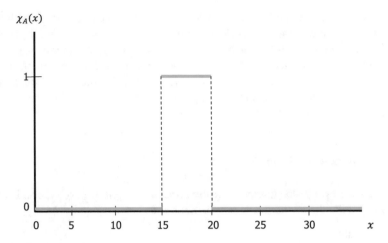

Fig. 9.16 The characteristic function $\chi_A(x)$ of the set $A = \{x \in \mathbb{R}^+ \mid 15 \leq x \leq 20\}$

$$\chi_A(x) = \begin{cases} 1 & \text{if } x \in A \\ 0 & \text{if } x \notin A \end{cases} \tag{9.60}$$

The *characteristic function* of the set A is depicted in Fig. 9.16.

As it can be observed from Fig. 9.16, the *characteristic function* is a *discontinuous function* which has two discontinuities at values 15 and 20. Intuitively, a *discontinuous function* is a function which at some given points the value of the function when you approximate to these points by the right or the left is not the same.

Imagine that the set A was representing the set of "*cool temperatures*". With this definition of the set and its characteristic function, we are saying that a temperature of 15 °C is cool ($\chi_A(15) = 1$), one of 17 °C is cool ($\chi_A(17) = 1$) and that a temperature of 20 °C is also cool ($\chi_A(20) = 1$). However, a temperature of 14.9 °C is not belonging to the set ($\chi_A(14.9) = 0$), and thus is not a "cool temperature". The same happens with a temperature of 20.1 °C, which does not belong to the set ($\chi_A(20.1) = 0$), and hence is not cool.

Is this situation representing the reality? The difference between 14.9 and 15 is so huge that makes that the former is not cool and the latter is?

The idea of a *fuzzy set* is that each value of the set has a membership degree value of belonging to the set. Following the original definition from Zadeh (1965):

Given a space of points or objects usually named as the *universe of discourse U*, which commonly is assumed to be discrete, i.e.

$$U = \{x_1, x_2, \ldots, x_n\}$$

Then, a *fuzzy set A* in U is defined as a set of ordered pairs:

$$A = \{(x, \mu_A(x)) \mid x \in U\} \tag{9.61}$$

where $\mu_A(x)$ is the *membership function* of the set A or *degree of membership* of x to A. The *membership function* is defined as:

$$\mu_A : X \to M$$
$$x \mapsto \mu_A(x)$$

Where the *range of the membership space M* is a subset of the nonnegative real numbers with a supremum value which is finite.[17] If $\sup_x\{\mu_A(x)\} = 1$, the fuzzy set A is called *normal*.

A non-empty fuzzy set A can be always normalized:

$$\mu_{A\,norm}(x) = \frac{\mu_A(x)}{\sup_x\{\mu_A(x)\}} \tag{9.62}$$

Then, all the membership degree values are between 0 and 1. From now on, we will assume that the fuzzy sets are normalized. Usually, the elements with a zero degree of membership are not listed.

Let us consider the universe of discourse of the set of "normal" monthly salaries, $U = \{0, 250, 500, 1000, 1500, 2000, 2500, 3000\}$. Then an example of a fuzzy set of "fair and reasonable salary" could be:

$$A = \{(500, 0.1), (1000, 0.2), (1500, 0.6), (2000, 0.8), (2500, 1), (3000, 1)\}$$

Other ways to describe a fuzzy set used in the literature are:

$$A = \mu_A(x_1)/x_1 + \ldots + \mu_A(x_n)/x_n = \sum_{i=1}^{n} \mu_A(x_i)/x_i \tag{9.63}$$

where the symbol + should be understood not as a sum, but as an aggregation or union operation. In the continuous case:

$$A = \int_x \mu_A(x_i)/x_i \tag{9.64}$$

Another way of describing a fuzzy set when the membership function is a continuous function is describing the expression of the membership function. For instance, the corresponding fuzzy set for representing the concept "*real numbers close to 0*", could be:

---

[17] More generally, this range could be a suitable partial ordered set P.

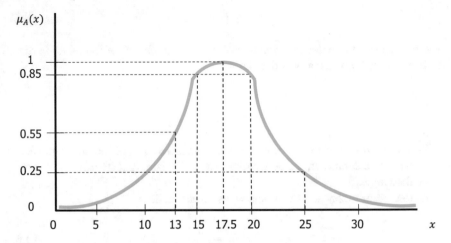

Fig. 9.17 The membership function of the fuzzy set "cool temperatures"

$$A = \{(x, \mu_A(x)) \mid \mu_A(x) = \frac{1}{1 + x^2}\}$$

Let us consider again the task of representing the set of "*cool temperatures*". One possibility would be the crisp set defined with the characteristic function of Fig. 9.16. However, as pointed above it has some drawbacks to actually represent the real situation. We could think that a temperature of 17.5 °C is our ideal "cool temperature" and its degree of membership to the set is maximum (i.e. $\mu_A(17.5) = 1$) and that all the other ones are not so good to be considered as "cool temperatures". Of course, temperatures near to 17.5 °C, for example, those ones between 15 °C ($\mu_A(15) = 0.85$) and 20 °C ($\mu_A(20) = 0.85$) would have a great degree of possibility of being "cool temperatures". On the other hand, as the temperature value starts increasing from 20 °C or decreasing from 15 °C, the temperatures have less degree of possibility to be considered "cool temperatures", and hence its degree of membership to the fuzzy set of "cool temperatures" decreases, like $\mu_A(13) = 0.55$ or $\mu_A(25) = 0.25$. This membership function for the fuzzy set "cool temperatures" is depicted in Fig. 9.17.

Hence, a *fuzzy set* is a generalization of an ordinary set, and its *membership function* is a generalization of the characteristic function of an ordinary set. As it will be analyzed afterwards, the membership functions could be of different types: differentiable functions like a bell curve function, continuous functions only segmentally differentiable like triangular or trapezoidal functions, etc.

Some basic definitions over fuzzy sets and related concepts are the following:

- *Empty* set: a fuzzy set $A \subseteq U$ is empty (denoted $A = \varnothing$) $\Leftrightarrow$[18] $\mu_A(x) = 0, \ \forall x \in U$
- *Equality* of two Fuzzy sets: Two fuzzy sets $A, B \subseteq U$ are *equal* or *equivalent* $\Leftrightarrow$ $\mu_A(x) = \mu_B(x), \ \forall x \in U$

---

[18] The double implication symbol $\Leftrightarrow$ in formal definitions in mathematics or philosophy can be read as *if and only if*, which sometimes is depicted as *iff*.

- *Subset* of a Fuzzy set: A fuzzy set $A \subseteq U$ is a *subset* or it is *included* $(A \subseteq B)$ in another fuzzy set $B \subseteq U \Leftrightarrow \mu_A(x) \leq \mu_B(x), \quad \forall x \in U$
- *Height* of a Fuzzy set: The height of a fuzzy set $A \subseteq U$ is defined as height $(A) = \sup \{\mu_A(x) \mid \forall x \in U\} = \sup (\mu_A(U))$
- *Normal* fuzzy set: A fuzzy set $A \subseteq U$ is said to be *normal* or *normalized* $\Leftrightarrow$ height$(A) = 1$
- *Support*: The *support* of a fuzzy set $A \subseteq U$ is a crisp set defined as Supp $(A) = A^{>0} = \{x \in U \mid \mu_A(x) > 0\}$. Thus, $\emptyset \subseteq \text{Supp}(A) \subseteq U$
- *Core* of a fuzzy set: The *core* or *kernel* of a fuzzy set $A \subseteq U$ is a crisp set defined as $\text{Core}(A) = \text{Kern}(A) = A^{=1} = \{x \in U \mid \mu_A(x) = 1\}$. Thus, $\emptyset \subseteq A^{=1} \subseteq U$
- *α-cut* of a Fuzzy set: The *α-cut* of a fuzzy set $A \subseteq U$ is a crisp set defined as $A^{\geq \alpha} = A_\alpha = \{x \in U \mid \mu_A \geq \alpha\}$. Thus, $\emptyset \subseteq A^{\geq \alpha} \subseteq U$
- *Strong α-cut* of a Fuzzy set: The *strong α-cut* of a fuzzy set $A \subseteq U$ is a crisp set defined as $A^{>\alpha} = A'_\alpha = \{x \in U \mid \mu_A > \alpha\}$. Thus, $\emptyset \subseteq A^{>\alpha} \subseteq U$
- *Width* of a Fuzzy Set: The *width* of a fuzzy set defined over real numbers $A \subseteq U \subseteq \mathbb{R}$ with a finite *support* set, i.e. bounded, is defined as follows: width$(A) = \sup \{\text{Supp}(A)\} - \inf \{\text{Supp}(A)\}$
- *Convex* Fuzzy set: A real fuzzy set $\subseteq U \subseteq \mathbb{R}$ is said to be a convex fuzzy set $\Leftrightarrow$ $\forall x, y \in U, \quad \forall \lambda \in [0, 1] \Rightarrow \mu_A(\lambda * x + (1 - \lambda) * y)) \geq \min (\mu_A(x), \mu_A(y))$
- *Cardinality* of a Fuzzy set: The *cardinality* of a fuzzy set $A \subseteq U$ is defined as $|A| = \sum_{x \in U} \mu_A(x)$ for finite U or $|A| = \int_x \mu_A(x)dx$ for infinite U, even it is not guaranteed to always exist in the latter case.

The basic definition of a fuzzy set provided above assumed that the membership space is the space of real numbers, and concretely, membership functions were crisp functions. This basic definition can be referred as *type 1 fuzzy set definition*. Some extensions and generalizations on this definition are possible. One of the most common is the generalization of the membership function to be a fuzzy set itself, instead of being a crisp ordinary set.

Thus, a *type 2 fuzzy set* in U can be defined as a fuzzy set whose membership values are *type 1 fuzzy sets* on $[0,1]$.

Furthermore, this definition can be generalized to a *type m fuzzy set* in U, which can be defined as a fuzzy set whose membership values are *type $m - 1$ $(m > 1)$ fuzzy sets* on $[0, 1]$.

9.2.5.2 Possibility Theory

The observation that natural language was inherently vague and imprecise led L. A. Zadeh to the introduction of the *theory of possibility* (Zadeh, 1978, 1982), which was extensively described in Dubois and Prade (1980, 1988). He coined this name but the *possibilistic* term was borrowed from Gaines and Kohout (1975). In fact, several pioneering authors have referred to similar concepts. In the 1950s, G. L. S. Shackle proposed the *min/max algebra* to describe *degrees of potential* surprise of events which were referring to *degrees of impossibility* (Shackle, 1961). In 1973, David Lewis considered a relation between possible worlds he called *comparative possibility* (Lewis, 1973). L. J. Cohen considered the problem of legal reasoning, and the concept of *degrees of provability*, which matches with *necessity measures* (Cohen, 1977).

The idea of Zadeh relied in the fact that possibility distributions were meant to provide a graded semantics to natural language statements. He proposed an interpretation of membership functions of fuzzy sets as possibility distributions encoding *flexible constraints* induced by natural language statements.

Before defining the *possibility distribution* concept in the framework of fuzzy sets, the concept of a *fuzzy restriction* must be explained. As defined by Zadeh in 1978, page 5:

Let be X a variable which takes values on the universe of discourse U, with the generic element of U denoted by u, and:

$$X = u$$

signifying that X is assigned the value u, $u \in U$

Given F, which is a *fuzzy set* of the universe *characterized* by a membership function $\mu_F(u)$,

Then, F is a *fuzzy restriction* on the variable X (or associated with X) if F acts as an elastic constraint on the values that may be assigned to X, in the sense that the assignment of the values u to X has the form:

$$X = u : \mu_F(u) \tag{9.65}$$

Where $\mu_F(u)$ is interpreted as the degree to which the constraint represented by F is satisfied when u is assigned to X. This equivalently implies that $1 - \mu_F(u)$ is the degree to which the constraint must be stretched in order to allow the assignment of u to X. Of course, a *fuzzy set* can be considered as a *fuzzy restriction* if it acts as a constraint on the values of a variable, which may take the form of a *linguistic term* or a *classical variable*.

Let $R(X)$ be a *fuzzy restriction* associated with X. Then, to express that F plays the role of a *fuzzy restriction* in relation to X, it is expressed as follows:

$$R(X) = F \tag{9.66}$$

This equation is called a *relational assignment equation*, which represents the assignment of the *fuzzy set F* to the *fuzzy restriction* associated with X, i.e. $R(X)$.

Consider a *proposition* or a *fact*[19] of the form: "*X* is *F*", where X is the name of an object, a variable or a proposition, $A(X)$ is an implied attribute of the variable X, and F is the name of a *fuzzy set* of U, then the proposition "*X* is *F*" can be expressed as:

---

[19] In the sense of a *fact* belonging to a *fact base* of a *rule-based reasoning* system.

$$R(A(X)) = F \qquad (9.67)$$

For instance, propositions such as "Sharon is very intelligent", "Ramon is young" or "X is a small number", can be expressed this way assuming that:

- "Sharon is very intelligent": $A(X) = Intelligence(X)$, $X = Sharon$, F is the fuzzy set "very intelligent". Thus:

$$R(\text{Intelligence}(\text{Sharon})) = \text{"very intelligent"}$$

- "Ramon is young": $A(X) = Age(X)$, $X = Ramon$, F is the fuzzy set "young". Thus:

$$R(\text{Age}(\text{Ramon})) = \text{"young"}$$

- "X is a small number": $A(X) = X$, X is a non-instanced variable, F is the fuzzy set "small number". Thus:

$$R(X) = \text{"small number"}$$

Now, the *possibility distribution* term can be defined as done by Zadeh (1978), page 6:

Let F be a *fuzzy set* in a universe of discourse U which is characterized by its *membership function* μ_F, with the *grade of membership*, $\mu_F(u)$, interpreted as the compatibility of u with the concept labelled F.

Let X be a variable taking values in U, and let F act as a *fuzzy restriction*, $R(X)$, associated with X. Then:

The proposition "X is F", which translates into

$$R(X) = F$$

associates a *possibility distribution*, Π_X, with variable X which is postulated to be equal to $R(X)$. Thus:

$$\Pi_X = R(X) \qquad (9.68)$$

Correspondingly, the *possibility distribution function associated with X* (or the *possibility distribution function* of Π_X) is denoted by π_X and is defined to be numerically equal to the membership function of F. Thus:[20]

---

[20] The mathematical symbol $\triangleq$ usually means "is defined as" or "denotes".

$$\pi_X \triangleq \mu_F \qquad (9.69)$$

Thus, $\pi_X(u)$, the possibility that $X = u$, is postulated to be equal to $\mu_F(u)$.

The *relational assignment equation* $R(X) = F$, taking into account that $\Pi_X = R(X)$ may be expressed equivalently in the form:

$$\Pi_X = F \qquad (9.70)$$

outlining that the proposition $p \triangleq X$ *is* F has the effect of associating X with a *possibility distribution* Π_X, which is equal to F. When expressed in the form of $\Pi_X = F$, a *relational assignment equation* will be referred to as a *possibility assignment equation*, with the understanding that Π_X is induced by p.

Let as consider an example. Assume that the universe of discourse U is formed by all positive integers, and let F be the fuzzy set of "small even integers", defined as:

$$F = \{(2,1),(4,0.8),(6,0.4)\}$$

The proposition "X is a small even integer" associates with variable X, the possibility distribution:

$$\Pi_X = F$$

so that a term such as (6,0.4) means that the possibility that X *is* 6, given that X is a small even integer, is 0.4.

As Zadeh outlined in Zadeh (1978) in page 7: An important consequence is that the *possibility distribution* Π_X may be regarded as an *interpretation of the concept of a fuzzy restriction* and, consequently, that the mathematical apparatus of the theory of fuzzy sets, and, especially, the *calculus of fuzzy restrictions*, provides a basis for the manipulation of *possibility distributions* by the rules of this calculus.

Furthermore, the definition of $\pi_X(u)$ implies that the *degree of possibility* may be any number in the interval [0,1] rather than just 0 or 1. In this connection, it should be noted that the existence of *intermediate degrees of possibility* is implicit in such commonly encountered propositions in natural language as "There is a slight possibility that Marilyn is very rich", "It is quite possible that Jean-Paul will be promoted", or in "It is almost impossible to find a needle in a haystack".

Let us define the concept of a *possibility measure*. As defined by Zadeh (1978), page 9:

Let A be a *non-fuzzy* set, $A \subseteq U$, and let π_X be a possibility distribution associated with a variable X which takes values in U. Then, the *possibility measure*, $\pi(A)$, of A is defined as a number in [0, 1] given by:

$$\pi(A) \triangleq \sup_{u \in A}\{\pi_X(u)\} \qquad (9.71)$$

where $\pi_X(u)$ is the possibility distribution function of π_X. This number, then, may be interpreted as the possibility that a value of X belongs to A, that is:

$$\text{Poss}(X \in A) \triangleq \pi(A) \triangleq \sup_{u \in A}\{\pi_X(u)\} \tag{9.72}$$

The general definition of *possibility measure extended for fuzzy sets* is the following, as defined in Zadeh (1978), page 9:

Let A be a *fuzzy set* of U and let π_X be a possibility distribution associated with a variable X which takes values in U. Then, the *possibility measure*, $\pi(A)$, of A is defined in the possibility scale [0, 1] as:

$$\text{Poss}(X \text{ is } A) \triangleq \pi(A) \triangleq \sup_{u \in U}\{ \min \{\mu_A(u), \pi_X(u)\}\} \tag{9.73}$$

where $\pi_X(u)$ is the possibility distribution function of π_X and $\mu_A(u)$ is the membership function of A.

D. Dubois and H. Prade further contributed to the development of a *general possibility theory* (Dubois & Prade, 1980, 1988) as a new general uncertainty theory devoted to the modelling of incomplete information. Let us define the possibility theory basics as defined by Dubois and Prade (1988):

Let S be a *set of states of affairs* or *states of the world* or *frame of discernment*. This set can be of different nature: unidimensional like the domain of an attribute; multidimensional like the Cartesian product of attribute domains, the set of interpretations of a propositional language, etc.

Let L be a *possibility scale* or *plausibility scale*. Thus, it is a totally ordered set of plausibility levels with top denoted by 1 and bottom by 0.

Then, a *possibility distribution* is a mapping π, from S to a totally ordered scale L:

$$\pi : \quad S \to L$$

$$s \mapsto \pi(s)$$

such that $\exists s \in S, \ \pi(s) = 1$ (normality condition).

The *possibility scale L* can be the unit interval as suggested by Zadeh, or generally any finite chain, or even the set of non-negative integers.

In the finite case:

$$L = \{1 = \lambda_1 > \lambda_2 > \ldots > \lambda_n > \lambda_{n+1} = 0\} \tag{9.74}$$

Usually, it is assumed that L is provided with an order-reversing mapping ν:

$$\nu : \quad \lambda \mapsto 1 - \lambda$$

The function π represents the state of knowledge of an agent (about the actual state of affairs), also called an *epistemic state* distinguishing what is plausible from what is less plausible, what is the normal course of things from what is not, what is surprising from what is expected. It represents a flexible restriction on what is the actual state with the following conventions:

- $\pi(s) = 0$ means that state s is rejected as impossible
- $\pi(s) = 1$ means that state s is totally possible or plausible

The larger $\pi(s)$, the more possible, i.e. plausible the state s is. Formally, the mapping π is the membership function of a fuzzy set, where membership grades are interpreted in terms of plausibility.

The extension of the possibility concept to a set of states or events A, $A \subseteq S$, leads the *possibility theory* to the definition of two dual functions *possibility measure* (expressing *feasibility* or *plausibility*) and *necessity measure* (expressing *certainty*) of an event or a set A, defined as:

$$\text{Possibility measure} : \Pi(A) = \sup_{s \in A} \{\pi(s)\} \tag{9.75}$$

$$\text{Necessity measure} : N(A) = \inf_{s \notin A} \{1 - \pi(s)\} \tag{9.76}$$

$\Pi(A)$ evaluates to what extent A is consistent with π (possible), while $N(A)$ evaluates to what extent no element outside A is possible (to what extent A is certainly implied by π). The possibility–necessity duality is expressed by:

$$N(A) = 1 - \Pi(A^C) \tag{9.77}$$

where A^C is the complement set of A.

Basic axioms of possibility and necessity measures are:

- $\Pi(\varnothing) = 0$
- $\Pi(A \cup B) = \max \{\Pi(A), \Pi(B)\}$ (maxitivity axiom)
- $N(A \cap B) = \min \{N(A), N(B)\}$ (dual axiom)

Generally, as π is normalized to 1, then:

- $\Pi(S) = 1$
- $N(\varnothing) = 1 - \Pi(S) = 0$
- $N(S) = 1 - \Pi(\varnothing) = 1$
- $\max\{\Pi(A), \Pi(A^c)\} = 1$
- $\min \{N(A), N(A^c)\} = 0$

And hence, if $N(A) > 0 \Rightarrow \Pi(A) = 1$

Thus, the *possibility distribution* concept is a different but related concept to the *probability distribution*. Both are based on the set functions. However, *possibility theory* differs by using a pair of dual set functions, i.e. *possibility and necessity measures*, instead of only one *probability measure* in *probability theory*.

Furthermore, the *probability measure* in *probability theory* is additive regarding the union, i.e. $p(A \cup B) = p(A) + p(B)$, if $A \cap B = \varnothing$, in contrast to the maxitivity axiom in the *possibility theory* which is not. Finally, the *possibility theory* is appropriate for ordinal structures.

9.2.5.3 Representation of Membership Functions and Linguistic Variables

There are several ways to represent *membership functions* in the context of fuzzy sets. The most genuine membership functions are differentiable functions like a *Gaussian*[21] *function*, also called the "bell curve":

$$\mu_A(x) = a * e^{-\frac{(x-b)^2}{2c^2}} \qquad a,b,c \in \mathbb{R}, \quad c \neq 0 \tag{9.78}$$

where the parameter a controls the height of the peak's curve, the parameter b indicates the position of the centre of the peak, and the parameter c regulates the width of the "bell".

In Fig. 9.18 a *Gaussian membership function* is depicted for the *fuzzy set A*= "real numbers around 5".

However, for practical reasons related to computational aspects when some operations between two membership functions must be done, the *original membership functions* are commonly represented by a simplified version of them. The original functions are substituted by a *polygonal approximation of the original membership function*. These *polygonal membership functions* are differentiable by

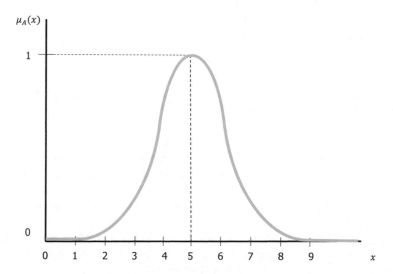

Fig. 9.18 A *Gaussian membership function* for the *fuzzy set A*= "real numbers around 5"

---

[21] A *Gaussian function* or "*bell curve*" *function* or *normal distribution function* in probability $(a = 1/(c\sqrt{2\pi}), b = \mu$ and $c = \sigma$, being μ the mean and σ the standard deviation of the distribution) was named to honour the great German mathematician and physicist Johan Carl Friedrich Gauss (1777–1855).

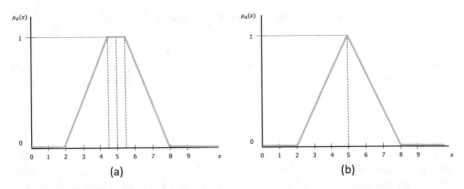

Fig. 9.19 A *trapezoidal membership function* (**a**) and a *triangular membership function* (**b**) for the *fuzzy set A*="real numbers around 5"

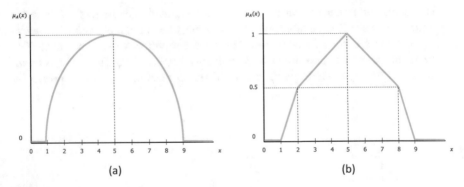

Fig. 9.20 A *parabolic membership function* (**a**) and its corresponding *approximated polygonal membership function* (**b**) for the *fuzzy set A*="real numbers around 5"

the straight segments which approximate the original curve. In the case of a *Gaussian function,* it is usually represented by the so-called *triangular membership functions* or *the trapezoidal membership functions*. In Fig. 9.19, a *trapezoidal membership function* and a *triangular membership function* approximating the original Gaussian membership function for the *fuzzy set A*= "real numbers around 5" are depicted.

Other used membership functions are *parabolic membership functions*, like the one depicted in Fig. 9.20 for the *fuzzy set A*= "real numbers around 5" and the corresponding *approximating polygonal membership function*.

Some other continuous functions but not differentiable like a composition of different functions are also used like the *composed membership function* and its corresponding *approximate polygonal membership function* in Fig. 9.21 for the *fuzzy set A*= "real numbers around 5".

Another important concept in *fuzzy set theory* because they are the basis for a fuzzy inference process or approximate reasoning are the *linguistic variables* (Zadeh, 1973a).

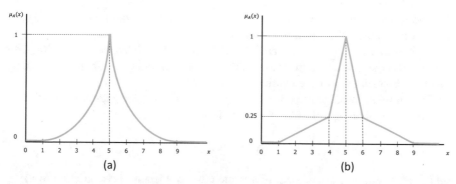

Fig. 9.21 A *composed membership function* (**a**) and its corresponding *approximated polygonal membership function* (**b**) for the *fuzzy set A*="real numbers around 5"

Let us first define a conventional (*non-fuzzy variable*) as defined in Zadeh (1973a), p. 14:

A *variable* is characterized by a triple $\langle X, U, R(X; u)\rangle$, in which X is the *name* of the variable; U is a universe of discourse (finite or infinite set); u is a generic name for the elements of U; and $R(X; u)$ is a subset of U which represents a *restriction* or *range* on the values of u imposed by X.

For convenience, $R(X; u)$ is abbreviated to $R(X)$ or $R(u)$ or $R(x)$, where x denotes a generic name for the values of X and will refer to $R(X)$ simply as the restriction on u or the restriction imposed by X.

In addition, a variable is associated with an *assignment equation*:

$$x = u : R(X) \tag{9.79}$$

or equivalently:

$$x = u, \quad u \in R(X)$$

which represents the assignment of a value u to x subject to the restriction $R(X)$. Thus, the assignment equation is satisfied $\Leftrightarrow u \in R(X)$.

For example, if we had a variable named *Age*. Then, U would be the set of integers, i.e. $U = \{0, 1, 2, 3, \ldots\}$, and $R(X)$ would be the range: 0, 1, 2, ..., 100. Thus $R(X) = \{0, 1, 2, \ldots, 100\}$.

Hence, a valid assignment equation would be expressed as:

$$age = u, \ u \in R(\text{Age}) = \{0, 1, 2, \ldots, 100\}. \text{ And for instance, } u = 56.$$

Now, a *fuzzy variable* can be defined (Zadeh, 1973a), p. 59:

A *fuzzy variable* is characterized by a triple $\langle X, U, R(X; u)\rangle$, in which X is the *name* of the variable; U is a universe of discourse (finite or infinite set); u is a generic name for the elements of U; and $R(X; u)$ is a *fuzzy set* over U which represents a *fuzzy restriction* on the values of u imposed by X.

As in the case of non-fuzzy variables, $R(X; u)$ is abbreviated to $R(X)$ or $R(u)$ or $R(x)$, where x denotes a generic name for the values of X and will refer to $R(X)$ simply as the restriction on u or the restriction imposed by X.

The non-restricted non-fuzzy variable u constitutes the *base variable* for X.

The *assignment equation* for X has the form:

$$x = u : R(X)$$

which represents the assignment of a value u to x subject to the restriction $R(X)$.

The degree to which this equation is satisfied will be referred to as the *compatibility of u with R(X)* and will be denoted by $c(u)$. By definition:

$$c(u) = \mu_{R(X)}(u), \qquad u \in U \tag{9.80}$$

where $\mu_{R(X)}(u)$ is the grade of membership of u to the restriction $R(X)$, i.e. to the *fuzzy set* representing the restriction.

Now is time to define the concept of *linguistic variable*. An important facet of the concept of a linguistic variable is that it is a variable of a *higher order* than a *fuzzy variable*, because a *linguistic variable* takes *fuzzy variables* as its values.

As defined by Zadeh in 1973a, p. 3: A *linguistic variable* is a variable whose values are words or sentences in a natural or artificial language. The motivation for the use of words or sentences rather than numbers is that linguistic characterizations are, in general, less specific than numerical ones, and hence, can model imprecision.

Taking the typical example from Zadeh (1973a), p. 3: speaking of age, when we say "John is young", we are less precise than when we say, "John is 25". In this sense, the label *young* may be regarded as a *linguistic value* of the variable *Age*, with the understanding that it plays the same role as the numerical value 25 but is less precise and hence less informative. The same is true of the linguistic values *very young, not young, extremely young, not very young*, etc., as contrasted with the numerical values 20, 21, 22, 23, etc.

The formal definition of a *linguistic variable* was given by Zadeh in 1973a, p. 75: a *linguistic variable* is characterized by the quintuple

$$\langle \chi, T(\chi), U, G, M \rangle \tag{9.81}$$

in which:

χ is the *name* of the *linguistic variable*;

$T(\chi)$ (or simply T) denotes the *term set* of χ, that is, the set of names of *linguistic values* of χ;

Each *linguistic value* is a *fuzzy variable* denoted generically by X, which ranges over a universe of discourse U which is associated with the *base variable u*;

G is a *syntactic rule* (which usually has the form of a grammar) for generating the name X of values of χ;

M is a *semantic rule* for associating with each X its meaning, $M(X)$, which is a *fuzzy set* over U.

A particular X that is a *linguistic value name* generated by G, is called a *term*.

A *term* consisting of a word or words which function as a unit (i.e. always occur together) is called an *atomic term*. A term containing one or more atomic terms is a *composite term*. A concatenation of components of a composite term is a *subterm*.

The meaning $M(X)$ of a term X is defined to be the restriction $R(X)$ on the base variable u which is imposed by the *fuzzy variable* named X. Thus

$$M(X) = R(X) \qquad\qquad (9.82)$$

with the understanding that $R(X)$, and hence $M(X)$, may be viewed as a *fuzzy set* of U carrying the name X.

Let us consider the same example than before and consider a *linguistic variable* named *Age*, and a possible *linguistic value* X ranging over $U = [0, 100]$.

A linguistic value of *Age* might be named *old*, with *old* being an *atomic term*. Another value might be named *very old*, in which case *very old* would be a *composite term* containing *old* as an *atomic component* and having *very* and *old* as *subterms*. The value of *Age* named *more or less young* would be a *composite term* containing *young* as an *atomic term* and in which *more or less* is a *subterm*.

The *term set* associated with *Age* could be expressed as:

$$T(\text{Age}) = \{\text{old, very old, not old, more or less young, quite young,}$$
$$\text{young, not very old and not very young,} \ldots\}$$

in which each *term* is the name of a *fuzzy variable* in the universe of discourse $U = [0, 100]$. The base variable u is the *age* in years of life. The restriction imposed by a *term*, like for instance $R(\text{old})$, constitutes the meaning of old:

$$R(\text{old}) = M(\text{old})$$

And thus, it is the fuzzy set:

$$M(\text{old}) = \{(u, \mu_{\text{old}}(u)) \mid u \in [0, 100]\}$$

where,

$$\mu_{old}(u) = \begin{cases} 0 & u \in [0, 50] \\ \dfrac{1}{1 + \left(\frac{5}{u-50}\right)^2} & u \in (50, 100] \end{cases}$$

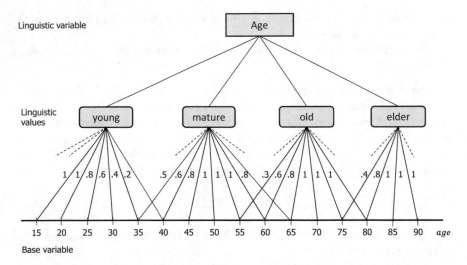

Fig. 9.22 Hierarchical relationship between the *linguistic variable* Age, its possible *linguistic values X*, i.e. the *term set T*(Age) = {young, mature, old, elder}, and the numeric *base variable age*. (Adapted from Zadeh 1973a)

Figure 9.22 shows the connection between a *linguistic variable* χ, the possible *linguistic values X*, and the *base variable u* for the case of the *linguistic variable Age*.

Thus, *linguistic variables* provide with a nice artefact to model the imprecision of several magnitudes or variables, that can be represented with different *linguistic values* or *terms*, which in turn are *fuzzy sets*. For instance, the concept of temperature can be expressed as:

- Temp: linguistic variable
- T(Temp) = {cold, nice, hot}, the *term set*
- *temp*: *base variable* over the universe of real values, $U = \mathbb{R}$, representing the outside temperature in °C

This way, the different *linguistic values* or *terms*, which are really *fuzzy variables*, can be graphically represented by their membership functions like depicted in Fig. 9.23.

The corresponding membership functions for each *linguistic value* or *label* are the following ones:

$$\mu_{\text{cold}}(u) = \begin{cases} 1 & u \in (-\infty, 5] \\ \dfrac{15 - u}{10} & u \in [5, 15] \\ 0 & u \in [15, +\infty) \end{cases}$$

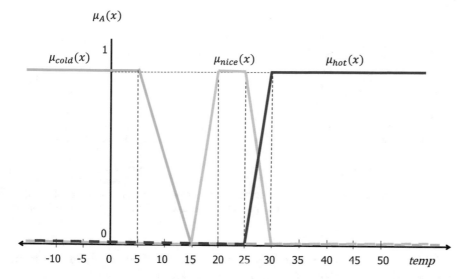

Fig. 9.23 Graphical representation of the *membership functions* of the *terms* or *fuzzy variables* (*cold, nice,* and *hot*) of the *linguistic variable Temp*

$$
\mu_{\text{nice}}(u) = \begin{cases}
0 & u \in (-\infty, 15] \\
\dfrac{u}{5} - 3 & u \in [15, 20] \\
1 & u \in [20, 25] \\
6 - \dfrac{u}{5} & u \in [25, 30] \\
0 & u \in [30, +\infty)
\end{cases}
$$

$$
\mu_{\text{hot}}(u) = \begin{cases}
0 & u \in (-\infty, 25] \\
\dfrac{u}{5} - 5 & u \in [25, 30] \\
1 & u \in [30, +\infty)
\end{cases}
$$

A common and practical way of representing a *trapezoidal membership function* characterizing a *fuzzy set* or *fuzzy variable* is the following:

$$\text{Term} : \langle a, b, c, d \rangle$$

where, a, b, c, d are the x-axis coordinates of the vertices of the *trapeze*, $\{(a, 0), (b, 1), (c, 1), (d, 0)\}$, which represents the *membership function* of the corresponding *linguistic value* or *term*.

Note that the corresponding y-axis coordinates are always the same, and hence are not needed: (0, 1, 1, 0). Figure 9.24a depicts this situation.

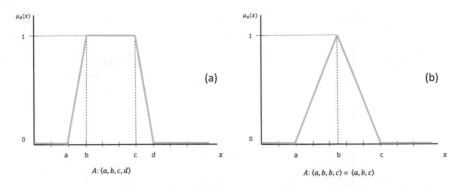

Fig. 9.24 Graphical representation of a general *trapezoid membership function* (**a**) and of a general *triangular membership function* (**b**) of a given *term* for a given linguistic variable through its vertices

In the specific case of *triangular membership functions*, a common way of representing a *fuzzy set* or *fuzzy variable* is the following:

$$\text{Term} : \langle a, b, b, c \rangle \equiv \langle a, b, c \rangle$$

where, *a, b, c*, are the *x*-axis coordinates of the vertices of the *triangle*, $\{(a, 0), (b, 1), (c, 0)\}$, which represents the *membership function* of the corresponding *linguistic value* or *term*.

Note that the second and third coordinates of the fuzzy set characterization are the same, indicating that the *trapeze* has an empty higher basis, i.e. it is a *triangle*. Hence, sometimes a reduced characterization of the *fuzzy set* uses only three coordinates instead of the four-coordinate scheme.

In addition, the corresponding *y*-axis coordinates are always the same, and hence are not needed: (0, 1, 1, 0). Sometimes Fig. 9.24b depicts this situation.

9.2.5.4   Fuzzy Logic and Fuzzy Connectives

Classic set theory is inherently related to the first-order logic.[22] As the fuzzy sets are some kinds of extension of the classic sets, there is a question in the air. Is there any somehow equivalent extension of the first-order logic matching the *fuzzy sets*?

In other words, can exist some "*Fuzzy logic*" related to the *fuzzy sets* as the classic sets are related to the first-order logic?

---

[22] *First-order logic* or *predicate logic*, is a formal system used in several sciences, which generalizes *propositional logic* allowing the parametrization of logical propositions with variables, the use of quantification operators (i.e., $\forall$, $\exists$) over these variables and the use of relations among variables (i.e., functions).

Consider that any logic theory as the formal basis for any kind of reasoning is composed by three main elements: the basic objects (the *truth-values*), the *logic connectives* or operators to generate more complex objects, and the *reasoning mechanism* to derive by inference new objects, which could be based on *tautologies*[23] or *syllogisms*.[24]

For instance, in classic *Boolean logic* (two-valued logic), the *truth-values* are true (1) and false (0). The operators are the *classic logic connectives*: negation ($\neg$), conjunction ($\wedge$), disjunction ($\vee$), and implication ($\rightarrow$). Finally, the *reasoning mechanism* is based on the tautologies like the *modus ponens, modus tollens, syllogisms*, etc.

As described by Zadeh in 1973a, pp. 97–99: the degree of truth of a proposition or statement can be characterized by natural language expressions such as *very true, quite true, more or less true, essentially true, false, quite false, completely false*, etc. This fact suggests that these values can be represented with the values of a linguistic variable, and hence, to treat *truth* as a linguistic variable (*Truth*). This way, the basic values of *false* and *true* are two of the *atomic terms* in its *term set* of *truth-values*. Therefore, it can be represented as:

- *Truth*: linguistic variable
- *T(Truth)* = {*completely true, true, quite true, essentially true, ..., more or less true, ..., not true, essentially false, quite false, false, completely false*}: The *term set* of *linguistic truth-values*
- *truth-values*: numeric *truth-values* over the universe of [0, 1].

As an example of the definition of some *linguistic truth-values*, Zadeh suggested the following membership functions for the *fuzzy set* representing the terms *true* and *false*:

$$\mu_{\text{true}}(v) = \begin{cases} 0 & v \in [0, a] \\ 2 \cdot \left(\frac{v-a}{1-a}\right)^2 & v \in \left[a, \frac{a+1}{2}\right] \\ 1 - 2 \cdot \left(\frac{v-1}{1-a}\right)^2 & v \in \left[\frac{a+1}{2}, 1\right] \end{cases} \tag{9.83}$$

where,

$$v = \frac{a+1}{2}$$

---

[23] A *tautology* is a logical formula that is true in every possible interpretation of its propositional variables.

[24] A *syllogism* is a specific logical argument that through the application of deductive reasoning arrives at a conclusion based on two or more predicates that are assumed to be true.

is the *crossover point* of the membership function (i.e. the point v which $\mu_{\text{true}}(v) = 0.5$) and $a \in [0, 1]$ is a parameter indicating the subjective judgement about the minimum value of v in order to consider a statement as *true*.

And,

$$\mu_{\text{false}}(v) = \mu_{\text{true}}(1 - v) \qquad 0 \le v \le 1 \tag{9.84}$$

Treating truth as a linguistic variable leads to a *fuzzy linguistic logic*, or simply *fuzzy logic*. *Fuzzy logic* may offer a more realistic framework for human reasoning than the traditional two-valued logic by providing a basis for *approximate reasoning*, which is a mode of reasoning in which the *truth-values* and the rules of inference are fuzzier than precise (Zadeh, 1973a).

Therefore, to construct a basis for *fuzzy logic*, in addition to the *linguistic truth-values*, it is necessary to extend the meaning of such logical operations as negation, disjunction, conjunction, and implication to *fuzzy logic connectives* operating with *linguistic truth-values*, which are *fuzzy sets*. Therefore, there is the need to define how to compute the *complementary or negation* of a *fuzzy set*, the *union*, the *intersection,* and the *implication* of *fuzzy sets*.

Zadeh (1973a) proposed the application of the *extension principle* to define operators for the *fuzzy connectives*. These early operators were *max* for the intersection, the operator *min* for the *union*, and the operator $1 - x$ for the *complementary* or *negation*.

The *extension principle* (Zadeh, 1965, 1973a; Dubois & Prade, 1980) can be formulated as follows:

Given,

      a Cartesian product of universes, $X = X_1 \times X_2 \times \ldots \times X_r$
   $A_1, A_2, \ldots, A_r$, which are r *fuzzy sets* in $X_1, X_2, \ldots, X_r$ respectively
   f, which is a *mapping* from X to a universe Y: $y = f(x_1, x_2, \ldots, x_r)$
   The *extension principle* allows to define a *fuzzy set* B in Y as:

$$B = \{(y, \mu_B(y)) \mid y = f(x_1, x_2, \ldots, x_r), (x_1, x_2, \ldots, x_r) \in X\} \tag{9.85}$$

where,

$$\mu_B(y) = \begin{cases} \sup_{(x_1, \ldots, x_r) \in f^{-1}(y)} \{\min\{\mu_{A_1}(x_1), \ldots, \mu_{A_r}(x_r)\}\} & \text{if } f^{-1}(y) \ne \emptyset \\ 0 & \text{otherwise} \end{cases}$$

$$\tag{9.86}$$

and f^{-1} is the inverse function of f.

In the special case when $r = 1$, the *extension principle* is simplified to:

$$B = f(A) = \{(y, \mu_B(y)) \mid y = f(x), x \in X\} \tag{9.87}$$

where,

$$\mu_B(y) = \begin{cases} \sup_{x \in f^{-1}(y)}\{\mu_A(x)\} & \text{if } f^{-1}(y) \neq \emptyset \\ 0 & \text{otherwise} \end{cases} \tag{9.88}$$

However, in the literature, there have been many operators suggested for the interpretation of these operations (Menger, 1942; Schweizer & Sklar, 1961; Zadeh, 1965). They differ in the *degree of generality* or *adaptability* (some of the are parametrized and others are more specific) and the *degree of justification*, ranging from intuitively justifiable to some that are axiomatically justifiable. Usually, taking into account the domain of application can help to decide using one operator or another one due to its different appropriateness to handle the real system being modelled.

Main operators for modelling the *intersection* of *fuzzy sets* are the so-called *triangular norms* or *T-norms*. Similarly, the usual operators for modelling the *union* of *fuzzy sets* are the *triangular conorms* or *T-conorms* or also named as *S-norms* (Kruse et al., 1994; Klement et al., 2000).

Fuzzy logic connectives, i.e. operations with *fuzzy sets*, are defined as continuous functions in the interval [0, 1], which generalize the classic logic connectives. Several proposals for these operators have been done in the literature such as the suggestions of Zadeh (1965), Giles (1976), Hamacher (1978), Dubois and Prade (1980), Yager (1980), Mizumoto and Tanaka (1981), Mizumoto (1981), Dubois and Prade (1988).

In Table 9.3, there is the correspondence among set operations, Boolean logical connectives, and *fuzzy logic connectives*.

Let us consider the different possible operators to model the different *fuzzy logic connectives*. Each operator needs to satisfy several properties, and there are different examples of operators proposed in the literature.

The *fuzzy negation connective* or *complementary fuzzy set* is modelled through the *negation functions*:

$$N : [0, 1] \rightarrow [0, 1]$$

$$x \mapsto N(x)$$

- Properties:

  - $N(0) = 1$ and $N(1) = 0$ *context conditions*
  - if $p \leq q$ then $N(p) \geq N(q)$ *non-increasing*

Table 9.3 Correspondence among set operations, Boolean logic connectives, and *fuzzy logic connectives*

	Set operations	Boolean logic connectives	Fuzzy logic connectives
Complementary	$\bar{P}$ or P^C	$\neg P$	Negation(P)
Intersection	$P \cap Q$	$P \wedge Q$	$T -$ norm (P, Q)
Union	$P \cup Q$	$P \vee Q$	$T -$ conorm(P, Q) or $S -$ norm (P, Q)
Inclusion	$P \subset Q$	$P \rightarrow Q$	Implication(P, Q)

All functions satisfying the above properties are *fuzzy negation functions*. However, if some additional properties are satisfied by these functions, then they are classified into *strict negation functions* and *strong negation functions*.

- Additional properties for *strict negation functions*:

  - if $p < q$ then $N(p) > N(q)$ *strictly non-increasing*
  - N is continuous   *continuity*

- Additional property for *strong negation functions*:

  - $N(N(p)) = p$ *involution*

- Example of a *strict negation function*:

  - $N(x) = 1 - x^2$

- Examples of *strong negation functions*:

  - $N(x) = 1 - x$ *standard negation*
  - $N(x) = \frac{1}{2}(1 + \cos(\pi x))$ *cosine negation*
  - $N_w(x) = (1 - x^w)^{1/w} \ \forall \ w > 0$ *Yager Family functions*
  - $N_t(x) = (1 - x)/(1 + t \cdot x) \ \forall \ t > -1$ *Sugeno Family functions*

As the *involution* property is corresponding to its Boolean logic counterpart ($\neg(\neg p) = p$), usually *strong negation functions* are used to implement the *fuzzy negation* operation with *fuzzy sets*.

The *fuzzy conjunction connective* or *fuzzy set intersection* is modelled through the *functions* called *T-norms*:

$$T : [0, 1] \times [0, 1] \rightarrow [0, 1]$$

$$x, y \ \mapsto T(x, y)$$

- Properties:

  - $T(1, p) = T(p, 1) = p$ *identity law*
  - $T(p, q) = T(q, p)$ *commutativity*
  - $T(p, T(q, r)) = T(T(p, q), r)$ *associativity*
  - if $q \leq r$ then $T(p, q) \leq T(p, r)$ *monotonicity*

- Derived Properties:
  From the basic properties, some new properties can be derived like the following ones:

  - $T(0, p) = T(p, 0) = 0$ *absorbent element*
  - $T(p, q) \leq \min \{p, q\}$ *upper bound*

- Examples:

  - $T(x, y) = \min \{x, y\}$ *minimum*
  - $T(x, y) = x * y$ *algebraic product*

- $T(x, y) = \max \{0, x + y - 1\}$ *bounded difference (Lukasiewicz t-norm)*
- $T(x, y) = x * y/(\alpha + (1 - \alpha) * (x + y - x * y)), \alpha \geq 0$
                                          Hamacher product family
- $T(x, y) = x * y/(x + y - x * y)$ *Hamacher product* $(\alpha = 0)$

The *fuzzy disjunction connective* or *fuzzy set union* is modelled through the following functions called *T-conorms* or *S-norms*:

$$S : [0, 1] \times [0, 1] \rightarrow [0, 1]$$

$$x, y \mapsto S(x, y)$$

- Properties:

    - $S(0, p) = S(p, 0) = p$ *identity law*
    - $S(p, q) = S(q, p)$ *commutativity*
    - $S(p, S(q, r)) = S(S(p, q), r)$ *associativity*
    - if $q \leq r$ then $S(p, q) \leq S(p, r)$ *monotonicity*

- Derived Properties:
    From the basic properties, some new properties can be derived like the following ones:

    - $S(1, p) = S(p, 1) = 1$ *absorbent element*
    - $\max \{p, q\} \leq S(p, q)$ *lower bound*

- Examples:

    - $S(x, y) = \max \{x, y\}$ *maximum*
    - $S(x, y) = x + y - x * y$ *algebraic sum*
    - $S(x, y) = \min \{x + y, 1\}$ *bounded sum/Lukasiewicz t-conorm*
    - $S(x, y) = (x + y + (\beta - 1) * x * y)/(1 + \beta * x * y), \beta \geq -1$
                                          Hamacher sum family
    - $S(x, y) = (x + y - 2x * y)/(1 - x * y)$ *Hamacher sum* $(\beta = -1)$

T-conorms are dual to *T-norms* in a logic sense, according to the De Morgan laws, under a suitable *strong negation function N*:

$$N(T(x, y)) = S(N(x), N(y)) \tag{9.89}$$

$$N(S(x, y)) = T(N(x), N(y)) \tag{9.90}$$

Common pairs of dual *T-norms* and *T-conorms* are the following ones:

- $T(x, y) = \min \{x, y\}$ *minimum*
  $S(x, y) = \max \{x, y\}$ *maximum*
- $T(x, y) = x * y$ *algebraic product*
  $S(x, y) = x + y - x * y$ *algebraic sum*

- $T(x, y) = \max \{0, x + y - 1\}$ *bounded difference*
  $S(x, y) = \min \{x + y, 1\}$ *bounded product (Lukasiewicz t-norm and t-conorm)*
- $T(x, y) = x * y/(x + y - x * y)$ *Hamacher product*
  $S(x, y) = (x + y - 2x * y)/(1 - x * y)$ *Hamacher sum*

Now that formal definitions of the *fuzzy connectives* are modelled through the *strong negation functions, t-norms* and *t-conorms*, let us consider how these *fuzzy logic connectives* are computed, i.e. how the final *fuzzy set* resulting from the operation is computed.

Fuzzy Connectives Operating on the Same Universe

Remember that previously it was stated that a *proposition* or a *fact* of the form: "*X* is *F*", where *X* is the name of an object, a variable, or a proposition, $A(X)$ is an implied attribute of the variable *X*, and *F* is the name of a *fuzzy set* of *U*, then the proposition "*X* is *F*" can be expressed as the following fuzzy restriction:

$$R(A(X)) = F$$

The following notation for representing the proposition "*X* is *L*", where *X* is a *linguistic variable* and *L* is a *linguistic label term*, *i.e.* a *fuzzy set*, will be used:

$$[X \text{ is } L] \equiv {}^{"}X \text{ is } L^{"}$$

Let us consider the following situation:

If $F = [X \text{ is } A]$ and $G = [X \text{ is } B]$ are two propositions bind to the corresponding fuzzy sets *A* and *B* with the corresponding possibility distributions of variable *X*, $\Pi_A(u) = \mu_A(u)$, and $\Pi_B(u) = \mu_B(u)$ defined over the same universe *U*, then the *fuzzy connective computation* is done as follows:

$$F \wedge G \equiv [X \text{ is } A \wedge B] \text{ with } \Pi_{A \cap B}(u) = T(\Pi_A(u), \Pi_B(u)) \tag{9.91}$$

where $u \in U$ and $T(\Pi_A(u), \Pi_B(u))$ is a suitable *T-norm*.

$$F \vee G \equiv [X \text{ is } A \vee B] \text{ with } \Pi_{A \cup B}(u) = S(\Pi_A(u), \Pi_B(u)) \tag{9.92}$$

where $S(\Pi_A(u), \Pi_B(u))$ is a suitable T-*conorm*.

$$\neg F \equiv [X \text{ is} \neg A] \text{ with } \Pi_{\bar{A}}(u) = N(\Pi_A(u)) \tag{9.93}$$

where $N(\Pi_A(u))$ is a suitable *strong negation function*.

To illustrate this computation, which is especially nice in a graphical way, the following *functions* will be considered to implement the *fuzzy connectives*:

- $N(x) = 1 - x$
- $T(x, y) = \min\{x, y\}$
- $S(x, y) = \max\{x, y\}$

As an example, the following *fuzzy connectives* can be computed:

$$[\text{Temp is nice}] \wedge [\text{Temp is hot}] \equiv [\text{Temp is nice} \wedge \text{hot}]$$

$$[\text{Temp is nice}] \vee [\text{Temp is hot}] \equiv [\text{Temp is nice} \vee \text{hot}]$$

$$\neg[\text{Temp is nice}] \equiv [\text{Temp is}\neg\text{nice}]$$

with the corresponding *membership functions* $\mu_{\text{nice}}(\text{temp})$ and $\mu_{\text{hot}}(\text{temp})$ defined above in Sect. 9.2.5.3.

Figure 9.25 illustrates the computation of the *fuzzy intersection of the fuzzy sets nice and hot*. Figure 9.26 depicts the *fuzzy union of the fuzzy sets nice and hot*. Finally, the *fuzzy negation* of the fuzzy set nice is shown in Fig. 9.27. These *fuzzy sets* (nice, hot) are *linguistic term values* of the *linguistic variable* Temp, which expresses the outside temperature in °C.

Fuzzy Connectives Operating on Different Universes

If $F = [X \text{ is } A]$ and $G = [Y \text{ is } B]$ are two propositions bind to the corresponding fuzzy sets A and B with the corresponding possibility distributions $\Pi_A(u) = \mu_A(u)$ defined

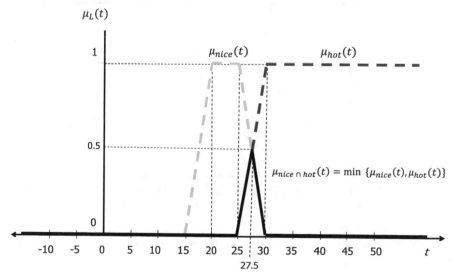

Fig. 9.25 Graphical representation of the *membership function* of the *fuzzy intersection of nice* and *hot fuzzy sets*

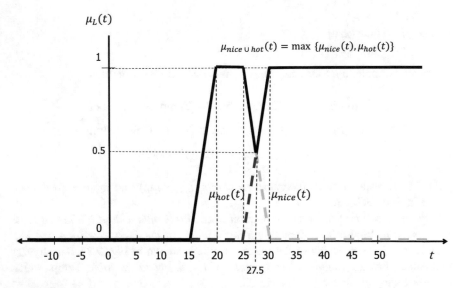

Fig. 9.26 Graphical representation of the *membership function* of the *fuzzy union* of *nice* and *hot* *fuzzy sets*

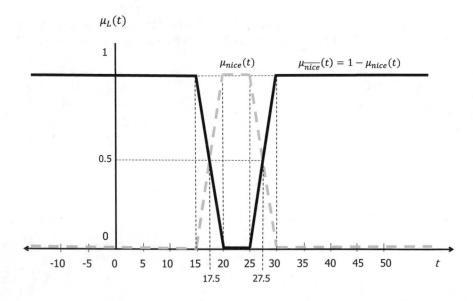

Fig. 9.27 Graphical representation of the *membership function* of the *fuzzy negation* of *nice fuzzy set*

over a universe U and $\Pi_B(v) = \mu_B(v)$ defined over a universe V, where $U \neq V$, then the *fuzzy connective computation* in two different universes is done as follows:

$$F \wedge G \equiv [X \text{ is } A] \wedge [Y \text{ is } B] \quad \text{with} \quad \Pi_{A \cap B}(u, v) = T(\Pi_A(u), \Pi_B(v)) \qquad (9.94)$$

where $u \in U$, and $v \in V$ and $T(\Pi_A(u), \Pi_B(v))$ is a suitable *T-norm*.

$$F \vee G \equiv [X \text{ is } A] \vee [Y \text{ is } B] \quad \text{with} \quad \Pi_{A \cup B}(u, v) = S(\Pi_A(u), \Pi_B(v)) \qquad (9.95)$$

where $S(\Pi_A(u), \Pi_B(u))$ is a suitable *T-conorm*.

To illustrate this computation in a graphical way, the following *functions* implementing the *fuzzy connectives* will be considered:

- $T(x, y) = \min\{x, y\}$
- $S(x, y) = \max\{x, y\}$

As an example, let us compute the following *fuzzy connectives*:

$$[\text{Price is expensive}] \wedge [\text{Speed is low}]$$

$$[\text{Price is expensive}] \vee [\text{Speed is low}]$$

with the corresponding membership functions $\mu_{\text{expensive}}(\text{price})$ and $\mu_{\text{low}}(\text{speed})$ defined as follows:

$$\mu_{\text{expensive}}(p) = \begin{cases} 0 & p \in [0, 25] \\ \dfrac{p - 25}{15} & p \in [25, 40] \\ 1 & p \in [40, +\infty) \end{cases}$$

$$\mu_{\text{low}}(s) = \begin{cases} 1 & s \in [0, 60] \\ \dfrac{120 - s}{60} & s \in [60, 120] \\ 0 & s \in [120, +\infty) \end{cases}$$

Figure 9.28 illustrates the computation of the *fuzzy intersection* of the *fuzzy sets expensive* and low. The *fuzzy set expensive* is a *linguistic term value* of the *linguistic variable Price*, which expresses the price of a car in thousands of euros (€) and the *fuzzy set low* is a *linguistic term value* of the *linguistic variable Speed* which expresses the speed of a car expressed in km/h.

Figure 9.29 depicts the *fuzzy union* of the *fuzzy sets expensive* and *low*.

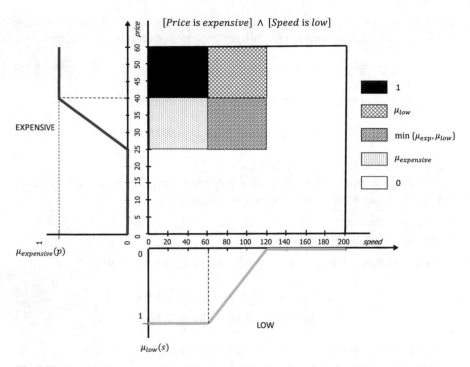

Fig. 9.28 Graphical representation of the *membership function* of the *fuzzy intersection of expensive* and *low fuzzy sets*

9.2.5.5   Approximate Reasoning and Fuzzy Inference in a Rule-Based System

As defined by Zadeh, *the approximate reasoning* is the reasoning mechanism of *fuzzy logic*. *Approximate reasoning* must hold *linguistic truth-values* rather than numerical truth-values.

To undertake this reasoning, once the *linguistic truth-values* have been considered and the *basic fuzzy connective* operators have been defined, the *fuzzy implication connective* must be extended from the classic *implication* operator.

The implication operator is the basis of the most used crisp reasoning mechanisms: *modus ponens*, *modus tollens*, and *hypothetical syllogism*. The *modus ponens*[25] is the basis for *forward reasoning* in rule-based reasoning systems:[26]

---

[25] *Modus Ponendo Ponens* or abbreviated *Modus Ponens* is the name of the Latin expression for a deductive argument form and inferential reasoning mechanism in propositional logic meaning: "mode that by affirming, affirms".

[26] The symbol ⊢ in classic crisp logic means "then it is deduced or implied".

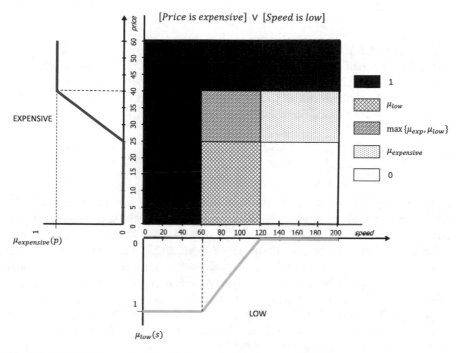

Fig. 9.29 Graphical representation of the *membership function* of the *fuzzy union of expensive* and *low fuzzy sets*

$$A \rightarrow B, A \vdash B$$

which means that if it is known that if A is true then B is also true, and in addition, that A is true, it can be concluded that B is true. The *modus tollens*,[27] which can be considered as the basis for the *backward reasoning* in rule-based reasoning systems:

$$A \rightarrow B, \ \neg B \vdash \neg A$$

which means that if it is known that if A is true then B is also true, and in addition, that B is false, it can be concluded that A is *false*. Finally, the *hypothetical syllogism*:[28]

---

[27] *Modus Tollendo Tollens* or abbreviated *Modus Tollens* is the name of the Latin expression for a deductive argument form and inferential reasoning mechanism in propositional logic meaning: "mode that by denying, denies".

[28] *Hypothetical syllogism* is the name of a valid inferential reasoning mechanism or rule of inference also called the chain argument or chain rule. It states the principle of transitivity of implication.

$$A \rightarrow B, B \rightarrow C \vdash A \rightarrow C$$

which means that if it is known that if A is true then B is also true, and in addition, that if B is true then C is also true, it can be concluded that if A is true then C is also true.

The *fuzzy implication connective* or *fuzzy set implication* is described through the following functions called *fuzzy implication functions* or *implication functions*:

$$I : [0, 1] \times [0, 1] \rightarrow [0, 1]$$

$$x, y \mapsto I(x, y)$$

Commonly, it is assumed that any fuzzy concept must generalize the corresponding crisp concept. Therefore, depending on the interpretation of the crisp *logical implication*, this leads to different families of *implication functions* (Mas et al., 2007). Notwithstanding, all of them satisfy that true $\rightarrow q \equiv q$ and false $\rightarrow q \equiv$ true and some increasing/decreasing relationship related to both arguments of the function. Most of the *fuzzy implication functions* in the literature belong to some of the families below described, even though there are some implication functions that do not belong to any family. On the contrary, the *Lukasiewicz implication* belongs to the four families. A *fuzzy implication function* should satisfy the following three properties:

- Properties:

    - if $p \leq q$ then $I(p, r) \geq I(q, r)$ $\forall r \in [0, 1]$ *non-increasing in the first variable*
    - if $p \leq q$ then $I(r, p) \leq I(r, q) \forall r \in [0, 1]$ *non-decreasing in the second variable*
    - $I(0, 0) = I(1, 1) = 1$ and $I(1, 0) = 0$

- Observation:

    - From the above properties it follows that $I(p, 1) = 1$ and $I(0, q) = 1$

- Families:

    - *S-implications*: they derive from the interpretation of the implication as in Boolean logic, where: $p \rightarrow q \equiv \neg p \vee q$.
      This means that:

$$I(x, y) = S(N(x), y) \tag{9.96}$$

where $S(x, y)$ is a *T-conorm* function, and $N(x)$ is a *strong negation* function. Due to this fact, they are also called *Strong implications*.

Examples using as a *strong negation function* $N(x) = 1 - x$, and as *T-conorms* or *S-norms*, $S(x, y) = \max \{x, y\}$ (*Kleene-Dienes*), $S(x, y) = x + y - x * y$ (*Reichenbach*), $S(x, y) = \min \{x + y, 1\}$ (*Lukasiewicz*):

$$I(x, y) = \max\{1 - x, y\} \qquad \textit{Kleene-Dienes}$$
$$I(x, y) = 1 - x + x * y \qquad \textit{Reichenbach}$$
$$I(x, y) = \min\{1 - x + y, 1\} \quad \textit{Lukasiewicz}$$

- *R-implications*: any implication satisfying the following lead to an *R-implication*:

$$I(x, y) = \sup\{z \in [0, 1] \mid T(x, z) \le y\}, \quad x, y \in [0, 1] \qquad (9.97)$$

R-implications derive from residuated lattices based on the residuation property (hence, the *R* in the name) that in the case of T-norms can be written as:

$$T(x, y) \le z \Leftrightarrow I(x, z) \ge y, \quad \forall x, y, z \in [0, 1]$$

Examples using as *T-norms min* $\{x, y\}$ *(Göedel), product T-norm (Goguen),* and *Lukasiewicz T-norm (Lukasiewicz)*:

$$I(x, y) = \begin{cases} 1 & \text{if } x \le y \\ y & \text{if } x > y \end{cases} \qquad \textit{Göedel}$$

$$I(x, y) = \begin{cases} 1 & \text{if } x \le y \\ \frac{y}{x} & \text{if } x > y \end{cases} \qquad \textit{Goguen}$$

$$I(x, y) = \min\{1 - x + y, 1\} \qquad \text{Lukasiewicz}$$

- *QL-implications*: They derive from the interpretation of the implication in *Quantum Logic*, where: $p \to q \equiv \neg p \vee (p \wedge q)$.
  This means that:

$$I(x, y) = S(N(x), T(x, y)), \quad x, y \in [0, 1] \qquad (9.98)$$

where $S(x, y)$ is a *T-conorm* function, $T(x, y)$ is a *T-norm* function, and $N(x)$ is a *strong negation* function.

Examples using as *T-norms* and *T-conorms* the algebraic product and the algebraic sum, *Lukasiewicz T-norm* and *Lukasiewicz T-conorm (Kleene-Dienes)*, all with the usual *negation function* leads to the following:

$$I(x, y) = 1 - x + x^2 * y \qquad \textit{algebraic}$$
$$I(x, y) = \max\{1 - x, y\} \qquad \textit{Kleene-Dienes}$$
$$I(x, y) = 1 - x + x * y \qquad \textit{Reichenbach}$$
$$I(x, y) = \min\{1 - x + y, 1\} \qquad \textit{Lukasiewicz}$$

- *D-implications*: They derive from the interpretation of the implication as Dishkant arrow in orthomodular lattices: $p \to q \equiv q \vee (\neg p \wedge \neg q)$. They are the contraposition ($I(x, y) = I(N(y), N(x))$) with respect to *N* of *QL-implications*.
  This means that:

$$I(x, y) = S(T(N(x), N(y)), y), \quad x, y \in [0, 1] \qquad (9.99)$$

where $S(x, y)$ is a *T-conorm* function, $T(x, y)$ is a *T-norm* function, and $N(x)$ is a *strong negation* function.

Examples:

$$I(x, y) = \ \max\ \{1 - x, y\} \qquad\qquad\qquad Kleene\text{-}Dienes$$
$$I(x, y) = 1 - x + x * y \qquad\qquad\qquad\qquad Reichenbach$$
$$I(x, y) = \ \min\ \{1 - x + y, 1\} \qquad\qquad\quad Lukasiewicz$$

- *Non-family Implications*: There are some fuzzy implication operators which satisfy the three properties but do not belong to any of the families like:

$$I(x, y) = \begin{cases} 1 & \text{if } x = y = 0 \\ y^x & \text{otherwise} \end{cases} \qquad Yager$$

$$I(x, y) = \begin{cases} 1 & \text{if } x \leq y \\ 0 & \text{if } x > y \end{cases} \qquad Gaines\text{-}Rescher$$

- *Pseudo-implication* Operators: There are some operators proposed in the literature that even though do not satisfy the necessary above three properties for being a *fuzzy implication function* have been successfully used as *fuzzy implications* especially in the *fuzzy control* field. Some of them are the following:

$$I(x, y) = \ \max\ \{1 - x, \min\ \{x, y\}\} \quad early\ Zadeh$$
$$I(x, y) = \ \min\ (x, y) \qquad\qquad\qquad\quad Mamdani$$
$$I(x, y) = x * y \qquad\qquad\qquad\qquad\quad\ Larsen$$

In addition to the basic three properties listed above, there are other formal properties that *fuzzy implication operators* should satisfy to match the usual formal logic properties of classic implication like:

- $I(1, x) = x \ \forall\ x \in [0, 1]$ *neutrality of true*
- $I(x, I(y, z)) = I(y, I(x, z)) \ \forall\ x, y, z \in [0, 1]$ *exchange principle*
- $I(x, y) = I(N(y), N(x)) \ \forall\ x, y \in [0, 1]$ being N a *strong negation* funct. contraposition regarding N
- $I(x, y) = 1 \Leftrightarrow x \leq y$ *ordering property* or *boundary condition*
- I is a continuous function   *continuity*

All the different *fuzzy implication* functions satisfy several of those properties. This variety of *fuzzy implication* operators available, means that depending on the context and the application domain, some of them can be more appropriate than others.

Fuzzy Inference

In addition to the definition of the *fuzzy implication* operator, an additional extension must be done for the *approximate* or *fuzzy inference reasoning*. The most used inference mechanism is the *modus ponens*. Therefore, the classical *modus ponens* reasoning mechanism must be generalized and extended. The classic *modus ponens* can be schematized as follows:

if *A* is *true*, then *B* is *true*
A is *true*

B is *true*

The generalization or relaxation of the classic *logic propositions* for what we will name as *fuzzy propositions* or *fuzzy facts*, such as:

$$[X \text{ is } A]$$

$$[Y \text{ is } B]$$

Will lead to a basic *fuzzy version of modus ponens*:

if $[X \text{ is } A]$, then $[Y \text{ is } B]$
$[X \text{ is } A]$

$[Y \text{ is } B]$

Thus, in fuzzy logic, the *fuzzy modus ponens* after interpreting the truth-values of the *fuzzy propositions* becomes semantically interpreted as: the truth-value of the conjunction of *A* and *A* → *B* must be less or equal than the truth-value of *B*. Thus, getting the same interpretation than in the crisp modus ponens: *B* is at least as true as *A* and *A* → *B* are.

This can be expressed in the following way:

$$M(x, I(x, y)) \leq y \quad \forall x, y \in [0, 1] \tag{9.100}$$

Where *M* is named, in general, as a *Modus Ponens Generating Function*, which in addition must verify at least:

- $M(1, 1) = 1$ *preservation of the crisp modus ponens*

And usually, these other properties should be also verified:

- $M(0, I(x, y)) = I(x, y)$ *from false all can be deduced*
- If $x \leq x'$ then $M(x, y) \leq M(x', y)$ *stability in chained reasoning*

When an implication function $I(x, y)$ satisfies the above condition $M(x, I(x, y)) \leq y \ \forall x, y \in [0, 1]$, then it is called an *MP-implication* for the *M Modus Ponens generation function*, and then both can be used to implement the *fuzzy modus ponens*.

A particular case of the *Modus Ponens generation functions* are the *t-norms*. This case has been thoroughly studied in the literature (Trillas et al., 2004), showing interesting results regarding which implication functions are *MP-implications* given a concrete *t-norm* function. Taking into account particular *M* functions as *t-norm* functions, the *inequality functional condition* for implementing a *fuzzy modus ponens* is:

$$T(x, I(x, y)) \leq y \quad \forall x, y \in [0, 1] \tag{9.101}$$

where $T(x, y)$ is a *t-norm* performing the conjunction and $I(x, y)$ is an *implication function* performing the conditional. If the implication function $I(x, y)$ satisfies the above inequality regarding a given t-*norm* $T(x, y)$, it is said that $I(x, y)$ is a *T-conditional*, i.e., a particular naming for an *MP-implication* bind to a *t-norm*.

In Trillas et al. (2004), it was analyzed which implications $I(x, y)$ and *t-norms* $T(x, y)$ satisfy the *T-conditionality*. Several classes of implications were studied.

One of the known results is that if the implication function $I_{T_1}(x, y)$ is an *R-implication* formed from a left continuous *t-norm* T_1, i.e.:

$$I_{T_1}(x, y) = \sup \{z \in [0, 1] \mid T_1(x, z) \leq y\} \tag{9.102}$$

is always T_1-conditional

Particularly, if $T_1(x, y) = \min (x, y)$, the corresponding $I_{T_1}(x, y)$ is *T-conditional* for any *t-norm* T.

And a more general result states that for a given continuous *t-norm* T, an *implication function* $I(x, y)$ is an *MP-implication* or *T-conditional* if and only if $I(x, y) \leq I_T(x, y)$.

The *Mamdani-Larsen's operators*, which are *pseudo-implication functions* as described before, are those represented by a function:

$$I_{ML}(x, y) = T_1(\varphi_1(x), \varphi_2(y)) \tag{9.103}$$

where T_1 is a continuous *t-norm*, φ_1 is an order automorphism and, $\varphi_2 : [0, 1] \to [0, 1]$ is a non-null contractive mapping, i.e. it verifies $\varphi_2(x) \leq x$ for any $x \in [0, 1]$ and $\varphi_2(x) > 0$ for some $x \neq 0$.

- *Mamdani Implication:*[29] when[30] $\varphi_1 = \varphi_2 = Id$ and $T_1(x, y) = \min \{x, y\}$, then $I_{ML}(x, y) = T_1(\varphi_1(x), \varphi_2(y)) = \min \{x, y\}$. Thus:

$$I_M(x, y) = \min \{x, y\} \tag{9.104}$$

---

[29] As previously told, both *Mamdani operator* and *Larsen operator* are pseudo-implication functions, but they will be referred as they were actual implication functions.

[30] Id function is the identity function, i.e., a function mapping the same value to each argument, $\varphi(x) = x$.

- *Larsen Implication*: when $\varphi_1 = \varphi_2 =$ Id and $T_1(x, y) = x * y$, then $I_{ML}(x, y) = T_1(\varphi_1(x), \varphi_2(y)) = x * y$. Thus:

$$I_L(x, y) = x * y \qquad (9.105)$$

A general result states that given two continuous *t-norms* T and T_1, φ_1 an order automorphism and, $\varphi_2 : [0, 1] \to [0, 1]$ a non-null contractive mapping, a *Mamdani-Larsen operator* defined as $I_{ML}(x, y) = T_1(\varphi_1(x), \varphi_2(y))$, is an *MP-implication* or *T-conditional* for any *t-norm* T.

Therefore, *Mamdani-Larsen's operators* are *MP-implications* or *T-conditionals* for any *t-norm* T, and, in particular, for $T(x, y) = \min \{x, y\}$.

Usually, *fuzzy implications functions* satisfy the axiom $I(0, y) = 1$, which means that from a *false* antecedent a *false* or *true* consequent can be deduced, which is not always a good and desired fact. On the contrary, *Mamdani-Larsen's operators* satisfy $I_{ML}(0, y) = 0$, which is usually more convenient to model real-life situations in applications. This fact jointly with their simplicity of computation and that are always *MP-implications* or *T-conditionals* for any *t-norm* T are the reasons why they have been commonly used in *fuzzy reasoning* for control systems.

With this extension of *modus ponens* to a *fuzzy modus ponens*, with a convenient *t-norm* T and a suitable *implication function* I being a *MP-implication* or *T-conditional*, a fuzzy reasoning procedure can be applied.

Therefore, a *rule-based reasoning system* can be created, and provided that all the facts appearing in the knowledge base of the system would be *fuzzy facts* or *fuzzy propositions*, the system can reach *fuzzy conclusions*, which is a very nice accomplishment.

The *imprecision* of a given domain or problem can be captured this way, and incorporated in an IDSS.

Summarising, if all the inference rules of our rule-based reasoning systems are like the below rule:

$$R : \text{if } [X \text{ is } A] \text{ then } [Y \text{ is } B]$$

where,

X is an input linguistic variable,
A is an attribute, i.e. a *linguistic value or term for the variable X*, which is a *fuzzy variable* modelled with a *fuzzy set* with a *membership function* $\mu_A(x)$,
Y is the output linguistic variable,
B is an attribute, i.e. a *linguistic value or term for the variable Y*, which is a *fuzzy variable* modelled with a *fuzzy set* with a membership function $\mu_B(y)$

And the conditional statement expressed in the rule R is modelled through an implication function $I(\mu_A(x), \mu_B(y))$. To implement the inferential process of *modus ponens*, each one of the rules R of the rule-based reasoning system, must satisfy the deduction rule of *modus ponens*, i.e.:

$$\text{if } [X \text{ is } A] \text{ then } [Y \text{ is } B]$$
$$[X \text{ is } A]$$

$$\overline{}$$

$$[Y \text{ is } B]$$

Which as previously explained will be satisfied when a corresponding continuous *t-norm T* is selected according to an implication function *I* which is a *MP-implication* or *T-conditional* regarding *T*, such that the *modus ponens* inequality is satisfied:

$$T(\mu_A(x), I(\mu_A(x), \mu_B(y))) \le \mu_B(y) \qquad \forall x \in X, \forall y \in Y \tag{9.106}$$

The truth-value of the consequent *Y* is *B* will be evaluated by the truth-value of "*X* is *A* and *X* is $A \to Y$ is *B*".

This process can be repeated for each rule, in a *forward strategy* until the inference engine will stop when no more rule applications can be done or the problem under consideration would be solved.

However, many times this basic *fuzzy relaxation* of *modus ponens* will not be enough to capture the inherent *approximate reasoning* of humans. Usually, humans would try to apply a basic *fuzzy rule*, but not using exactly the same identical *fuzzy propositions* or *fuzzy facts*, but using a slightly different *fuzzy propositions*.

Thus, the *fuzzy modus ponens* must be generalized. Thus, the idea is to extend it to a *generalized modus ponens* to solve the following general scheme:

$$\text{if } [X \text{ is } A] \text{ then } [Y \text{ is } B]$$
$$[X \text{ is } A']$$

$$\overline{}$$

$$[Y \text{ is } B']$$

Where A' is similar to A, i.e. A' it is an approximation in some sense to A. Then, human reasoning would expect that from the conditional if X is A, then Y is B, even the premise A' is not exactly the same than the antecedent of the conditional A but similar, could obtain some approximate inference over B, like B', which would be an approximation to B.

For instance, let us imagine the following inference rule regarding a strawberry S and its degrees of ripeness:

$$\text{if } [S \text{ is } Red] \text{ then } [S \text{ is } Ripe]$$
$$[S \text{ is } very \ Red]$$

$$\overline{}$$

$$[S \text{ is } very \ Ripe]$$

Where *very Red* is a different attribute (linguistic term) of the linguistic variable S, quite similar to *Red*, but not exactly the same. Furthermore, *very Ripe* is a similar attribute (linguistic term) to *Ripe* for the linguistic variable S. Even though humans would expect that this reasoning can be done, the *fuzzy modus ponens* by itself does not allow to obtain the desired conclusions from not exactly premises. To be used, some knowledge or procedure should be done to modify or transform the premises (from *very Red* to *Red*) and their consequences (from *Ripe* to *very Ripe*).

Therefore, a more powerful mechanism is needed. This mechanism is the *compositional rule of inference* proposed by Zadeh (1973a). Before that, some related concepts proposed must be introduced like a *fuzzy relation* and the *composition of relations*.

Let us consider n universes of discourse: $U_1, \ldots, U_n$

Let $U : U_1 \times \ldots \times U_n$ be the Cartesian product of $U_1, \ldots, U_n$

Then, an *n-ary fuzzy relation*, $R(u_1, \ldots, u_n)$ on U is a *fuzzy subset* of U, which may be expressed as the union of its constituent fuzzy singletons $\mu_R(u_1, \ldots, u_n)/(u_1, \ldots, u_n)$ as follows:

$$R(u_1, \ldots, u_n) = \int_{U_1 \times \ldots \times U_n} \mu_R(u_1, \ldots, u_n)/(u_1, \ldots, u_n) \qquad (9.107)$$

For the case where $U_1, \ldots, U_n$ are infinite, or for the case where $U_1, \ldots, U_n$ are discrete:

$$R(u_1, \ldots, u_n) = \{((u_1, \ldots, u_n), (\mu_R(u_1, \ldots, u_n))) \mid (u_1, \ldots, u_n) \in U_1 \times \ldots \times U_n \} \qquad (9.108)$$

where $\mu_R(u_1, \ldots, u_n)$ is the membership function of $R(u_1, \ldots, u_n)$.

As a concrete case, a *binary fuzzy relation* or simply a *fuzzy relation* is the particularization for only two universes X, Y.

Given both universes X, Y and its Cartesian product: $X \times Y$, a *fuzzy relation* $R(x, y)$ on $X \times Y$ is described as the following *fuzzy subset* of $X \times Y$:

$$R(x, y) = \int_{X \times Y} \mu_R(x, y)/(x, y) \qquad (9.109)$$

or

$$R(x, y) = \{((x, y), \mu_R(x, y)) \mid (x, y) \in X \times Y \} \qquad (9.110)$$

depending on the nature of the universes, and being $\mu_R(x, y)$ the membership function of $R(x, y)$. Note that universes could be both *crisp sets* or *fuzzy sets*. Former generates *fuzzy relations on sets*, and the latter ones produce *fuzzy relations on fuzzy sets*.

Some usual examples of *binary fuzzy relations* could be "*much greater than*", "*less than*", "*resembles*", "*is relevant to*", "*is close to*", "*approximately equal to*", etc.

In the case that universes are discrete, sometimes the *fuzzy relation* is expressed through a *relation matrix*. For instance, let us suppose that $X = \{1,2,3\}$ and $Y = \{1,2,3,4\}$, then, the relation "X is very close to Y" can be described as:

$$R(x,y) = \begin{pmatrix} 1 & 0.8 & 0.6 & 0.2 \\ 0.8 & 1 & 0.8 & 0.6 \\ 0.6 & 0.8 & 1 & 0.8 \end{pmatrix}$$

Fuzzy relations in different Cartesian products can be combined with others by the *composition of relations* operation. In the literature several versions of the *composition* operation have been defined (Zadeh, 1965, 1971; Kaufmann, 1975; Rosenfeld, 1975; Zadeh et al., 1975). They differ in their results and in the mathematical properties. Among all of them, the *max-min composition* is the most-frequently used, but others like *max-product* or *max-average* have been defined.

Given a *fuzzy relation* $R_1(x,y) = \{((x,y), \mu_{R_1}(x,y)) | (x,y) \in X \times Y\}$ defined on $X \times Y$, and another *fuzzy relation* $R_2(y,z) = \{((y,z), \mu_{R_2}(y,z)) | (y,z) \in Y \times Z\}$ defined on $Y \times Z$.

The (max–min) *composition* of two fuzzy relations R_1 and R_2, denoted by $R_1 \circ R_2$, is another *fuzzy relation* on $X \times Z$, i.e. a *fuzzy set*, defined as follows:

$$R_1 \circ R_2 = \{((x,z), \mu_{R_1 \circ R_2}(x,z)) \mid x \in X, z \in Z\} \tag{9.111}$$

where,

$$\mu_{R_1 \circ R_2}(x,z) = \max_{y \in Y} \left\{ \min_{x,y,z} \{\mu_{R_1}(x,y), \mu_{R_2}(y,z)\} \right\} \tag{9.112}$$

is the *membership function* of the *composition of the two relations* $R_1 \circ R_2$. The *maximum* can be generalized by the *supremum* for the infinite case.

The interpretation of this membership function is straightforward. The degree of membership of the relationship $R_1 \circ R_2(x,z)$ will be the *conjunction* of the degree of membership of the intermediate relationships $R_1(x,y)$ and $R_2(y,z)$. Thus, a *t-norm* could be used. In the original formulation of Zadeh, the *min* conjunction operator was proposed. However, as the relationship $R_1 \circ R_2(x,z)$ could be composed by alternative intermediate relationships connecting (x,z) through different y values, the *aggregation (disjunction)* of all these different degrees of memberships obtained by different intermediate paths through different y values must be done. Thus, a *t-conorm* could be used. In the original formulation of Zadeh, the max (x,y) operator was proposed.

Notwithstanding, other *t-norms, t-conorms* and *aggregation functions* have been proposed in the literature as explained above.

The *compositional rule of inference*, as defined by Zadeh (1973a) in p. 148 and in Zadeh (1975) in p. 59 states:

Let U and V be two universes of discourse with base variables u and v respectively.

Let $R(u)$, $R(u, v)$ and $R(v)$ denote restrictions on u, (u, v) and v respectively, with the understanding that $R(u)$, $R(u, v)$, and $R(v)$ are fuzzy relations in U, $U \times V$, and V

Let A and F denote particular *fuzzy subsets* of U, $U \times V$

Then, the *compositional rule of inference* asserts that the solution of the *relational assignment equations*:

$$R(u) = A \qquad (9.113)$$

and

$$R(u, v) = F \qquad (9.114)$$

is given by

$$R(v) = A \circ F \qquad (9.115)$$

where $A \circ F$ is the composition of A and F.

In this sense, it can be inferred that $R(v) = A \circ F$ from $R(u) = A$ and $R(u, v) = F$

As we will illustrate with an example, the *compositional rule of inference* provides a solution to the approximate inference reasoning using *modus ponens* when the fuzzy propositions are not exactly the same than is in the conditional of a fuzzy rule in a knowledge base.

As an example, let us consider the following example extracted from Zadeh (1975):

Let us assume that we have,

$$U = V = \{1, 2, 3, 4\} \text{ as the universes,}$$

let be A the corresponding *linguistic value* "*small*", which is a *fuzzy set*:

$$A = \{(1, 1), (2, 0.6), (3, 0.2), (4, 0)\}$$

and let be F the corresponding *linguistic value* "*approximately equal*", which is a fuzzy set:

$$F = \{((1, 1), 1), ((1, 2), 0.5), ((2, 1), 0.5), ((2, 2), 1), ((2, 3), 0.5),$$
$$((3, 2), 0.5), ((3, 3), 1), ((3, 4), 0.5), ((4, 3), 0.5), ((4, 4), 1)\}$$

Therefore,

A is a unary fuzzy relation in U named "*small*"
F is a binary fuzzy relation in $U \times V$ named "*approximate equal*"

The relational assignment equations in this case are:

$$R(u) = \text{small}$$

$$R(u, v) = \text{approximate equal}$$

and hence applying the compositional rule of inference, we have:

$$R(v) = \text{small} \circ \text{approximate equal}$$

To compute $R(v)$ we must compute the composition of the two fuzzy relations. It can be done using the *max–min composition*. For an easier computation, the two fuzzy relations can be expressed in *relation matrix* form. This way, we have:

$$R(v) = \text{small} \circ \text{approximate equal}$$

$$= (\,1 \quad 0.6 \quad 0.2 \quad 0\,) \circ \begin{pmatrix} 1 & 0.5 & 0 & 0 \\ 0.5 & 1 & 0.5 & 0 \\ 0 & 0.5 & 1 & 0.5 \\ 0 & 0 & 0.5 & 1 \end{pmatrix}$$

$$= (\,1 \quad 0.6 \quad 0.5 \quad 0.2\,)$$

Thus,

$$R(v) = (\,1 \quad 0.6 \quad 0.5 \quad 0.2\,)$$

This *fuzzy set*, which is a restriction on the values on V could be approximated by a corresponding *linguistic term*, that in this case could be *"more or less small"*.

Zadeh proposed the mechanism of *linguistic approximation* to the solution of the simultaneous equations, converting a *fuzzy set* described by its membership degree values to some *linguistic term*. This *linguistic term* would be the most similar *linguistic term* of the corresponding *linguistic variable* fitting the membership distribution values. Those *truth values* for the different *linguistic terms* can be computed and stored as *truth values tables* with its corresponding labels like *"more or less small"*, *"more or less large"*, *"large"*, etc.

Therefore,

$$R(v) = (\,1 \quad 0.6 \quad 0.5 \quad 0.2\,) \text{ computed exactly}$$

and,

$$R(v) = \text{more or less small as a linguistic approximation}$$

The *relational assignment equations* solved corresponds to the following inference reasoning process:

> *u* is *small*
>
> *u* and *v* are *approximate equal*
> _____
>
> *v* is *more or less small*

In fact, Zadeh (1973a) proposed the idea that the *generalized fuzzy modus ponens*
can be viewed as a special case of the *compositional rule of inference*. The compo-
sitional rule of inference is a general mechanism wrapping the *generalized fuzzy
modus ponens* because the implication or conditional is a *fuzzy binary relation.*
Thus:

Considering that $A_1 \subset U$, $A_2 \subset U$, and $B, B' \subset V$ are fuzzy subsets of the
universes U and V. Let us consider the following generalized fuzzy modus ponens:

> if $[X$ is $A_2]$ then $[Y$ is $B]$
> $[X$ is $A_1]$
> _____
>
> $[Y$ is $B']$

A_1 is a restriction on the values of U, B, and B' are restrictions on the values of V,
and that the relation $A_2 \rightarrow B$ is a restriction on the values of $U \times V$, we have:

$$R(u) = A_1 \tag{9.116}$$

$$R(u, v) = A_2 \rightarrow B \tag{9.117}$$

Applying the *compositional rule of inference*, these relational assignment equations can be solved for the restriction on v, as follows:

$$R(v) = A_1 \circ (A_2 \rightarrow B) \tag{9.118}$$

And it can be solved using the *max–min composition* to compute the *membership
function* of B' when a corresponding continuous *t-norm* T is selected according to an
implication function I which is a *MP-implication* or *T-conditional* regarding T:

$$\mu_{B'}(y) = \sup_{x \in X} \left\{ T\left(\mu_{A_1}(x), I\left(\mu_{A_2}(x), \mu_B(y)\right)\right) \right\} \quad \forall y \in Y \tag{9.119}$$

This way, from a set of *fuzzy inference rules* several fuzzy propositions can be
deduced in *forward reasoning* at each application of the rules.

Symmetrically, *modus tollens* can be used to make deductions based on the
backward reasoning in a fuzzy rule-based system.

$$\text{if } [X \text{ is } A] \text{ then } [Y \text{ is } B]$$
$$[Y \text{ is } \neg B]$$

$$[X \text{ is } \neg A]$$

Which will be satisfied when a corresponding continuous *t-norm T* and a strong negation function *N* are selected according to an implication function *I* which is named as *MT-implication* or *T-conditional* regarding *T* and *N*, such that the *modus tollens* inequality is satisfied:

$$T(N(y), I(x, y)) \leq N(x) \qquad \forall x \in X, \forall y \in Y \qquad (9.120)$$

Considering that the *modus tollens* reasoning is expressed as follows:

$$A \rightarrow B, \neg B \vdash \neg A$$

And this reasoning is equivalent to:

$$\neg B \rightarrow \neg A, \neg B \vdash \neg A$$

because $A \rightarrow B \equiv \neg B \rightarrow \neg A$ and hence, it can be implemented using *modus ponens*, using the same application of the *composition rule of inference* as follows:

$$\mu_{\neg A'}(x) = \sup_{y \in Y} \{T(\mu_{\neg B'}(y), I(\mu_{\neg B}(y), \mu_{\neg A}(x)))\} \qquad \forall x \in X \qquad (9.121)$$

Fuzzy Inference with Precise Data: Fuzzy Control

In the previous subsection the fuzzy reasoning mechanism, i.e., *approximate reasoning*, based on *fuzzy logic* and its *fuzzy connectives* have been analyzed. It has been explained how to make the inferences from *fuzzy inference rules* where the propositions or facts are *fuzzy sets*, and both the input information provided to the inference process and the output information obtained from the inference process are *fuzzy propositions* or *fuzzy sets*.

However, in many real-world situations both the input information and the output information must be *precise* o *crisp data* to be used to solve a concrete problem like the automatic control of a car, the regulation of a temperature through a thermostat, etc., from a knowledge base of *fuzzy rules*. In those kinds of systems, a *fuzzy set* as the output of the reasoning mechanism is not useful, because it cannot be directly applied to any real computer system which needs digital numeric values. In these situations, the output values must be *precise* or *crisp values* that could be sent to the regulation and controlling mechanism of some systems.

These real-world situations where a fuzzy inference process can be performed to support some decisions which are usually made in a rather automatic way fall in the realm of *control systems*, even though any real-world application of *fuzzy rules* with precise or crisp input data needing precise or crisp output data can be managed in the same way. Commonly, these systems aim to control complex processes, which usually are controlled through the use of human experience and expertise in a given domain.

The application of *fuzzy logic* to *control systems* created a new application field named as *fuzzy logic control* or simply *fuzzy control*. The foundations of fuzzy control were given by Zadeh in 1973b. The first application of *fuzzy set theory* to the control of systems was described in Mamdani and Assilian (1975) regarding the control of a laboratory model steam engine. And the first industrial application of fuzzy control was the control of a cement kiln in Denmark (Holmblad & Ostergaard, 1982).

Fuzzy controllers have been used with success in nonlinear control problems. In fact, in Castro (1995) it was proved that fuzzy controllers are *universal approximators*. This means that any bounded continuous function may be approximated by a fuzzy controller with as much accuracy as desired.

Control systems aim to lead and maintain a process to a desired state/s. To get that, they continually compare some output measures of the evaluated state of the process against the desired state/s to be reached, i.e. set-points, and then, they adjust the input values to the process. In the *fuzzy control* field, most common approaches proposed in the literature are:

- The Mamdani fuzzy control model (Mamdani & Assilian, 1975)
- The Sugeno/Takagi–Sugeno fuzzy control model (Sugeno, 1985; Sugeno & Nishida, 1985; Takagi & Sugeno, 1985)

The Mamdani Fuzzy Control Method

The *Mamdani fuzzy control* method is based on the use of *fuzzy inference rules*, which had *fuzzy facts*, i.e. *linguistic variables*, which can be combined with the usual *fuzzy connectives* (disjunction, conjunction, and negation) in the antecedent of the rule. The consequent of a *fuzzy rule* is also a *fuzzy fact*, i.e. a *linguistic variable* representing the output variable. The base variables of these linguistic variables will be the measured values from the process or the output values of the controller.

For each control application, the number of rules needed to control a system which should be written depends on the number of linguistic variables and the different terms of each linguistic variable.

The rules according to the Mamdani fuzzy control model can be formally described as follows:

R_1 : if $\left[X_1 \text{ is } A_1^1\right]$ and $\left[X_2 \text{ is } A_2^1\right]$ and ... and $\left[X_n \text{ is } A_n^1\right]$ then $\left[Y \text{ is } B^1\right]$

R_2 : if $\left[X_1 \text{ is } A_1^2\right]$ and $\left[X_2 \text{ is } A_2^2\right]$ and ... and $\left[X_n \text{ is } A_n^2\right]$ then $\left[Y \text{ is } B^2\right]$

...

R_m : if $\left[X_1 \text{ is } A_1^m\right]$ and $\left[X_2 \text{ is } A_2^m\right]$ and ... and $\left[X_n \text{ is } A_n^m\right]$ then $\left[Y \text{ is } B^m\right]$

where,

$X_1, \ldots, X_n$ are the input variables

$A_1^i, \ldots, A_n^i$ are the linguistic term values of the input variables $(X_1, \ldots, X_n)$ in the rule R_i

Y is the output variable

B^i is the linguistic term value of the output variable Y in the rule R_i

The whole process of a Mamdani fuzzy control method follows the next steps starting from the crisp values $(x_1, \ldots, x_n)$ of the input variables $(X_1, \ldots, X_n)$ for the set of rules $\{R_1, \ldots, R_m\}$ related to the same output variable Y:

- Evaluation of the antecedent for each rule R_i $i = 1, \ldots, m$: for each crisp value (x_j) of the corresponding input variable (X_j) and its corresponding linguistic term value (i.e. the fuzzy set A_j^i) from the antecedent of the rule, the degree of membership of x_j to A_j^i is computed $(\mu_{A_j^i}(x_j))$, and the obtained values are combined according to the interpretation of the *fuzzy logic connectives* appearing (conjunction, disjunction, negation). As the rules usually have the *conjunction connective*, it is implemented through the classic *t-norm,* the min operator:

$$\mu_{\text{Ant}(R_i)}(x_1, \ldots, x_n) = \min\left(\mu_{A_1^i}(x_1), \ldots, \mu_{A_n^i}(x_n)\right) \qquad (9.122)$$

- Evaluation of the consequent of each rule R_i $i = 1, \ldots, m$: the consequent of each rule R_i is evaluated through the application of the *fuzzy modus ponens* using as an *implication function* the Mamdani implication operator, $I_M(x, y) = \min(x, y)$, as an *MP-implication* and as a *t-norm* for implementing the *modus ponens generation function*, the $T(x, y) = \min(x, y)$ too. Thus:

$$\mu_{\text{Con}(R_i)}(y) = T\left(\mu_{\text{Ant}(R_i)}(x_1 \ldots, x_n), I_M\left(\mu_{\text{Ant}(R_i)}(x_1 \ldots, x_n), \mu_{B^i}(y)\right)\right) \qquad (9.123)$$

$$\mu_{\text{Con}(R_i)}(y) = \min\left(\mu_{\text{Ant}(R_i)}(x_1 \ldots, x_n), \min\left(\mu_{\text{Ant}(R_i)}(x_1 \ldots, x_n), \mu_{B^i}(y)\right)\right) \qquad (9.124)$$

and finally:

$$\mu_{\text{Con}(R_i)}(y) = \min\left(\mu_{\text{Ant}(R_i)}(x_1 \ldots, x_n), \mu_{B^i}(y)\right) \qquad (9.125)$$

This computation has a clear interpretation. If the degree of membership of the antecedent is completely true, $\mu_{\text{Ant}(R_i)}(x_1 \ldots, x_n) = 1$, then the membership function of the consequent will be the same appearing in the rule: $\mu_{B^i}(y)$. On the contrary, when the degree of membership of the antecedent is not completely true, $\mu_{\text{Ant}(R_i)}(x_1 \ldots, x_n) < 1$, the membership function of the consequent will be less than the membership function on the rule. The membership function of the consequent will be clipped to the value of the degree of membership of the antecedent. In the case that the degree of membership of the antecedent is completely false, $\mu_{\text{Ant}(R_i)}(x_1 \ldots, x_n) = 0$, the membership function of the consequent will be $\mu_{\text{Con}(R_i)}(y) = 0$.

- Combination of the consequents of each rule R_i $i = 1, \ldots, m$: the final membership function of the consequent of the set of all rules will be computed as the disjunctive aggregation (union) of the different membership functions of the consequents of all the rules. The disjunction will be implemented through a *t-conorm*, usually the max operator. Therefore:

$$\mu_{\text{Con}}(y) = \max\left(\mu_{\text{Con}(R_1)}(y), \ldots, \mu_{\text{Con}(R_m)}(y)\right) \tag{9.126}$$

This combination of all the rules is the result of imaging as we had a fictitious rule R, formed by several related rules with the same output variable, $R = \{R_1, R_2, \ldots, R_m\}$. The application or obtaining of a conclusion from the rule R is the result of obtaining a conclusion from the rule R_1 or from the rule R_2 or $\ldots$ or from the rule R_m. Thus, to obtain the union of the conclusions a *t-conorm* should be used, and commonly the *max* function is used.

- *Defuzzification* of the Final Consequent: The final membership function of the consequent of the set of all rules is a *fuzzy set*. As the output variable require a precise crisp value, the *fuzzy set* must be transformed to a crisp value. This transformation process of a fuzzy set to a precise value is named as a *defuzzification* process.

There are several *defuzzification* strategies. The crisp value to be obtained usually it is an element belonging to the supports of the different fuzzy sets aggregated into the final consequent. Major strategies can be divided into two groups:

- *Extreme Value Strategies*: these *defuzzification* strategies use extreme values of the membership functions, usually the maxima, to define the crisp precise value. If you remind the definition of the core of a fuzzy normalized set as:

$$\text{Core}(A) = \{x \in U \mid \mu_A(x) = 1\}$$

which for a non-normalized fuzzy set is:

$$\text{Core}(A) = \{x \in U \mid \nexists y \in U \text{ such that } \mu_A(y) > \mu_A(x)\} \tag{9.127}$$

Major used strategies are:

Left of maximum (LoM): the left value of the *Core* of the final consequent of the rules is used:

$$y_{\text{LoM}} = \min \ \{y \mid y \in \text{Core}(\text{Con})\} \tag{9.128}$$

Right of maximum (RoM): the right value of the *Core* of the final consequent of the rules is used:

$$y_{\text{RoM}} = \max \ \{y \mid y \in \text{Core}(\text{Con})\} \tag{9.129}$$

Centre of maximum (CoM): the centre value of the *Core* of the final consequent of the rules is used:

$$y_{\text{CoM}} = \frac{y_{\text{RoM}} - y_{\text{LoM}}}{2} \tag{9.130}$$

- *Area Strategies*: These *defuzzification* strategies use the idea of computing the centroids of some surfaces or areas. Most commonly used are:

Centre of Area (CoA): This method selects the value corresponding to the centre of the area with membership values greater than zero. Thus, it is the support value that divides the area below a continuous membership function into two equal parts. Therefore:

$$\int_{y_{\min}}^{y_{\text{CoA}}} \mu_{\text{Con}}(y) \ \text{d}y = \int_{y_{\text{CoA}}}^{y_{\max}} \mu_{\text{Con}}(y) \ \text{d}y \tag{9.131}$$

Centre of Gravity (CoG): This method computes the value corresponding to the centre of gravity or centre of mass of the area formed by the continuous membership function. It can be obtained as:

$$y_{\text{CoG}} = \frac{\int_y y * \mu_{\text{Con}}(y) \ \text{d}y}{\int_y \mu_{\text{Con}}(y) \ \text{d}y} \tag{9.132}$$

Notwithstanding, the selection of the best *defuzzification* strategy must be done depending on the consequent membership function. Several aspects must be taken into account such as whether it is *unimodal*, i.e. it has only one maximum value, or not, or there not exist a core of the *fuzzy set* but separate different maxima values, or whether the fuzzy set is compact or it is presenting some intervals of the output variable which do not belong to the support of the *fuzzy set*, and no value should be selected from this zone.

This process can be summarized in the following algorithm:

---

algorithm Mamdani fuzzy control method

input $x_1, \dots, x_n$: the set of input values of the input variables $(X_1, \dots, X_n)$
             to the fuzzy rules
          $R_1, \dots, R_m$: the set of fuzzy rules
          $A_1^i, \dots, A_n^i$: the linguistic term values of the input variables for each rule R_i
          B^i: is the linguistic term value of the output variable Y for each rule R_i

output y: the output value of the output variable Y

begin
 for each i **in** 1, ..., m **do**
   // Rule antecedent evaluation
   $\mu_{Ant(R_i)}(x_1, \dots, x_n) \leftarrow min(\mu_{A_1^i}(x_1), \dots, \mu_{A_n^i}(x_n))$
   // Rule consequent evaluation
   $\mu_{Con(R_i)}(y) \leftarrow min(\mu_{Ant(R_i)}(x_1 \dots, x_n), \mu_{B^i}(y))$
 endforeach
   // Rules' consequent combination
   $\mu_{Con}(y) \leftarrow max(\mu_{Con(R_1)}(y), \dots, \mu_{Con(R_m)}(y))$
   // Deffuzzification of the final fuzzy set using the CoG strategy, for instance
   $y_{CoG} \leftarrow \dfrac{\int_y y * \mu_{Con}(y)\, dy}{\int_y \mu_{Con}(y)\, dy}$

 $y_M \leftarrow y_{CoG}$
 return y_M
end

The Sugeno/Takagi–Sugeno Fuzzy Control Method

The *Mamdani fuzzy control* method can be modified in several ways. One of the most commonly used in the literature and in the *fuzzy control* field was proposed by Sugeno (1985) and Sugeno and Nishida (1985). It uses *fuzzy inference rules*, which had *fuzzy facts*, i.e. *linguistic variables*, which can be combined with the usual *fuzzy connectives* (disjunction, conjunction, and negation) in the antecedent of the rule, such as in the Mamdani controller. However, these rules have crisp consequences, which are either constant values or functions of the input variables. Usually, the consequence function, which depends on the input variables, is linear, but any other function can be used.

The rules according to the Sugeno/Takagi–Sugeno fuzzy control model can be formally described as follows:

R_1 :   if $\left[X_1 \text{ is } A_1^1\right]$ and $\left[X_2 \text{ is } A_2^1\right]$ and ... and $\left[X_n \text{ is } A_n^1\right]$ then Y is $f_{R_1}(X_1, \ldots, X_n)$

R_2 :   if $\left[X_1 \text{ is } A_1^2\right]$ and $\left[X_2 \text{ is } A_2^2\right]$ and ... and $\left[X_n \text{ is } A_n^2\right]$ then Y is $f_{R_2}(X_1, \ldots, X_n)$

...

R_m :   if $\left[X_1 \text{ is } A_1^m\right]$ and $\left[X_2 \text{ is } A_2^m\right]$ and ... and $\left[X_n \text{ is } A_n^m\right]$ then Y is $f_{R_m}(X_1, \ldots, X_n)$

where,

$X_1, \ldots, X_n$ are the input variables
$A_1^i, \ldots, A_n^i$ are the linguistic term values of the input variables $(X_1, \ldots, X_n)$ in the rule
  R_i
Y is the output variable
$f_{R_i}(X_1, \ldots, X_n)$ is the consequence function to compute the value for the output
  variable Y in the rule R_i

The whole process of a Sugeno/Takagi–Sugeno fuzzy control method follows the next steps starting from the crisp values $(x_1, \ldots, x_n)$ of the input variables $(X_1, \ldots, X_n)$ for the set of rules $\{R_1, \ldots, R_m\}$ related to the same output variable Y:

- Evaluation of the antecedent for each rule R_i $i = 1, \ldots, m$: for each crisp value (x_j) of the corresponding input variable (X_j) and its corresponding linguistic term value (i.e. the *fuzzy set* A_j^i) from the antecedent of the rule, the degree of membership of x_j to A_j^i is computed $(\mu_{A_j^i}(x_j))$, and the obtained values are combined according to the interpretation of the *fuzzy logic connectives* appearing (conjunction, disjunction, negation). As the rules usually have the *conjunction connective*, it is implemented through the classic *t-norm*, the *min* operator:

$$\mu_{\text{Ant}(R_i)}(x_1, \ldots, x_n) = \min\left(\mu_{A_1^i}(x_1), \ldots, \mu_{A_n^i}(x_n)\right) \qquad (9.133)$$

- Computation of the output value for the output variable Y: all output values from all rules R_i, which are the output values of the corresponding functions $f_{R_i}(X_1, \ldots, X_n)$ are aggregated in an average weighted by the different degrees of membership of the antecedents of the rules. The final value is computed as follows:

$$y_{S/TS} = \frac{\sum\limits_{i=1}^{m} \left(\mu_{\text{Ant}(R_i)}(x_1, \ldots, x_n) * f_{R_i}(x_1, \ldots, x_n)\right)}{\sum\limits_{i=1}^{m} \mu_{\text{Ant}(R_i)}(x_1, \ldots, x_n)} \qquad (9.134)$$

The interpretation of this process is that each value of the output function of each rule is just considered with a weight equal to the degree of membership of the antecedent of the rule. A higher degree of membership on the antecedent, means more confidence on the output value of the function of the corresponding rule and vice versa.

This process can be summarized in the following algorithm:

---

algorithm Sugeno/Takagi-Sugeno fuzzy control method

input $x_1, \dots, x_n$: the set of input values of the input variables $(X_1, \dots, X_n)$
          to the fuzzy rules
    $R_1, \dots, R_m$: the set of fuzzy rules
    $A_1^i, \dots, A_n^i$: the linguistic term values of the input variables for each rule R_i
    $f_{R_i}(X_1, \dots, X_n)$: the consequence function to compute the output value for
            the output variable Y in the rule R_i

output y: the output value of the output variable Y

begin
  for each i **in** 1, ..., m **do**
    // Rule antecedent evaluation
    $\mu_{Ant(R_i)}(x_1, \dots, x_n) \leftarrow min(\mu_{A_1^i}(x_1), \dots, \mu_{A_n^i}(x_n))$
  endforeach
  // Computing of the output value
  $y_{S/TS} = \dfrac{\sum_{i=1}^{m}\left(\mu_{Ant(R_i)}(x_1,\dots,x_n) * f_{R_i}(x_1,\dots,x_n)\right)}{\sum_{i=1}^{m}\mu_{Ant(R_i)}(x_1,\dots,x_n)}$
  return $y_{S/TS}$
end

An Example Using the Mamdani Fuzzy Control Method

As an illustration of the application of the *Mamdani fuzzy control method*, which is more complex than the Sugeno/Takagi– Sugeno method, let us consider the following scenario: Imagine that we had a basic prototype of a *control system for an autonomous or self-driving car*. Let us think that our control variable that should be adjusted with the automatic intelligent control system is the "*Steering Wheel Rotation (SWR)*" to avoid obstacles and collisions against them, and that the only two variables considered in this simplified scenario are: the "*Obstacle Disposition (ODP)*" and "*Obstacle Distance (OD)*". If this scenario must be modelled through *fuzzy sets* and *fuzzy reasoning*, the different variables should be considered as linguistic variables, and its corresponding *linguistic term* values that will be *fuzzy sets* must be defined.

Thus, the three *linguistic variables* will be:

- *"Steering Wheel Rotation (SWR)"*: It expresses the rotation in angle degrees that must be applied to the steering wheel at each moment to avoid collisions with possible near obstacles and it is measured on the base variable of positive and negative angles in degrees belonging to $[-90°, +90°]$. In addition, it has five possible *linguistic terms* such as *"A Bit on the Right (BR)"*, *"Much on the Right (MR)"*, *"Approximately Zero (AZ)"*, *"A Bit on the Left (BL)"*, and *"Much on the Left (ML)"*.
- *"Obstacle Distance (OD)"*: It expresses the distance from the car to the nearest obstacle detected and is measured on the base variable of longitudinal distance expressed in meters belonging to $[0,+\infty)$. Furthermore, it has three possible *linguistic terms* such as *"Short (SH)"*, *"Medium (MED)"*, and *"Large (LRG)"*.
- *"Obstacle Disposition (ODP)"*: It expresses the angle deviation in degrees described by the straight line joining the centre of the obstacle and the centre of the car angle regarding the orthogonal line to the front of the car and is measured on the base variable of positive and negative angle degrees belonging to $[-60°, +60°]$. Furthermore, it has three possible *linguistic terms* such as *"Left (LFT)"*, *"Centred (CNT)"*, and *"Right (RGT)"*.

Each *linguistic term* is a *fuzzy set* that has its corresponding *membership functions* defined as follows for the *terms* of *"Steering Wheel Rotation (SWR)"* linguistic *variable*:

$$
\mu_{ML}(y) = \begin{cases} 0 & y \in (-\infty, -90] \\ \dfrac{y+90}{30} & y \in [-90, -60] \\ \dfrac{-y-30}{30} & y \in [-60, -30] \\ 0 & y \in [-30, +\infty) \end{cases}
$$

$$
\mu_{BL}(y) = \begin{cases} 0 & y \in (-\infty, -60] \\ \dfrac{y+60}{30} & y \in [-60, -30] \\ \dfrac{-y}{30} & y \in [-30, 0] \\ 0 & y \in [0, +\infty) \end{cases}
$$

$$
\mu_{AZ}(y) = \begin{cases} 0 & y \in (-\infty, -30] \\ \dfrac{y+30}{30} & y \in [-30, 0] \\ \dfrac{30-y}{30} & y \in [0, 30] \\ 0 & y \in [30, +\infty) \end{cases}
$$

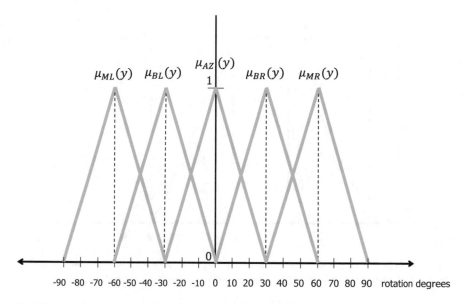

Fig. 9.30 Graphical representation of the *membership functions* of the *linguistic terms A Bit on the Right (BR), Much on the Right (MR), Approximately Zero (AZ), A Bit on the Left (BL)* and *Much on the Left (ML)* of the *linguistic variable Steering Wheel Rotation (SWR)*

$$
\mu_{BR}(y) = \begin{cases} 0 & y \subset (-\infty, 0] \\ \dfrac{y}{30} & y \in [0, 30] \\ \dfrac{60 - y}{30} & y \in [30, 60] \\ 0 & y \in [60, +\infty) \end{cases}
$$

$$
\mu_{MR}(y) = \begin{cases} 0 & y \in (-\infty, 30] \\ \dfrac{y - 30}{30} & y \in [30, 60] \\ \dfrac{90 - y}{30} & y \in [60, 90] \\ 0 & y \in [90, +\infty) \end{cases}
$$

These membership functions are graphically depicted in Fig. 9.30.

The corresponding *membership functions* for the *terms* of *"Obstacle Distance (OD)"* linguistic variable are defined as follows:

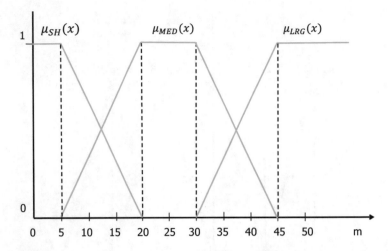

Fig. 9.31 Graphical representation of the *membership functions* of the *linguistic terms Short (SH)*, *Medium (MED)* and *Large (LRG)* of the *linguistic variable Obstacle Distance (OD)*

$$\mu_{SH}(x) = \begin{cases} 1 & x \in [0,5] \\ \dfrac{20-x}{15} & x \in [5,20] \\ 0 & x \in [20,+\infty) \end{cases}$$

$$\mu_{MED}(x) = \begin{cases} 0 & x \in [0,5] \\ \dfrac{x-5}{15} & x \in [5,20] \\ 1 & x \in [20,30] \\ \dfrac{45-x}{15} & x \in [30,45] \\ 0 & x \in [45,+\infty) \end{cases}$$

$$\mu_{LRG}(x) = \begin{cases} 0 & x \in [0,30] \\ \dfrac{x-30}{15} & x \in [30,45] \\ 1 & x \in [45,+\infty) \end{cases}$$

These membership functions are graphically depicted in Fig. 9.31.

The corresponding *membership functions* for the *terms* of "*Obstacle Disposition (ODP)*" linguistic variable are defined as follows:

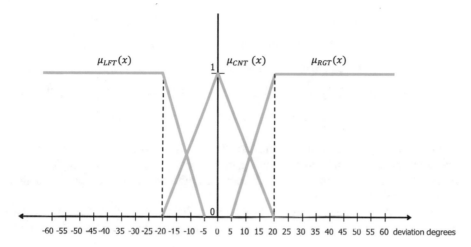

Fig. 9.32 Graphical representation of the *membership functions* of the *linguistic terms Left (LFT)*, *Centred (CNT)*, and *Right (RGT)* of the *linguistic variable Object Disposition (ODP)*

$$\mu_{\text{LFT}}(x) = \begin{cases} 1 & x \in (-\infty, -20] \\ \dfrac{-x-5}{15} & x \in [-20, -5] \\ 0 & x \in [-5, +\infty) \end{cases}$$

$$\mu_{\text{CNT}}(x) = \begin{cases} 0 & x \in (-\infty, -20] \\ \dfrac{x+20}{20} & x \in [-20, 0] \\ \dfrac{20-x}{20} & x \in [0, 20] \\ 0 & x \in [20, +\infty) \end{cases}$$

$$\mu_{\text{RGT}}(x) = \begin{cases} 0 & x \in (-\infty, 5] \\ \dfrac{x-5}{15} & x \in [5, 20] \\ 1 & x \in [20, +\infty) \end{cases}$$

These membership functions are graphically depicted in Fig. 9.32.

Furthermore, let us consider that the expert rules for controlling the output variable *Steering Wheel Rotation (SWR)* are the following five ones:

R_1 : if [OD is SH] and ([ODP is RGT] or [ODP is CNT]) then [SWR is ML]

$$R_2 : \text{ if [OD is SH] and [ODP is LFT] then [SWR is MR]}$$

$$R_3 : \text{ if ([OD is MED] or [OD is LRG]) and [ODP is RGT] or}$$

$$([OD \text{ is MED] and [ODP is CNT]) then [SWR is BL]}$$

$$R_4 : \text{ if ([OD is MED] or [OD is LRG]) and [ODP is LFT] then [SWR is BR]}$$

$$R_5 : \text{ if [OD is LRG] and [ODP is CNT] then [SWR is AZ]}$$

Let us suppose that in a real situation the car is at 40 m of distance from an obstacle, and the obstacle disposition regarding the car has a deviation angle of 15° (degrees). Then, what should be the rotation angle that must be applied to the steering wheel?

This situation can be summarized as $< OD = 40, ODP = 15, SWR = ? >$. To apply the *Mamdani fuzzy control* method the following steps must be undertaken:

1. Evaluation of the antecedent for each rule R_i $i = 1, \ldots, 5$.

The *min* operator will be used as *t-norm* for modelling the conjunction and the *max* operator will be used as *t-conorm* for modelling the disjunction. First, let us compute the degree of membership of the values of the input variables regarding the different linguistic terms, according to the above function definitions:

$$\mu_{\text{SH}}(40) = 0$$

$$\mu_{\text{MED}}(40) = \frac{45 - 40}{15} = \frac{5}{15} = \frac{1}{3}$$

$$\mu_{\text{LRG}}(40) = \frac{40 - 30}{15} = \frac{10}{15} = \frac{2}{3}$$

$$\mu_{\text{RGT}}(15) = \frac{15 - 5}{15} = \frac{10}{15} = \frac{2}{3}$$

$$\mu_{\text{CNT}}(15) = \frac{20 - 15}{20} = \frac{5}{20} = \frac{1}{4}$$

$$\mu_{\text{LFT}}(15) = 0$$

Thus, for each rule we have:

$$\mu_{\text{Ant}(R_1)}(40, 15) = \min \left(\mu_{\text{SH}}(40), \max \left(\mu_{\text{RGT}}(15), \mu_{\text{CNT}}(15) \right) \right)$$

$$= \min \left(0, \max \left(\tfrac{2}{3}, \tfrac{1}{4} \right) \right) = \min \left(0, \tfrac{2}{3} \right) = 0$$

$$\mu_{\text{Ant}(R_2)}(40, 15) = \min \left(\mu_{\text{SH}}(40), \mu_{\text{LFT}}(15) \right) = \min \left(0, 0 \right) = 0$$

$$\mu_{Ant(R_3)}(40, 15) = \max\left(\min\left(\max\left(\mu_{MED}(40), \mu_{LRG}(40)\right), \mu_{RGT}(15)\right),\right.$$
$$\left.\min\left(\mu_{MED}(40), \mu_{CNT}(15)\right)\right)$$
$$= \max\left(\min\left(\max\left(1/3, 2/3\right), 2/3\right), \min\left(1/3, 1/4\right)\right)$$
$$= \max\left(\min\left(2/3, 2/3\right), 1/4\right)$$
$$= \max\left(2/3, 1/4\right) = 2/3$$

$$\mu_{Ant(R_4)}(40, 15) = \min\left(\max\left(\mu_{MED}(40), \mu_{LRG}(40)\right), \mu_{LFT}(15)\right)$$
$$= \min\left(\max\left(1/3, 2/3\right), 0\right) = \min\left(2/3, 0\right) = 0$$

$$\mu_{Ant(R_5)}(40, 15) = \min\left(\mu_{LRG}(40), \mu_{CNT}(15)\right) = \min\left(2/3, 1/4\right)$$
$$= 1/4$$

2. Evaluation of the consequent for each rule R_i $i = 1, \ldots, 5$.

$$\mu_{Con(R_1)}(y) = \min\left(\mu_{Ant(R_1)}(40, 15), \mu_{ML}(y)\right)$$
$$= \min\left(0, \mu_{ML}(y)\right) = 0$$

$$\mu_{Con(R_2)}(y) = \min\left(\mu_{Ant(R_2)}(40, 15), \mu_{MR}(y)\right)$$
$$= \min\left(0, \mu_{MR}(y)\right) = 0$$

$$\mu_{Con(R_3)}(y) = \min\left(\mu_{Ant(R_3)}(40, 15), \mu_{BL}(y)\right) = \min\left(2/3, \mu_{BL}(y)\right)$$

$$\mu_{Con(R_4)}(y) = \min\left(\mu_{Ant(R_4)}(40, 15), \mu_{BR}(y)\right)$$
$$= \min\left(0, \mu_{BR}(y)\right) = 0$$

$$\mu_{Con(R_5)}(y) = \min\left(\mu_{Ant(R_5)}(40, 15), \mu_{AZ}(y)\right) = \min\left(1/4, \mu_{AZ}(y)\right)$$

Therefore, the rules R_1, R_2, and R_4 generates *empty fuzzy sets* as consequents, i.e. $\mu_{Con(R_1)}(y) = \mu_{Con(R_2)}(y) = \mu_{Con(R_4)}(y) = 0$ $\forall y$.

The rules R_3 and R_5 generates two *consequent fuzzy sets* described by the following membership functions:

$$\mu_{Con(R_3)}(y) = \min\left(2/3, \mu_{BL}(y)\right)$$
$$\mu_{Con(R_5)}(y) = \min\left(1/4, \mu_{AZ}(y)\right)$$

These consequent fuzzy sets will derive from the trimming of the corresponding original consequent fuzzy sets, $\mu_{BL}(y)$ and $\mu_{AZ}(y)$, with the corresponding values from the antecedent of each rule, $2/3$ and $1/4$, respectively.

These sets are shown in Figs. 9.33 and 9.34, respectively.

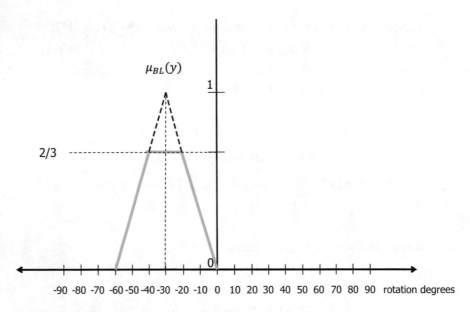

Fig. 9.33 Graphical representation of the *membership function* of the *consequent fuzzy set* $\mu_{BL}(y)$ of the rule R_3 regarding the output variable *Steering Wheel Rotation (SWR)*

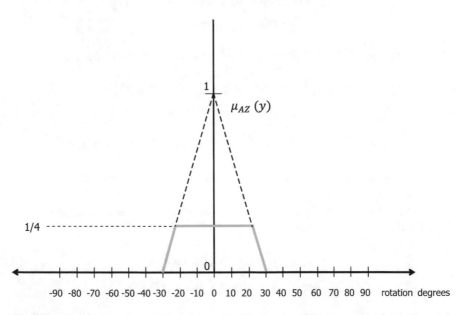

Fig. 9.34 Graphical representation of the *membership function* of the *consequent fuzzy set* $\mu_{AZ}(y)$ of the rule R_5 regarding the output variable *Steering Wheel Rotation (SWR)*

3. Combination of the consequents of all rules.

The membership function of the consequent of the set of all rules is the disjunctive aggregation of the different membership functions of the consequents

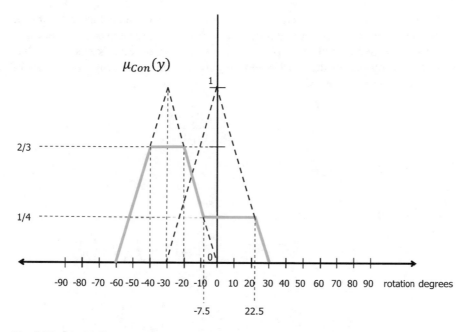

Fig. 9.35 Graphical representation of the *membership function* of the *aggregated fuzzy set* of the conclusion of all the rules regarding the output variable *Steering Wheel Rotation (SWR)*

of all the rules. The disjunction is implemented through a *t*-conorm. The max operator will be used.

$$\mu_{\mathrm{Con}}(y) = \max\left(\mu_{\mathrm{Con}(R_1)}(y), \ldots, \mu_{\mathrm{Con}(R_5)}(y)\right)$$
$$= \max\left(0, 0, \min\left(\tfrac{2}{3}, \mu_{\mathrm{BL}}(y)\right), 0, \min\left(\tfrac{1}{4}, \mu_{\mathrm{AZ}}(y)\right)\right)$$
$$= \max\left(\min\left(\tfrac{2}{3}, \mu_{\mathrm{BL}}(y)\right), \min\left(\tfrac{1}{4}, \mu_{\mathrm{AZ}}(y)\right)\right)$$

The final fuzzy set will be the aggregation of the two consequents of the rules R_3 and R_5. Graphically, the final aggregated fuzzy set is depicted in Fig. 9.35.

The membership function of the final consequent set after the aggregation can be described analytically as follows:

$$\mu_{\mathrm{CON}}(y) = \begin{cases} 0 & y \in (-\infty, -60] \\ \dfrac{y+60}{30} & y \in [-60, -40] \\ \dfrac{2}{3} & y \in [-40, -20] \\ \dfrac{-y}{30} & y \in [-20, -7.5] \\ \dfrac{1}{4} & y \in [-7.5, 22.5] \\ \dfrac{30-y}{30} & y \in [22.5, 30] \\ 0 & y \in [30, +\infty) \end{cases}$$

4. **Defuzzification of the final consequent.** In order to get the final crisp value for the output variable *Steering Wheel Rotation (SWR)*, the aggregated consequent fuzzy set must be defuzzified. As explained before there are several methods available. Let us apply one of the most commonly used such as the centre of gravity or centre of mass (Eq. 9.132):

$$y_{CoG} = \frac{\int_y y * \mu_{Con}(y)\, dy}{\int_y \mu_{Con}(y)\, dy}$$

Thus, in our example:

$$\int_y y * \mu_{Con}(y)\, dy = \int_{-60}^{30} y * \mu_{Con}(y)\, dy$$

$$= \int_{-60}^{-40} y * \left(\frac{y+60}{30}\right) dy + \int_{-40}^{-20} y * \left(\frac{2}{3}\right) dy + \int_{-20}^{-7.5} y * \left(\frac{-y}{30}\right) dy$$

$$+ \int_{-7.5}^{22.5} y * \left(\frac{1}{4}\right) dy + \int_{22.5}^{30} y * \left(\frac{30-y}{30}\right) dy$$

$$= \frac{1}{30} * \left[\frac{y^3}{3} + \frac{60y^2}{2} + c_1\right]_{-60}^{-40} + \frac{2}{3} * \left[\frac{y^2}{2} + c_2\right]_{-40}^{-20} - \frac{1}{30} * \left[\frac{y^3}{3} + c_3\right]_{-20}^{-7.5}$$

$$+ \frac{1}{4} * \left[\frac{y^2}{2} + c_4\right]_{-7.5}^{22.5} + \frac{1}{30} * \left[\frac{30y^2}{2} - \frac{y^3}{3} + c_5\right]_{22.5}^{30}$$

$$= \frac{1}{30} * \left[\left(\frac{(-40)^3}{3} + \frac{60*(-40)^2}{2} + c_1\right) - \left(\frac{(-60)^3}{3} + \frac{60*(-60)^2}{2} + c_1\right)\right]$$

$$+ \frac{2}{3} * \left[\left(\frac{(-20)^2}{2} + c_2\right) - \left(\frac{(-40)^2}{2} + c_2\right)\right]$$

$$- \frac{1}{30} * \left[\left(\frac{(-7.5)^3}{3} + c_3\right) - \left(\frac{(-20)^3}{3} + c_3\right)\right]$$

$$+ \frac{1}{4} * \left[\left(\frac{(22.5)^2}{2} + c_4\right) - \left(\frac{(-7.5)^2}{2} + c_4\right)\right]$$

$$+ \frac{1}{30} * \left[\left(\frac{30*30^2}{2} - \frac{30^3}{3} + c_5\right) - \left(\frac{30*(22.5)^2}{2} - \frac{(22.5)^3}{3} + c_5\right)\right]$$

$$= \frac{1}{30} * \left[\frac{160000}{6} - \frac{216000}{6}\right] + \frac{2}{3} * \left[\frac{400}{2} - \frac{1600}{2}\right]$$

$$- \frac{1}{30} * \left[\frac{-421.875}{3} - \frac{8000}{3}\right] + \frac{1}{4} * \left[\frac{506.25}{2} - \frac{56.25}{2}\right]$$

$$+ \frac{1}{30} * \left[\frac{27000}{6} - \frac{22781.25}{6}\right]$$

$$= -\frac{56000}{180} - \frac{1200}{3} - \frac{7578.125}{90} + \frac{450}{8} + \frac{4218.75}{180}$$

$$= -\frac{257625}{360} = -715.625$$

and,

$$\int_y \mu_{\text{Con}}(y) \, dy = \int_{-60}^{30} \mu_{\text{Con}}(y) \, dy$$

$$= \int_{-60}^{-40} \left(\frac{y+60}{30}\right) dy + \int_{-40}^{-20} \frac{2}{3} \, dy + \int_{-20}^{-7.5} \frac{-y}{30} \, dy$$

$$+ \int_{-7.5}^{22.5} \frac{1}{4} \, dy + \int_{22.5}^{30} \frac{30-y}{30} \, dy$$

$$= \frac{1}{30} * \left[\frac{y^2}{2} + 60y + c_1\right]_{-60}^{-40} + \left[\frac{2}{3} * (y + c_2)\right]_{-40}^{-20} - \frac{1}{30} * \left[\frac{y^2}{2} + c_3\right]_{-20}^{-7.5}$$

$$+ \left[\frac{1}{4} * (y + c_4)\right]_{-7.5}^{22.5} + \frac{1}{30} * \left[30y - \frac{y^2}{2} + c_5\right]_{22.5}^{30}$$

$$- \frac{1}{30} * \left[\left(\frac{(-40)^2}{2} + 60 * (-40) + c_1\right) - \left(\frac{(-60)^2}{2} + 60 * (-60) + c_1\right)\right]$$

$$+ \frac{2}{3} * [(-20 + c_2) - (-40 + c_2)]$$

$$- \frac{1}{30} * \left[\left(\frac{(-7.5)^2}{2} + c_3\right) - \left(\frac{(-20)^2}{2} + c_3\right)\right]$$

$$+ \frac{1}{4} * [(22.5 + c_4) - (-7.5 + c_4)]$$

$$+ \frac{1}{30} * \left[\left(30 * 30 - \frac{30^2}{2} + c_5\right) - \left(30 * 22.5 - \frac{(22.5)^2}{2} + c_5\right)\right]$$

$$= \frac{1}{30} * \left[\left(\frac{1600}{2} - 2400\right) - \left(\frac{3600}{2} - 3600\right)\right]$$

$$+ \frac{2}{3} * [-20 - (-40)] - \frac{1}{30} * \left[\frac{56.25}{2} - \frac{400}{2}\right]$$

$$+ \frac{1}{4} * [22.5 - (-7.5)]$$

$$+ \frac{1}{30} * \left[\left(900 - \frac{900}{2}\right) - \left(675 - \frac{506.25}{2}\right)\right]$$

$$= \frac{200}{30} + \frac{40}{3} + \frac{343.75}{60} + \frac{30}{4} + \frac{28.125}{30}$$

$$= \frac{2050}{60} = 34.1666$$

Therefore, finally the crisp output value inferred from the set of rules and from the input variable values is:

$$y_{\text{CoG}} = \frac{\int_{-60}^{30} y * \mu_{\text{Con}}(y) \, dy}{\int_{-60}^{30} \mu_{\text{Con}}(y) \, dy} = \frac{-715.625}{34.1666} = -20.95$$

This means that given the scenario where the autonomous car is at 40 meters of an obstacle (*Obstacle Distance*, OD = 40), and that the disposition of the obstacle has a deviation of 15° (*Obstacle Disposition*, ODP = 15), the *Steering Wheel*

Rotation (SWR) to be applied must be of $-20.95°$. This angle means that the steering wheel would be slightly rotated to the left to avoid the collision with the obstacle.

9.3   Temporal Reasoning Issues

Interest in the area of temporal reasoning, as well as in spatial reasoning, is growing within the artificial intelligence field. In many application domains, both the temporal information or/and the spatial information must be managed (Renz & Guesguen, 2004; Sànchez-Marrè et al., 2006, 2008). Reasoning about time is a need emerging in many AI applications, such as in *medicine/health-care applications, sustainable/ environmental system management, industrial process supervision, robot planning navigation, or natural language understanding* applications. In all of them, the temporal dimension is important, and has a clear influence on the reasoning mechanisms to solve problems in those domains.

In any IDSS for a dynamic real-world problem, the concept of *time* is fundamental to reason about *change* and *actions*. When some concept, variable, fact, proposition, etc., changes, it means that some *states* or *situations* exist and they change and evolve. This launches the issue of how are related the different states or situations, how they evolve along time, and how important they are to influence reasoning mechanisms used in the different AI and statistical models integrated in an IDSS.

Remembering the three-layer cognitive architecture proposed in Chap. 4 for the deployment of an IDSS, the *analytical tasks* and the *synthetic tasks* are the main issues where the temporal aspects must be integrated to produce a better reasoning process. Therefore, both in the *diagnostic process* of an IDSS and in the *solution-generation process* of an IDSS, the temporal aspects must be used in the new temporal reasoning approaches to be developed to cope with the time dimension. In the *prognostic tasks*, usually the time component is already included as they produce an estimation for the future time of several aspects of the knowledge in a concrete domain.

9.3.1   The Temporal Reasoning Problem

A temporal reasoning approach should provide a formalism to model the notion of time and/or its effects and to represent and reason about the temporal aspects of knowledge about a domain. Thus, usually two components must be specified (Ligozat et al., 2004):

- A new *formalism* or *representation language* or an extension of an old formalism or language to represent the temporal aspects of the domain knowledge. This incorporation of the time dimension can be done in several ways depending on how the knowledge about the domain is represented. It could be the incorporation of a new parameter in a logic predicate within a logic reasoning mechanism, or the integration of new edges and/or nodes in an existing network (*Bayesian Network, Artificial Neural Network*, etc.) to represent temporal aspects or the incorporation of the concept of temporal episodes in a *Case-Based Reasoning* framework, etc.

- A *method for reasoning* using the new extended knowledge formalisms that integrate the above temporal aspects. It includes techniques and methods for managing and reasoning about the knowledge. It could be a new reasoning mechanism taking into account the temporal arguments of logic predicates in a rule-based reasoning system, or a new algorithm or method for making inferences in a new kind of network with temporal elements, or could be a new algorithm/s or method/s for reasoning using temporal episodes, etc.

Continuous or dynamic or time-dependent or temporal domains commonly involve a set of features, which make them really difficult to work with, such as:

- A large amount of new valuable knowledge and/or experiences is continuously generated
- The *current state* or *situation* of the domain depends on previous temporal states or situations of the domain
- *States* can have *multiple diagnoses*.

Taking into account their major characteristics, temporal domains could be theoretically defined as those domains where it is satisfied the following property at least by one assertion at a given instant time:

$$
\operatorname*{truth}_{\substack{1\leq k\leq |la_{t_i}|}} (a_{k,t_i}) = f\left(\operatorname*{truth}_{\substack{1\leq h_i\leq |la_{t_i}| \\ h_i\neq k}} (a_{h_i,t_i}), \operatorname*{truth}_{\substack{1\leq h_j\leq |la_{t_j}| \\ t_j<t_i}} (a_{h_j,t_j}) \right) \tag{9.135}
$$

where,

t_i is the current instant time
t_j is any previous instant time to the current one $(t_j < t_i)$
la_{t_i} is the list of assertions at instant time t_i, being $|la_{t_i}|$ the cardinal of this set
la_{t_j} is the list of assertions at instant time t_j, being $|la_{t_j}|$ the cardinal of this set
a_{k,t_i} is a concrete assertion k at current instant time t_i $(a_{k,t_i} \in la_{t_i})$
a_{h_i,t_i} is whatever assertion h_i in time t_i $(a_{h_i,t_i} \in la_{t_i})$
a_{h_j,t_j} is whatever assertion h_j in time t_j $(a_{h_j,t_j} \in la_{t_j})$

The above property means that the truth of a logic assertion k at current instant time t_i (a_{k,t_i}) depends both on the truth of other logic assertions at current instant time t_i $(a_{h_i,t_i} \in la_{t_i}, h_i \neq k)$, and on the truth of logic assertions at any previous instant time t_j, $(a_{h_j,t_j}, t_j < t_i)$.

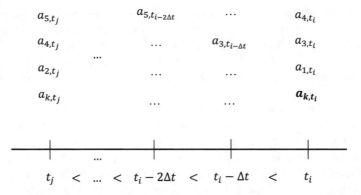

Fig. 9.36 Graphical representation of an example of a temporal dependency of assertions

Imagine that an assertion a_{k,t_i} at the current instant t_i depends on the assertions $a_{1,t_i}, a_{3,t_i}, a_{4,t_i}$ at the current instant t_i, on the assertion $a_{3,t_{i-\Delta t}}$ at the instant time $t_{i-\Delta t}$, on the assertion $a_{5,t_{i-2\Delta t}}$ at the instant time $t_{i-2\Delta t}$, and on the assertions $a_{k,t_j}, a_{2,t_j}, a_{4,t_j}, a_{5,t_j}$ at the instant time t_j $(t_j < t_i)$. The minimum decrement/increment of an instant time t is Δt. Therefore:

$$\text{truth}(a_{k,t_i}) = f\left(\text{truth}(a_{1,t_i}, a_{3,t_i}, a_{4,t_i}), \text{truth}\left(a_{3,t_{i-\Delta t}}, a_{5,t_{i-2\Delta t}}, a_{k,t_j}, a_{2,t_j}, a_{4,t_j}, a_{5,t_j}\right)\right)$$

This condition on the assertion a_{k,t_i} is graphically illustrated in Fig. 9.36 regarding the timeline.

9.3.2   Approaches to Temporal Reasoning

In Computer Science and Artificial Intelligence field, there are many techniques or formalisms which have been developed to deal with temporal reasoning including non-monotonic logics, modal logics, circumscription methods, chronological minimization methods, relation algebras, and applications of constraint-based reasoning, but a generalized understanding across different domains of time/space does not exist. No formal general-purpose methodology has been developed and proven to be useful for different spatio-temporal calculi methods (Renz & Guesguen, 2004). In fact, each one of the methodologies is commonly oriented to slightly different features of the time/space problem. This is why temporal reasoning within IDSS is an advanced challenge to be deeply studied and analyzed.

Formalisms developed to manage temporal reasoning could be grouped as follows:

- *Theoretically oriented models*, which are basically, inspired by certain kinds of *logic* or *relation algebras*. The different theories can be divided into two major models:

  - *Reasoning About Actions and Change*: Major formalisms are the *Situation Calculus* (McCarthy & Hayes, 1969) that is based on the concepts of situation and action, the *Event Calculus* was originally formulated by Kowalski and Sergot (1986) and an axiomatization in classical logic, using *circumscription* as a method for default reasoning to solve the *frame problem* and other related ones was summarized in Miller and Shanahan (1999). Other approaches are the *Features and Fluents formalism* proposed by Sandewall (1994) and the *Fluent Calculus* (Hölldobler, 1997; Thielscher, 1998).
  - *Reasoning About Temporal Constraints*: These approaches are focussed on the management of relations among temporal entities, such as points, intervals, etc., using some properties of the underlying temporal domain, such as discreteness, metric, or ordering. Thus, temporal relations are interpreted as constraints on temporal entities. *Qualitative* and *quantitative temporal constraints* have been formulated in the literature. Outstanding models for qualitative temporal constraints are the *temporal Interval Algebra* by Allen (1983), the *Point Algebra* (Vilain, 1982; Vilain & Kautz, 1986), the Generalized Interval Calculus (Ligozat, 1991), and the *ORD-Horn Algebra* (Nebel & Bürckert, 1995). Major work on quantitative temporal constraints was studied in *Distance Algebra* (Dechter et al., 1991), which models distances between points as well as duration of intervals.

- *Adaptation-based models*, which are methods based on the adaptation or generalization of a known method in Artificial Intelligence to accommodate the temporal dimension. Commonly, they are more inspired by the practical use of the models. Most representative approaches are *Dynamic Bayesian Networks (DBN)*, which are a generalization of *Bayesian Networks* to incorporate the time dimension (Dean & Kanazawa, 1989; Dagum et al., 1991, 1995), *Temporal Artificial Neural Networks* like *Time-Delay Neural Networks* (TDNN) (Waibel et al., 1989) or *Recurrent Artificial Neural Networks (RNNs)* (Rumelhart et al., 1986; Jordan, 1997; Elman, 1990; Cleeremans et al., 1989; Hochreiter & Schmidhuber, 1997), *Temporal Case-based Reasoning* approaches like *Episode-Based Reasoning (EBR)*, that is a generalization of Case-Based Reasoning (Sànchez-Marrè et al., 2005) and *Data Stream Mining* or *Incremental/Dynamic Machine Learning models* (Gaber et al., 2005; Aggarwal, 2007; Gama, 2010), which adapt classic static machine learning models to incrementally integrate the time evolution. *Adaptation-based models* are more concerned on the efficiency and accuracy of the problems solved in the concrete domain than on the logic properties and consistency of the modelling.

The huge complexity of real-world systems makes IDSS modelling difficult with *theoretically oriented models* because many logic assertions should be stated and demonstrated before some reasoning mechanisms can be applied. On the other hand, *adaptation-based models* are mainly concerned with allowing effective and accurate reasoning capabilities in order to make the appropriate decisions about the real-world system. Next, main adaptation-based models will be described.

9.3.2.1 Dynamic Bayesian Networks

A *Dynamic Bayesian Network* (*DBN*) or a *Temporal Bayesian Network* is a *Bayesian Network* (*BN*) expressing the relation among variables over consecutive time steps (Dean & Kanazawa, 1989; Dagum et al., 1991, 1995). *Dynamic Bayesian Networks* (*DBNs*) are directed graphical models of stochastic processes. They generalize *Hidden Markov Models* (*HMMs*) and *Linear Dynamical Systems* (*LDSs*), such as the *Kalman Filters* (*KFs*) by representing the hidden observed state of a dynamical process in terms of *state variables*, which can have complex interdependencies, and some *observed variables* or *perception variables* or *sensor variables*. They are named as *sensor variables* because in many real-world processes, the observations came into the system through the use of sensors. The graphical structure provides an easy way to specify these conditional independencies, and hence to provide a compact parameterization of the model.

A *DBN* models a temporal domain or process as a series of instant snapshots, named *time slices*. Each *time slice* contains a set of random variables. Some of these are unobservable or hidden variables which are the *state variables* at time t, X_t. The other variables are the *observable* or *sensor* or *evidence variables* at time t, E_t. Therefore, at each time t, the concrete evidences observed, e_t, form the observation E_t. Figure 9.37 depict a time slice of a *DBN* with general nodes representing the *state variables* (State$_t$) and the *observation variables* (Observation$_t$). Of course, for each concrete problem the *state variables* can be several variables, i.e. nodes, with its conditional dependencies, and the *observation variables* can be several observable variables. The darker coloured node depicts the evidence values accumulated so far until the current time.

For instance, let us consider again the problem of the *autonomous car driving* problem analyzed in previous Sect. 9.2.5.5, where there are the following three *state variables*: the controlled variable "*Steering Wheel Rotation (SWR)*" and the two variables considered in that simplified scenario: "*Obstacle Disposition (ODP)*" and

Fig. 9.37 A time slice in a Dynamic Bayesian Network (DBN)

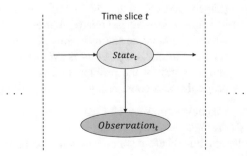

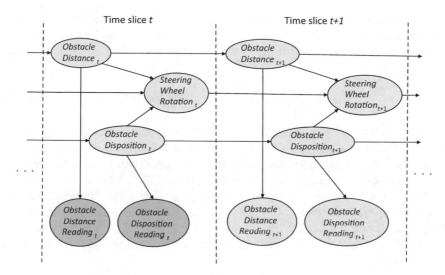

Fig. 9.38 A two-slice fragment of a *DBN* for the monitoring and control of the steering wheel rotation of an autonomous car driving scenario

"*Obstacle Distance (OD)*". Of course, the system needs some observed variable measures from both the obstacle distance, "*Obstacle Distance Reading (ODR)*", and from the Obstacle disposition, "*Obstacle Disposition Reading (ODPR)*". This scenario is described in Fig. 9.38. The darker coloured nodes depict the evidence values accumulated so far until the current time.

Once the set of state variables and the set of observation variables are defined, the evolution of the domain must be specified. The evolution of the state variables is named as the *transition model*, and how the environment generates the evidence values is named as the *sensor model or observation model*.

The *transition model* determines the probability distribution of the state variables at a concrete time given the values of the variables in previous instant times, i.e. $p(X_t| X_0, X_1, \ldots, X_{t-1})$. To solve this unbounded computation, *DBNs* assume the Markov assumption that the current state depends only on the previous state (a first-order Markov process). Thus:

$$p(X_t|X_0, X_1, \ldots, X_{t-1}) = p(X_t| X_{t-1}) \tag{9.136}$$

Furthermore, as there could be infinitely many possible values of t, a different distribution would need to be specified for each time step. Notwithstanding, this drawback is solved with the assumption that the temporal process is a *stationary process*. This means that:

$$p(X_t| X_{t-1}) \text{ is the same } \forall t \tag{9.137}$$

Regarding the *sensor* or *observation model*, the evidence or observed variables at time t, E_t, could depend on variables at previous time instants, both evidence or state variables, as well as the current state variables. However, a *Markov sensor assumption* is made:

$$p(E_t|E_0, E_1, \ldots, E_t, X_0, X_1, \ldots, X_t) = p(E_t|X_t) \qquad (9.138)$$

Due to previous Markov assumptions, *DBNs* are sometimes referred as a *Two-Time slice Bayesian Network (2TBN)* because at any point in time t, the value of a *state variable* can be calculated from the other conditioning variables and the immediate prior value of the same variable at time $t - 1$.

In addition to previous *transition* and *sensor models* which are described with conditional probability tables, the *prior probability distribution at initial time 0* must be specified: $p(X_0)$. Then, the complete joint distribution of the whole *DBN*, using the conditional independence relationships among the variables in the network we would have:

$$p(X_0, X_1, \ldots, X_t, E_0, E_1, \ldots, E_t) = p(X_0) * \prod_{i=1}^{t} p(X_i|X_{i-1}) * p(E_i|X_i) \quad (9.139)$$

The basic *inference reasoning processes* in the temporal DBN model to be undertaken are basically *state estimation* or *filtering*, and *prediction* of the future state. Some other ones are *smoothing, finding the most likely explanation,* and *learning of the DBN structure.* Let us describe how to manage the two basic inference processes:

- *State Estimation*: The computation of the current state means to calculate the posterior distribution of the current time t given all the evidence values up to the current time. This step is also named as filtering. Thus, it must be computed the following distribution:

$$p(X_t|E_1 = e_1, \ldots, E_t = e_t) = p(X_t|e_1, \ldots, e_t)$$

This calculation can be done in a recursive manner due to the Markovian hypothesis, which is usually named as a recursive estimation process, because:

$$p(X_t|e_1, \ldots, e_t) = f(e_t, p(X_{t-1}|e_1, \ldots, e_{t-1})) \qquad (9.140)$$

The distribution can be computed applying several algebraic assumptions, and using the Bayes rule and the sensor Markov assumption as follows:

$$
\begin{aligned}
p(X_t|e_1, \ldots, e_t) &= p(X_t|e_1, \ldots, e_{t-1}, e_t) \\
&= \alpha * p(e_t|X_t, e_1, \ldots, e_{t-1}) * p(X_t|e_1, \ldots, e_{t-1}) \\
&= \alpha * p(e_t|X_t) * p(X_t|e_1, \ldots, e_{t-1})
\end{aligned}
$$

Therefore:

$$p(X_t|e_1, \ldots, e_t) = \alpha * p(e_t|X_t) * p(X_t|e_1, \ldots, e_{t-1}) \tag{9.141}$$

where,

α is a normalizing factor for the probability distribution

$p(e_t|X_t)$ is a probability from the sensor model

$p(X_t|e_1, \ldots, e_{t-1})$ is a one-step prediction of the current state

Finally, the one-step prediction of the current state can be obtained by conditioning on the past state X_{t-1}:

$$p(X_t|e_1, \ldots, e_t) = \alpha * p(e_t|X_t) * \sum_{x_{t-1}} p(X_t|x_{t-1}, e_1, \ldots, e_{t-1}) * p(x_{t-1}|e_1, \ldots, e_{t-1})$$

And applying the Markov assumption again, finally we have:

$$p(X_t|e_1, \ldots, e_t) = \alpha * p(e_t|X_t) * \sum_{x_{t-1}} p(X_t|x_{t-1}) * p(x_{t-1}|e_1, \ldots, e_{t-1}) \tag{9.142}$$

- *Prediction of the Future State*: The goal is to compute the posterior distribution of a *future state*, given all the evidence values until to the current time. It must be computed the following distribution:

$$p(X_{t+k}|e_1, \ldots, e_t), \text{for some } k > 0 \tag{9.143}$$

In fact, the prediction task can be figured out as a state estimation without the addition of new evidence information, because it already integrates a one-step prediction in its computation. Let us define how to compute the state prediction at time $t + k + 1$ from the prediction at time $t + k$:

$$p(X_{t+k+1}|e_1, \ldots, e_t) = \sum_{x_{t+k}} p(X_{t+k+1}|x_{t+k}) * p(x_{t+k}|e_1, \ldots, e_t) \tag{9.144}$$

See Russell and Norvig (2010 Chap. 15, Sect. 15.2) for a summary on the other inference processes.

Summarizing, given a temporal sequence of observations, the full *Bayesian Network* representation of a *DBN* can be constructed by replicating slices until the generated network is large enough to host all the observations. This process is commonly referred as *unrolling* the *DBN*. In Fig. 9.39 there is a general schema of a *DBN* unrolled to fit the corresponding observations until current time t plus the two next future time slices. As before, the darker coloured nodes depict the evidence values accumulated so far until the current time.

To compute the probability distributions of the inference processes such as the state estimation or a prediction defined above, as a *Dynamic Bayesian Network* (*DBN*) is in fact a *Bayesian Network*, any *exact inference method* or *approximate*

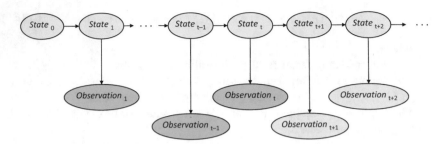

Fig. 9.39 A generic structure of a *Dynamic Bayesian Network* (*DBN*) after the corresponding *unrolling* process to fit all the observations plus the two next time slices

method for *Bayesian Networks* like the ones described in Sect. 9.2.4.2 can be applied.

However, as the *unrolling* process of a *DBN* when the number of observed evidences is huge could not be computationally adequate, usually the approximate methods are used. Concretely, the *likelihood weighting method* (described also in Sect. 9.2.4.2) can be more easily adapted to fit a *DBN*.

9.3.2.2   Temporal Artificial Neural Networks

In classic and static ANNs, the input values to the ANN are provided altogether. However, when there is a temporal component in the input values, a temporal sequence must be learned or processed, in the same way than in a *Dynamic Bayesian Network*. These ANNs that cope with a temporal dimension of the input data, and process/learn a temporal sequence can be generally denominated as Dynamic ANNs.

Among them, the first attempt to manage a temporal sequence in the data was the so-called *Time-Delay Neural Network* (*TDNN*) model (Waibel et al., 1989) proposed by Waibel et al. A *Time-Delay Neural Network* (*TDNN*) is a Multi-layer Feedforward ANN, where the previous *k* input values, for a given *temporal window of length k*, are delayed in time until the final input is available, synchronizing all the input elements of the window. When all the input elements are available, they are provided to the *TDNN*, which can be trained as a usual Multi-layer feedforward ANN with the backpropagation algorithm. Figure 9.40 shows a general scheme of a *TDNN*.

One of the major drawbacks of *Time-Delay Neural Networks* (*TDNNs*) is that the size of the time window must be determined a priori.

Recurrent Artificial Neural Networks (RNNs) (Rumelhart et al., 1986; Jordan, 1997; Elman, 1990; Cleeremans et al., 1989) on the contrary of Feedforward ANNs, where each neuron can only be connected to neurons on the next forward layer, they allow connections to neurons in the *previous layers* or even *self-connections* in the same layer. These *recurrent connections* act as a *short-term memory* and allows the network to remember what happened in the past exhibiting a temporal dynamic behaviour.

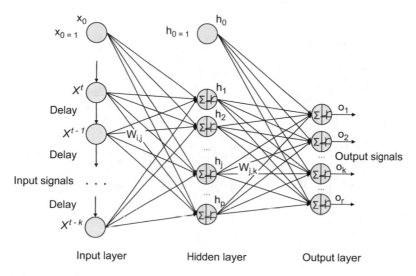

Fig. 9.40 A generic structure of a *Time-Delay Neural Network* (TDNN) with a generic time window of length *k* at the input layer. A node *X* represents all the input vector signals

In *RNNs*, connections between nodes form a directed graph along a temporal sequence. These closed loop connections provide feedback in the *RNN*. *RNNs* can use their internal state (memory) to process variable length sequences of inputs. They have been used in handwritten pattern recognition, speech processing, music composition, non-Markovian control, etc. Usually, the hidden neurons or units in the feedback/backward connections are named as the *context units*.

RNN architectures range from *Simply Recurrent Neural Networks* (*SRNNs*) where only some neurons or units are connected backward to *Fully Recurrent Neural Networks* (*FRNNs*), which connect the outputs of all neurons to the inputs of all neurons. Among the *SRNNs*, the most used in the literature are:

- *Elman Network* (Elman, 1990): An *Elman Network* is a three-layer network with the addition of a *context layer*, i.e. a set of *context units*. The first layer is the *input layer*. The second layer is a *hidden layer*, and the third one is the *output layer*. The *hidden layer* is connected to these *context units* with a fixed weight of one. At each time step, the input values are fed forward and a weight learning method is applied. The fixed back-connections save a copy of the previous values of the hidden units in the *context units* (since they propagate over the connections before the weight learning method is applied). Thus, the network can maintain a sort of state, allowing it to perform a time sequence prediction or classification. Figure 9.41 depicts a general scheme of an *Elman Network*.
- *Jordan Network* (Jordan, 1997): A *Jordan Network* is similar to an *Elman Network*. It has three-layers too, and an additional *context layer*: the *input layer*, the *hidden layer*, the *output layer*, and the *context layer*. Main difference is that the *context units* are fed from the *output layer* instead of the *hidden layer*,

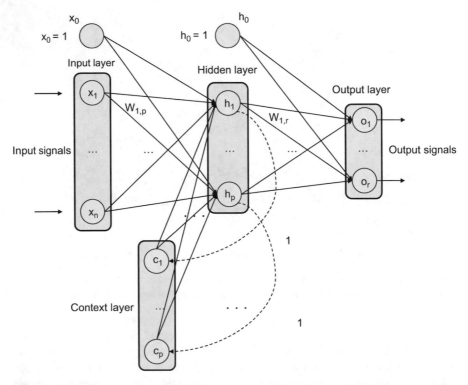

Fig. 9.41 A generic structure of an *Elman Network*. Connections between the *hidden layer* and the *context layer* are one-to-one connections with fixed weight of 1, represented with dashed lines

and that they have a recurrent connection to themselves. The context units in a *Jordan Network* are usually referred as the *state layer*. A general scheme of a *Jordan Network* is depicted in Fig. 9.42.

In case that the time sequences had a small maximum length, the *RNN* can be *unfolded in time*, which is a similar idea to the unrolling in *DBNs*, converting a *RNN* into an equivalent feedforward ANN. Then, a separate unit and connection is generated for copies at different times. The obtained multi-level feedforward ANN can be trained using classic backpropagation with the added constraint that all copies of each connection must remain equal. The final solution for learning the weights is to sum-up the different weight changes in time and update the weight by the average. This process is knowns as *Backpropagation Through Time* (*BTT*) (Rumelhart et al., 1986). If the length of the sequence is large the memory consumption can be really large.

Another approach for the training of an *RNN* is the *Real-Time Recurrent Learning* (*RTRL*) (Williams & Zipser, 1989) which does not use unfolding in time and can hold sequences of larger length.

Notwithstanding, as pointed in Hochreiter and Schmidhuber (1997), with conventional *Backpropagation Through Time* (*BPTT*) (Williams & Zipser, 1992;

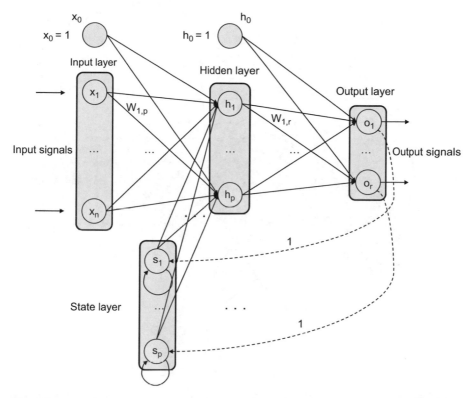

Fig. 9.42 A generic structure of a *Jordan Network*. Connections between the *output layer* and the *state layer* are one-to-one connections with fixed weight of 1, represented with dashed lines. State layer units have a recurrent connection to themselves

Werbos, 1988) or *Real-Time Recurrent Learning (RTRL)* (Robinson & Fallside, 1987), error signals going backwards in time tend to either blowing up causing an oscillation to the weights or vanishing because the temporal evolution of the back-propagated error exponentially depends on the size of the weights, and therefore, learning to bridge long time lags can take a prohibitive amount of time, or even not working properly. This problem is commonly known as the *vanishing gradient problem*.

One of the proposed powerful models to solve those problems is the *Long Short-Term Memory (LSTM) model* (Hochreiter & Schmidhuber, 1997). A *Long Short-Term Memory (LSTM) is a RNN* using a convenient and appropriate gradient-based learning algorithm avoiding the *exploding* or *vanishing gradient problems*. This is achieved by an efficient gradient-based algorithm enforcing a constant error flow through *internal states of special units*. The gradient computation is truncated at certain specific points in the network architecture.

A LSTM network has the following topology: There is one *input layer*, one *hidden layer*, and one *output layer*. The fully self-connected *hidden layer* contains *memory*

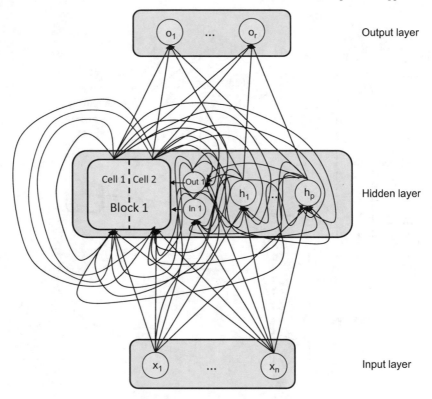

Fig. 9.43 A general scheme of a fully connected LSTM network with a memory block of two memory cells. (Adapted from Hochreiter and Schmidhuber 1997)

cells and corresponding *gate units* which are considered as *hidden units* located in the *hidden layer*. The *hidden layer* may also contain *conventional hidden units* providing inputs to *gate units* and *memory cells*. All units except *gate units* in all layers have directed connections, serving as inputs, to all units in higher layers.

S memory cells sharing one *input gate* and one *output gate* form a *memory cell block of size S*. They provide information storage as conventional neural nets. It is not so easy to code a distributed input within a single cell. Since each *memory cell block* has as many gate units as a *single memory cell*, namely two, the block architecture can be even slightly more efficient. A *memory cell block of size 1* is just a *simple memory cell*. A general scheme of a LSTM network with *one memory block* formed with *two memory cells* is depicted in Fig. 9.43. It is assumed that in general any *hidden unit* of the *hidden layer* as well as the *input gate*, the *output gate*, and the *memory cells* of the *memory block* are *fully connected*. Thus, each gate unit, memory cell unit or hidden unit receives all the non-output unit values.

The general scheme of a *memory cell c_j* is depicted in Fig. 9.44. The linear unit *cell c_j* (the light-red unit in Fig. 9.44) with a fixed self-connection keeps the internal *state* of the *memory cell*, which is usually named as the *Constant Error Carrousel* (CEC). It stores an error signal which is protected unless it is worth to be updated.

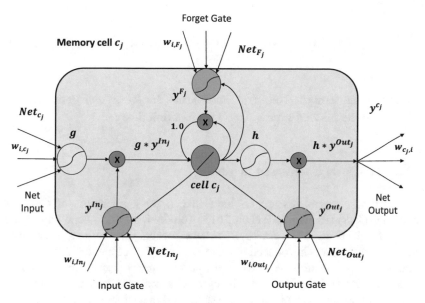

Fig. 9.44 A single memory cell scheme of an LSTM network, adapted from Graves et al. (2009). A weight $w_{i,j}$ represents the weight of the connection from unit i to unit j

A *memory cell* has three *gate units*: *input gate* (In$_j$), *output gate* (Out$_j$), and *forget gate* (F_j). *Gate units* are represented by the green units in Fig. 9.44. The three gates gather the input from the rest of the connected network (Net$_{In_j}$, Net$_{Out_j}$, Net$_{F_j}$), and control the cell using *multiplicative units* (the small blue units in Fig. 9.44). The *input* and *output gates* scale and control the input and output of the memory cell, while the *forget gate* controls the recurrent connection of the cell c_j. The cell c_j has a recurrent connection with fixed weight 1.0, which allows the network to keep error information with a delay of one unit of time. In fact, the network uses the *input gate* In$_j$ to decide whether to keep or override information in the cell c_j, and the *output gate* Out$_j$ to decide whether to use the content of cell c_j to be propagated to other units or to prevent other units from being perturbed by c_j. The functions g and h squashes or scales the information before and after the cell. The internal connections from the cell c_j to the three gates are usually known as the *peephole weights*.

The basic time-step operation of a memory cell can be summarized as follows: at time t, the memory cell c_j output $y^{c_j}(t)$ is computed this way:

$$y^{c_j}(t) = y^{Out_j}(t) * h\big(s_{c_j}(t)\big) \tag{9.145}$$

where $s_{c_j}(t)$ represents the *internal state* of the cell, and it is computed as follows:

$$s_{c_j}(t) = \begin{cases} 0 & t = 0 \\ y^{F_j}\left(\text{Net}_{F_j}(t)\right) * s_{c_j}(t-1) + y^{\text{In}_j}\left(\text{Net}_{\text{In}_j}(t)\right) * g\left(\text{Net}_{c_j}(t)\right) & t > 0 \end{cases}$$

$$(9.146)$$

Note that the term $y^{F_j}\left(\text{Net}_{F_j}(t)\right)$ implements the *forget gate* activating (1) or inhibiting (0) the *internal state* $s_{c_j}(t-1)$ at past unit time $t-1$.

9.3.2.3   Temporal Case-Based Reasoning

In recent years, several researchers have studied the applicability of *Case-Based Reasoning (CBR)* (Richter & Weber, 2013; Kolodner, 1993) to cope with dynamic or temporal domains. *Case-Based Reasoning* mechanism, both as a classifier or predictive method, was introduced in Chap. 6, Sects. 6.4.2.1 and 6.4.2.2. In temporal domains, the *current state* depends on the *past temporal states*. This feature really makes difficult to cope with them. Therefore, classical *isolated case* retrieval is not very accurate, as the dynamic domain is structured in a temporally related *stream of cases* rather than in single cases.

As it is depicted in Fig. 9.45, a *stream of consecutive cases in time* has temporal dependencies, which cannot be detected and used if the cases are managed as *isolated individual cases*.

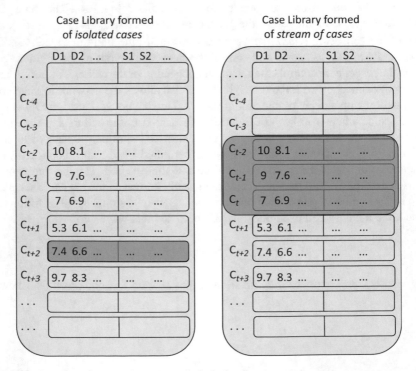

Fig. 9.45   A stream of temporal cases versus the isolated cases management

Thus, Case-Based Reasoning is becoming a promising framework to deal with temporal domains (Bichindaritz & Conlon, 1996). The main reason is that CBR itself operates by retrieving similar solutions within the realm of past experiences (past time actions) to solve a new unseen problem. Thus, it could be easier to incorporate the temporal component in this kind of system.

Main temporal CBR extensions to incorporate the *time dimension* in the CBR paradigm in the literature are:

- Branting and Hastings (1994; Branting et al., 1997) deployed CARMA, which is a system for rangeland pest management advising that uses *model-based matching and adaptation* to integrate case-based reasoning with model-based reasoning for prediction in rangeland ecosystems. It incorporates a method named *temporal projection*, which aligns two cases in time, by projecting a retrieved case forward or backwards in order to match on other parameters, such as the development stage of an insect.

- Ram and Santamaría (1993, 1997) proposed a new method for *continuous case-based reasoning*, and describe how it can be applied to the dynamic selection, modification, and acquisition of robot behaviours in autonomous navigation systems. The *case similarity metric* is based on the mean squared difference between each of the vector values of the case over a *trending window*, and the vector values of the environment. Case-based reasoning in such task domains requires a *continuous representation* of cases, in terms of the available features, that represent the time course of these features over *suitably chosen time windows*.

- Jaczynski (1997) proposed a method based on the so-called *time-extended situations*. Temporal knowledge is represented as temporal patterns, i.e. multiple streams of data related to time points. The representation holds cases as well as general knowledge, which both are taken into account during retrieval. Cases are represented in one of three different forms. *Abstract cases*, also called domain scripts, capture general, or prototypical combinations of data. *Concrete cases* are explicit situations that have been reused at least one time in the CBR-cycle. *Potential cases* are case templates whose contents have to be filled from records in the database. Potential cases are stored as concrete cases once they get activated and used. The retrieval method first tries abstract cases, then concrete cases, and finally, potential cases. This approach has been applied both to plant nutrition control system and prediction of user behaviour for web navigation system.

- Meléndez et al. suggest a method for supervising and controlling the sequencing of process steps that have to fulfil certain conditions (Meléndez et al., 2001). Their main application domain is the control of sets of recipes for making products, such as plastic or rubber pieces, from a set of ingredients. A case represents a recipe, and the temporal problem is the control of a set of recipes, i.e. *a batch*, in order to fulfil process conditions and achieve a production goal. A *deviation* from a normally operating condition is called an *event*, and consists of actions and reactions. Together, the events represent significant points in the history of a product. An *episode* contains information related to the behaviour between two consecutive *events*. The retrieval method first matches general conditions such as the initial and final sub-processes, and then the initial conditions of the corresponding episodes.

- Jære et al. in 2002 propose to use a method for representing temporal cases inside the knowledge-intensive CBR framework of their Creek system. They propose a *qualitative model* derived from the *temporal interval logic* from Allen's theory (Allen, 1983). Allen's temporal intervals are incorporated into the *semantic network* representation of the Creek system. They have been applying their approach in decision support for oil well drilling management tasks.

- Ma and Knight (2003) propose a theoretical framework to support historical CBR, which allows expression of both *relative* and *absolute temporal knowledge*. The formalism is founded on a *general temporal theory* that accommodates both *points* and *intervals* as *primitive time* elements. Similarity evaluation is based on two components: non-temporal similarity, based on elemental cases, and temporal similarity, based on graphical representations of temporal references. They used the concept of *fluents*, and introduced the new concepts of *elemental cases*, *time elements,* and *case histories*.

- In Martín and Plaza (2004), F. J. Martín and E. Plaza propose *Ceaseless Case-Based Reasoning*, a new model which on the one hand considering the CBR task as on-going rather than one-shot task, and on the other hand aiming at finding the best explanation of an *unsegmented sequence of alerts* with the purpose of pinpointing whether *undesired situations* have occurred or not and, if so, indicating the *multiple responsible sources* or at least which ones are the most plausible. It is closely related to the approach of *Continuous Case-Based Reasoning* (Ram & Santamaría, 1997). They applied this paradigm to an intrusion detection system managing an alert stream data.

- Montani and Portinale (2005) propose to represent temporal information at two levels: The *case level* and the *history level*. The *case level* is used whenever some features describe parameters varying within a period of time, which is the duration of the case, and are gathered as a time series. The *history level* is used whenever the evolution of the system can be reconstructed by retrieving temporally related cases. They provide a framework for case representation and case retrieval, which is able to take into account the temporal dimension in any time dependent domain. They applied the framework in RHENE, which is a system for managing patients in a haemodialysis regimen.

- Sànchez-Marrè et al. (2005) propose a new approach for *Temporal Case-Based Reasoning*, named *Episode-Based Reasoning (EBR)*. It is based on the *abstraction of temporal sequences of cases*, which are named as *episodes*. In temporal domains, it is really important to detect similar *temporal episodes of cases*, rather than *similar isolated* cases. Thus, a more accurate diagnosis and problem solving of the dynamic domain could be achieved, taking into account such *temporal episodes of cases* rather than analyzing only the current *isolated case*, as depicted in Fig. 9.46. Furthermore, the problem of multiple-diagnoses in the temporal domain, which is quite usual, can be satisfactorily managed by this approach.

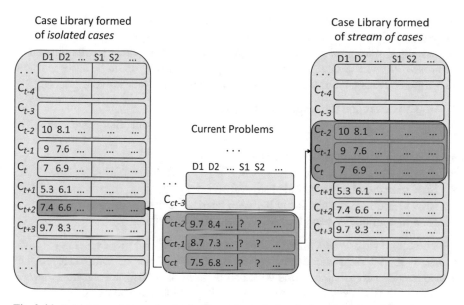

Fig. 9.46 Retrieving a *temporal episode of cases* instead of an *isolated case* for solving the current problem

This framework has been applied to the online supervision of a wastewater treatment plant.

- More recent works in the literature have proposed similar Temporal CBR approaches to previous ones applied to concrete domains, specifically to *healthcare* and *medical domains*. For instance, the description of temporal cases using the *event* concept as *temporal event sequences* for monitoring elderly people at home in Lupiani et al. (2017) is similar to Meléndez et al. (2001) and Montani and Portinale (2005) formulations. Furthermore, in Brown et al. (2018), for Type 1 Diabetes Mellitus Bolus Insulin decision support they use the concept of *temporal case sequence* and isolated cases which are the same concepts than an *episode and isolated cases* as defined in Sànchez-Marrè et al. (2005).

Next, the *Episode-Based Reasoning* (*EBR*) approach will be detailed, as it has been successfully applied to supervise some environmental systems, and was proposed by the author of this book, and hence, it can be detailed and explained in a deeper level of detail.

Episode-Based Reasoning (EBR)

Sànchez-Marrè et al. (2005) propose a new framework for the development of temporal CBR systems: the *Episode-Based Reasoning* (*EBR*) model. It is based on the *abstraction of temporal sequences of cases*, named as *episodes*. Similar temporal episodes of cases can be better past examples of similar situations to the current episode, rather than similar isolated cases which neglect the past history of the different attributes characterizing a case. Using episodes more accurate detection of actually similar situations can be obtained and a more reliable actuation can be proposed. Notwithstanding, working with episodes instead of single cases is useful in temporal domains, but also raises some difficult tasks to be solved, such as:

- How to determine the length of an episode?
- How to represent the episodes, taking into account that they could be overlapping?
- How to represent the isolated cases?
- How to relate them to form episodes?
- How to undertake the episode retrieval?
- How to evaluate the similarity between temporal episodes of cases?
- How to continually learn and solve new episodes?

This approach answers almost all of these questions, and proposes a new framework to model temporal dependencies by means of the episode concept. The Episode-Based Reasoning framework can be used as a basis for the development of temporal CBR systems. This framework provides mechanisms to represent temporal episodes, to retrieve episodes, and to learn new episodes. An experimental evaluation has shown the potential of this new framework for temporal domains (Martínez, 2006; Sànchez-Marrè et al., 2005).

Basic Terminology for EBR

An *isolated case*, or simply *a case* describing several features of a temporal domain at a given time t, is defined as a structure formed by the following components:

$$C_t = \langle \text{CI}, t, \text{CD}, \text{CDL}, \text{CS}, \text{CE} \rangle \qquad (9.147)$$

where,

CI is the *Case Identifier*,
t is the temporal identifier or time stamp that could be measured in any unit of time (month, day, hour, minute, second, etc.)
CD is the *Case* Situation *Description*
CDL is the *Case Diagnostics List*
CS is the *Case Solution* Plan
CE is the *Case* Solution *Evaluation*

The description of the domain situation at a given moment, CD, is a snapshot of the state of the domain, which will consist of the list of values V_{A_k} for the different attributes A_k characterizing the system:

$$\text{CD} = [\langle A_1, V_{A_1} \rangle, \langle A_2, V_{A_2} \rangle, \ldots, \langle A_N, V_{A_N} \rangle] \qquad (9.148)$$

In the temporal domains being addressed by *EBR* proposal, the basic data stream describing the domain can be structured as a feature vector. This hypothesis is not a hard constraint, since most of real temporal systems use this formalism, and also because other structured representations can be transformed into a vector representation. Notwithstanding, some information loss can occur with this transformation process.

For instance, an isolated case in the domain of volcanic and seismic prediction domain, could be described as follows:

$$C_t = < CI = CASE - 134,$$
$$t = 27 - 11 - 2004,$$
$$CD = [< SEISMIC - ACT, Invaluable >,$$
$$< DEFORMATIONS, mean - value >,$$
$$< GEOCHEMICAL - EVOL, normal >,$$
$$< ELECT - PHEN, level - 1 >]$$
$$CDL = [No - eruption, Seismic - pre - Alert]$$
$$CS = [Alert - Emergency - Services]$$
$$CE = correct >$$

A *temporal episode of cases* of length l starting at time t, which is a sequence of l consecutive cases in time, is a structure described as follows:

$$E_{t,l}^d = \langle EI, t, l, d, ED, ES, EE, LoC_{t,l} \rangle \tag{9.149}$$

where,

EI is the *Episode Identifier*,
t is the initial time where the episode starts
l is the length of the episode, i.e. the number of consecutive cases forming it
d is the *Episode Diagnosis*
ED is the *Episode Description*
ES is the *Episode Solution* Plan
EE is the *Episode* Solution *Evaluation*
$LoC_{t, l}$ is the *List* of the l consecutive *Cases* forming the episode

The list of l consecutive cases starting at time t, LoC is:

$$LoC_{t,l} = [C_t, C_{t+1}, \ldots, C_{t+l-1}] \tag{9.150}$$

Episode-Based Reasoning Memory Model

Main outstanding features considered to organise the structure memory of the Episode-Based Reasoning (EBR) system are the following:

- The same *case* could belong to different *episodes*.
- The description, or state depicted by a *case* could correspond to several situations or problems, i.e. multiple diagnostics, at the same time, and not only one, as it is assumed by most CBR system models.
- *Episodes* could overlap among them, and this fact should not imply a case base representation redundancy of the common cases overlapped by the episodes.
- *Episode retrieval* for each case belonging to an episode, should be as efficient as possible.

Taking into account the above facts, the memory proposal integrates hierarchical formalisms to represent the *episodes*, and flat representations for the *cases*. This representation model will set an abstraction process that allows splitting the *temporal episode concept* and the *real case* of the domain. Discrimination trees for the episodes (Episode Base or EpB), and a flat structure for cases (Case Base or CsB) are used. The discrimination tree enables to search which episodes should be retrieved, according to the feature values of the current episode description. Episodes have the appropriate information to retrieve all cases belonging to them.

This structure of the memory of the EBR system allows one case to belong to more than one episode. In addition, it allows the overlapping of episodes, and even though the extreme scenario, which is very common in complex temporal real domains, where the exactly same cases form several different episodes.

To increase even more the efficiency and accuracy of the retrieval step, the use of the mechanism of *episode abstraction* by means of *episode prototypes* or *meta-episodes* is proposed. This technique was originally proposed in Sànchez-Marrè et al. (2000) for a case base. Here, it is used for *episode* categorization instead.

The *meta-episodes* and induced *episode bases* are semantic patterns containing aspects considered as relevant. These relevant aspects (features and feature ordering) constitute the basis for the biased search in the *general episode base*. This new step adds the use of domain knowledge-intensive methods to understand and bias the new problem within its context.

The setting of several *meta-episodes* induces the splitting of the *general episode base* into several *episode bases* with different hierarchical structures. Each *episode base* will store similar episodes that can be characterized with the same set and order of predictive features.

In the retrieval step, first, the *EBR* system will search within the previously established classification to identify which kind of *episode* it is coping with. For each established class (*meta-episode*) there will be a possible different set of specific discriminating features and a different *episode base*. Then, the retrieval will continue in the *episode base/s* induced by the *meta-episode/s* best matching the current episode.

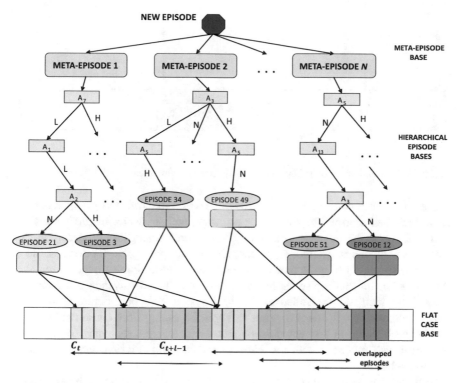

Fig. 9.47 *Hierarchical three-layered* memory structure

The memory model of the approach is composed by a set of meta-episodes, which will constitute the Meta-Episode Base (MEpB). For each meta-episode there exists a hierarchical Episode Base (EpB). Cases are organized within a flat case base (CsB). Also there exists a Meta-Case Base (MCsB) for the diagnostic list computation of a case. This *hierarchical three-layered structure* allows a more accurate and faster retrieval of similar episodes to the current episode/s. Figure 9.47 shows this memory structure.

Episode Retrieval and Learning

Retrieval task of episodes is activated each time the EBR system receives a new *current case* (C_{ct}) with data gathered from the domain, at the current time (ct). First step is *getting the possible diagnostics of the current case.*

This label list can be obtained by different ways. For example, using a *set of inference rules*, which can diagnose the state or situation of the domain from the relevant features. These classification rules could be directly collected from domain experts or could be induced from real data. Another way is using the *meta-cases* technique, and to evaluate the similarity between the current case (C_{ct}) and the meta-cases. The current case is labelled with the diagnostic labels of most similar meta-

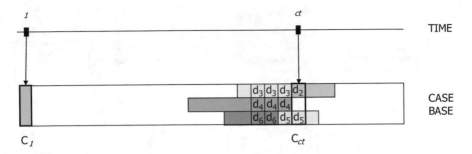

Fig. 9.48 New and/or continued episodes arising from the current case

cases. Meta-cases can be obtained, in the same way as the rules: from experts or from an inductive clustering process. In our proposal, meta-cases technique is used.

Next step is the *generation of possible episodes* arising from the current case. This means to check whether some episodes are continuing from prior cases to the current case, and/or to build new episodes, which are starting from the current case. At this time, finished episodes are detected, and the *EBR* system can *learn new episodes*, which will be added to the *EBR* system memory. Figure 9.48 depicts several alternative episode formation and episode ending from current case.

For each *possible current episode*, most similar episodes must be retrieved. Retrieval task proceeds with the *hierarchical three-layered memory structure* as explained before.

For each one of the *retrieved episodes* and the corresponding *current episode*, a degree of similarity is computed. This value is computed through an *episode similarity measure*. Each retrieved episode is added to a sorted list of episodes by decreasing degree of similarity. Thus, at the end of the process, the *first episode* of the list is the episode with a higher similarity value to a possible current episode. The *EBR* system will use this episode to solve the domain problem, but other policies, such as using the most similar episode for each possible diagnostic, or making a user-dependent choice, can be done.

Episodic similarity between two episodes is computed based on the aggregation of the similarity values among cases belonging to each episode. Episodes are compared based on a left alignment of cases holding all the same importance, which could be changed to give different degree of importance to the different cases of the episode. This episodic similarity measure can be formalized as:

$$
\text{Sim}_{\text{Ep}}\left(E_{t_1,l_1}^d, E_{t_2,l_2}^d\right) =
\begin{cases}
\dfrac{1}{l} \displaystyle\sum_{i=1}^{l} \text{Sim}_C(C_{t_1+i-1}, C_{t_2+i-1}) & l_1 = l_2 = l \\[3ex]
\dfrac{1}{\max(l_1, l_2)} \displaystyle\sum_{i=1}^{\min(l_1,l_2)} \text{Sim}_C(C_{t_1+i-1}, C_{t_2+i-1}) & l_1 \neq l_2
\end{cases}
$$

$$
(9.151)
$$

where Sim_C can be computed with any case similarity measure. In the original approach, *L'Eixample* measure (Sànchez-Marrè et al., 1998) was proposed, because some performance tests done showed it as a good measure.

9.3.2.4   Incremental Machine Learning Techniques and Data Stream Mining

Most inductive Data Mining models, and particularly, Machine Learning models are static. This means that the models are obtained from a static set of data, i.e. from a *batch of data*, and cannot take into account any variation in the data distribution that could happen in new data available from the same domain. This phenomenon happens both to unsupervised data mining models and supervised data mining models.

The only way these static learning methods have to update their own generated models is to induce again a new model from the old data plus a new batch data, which is not very efficient form the computational point of view. Figure 9.49 shows this batch learning scheme.

This problem that can happen in static domains could be more frequent in temporal domains because the temporal variability of data more often can produce different data distribution and the appearance or modification of existing *concepts*.

The solution to this temporal variability, and to capture the possible *concept modification* in the data is the deployment of *incremental data mining methods*. These approaches need to incrementally update the previously induced model with the new data available to integrate the *concept shift* or *concept drift*, without needing to induce again a new model from the old batch plus the new instances or observations. This incremental dynamic learning approach is depicted in Fig. 9.50.

The extreme situation of incremental dynamic learning happens when the new models must be considered to be updated at each new instance arriving from the data

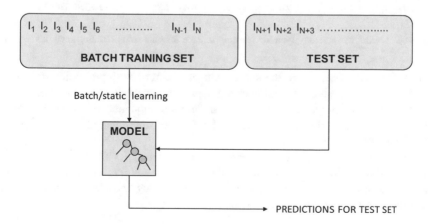

Fig. 9.49 Batch learning scheme for static machine learning methods

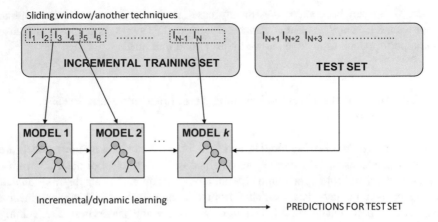

Fig. 9.50 Incremental dynamic learning scheme for incremental machine learning methods

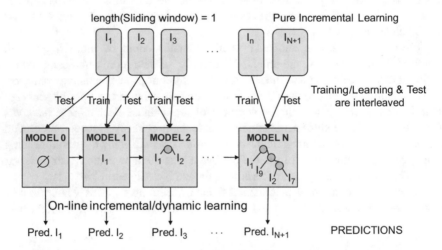

Fig. 9.51 Online incremental dynamic learning scheme for data stream mining methods

source. This situation where the source data are continuously producing data units, which arrives to the incremental learning approach is named as a *data stream*.

More technically, a *data stream* is an ordered sequence of instances, which commonly can be read just once or a small number of times using limited computing and storage capabilities.

Data stream learning or *data stream mining* (Gaber et al., 2005; Aggarwal, 2007; Gama & Gaber, 2007; Gama, 2010), also named as *online incremental dynamic learning*, or simply *online learning*, consists in extracting knowledge structures from continuous, rapid data records, or instances.

This process is extremely complex, as illustrated in Fig. 9.51 because it must cope with several problems such as the data volume could be impossible to store on a disk, or even when data could be stored, examples must be processed just once. Furthermore, as a *data stream* flows, the underlying concepts or/and data distribution can be modified, i.e. *concept drift* or *concept shift* can appear (Hulten et al., 2001; Klinkenberg & Joachims, 2000; Maloof & Michalski, 2000; Kubat & Widmer, 1995), and hence, the previously induced models should be changed accordingly. It is remarkable that in data stream mining process the learning and test phases are interleaved, as shown in Fig. 9.51.

The *data stream mining algorithms* must meet the following properties as much as possible (Hulten et al., 2001): incremental learning process, constant time to process each example, a single scan over the training set, taking *concept drift* into account, and maintain the data volume under control.

The *concept drift* or *concept shift* affects the different methods in different ways. In *predictive methods*, the predictive patterns can change as well, and the approximation function being estimated by the predictive method changes according to the new information. Previously, the *Dynamic ANNs* approaches have been reviewed, which can cope with temporal domains and *concept shift/drift* in *predictive methods*.

In *clustering algorithms*, the distribution of the clusters can change with time, the bounds of the clusters can change, new clusters can appear or even some of them could disappear or being subsumed by other ones.

Some efforts have been done in the literature regarding *dynamic clustering* methods such as:

- COBWEB (Fisher, 1987, 1996), which is an incremental clustering technique that keeps a hierarchical clustering model in the form of a classification tree. Each new instance is located in the node optimizing a category utility function.
- BIRCH (Zhang et al., 1996), which builds a hierarchical data structure to incrementally cluster the incoming points using the available memory and minimizing the amount of input/output required.
- STREAM (Guha et al., 2000), which can solve the k-median problem on a data stream in a single pass.
- Sequential k-means clustering method, Sequential nearest-neighbour clustering method and Sequential agglomerative clustering methods described in Ackerman and Dasgupta (2014)
- Stochastic learning for dynamic clustering, which build-ups dynamic clusters for guiding a dynamic CBR approach (Orduña Cabrera & Sànchez-Marrè, 2018).

In *classifier or discriminant methods*, the classifier patterns can change as a consequence of the temporal evolution of the data, and new patterns can be detected, others can change and some others could disappear or decrease its importance.

Next, an overview of the *dynamic classifier methods* is done grouping the different type of dynamic classifiers in three main groups (*adaptive base learners, learners modifying the training set,* and *ensemble techniques*):

- *Adaptive Base Learners*: They are able to *dynamically adapt to new training data* that contradicts a learned concept. Depending on the *base learner* employed, this adaptation can take on many forms, but usually relies on *restricting or expanding the data that the classifier uses* to predict new instances in some region of the feature space. Some approaches according to the different base learners used are:

  – *Decision Tree-based methods,* like the following approaches:

    Very Fast Decision Trees (VFDT) (Domingos & Hulten, 2000) which make use of *Hoeffding bounds*[31] to grow decision trees in streaming data.

    Concept-Adapting Very Fast Decision Tree (CVFDT) (Hoeglinger & Pears, 2007), which uses a *sliding window* of instances retained in a short-term memory.

    Hoeffding Option Tree (HOT) (Pfahringer et al., 2007) and a modified version, Adaptive Hoeffding Option Tree (AHOT) using an adaptive window (Bifet et al., 2009).

    Hoeffding Window Tree (HWT) and Hoeffding Adaptive Tree (HAT) (Bifet & Gavaldà, 2009), which are approaches similar to CVFDT, and the last one using an adaptive window to detect change

  – *Similarity-based methods,* like the following ones:

    k-NN based methods (Alippi & Roveri, 2007, 2008), which are adaptation of the *k-NN classifier* when *concept drift* is detected, removing obsolete instances from the memory.

    A *Dynamic CBR classifier* (Orduña-Cabrera & Sànchez-Marrè, 2009, 2016), which is an adaptation of CBR for dynamic data. It uses a dynamic multi-library with incremental creation of prototypes and associated discriminant trees.

  – *Fuzzy ARTMAP-based methods* (Carpenter et al., 1992). ARTMAP attempts to generate a new "cluster" for each pattern that it finds in the dataset, and then maps the cluster to a class. If a new pattern is found that is sufficiently different (defined via a vigilance parameter), then the new pattern is added with its corresponding class. It uses Self-Organising Artificial Neural Networks.

---

[31] Wassily Hoeffding proved (Hoeffding, 1963), the so-called *Hoeffding's inequality* which provides an upper bound on the probability that the sum of bounded independent random variables deviates from its expected value by more than a certain amount.

- *Learners Modifying the Training Set*: Another popular approach of addressing *concept drift* is by modifying the training set seen by the classification algorithm. The most common approaches employed are *windowing*, where only a subset of previously seen instances is used, and *instance weighting*. One of the strengths of the modification approach over the adaptive base learners approach is that the modification strategies are independent of the classifier used. Proposals in the literature can be divided in:

  - *Windowing Techniques*: They use window/s to determine the instances forming the training set of the classifier. The basic naïve approach keeps a window over the newest instances. Several approaches in the literature are:

    FLORA3 (Widmer & Kubat, 1996), which is an extension of previous FLORA and FLORA2 versions. It introduces an *adaptive window* which attempts to vary its size to fit the current concept.
    Based on SVMs (Klinkenberg & Joachims, 2000). They use an estimator to compute a bound on the error rate of the SVM, and *choose the window size* corresponding to the minimum estimated error.
    Method based on the learner's error rate over time (Gama et al., 2004).
    Multiple Windows method (Lazarescu et al., 2004), which used several windows to fit better the concept drift.
    Adaptive Window (ADWIN) and ADWIN2 (Bifet & Gavaldà, 2007), which uses an adaptive window to determine the best length of the window.

  - *Weighting Techniques*: They use weights to determine the instances that should be part of the training dataset to induce the classifier model. One of the well-known approaches is:

    Adaptive weighted k-NN (Alippi et al., 2009). The instance weight expresses the likelihood of an instance to be from the current concept.

- *Incremental Learners Through Ensemble Techniques*: They can deal with *reoccurring concepts* (Hoens et al., 2012). A *reoccurring concept* is a concept that was formed from past data, but was discontinued with new data until it appears again. Since *ensembles* can contain *models built from past data*, such models can be reused to classify new instances if they are drawn from a reoccurring concept. Other approaches often discard historical data in order to learn the new concepts. When combined in an *ensemble*, however, the multiple batch learners can be trained on different subsets of the data to create an *incremental learner*.

  - *Accuracy weighted ensembles*, like the next ones:

    Streaming Ensemble Algorithm (SEA) (Street & Kim, 2001), which proposes pruning the worst classifier if the ensemble is full.
    Dynamic Weighted Majority (DWM) (Kolter & Maloof, 2003, 2005, 2007)
    Weighting ensemble techniques (Wang et al., 2003; Becker & Arias, 2007)

- *Bagging and Boosting methods* like the following ones:

    Learn++ (Polikar et al., 2001), which is an incremental learner for learning ANN classifiers in a data stream based on Adaboost.

    Adaptive Boosting Ensemble (ABE) (Chu & Zaniolo, 2004).

    Learn++.NSE (Elwell & Polikar, 2011; Muhlbaier et al., 2009), which uses weighting votes of classifiers. High weight is given to classifiers performing well in the *current concept*.

    ADWIN bagging (Bifet et al., 2009), which uses adaptive classifiers in the ensemble.

- *Concept locality-based approaches* which assume that *concept drift* does not need to occur on a global scale, but on a local one. Some approaches are the following:

    Local Concept drift strategy (Tsymbal et al., 2006, 2008), where the dynamic integration of the classifiers in the ensemble is based on the local accuracy of each classifier.

    Local Concept drift which is based on feature space partition (Wang et al., 2006).

    Adaptive Classifier Ensemble (ACE) (Nishida et al., 2005), which combines batch learners (*long-term memory*) and an online learner (*short-term memory*), and a *drift detection* mechanism.

    CDC (Stanley, 2003), which undertakes the learning of the *concept drift* with a committee of Decision Trees.

9.4 Spatial Reasoning Issues

According to Timpf and Frank (1997) a possible definition of spatial reasoning is the following: "Any deduction of information from a representation of a spatial situation". Space perception, representation, and reasoning like the time dimension, is one important human cognition ability. *Space* and *spatial representation and reasoning* (Davis, 1990; Freksa, 1991; McDermott, 1992; Freksa & Rohrig, 1993; Hernández, 1994; Vieu, 1997) are formed of entities, concepts, and abilities very related to Cognitive Science, and specifically to vision, perception, motion, touch, etc. However, space is more complex than time as it is multidimensional. Many real-world systems have to be managed as two-dimensional or three-dimensional processes. Many real-world applications to show a reliable performance need to use *spatial representations* and *spatial reasoning* abilities such as in *robotics for navigation planning to avoid obstacles*, in *computer vision for suitable scene identification and interpretation*, in *physical reasoning coping with real objects*, in *autonomous car driving for a safe navigation*, in *natural language understanding for a correct interpretation and mining of spatial relationships among entities appearing in the text*, in *computer-aided design of spatial systems like a microprocessor chip or an architectural design*, and in *Geographic Information Systems*

(*GIS*), which are common components of an IDSS, especially in environmental, sustainability, economical, biological or geographical applications.

In any IDSS for a dynamic real-world problem, the concept of *space* is fundamental to reason about *spatial relations* and *properties* like the location of objects and about the relationships among objects. Common grounding concepts in space are notions coming from *Topology*, and some concepts taken from the *Geometry* field like the concept of *orientation* and the concept of *distance*.

Furthermore, it should be noted that space can be *quantitatively represented with numerical magnitudes*, and those magnitudes are processed by numerical algorithmic methods. Usually, this area is studied in *computational geometry* field. On the other hand, space can be *qualitatively represented with qualitative measures*, and those qualitative measures, i.e. qualitative quantity spaces (see Chap. 5, Sect. 5.5.3.2 where qualitative reasoning representation models were explained) are processed by *symbolic reasoning* mechanism. Therefore, what is usually named as spatial representation and reasoning in Artificial Intelligence, is mostly a *qualitative spatial reasoning*.

In the same way that the time problem, regarding the three-layer cognitive architecture proposed in Chap. 4 for the deployment of an IDSS, the main issues where the *spatial reasoning* aspects must be integrated to produce a better reasoning process must be taken into account in both the *analytical tasks*, the *synthetic tasks*, and the *prognostic* tasks. Therefore, in all *processes* of an IDSS, the spatial aspects must be used in the new spatial representation and reasoning approaches to be developed to cope with the multidimensional space.

There are different conceptualizations of *space* according to different views coming from disciplines like Philosophy, Mathematics, Physics, or Psychology. Some of them argue that space is the *physical space where humans live* that can be measured, modelled and reasoned upon it. Others says that it is the *cognitive perception and representation* of the physical space. Finally, others think that it is a *mathematical abstract artefact* built for modelling the above-mentioned space conceptualizations.

In the different *spatial representation and reasoning* mechanisms developed in the literature, there are approaches modelling the *physical space* or the *common-sense cognitive space*.

In addition, there are two contrary options to consider space: an *absolute space* and a *relative space*. The conception of an *absolute space* is like an empty container, existing beforehand, independently of the physical, or mental objects to be located in it. A *relative space* is conceived as a construct induced by spatial relations over non-purely spatial entities, i.e. material bodies in the case of physical space and mental entities with more properties than just spatial ones in the cognitive space.

An *absolute space* can be further classified as *global* or *local*. In a *global absolute space*, each spatial entity is a location in a general reference frame, whose relative position respect to all other spatial entities must be completely determined. In a *local absolute space*, a spatial entity is situated through a number of explicit spatial relations with some other spatial entities, but not mandatory with all.

Most approaches of *qualitative spatial reasoning* in the literature consider the space as being a *local absolute space*.

9.4.1   The Spatial Reasoning Problem

A *spatial reasoning approach* should provide a formalism to *model the notion* of *space* and *how to represent* it, and in addition, the corresponding *spatial reasoning mechanisms*. Notwithstanding, in the Computer Science and Artificial Intelligence community, more efforts have been done in the *spatial reasoning problem* than in the *spatial representation*, which have been faced more recently.

The different *spatial reasoning tasks* raising its own *problems* can be grouped in the following ones, as suggested in Vieu (1997):

- *Deduction* of facts from a given spatial representation: This means the exploitation of the spatial knowledge and spatial information gathered in a spatial representational framework.

   This deduction could refer to the simple process of making explicit a given fact that was implicit in the spatial representation by using properties of spatial relations and entities, combining several facts. It is a *deductive reasoning process*.

   Furthermore, it can refer to the process of inferring possible true facts on the basis of some hypotheses on the structure of the space and spatial relationships and entities. Usually, information or knowledge can be uncertain or incomplete.
- *Transformation* of a given spatial representation formalism to another one: This task means to convert information and knowledge represented in one spatial representation framework into a different spatial representation framework.
- *Design* of spatial configurations satisfying some constraints: This task refers to the construction of spatial or even spatiotemporal configurations satisfying certain requirements. It implies obtaining new spatial representations with new entities.

In many situations in a deployment of an IDSS regarding the *spatial reasoning*, different *spatial relationships and dependencies* can be exploited to make the *deductions* or *inferences* more reliable, like the situation illustrated in Fig. 9.52, where in a *spatial grid* of $m \times n$ cells, some spatial properties such as location or other assertions at cell C_{ij} can depend on the same spatial properties or assertions in the neighbour cells of C_{ij}, which are within the neighbour red rectangles.

For instance, a key aspect of complex *spatial representation* of GIS raster-based models is controlling how adjacent cells interact. Does the value of one cell depend on the value of adjacent cells? The concept of a *spatial episode of cells,* which is the equivalent to the *temporal episode of cases* concept within the *Episode-Based Reasoning (EBR)* framework (Sànchez-Marrè et al., 2005) can be extrapolated to manage this dependency. The concept of a *moving window* has been commonly used in many environmental applications ranging from wildlife habitat models to soil science to estimate land use change.

Taking into account these spatial dependencies, some spatial domains satisfy the following dependency property by some assertion at a given spatial entity or region:

Fig. 9.52 Spatial dependency of assertions in cell or region C_{ij} on the neighbour cells of a spatial grid

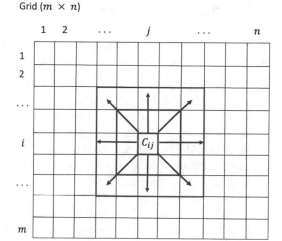

Grid $(m \times n)$

$$\underset{\substack{1 \le k \le |la_{C_{ij}}|}}{\text{truth}} \left(a_{k,C_{ij}} \right) = f \left(\underset{\substack{1 \le h_l \le |la_{C_{ij}}| \\ h_l \ne k}}{\text{truth}} \left(a_{h_l,C_{ij}} \right), \underset{\substack{1 \le h_r \le |la_{C_{pq}}| \\ C_{pq} \in \text{Neib}(C_{ij})}}{\text{truth}} \left(a_{h_r,C_{pq}} \right) \right) \tag{9.152}$$

where,

C_{ij} is the given spatial cell or region
C_{pq} is any cell or region belonging to the neighbourhood of the given one $(C_{pq} \in \text{Neib}(C_{ij}))$
$\text{Neib}(C_{ij})$ is the set of neighbour cells to cell or region C_{ij}
$la_{C_{ij}}$ is the list of assertions at cell C_{ij}, being $\left| la_{C_{ij}} \right|$ the cardinal of this set
$la_{C_{pq}}$ is the list of assertions at cell C_{pq}, $\forall C_{pq} \in \text{Neib}(C_{ij})$ being $\left| la_{C_{pq}} \right|$ the cardinal of this set
$a_{k,C_{ij}}$ is a concrete assertion k at given cell or region C_{ij} $\left(a_{k,C_{ij}} \in la_{C_{ij}} \right)$
$a_{h_l,C_{ij}}$ is whatever assertion h_l in cell or region C_{ij} $\left(a_{h_l,C_{ij}} \in la_{C_{ij}} \right)$
$a_{h_r,C_{pq}}$ is whatever assertion h_r in cell or region C_{pq}, $\forall C_{pq} \in \text{Neib}(C_{ij})$ $\left(a_{h_r,C_{pq}} \in la_{C_{pq}} \right)$

The above property means that the truth of a logic assertion k at a given cell or region C_{ij} $\left(a_{k,C_{ij}} \right)$ depends both on the truth of other logic assertions at the same cell or region C_{ij} $(a_{h_l,C_{ij}} \in la_{C_{ij}}, h_l \ne k)$, and on the truth of logic assertions at any other neighbour cell or region of C_{ij}, C_{pq}, $\left(a_{h_r,C_{pq}}, C_{pq} \in \text{Neib}(C_{ij}) \right)$.

9.4.2   Approaches to Spatial Representation and Reasoning

The different approaches existing in the Artificial Intelligence literature to the *spatial representation* problem can be grouped as follows as described in Vieu (1997):

- *Point-Based Spaces*: these approaches are based on points and defining local spaces, typically focussing on orientation and distance concepts, since dealing with topology would require going to higher order predicates. Among these, there are different approaches such as the works of Hernández (1994), Freksa (1992), Ligozat (1993) which use local reference frames. Other approaches use distance and orientation concepts, such as in Vieu (1993), Zimmermann (1993), Frank (1992), Hernández et al. (1995).
- *Space as Arrays*: In computer vision and in spatial databases *occupancy arrays* are commonly used. They are a discrete coordinate system. These arrays have been modified in the so-called *symbolic arrays* approach (Glasgow & Papadias, 1992; Glasgow, 1993).
- *Interval-Based Spaces*: The temporal Allen's interval calculus have been extended to a multidimensional domain like the spatial domain such as in Güsgen (1989), Mukerjee (1989), Mukerjee and Joe (1990). In those approaches, spatial regions are represented by means of tuples of intervals being the projections of the regions on the axes of a given absolute reference frame.
- *Region-Based Spaces*: regions are extended entities of any shape, which makes them closer to spatial cognition. In addition, they are first-order theories dealing properly with topological concepts. Most of the works are originated from the research on mereotopologies[32] in formal ontology by Whitehead (1929), Clarke (1981, 1985). Different approaches are based on regions (Fleck, 1996) and other works on non-differentiated regions such as RCC theory or calculus (Randell et al., 1992; Gotts et al., 1996), and other similar approaches (Egenhofer & Franzosa, 1991) and others based on adjacency structure on cells and fuzzy spaces (Fleck, 1987). Some other works use orientation such as the approaches of (Randell & Cohn, 1989; Borgo et al., 1996a; Hernández, 1994). Finally, other techniques use the concept of non-qualitative distance in a pseudo-metric space (Gerla, 1990).
- *Spaces with Mixed Ontologies*: Some other approaches are based on entities of different dimensionality. Their spatial ontologies are a mixture of extended and non-extended entities. Some approaches use *mereotopology on regions* such as Gotts (1996) and Galton (1996). Other techniques use geometrical concepts in a qualitative segmentation of the Cartesian space, such as the works of Davis (1988, 1990), Forbus (1983, 1995).

---

[32]*Mereotopology*, in ontological computer science, refers to a first-order theory, embodying *mereological* (i.e., relative to the study of parts and the wholes they form), and *topological* concepts, of the relations among wholes, parts, parts of parts, and the boundaries between parts.

The different *spatial reasoning* approaches, following the previous classification can be grouped according to the tasks they solve:

- *Deduction* of facts from a given spatial representation: simple deduction reasoning mechanisms were using relational algebras and transitivity or compositional tables to previously store all possible composition of relations. This way theorem proving was not needed and simply replaced by *looking-up in a table*. These methods are applied in Randell and Cohn (1989), Egenhofer (1991), Freksa (1992), Randell et al. (1992), Zimmermann (1993), Hernández (1994), Grigni et al. (1995).

  Other approaches applied *constraint satisfaction methods* as in Hernández (1994), Grigni et al. (1995).

  Dealing with more complex situations with uncertainty in space—time dimensions has accounted for some works exploiting also physical theories to predict the result of motion. They are called *qualitative kinematics*, based on principle from qualitative physics such as Forbus (1983, 1995), Faltings (1990), Forbus et al. (1991).

  Other approaches are based on the *classic theorem-proving method* within the logical framework such as the of a logical model using topological, geometrical and mechanic concepts, such as the works of Davis (1988), Nielsen (1988).

  And other works manage the *continuity of motion* through neighbourhood structures and continuity or transition graphs, such as the works of Cui et al. (1992), Hernández (1993), Galton (1993).

- *Transformation* of a given spatial representation formalism to another one: some approaches worked on the *translation between vision and spatial expressions in natural language*. Most of them used a *global numeric representation of space* and usual computational geometric methods such as in Marr (1982), Ballard and Brown (1982), Chen (1990).

  Other works are based on the needed *transformations according to the different perspective or granularity*. Some works proposed regarding space as global are Hobbs (1985), Hobbs et al. (1987) and Borgo et al. (1996b) regarding space as local.

  Some approaches manage *granularity in an implicit way* by using operations to switch from alternative representations of different granularity, such as some *hierarchical structures* in Samet (1984, 1989), Glasgow (1993) or using a *modal operator of refinement* such as in Asher and Vieu (1995).

- *Design* of spatial configurations satisfying some constraints: The approaches to solve this reasoning problem are focussed on the *route-finding problem*, usually for a robot. A pioneer work by Lozano-Pérez (1983) is based on using a *dense coordinate space*, and the robot position is determined to be a concrete point by appropriate transformations of the spatial environment.

  Others (Slack & Miller, 1987; Fujimura & Samet, 1989) use *discrete global spaces and local spaces* approach based on cells and adjacency relation.

  Other well-known approaches such as Kuipers (1978), Kuipers and Levitt (1988), Levitt and Lawton (1990) use *cognitive maps*, which are local spatial representations based on a mixed ontological representation combining regions, lines and points into a graph, coming from cognitive psychology field.

Finally, some approaches used *constraint satisfaction methods* are applied in local spaces to spatial planning in kitchen design (Baykan & Fox, 1987), apartment design (Baykan & Fox, 1997) and on a hyper-frequency electronic equipment design (du Verdier, 1993).

9.4.3   Geographic Information Systems (GISs)

A Geographic Information System (GIS) (Tomlinson, 1969) is a straightforward application where *spatial reasoning* can be applied. According to the encyclopedic entry of National Geographic (NatGeo, 2021) for a GIS system, it can be defined as follows: "a Geographic Information System (GIS) is a computer system for capturing, storing, checking, and displaying data related to positions on Earth's surface".

It can be said that effectively a GIS can capture, store, check and display *spatial data* organized in several *layers* that could be merged as required and visualized over a *map*. There are two basic file formats used to store data in a GIS: *raster files* and *vector files*. *Raster formats* are grids of cells or pixels. Raster formats are useful for storing varying GIS data, such as elevation or satellite imagery. *Vector formats* are usually geographical features considered as geometric shapes such as points, lines, or polygons. Vector formats are useful for storing GIS data with firm borders, like wells, rivers, or lakes. Data visualized can be of different nature, such as vegetation, population, streets, forest trails, roads, transport lines, economic measures, health services, disease level, etc., but always is related to a concrete *space region*, which is usually named as a *cell* of the whole *map* or *grid*. This data layer merging generates an aggregated information which makes easier the *spatial data analysis* and *spatial knowledge patterns extraction* from spatial data.

GIS allows to relate information, through the use of *location* as the key variable. *Locations* and *extents* that are found in the Earth's *spacetime,*[33] are able to be recorded through the *date* and *time* of occurrence, along with x, y, and z coordinates. Coordinates represents: *longitude* (x), *latitude* (y), and *elevation* (z). All Earth-based, *spatial–temporal location* and *extent* references, can be related to one another, and ultimately, to a real physical *location* or *extent*.

One of the first and original issues of a GIS was spatial data visualization on geographical maps, where the maps have been discretized in some *space regions* or *cells*. In addition, a GIS allows the user to create interactive queries to consult and visualize different layers and desired combinations over a corresponding map.

Notwithstanding of GIS popularity, the major problem is about interaction. Current GISs do not support intuitive or common-sense *spatially oriented* human–

---

[33] In physics, *spacetime* refers to any mathematical model which fuses the three dimensions of space and the one dimension of time into a single four-dimensional manifold.

computer interaction. Users may wish to abstract away from the mass of numerical data and specify a query in a way which is essentially, a *qualitative spatial reasoning* query. Therefore, an extended spatial query language for GIS is required. Bettini and Montanari (2002) provide a summary of the related research needs and promote the linkage between GIS and AI. In fact, GIS is one common complementary component of an IDSS for many environmental, geographical, life sciences, medical or sustainability domains.

For interested readers in GIS, there are several open-source GIS software like QGIS,[34] GeoDa,[35] GRASS GIS[36] (Geographic Resources Analysis Support System), gvSIG,[37] SAGA GIS,[38] or OrbisGIS.[39]

Finally, let us remark that *spatial* and *temporal reasoning* share many commonalities, and often, spatial problems must be represented in time steps or some other temporal framework.

9.5 Recommender Systems

With the great spread of internet around the world in last decades, with the corresponding overflow of information for people, and with the explosion of social networks over internet, a new special kind of Intelligent Decision Support Systems have appeared. They are the so-called *Recommender Systems* (Goldberg et al., 1992; Resnick & Varian, 1997; Jannach et al., 2010; Ricci et al., 2015), which emerged in the middle 90s. As it will be described in this section, they are a particularization of an IDSS, where always the possible alternatives in the decision problem at hand are different *items* or *list of items* to be suggested to a *user*.

Thus, the particular problem relies on suggesting the *best items* to be used or consumed by a *concrete user* given some *knowledge regarding users' preferences* over the *items* in a concrete domain.

People face numerous decision-making situations related to selection of items, such as products, films, cars, documents, movies, songs, etc. In all these situations, there are several common characteristics describing the decision-making scenario, such as the following ones:

- Usually, there are numerous, and often a huge number of *alternatives* to cover the same needs, which are impossible to analyze and evaluate by the user herself or himself.

---

[34] QGIS (https://www.qgis.org).

[35] GeoDA (http://geodacenter.github.io/).

[36] GRASS GIS (https://grass.osgeo.org).

[37] gvSIG (http://www.gvsig.com).

[38] SAGA GIS (http://www.saga-gis.org).

[39] OrbisGIS (http://orbisgis.org/).

- There is a *huge amount of information*, especially in internet, from different types, leading the user to an overwhelming and a frustrating situation.
- Users seek to obtain the maximum *utility* from their transaction, i.e. from their *item* or *items* finally selected.

A *Recommender System* (RS) is an Intelligent Decision Support System that provides support in form of suggestions about which are the *best useful items* being of interest to a *given user or users* in a concrete domain.

There are many domains where RSs have emerged, especially in internet websites of companies selling different kind of items. Books recommended to be read by a user, songs suggested to be listened for a given user in a radio station or radio podcast internet application, news recommended to be read in an online media or movies suggested to be watched by a given user in a subscription-based streaming internet are examples of this recommendation process.

9.5.1   Formulation of the Problem

The problem to be solved by a *Recommender System* can be mathematically formulated as follows:

Given,

$U = \{u_1, u_2 \ldots, u_k\}$ that is the set of k *users* identified in the system,

$I = \{i_1, i_2, \ldots, i_l\}$ that is the set of all l available *items*,

$T = \{t_1, t_2, \ldots, t_m\}$ is the set of m recorded *transactions* between the *users* and the system

and, a *utility function R* defined on real values that a *user* $u \in U$ gets by using an *item* $i \in I$, which is defined as:

$$R: \quad U \times I \to \mathbb{R}$$

$$u, i \longmapsto R(u, i)$$

such that the *utility of an item* for a given user, $R(u, i)$ can be approximated or estimated by the *rating* $r_{p, q}$ that a user u_p assigns to an item i_q, showing the level of the *item* preference for the *user*, in the interaction of the user with the system recorded in a given *transaction* t:

$$R\left(u_p, i_q\right) \approx r_{p,q} = \widehat{R}\left(u_p, i_q\right) \tag{9.153}$$

then, these *ratings* on the different *items* given by the *users* collected from the *transactions* between the *users* and the system, can be expressed in the so-called *user-item matrix* or *rating matrix* (Table 9.4).

Table 9.4 Structure of a
*user-item matrix or rating
matrix*

	i_1	i_2	$\ldots$	i_l
u_1	$r_{1,1}$	$r_{1,2}$	$\ldots$	$r_{1,l}$
u_2	$r_{2,1}$	$r_{2,2}$		$r_{2,l}$
$\ldots$	$\ldots$	$\ldots$	$\ldots$	$\ldots$
u_k	$r_{k,1}$	$r_{k,2}$	$\ldots$	$r_{k,l}$

The goal of a *Recommender System* is to recommend the *item* $i_{\max}$ or the *set of N items* $\{i_{l_1}, i_{l_2}, \ldots, i_{l_N}\}$ maximizing the *predicted* or *estimated utility* $\widehat{R}(u_p, i_q)$ for the user u_p. Thus:

$$\forall i_q \subseteq I, \quad i_{\max} = \operatorname*{argmax}_{1 \leq q \leq l} \widehat{R}(u_p, i_q) \tag{9.154}$$

or

$$\forall i_q \in I, \quad \{i_{l_1}, i_{l_2}, \ldots, i_{l_N}\} = N - \operatorname*{argmax}_{1 \leq q \leq l} \widehat{R}(u_p, i_q) \tag{9.155}$$

The *rating matrix* is built as the result of the interaction between a *user* and the system, once the *user* has been received a recommendation for an *item* or set of *items*, and then, the *user* gives a feedback for the recommended *item* or *items* in form of a *rating* for each *item recommended*, which is the estimation of the *utility* of a given *recommended item* for the *user*.

Ratings can be expressed in many *ordered scoring real values*. In internet-based recommender systems, it is very popular the *five-star rating system*, where a user gives a value of *one star* if she or he is rather unsatisfied with the recommended item, and a value of *five stars* if the satisfaction is maximum. Of course, this five-star rating is transformed internally into a *real-valued rating* belonging to the interval [1,5].

The *rating matrix* commonly is a huge *sparse matrix* because a system could have a huge number of items, and of course, a given user has not used or consumed or bought all the possible items, and hence, just a *few ratings* will appear in the corresponding row of that user in the *rating matrix*. These means that there a lot of null values for many items that are not evaluated by a user.

A typical example is in an internet subscription-based streaming service, where the system can have millions of movies, but a user probably only will be able to watch and rate just hundreds of movies at most.

Therefore, *Recommender Systems* commonly must use some efficient way to represent, retrieve and update the *sparse rating matrix*.

This *explicit gathering of feedback* through *ratings* is one possibility to collect information about the users' behaviour related to the system. However, many real applications can collect information about the user with other *complementary explicit techniques* such as:

- Asking a user to search, and observing what are the queries of the user to estimate her/his preferences.
- Asking a user to rank a collection of items from the most favourites to the least favourites.
- Presenting two items to a user and asking her or him to choose the one that it is preferred over the other. Usually, this process is like a pairwise comparison done in several steps that allows to get information about preferences of the user.
- Asking a user to create a list of items that she or he likes, which let gathers the *harmonization* and *similarity* of several items when a list of items must be recommended

Other *Recommender Systems* use an *implicit data collection* procedure, which should be known by the user according to the data protection regulation laws existing in many countries, like, for example, in the European Union, where the General Data Protection Regulation (GDPR) (EU 2016/679), which entered into force in 2016, is mandatory. These systems gather information and knowledge about the user interaction with the system by observing her/his behaviour with some *explicit collection techniques*, such as:

- Observing the items that a user views in an online system an obtaining information about the preferences of the users that could allow the system to create some profile of the user.
- Accounting the time that a user spends viewing different items in the system. The spent time can be correlated with the interest of the item for the user.
- Keeping a record of the items that a user has purchased or listened to or watched in the internet application. This allows the system to build a profile of the user with her/his preferences.
- Analyzing the user's social network to discover similar items or related information that are liked or disliked by the user that helps to complement her/his profile.

9.5.2 General Architecture of a Recommender System

The solution generated by a *Recommender System* is based on two great assumptions. Users need intelligent systems being able to support them in items' search and selection due to the limited cognitive abilities of humans, and users rely on the opinion of their *social network* like friends, connected people, and of specialists to be supported in their final selection of items.

Recommender Systems usually recommend items similar to those that a user has liked in the past or items that similar users to the user have liked in the past. These means that this kind of systems need a great interaction between the recommendation system and the users to capture the feedback and real utility from the suggested items to a user.

The main goals of a *Recommender System* are commonly:

- The generation of a *list of the most useful items*, i.e. a kind of *top-N items*, according to an estimated utility of the *items* given that the system has an historical record of users and items interaction.
- The prediction whether a *specified item* will be of interest for a *user* given that the system has an historical record of users and items interaction.

Recommender Systems are developed assuming the following hypothesis are true, and can show a good performance while these assumptions are held:

- Similar *users* have a similar *behaviour*. This is a fundament hypothesis of *Recommender Systems*, especially from the first ones developed in the field, which were based on the similarity among users.
- Users select the *items* maximizing their *utility*. This matches the general *principle of rationality* in Artificial Intelligence, formulated by Newell (1982) in the context of knowledge-based systems: "*if an agent has knowledge that one of its actions will lead to one of its goals, then the agent will select that action*". As the user has the goal of maximizing the utility of the transaction, then recommender system will recommend the item maximizing the utility of the transaction.
- *Users' behaviour* and *users' preferences* remain stable upon time. Of course, this hypothesis could not be completely true along time, and probably the *behaviour* and *preferences* of a user can be dynamic.

The architecture of *Recommender Systems* is a particularization of the general architecture of an Intelligent Decision Support System. The Fig. 9.53 depicts this general architecture for a *Recommender System*.

Recalling the general architectural framework for an IDSS proposed in Chap. 4, Sects. 4.6 and 4.7, and concretely the charts depicted in Figs. 4.12 and 4.13, it can be easily appreciated that both architectures perfectly match.

A recommender system has four main components:

- *User Interface*: The user interface is in charge to inform the user with the *list of recommended items* ordered by the utility estimation of the system, *i.e*, the *rating*.
    Of course, the user can input whatever information required to start or continue the recommendation process.
- *Recommendation Process Unit*: This is the *core* part where the different methods and techniques can be used to recommend the *list of items*. As it will be analyzed

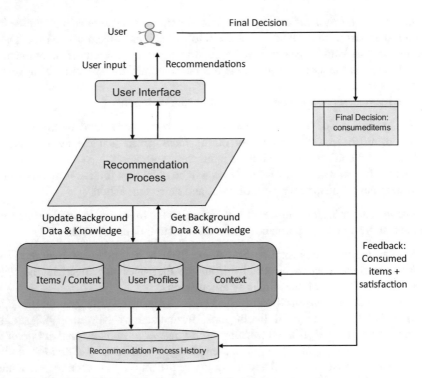

Fig. 9.53 General architecture of a Recommender System

in the next section several different techniques can be used such as *collaboration filtering techniques*, *content-based techniques*, etc.

- *Background Data and Knowledge Repository*: This component is where the different data structures and databases are used to keep the required information about the *user profiles*, the *items* characteristics, the *user-item matrix* or *rating matrix*, and other contextual information.

- *Recommendation Process History*: The recommendation process history is the transaction database, where the *transactions* between the *users* and the system are stored for further processing or analysis. For instance, this analysis can provide knowledge for building or updating the *users' profile*.

It is worth to note that in most *Recommender Systems* is very important the feedback provided by the user after each interaction with the system to associate the consumed/used *items* and their satisfaction, which will cause the updating of the corresponding data structures like the *rating matrix*, etc. The user asking for a recommendation is usually known as the active user, u_a.

9.5.3 *Recommender System Techniques*

A *Recommender System* can use several models and techniques. The most used are *collaborative filtering* and *content-based techniques*. Furthermore, some other recommendation approaches have appeared in the field, such as *knowledge-based*, *demographic-based* and *community-based* (Ricci et al., 2015; Burke, 2007), and of course several *hybrid techniques*. These approaches are analyzed in the next sections.

9.5.3.1 Collaborative Filtering

Collaborative Filtering was used in the first recommender system known in the literature (Goldberg et al., 1992). This technique recommends to the active user u_a, items that other users with similar preferences to the active user, liked in the past (Schafer et al., 2007).

The similarity among the active user u_a, and the other users u_p is computed based on the characterization of each user with her/his preferences, i.e. with the *ratings*, given by the users to the items consumed in the past. This means that the active user u_a, characterized with some ratings is the vector:

$$u_a = [r_{a1}, ?, r_{a3}, ?, ?, r_{a6}, \ldots, ?, r_{al}] \qquad (9.156)$$

and that the similarity between u_a and all the other users $u_p \in U$, $p \neq a$, must be computed using any similarity measure:

$$\text{sim}(u_a, u_p)$$

Afterwards, the items not yet consumed by u_a, with good ratings for similar users will be recommended. Other option is to estimate the *rating* of those items not yet consumed by u_a aggregating the ratings given to them by the similar users to u_a.

algorithm Collaborative Filtering

input U: the set of users $(u_1, \dots, u_k)$
    I: the set of items $(i_1, \dots, i_l)$
    R: the ratings matrix or user-item matrix, formed by each rating r_{pq}
    u_a: the active user

output $RecItems$: the list of recommended items

begin
  // Collaborative step
  $LUsSim \leftarrow \emptyset$
  for each $u_p \in U\ (p \neq a)$ **do**
    // Compute the similarity between all users u_p and the u_a
    $LUsSim \leftarrow \text{add}(\langle u_p, sim(u_a, u_p)\rangle, LUsSim)$
  endforeach
  $LUsOrdSim \leftarrow \text{order}(LUsSim)$ // The list of users ordered by similarity
  $U_{MS} \leftarrow \text{mostSimilarUsers}(\sigma, LUsOrdSim)$ // Users u_p with $sim(u_a, u_p) \geq \sigma$
  $I_{CAND} \leftarrow \emptyset$ // The list of candidate items
  // From U_{MS} compute items i_{CAND} not yet used by u_a and of a minimum interest
  for each $u_{MS} \in U_{MS}$ **do**
    for each i_{CAND} rated by u_{MS} and not yet used by u_a and $(r_{MS,i_{CAND}} \geq \rho)$ **do**
      $I_{CAND} \leftarrow \text{add}(i_{CAND}, I_{CAND})$
    endforeach
  endforeach
  $RecItems \leftarrow I_{CAND}$
  // If the estimation of ratings for the candidate items is required, then continue
  $IRat_{CAND} \leftarrow \emptyset$ // The list of candidate items with its rating
  for each $i_{CAND} \in I_{CAND}$ **do**
    // Compute the estimation of the ranking $r_{a,i_{CAND}}$ for each candidate item i_{CAND}
    $r_{a,i_{CAND}} \leftarrow \underset{p \in U_{MS}}{\text{aggreg}}(r_{p,i_{CAND}})$
    // A usual aggregation function is the average rating weighted by $sim(u_a, u_p)$
    // $\underset{p \in U_{MS}}{\text{aggreg}}(r_{p,i_{CAND}}) = \frac{1}{|U_{MS}|}\sum_{p=1}^{|U_{MS}|} sim(u_a, u_p) * r_{u_p, i_{CAND}}$
    $IRat_{CAND} \leftarrow \text{addRating}(\langle i_{CAND}, r_{a,i_{CAND}}\rangle, IRat_{CAND})$
  endforeach
  // Filtering step
  $RecItems \leftarrow \text{topItems}(n, \text{order}(IRat_{CAND}))$ // The top n items are recommended
  return $RecItems$
end

This *collaborative filtering* procedure can be summarized in the above detailed algorithm.

This *collaborative filtering* approach has been also named as *memory-based* or *neighbourhood-based* in contraposition with other variations named as *model-based*. *Memory-based* approaches generate the recommended items directly from the exploitation of the information from the *rating matrix*.

Model-based techniques use the *rating matrix* to learn a *model* or *models*, which will be used to generate recommendations for the active user. Some models applied are Bayesian Networks, Latent Factor models or Clustering models. For instance, using a clustering process, the different users can be clustered in several classes. When recommendations are searched for the active user, the cluster to which she/he belongs is determined. Then, the items liked by other users of the same cluster are taken into account for generating the recommended items.

One of the typical problems that *collaborative filtering* techniques suffer is the so-called *cold-start* problem and the *popularity-bias* problem. The *cold-start* problem is produced when new users access the recommender system for the first time/s and there is no information about her/his preferences over the items, and then the similarity to other users cannot be computed. Also, it happens when new items are added to the system, because they cannot be recommended because nobody has rated them yet.

The *popularity-bias* problem means that the items that are more popular, i.e. the items having been rated more times, tend to be recommended with higher frequency than the other items less popular. This generates a *long tail* distribution of items with low popularity, which rarely will be recommended.

9.5.3.2 Content-Based

Content-based techniques or *Content-based Filtering* techniques are based on obtaining the recommended items by analyzing the items liked by the active user u_a, and then, searching for similar items to these ones. This approach needs that the items be described by several features characterizing them, and this *item information* must be available.

For instance, in a film recommender system, the features characterizing a film could be the language of the film, the genre type of the film (western, comedy, drama, mystery, etc.), the year of the film, the director of the film, the main actress, and/or main actor of the film. The idea is that if a user has liked a film with some given characteristics, the *Recommender System* will search for films of similar characteristics, i.e. similar genre/s or/and language or/and year and/or director and/or main actress and/or main actor, etc.

This technique can be summarized in the following algorithm:

---

algorithm Content-based

input U: the set of users $(u_1, \ldots, u_k)$
    I: the set of items $(i_1, \ldots, i_l)$
    R: the ratings matrix or user-item matrix, formed by each rating r_{pq}
    u_a: the active user

output *RecItems*: the list of recommended items

begin
  // Content-based step
  IteLk ← getLikedItems(u_a, ρ)   // The items liked by the u_a: the ones with $r_{ai} \geq \rho$
  LIteSim ← ∅ // The list of items i_q with its similarity to items liked by the u_a
  for each $i_l \in$ *IteLk* **do**
    // Compute the similarity between the i_l and all other items i_q not yet used
    for each $i_q \in I$ $(i_q \neq i_l, \forall i_l \in$ *IteLk*$)$ **do**
      LIteSim ← add($\langle i_q, sim(i_l, i_q)\rangle$, *LIteSim*)
    endforeach
  endforeach
  LIteOrdSim ← Order, (*LIteSim*)   // The list of items ordered by similarity
  I_{MS} ← mostSimilarItems(σ, *LIteOrdSim*) // Items with $sim(i_l, i_q) \geq \sigma$
  // Filtering step
  RecItems ← topItems(n, I_{MS})   // The top n items are recommended
  return *RecItems*
end

When the top n items are computed, it must be taken into account that the *same item* can appear more than once in the list of ordered items, because they could be similar to more than one item of those belonging to the liked ones by the active user. This *frequency of repetition* of an item can be used to gain positions in the list, and thus, increase its probability of being recommended.

The most common drawback of content-based techniques is the *overspecialization* problem. The *overspecialization* problem can be produced when most recommended items have a very *limited diversity*, because are too much closely related to items previously liked by the users.

9.5.3.3   Other Techniques

Other recommendation techniques used in recommendation applications are:

- *Demographic-based*: these techniques are based on the exploitation of the demographic characteristics of users (Pazzani, 1999). The *demographic profile* of users is described with several features like the native language of the user, the country where they live, the size of the city they live, the age they have, the sex, the profession they have, education, and other characteristics.

Some existing approaches have been developed specifically in the marketing area. Many times, these systems take into account just one characteristic, and for instance according to the language or/and the country of the active user, they are redirected to a concrete website.

Other variants probably more rational can be to obtain different profiles of the users from its demographic data through a clustering process. Then, for each active user, her/his demographic profile is determined, through some similarity measure computed between the profile prototypes and the active user or using some expert inference rules. Once the profile is known, the possible items well rated by the users belonging to the same profile are considered to be recommended.

- *Knowledge-Based*: These techniques use *knowledge models* for the recommendation process. They usually recommend items based on a *specific domain knowledge* about how certain item features satisfy users' needs and preferences.

  One common knowledge-based recommendation techniques are *case-based recommendation* approaches (Bridge et al., 2006; Gatzioura & Sànchez-Marrè, 2015). In this kind of approaches, the similarity function of CBR systems assesses how much the user preferences can be satisfied with some items. The user preferences are described within the descriptive part of a case, and the possible items to be recommended constitute the solution part of cases. Other techniques are *constraint-based recommenders*. Constraint-based techniques use predefined knowledge bases containing inference rules to link user requirements to item characteristics.

- *Community-based*: this type of recommendation techniques is based on the preferences of the *social community* of the users. Some studies and practical evidence confirm that people rely more on recommendations from their friends than on recommendations from anonymous people, even being similar from the point of view of preferences. These systems browse and acquire information about the social connections of the users and about the preferences of their friends. The recommended items are based on the ratings of the user's friends. These kinds of community-based systems are also named *social recommender systems* (Golbeck, 2006).

- *Hybrid Systems*: a hybrid technique is whatever approach combining different kinds of recommendation techniques, such as the other ones described above. Of course, the goal of this approach is to take profit of the advantages of some recommendation techniques and to avoid the drawbacks of other recommendation techniques (Burke, 2007), such as the *cold-start, popularity-bias,* or *overspecialization* problems. There are several ways of combining the different recommendations. See Chap. 6, Sect. 6.4.2.1 on *Ensemble methods*.

  One strategy is to combine the different ratings coming from the different recommendation techniques through a (weighted) aggregation process, i.e. a *voting* scheme of the *ensemble* of techniques. Another approach is based on the selection of the best recommendation technique for a given problem based on a *meta-classifier*, i.e. a *gating* approach. Collecting together all the recommendations provided by all the recommender techniques is another common hybrid approach.

Analyzing the different recommendation techniques, it is easy to observe that the *core mechanism* in all techniques is the *similarity assessment* between several entities such as users' ratings, items description, users' profiles, etc. Considering that this similarity assessment between entities is one main step in the retrieval stage in the Case-Based Reasoning approach, and taking a close look to all recommendation techniques, it can be outlined that most recommendation techniques and the problem at hand can be formulated as a Case-Based Reasoning problem. Sometimes the Case Base is formed by users described with its corresponding ratings. Other ones the Case Base is formed by items described by its main characteristics, other ones the Case Base could be formed of users' profile, etc.

9.5.4   Evaluation of Recommender Systems

Recommender systems as any computational system, and specially as an IDSS must be evaluated (Wang & Kobsa, 2009). As any complex system such as an IDSS, two types of evaluation must be done. One regarding its *components*, and the other regarding the *whole system* performance.

The evaluation of its *components* is mainly focussed in the evaluation of the *different recommendation techniques* used by the system. Thus, it is an offline evaluation carried out in the design/implementation stage (Nguyen & Ricci, 2007). The different recommendation techniques are compared applying them on the same datasets containing the user interactions, i.e. *the ratings*, and assessing their performance. Usually, the comparison is done using some public benchmark data or with particular collected data in a given domain.

Furthermore, the *whole recommender system* must be evaluated once it is available for end-users, because the global performance of the system and its usability by end-users must be assessed. When possible, this evaluation is performed online with open uncontrolled real users of the system. The system logs are analyzed to improve the performance of the system.

However, when it is not possible to carry out this online open evaluation with real users, a controlled testing is organized observing the behaviour of a reduced group of users. These users are asked to perform different tasks with different releases of the recommender system. Afterwards, the users' feedback about the performance and suitability of the system is analyzed.

In any type of evaluation, either the *components* or of the *whole system*, the performance evaluation can be done both from a *quantitative point of view* and from a *qualitative perspective*.

The *quantitative evaluation* is usually undertaken using classic accuracy measures like the *Precision, Recall,* or *F-measure*:

- *Precision* is the percentage of recommended or returned items that are relevant or interesting for the active user:

$$\text{Precision} = \frac{\#\text{Relevant} - \text{Recommended} - \text{Items}}{\#\text{Recommended} - \text{Items}} \qquad (9.157)$$

- *Recall* is the percentage of relevant or interesting items that were recommended or returned to the active user:

$$\text{Recall} = \frac{\#\text{Relevant} - \text{Recommended} - \text{Items}}{\#\text{Relevant} - \text{Items}} \qquad (9.158)$$

- *F-measure* is the harmonic mean of *Precision* and *Recall*:

$$F - \text{measure} = \frac{2 * \text{Precision} * \text{Recall}}{\text{Precision} + \text{Recall}} \qquad (9.159)$$

The *qualitative evaluation* is performed using other measures not related to the *accuracy*. Commonly, they are called *beyond accuracy measures*. Some measures applicable to the general recommendation process are *Diversity* and *F-measure of Precision and Diversity*:

- *Diversity* or *Intra List Diversity* (*ILD*) is computed as the average aggregated pairwise dissimilarity of the items in the list. Given a similarity or a distance metric between two items, related as $d(i,j) = 1 - \text{sim}(i,j)$, the diversity of a list or set of size |*Lst*| can be calculated as proposed in (Adomavicius & Kwon, 2012; Castells et al., 2015):

$$\text{ILD} = \frac{1}{|\text{Lst}| * |\text{Lst} - 1|} \sum_{i \in \text{Lst}} \sum_{\substack{j \in \text{Lst} \\ j \neq i}} d(i,j) \qquad (9.160)$$

- *F-measure* of *Precision* and *Diversity* (Borràs et al., 2017) could be used to evaluate the overall performance of the algorithms related to both accuracy and diversity concepts. It is defined as:

$$F_{\text{d}} - \text{measure} = \frac{2 * \text{Precision} * \text{Diversity}}{\text{Precision} + \text{Diversity}} \qquad (9.161)$$

9.5.5   Applications of Recommender Systems

Some well-known applications of *Recommender Systems* are the following:

- In 1992, the first commercial RS, called Tapestry, was developed at the Xerox Palo Alto Research Center, in order to handle a large volume of data and to recommend documents to collections of users using a "social" *collaborative filtering* approach. Tapestry was motivated by the increase of the amount of electronic mail people receive and their possibility to subscribe to newsgroups thus receiving only documents of interest to them (Goldberg et al., 1992).
- Another recommendation system based on *collaborative filtering* techniques, designed to handle problems of information overload, was implemented in 1994, by GroupLens for Usenet newsgroups. Newsgroups is a high turnover discussion lists service on the Internet. This distributed system enables users in finding articles of interest to them. It used past users' ratings to predict other users' interest in articles based on their previous subjective evaluations of articles (Konstan et al., 1997; Resnick et al., 1994).
- Netflix,[40] an online streaming video and DVD rental service, recognizing the importance of an effective and accurate RS announced in 2006, a competition for the implementation of the best *collaborative filtering algorithm* with a high prize for the winner in order to improve the recommendation algorithm that was using. The aim was to improve the prediction accuracy and produce a reduction of at least 10% in the RMSE in comparison to Cinematch, the RS Netflix was using by then. Netflix released for the competition a dataset containing about 100 million ratings from 480,000 users on approximately 18,000 movies (Takács et al., 2009; Schafer et al., 1999).

Recommender Systems have frequently been integrated in e-commerce applications and online marketing activities. The managers of those companies have great expectations as efficient personalized RSs increase the possibility of a user purchasing an item, or a touristic trip, or reading news in an internet media (Takács et al., 2009). Some of more recent RSs applications have been integrated in many well-known internet-activity companies, such as Amazon,[41] YouTube,[42] Ebay,[43] CDNow, Moviefinder, Last.fm,[44] or IMDb:[45]

- The Amazon website, in its section devoted to books selling, and concretely in the information page of a selected book, displays three recommendation hints:

---

[40] Netflix, Inc. (https://www.netflix.com).

[41] Amazon Inc. (©1996–2021) (https://www.amazon.com/).

[42] © 2021 Google LLC YouTube (https://www.youtube.com/).

[43] Copyright © 1995–2021 eBay Inc. (https://www.ebay.com/).

[44] CBS Interactive © 2021 Last.fm Ltd (https://www.last.fm/).

[45] © 1990-2021 by IMDb.com, Inc. (https://www.imdb.com/).

- *"Customers who viewed this item also viewed"* list that suggests books that are usually viewed by customers who viewed the concrete book selected.
- *"Customers who bought this item also bought"* list that provides recommendation books that are usually purchased by customers who bought the concrete book.
- *"More items to explore"* list that suggests books that should be interesting for the user.

In general, Amazon uses recommendation algorithms to personalize the market experience for each customer based on his personal interests. An *item-to-item scalable collaborative filtering* approach is used in order to provide recommendations of high quality in real time (Linden et al., 2003; Schafer et al., 2001).

- In CDNow, Inc. (1994–2013), which was a dot-com company operating an online shopping website selling compact discs and music-related product, the album advisor recommender worked in two different modes: the *single album* and the *multiple artist mode* (Schafer et al., 2001).

  - The *single album mode* generated a list of ten albums that may be of interest to the user based on an album already selected.
  - The *multiple artist mode* produced a list of ten recommended albums based on the user's selection of up to three artists.

- Moviefinder was the movie site maintained by E! Online. The Moviefinder's Match Maker enabled users in finding movies. Recommendations were generated according to three lists (Schafer et al., 2001):

  - One contained the recommended films with similar mood, theme or cast to a concrete movie.
  - The other one contained links to other films by the same director and/or key actors of the film.
  - In addition, there was the section *"We Predict"* which was a recommender system based on a customer's previous indicated ratings.

9.5.6   Future Trends in Recommender Systems

Recommender Systems is a fast-evolving field, and many new trends are emerging, constituting the open challenges for researchers (Adomavicius & Tuzhilin, 2005; Anand & Mobasher, 2005). Some of these trends can be enumerated, such as:

- *User group recommendation*, which is addressing the problem of recommending items that should be of interest to all members in a group, which is a more difficult task than a single user recommendation process.
- *Incremental recommender systems*. The incrementality is a typical problem in all machine learning techniques, which should address the task of updating the models or techniques previously induced or built up from data, when new data is available.

- *Context-aware recommender systems*, which are recommender systems taking inti account the context of the problem or/and the user to improve the performance of the recommendations. For instance, the user context such as her/his current *state of mind* or *feelings* can be taken into account to make recommendations about a song or a movie or a book. This different *mood* of the user must guide the search for recommended items where the user had the same or similar mood to be more accurate.
- *Scalable recommender systems*. When applying a recommender system to a real on-line dynamic system where a lot of data is being continuously generated, the scalability of the different recommendation techniques to huge datasets must be addressed.
- *Beyond accuracy recommender systems*. Most of the evaluation techniques are based on quantitative evaluation of the recommendation techniques, using accuracy-related measures. However, other beyond accuracy measures should be used to evaluate more precisely the quality of the recommended items. Often, a good recommender system can suggest new items, not very popular, which can be of great interest to the users, but as they are not already rated, the accuracy-related measures will not be able to capture and assess its interest for the users. Concepts like *diversity* or *novelty* of the items recommended or *trust* on the recommendation process are some examples.
- *Recommending a list or sequence of items* rather than just an item or a set of items. Common recommender systems suggest one or more unrelated items being of interest by users. In some application domains, such as in recommending song playlists, all the sequence or list of items recommended has an underlying musical concept which is shared by all the items in the list. Thus, the recommendation process cannot be based as the union of some unrelated items but aiming to find a *coherent* and *related* list of items.
- *Robust recommender systems*, which should be more robust and secure recommender systems to be able to protect users of being maliciously influenced by other users.

References

Ackerman, M., & Dasgupta, S. (2014). Incremental clustering: The case for extra clusters. In *Proceedings of the 27th International Conference on Neural Information Processing Systems (NIPS'14)* (Vol. 1, pp. 307–315)

Adomavicius, G., & Kwon, Y. (2012). Improving aggregate recommendation diversity using ranking-based techniques. *IEEE Transactions on Knowledge and Data Engineering, 24*(5), 896–911.

Adomavicius, G., & Tuzhilin, A. (2005). Toward the next generation of recommender systems: A survey of the state-of-the-art and possible extensions. *IEEE Transactions on Knowledge and Data Engineering, 17*(6), 734–749.

Aggarwal, C. C. (Ed.). (2007). *Data streams. Models and algorithms* (Series advances in database systems) (Vol. 31). Springer.

Alippi, C., Boracchi, G., & Roveri, M. (2009). Just in time classifiers: Managing the slow drift case. In *IEEE International Joint Conference on Neural Networks, (IJCNN 2009)* (pp. 114–120). IEEE.

Alippi, C., & Roveri, M. (2007). Just-in-time adaptive classifiers in non-stationary conditions. In *IEEE International Joint Conference on Neural Networks, (IJCNN 2007)* (pp. 1014–1019). IEEE.

Alippi, C., & Roveri, M. (2008). Just-in-time adaptive classifiers part ii: Designing the classifier. *IEEE Transactions on Neural Networks, 19*(12), 2053–2064.

Allen, J. F. (1983). Maintaining knowledge about temporal intervals. *Communications of the ACM, 26*(11), 832–843.

Anand, S. S., & Mobasher, B. (2005). Intelligent techniques for web personalization. In *Intelligent Techniques for Web Personalization* (pp. 1–36). Springer.

Asher, N., & Vieu, L. (1995). Toward a geometry of common sense: A semantics and a complete axiomatization of mereotopology. In *Proceedings of the 14th International Joint Conference on Artificial Intelligence (IJCAI-95)*, San Mateo, CA (pp. 846–852).

Ballard, D., & Brown, C. (1982). *Computer vision*. Prentice-Hall.

Baykan, C., & Fox, M. (1987). An investigation of opportunistic constraint satisfaction in space planning. In *Proceedings of the 10th International Joint Conference on Artificial Intelligence (IJCAI-87)*, Los Altos, CA (pp. 1035–1038).

Baykan, C., & Fox, M. (1997). Spatial synthesis by disjunctive constraint satisfaction. *Artificial Intelligence for Engineering Design, Analysis and Manufacturing, 11*, 245–262.

Becker, H., & Arias, M. (2007). Real-time ranking with concept drift using expert advice. In *Proceedings of the 13th ACM SIGKDD International Conference on Knowledge Discovery and Data Mining (KDD'07)*, New York, 2007 (pp. 86–94).

Bettini, C., & Montanari, A. (2002). Research issues and trends in spatial and temporal granularities. *Annals of Mathematics and Artificial Intelligence, 36*(1–2), 1–4.

Bichindaritz, I., & Conlon, E. (1996). Temporal knowledge representation and organization for case-based reasoning. In *Proceedings of 3rd International Workshop on Temporal Representation and Reasoning (TIME'96)* (pp. 152–159).

Bifet, A., & Gavaldà, R. (2007) Learning from time-changing data with adaptive windowing. In *Proceedings of the 2007 SIAM International Conference on Data Mining* (pp. 443–448).

Bifet, A., & Gavaldà, R. (2009). Adaptive learning from evolving data streams. In *Proceedings of the 8th International Symposium on Intelligent Data Analysis: Advances in Intelligent Data Analysis VIII (IDA'09)* (pp. 249–260). Springer.

Bifet, A., Holmes, G., Pfahringer, B., Kirkby, R., & Gavaldà, R. (2009). New ensemble methods for evolving data streams. In *Proceedings of the 15th ACM SIGKDD International Conference on Knowledge Discovery and Data Mining (KDD'09)*, New York (pp. 139–148).

Borgo, S., Guarino, N., & Masolo, C. (1996a). A pointless theory of space based on strong connection and congruence. In L. C. Aiello, & S. Shapiro (Eds.), *Proceedings of KR'96, Principles of Knowledge Representation and Reasoning*, San Mateo (CA) (pp. 220–229).

Borgo, S., Guarino, N., & Masolo, C. (1996b). Towards an ontological theory of space and matter. In *CESA'96, Proceedings of IMACS-IEEE/SMC Conference on Computational Engineering in Systems Applications*, Lille.

Borràs, J., Moreno, A., & Valls, A. (2017). Diversification of recommendations through semantic clustering. *Multimedia Tools and Applications, 76*(22), 24165–24201.

Branting, L. K., & Hastings, J. D. (1994). An empirical evaluation of model-based case matching and adaptation. In *Proceedings of the Workshop on Case-Based Reasoning, AAAI-94*, Seattle, Washington, July 1994 (pp. 72–78).

Branting, L. K., Hastings, J. D., & Lockwood, J. A. (1997). Integrating cases and models for prediction in biological systems. *AI Applications, 11*(1), 29–48.

Bridge, D., Göker, M., McGinty, L., & Smyth, B. (2006). Case-based recommender systems. *The Knowledge Engineering Review, 20*(3), 315–320.

Brown, D., Aldea, A., Harrison, R., Martin, C., & Bayley, I. (2018). Temporal case-based reasoning for type 1 diabetes mellitus bolus insulin decision support. *Artificial Intelligence in Medicine, 85*, 28–42.

Burke, R. (2007). Hybrid web recommender systems. In   *The adaptive web* (pp. 377–408). Springer.

Carpenter, G., Grossberg, S., Markuzon, N., Reynolds, J., & Rosen, D. (1992). Fuzzy ARTMAP: A neural network architecture for incremental supervised learning of analog multidimensional maps. *IEEE Transactions on Neural Networks, 3*(5), 698–713.

Castells, P., Hurley, N. J., & Vargas, S. (2015). Novelty and diversity in recommender systems. In *Recommender Systems Handbook* (pp. 881–918).

Castillo, E., Gutierrez, J. M., & Hadi, A. S. (1997). *Expert systems and probabilistic network models*. Springer-Verlag.

Castro, J. L. (1995). Fuzzy logic controllers are universal approximators. *IEEE Transactions on Systems, Man and Cybernetics, 25*, 629–635.

Chen, S. (Ed.). (1990). *Advances in spatial reasoning* (Vol. 1). Ablex.

Chu, F., & Zaniolo, C. (2004). Fast and light boosting for adaptive mining of data streams. In H. Dai, R. Srikant, & C. Zhang (Eds.), *Advances in knowledge discovery and data mining (PAKDD 2004)* (Lecture notes in computer science) (Vol. 3056, pp. 282–292). Springer.

Clarke, B. (1981). A calculus of individuals based on "connection". *Notre Dame Journal of Formal Logic, 22*(3), 204–218.

Clarke, B. (1985). Individuals and points. *Notre Dame Journal of Formal Logic, 26*(1), 61–75.

Cleeremans, A., Servan-Schreiber, D., & McLelland, J. M. (1989). Finite state automata and simple recurrent networks. *Neural Computation, 1*, 372–381.

Cohen, L. J. (1977). *The probable and the provable*. Clarendon.

Cui, Z., Cohn, A. G., & Randell, D. (1992). Qualitative simulation based on a logical formalism of space and time. In *Proceedings of the 10th National Conference of the American Association for Artificial Intelligence (AAAI-91)*, Menlo Park, California (pp. 679–684).

Dagum, P., Galper, A., & Horvitz, E. (1991). *Temporal probabilistic reasoning: Dynamic network models for forecasting, Knowledge Systems Laboratory. Section on medical informatics.* Stanford University.

Dagum, P., Galper, A., Horvitz, E., & Seiver, A. (1995). Uncertain reasoning and forecasting. *International Journal of Forecasting, 11*(1), 73–87.

Davis, E. (1988). A logical framework for commonsense predictions of solid object behaviour. *Artificial Intelligence in Engineering, 3*(3), 125–140.

Davis, E. (1990). *Representations of commonsense knowledge*. Morgan Kaufmann.

Dean, T., & Kanazawa, K. (1989). A model for reasoning about persistence and causation. *Computational Intelligence, 5*(3), 142–150.

Dechter, R., Meiri, I., & Pearl, J. (1991). Temporal constraint networks. *Artificial Intelligence, 49*, 61–95.

Dempster, A. P. (1967). Upper and lower probabilities induced by a multi-valued mapping. *The Annals of Mathematical Statistics, 38*(2), 325–339.

Domingos, P., & Hulten, G. (2000). Mining high-speed data streams. In *Proceedings of the 6th ACM SIGKDD International Conference on Knowledge Discovery and Data Mining (KDD '00)*, New York (pp. 71–80).

du Verdier, F. (1993). Solving geometric constraint satisfaction problems for spatial planning. In *Proceedings of the 13th International Joint Conference of Artificial Intelligence (IJCAI-93)* (pp. 1564–1569).

Dubois, D., & Prade, H. (1980). *Fuzzy sets and systems: Theory and applications*. Academic Press.

Dubois, D., & Prade, H. (1988). *Possibility theory: An approach to computerized processing of uncertainty*. Plenum Press.

Duda, R. O., Hart, P. E., Nilsson, Reboh, R., Slocum, J., & Sutherland, G. L. (1977). Development of a computer-based consultant for mineral exploration. *Annual Report SRI Projects 5821 and 6415*, October 1976-September 1977, Menlo Park, California.

Egenhofer, M. (1991). Reasoning about binary topological relations. In O. Gunther, & H. Schek (Eds.), *Proceedings of Advances in Spatial Databases (SSD'91)*, Berlin (pp. 143–160).

Egenhofer, M., & Franzosa, R. (1991). Point-set topological spatial relations. *International Journal of Geographical Information Systems, 5*(2), 161–174.

Elman, J. L. (1990). Finding structure in time. *Cognitive Science, 14*(2), 179–211.

Elwell, R., & Polikar, R. (2011). Incremental learning of concept drift in nonstationary environments. *IEEE Transactions on Neural Networks, 22*(10), 1517–1531.

Faltings, B. (1990). Qualitative kinematics in mechanisms. *Artificial Intelligence, 44*(1–2), 89–119.

Fisher, D. (1987). Knowledge acquisition via incremental conceptual clustering. *Machine Learning, 2*, 139–172.

Fisher, D. (1996). Iterative optimization and simplification of hierarchical clusterings. *Journal of Artificial Intelligence Research, 4*, 147–179.

Fisher, M., Gabbay, D. M., & Vila, L. (2005). Handbook of temporal reasoning in artificial intelligence. In *Foundations of artificial intelligence book series* (Vol. 1, 1st ed.). Elsevier.

Fleck, M. (1987). Representing space for practical reasoning. In *Proceedings of 10th International Joint Conference on Artificial Intelligence (IJCAI-87)* (pp. 728–730).

Fleck, M. (1996). The topology of boundaries. *Artificial Intelligence, 80*, 1–27.

Forbus, K. (1983). Qualitative reasoning about space and motion. In D. Gentner & A. Stevens (Eds.), *Mental models* (pp. 53–73). Erlbaum.

Forbus, K. (1995). Qualitative spatial reasoning. Framework and frontiers. In J. Glasgow, N. Narayanan, & B. Chandrasekaran (Eds.), *Diagrammatic reasoning. Cognitive and computational perspectives*. AAAI Press/MIT Press.

Forbus, K., Nielsen, P., & Faltings, B. (1991). Qualitative spatial reasoning: The CLOCK project. *Artificial Intelligence, 51*, 417–471.

Frank, A. (1992). Qualitative spatial reasoning about distances and directions in geographic space. *Journal of Visual Languages and Computing, 3*, 343–371.

Freksa, C. (1991). Qualitative spatial reasoning. In D. Mark & A. Frank (Eds.), *Cognitive and linguistic aspects of geographic space* (pp. 361–372). Kluwer.

Freksa, C. (1992). Using orientation information for qualitative spatial reasoning. In A. Frank, I. Campari, & U. Formentini (Eds.), *Theories and methods of spatiotemporal reasoning in geographic space. Proceedings of the International Conference GIS - From Space to Territory* (pp. 162–178). Springer.

Freksa, C., & Rohrig, R. (1993). Dimensions of qualitative spatial reasoning. In N. P. Carreté & M. Singh (Eds.), *Qualitative reasoning and decision technologies* (pp. 482–492). CIMNE.

Fujimura, K., & Samet, H. (1989). A hierarchical strategy for path planning among moving obstacles. *IEEE Transactions on Robotics and Automation, 5*(1), 61–69.

Fung, R., & Chang, K. C. (1989). Weighting and integrating evidence for stochastic simulation in Bayesian networks. In *Proceedings of 14th Conference on Uncertainty in Artificial Intelligence (UAI'98)* (pp. 209–220).

Gaber, M. M., Zaslavsky, A., & Krishnaswamy, S. (2005). Mining data streams: A review. *ACM SIGMOD Record, 34*(2), 18–26.

Gaines, B. R., & Kohout, L. (1975). Possible automata. In *Proceedings of International Symposium on Multiple-Valued Logics*, Bloomington, Indiana (pp. 183–196).

Galton, A. (1993). Towards an integrated logic of space, time and motion. In R. Bajcsy (Ed.), *Proceedings of the 13th International Joint Conference on Artificial Intelligence (IJCAI-93)*, San Mateo, CA (pp. 1550–1555).

Galton, A. (1996). Taking dimension seriously in qualitative spatial reasoning. In W. Wahlster (Ed.), *Proceedings of the 12th European Conference on Artificial Intelligence (ECAI'96)*, Chichester (pp. 501–505).

Gama, J. (2010). *Knowledge discovery from data streams. Data mining and knowledge discovery*. Chapman and Hall.

Gama, J., Medas, P., Castillo, G., & Rodrigues, P. (2004). Learning with drift detection. In A. L. C. Bazzan & S. Labidi (Eds.), *Advances in artificial intelligence (SBIA 2004)* (Lecture notes in computer science) (Vol. 3171, pp. 66–112). Springer.

Gama, J., & Gaber, M. M. (Eds.). (2007). *Learning from data streams: Processing techniques in sensor networks*. Springer.

Gatzioura, A., & Sànchez-Marrè, M. (2015). A case-based recommendation approach for market basket data. *IEEE Intelligent Systems, 30*(1), 20–27.

Geman, S., & Geman, D. (1984). Stochastic relaxation, Gibbs distributions, and Bayesian restoration of images. *IEEE Transactions on Pattern Analysis and Machine Intelligence, 6*(6), 721–741.

Gerla, G. (1990). Pointless metric spaces. *The Journal of Symbolic Logic, 55*(1), 207–219.

Giles, R. (1976). Lukasiewicz logic and fuzzy theory. *International Journal of Man-Machine Studies, 8*, 313–327.

Glasgow, J. (1993). The imagery debate revisited: A computational perspective. *Computational Intelligence, 9*(4), 309–333.

Glasgow, J., & Papadias, D. (1992). Computational imagery. *Cognitive Science, 16*(3), 355–394.

Golbeck, J. (2006). Generating predictive movie recommendations from trust in social networks. In *Proceedings of 4th International Conference on Trust Management (iTrust 2006)* (pp. 93–104).

Goldberg, D., Nichols, D., Oki, B. M., & Terry, D. (1992). Using collaborative filtering to weave an information tapestry. *Communications of the ACM, 35*(12), 61–70.

Gotts, N. (1996). Formalizing commonsense topology: The INCH calculus. In *Proceedings of the 4th International Symposium on Artificial Intelligence and Mathematics (AI/MATH'96), Fort Lauderdale (FL)* (pp. 72–75).

Gotts, N. M., Gooday, J. M., & Cohn, A. G. (1996). A connection based approach to common-sense topological description and reasoning. *The Monist, 79*(1), 51–75.

Graves, A., Liwicki, M., Fernández, S., Bertolami, R., Bunke, H., & Schmidhuber, J. (2009). A novel connectionist system for unconstrained handwriting recognition. *IEEE Transactions on Pattern Analysis and Machine Intelligence, 31*(5), 855–868.

Grigni, M., Papadias, D., & Papadimitriou, C. (1995). Topological inference. In *Proceedings of the 14th International Joint Conference on Artificial Intelligence (IJCAI-95), San Mateo, CA* (pp. 901–906).

Guha, S., Mishra, N., Motwani, R., & O'Callaghan, R. (2000). Clustering data streams. In *Proceedings of IEEE Symposium on Foundations of Computer Science* (pp. 359–366).

Güsgen, H. (1989). Spatial reasoning based on Allen's temporal logic. *Report ICSI TR-89-049*. International Computer Science Institute.

Hamacher, H. (1978). *Über logische Aggregationen nicht-binär expliziter Entscheidungskriterien* (In German). Frankfurt/Main

Hastings, W. K. (1970). Monte Carlo sampling methods using Markov chains and their applications. *Biometrika, 57*(1), 97–109.

Heckerman, D. (1995). A tutorial on learning Bayesian networks. *Technical Report MSR-TR-95-06*. Microsoft Research.

Henrion, M. (1988). Propagation of uncertainty by logic sampling in Bayes' networks. In J. F. Lemmer & L. N. Kanal (Eds.), *Uncertainty in artificial intelligence* (Machine intelligence and pattern recognition) (Vol. 5, pp. 149–164). North Holland.

Hernández, D. (Ed.). (1994). *Qualitative representation of spatial knowledge* (Lecture notes in artificial intelligence) (Vol. 804). Springer.

Hernández, D. (1993). Reasoning with qualitative representations: Exploiting the structure of space. In N. Piera Carreté & M. Singh (Eds.), *Qualitative reasoning and decision technologies* (pp. 493–502). CIMNE.

Hernández, D., Clementini, E., & Di Felice, P. (1995). Qualitative distances. In A. Frank & W. Kuhn (Eds.), *Proceedings of European Conference on Spatial Information Theory—A Theoretical Basis for GIS (COSIT'95), Berlin* (pp. 45–57).

Hobbs, J. (1985). Granularity. In *Proc. of the 9th Int. Joint Conference on Artificial Intelligence (IJCAI-85)* (pp. 432–435).

Hobbs, J. (1987). Granularity. In *Proceedings of the 9th International Joint Conference on Artificial Intelligence (IJCAI-87)* (pp. 432–435).

Hobbs, J., Croft, W., Davies, T., Edwards, D., & Laws, K. (1987). Commonsense metaphysics and lexical semantics. *Computational Linguistics, 13*(3–4), 241–250.

Hochreiter, S., & Schmidhuber, J. (1997). Long short-term memory. *Neural Computation, 9*(8), 1735–1780.

Hoeffding, W. (1963). Probability inequalities for sums of bounded random variables. *Journal of the American Statistical Association, 58*(301), 13–30.

Hoeglinger, S., & Pears, R. (2007). Use of Hoeffding trees in concept based data stream mining. In *Proceedings of 3rd International Conference on Information and Automation for Sustainability (ICIAFS 07)* (pp. 57–62).

Hoens, T. R., Polikar, R., & Chawla, N. V. (2012). Learning from streaming data with concept drift and imbalance: An overview. *Progress in Artificial Intelligence, 1*, 89–101.

Hölldobler, S. (1997). Situations, actions, and causality in the fluent calculus. *Technical Report WV 97-01*. Knowledge Representation and Reasoning Group, AI Institute, Department of Computer Science, Dresden University of Technology.

Holmblad, L. P., & Ostergaard, J. J. (1982). Control of cement kiln by fuzzy logic. In M. M. Gupta & E. Sanchez (Eds.), *Approximate reasoning in decision analysis* (pp. 389–400). North Holland.

Hulten, G., Spencer, L., & Domingos, P. (2001) Mining time-changing data streams. In *Proceedings of the 7th ACM SIGKDD International Conference on Knowledge Discovery and Data Mining (KDD'01)*, New York (pp. 97–106).

Jaakkola, T., & Jordan, M. I. (1996). Computing upper and lower bounds on likelihoods in intractable networks. In *Proceedings of the 12th International Conference on Uncertainty in Artificial Intelligence (UAI'96)* (pp. 340–348). Morgan Kaufmann.

Jaczynski, M. (1997). A framework for the management of past experiences with time-extended situations. In *Proceedings of 6th International Conference on Information and Knowledge Management (CIKM'97)*, Las Vegas, Nevada, USA, November 1997 (pp. 32–39).

Jære, M., Aamodt, A., & Shalle, P. (2002). Representing temporal knowledge for case-based reasoning. In *Proceedings of the 6th European Conference on Case-Based Reasoning (ECCBR 2002)*, Aberdeen, Scotland, UK, September 2002 (pp. 174–188).

Jannach, D., Zanker, M., Felfernig, A., & Friedrich, G. (2010). *Recommender systems. An introduction*. Cambridge University Press.

Jensen, F. V. (1989). Bayesian updating in recursive graphical models by local computations. *Technical Report R-89-15*. Department of Mathematics and Computer Science, University of Aalborg.

Jensen, F. V., Lauritzen, S. L., & Olesen, K. G. (1990). Bayesian updating in causal probabilistic networks by local computations. *Computational Statistics Quarterly, 5*(4), 269–282.

Jensen, F. V., & Nielsen, T. D. (2007). *Bayesian networks and decision diagrams* (2nd ed.). Springer.

Jordan, M. I. (1997). Serial order: A parallel distributed processing approach. *Advances in Psychology, 121*, 471–495.

Kaufmann, A. (1975). *Introduction to the theory of fuzzy subsets* (Vol. I). Academic.

Kim, J. H., & Pearl, J. (1983). A computational model for combined causal and diagnostic reasoning in inference systems. In *Proceedings of the 8th International Joint Conference on Artificial Intelligence (IJCAI-83)* (pp. 190–193). Morgan Kaufmann.

Klaua, D. (1965). Über einen Ansatz zur mehrwertigen Mengenlehre. Monatsb. *Deutsch Akad. Wiss. Berlin 7*, 859–876, (in German), 1965. A recent in-depth analysis of this paper has been provided by Gottwald, S. An early approach toward graded identity and graded membership in set theory. *Fuzzy Sets and Systems* 161(18):2369–2379, 2010.

Klement, E. P. L., Mesiar, R., & Pap, E. (Eds.). (2000). *Triangular norms*. Kluwer.

Klinkenberg, R., & Joachims, T. (2000). Detecting concept drift with support vector machines. In *Proceedings of 17th International Conference on Machine Learning (ICML'00)* (pp. 487–494).

Kolodner, J. (1993). *Case-based reasoning*. Morgan Kaufmann.

Kolter, J., & Maloof, M. (2003). Dynamic weighted majority: A new ensemble method for tracking concept drift. In *Proceedings of 3rd IEEE International Conference on Data Mining (ICDM'03)*, New York (pp. 123–130).

Kolter, J., & Maloof, M. (2005). Using additive expert ensembles to cope with concept drift. In *Proceedings of 22nd International Conference on Machine Learning (ICML'05)*, New York (pp. 449–456).

Kolter, J., & Maloof, M. (2007). Dynamic weighted majority: An ensemble method for drifting concepts. *Journal of Machine Learning Research, 8*, 2755–2790.

Konstan, J. A., Miller, B. N., Maltz, D., Herlocker, J. L., Gordon, L. R., & Riedl, J. (1997). Grouplens: Applying collaborative filtering to Usenet news. *Communications of the ACM, 40* (3), 77–87.

Kowalski, R., & Sergot, M. (1986). A logic-based calculus of events. *New Generation Computing, 4*, 67–95.

Kruse, R., Gerhardt, J., & Klawonn, F. (1994). *Foundations of fuzzy systems*. John Wiley.

Kubat, M., & Widmer, G. (1995). Adapting to drift in continuous domain. In *Proceedings of the 8th European Conference on Machine Learning (ECML'95)* (pp. 307–310). Springer-Verlag.

Kuipers, B. (1978). Modelling spatial knowledge. *Cognitive Science, 2*(2), 129–154.

Kuipers, B., & Levitt, T. (1988). Navigation and mapping in a large-scale space. *AI Magazine, 9*(2), 25–43.

Lauritzen, S. L., & Spiegelhalter, D. J. (1988). Local computations with probabilities on graphical structures and their application to expert systems. *Journal of the Royal Statistical Society B, 50* (2), 157–224.

Lazarescu, M., Venkatesh, S., & Bui, H. (2004). Using multiple windows to track concept drift. *Intelligent Data Analysis, 8*(1), 29–59.

Levitt, T., & Lawton, D. (1990). Qualitative navigation for mobile robots. *Artificial Intelligence, 44*, 305–360.

Lewis, D. L. (1973). *Counterfactuals*. Basil Blackwell.

Ligozat, G. (1991). On generalized interval calculi. In *Proceedings of the 9th National Conference of the American Association for Artificial Intelligence (AAAI'91)* (pp. 234–240). AAAI Press/ MIT Press.

Ligozat, G. (1993). Qualitative triangulation for spatial reasoning. In A. Frank & I. Campari (Eds.), *Proceedings of European Conference on Spatial Information Theory—A Theoretical Basis for GIS (COSIT'93)*, Berlin (pp. 54–68).

Ligozat, G., Mitra, D., & Condotta, J. F. (2004). Spatial and temporal reasoning: Beyond Allen's calculus. *AI Communications, 17*(4), 223–233.

Linden, G., Smith, B., & York, J. (2003). Amazon.com recommendations: Item-to-item collaborative filtering. *IEEE Internet Computing, 7*(1): 76–80..

López de Mántaras, R. (1990). *Approximate reasoning models. Ellis Horwood series in artificial intelligence*. Ellis Horwood/Halsted Press.

Lozano-Pérez, T. (1983). Spatial planning: A configuration space approach. *IEEE Transactions on Computers, C-32*(2), 108–120.

Lupiani, E., Juarez, J. M., Palma, J., & Marin, R. (2017). Monitoring elderly people at home with temporal case-based reasoning. *Knowledge-Based Systems, 134*, 116–134.

Ma, J., & Knight, B. (2003). A framework for historical case-based reasoning. In *Proceedings of 5th International Conference on Case-Based Reasoning (ICCBR'2003)* (Lecture notes in computer science (LNCS-2689)) (pp. 246–260). Springer.

Maloof, M., & Michalski, R. (2000). Selecting examples for partial memory learning. *Machine Learning, 41*, 27–52.

Mamdani, E. H., & Assilian, S. (1975). An experiment in linguistic synthesis with a fuzzy logic controller. *International Journal of Man-Machine Studies, 7*(1), 1–13.

Marr, D. (1982). *Vision: A computational investigation into the human representation and processing of visual information*. Freemann.

Martín, F. J., & Plaza, E. (2004). Ceaseless case-based reasoning. In *Proceedings of 7th European Conference on Case-Based Reasoning (ECCBR 2004)* (Lecture notes in artificial intelligence (LNAI-3155)) (pp. 287–301). Springer.

Martínez, M. (2006). *A dynamic knowledge-based decision support system to handle solids separation problems in activated sludge systems: development and validation*. Ph.D. thesis, Universitat de Girona.

Mas, M., Monserrat, M., Torrens, J., & Trillas, E. (2007). A survey on fuzzy implication functions. *IEEE Transactions on Fuzzy Systems, 15*(6), 1107–1121.

McCarthy, J., & Hayes, P. J. (1969). Some philosophical problems from the standpoint of artificial intelligence. In B. Meltzer & D. Michie (Eds.), *Machine intelligence* (Vol. 4, pp. 463–502). Edinburgh University Press.

McDermott, D. (1992). Reasoning, "spatial" entry. In E. Shapiro (Ed.), *Encyclopaedia of artificial intelligence* (2nd ed., pp. 1322–1334). Wiley.

Meléndez, J., Macaya, D., & Colomer, J. (2001). Case based reasoning methodology for supervision. In *Proceedings of 2001 European Control Conference (ECC)* (pp. 1600–1605).

Menger, K. (1942). Statistical metrics. *Proceedings of National Academy U S A, 8*, 535–537.

Metropolis, N., Rosenbluth, A. W., Rosenbluth, M. N., Teller, A. H., & Teller, E. (1953). Equation of state calculations by fast computing machines. *Journal of Chemical Physics, 21*(6), 1087–1092.

Miller, R., & Shanahan, M. (1999). The event calculus in classical logic—Alternative axiomatizations. *Linköping Electronic Articles in Computer and Information Science, 4*(16).

Mizumoto, M. (1981). Fuzzy sets and their operations, II. *Information Control, 50*, 160–174.

Mizumoto, M., & Tanaka, K. (1981). Fuzzy sets and their operations. *Information Control, 48*, 30–48.

Montani, S., & Portinale, L. (2005) Case based representation and retrieval with time dependent features. In *Proceedings of 6th International Conference on Case-Based Reasoning (ICCBR'2005)*. Lecture Notes in Artificial Intelligence (Vol. 3620, pp. 353–367). Springer.

Muhlbaier, M., Topalis, A., & Polikar, R. (2009). Learn++. nc: Combining ensemble of classifiers with dynamically weighted consult-and-vote for efficient incremental learning of new classes. *IEEE Transactions on Neural Networks, 20*(1), 152–168.

Mukerjee, A. (1989). A representation for modeling functional knowledge in geometric structures. In Ramani, Chandrasekan, and Anjancluyu (Eds.), *Proceedings of KBCS'89* (pp. 192–202).

Mukerjee, A., & Joe, G. (1990). A qualitative model for space. In *Proceedings of the 8th National Conference of the American Association for Artificial Intelligence (AAAI'90)* (pp. 721–727).

Murphy, K., Weiss, Y., & Jordan, M. I. (1999). Loopy belief propagation for approximate inference: An empirical study. In *Proceedings of 15th Conference on Uncertainty in Artificial Intelligence (UAI'99)* (pp. 467–475).

Murphy, K. A. (1998). *Brief introduction to graphical models and Bayesian Networks*. Tutorial. http://www.cs.berkeley.edu/~murphyk/Bayes/bayes.html

National Geographic (2021). *GIS (Geographic Information System)*. Encyclopaedic Entry. https://www.nationalgeographic.org/encyclopedia/geographic-information-system-gis/

Neapolitan, R. E. (1990). *Probabilistic reasoning in expert systems: Theory and algorithms*. John Wiley & Sons.

Nebel, B., & Bürckert, H. J. (1995). Reasoning about temporal relations: A maximal tractable subclass of Allen's Interval Algebra. *Journal of the ACM, 42*(1), 43–66.

Newell, A. (1982). The knowledge level. *Artificial Intelligence, 18*(1), 87–127.

Nguyen, Q. N., & Ricci, F. (2007). Replaying live-user interactions in the off-line evaluation of critique based mobile recommendations. In *Proceedings of the 2007 ACM conference on Recommender Systems (RecSys'07)* (pp. 81–88). ACM Press.

Nielsen, P. (1988). A qualitative approach to mechanical constraint. In *Proceedings of the 7th National Conference of the American Association for Artificial Intelligence (AAAI-88)* (pp. 270–274).

Nishida, K., Yamauchi, K., & Omori, T. (2005). ACE: Adaptive classifiers-ensemble system for concept-drifting environments. In *Proceedings of the 6th International Conference on Multiple Classifier Systems (MCS'05)* (pp. 176–185).

Orduña Cabrera, F., & Sànchez-Marrè, M. (2018). Environmental data stream mining through a case-based stochastic learning approach. *Environmental Modelling and Software, 106,* 22–34.

Orduña-Cabrera, F., & Sànchez-Marrè, M. (2009). Dynamic adaptive case library for continuous domains. In *Proceedings of 12th International Conference of the Catalan Association of Artificial Intelligence (CCIA'2009). Frontiers in artificial intelligence and applications series* (Vol. 202, pp. 157-166).

Orduña-Cabrera, F., & Sànchez-Marrè, M. (2016). Dynamic learning of cases from data streams. In *Electronic Proceedings of International Workshop on Advances and Applications of Data Science & Engineering (IWAADS&E)* (pp. 101–106). Real Academia de Ingeniería.

Pazzani, M. J. (1999). A framework for collaborative, content-based and demographic filtering. *Artificial Intelligence Review, 13,* 393–408.

Pearl, J. (1982). Reverend Bayes on inference engines: A distributed hierarchical approach. In *Proceedings of the 2nd National Conference of the American Association for Artificial Intelligence (AAAI-82)* (pp. 133–136).

Pearl, J. (1988). *Probabilistic reasoning in intelligent systems: Networks of plausible inference.* Morgan Kaufmann.

Pearl, J., & Russell, S. (2000). Bayesian networks. *Technical Report R-277.* UCLA Cognitive Systems Laboratory.

Pearl, J. A. (1986). Constraint-propagation approach to probabilistic reasoning. In L. N. Kanal & J. F. Lemmer (Eds.), *Uncertainty in artificial intelligence* (Machine intelligence and pattern recognition) (Vol. 4, pp. 357–369). North Holland.

Peot, M. A., & Shachter, R. D. (1991). Fusion and propagation with multiple observations in belief networks. *Artificial Intelligence, 48*(3), 299–318.

Pfahringer, B., Holmes, G., & Kirkby, R. (2007). New options for Hoeffding trees. In *Proceedings of 22nd National Conference of the American Association for Artificial Intelligence (AAAI-07)* (pp. 90–99).

Polikar, R., Upda, L., Upda, S. S., & Honavar, V. (2001). Learn++: An incremental learning algorithm for supervised neural networks. *IEEE Transactions on Systems, Man, and Cybernetics Part C, 31*(4), 497–508.

Ram, A., & Santamaría, J. C. (1993). Continuous case-based reasoning. In *Proceedings of case-based reasoning workshop at AAAI-93* (pp. 86–93).

Ram, A., & Santamaría, J. C. (1997). Continuous case-based reasoning. *Artificial Intelligence, 90,* 25–77.

Randell, D., & Cohn, A. G. (1989). Modelling topological and metrical properties of physical processes. In R. Brachman, H. Levesque, & R. Reiter (Eds.), *Proceedings of the 1st International Conference on the Principles of Knowledge Representation and Reasoning* (pp. 55–66).

Randell, D., Cui, Z., & Cohn, A. G. (1992). A spatial logic based on regions and connection. In *Proceedings of the 3rd International Conference on Principles of Knowledge Representation and Reasoning (KR'92)* (pp. 165–176).

Renz, J., & Guesguen, H. W. (2004). Guest editorial: Spatial and temporal reasoning. *AI Communications, 17*(4), 183–184.

Resnick, P., Iacovou, N., Suchak, M., Bergstrom, P., & Riedl, J. (1994). Grouplens: An open architecture for collaborative filtering of netnews. In *Proceedings of ACM Conference on Computer-Supported Cooperative Work* (pp. 175–186).

Resnick, P., & Varian, H. R. (1997). Recommender systems. *Communications of the ACM, 40*(3), 56–58.

Ricci, F., Rokach, L., & Shapira, B. (2015). Recommender systems: Introduction and challenges. In *Recommender systems handbook* (pp. 1–34). Springer US.

Richter, M. M., & Weber, R. O. (2013). *Case-based reasoning. A textbook*. Springer.

Robinson, A. J., & Fallside, F. (1987). The utility driven dynamic error propagation network. *Technical Report CUED/F-INFENG/TR.1*. Cambridge University Engineering Department.

Rosenfeld, A. (1975). Fuzzy graphs. In Zadeh et al. (Eds.), *Fuzzy sets and their applications to cognitive and decision processes*, New York, London, (pp. 77–96).

Rumelhart, D. E., Hinton, G. E., & Williams, R. J. (1986). Learning internal representations by error propagation. In D. E. Rumelhart, McClelland and the PDP Research Group (Eds.), *Parallel distributed processing* (pp. 318–362).

Russell, S. J., & Norvig, P. (2010). *Artificial intelligence. A modern approach* (3rd ed.). Prentice-Hall, Pearson Education.

Salii, V. N. (1965). Binary L-relations. Izv. Vysh. Uchebn. Zaved. *Matematika, 44*(1), 133–145. (in Russian).

Samet, H. (1984). The quadtree and related hierarchical data structures. *Computing Surveys, 16*(2), 187–260.

Samet, H. (1989). *The design and analysis of spatial data structures*. Addison Wesley.

Sànchez-Marrè, M., Cortés, U., Martínez, M., Comas, J., & Rodríguez-Roda, I. (2005). An approach for temporal case-based reasoning: episode-based reasoning. In *Proceedings of 6th International Conference on Case-Based Reasoning (ICCBR'2005)* (Lecture notes in artificial intelligence 3620) (pp. 465–476). Springer-Verlag.

Sànchez-Marrè, M., Cortés, U., Roda, I. R., & Poch, M. (1998). L'Eixample Distance: a new similarity measure for case retrieval. In *Proceedings of 1st Catalan Conference on Artificial Intelligence (CCIA'98)*. ACIA Bulletin 14-15:246–253.

Sànchez-Marrè, M., Cortés, U., Roda, I. R., & Poch, M. (2000). Using meta-cases to improve accuracy in hierarchical case retrieval. *Computación y Sistemas, 4*(1), 53–63.

Sànchez-Marrè, M., Gibert, K., Sojda, R., Steyer, J. P., Struss, P., & Rodríguez-Roda, I. (2006). Uncertainty management, spatial and temporal reasoning and validation of intelligent environmental decision support systems. In *Proceedings of 3rd International Congress on Environmental Modelling and Software (iEMSs'2006)*, Burlington, VT, USA (pp. 1352–1377).

Sànchez-Marrè, M., Gibert, K., Sojda, R., Steyer, J. P., Struss, P., Rodríguez-Roda, I., Comas, J., Brilhante, V., & Roehl, E. A. (2008). Intelligent environmental decision support systems. In A. J. Jakeman, A. Rizzoli, A. Voinov, & S. Chen (Eds.), *Environmental modelling, software and decision support. State of the art and new perspectives* (pp. 119–144). Elsevier Science.

Sandewall, E. (1994). *Features and fluents. A systematic approach to the representation of knowledge about dynamical systems*. Oxford University Press, Oxford.

Sangüesa, R., & Cortés, U. (1997). Learning causal networks from data: A survey and a new algorithm for recovering possibilistic causal networks. *AI Communications, 10*, 31–61.

Saul, L. K., Jaakkola, T., & Jordan, M. I. (1996). Mean field theory for sigmoid belief networks. *Journal of Artificial Intelligence Research, 4*, 61–76.

Schafer, J. B., Frankowski, D., Herlocker, J., & Sen, S. (2007). Collaborative filtering recommender systems. In *The adaptive web* (pp. 291–324). Springer.

Schafer, J. B., Konstan, J. A., & Riedl, J. (1999). Recommender systems in E-commerce. In *Proceedings of E-Commerce 99*, Denver, Colorado (pp. 158–166).

Schafer, J. B., Konstan, J. A., & Riedl, J. (2001). E-commerce recommendation applications. *Data Mining and Knowledge Discovery, 5*(1–2), 115–153.

Schweizer, B., & Sklar, A. (1961). Associative functions and statistical triangle inequalities. *Publicationes Mathematicae Debrecen, 8*, 169–186.

Shachter, R. D. (1988). Probabilistic inference and influence diagrams. *Operations Research, 36*, 589–605.

Shachter, R. D. (1990). Evidence absorption and propagation through evidence reversals. In M. Henrion, R. D. Shachter, L. N. Kanal, & J. F. Lemmer (Eds.), *Uncertainty in artificial intelligence* (Machine intelligence and pattern recognition) (Vol. 10, pp. 173–190). North Holland.

Shachter, R. D., & Peot, M. (1989). Simulation approaches to general probabilistic inference on belief networks. *Proc. of the Fifth Conference on Uncertainty in Artificial Intelligence (UAI-1989)* (pp. 311–318).

Shackle, G. L. S. (1961). *Decision order and time in human affairs* (2nd ed.). Cambridge University Press.

Shafer, G. (1976). *A mathematical theory of evidence*. Princeton University Press.

Shafer, G., & Shenoy, P. (1990). Probability propagation. *Annals of Mathematics and Artificial Intelligence, 2*, 327–352.

Shortliffe, E. H., & Buchanan, B. G. (1975). A model of inexact reasoning in medicine. *Mathematical Biosciences, 23*(3-4), 351–379.

Slack, M., & Miller, D. (1987). Path planning through time and space in dynamic domains. In *Proceedings of the 10th International Joint Conference on Artificial Intelligence (IJCAI-87)* (pp. 1067–1070).

Stanley, K. (2003). Learning concept drift with a committee of decision trees. *Technical Report AI-03-302*. Computer Science Department, University of Texas-Austin.

Stock, O. (Ed.). (1997). *Spatial and temporal reasoning*. Springer-Science+Business Media.

Street, W., & Kim, Y. (2001). A streaming ensemble algorithm (SEA) for large-scale classification. In *Proceedings of the 7th ACM SIGKDD International Conference on Knowledge Discovery and Data Mining (KDD 2001)*, New York (pp. 377–382).

Suermondt, H. J., & Cooper, G. F. (1991). Initialization for the method of conditioning in Bayesian belief networks. *Artificial Intelligence, 50*, 83–94.

Sugeno, M. (1985). An introductory survey of fuzzy control. *Information Science, 36*, 59–83.

Sugeno, M., & Nishida, M. (1985). Fuzzy control of model car. *Fuzzy Sets and Systems, 16*, 103–113.

Takács, G., Pilászy, I., Németh, B., & Tikk, D. (2009). Scalable collaborative filtering approaches for large recommender systems. *Journal of Machine Learning Research, 10*, 623–656.

Takagi, T., & Sugeno, M. (1985). Fuzzy identification of systems and its applications to modeling and control. *IEEE Transactions on Systems, Man, and Cybernetics, 15*(1), 116–132.

Thielscher, M. (1998). Introduction to the fluent calculus. *Linköping Electronic Articles in Computer and Information Science, 3*(14).

Timpf, A., & Frank, A. U. (1997). Using hierarchical spatial data structure for hierarchical spatial reasoning. In S. C. Hirtle & A. U. Frank (Eds.), *Proceedings of European Conference on Spatial Information Theory—A Theoretical Basis for GIS (COSIT'97)* (Lecture notes in computer science) (Vol. 1329, pp. 69–83). Springer.

Tomlinson, R. F. (1969). A geographic information system for regional planning. *Journal of Geography, 78*(1), 45–48.

Trillas, E., Alsina, C., & Pradera, A. (2004). On MPT-implication functions for fuzzy logic. *Revista de la Real Academia de Ciencias Exactas, Físicas y Naturales. Serie A: Matemáticas (RACSAM), 98*, 259–271.

Tsymbal, A., Pechenizkiy, M., Cunningham, P., & Puuronen, S. (2006). Handling local concept drift with dynamic integration of classifiers: domain of antibiotic resistance in nosocomial infections. In *19th IEEE Symposium on Computer-Based Medical Systems (CBMS'06)* (pp. 679–684).

Tsymbal, A., Pechenizkiy, M., Cunningham, P., & Puuronen, S. (2008). Dynamic integration of classifiers for handling concept drift. *Information Fusion, 9*(1), 56–68.

van Asselt, M. B. A., & Rotmans, J. (2002). Uncertainty in integrated assessment modelling: From positivism to pluralism. *Climatic Change, 54*, 75–105.

Vieu, L. (1993). A logical framework for reasoning about space. In A. Frank & I. Campari (Eds.), *Proceedings of European Conference on Spatial Information Theory—A Theoretical Basis for GIS (COSIT'93)*, Berlin (pp. 25–35).

Vieu, L. (1997). Spatial representation and reasoning in artificial intelligence. In O. Stock (Ed.), *Spatial and temporal reasoning*. Springer.

Vilain, M. B. (1982). A system for reasoning about time. In *Proceedings of the 2nd National Conference of the American Association for Artificial Intelligence (AAAI-82)* (pp. 197–201). AAAI Press.

Vilain, M. B., & Kautz, H. (1986). Constraint propagation algorithms for temporal reasoning. In *Proceedings of the 5th National Conference of the American Association for Artificial Intelligence (AAAI-86)*, Philadelphia, PA (pp. 377–382). AAAI Press.

Waibel, A., Hanazawa, T., Hinton, G., Shikano, K., & Lang, K. J. (1989). Phoneme recognition using time-delay neural network. *IEEE Transactions on Acoustics, Speech and Signal Processing, 37*(3), 328–339.

Wainwright, M. J., & Jordan, M. I. (2008). Graphical models, exponential families, and variational inference. *Machine Learning, 1*(1–2), 1–305.

Walker, W. E., Harremoës, P., Rotmans, J., van der Sluijs, J. P., van Asselt, M. B. A., Janssen, P., & Krayer von Krauss, M. P. (2003). Defining uncertainty—A conceptual basis for uncertainty management in model-based decision support. *Integrated Assessment, 4*(1), 5–17.

Walley, P. (1996). Measures of uncertainty in expert systems. *Artificial Intelligence, 83*, 1–58.

Wang, H., Fan, W., Yu, P., & Han, J. (2003). Mining concept-drifting data streams using ensemble classifiers. In *Proceedings of KDD 2003*, New York (pp. 226–235).

Wang, H., Yin, J., Pei, J., Yu, P., & Yu, J. (2006) Suppressing model overfitting in mining concept-drifting data streams. In. *Proceedings of 12th ACM SIGKDD International Conference on Knowledge Discovery and Data Mining (KDD 2006)*, New York (pp. 736–741).

Wang, Y., & Kobsa, A. (2009). Performance evaluation of a privacy-enhancing framework for personalized websites. In G. J. Houben, G. I. McCalla, F. Pianesi, & M. Zancanaro (Eds.), *Proceedings of International Conference on User Modeling, Adaptation, and Personalization (UMAP 2009)* (Lecture notes in computer science) (Vol. 5535, pp. 78–89). Springer.

Weiss, Y. (2000). Correctness of local probability propagation in graphical models with loops. *Neural Computation, 12*(1), 1–41.

Werbos, P. J. (1988). Generalization of backpropagation with application to a recurrent gas market model. *Neural Networks, 1*(4), 339–356.

Whitehead, A. (1929). *Process and reality: An essay in cosmology*. The MacMillan Company.

Widmer, G., & Kubat, M. (1996). Learning in the presence of concept drift and hidden contexts. *Machine Learning, 23*(1), 69–101.

Williams, R. J., & Zipser, D. (1989) A learning algorithm for continually running fully recurrent networks. *Neural Computation, 1*, 270–280.

Williams, R. J., & Zipser, D. (1992). Gradient-based learning algorithms for recurrent networks and their computational complexity. In *Back-propagation: Theory, architectures and applications*. Erlbaum.

Wright, S. (1921). Correlation and causation. *Journal of Agricultural Research, 20*, 557–585.

Yager, R. R. (1980). On a general class of fuzzy connectives. *Fuzzy Sets and Systems, 4*, 235–242.

Zadeh, L. A. (1965). Fuzzy sets. *Information and Control, 8*, 338–353.

Zadeh, L. A. (1971). Similarity relations and fuzzy orderings. *Information Sciences, 3*(2), 177–200.

Zadeh, L. A. (1973a). The concept of a linguistic variable and its application to approximate reasoning. *Memorandum/Technical Report UCB/ERL M411*. EECS Department, University of California.

Zadeh, L. A. (1973b). Outline of a new approach of the analysis of complex systems and decision processes. *IEEE Transactions on Systems, Man and Cybernetics, 3*, 28–44.

Zadeh, L. A. (1975). The concept of a linguistic variable and its application to approximate reasoning-III. *Information Sciences, 9*, 43–80.

Zadeh, L. A. (1978). Fuzzy sets as the basis for a theory of possibility. *Fuzzy Sets and Systems, 1*, 3–28.

Zadeh, L. A. (1982). Possibility theory and soft data analysis. In L. Cobb & R. Thrall (Eds.), *Mathematical frontiers of social and policy sciences* (pp. 69–129). Westview.

Zadeh, L. A., Fu, K. S., Tanaka, K., & Shimura, M. (Eds.). (1975). *Fuzzy sets and their applications to cognitive and decision processes*. Academic Press.

Zhang, N. L., & Poole, D. (1994). A simple approach to Bayesian network computations. In *Proceedings of the 10th Canadian Conference on Intelligence* (pp. 171–178).

Zhang, N. L., & Poole, D. (1996). Exploiting causal independence in Bayesian network inference. *Journal of Artificial Intelligence Research, 5*, 301–328.

Zhang, T., Ramakrishnan, R., & Livny, M. (1996). BIRCH: An efficient data clustering method for very large databases. *ACM SIGMOD Record, 25*(2), 103–114.

Zimmermann, H.-J. (2000). An application-oriented view of modeling uncertainty. *European Journal of Operational Research, 122*(2), 190–198.

Zimmermann, K. (1993). Enhancing qualitative spatial reasoning—Combining orientation and distance. In A. Frank & F. Campari (Eds.), *Proceedings of European Conference on Spatial Information Theory—A Theoretical Basis for GIS (COSIT'93)*, Berlin (pp. 69–76).

Further Reading

Cowell, R. G., Dawid, A. P., Lauritzen, S. L., & Spiegelhalter, D. J. (1999). *Probabilistic networks and expert systems*. Springer-Verlag.

Fisher, M., Gabbay, D. M., & Vila, L. (2005). Handbook of temporal reasoning in artificial intelligence. In *Foundations of artificial intelligence* (Vol. 1, 1st ed.). Elsevier.

Klir, G., & Yuan, B. (1995). *Fuzzy sets and fuzzy logic: Theory and applications*. Prentice-Hall.

Koller, D., & Friedman, N. (2009). *Probabilistic graphical models: Principles and techniques*. The MIT Press.

Neapolitan, R. E. (2019). *Learning Bayesian Networks*. Pearson.

Pearl, J. (2013). *Causality: Models, reasoning and inference* (2nd ed.). Cambridge University Press.

Ricci, F., Rokach, L., & Shapira, B. (Eds.). (2015). *Recommender systems handbook* (2nd ed.). Springer.

Zimermann, H.-J. (2001). *Fuzzy sets theory and its applications* (4th ed.). Kluwer Academic.

Chapter 10
Summary, Open Challenges, and Concluding Remarks

10.1 Summary

This book has intended to provide a thorough insight into the world of *Intelligent Decision Support Systems*, making a special emphasis on the computer science, and artificial intelligence aspects.

In the first part of the book, the *fundamentals of IDSSs* were studied, starting from the humans' need for decision support tools to cope with the complexity of real-world systems.

Afterwards, an immersion into *decisions*, its nature and typology was performed. The classic *Decision Theory* was analyzed to provide the readers with some insights on how to manage the decision process, from a formal way. The analysis of the different *decision-making scenarios*, such as decision-making under certainty, decision-making under risk, and decision-making under uncertainty was done. As a consequence of this analysis, it was made clear that for solving complex decision-making problems, some powerful computational tools are required: the IDSSs. Most used formalisms to model multiple sequential decisions were presented: *Decision Trees* and *Influence Diagrams*.

The *historical evolution of IDSSs* was outlined, and how they were originated from the pioneering Management Information Systems in the late 60s and initial 70s has been described. IDSSs are the evolution of former *Decision Support Systems* (*DSS*). DSSs were born in the Business Management field in the middle 70s. The progress of the DSSs was parallel to the evolution of computers and their computational environment. New computer languages appeared and new and more powerful computer hardware architectures appeared providing interactivity with end users. The *Advanced DSSs* appeared in the 80s, integrating business first-principles models and mathematical optimization models or linear programming techniques. Furthermore, Advanced DSS incorporated an important and characteristic feature of DSSs: the "what-if" scenario analysis. This capability provided forecasting abilities to assess the different alternatives for the decision-making problem. Finally, the

M. Sànchez-Marrè, *Intelligent Decision Support Systems*,
https://doi.org/10.1007/978-3-030-87790-3_10

integration of *data-driven models* from the inductive Machine Learning field and *model-driven techniques* from Artificial Intelligence (AI) originated the *Intelligent Decision Support Systems* (*IDSSs*) in the 90s. The DSSs in general, and the IDSSs in particular, were classified in two major categories, *model-driven IDSSs* and *data-driven IDSSs*, according to the knowledge source origin of its models or methods. Notwithstanding, most IDSSs are hybrid because they use both model-driven methods and data-driven methods combined in the same IDSS.

After the fundamentals of IDSSs have been presented, *Intelligent Decision Support Systems* (*IDSSs*) are analyzed in-depth, in the second part of the book. A brief introduction to *Artificial Intelligence* (AI) was done to situate the reader in the more general context of AI. Afterwards, different features of IDSSs were explored such as different *typology of IDSSs* according to both the nature of the application domain of the IDSS, such as *static IDSS*, *dynamic IDSS* and *hard-constrained dynamic IDSS*, and the knowledge source of their methods: *model-driven IDSS* and *data-driven IDSS*. Furthermore, the *conceptual components of an IDSS* were detailed. Other features of IDSSs, such as the main *requirements for an IDSS*, and its analysis and design were discussed. A *proposal for a general architecture of an IDSSs* has been made as well as a *proposal for a cognitive framework for IDSSs development*. A first case study of an IDSS for customer loyalty analysis was illustrated.

In the next chapters, different methods, models, and reasoning components of an IDSSs were explained, both belonging to model-driven methods or to data-driven methods. Regarding *model-driven techniques*, *agent-based simulation models*, *expert-based models*, *model-reasoning techniques* and *qualitative reasoning models* are explained in detail. Examples for all methods are provided.

Data-driven models were explained starting with the whole *data mining process*: *pre-processing* techniques, *data mining* models, and *post-processing* techniques. Data mining methods reviewed range from *descriptive models*, such as *clustering techniques*; *associative models* such as *association rules*; *discriminant or classifier methods*, such as *decision trees*, *case-based discriminant models*, and *ensemble methods*; *numeric predictive methods* such as *artificial neural networks*, *case-based predictor models*, *multiple linear regression models*, *regression trees* and *numeric predictor ensembles*; and *second-order data mining models*, such as *genetic algorithms*.

Finally, to illustrate the *practical application of above-mentioned models and techniques in IDSSs*, many case studies were detailed for most kinds of methods. Case studies ranged from a river basin management IDSS, an IDSS for garden design, a combinatorial circuit fault location IDSS, an IDSS for sustainable action planning, a crime analysis, control and prevention IDSS, an IDSS for market basket analysis, a manufacturing industry quality control IDSS to an IDSS for sales planning in a company.

The third and last part of the book was devoted to the *development and applications of IDSSs*. Different *tools available for IDSS development* were analyzed. Some tools for data-driven models and model-driven techniques were reviewed. Most of them are open-source tools, and just a few are commercial ones. As explained, to the best of the author's knowledge, there is not a *general development environment* to develop an IDSS, but both the *R* and *Python* languages along with the high quantity of software packages available are probably the best options to develop an IDSS.

Furthermore, some *advanced topics in IDSSs* were presented. *Uncertainty management* and its main approaches were reviewed and detailed. The *pure probabilistic method*, the *certainty factor model*, and a deep study of the *Bayesian Network model* and of the *Fuzzy Set Theory* were undertaken. In addition, *Temporal Reasoning issues* were explained, and some approaches, such as *Dynamic Bayesian Networks*, *Temporal Artificial Neural Networks*, *Temporal Case-Based Reasoning*, and *Incremental Machine Learning* techniques and data stream mining were described. *Spatial Reasoning* issues was another advanced topic explored. Moreover, a particular successful application of IDSSs for selecting items for a given user, i.e. *Recommenders Systems* is detailed. They have been frequently used in internet commercial applications in the last decades.

Finally, in this last chapter, some open challenges in IDSSs are reviewed, and some proposals to manage them are made. *Interoperability of methods, models and components in IDSSs, evaluation of IDSSs*, and *a general framework for IDSSs* development are addressed. Furthermore, some conclusions and final remarks are drawn up.

Now, the readers should be able to recognize a complex real-world domain encompassing some decision/s requiring the development of an IDSS. They should have a clear knowledge of what an IDSS is, what components it should have, which models and methods can be used, and how they work. Moreover, readers must know what software tools or/and integrated development tools can use for the development of an IDSS.

10.2 Open Challenges in IDSS

In the previous Chap. 9 some advanced topics in IDSSs were inspected, such as *uncertainty management, temporal reasoning*, and *spatial reasoning*. Even though these topics are advanced, and not present in all IDSSs, some approaches to solve them have been proposed in the literature.

Notwithstanding, there are some other advanced topics that in the opinion of the author are not still completely achieved and fully implemented in most of the IDSSs deployed in different application areas. These open challenges are the *lack of automated integration and interoperation of models/methods* in IDSSs, the *absence of a general framework and associated tool for the development* of IDSSs, and finally the need for a *general standardized protocol for the evaluation* of IDSSs.

In this section, they will be reviewed, and some proposals for coping with them will be presented. Of course, as they are open challenges, there could be other possible ways to cope with them. Approaches presented here, are just some author's ideas that could be useful for the practitioners in the IDSS field.

10.2.1 Integration and Interoperation of Models in IDSS

As explained in this book, the core of the reliability and high performance of an IDSS is strictly related to the methods or models integrated within it. As already described in Chaps. 4, 5, and 6, different type of models or methods exists, which can be classified as *data-driven models* and *model-driven techniques*.

A *model* is a description of a system, usually a simplified description less complex than the actual system, i.e. an abstraction, designed to help an observer to understand how it works and to predict its behaviour. Historically in science and technology, *models* have been divided into *mechanistic models* and *empirical models*.

Mechanistic models are based on an understanding of the behaviour of a system's components, analyzing the system from its *first-principles*. Usually, these mechanistic models are expressed as set of mathematical formulas and equations (differential equations, etc.), such as an econometric model for forecasting the sales benefit of one company or a physical model describing the movement, velocity and acceleration of a body. The first DSS were using only mechanistic models.

Notwithstanding, taking into account that usually a huge amount of data gathered from the system's being managed were available, some new *empirical models* were started to be used. *Empirical models* are based on direct observation, measurement, and extensive data records. The first *empirical models* used were *mathematical and statistical methods* like Multiple Linear Regression models, Principal Component Analysis models, etc. and *expert-based models* formed of knowledge compiled by the experts through their own observation and experience in problem solving. The success of several *inductive machine learning* techniques, within the Artificial Intelligence area, lead to the use of another kind of *empirical models*: the *intelligent data analysis models*. Some instances are the Association Rules model, Decision Tree models, Artificial Neural Networks models, Case-Based Reasoning models, etc., which have been extensively analyzed in this book.

Since the 80s, both the former *mathematical/statistical empirical models* and the later *machine learning empirical models* have been named as *data mining methods*, because the models constructed are the result of a mining process among the data.

With the use of *data mining models* and other reasoning techniques coming from the Artificial Intelligence field, the DSS has been evolved to the Intelligent Decision Support Systems (IDSS). IDSS is built using some Artificial Intelligence model or integrating several artificial intelligence models to be more powerful, jointly with geographical information system components, mathematical or statistical models, and environmental/health ontologies, and some minor economic components as explained in Chap. 4.

IDSS can integrate both *model-driven techniques* such as expert-based models, model-based reasoning methods, qualitative reasoning models, or agent-based simulation models and *data-driven models* such as descriptive models, associative models, discriminant models, or predictive models. This *integration* is the key point to provide IDSSs with high reliability and increased performance, taking profit from the advantages of several models and methods.

Major drawback in IDSS development is that this integration of models is usually built-up on a *manual ad-hoc basis*, and no standardization and automation of the communication and integration of models or methods is not yet clearly established in the IDSS scientific community.

Single AI models or numerical/statistic models provide a solid basis for the construction of reliable and real IDSS applications, but the *standardized* and *automated interoperability* of whatever AI models or numerical/statistical models or reasoning mechanisms is one of the main open challenges in this field.

10.2.1.1 Interoperation and Model Interoperability

The Institute of Electrical and Electronics Engineers (IEEE) defined *interoperability* as "the ability of two or more systems or components to exchange information and to use the information that has been exchanged" (IEEE, 1990). *Semantic interoperability* is achieved when, in addition, the components share a common understanding of the information model behind the data being interchanged (Manguinhas, 2010).

Semantic integration or *semantic interoperation* has been the focus of some research works in different IDSS application domains such as in medical domains, where some works used service-oriented architectures (Komatsoulis et al., 2008) or other works aimed at performing data interoperability on any Health Level Seven's[1] standard (Dolin & Alschuler, 2011).

In the information systems field, several works were done in semantic integration of business components (Elasri & Sekkaki, 2013; Kzaz et al., 2010). Other interesting works focussed on the semantic interoperability through service-oriented architectures (Vetere & Lenzerini, 2005).

In environmental science domains, some pioneer work was done in semantic integration of environmental models for application to global information systems and decision making, especially related to GIS components and models (Mackay, 1999; Wesseling et al., 1996). Other works in environmental IDSS, the so-called Intelligent Environmental Decision Support Systems, were done in model and data integration and reuse (Rizzoli et al., 1998) and an overview of model integration was performed in Argent (2004). More recently, an interesting work was presented in Sottara et al. (2012) using the Drools Rule-based integration platform using a unified data model and execution environment.

Although there are some architecture proposals in the literature to combine some of these models, there is not a *common framework* to be taken into account as the first guideline to deploy *interoperable IDSS* providing an easy way to integrate and (re)use several AI models or statistical/numerical models in a single IDSS.

---

[1] Health Level Seven International (HL7, http://www.HL7.org), founded in 1987, is a not-for-profit, ANSI-accredited standards developing organization dedicated to providing a comprehensive framework and related standards for the exchange, integration, sharing and retrieval of *electronic health information* that *supports clinical practice* and the *management, delivery and evaluation of health services*.

As outlined above, most of the interoperability of the models is achieved by a *manual ad hoc* model interaction.

In Sànchez-Marrè (2014), a research work provided a *systematic approach* to *interoperate* different models at different steps in the IDSS solving process. Now, that approach will be reviewed, and some extensions and modifications to the original proposal will be provided.

In an IDSS, regarding the models used, there are two different steps: the *model production* and the *model use*. The *model use* is commonly carried through the *model visualization* or *model execution*. Models are basically *produced* in a data mining process extracting the model from data (data-driven model) or directly from a domain expert who provides the knowledge body of the model (inference rules, probability values, etc.), or a first-principles model, expressing a well-established theory in one domain (a set of differential equations, etc.), or other model-based methods.

After that step, the models can be *used* for *visualization purposes* or for *execution purposes*. The visualization of a decision tree, the visualization of a Bayesian network, the visualization of a rule set, or the visualization of a set of equations aims at validating and/or interpreting the model by the experts. The *execution of a model* produces some results such as the diagnosis of the current state of a dynamic system, the prediction of values for numerical variables, the class discrimination of new instances, or the simulation of what-if scenarios, among others.

This means that in a general IDSS, at least three different kinds of processes, related to models, exist:

- *Model Production*, like for instance a decision tree *mining* algorithm which produces a decision tree model from training data, or a rule-based *interface that allows to supply* a set of inference rules model by an expert.
- *Model Visualization*, such as a tree *visualizer* algorithm that produces a tree shape image of the nodes and the leaves of the tree, from a decision tree model, or a *visualizer* algorithm that shows a set of inference rules in an understandable and easily interpretable way for humans, from an expert-based model expressed as a set of inference rules.
- *Model Execution* or *Interpretation*, such as a decision tree *interpreter* which discriminates the class to which belongs a new instance or example, from a decision tree model, or a set of rules *interpreter*, i.e. an inference engine, which makes new inferences based on the set of rules and set of facts.

All three model tasks, production, visualization, and execution, must be able to *share the models* and have a clear understanding of them. Sharing the models can be accomplished if there is a *model interchange format* being able to codify different kinds of models, with semantic expressivity. This interchange process is depicted in Fig. 10.1.

In the scientific community of data mining, there have been some efforts to adopt a standard model interchange format. A model interchange format will achieve the *inter-task interoperation* between these three model tasks. Thus, for instance, as shown in Fig. 10.1, the "Empirical Tree Model 1", once it has been generated by the

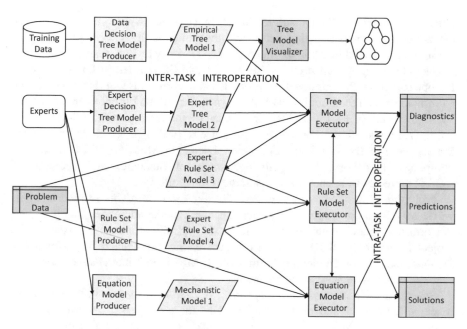

Fig. 10.1 Inter-task and intra-task interoperation

"Data Decision Tree Model Producer", can be afterwards be visualized by the "Tree Model Visualizer" and executed/interpreted by the "Tree Model Executor". In addition, this interoperability through the interchange of models must be achieved inside the same task category, i.e. *intra-task interoperation*. For example, model executors can generate new models adopting the role of model producer, such as the tree model executor depicted in the Fig. 10.1, which produces the "Expert Rule Set Model 3", from the "Expert Tree Model 2". Furthermore, model producers can activate other model executors, such as the "Rule Set Model Executor" which calls both the "Tree Model Executor" and the "Equation Model Executor" because of some rule inference action, as illustrated in Fig. 10.1. Thus, both *intra-task interoperability* and *inter-task interoperability* must be accomplished using the same model interchange format plus the possibility to activate/call other model executors from a model executor.

10.2.1.2   Tools and Techniques for Achieving Model Interoperability

Some organizations in the world have faced *semantic interoperability*, from an eGovernment, eBusiness, or data exchange perspective such as the Network Centric Operations Industry Consortium[2] (NCOIC), which is an international not-for-profit

---

[2]https://www.ncoic.org/home/.

consortium created in 2004, chartered in the United States, whose goal is to facilitate the adoption of cross-domain interoperability standards. In Europe, the European Commission created in 2004, the Interoperable Delivery of European eGovernment Services to public Administrations, Businesses, and Citizens (IDABC), which was a European Union Programme that promoted the correct use of Information and Communication Technologies. IDABC program issued the European Interoperability Framework[3] (EIF). IDABC was succeeded by the Interoperability Solutions for public Administrations, businesses, and citizens (ISA[4]) program. In the United States, there is the National Information Exchange Model[5] (NIEM) work and component repository. The National Institute of Standards and Technology[6] (NIST) serves as an agency for measurement standards.

From the point of view of concrete *semantic interoperability of models*, some works have been done, such as a queueing network model interoperability through a Performance Model Interchange Format (PMIF) (Smith et al., 2010), a JavaScript Object Interchange (JSOI) (Hoyos-Rodríguez & Sánchez-Piña, 2019), which is a JSON-based Eclipse Modelling Framework (EMF) model interchange format for efficient object interchange management.

Specifically for *semantic interoperability* of *predictive models* in the context of machine learning or/and data mining models, there are just a few works. *Open Neural Network Exchange[7] (ONNX)* is an open ecosystem for interoperable AI models, focussed at the beginning on deep learning models based on the Convolutional Neural Networks, but enlarged to other machine learning models. It emerged from Microsoft and Facebook in 2017, but now it is an open community project.

Another approach is the *Predictive Model Markup Language (PMML)* which is an XML-based predictive model interchange format conceived in 1997 by Prof. Robert Lee Grossman in National Center for Data Mining (NCDM), the University of Illinois at Chicago, and continued by a technical working group led by himself when he was the chair of the Data Mining Group[8] (DMG) (1998–2010). PMML provides a way for analytic applications to describe and exchange predictive models produced by data mining and machine learning algorithms. New versions of PMML have been developed, being the current version PMML 4.4.1. Furthermore, DMG has developed a *Portable Format for Analytics (PFA)*, which combines the ease of portability across systems with algorithmic flexibility: models, pre-processing, and post-processing are all functions that can be arbitrarily composed, chained, or built into complex workflows.

---

[3] https://ec.europa.eu/isa2/eif_en.

[4] https://ec.europa.eu/isa2/home_en.

[5] https://www.niem.gov/.

[6] https://www.nist.gov/.

[7] https://onnx.ai/index.html.

[8] The Data Mining Group (DMG) is an independent, vendor led consortium that develops data mining standards (http://dmg.org/).

Interoperation of models both at inter-task level and at intra-task level can be achieved by means of using a *model interchange format*. Furthermore, the use of *workflows* can help with the interoperation of models.

The selected model interchange format should be flexible enough to store both data mining models, model-driven techniques, reasoning mechanisms, etc., to guarantee more flexibility on the deployment of the IDSS.

XML (eXtensible Markup Language) is an effective way to interchange information between several software components and to share the semantics associated with the information. XML is a meta-language intended to supplement HTML's presentation features with the ability to describe the nature of the information being presented (Erl, 2004). XML adds a layer of intelligence to information being interchanged, providing meta-information, which is encoded as self-descriptive labels for each piece of text that go wherever the document goes. XML is implemented as a set of elements, which can be customized to represent data in unique contexts. A set of related XML elements can be classified as a vocabulary. An instance of a vocabulary is an XML document. Vocabularies can be defined formally using a schema definition language like Document Type Definition (DTD) or XML Schema Definition Language (XSD).

Therefore, taking into account the XML flexibility and that PMML, as described above, is based on XML, it seems a good choice for implementing the model interchange format.

PMML uses XML to represent data mining models. An XML Schema describes the structure of the models. One or more mining models can be contained in a PMML document. A PMML document is an XML document with a root element of type PMML. A PMML document can contain more than one model, and most common data mining models are supported (AssociationModel, BayesianNetworkModel, ClusteringModel, NeuralNetwork, RegressionModel, RuleSetModel, TimeSeriesModel, TreeModel, etc.). The general structure of a PMML document is:

```
<?xml version="1.0"?>
<PMML version="4.4"
xmlns="http://www.dmg.org/PMML-4_4"
xmlns:xsi="http://www.w3.org/2001/XMLSchema-instance">

<Header copyright="Example.com"/>
<DataDictionary> ... </DataDictionary>
<TreeModel ... >

 . . . A model . . .

</TreeModel>

 . . . Other models . . .

</PMML>
```

Using PMML for the data mining models means that one model could be *produced* by whatever data mining tool, and could be *visualized* by another visualizing tool, and can be *executed* by another model interpreter tool, given that all are compliant with the PMML model interchange format. Major commercial and open-source data mining tools supports PMML, such as *IBM SPSS Modeler*, *SAS Enterprise Miner*, *RapidMiner*, *KNIME,* or *WEKA*. Furthermore, both *R* has a *PMML package*, and *Python* has a *sklearn-pmml-model package* available. Therefore, for the building of an IDSS, some of the models used, could be obtained from data or experts in some data mining tool, and then incorporated in the IDSS. This fact provides great flexibility in the development step of the IDSS, in addition to the interoperation ability.

As mentioned before, the use of *workflows* can make easier the development stage of an IDSS facilitating the interoperation and reuse of models. The use of *workflows* can be very useful for specifying the workflow involving the raw data, the models produced, the model visualizers, the model executors, and auxiliary processes. Furthermore, the possible automated execution of *workflows* is an added advantage of its use.

Workflows are graphical notations, which were introduced in business management area to model and describe business processes. They let the design and specification of a *workflow* involving several elements such as:

- Data in whatever form: dataset, database, etc.
- Models described in an interchange model format
- Model producers
- Model visualizers
- Model executors
- Solution combiners
- Current problem specification
- Other components

An example of a *workflow* is depicted in the Fig. 10.1.

In the business field, there exist *jBPM*,[9] which is an open-source *workflow engine* written in Java that can execute business processes described in Business Process Model and Notation (BPMN) 2.0 standard, providing a graphical workflow editor. Other open-source *workflow engines* exist in the literature such as *Apache Airflow*,[10] which is coded in Python and is a platform to programmatically author, schedule and monitor workflows or *Prefect Core*,[11] which is another open-source *workflow* engine written in Python.

---

[9] https://www.jbpm.org/.

[10] https://airflow.apache.org/.

[11] https://www.prefect.io/core.

The idea is that the solving process of the IDSS through its different tasks and layers (data mining, diagnosis tasks, solution-generation tasks and predictive tasks) will be described through the *workflow*, and the *workflow* will be directly executed.

10.2.2   An Interoperable Framework for IDSS Development

As it was outlined in Chap. 8, there are no comprehensive and thorough software tools to develop a whole IDSS. Furthermore, as explained in Chap. 4, there are not a common-agreed architectural framework for the development of IDSS, and each IDSS is deployed in an *ad-hoc way*. In Chap. 4, the first approach for a general framework in the development of IDSS was proposed. Starting from the *dynamic Intelligent Decision Support System architecture* proposed and depicted in Fig. 4.12, and the *cognitive-oriented approach for IDSS development* depicted in Fig. 4.13, a new *interoperable framework for the development of IDSS* will be proposed.

In addition, to make easier the achievement of model/method interoperability, it would be interesting to have an *ontology* of all the possible components, models, methods, and operators involved in the solving process of the IDSS.

The ontology must contain the possible AI and numeric/statistic models/methods to be used in the development of an IDSS. In addition, *reasoning methods* such a rule-based reasoning, case-based reasoning, model-based reasoning, or qualitative reasoning must be also in the ontology, and other possible components, such as combination/aggregation operators, data validation operators, etc. An *ontology* is a formal description and characterization of the knowledge about some domain. It defines the important concepts and entities, their relevant features, and their relationships. In our case, the domain is the myriad of all possible components of an IDSS. Each possible method, model, reasoning approach, or operators must be described enumerating its preconditions to be used, its goals, its input data, the output generated, etc. All this information and knowledge will help in the building, verification, and execution of the *workflows*.

Furthermore, at the *prognosis layer*, several alternative solutions must be evaluated to give the corresponding feedback to the IDSS user. Our approach proposes that numerical simulation models or/and qualitative simulation models can be used, and in general, all of them could be encompassed in an *agent-based simulation tool* to model and evaluate the complex scenarios that must be assessed in many IDSS such as in an environmental domain as in Rendón-Sallard et al. (2006).

Any of the proposed multi-agent platform tools detailed in Chap. 8, Sect. 8.3.1., such as *Jadex*, *Ascape*, or *FLAME* written in Java, or such as *Mesa* or *dworp* tools, written in Python, or *NetlogoR*, which is written in R, could be used.

Therefore, the basis of our approach relies on the *cognitive-oriented approach for IDSS development*, which was a three-layer architecture distinguishing the main three tasks performed in the problem-solving of a IDSSs: analytical tasks, synthesis tasks, and prognosis tasks.

The separation between *model production* and *model execution* is an important feature to achieve interoperability. Thus, a new *four-layer cognitive-oriented approach for IDSS development* is proposed, based on previous work (Sànchez-Marrè, 2014). From those four layers, three correspond to the three cognitive tasks mentioned before. This four-layer architecture is described as follows:

- *Model Production Layer*: At this stage, which is usually done offline, the different models are produced. Both data gathering and knowledge discovery processes are run using some data mining techniques to build *data-driven models*. In addition, some knowledge acquisition processes acquire expert knowledge to build expert-based models, i.e. *model-driven methods*. This way the different kinds of models are obtained: *diagnostic models*, *solution-generation models*, and *predictive models*. These models will be used online at the corresponding task layers.
- *Model Execution Layer*: This layer, where the different models and methods will be executed online, includes the analysis tasks, synthesis tasks, and prognosis tasks. It is composed by three sub-layers:

  - *Analysis Task Layer*: In this layer is where most of the interpretative processes are run. The *diagnosis models* are executed here to provide the IDSS with strong *analytical power* to get an insight into what is going on in the system/entity being supervised or managed in real-time or on an offline basis. The system could be either an industrial process or an individual patient in a clinical IDSS.
  - *Synthesis Task Layer*: This layer wraps all the work necessary to synthesize possible *alternative solutions* for the different diagnostics found in the previous step. This synthetization can be done through the execution of several *solution-generation models* constructed on the *model production layer*. Those models and methods could be AI models, mathematical/statistical models, expert-based models or other model-driven methods, reasoning engine components, and even mechanistic models. The integration and interoperability of different nature models and components would enhance the problem-solving ability of the IDSS.
  - *Prognosis Task Layer*: At this upper layer lies the inherent ability of IDSS to decision support and or recommendation tasks. At this level, the *predictive models* constructed at the *model production layer*, which can be numerical (mostly simulations) or rather qualitative (qualitative reasoning or qualitative simulations) or agent-based (agent-based simulation), are used to estimate and forecast the consequences of several actions proposed in the previous layer by the solution-generation models. These "what if" models let the final user/decisor to make a rational decision based on the evaluation of several possible alternatives.

At any stage, depending on the nature of the IDSS application domain, the *temporal* and *spatial features*, as well as the *uncertainty management* of the information and knowledge used could be very important for obtaining reliable models. As much reliable are the models better performance and quality will get the IDSS.

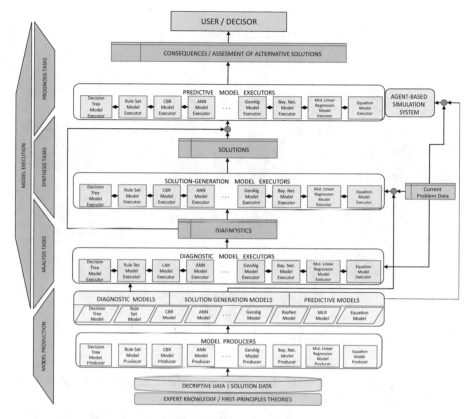

Fig. 10.2 Interoperable framework for IDSS development

The interoperable framework proposal for the development of an IDSS is depicted in Fig. 10.2.

10.2.3 IDSS Evaluation

Intelligent Decision Support Systems (IDSS) are software systems. Ensuring the correctness, completeness, consistency, accuracy, and performance of software systems is a quite hard task (Andriole, 1986). This task is even more complex when the software system, such as an IDSS, is a complex software system facing an unstructured or semi-structured problem.

In the IDSS literature, there are not many common frameworks for the assessment of the correctness of an IDSS. Usually, these evaluation processes are deployed for specific application domains such as in Papamichail and French (2005), where the evaluation of an IDSS for nuclear emergencies is described. In Sànchez-Marrè et al. (2008), a general evaluation approach was presented for Intelligent Environmental

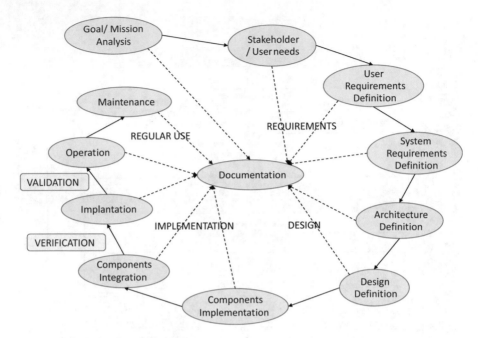

Fig. 10.3 Software development life cycle

Decision Support Systems (IEDSS), and in Sojda (2007), an empirical evaluation of Decision Support Systems was analyzed.

In this section, a general and standard procedure for evaluating an IDSS will be described. First, the general evaluation task of a software system is analyzed.

The usual *life cycle* of a software system development is composed of several activities: *goal/mission analysis, stakeholder/user needs analysis and requirements definition, system requirements definition, architecture definition, design definition, components implementation, components integration, implantation of the system, operation,* and *maintenance*. In addition, there is the *documentation* activity, which must be done throughout all the development process. This cycle is depicted in Fig. 10.3.

The process of ensuring the quality and the reliability of a software system aims to provide with an objective assessment of a software system correctness along its development *life cycle*. This process is composed of two procedures: *verification* and *validation*. Notwithstanding, both terms are commonly confused or used interchangeably. Some people simply talk about *software validation*, even though they refer to the *global software evaluation* process.

As described in the IEEE Standard for System, Software, and Hardware Verification and Validation (IEEE, 2017), *verification* is defined as "The process of evaluating a system or component to determine whether the products of a given development phase satisfy the conditions imposed at the start of that phase". Thus, *verification* assess whether each *life cycle step* has been successfully completed, and all the specifications are satisfied for the next life cycle steps or not. In other words, it is in charge to evaluate whether *the system is built-up in a correct way* or not.

On the other hand, *validation* is defined as "The process of evaluating a system or component during or at the end of the development process to determine whether it satisfies specified requirements". Thus, validation assess whether the system solves the right problem or not. In other words, it is in charge to evaluate whether *the correct system is built-up* or not.

Verification process evaluates the software system to determine whether the product/s or output of a given development phase satisfy the conditions at the beginning of that phase. It certifies the correctness of the transformation process for each life cycle step, i.e. the output conforms to the input, starting from the initial requirements to the final system, but it does not ensure that the output of each step or the final output is actually correct. *Validation process* evaluates whether the software product meets the needs of stakeholders, i.e., whether the requirements specification correctly expressed the needs of users. *Validation* could happen during development process or at the end, but commonly it is done at the end of the software development cycle, after the implantation of the software at the users' facility and before using it in regular operation (see Fig. 10.3). *Verification* of the different steps, is usually done after all the software code is integrated as shown in Fig. 10.3.

Verification tasks take the results of a given step in the development life cycle and verify that they match with the corresponding input specification of that step. For instance, it checks whether the proposed architecture design matches with the user and system requirements or whether the design definition proposed matches with the architecture definition.

The *verification methods* are different depending on the development cycle step:

- The development steps related to the *requirements stage* on the upper right zone of Fig. 10.3 or the *design stage* on the lower right part of Fig. 10.3 must use *static testing approaches* not involving the execution or operation of the system because it is not already implemented. Thus, usual techniques in those stages are based on the *manual revision of documentation reports* of the output of each step to verify that satisfy the requirements of the previous step.
- However, the verification techniques used in the *implementation stage* on the lower left part of Fig. 10.3 commonly use *dynamic testing approaches* involving the execution of a software component or the whole software system. Both *functional testing* and *structural testing* approaches are used.

There are two basic *validation methods* at the end of the development life cycle:

- *Internal Validation*: This approach assumes that stakeholder needs were correctly understood and described in the requirements in a right way. Thus, just it needs to validate whether the software meets the requirements specification. If so, then the software is validated.
- *External Validation*: This approach is based on asking to stakeholders whether the software meets their needs. Therefore, this means that stakeholders should test the system through some use cases, run the software system and afterwards

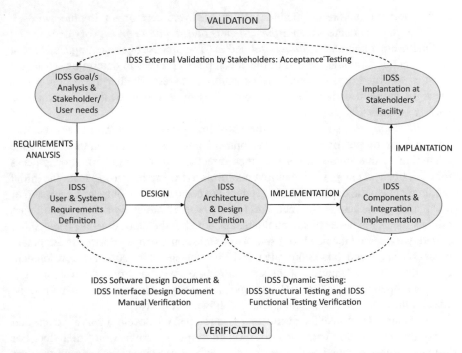

Fig. 10.4 IDSS development life cycle and its corresponding verification and validation strategies

analyze the results. The validation results can be successful or some anomalies can be detected. This approach is usually named *acceptance testing*.

Once the general software verification and validation processes have been described, and taking into account the particular architecture of IDSS, as specified in Chap. 4, Fig. 4.12, a *general standard protocol for the evaluation of an IDSS* is proposed.

This protocol is summarized in Fig. 10.4. As it can be observed in Fig. 10.4, all the development steps can be collapsed into four general steps: IDSS requirements specification, IDSS design, IDSS implementation, and IDSS implantation at stakeholders' facility.

The *general standard protocol for the evaluation of an IDSS* can be described as follows as two sequential processes: the verification of the IDSS and the validation of the IDSS.

Verification of the IDSS: It is proposed to be carried out twofold according to the fact that the IDSS development process can be particularized, from the general one described in Fig. 10.3, to the one depicted in Fig. 10.4. There are two great stages in the IDSS development: The design step and the implementation step. Our proposal is to use a different verification method for each step:

- *IDSS Design Verification*: in the design verification, the IDSS architecture and the IDSS design, both the high-level design and the different components design, must be verified. They should match the user and system requirements. This process is a manual task that must be done by a knowledge engineer. The process analyses the IDSS Design Document (IDSS–DD) and the IDSS Interface Design Document (IDSS–IDD) to check whether they really correspond and match the IDSS Requirements Specification Document (IDSS–RSD) and the IDSS Interface Requirements Specification Document (IDSS–IRSD).

  The IDSS Design Document (IDSS–DD) details the design information of the IDSS. It contains the architectural design of the IDSS, making explicit all the designed components (AI models, statistical models, reasoning components, forecasting models, etc.) for all the three cognitive tasks (analytical, synthetic, and prognostic) of the IDSS with its functionalities, attributes, etc. It constitutes the conceptual model of the IDSS. The IDSS Interface Design Document (IDSS–IDD) describes the architecture and design of IDSS interfaces between system and components. Both the IDSS Requirements Specification Document (IDSS–RSD) and the IDSS Interface Requirements Specification Document (IDSS–IRSD) details the specification of the IDSS requirements such as functions, attributes, design constraints, and performance, the requirements for its external interfaces, and the requirements for the interfaces between the IDSS main system and its components.

  The output of this verification process is a *design evaluation report* and/or an *anomaly report*.

- *IDSS Implementation Verification*: The verification of the IDSS implementation can be done running the IDSS code implemented to check whether the implementation matches with the architecture and design defined. Therefore, this testing of the IDSS is a dynamic one. Each *test case* is formed of the *test input data*, the *expected test output*, and the *test output result*. The test results are compared with the expected ones to verify the right implementation of the IDSS. The two main aspects to be tested are the IDSS *structure* and the *functionality* of the IDSS. These two testing processes are:

  - *Structural Component-centred Testing*: This testing checks the operation of all the IDSS components. The process can be divided in the following steps:

    IDSS Component Identification: For each layer of the IDSS architecture the different components must be identified: data source (sensors, laboratories, observations, opinions) and data quality modules (data gathering components), model generation components (data-driven/data mining, model-driven), model use components (AI models, statistical models, etc.), reasoning components (rule-based reasoning, case-based reasoning, etc.), decision-support components (forecasting models, etc.), learning component, and additionally for dynamic IDSS: planning components and actuation components.

    IDSS Unit Component Verification: Each one of the previous identified components must be individually tested to check that the test output matches the expected test output to verify the correct operation of each component. Depending on the nature of the component the test must be designed accordingly. For instance, a test for a predictive data-driven model component must prepare some test dataset and the expected predictions to be verified when applying the data-driven model with the test dataset.

– *Functional Task-centred Testing*: This testing checks the functionality of the whole IDSS through the testing of all the *cognitive tasks* involved in the IDSS. This way both the functionalities of the IDSS and the interactions among the different components are tested. This testing process can be divided in the following steps:

IDSS Task Identification: The existing cognitive tasks are identified. Usually in an IDSS there are:

- Analytic Tasks: They are the tasks determining the state or situation of a domain or process, i.e. diagnosis, through the interpretation of the available information and knowledge.
- Synthetic Tasks: They are the tasks providing the solution-generation for the previous identified diagnostics of the system.
- Prognostic Tasks: They are the tasks aiming at providing the future evolution of the system if some alternative is selected. Thus, they provide the relevant information for evaluating the different alternatives.
- Recommendation Task: Is the final task providing the ordered list of alternatives to be recommended to the user, for the final decision making.

IDSS Functionality Verification: The different functions of the IDSS are tested throughout the testing of all cognitive tasks of the IDSS involved. This verification process comprises:

- *Definition of Use Cases*: Several *use cases* will be defined which activate different cognitive tasks as the interaction between a certain subset of the components of the system such as gathering current data from sensors and user's answers, using an expert-based model, initiate a case-based reasoning process, evaluating some qualitative simulation model, and recommend the more suitable alternatives to the user. Furthermore, the interfacing between the involved components is also evaluated.
- *Testing the Use Cases*: The different *use cases* are executed in the IDSS to obtain the recommended alternatives.
- *Analyzing the Results of the Use Cases*: The recommendations of each *use case* are analyzed and verified for its correctness, completeness, performance, etc.

Validation of the IDSS: the validation of the IDSS is proposed to be carried out as an external validation process by the stakeholders. This means that the IDSS must be implanted at the stakeholders' facility, and the IDSS must be validated to check whether the IDSS satisfy their needs. This validation process is usually named as the *acceptance test*. *Acceptance testing* is a very similar process to the *IDSS functionality verification*. Major difference is the people in charge of the evaluation. In the above *IDSS functionality verification*, the responsible people are the knowledge engineer itself, but in the *IDSS external validation process*, the responsible is the stakeholder. The process is quite similar and can be described as follows:

- *Definition of a set of Scenarios*: *Scenarios* are complex use cases defined by the stakeholders to check different possible situations that the IDSS must be able to cope with. The *set of scenarios* must be *representative* of all possible scenarios that can happen. If in the given domain of the IDSS some *benchmarks*[12] exist, they can be used for scenario testing. Scenarios should cover normal situations as well as extreme or difficult situations to ensure a right operation and performance of the IDSS.

  For instance, in a medical IDSS for recommending suitable therapies to patients suffering from a concrete disease, stakeholders should define several scenarios to assess the *analytic ability* of the IDSS such as scenarios, where the patient has the disease, and it should be clearly detectable, others where the patient has the disease, but it is difficult to be detected, others where the patient do not have the disease, and it should be clearly detectable, and others where the patient do not have the disease, but it is difficult to be detected. Furthermore, other scenarios were there are missing data, and even errors in data, etc., should be tested. Other scenarios must test the *synthetic ability* of the IDSS to generate possible solutions, like suggesting the more adequate treatments and associate doses of drugs to the patients in different situations. Other scenarios must be defined to test the *prognostic ability* of the IDSS to forecast the right possible consequences of a given treatment administered to a patient making temporal projections about the patient's state. Finally, other scenarios must be defined to specifically test the *recommendation ability* of the IDSS to make a suitable ordered list of recommendations of all possible alternatives generated and projected in time in previous cognitive steps. Of course, as more representative, diverse, and complete is the set of scenarios the validation process will be more reliable. Furthermore, scenarios could test several cognitive tasks at the same time, too.

- *Testing the Scenarios in the IDSS*: Testing a *scenario* is performed operating the IDSS system at the stakeholder facility. This means running the software code supplying the input data from the different scenarios' tests to the IDSS, asking for the IDSS to provide recommendations to the given scenarios and collecting the

---

[12] A *benchmark*, in the context of computer science, is a standardized problem or test serving as the basis for evaluation or comparison of a software system performance.

scenarios' tests results, which at least include the recommendations suggested, but also other intermediate results can be analyzed.

- *Analyzing the Scenario IDSS Results*: Main goal is to analyze and check whether the IDSS provide a good performance. A good performance can be identified with the capacity of the system to provide the *right recommendations* in front of a certain scenario. This means that the intermediate diagnostics, the generated alternatives, the forecasted consequences, and the final recommendations are analyzed according whether they include reasonable, appropriate or expected results and not include wrong, inappropriate or implausible results. In addition to the *correctness* of the system, other performance indicators are assessed such as the *completeness* of the possible alternatives explored, the suggested *ordering of the alternatives* according to some uncertainty or plausibility measure, the *sensitivity* or robustness of the IDSS to noisy data, the *time efficiency* or *high reliability* in case that the IDDS is a dynamic hard-constrained IDSS where time response or avoiding dangerous consequences is a relevant factor. In addition, the IDSS ability to provide users with *explanations and/or justifications* could be assessed. After all the analysis, stakeholders must accept or not the IDSS depending on whether the IDSS satisfy their needs or not.

10.3   Concluding Remarks

In addition to the open challenges reported in the previous section, there are some *trends* that have already started to be taken into account and will grow in importance in next future for the development of IDSSs:

- *Explainability*: There are many efforts in scientific and technological community of AI to generate explainable AI systems. IDSSs, as particular intelligent systems must satisfy also the explanation requirements. Main reason for explainability is generating *trust* on the end users by providing *transparency* to the AI system. As AI/IDSS are coping with complex problems which usually has complex solution procedures, if the recommendations and solutions provided by IDSSs are not well motivated and explained, end users could not trust on them due to some lack of *transparency* (Schmidt et al., 2020). And if end users do not trust on an IDSS and in the solutions proposed by the system, it will not be used. For this reason, the use of interpretable models in AI is an important feature. *Model-driven techniques*, as they derive from an expert knowledge or from a first-principles theory are interpretable by the end users. However, some data-driven models which are usually named as *black-box* or *grey-box models*, because the reasons why the results are like they are, are not clearly interpretable, such as an ANN. This contrasts specially with the so-called *white-box models* were a rational interpretation and justification for the solution can be provided, such as in a decision tree. Therefore, the transparency in the reasoning processes and how the solutions are obtained enhances the trust of end users in IDSSs.

- Another important feature is ensuring *equality* in the AI system output.

  The ultimate goal is to have *non-discriminant* and *egalitarian* AI systems, specially referred to possible biases either on gender or origin or whatever human feature (UNESCO, 2020). AI/IDSS must provide recommendations that must be *fair* for all. This problem of biased AI/IDSS systems have become apparent with some past experiences in *biased predictive AI models*. One example was when it was detected that one algorithm used in court systems predicted twice as many false positives for a defendant becoming a recidivist for black offenders (45%) than white offenders (23%). Another example was detected when one company realized that their algorithm used for hiring employees was found to be biased against women. The algorithm was based on the number of resumes submitted over the past years, and since most of the applicants were men, it was trained to favour men over women. These two examples are instances of the same problem: a bad quality of data, not balanced, and not representative at all. As data-driven methods are very sensitive to data and generalize the patterns in the data, data must be balanced and representative of all people and all entities described within data. If data is not representative of the whole domain and not balanced, then the induced data-driven models could be biased to the overrepresented examples in the training dataset. Thus, all models, and specially the data-driven models, must be induced very careful in AI/IDSS systems.
- *Ethics*: Technical developers in general, and AI/IDSS scientists in particular should be aware and comply with ethical behaviour, because in AI/IDSS Systems much sensitive information is managed. In addition, the IDSS issues and its suggested recommendations must be fair, and cannot harm any people at any level. Therefore, no misuse or malicious use of an IDSS must be allowed. Technology in general, an IDSS in particular, are neutral. What can be dangerous, harmful or bad is a concrete use of technology. Therefore, some ethical behaviour rules and protocol must be established and satisfied by any AI practitioner. Recently, some reports and regulations have arisen, such as a pioneering report generated in the "Barcelona declaration for the proper development and usage of artificial intelligence in Europe" in 2017 (Steels & López de Mántaras, 2018).
- *Data Protection*: In IDSSs, data-driven models use data to induce the models. These data could be personal data, which must be protected, in order to preserve the confidentiality of people. Following this idea, several efforts have been done to regulate and protect personal data over the world. For instance, as reported in Chap. 9, in European Union the General Data Protection Regulation (GDPR) (EU 2016/679), which entered into force in 2016, ensures the protection of natural persons with regard to the processing of personal data and on the free movement of such data.

Some *conclusions* and *remarks* about Intelligent Decision Support Systems which can be extracted from the subject we have been explaining and investigating in the book are the following:

- *Reliable* IDSS for a real-world complex and ill-structured domains can be constructed. Several examples and case studies have been illustrated this statement along the book.
- Multi-knowledge and *interoperability* of several problem-solving techniques (AI, statistical, numerical, control models) is the key to obtaining more powerful and reliable IDSSs.
- *Data mining* is a key step to get inductive models from historical data, and get hidden patterns from data (*data-driven models*).
- Capturing *expert knowledge* (expert-based model) or *first-principles theories* to get a central corpus of knowledge are interesting *model-driven techniques*, which can be used in an IDSS.
- Capturing *experiential knowledge* (case-based reasoning) about either normal situations, prototypical situations, or idiosyncratic situations, can help to build flexible and dynamic IDSSs.
- IDSSs can be *dynamic learning environments* that can learn from their own solved problems in the past.
- An experiential knowledge component can ensure the *generalization ability* in order that the IDSS be *portable* to similar real-world processes
- Incorporating *predictive knowledge* to offer prognosis skills, i.e. the classic "what if" scenario analysis, offers great support to users in making the final decision.
- *Effective* decision support for real problems can be achieved with IDSSs. Medical diagnosis and treatment, Sustainability/Environmental management, Industrial process supervision, and Business management are application domains where IDSSs have been successfully applied.

Now, readers have arrived at the end of a travel around the Intelligent Decision Support Systems world. We hope that this trip has been interesting, challenging and fruitful!

If readers have reached the conclusion that there are not much differences between a general Artificial Intelligence System and an Intelligent Decision Support System, is the actual proof they have really understood the subject of this book!

Actually, any AI System solves a complex problem trying to get an optimal or suboptimal solution, such as an AI System for the diagnosis of the presence or absence of a patient's disease, or an AI System for the supervision of an industrial process or an AI system for a real-time robot planning. However, after a closer analysis of any AI System, it can be noticed that in any of them there are one or more decisions to which the AI System provides the best output information, i.e. support, in order that the end users select the best alternative, given the recommendation provided.

References

Andriole, S. J. (Ed.). (1986). *Software validation, verification, testing, and documentation.* Petrocelli Books.

Argent, R. M. (2004). An overview of model integration for environmental applications-Components, frameworks and semantics. *Environmental Modelling and Software, 19,* 219–234.

Dolin, R. H., & Alschuler, L. (2011). Approaching semantic interoperability in health level seven. *Journal of the American Medical Informatics Association, 18,* 99–103.

Elasri, H., & Sekkaki, A. (2013). Semantic integration process of business components to support information system designers. *Internation Journal of Web & Semantic Technologies, 4*(1), 51–65.

Erl, T. (2004). *Service-oriented architecture. A field guide to integrating XML and web services.* Prentice-Hall.

Hoyos-Rodríguez, H., & Sánchez-Piña, B. (2019). JSOI: a JSON-based interchange format for efficient format management. In *2019 ACM/IEEE 22nd International Conference on Model Driven Engineering Languages and Systems Companion (MODELS-C)* (pp. 259–266). IEEE.

IEEE. (1990). *IEEE standard computer dictionary: A compilation of IEEE standard computer glossaries.* Author.

IEEE. (2017). IEEE standard for system, software, and hardware verification and validation. In *IEEE Std 1012-2016* (pp. 1–260). Author. (Revision of IEEE Std 1012-2012/Incorporates IEEE Std 1012-2016/Cor1-2017).

Komatsoulis, G. A., Warzel, D. B., Hartel, F. W., Shanbhag, K., Chilukuri, R., Fragoso, G., de Coronado, S., Reeves, D. M., Hadfield, J. B., Ludet, C., & Covitz, P. A. (2008). caCORE version 3: Implementation of a model driven, service-oriented architecture for semantic interoperability. *Journal of Biomedical Informatics, 41*(1), 1–29.

Kzaz, L., Elasri, H., & Sekkaki, A. (2010). A model for semantic integration of business components. *International Journal of Computer Science & Information Technology, 2*(1), 1–12.

Mackay, D. S. (1999). Semantic integration of environmental models for application to global information systems and decision-making. *ACM SIGMOD Record, 28*(1), 13–19.

Manguinhas, H. (2010). Achieving semantic interoperability using model descriptions. *Bulletin of IEEE Technical Committee on Digital Libraries, 6*(2).

Papamichail, K. N., & French, S. (2005). Design and evaluation of an intelligent decision support system for nuclear emergencies. *Decision Support Systems, 41,* 84–111.

Rendón-Sallard, T., Sànchez-Marrè, M., Aulinas, M., & Comas, J. (2006). Designing a multi-agent system to simulate scenarios for decision making in river basin systems. In M. Polit, T. Talbet, B. López, & J. Meléndez (Eds.), *Proc. of 9th Int. Conference of the Catalan Association of Artificial Intelligence (CCIA'2006), Frontiers in Artificial Intelligence and Applications Series* (Vol. 146, pp. 291–298).

Rizzoli, A. E., Davis, J. R., & Abel, D. J. (1998). Model and data integration and re-use in environmental decision support systems. *Decision Support Systems, 24,* 127–144.

Sànchez-Marrè, M. (2014). Interoperable intelligent environmental decision support systems: A framework proposal. In D. P. Ames, N. W. T. Quinn, & A. E. Rizzoli (Eds.), *7th Int. Congress on Environmental Modelling & Software (iEMSs 2014)* (Vol. 1, pp. 501–508).

Sànchez-Marrè, M., Comas, J., Rodríguez-Roda, I., Poch, M., & Cortés, U. (2008). Towards a framework for the development of intelligent environmental decision support systems. In *Proc. of 4th Int. Congress on Environmental Modelling and Software (iEMSs'2008)* (pp. 398–406).

Schmidt, P., Biessmann, F., & Teubner, T. (2020). Transparency and trust in artificial intelligence systems. *Journal of Decision Systems, 29*(4), 260–278.

Smith, C. U., Lladó, C. M., & Puigjaner, R. (2010). Performance model interchange format (PMIF 2): A comprehensive approach to queueing network model interoperability. *Performance Evaluation, 67*(7), 548–568.

Sojda, R. S. (2007). Empirical evaluation of decision support systems: Needs, definitions, potential methods, and an example pertaining to waterfowl management. *Environmental Modelling and Software, 22,* 269–277.

Sottara, D., Bragaglia, S., Mello, P., Pulcini, D., Luccarini, L., & Giunchi, D. (2012). Ontologies, rules, workflow and predictive models: Knowledge ASSETS for an EDSS. In *Proc. of 6th Int. Congress on Environmental Modelling and Software (iEMSs'2012)* (pp. 204–211).

Steels, L., & López de Mántaras, R. (2018). Barcelona declaration for the proper development and usage of artificial intelligence in Europe. *AI Communications, 31*(6), 485–494.

UNESCO. (2020). *Artificial intelligence and gender equality. Key findings of UNESCO's global dialogue.* Author.

Vetere, G., & Lenzerini, M. (2005). Models for semantic interoperability in service-oriented architectures. *IBM Systems Journal, 44*(4), 887–903.

Wesseling, C. G., Karssenberg, D., Burrough, P. A., & Van Deursen, W. P. A. (1996). Integrating dynamic environmental models in GIS: The development of a dynamic modelling language. *Transactions in GIS, 1*(1), 40–48.

Further Reading

Andriole, S. J. (Ed.). (1986). *Software validation, verification, testing, and documentation.* Petrocelli Books.

Jobin, A., Ienca, M., & Vayena, E. (2019). The global landscape of AI ethics guidelines. *Nature Machine Intelligence, 1*(9), 389–399.

Rehm, G., Galanis, D., Labropoulou, P., Piperidis, S., Welß, M., Usbeck, R., Köhler, J., Deligiannis, M., Gkirtzou, K., Fischer, J., Chiarcos, C., Feldhus, N., Moreno-Schneider, J., Kintzel, F., Montiel, E., Doncel, V. R., McCrae, J. P., Laqua, D., Theile, I. P., ... Lagzdiņš, A. (2020). Towards an interoperable ecosystem of AI and LT platforms: A roadmap for the implementation of different levels of interoperability. *arXiv,* 2004.08355.

Correction to: Intelligent Decision Support Systems

Correction to:
Chapter 5 and Chapter 7 in: M. Sànchez-Marrè,
Intelligent Decision Support Systems,
https://doi.org/10.1007/978-3-030-87790-3

The book was inadvertently published without including the contributor's name Franz Wotawa in Chapters 5 and 7 and Table of Contents.

The contributor's name Franz Wotawa has now been updated.

The updated original version of these chapters can be found at
https://doi.org/10.1007/978-3-030-87790-3_5
https://doi.org/10.1007/978-3-030-87790-3_7

Index

© Springer Nature Switzerland AG 2022
M. Sànchez-Marrè, *Intelligent Decision Support Systems*,
https://doi.org/10.1007/978-3-030-87790-3

Printed in the United States
by Baker & Taylor Publisher Services